Lecture Notes in Electrical Engineering

Volume 1011

The book series *Lecture Notes in Electrical Engineering* (LNEE) publishes the latest developments in Electrical Engineering—quickly, informally and in high quality. While original research reported in proceedings and monographs has traditionally formed the core of LNEE, we also encourage authors to submit books devoted to supporting student education and professional training in the various fields and applications areas of electrical engineering. The series cover classical and emerging topics concerning:

- Communication Engineering, Information Theory and Networks
- Electronics Engineering and Microelectronics
- Signal, Image and Speech Processing
- Wireless and Mobile Communication
- Circuits and Systems
- Energy Systems, Power Electronics and Electrical Machines
- Electro-optical Engineering
- Instrumentation Engineering
- Avionics Engineering
- Control Systems
- Internet-of-Things and Cybersecurity
- Biomedical Devices, MEMS and NEMS

For general information about this book series, comments or suggestions, please contact leontina.dicecco@springer.com.

To submit a proposal or request further information, please contact the Publishing Editor in your country:

China

Jasmine Dou, Editor (jasmine.dou@springer.com)

India, Japan, Rest of Asia

Swati Meherishi, Editorial Director (Swati.Meherishi@springer.com)

Southeast Asia, Australia, New Zealand

Ramesh Nath Premnath, Editor (ramesh.premnath@springernature.com)

USA, Canada

Michael Luby, Senior Editor (michael.luby@springer.com)

All other Countries

Leontina Di Cecco, Senior Editor (leontina.dicecco@springer.com)

** This series is indexed by EI Compendex and Scopus databases. **

Yashwant Singh · Chaman Verma · Illés Zoltán ·
Jitender Kumar Chhabra · Pradeep Kumar Singh
Editors

Proceedings of International Conference on Recent Innovations in Computing

ICRIC 2022, Volume 2

 Springer

Editors
Yashwant Singh
Central University of Jammu
Jammu, Jammu and Kashmir, India

Illés Zoltán
Department of Media and Educational
Informatics, Faculty of Informatics
Eötvös Loránd University
Budapest, Hungary

Pradeep Kumar Singh
KIET Group of Institutions
Ghaziabad, Uttar Pradesh, India

Chaman Verma
Department of Media and Educational
Informatics, Faculty of Informatics
Eötvös Loránd University
Budapest, Hungary

Jitender Kumar Chhabra
Department of Computer Engineering
National Institute of Technology
Kurukshetra
Kurukshetra, Haryana, India

ISSN 1876-1100   ISSN 1876-1119 (electronic)
Lecture Notes in Electrical Engineering
ISBN 978-981-99-0603-1   ISBN 978-981-99-0601-7 (eBook)
https://doi.org/10.1007/978-981-99-0601-7

This Springer imprint is published by the registered company Springer Nature Singapore Pte Ltd.
The registered company address is: 152 Beach Road, #21-01/04 Gateway East, Singapore 189721,
Singapore

Preface

The volume two of the fifth version of international conference was hosted on the theme Recent Innovations in Computing (ICRIC-2022), and the conference was hosted by the Eötvös Loránd University (ELTE), Hungary, in association with the Knowledge University, Erbil, WSG University in Bydgoszcz Poland, Cyber Security Research Lab (CSRL) India, and other academic associates, technical societies, from India and abroad. The conference was held from August 13 to 14, 2022. The conference includes the tracks: cybersecurity and cyberphysical systems, Internet of things, machine learning, deep learning, big data analytics, robotics cloud computing, computer networks and Internet technologies, artificial intelligence, information security, database and distributed computing, and digital India.

The authors are invited to present their research papers at the second volume of Fifth International Conference on Recent Innovations in Computing (ICRIC-2022) in six technical tracks. We express our thanks to authors who have submitted their papers during the conference.

The first keynote address was delivered by Dr. Zoltán Illés, Eötvös Loránd University (ELTE), Hungary. He speaks on real-time systems and its role in various applications including educational purposes as well. The second keynote speech was delivered by Dr. Kayhan Zrar Ghafoor, Knowledge University, Erbil; he speaks on understanding the importance of the ultra-low latency in beyond 5G networks. The third keynote was delivered by Dr. Yashwant Singh, Central University of Jammu, J&K, India; he speaks on the Internet of things (IoT) and intelligent vulnerability assessment framework. He discusses the importance of IoT and IoVT during his talk.

The organizing committee expresses their sincere thanks to all session chairs—Dr. Chaman Verma, Dr. Amit Sharma, Dr. Ashutosh Sharma, Dr. Mayank Agarwal, Dr. Aruna Malik, Dr. Samayveer Singh, Dr. Yugal Kumar, Dr. Vivek Sehgal, Dr. Anupam Singh, Dr. Rakesh Saini, Dr. Nagesh Kumar, Dr. Vandana, Dr. Kayhan Zrar Ghafoor, Dr. Zdzislaw Polkowski, Dr. Pljonkin Anton for extending their help during technical sessions.

We are also grateful to rectors, vice rectors, deans, and professors of ELTE, Hungary, for their kind support and advice from time to time. The organizing committee is thankful to all academic associates and different universities for extending their support during the conference. Many senior researchers and professors across the world also deserve our gratitude for devoting their valuable time to listen, giving feedback, and suggestion on the paper presentations. We extend our thanks to the Springer, LNEE Series, and editorial board for believing in us.

Jammu and Kashmir, India   Yashwant Singh
Budapest, Hungary   Chaman Verma
Budapest, Hungary   Illés Zoltán
Kurukshetra, India   Jitender Kumar Chhabra
Ghaziabad, India   Pradeep Kumar Singh

Contents

E-Learning, Cloud and Big Data

About the Editors

Dr. Yashwant Singh is a Professor and Head of the Department of Computer Science and Information Technology at the Central University of Jammu where he has been a faculty member since 2017. Prior to this, he was at the Jaypee University of Information Technology for 10 Years. Yashwant completed his Ph.D. from Himachal Pradesh University Shimla, his Post Graduate study from Punjab Engineering College Chandigarh, and his undergraduate studies from SLIET Longowal. His research interests lie in the area of Internet of Things, Vulnerability Assessment of IoT and Embedded Devices, Wireless Sensor Networks, Secure and Energy Efficient Routing, ICS/SCADA Cyber Security, ranging from theory to design to implementation. He has collaborated actively with researchers in several other disciplines of computer science, particularly Machine Learning, Electrical Engineering. Yashwant has served on 30 International Conference and Workshop Program Committees and served as the General Chair for PDGC-2014, ICRIC-2018, ICRIC-2019, ICRIC-2020, and ICRIC-2021. He currently serves as coordinator of Kalam Centre for Science and Technology (KCST), Computational Systems Security Vertical at Central University of Jammu established by DRDO. Yashwant has published more than 100 Research Papers in International Journals, International Conferences, and Book Chapters of repute that are indexed in SCI and SCOPUS. He has 860 Citations, i10-index 27 and h-index 16. He has Research Projects of worth Rs.1040.9413 Lakhs in his credit from DRDO and Rs. 12.19 Lakhs from NCW. He has guided four Ph.D., 24 M.Tech. students and guiding four Ph.Ds. and five M.Tech. He has visited eight countries for his academic visits e.g. U.K., Germany, Poland, Chez Republic, Hungary, Slovakia, Austria, Romania. He is Visiting Professor at Jan Wyzykowski University, Polkowice Poland.

Dr. Chaman Verma is an Assistant Professor at the Department of Media and Educational Informatics, Faculty of Informatics, Eötvös Loránd University, Hungary. He has completed his Ph.D. Informatics from Doctoral School of Informatics, Eötvös Loránd University, Hungary, with Stipendium Hunagricum scholarship funded by Tempus Public Foundation, Government of Hungary. During his Ph.D., he won the EFOP Scholarship, Co-founded by the European Union Social Fund and the

Government of Hungary, as a professional research assistant in a Real-Time System for 2018–2021. He also received the Stipendium Hungaricum Dissertation Scholarship of Tempus Public Foundation, Government of Hungary, in 2021–22. He has been awarded several Erasmus scholarships for conducting international research and academic collaboration with European and non-European universities. He also received the best scientific publication awards from the Faculty of Informatics, Eötvös Loránd University, Budapest, Hungary, in the years 2021, 2022, and 2023. He has also been awarded ÚNKP Scholarship for research by the Ministry of Innovation and Technology and the National Research, Development and Innovation (NRDI) Fund, Government of Hungary, in 2020–2021 and 2022–2023. He has around ten years of experience in teaching and industry. He has over 100 scientific publications in the IEEE, Elsevier, Springer, IOP Science, MDPI, and Walter de Gruyter. His research interests include data analytics, IoT, feature engineering, real-time systems, and educational informatics. He is a member of the Editorial Board and a reviewer of various international journals and scientific conferences. He was the leading Guest Editor of the Special Issue "Advancement in Machine Learning and Applications" in Mathematics, IF–2.25, MDPI, Basel, Switzerland, in 2022. He is co-editor in the series conference proceedings of ICRIC-2021 and ICRIC-2022 published by Springer, Singapore. He is a life member of ISTE, New Delhi, India.He reviews many scientific journals, including IEEE, Springer, Elsevier, Wiley, and MDPI. He has Scopus citations of 615 with an H-index of 15. He has Web of Science citations 81 with an H-index 8.

Dr. Illés Zoltán, Ph.D., Habil. has started higher education studies in Mathematics and Physics at Eötvös Loránd University. He later took up the Computer Science supplementary course, which was started at that time. He got a Hungarian's Republic scholarship based on his outstanding academic achievements during his university studies. He graduated in 1985, after which he started working at the Department of Computer Science of Eötvös Loránd University. He completed his Ph.D. dissertation entitled "Implementation of Real-Time Measurements for High-Energy Ion Radiations" in 2001. In 2004, at the request of Jedlik Publisher, he also wrote a textbook on the C# programming language. This book has a second, expanded edition in 2008. In 2007, he was awarded a scholarship by the Slovak Academy of Sciences, where he spent six months in researching and teaching at the Constantine the Philosopher University in Nitra. The NJSZT awarded the Rezső Tarján Prize in 2016 for the success of the joint work that has been going on ever since. He and his colleagues also researched the issue of mobile devices and applications in the framework of a TéT_SK tender won in 2014. Based on their research findings, he launched a pilot project to support real time, innovative performance management. The first results of this research are an integral part of his habilitation dissertation. He has been an invited speaker at several international conferences and a member of the Amity University Advisory Board since 2020.

Prof. (Dr.) Jitender Kumar Chhabra is a professor, Computer Engineering Department at National Institute of Technology, Kurukshetra India. He has published 120 papers in reputed International and National Journals and conferences including more than 40 publications from IEEE, ACM, Elsevier, and Springer, most of which are SCI/Scopus indexed. His research interest includes Software Metrics, Data Mining, Soft Computing, Machine Learning, Algorithms, and related areas. He is a reviewer for most reputed journals such as *IEEE Transactions ACM Transactions* Elsevier, Wiley, and Springer. He has Google Scholar citations 1122, H-index 16, and i-10 index 24.

Dr. Pradeep Kumar Singh is currently working as a professor and head in the department of Computer Science at KIET Group of Institutions, Delhi-NCR, Ghaziabad, Uttar Pradesh, India. Dr. Singh is a senior member of Computer Society of India (CSI), IEEE, and ACM and a life member. He is an associate editor of the *International Journal of Information System Modeling and Design (IJISMD)* Indexed by Scopus and Web of Science. He is also an associate editor of *International Journal of Applied Evolutionary Computation (IJAEC)* IGI Global USA, Security and Privacy, and Wiley. He has received three sponsored research project grants from Government of India and Government of Himachal Pradesh worth Rs 25 Lakhs. He has edited a total of 12 books from Springer and Elsevier. He has Google Scholar citations 1600, H-index 20, and i-10 index 49. His recently published book titled *Handbook of Wireless Sensor Networks: Issues and Challenges in Current Scenarios* from Springer has reached more than 12000 downloads in the last few months. Recently, Dr. Singh has been nominated as a section editor for *Discover IoT* a Springer Journal.

Artificial Intelligence, Machine Learning, Deep Learning Technologies

A Three-Machine *n*-Job Flow Shop Scheduling Problem with Setup and Machine-Specific Halting Times

T. Jayanth Kumar, M. Thangaraj, K. J. Ghanashyam, and T. Vimala

Abstract Generally, machines may be serviced due to multiple reasons. The most common cause for machine repairs is due to allowing them to function uninterruptedly. This can be avoided by introducing mandatory machine halts that play a crucial role in preventing machine failures. However, preventive maintenance has been extensively addressed in the literature but halting times, which are indeed important in an effective manufacturing process, have received limited attention. Due to its practical importance, this paper considers a three-machine *n*-job flow shop-scheduling problem (FSSP) with transportation time, setup and machine-specific halting times. Here, setup times are separated from the processing time, and this occurs due to several causes such as assembling, cleaning and evocation. The objective of this problem is to minimize the overall makespan and mean weighted flow time such that halts must be supplied to the machines whenever they operate continuously for specified times. To the best of the author's knowledge, the present FSSP model is not addressed before in the literature. Focusing on heuristics or metaheuristic algorithms is inevitable given that the FSSP with three or four machines is NP-hard. Thus, an efficient heuristic algorithm is developed to tackle this problem. Finally, the proposed algorithm is demonstrated with a numerical example through which various performance measures are calculated. The current study is a modest attempt to look into the implications of machine-specific halting, setup and conveyance times for production.

Keywords Halting time · Setup time · Transportation time · Heuristic algorithm · Job weights · Flow shop-scheduling problem

T. J. Kumar · M. Thangaraj (✉) · K. J. Ghanashyam · T. Vimala
Faculty of Engineering and Technology, Department of Mathematics, Jain Deemed-to-Be University, Kanakapura 562112, India
e-mail: m.thangaraj@jainuniversity.ac.in

Y. Singh et al. (eds.), *Proceedings of International Conference on Recent Innovations in Computing*, Lecture Notes in Electrical Engineering 1011,
https://doi.org/10.1007/978-981-99-0601-7_1

1 Introduction

Scheduling has now become significantly important due to technical advancements in modern production systems. The flow shop-scheduling problem (FSSP) is an NP-hard optimization problem [1] that finds several practical applications in quality control process in steel making process [2], garments [3], other production, manufacturing and service systems [4], etc. The FSSP is described as a continuous flow of work across several machines for specific operations under certain conditions and is significant for manufacturing and designing companies to improve productivity and optimize costs. The scheduling for the entire production line is defined by how the jobs are distributed among the machines and how the order of these jobs is established to maximize a given parameter. Because of its extensive practical applications, the studies on FSSP have received great attention from the research community. As the majority of the FSSP and its allied problems are highly NP-hard, it is extremely difficult to achieve the optimal solution within polynomial time. According to [5], exact techniques can be employed only for FSSP with less than 20 jobs. Therefore, developing heuristic and metaheuristic algorithms is necessary, and it has gained attention for FSSP and its allied problems.

Addressing the developments of FSSP algorithms, since [6] has proposed the first solution methodology for the two machine FSSP to minimize the makespan, various algorithms have been developed for solving FSSP and its variants. An efficient CDS algorithm has been developed by [7] which effectively solves n-job and m-machine sequencing problems. An efficient constructive heuristic called NEH algorithm with two main processes, namely creating the priority sequence and inserting the jobs, has been developed by [8]. The FSSP with sequential start-up durations on the machines was studied by [9] and presented an efficient metaheuristic algorithm for finding optimal makespan. The structural condition-based binding method has been proposed by [10] to solve constrained FSSP with job weights. A heuristic technique that works in polynomial time for two machine FSSP with release dates has been developed by [11]. The authors [12] have adopted the structural conditions addressed by [10], presented a heuristic algorithm for solving n-job m-machine FSSP and compared it against the benchmark Palmer's, CDS and NEH algorithms. The authors [13] have proposed a heuristic algorithm for solving the n-job 3-machine FSSP under a fuzzy environment. The no-wait permutation FSSP was investigated by [14] and developed an improved productive heuristic algorithm for minimizing the total elapsed time. The preventive maintenance time based on cumulative machine flow time with two machines has been studied by [15] and presented an exact branch and bound algorithm for obtaining the optimal solutions. The authors [16] have studied the FSSP with permutation and sequence-independent setup time and compared those results with Johnson's rule and NEH algorithm. The FSSP with blocking constraints has been studied and proposed an iterated greedy algorithm by [17] to minimize the makespan. In addition to the above, more recently, particle swarm optimization technique [18], recursive greedy algorithm [19], evolutionary algorithm [20], constructive heuristic

[21], binding method-based heuristic algorithm [22], iterative meta-heuristic [23] and several other algorithms were developed for solving FSSP and its allied problems.

Halting time is crucial for preventing machine failures. In FSSP, halting time is crucial in preventing machine failures. To achieve the optimal job sequence and reduce the makespan and mean weighted flow time, relaxation must be provided to the machines while they are operating constantly. The setup time for each task on each machine is also important because there may be instances where a machine requires setup before performing jobs. By understanding the importance of various practical constraints, this study addresses the FSSP combined with setup time, conveyance time, the weight of jobs and machine-specific halting times. To the author's best knowledge, the present model has not been addressed in the literature.

2 Problem Description

2.1 Assumptions

The following assumptions are used to define the current model:

- Only one job can be performed by a machine at once.
- Only one machine can handle each job at once.
- At time zero, all jobs are available.
- All jobs must be completed in the predetermined machine sequence.
- The times at which halting starts for each machine are deterministic.
- Halting time duration is also deterministic and predetermined.
- Once a job gets started processing, it cannot be stopped by halting.
- Machine idle time is also utilized for halting time.

2.2 Problem Statement

The n-job and three-machine FSSP with setup time and machine-specific halting times is defined as follows: Let $R_{ij}\{i = 1, 2, \ldots, n, j = 1, 2, 3\}$ be the machine processing time, and $S_{ij}\{i = 1, 2, \ldots, n, j = 1, 2, 3\}$ be the setup time for each machine. Let h_i and l_i denote the conveyance time between Machine 1 to Machine 2 and Machine 2 to Machine 3, respectively. Conveyance time is the time that carries something to one place to another. Let w_i be the weight of jobs, which are relative importance between the jobs. Let CF_j be the cumulative flow time for all the machines. Depending on the cumulative flow time, each machine is sent for halting. A machine is allowed for halting only when the cumulative flow time is continuous, and it exceeds specified time. Let HS_j and HE_j be the halt starting time and halt ending time for each machine. The times at which Machine 1, Machine 2 and Machine 3 are allowed for halting are deterministic, which are denoted by

HS_1, HS_2 and HS_3, respectively. Halting duration times for all the machines are represented by HE_1, HE_2 and HE_3, respectively. Here, idle times of machines are assumed to utilize for halting to optimize the overall elapsed time. The objective is to find the optimal job sequence such that the total elapsed time is minimum. Due to the NP-hard nature of the present problem, an efficient heuristic algorithm is presented to minimize the total makespan. Since there are no existing studies on the present model, comparative studies are not performed. The outline of the described problem is given in Table 1, which is also graphically shown (see Fig. 1).

3   Proposed Methodology

The systematic algorithm for solving the present problem is described as follows:

Step 1: Modify the provided three-machine problem into two-machine problem by using the following structural conditions:

$$\left. \begin{array}{l} \text{Min}\{S_{i1} + R_{i1}\} \geq \text{Max}\{h_i + S_{i2} + R_{i2}\} \\ \text{(or)}\quad \text{Min}\{l_i + S_{i3} + R_{i3}\} \geq \text{Max}\{h_i + S_{i2} + R_{i2}\} \end{array} \right\} \tag{1}$$

$$\left. \begin{array}{l} \text{Min}\{S_{i1} + R_{i1} + h_i\} \geq \text{Max}\{S_{i2} + R_{i2}\} \\ \text{(or)}\quad \text{Min}\{l_i + S_{i3} + R_{i3}\} \geq \text{Max}\{S_{i2} + R_{i2}\} \end{array} \right\} \tag{2}$$

$$\left. \begin{array}{l} \text{Min}\{S_{i1} + R_{i1} + h_i\} \geq \text{Max}\{S_{i2} + R_{i2} + l_i\} \\ \text{(or)}\quad \text{Min}\{S_{i3} + R_{i3}\} \geq \text{Max}\{S_{i2} + R_{i2} + l_i\} \end{array} \right\} \tag{3}$$

$$\left. \begin{array}{l} \text{Min}\{S_{i1} + R_{i1}\} \geq \text{Max}\{h_i + S_{i2} + R_{i2} + l_i\} \\ \text{(or)}\text{Min}\{S_{i3} + R_{i3}\} \geq \text{Max}\{h_i + S_{i2} + R_{i2} + l_i\} \end{array} \right\} \tag{4}$$

(a) If condition (1) is satisfied, then

$$\left. \begin{array}{l} M_i = S_{i1} + R_{i1} + h_i + S_{i2} + R_{i2} \\ N_i = h_i + S_{i2} + R_{i2} + l_i + S_{i3} + R_{i3} \end{array} \right\} \tag{5}$$

(b) If condition (2) is satisfied, then

$$\left. \begin{array}{l} M_i = S_{i1} + R_{i1} + h_i + S_{i2} + R_{i2} \\ N_i = S_{i2} + R_{i2} + l_i + S_{i3} + R_{i3} \end{array} \right\} \tag{6}$$

(c) If condition (3) is satisfied, then

$$\left. \begin{array}{l} M_i = S_{i1} + R_{i1} + h_i + S_{i2} + R_{i2} + l_i \\ N_i = S_{i2} + R_{i2} + l_i + S_{i3} + R_{i3} \end{array} \right\} \tag{7}$$

Table 1 Problem layout

Job (*i*)	Setup time (S_{i1})	Processing time on Machine 1 (R_{i1})	Conveyance time $R_{i1} \rightarrow R_{i2} (h_i)$	Setup time (S_{i2})	Processing time on Machine 2 (R_{i2})	Conveyance time $R_{i2} \rightarrow R_{i3} (l_i)$	Setup time (S_{i3})	Processing time on Machine 3 (R_{i3})	Weight of jobs (w_i)
1	S_{11}	R_{11}	g_1	S_{12}	R_{12}	l_1	S_{13}	R_{13}	w_1
2	S_{21}	R_{21}	g_2	S_{22}	R_{22}	l_2	S_{23}	R_{23}	w_2
…	…	…	…	…	…	…	…	…	…
N	S_{n1}	R_{n1}	g_n	S_{n2}	R_{n2}	l_n	S_{n3}	R_{n3}	w_n

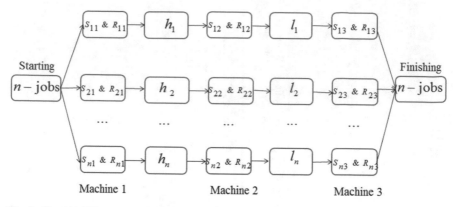

Fig. 1 Graphical illustration of the present problem

(d) If condition (4) is satisfied, then

$$M_i = S_{i1} + R_{i1} + h_i + S_{i2} + R_{i2} + l_i \left.\right\}$$
$$N_i = h_i + S_{i2} + R_{i2} + l_i + S_{i3} + R_{i3} \left.\right\}$$

$$(8)$$

where M_i and N_i are processing times of revised problem.

Step 2: Check the minimum value of M_i and N_i,

If $\min(M_i, N_i) = M_i$ then $M_i' = M_i - w_i$ and $N_i' = N_i$.

If $\min(M_i, N_i) = N_i$ then $M_i' = M_i$ and $N_i' = N_i + w_i$.

Step 3: Find the revised problem by computing M_i'' and N_i'',

where $M_i'' = \frac{M_i'}{w_i}$ and $N_i'' = \frac{N_i'}{w_i}$.

Step 4: Compute the optimal sequence for the revised problem using the Johnson's algorithm.

Step 5: Find the makespan for the original problem using the optimal sequence.

Step 6: Calculate the cumulative flow time (CF_i) for each machine:

(i) If the cumulative flow time of the machine j is continuous (no idle time between the jobs) and it exceeds the machine specified halting time (HS_j) of the machine j, then allow the jth machine for halting.

(ii) If the cumulative flow time of the machine j is not continuous (idle time between the jobs), then no need to halt the machine.

Every machine must be performing in this process.

Step 7: Identify the halting time affected jobs,

(i) If the job that requires halting time, then update the original table by adding the specified halting time duration with the processing times.

(ii) If the job that does not require halting time, then keep the same processing times in the original table.

Step 8: Calculate the performance measures such as total elapsed time, mean weighted flow time, flow time of each job, flow time of each machines and machine idle time for revised initial table.

4 Numerical Example

Consider the three-machine five-job FSSP with setup time, transportation time and weight of jobs with specific machine halting times. The objective of the given problem is to optimize the makespan and mean weighted flow time. The processing and conveyance times (in hours) and halting times (in hours) are given in Tables 2 and 3 as follows:

Step 1: The given problem satisfied the structural condition (2), the given problem can be reduced into two machine problem that is given in Table 4.

where $M_i = S_{i1} + R_{i1} + h_i + S_{i2} + R_{i2}$, $N_i = S_{i2} + R_{i2} + l_i + S_{i3} + R_{i3}$.

Steps 2–3: Apply Step 2 and Step 3 for the revised table, we get the following Table 5.

Table 2 Numerical example with setup time, conveyance time and weight of jobs

Job (i)	(S_{i1})	(R_{i1})	(h_i)	(S_{i2})	(R_{i2})	(l_i)	(S_{i3})	(R_{i3})	(w_i)
1	5	25	10	2	12	5	3	23	2
2	4	36	8	4	15	6	4	26	4
3	3	14	5	3	10	4	2	19	3
4	4	18	6	2	17	5	3	18	5
5	3	15	5	1	11	3	4	22	2

Table 3 Machine-specific halting times

Machines ($j = 1, 2, 3$)	Time (in hours) to start halting (HS_j)	Time (in hours) taking for halting (HE_j)
Machine 1	50	10
Machine 2	40	6
Machine 3	50	8

Table 4 Two-machine problem by adding setup and conveyance time

Job (i)	M_i	N_i	w_i
1	54	45	2
2	67	55	4
3	35	38	3
4	47	45	5
5	35	41	2

Table 5 Reduced two-machine problem

Job (i)	M_i''	N_i''	w_i
1	27	23.5	2
2	16.75	14.75	4
3	10.67	12.67	3
4	9.4	10	5
5	16.5	20.5	2

Table 6 Total elapsed time without halting times

Job (i)	(S_{i1})	(R_{i1})	(h_i)	(S_{i2})	(R_{i2})	(l_i)	(S_{i3})	(R_{i3})	(w_i)
4	4	4–22	6	2	30–47	5	3	55–73	5
3	3	25–39	5	3	50–60	4	2	75–94	3
5	3	42–57	5	1	63–74	3	4	**98–120**	2
1	5	**62–87**	10	2	99–111	5	3	123–146	2
2	4	91–127	8	4	139–154	6	4	164–190	4

Step 4: On applying Johnson's algorithm for the reduced two-machine problem, we obtain the following optimal sequence for all the jobs.

4	3	5	1	2

Step 5: Using the optimal sequence, the total elapsed time and halting time affected jobs for all the machines are computed and shown in Table 6.

The total makespan for the given optimal sequence is 190 h.

Step 6–7: Identify the halting time affected jobs by using cumulative flow time of each machine. The halting time affected jobs are Job 1 for Machine 1 and Job 5 for Machine 3. The Machine 2 flow time is not continuous, and then keeps the same processing times for Machine 2 in the original table, which is highlighted in bold as shown in Table 7.

Step 8: The summary results of total makespan, weighted mean flow time, flow time of each job, flow time of each machine and idle times of the machines are determined for reduced problem and shown in Tables 8 and 9. The final solution is demonstrated through Gantt chart (see Fig. 2).

From Table 6, it is observed that the total makespan due to absence of halting times is 190 h, whereas from the Table 8, it is seen that the total makespan is 200 h and increased by 10 h against the former makespan due to effect of halting times.

Table 7 Effect of processing time due to halting times

Job (i)	(S_{i1})	(R_{i1})	(h_i)	(S_{i2})	(R_{i2})	(l_i)	(S_{i3})	(R_{i3})	(w_i)
1	5	**35**	10	2	12	5	3	23	2
2	4	36	8	4	15	6	4	26	4
3	3	14	5	3	10	4	2	19	3
4	4	18	6	2	17	5	3	18	5
5	3	15	5	1	11	3	4	**30**	2

Table 8 Total makespan with the presence of halting time

Job (i)	(S_{i1})	(R_{i1})	(h_i)	(S_{i2})	(R_{i2})	(l_i)	(S_{i3})	(R_{i3})	(w_i)
4	4	4–22	6	2	30–47	5	3	55–73	5
3	3	25–39	5	3	50–60	4	2	75–94	3
5	3	42–57	5	1	63–74	3	4	98–128	2
1	5	62–97	10	2	109–121	5	3	131–154	2
2	4	101–137	8	4	149–164	6	4	174–200	4

Table 9 Summary of final results

(HS_j)	(HE_j)	Optimal sequence	Halting time affected jobs	Overall flow time of each job (in hrs)	Idle time of each machine (in hrs)	Overall flow time of each machines (in hrs)	Total elapsed time due to halting time (in hrs)	Mean weighted flow time (in hrs)
50, 40, 50	10, 6, 8	4–3–5–1–2	1 & 5	$f_4 = 73 f_3 = 60 f_5 = 72 f_1 = 95 f_2 = 103$	$M_1 = 0$ $M_2 = 77 M_3 = 132$	$M_1 = 137 M_2 = 77 M_3 = 132$	200	3.2034

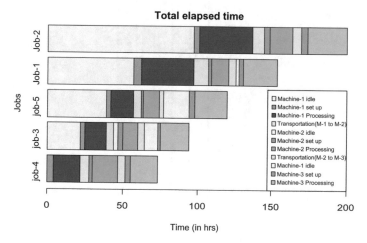

Fig. 2 Gantt chart for the solution

5 Conclusions

In this paper, we investigated the three-machine n-job FSSP with various constraints such as setup time, transportation time, job weights and machine-specific halting times. The problem is to minimize the overall makespan and mean weighted flow time such that halts must be supplied to the machines whenever they operate continuously for specified times. The present model is not addressed before in the literature. Due to the NP-hard nature of the present problem, an efficient heuristic algorithm is presented to minimize the total makespan. Since no existing studies on the considered model, comparative studies are not performed. However, the proposed algorithm is demonstrated with a numerical example through which various performance measures are computed. The current study is a modest attempt to look into the implications of machine-specific halting, setup and conveyance times for production.

References

1. Lian Z, Gu X, Jiao B (2008) A novel particle swarm optimization algorithm for permutation flow-shop scheduling to minimize makespan. Chaos, Solitons Fractals 35(5):851–861
2. Ren T, Guo M, Lin L, Miao Y (2015) A local search algorithm for the flow shop scheduling problem with release dates. Discret Dyn Nat Soc 2015:1–8
3. Erseven G, Akgün G, Karakaş A, Yarıkcan G, Yücel Ö, Öner A (2018) An application of permutation flowshop scheduling problem in quality control processes. In: The international symposium for production research. Springer, Cham, pp 849–860
4. Khurshid B, Maqsood S, Omair M, Sarkar B, Saad M, Asad U (2021) Fast evolutionary algorithm for flow shop scheduling problems. IEEE Access 9:44825–44839
5. Ruiz R, Maroto C, Alcaraz J (2006) Two new robust genetic algorithms for the flowshop scheduling problem. Omega 34(5):461–476
6. Johnson SM (1954) Optimal two-and three-stage production schedules with setup times included. Naval Res Logist Q 1(1):61–68
7. Campbell HG, Dudek RA, Smith ML (1970) A heuristic algorithm for the n job, m machine sequencing problem. Manage Sci 16(10):630–637
8. Nawaz M, Enscore E, Ham I (1983) A heuristic algorithm for the m-machine, n-job flow-shop sequencing problem. Omega 11(1):91–95
9. Ruiz R, Maroto C, Alcaraz J (2005) Solving the flowshop scheduling problem with sequence dependent setup times using advanced metaheuristics. Eur J Oper Res 165(1):34–54
10. Pandian P, Rajendran P (2010) Solving constrained flow-shop scheduling problems with three machines. Int J Contemp Math Sci 5(19):921–929
11. Aydilek H, Allahverdi A (2013) A polynomial time heuristic for the two-machine flowshop scheduling problem with setup times and random processing times. Appl Math Model 37(12–13):7164–7173
12. Gupta, D., Nailwal, K. K., Sharma, S.: A heuristic for permutation flowshop scheduling to minimize makespan. In: Proceedings of the Third International Conference on Soft Computing for Problem Solving, pp. 423–432, Springer, New Delhi (2014).
13. Thangaraj M, Rajendran P (2016) Solving constrained multi-stage machines flow-shop scheduling problems in fuzzy environment. Int J Appl Eng Res 11(1):521–528
14. Nailwal K, Gupta D, Jeet K (2016) Heuristics for no-wait flow shop scheduling problem. Int J Ind Eng Comput 7(4):671–680

15. Lee JY, Kim YD (2017) Minimizing total tardiness in a two-machine flowshop scheduling problem with availability constraint on the first machine. Comput Ind Eng 114:22–30
16. Belabid J, Aqil S, Allali K (2020) Solving permutation flow shop scheduling problem with sequence-independent setup time. J Appl Math 2020:1–11
17. Missaoui A, Boujelbene Y (2021) An effective iterated greedy algorithm for blocking hybrid flow shop problem with due date window. RAIRO-Oper Res 55(3):1603–1616
18. Abbaszadeh N, Asadi-Gangraj E, Emami S (2021) Flexible flow shop scheduling problem to minimize makespan with renewable resources. Sci Iranica 28(3):1853–1870
19. Ribas I, Companys R, Tort-Martorell X (2021) An iterated greedy algorithm for the parallel blocking flow shop scheduling problem and sequence-dependent setup times. Expert Syst Appl 184:115535
20. Sun J, Zhang G, Lu J, Zhang W (2021) A hybrid many-objective evolutionary algorithm for flexible job-shop scheduling problem with transportation and setup times. Comput Oper Res 132:105263
21. Shao W, Shao Z, Pi D (2021) Effective constructive heuristics for distributed no-wait flexible flow shop scheduling problem. Comput Oper Res 136:105482
22. Thangaraj M, Kumar T, Nandan K (2022) A precedence constrained flow shop scheduling problem with transportation time, breakdown times, and weighted jobs. J Project Manage 7(4):229–240
23. Liang Z, Zhong P, Liu M, Zhang C, Zhang Z (2022) A computational efficient optimization of flow shop scheduling problems. Sci Rep 12(1):1–16

Clinical Named Entity Recognition Using U-Net Classification Model

Parul Bansal and Pardeep Singh

Abstract Named entity recognition (NER) is a commonly followed standard approach in natural language processing. For example, properly recognize the names of people, sites and establishments in a sentence; or some domain-specific nouns, specific objects, etc. Consequently, the natural language processing domain can more effectively tackle complex tasks, such as question answering, information extraction and machine transformation. The deep learning models which are already proposed are quite complicated and need high execution time. In this research work, the clinical entity recognition technique is proposed which is based on the three phases: preprocessing, boundary detection and classification. The approach of U-Net is applied proposed for the classification for the named entity recognition. The proposed model is implemented in Python, and results are analyzed in terms of accuracy, precision and recall. The proposed model shows approx. 5% improvements in the results for the entity recognition.

Keywords Clinical entity recognition · Boundary detection · CNN · UNet · NLP · Deep learning

1 Introduction

Electronic medical records (EMRs) consist of plentiful health data and vital medicinal signs supporting clinical decision making and disease surveillance. But a great deal of non-structured medicinal texts substantially limits the search for knowledge and the appliance of EMRs. Discovering automated information retrieval techniques is the need of the current time to convert unstructured text into simple and machine-compatible structured data [1]. As a fundamental step in natural language processing

P. Bansal · P. Singh (✉)
National Institute of Technology, Hamirpur, India
e-mail: pardeep@nith.ac.in

P. Bansal
e-mail: 17MI559@nith.ac.in

© The Author(s), under exclusive license to Springer Nature Singapore Pte Ltd. 2023   15
Y. Singh et al. (eds.), *Proceedings of International Conference on Recent Innovations in Computing*, Lecture Notes in Electrical Engineering 1011,
https://doi.org/10.1007/978-981-99-0601-7_2

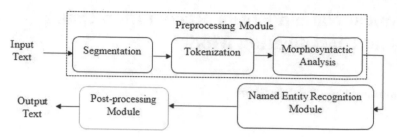

Fig.1 CNER component chain

(NLP), clinical named entity recognition (CNER) has been a well-known research topic on deriving all types of expressive information in non-structured clinical text. The purpose of the CNER is to discover and classify medical terminology in the EHR, for example, disease, symptoms, treatment, examination and body part [2]. However, this task becomes challenging for two reasons. One is the magnificence of the EHR; that is, the same word or sentence can represent multiple forms of named entities, and different types can define the entities with the same name. The second is the use of non-standard abbreviations or acronyms along with multiple forms of similar entities due to the large number of entities that rarely or do not occur in the training set. The development of the CNER component follows a modular design with a certain processing sequence (Fig. 1). This allows different modules belonging to the design sequence to be adaptive and independently instantiate a number of times.

The component's input contains clinical terms that are previously considered at the level of images and tables that may be present in a document. Here, the major objective is the identification and classification of sensitive clinical data, given its class. The processing chain is followed for this as depicted in the Fig. 1. The three main steps include pre-processing, named entity recognition and post-processing. All these tasks are performed in different modules. Each of these modules deals with solving a particular issue to access the output terminology [3]. The last module of the NER component chain is post-processing. This module aims at treating the results obtained from the former NER module and return the text (output) and entities discovered with the required format to the client. This facilitates the user to select to see the result in five dissimilar ways, as various forms of output are available for display based on their priority.

Machine learning (ML) being a form of data-driven artificial intelligence (AI) is able to learn about systems without clear programming. The algorithm based on ML formulates the clinical NER process as a chain labeling issue that is meant to discover the most optimal label chain (e.g., BIO) for a provided input chain (different words from the clinical text). Researchers have implemented several machines learning framework, including conditional random fields (CRFs), maximum entropy (ME), etc. Several prominent NER systems implemented the CRF model, which is one of the leading machine learning algorithms.

- Conditional random fields (CRFs): It is a sequence modeling model based on each node's setting token features and n-gram position model traits. A CRF consists of one exponential model for the combined probability of the whole chain of labels given the observed order. Hence, the weights of various traits can be swapped against each other in different cases. A CRF can be considered as a finite state model with unusual transition probabilities. However, in contrast to some other weighted finite-state solutions, CRFs provide a transparent and maximum likelihood or MAP estimation trained probability distribution over potential labeling. Furthermore, the loss function is convex, which guarantees convergence to the global optimum. CRFs readily generalize to similar stochastic context-free grammars suitable for problems such as RNA secondary structure prediction and natural language processing [4].
- Maximum entropy: E. T. JAYNES is the originator of the maximum entropy principle. Its fundamental assumption is that the distribution model avoids making assumptions about the hidden part in the case of a given finite amount of information. If only a given chunk of the knowledge of the unknown is known, it is desirable to pick up a probability distribution that is consistent with this knowledge but has maximal entropy. Some researchers used maximum entropy in natural language processing to stablish language models. Using the maximum entropy principle in natural language processing can solve text classification, pass tagging, phrase recognition problems, etc., and obtain fruitful results.

2   Literature Review

- Xing et al., discussed that various issues were occurred in the Chinese clinical domain, due to which the recognition of clinical named entities (CNEs) became challenging [5]. Thus, a glyph-based enhanced information model was introduced for which the convolutional neural network (CNN) and ALBERT were implemented for pre-training the language model so that the enhanced character information vector was attained. An attention mechanism was put forward on the BiLSTM-CNN-CRF (bidirectional long short-term memory network-CNN-conditional random field) structure for dealing with the issue related to extracting the attributes through the existing model without considering the importance and location information of words. The outcomes acquired on CCKS2018 dataset demonstrated the supremacy of the introduced model over others.
- Li et al., developed BiLSTM-Att-CRF algorithm in which bidirectional long-short time memory (BiLSTM) was integrated with an attention system for enhancing the efficiency to recognize the named entity in Chinese EMRs [6]. Hence, five kinds of clinical entities were recognized from CCKS2018. The outcomes indicated that the developed algorithm was performed well. Moreover, this algorithm yielded the F-score up to 85.79% without any additional attributes and 86.35% in the presence of extra attributes.

- Qiu et al., designed a residual dilated convolutional neural network with conditional random field (RD-CNN-CRF) algorithm in order to recognize the CNE [7]. Initially, the contextual information was captured using dilated convolutions, and the semantic and low-level attributes were deployed with the implementation of residual connections later on. In the end, the CRF was implemented as the output interface for attaining the optimal sequence of tags over the entire sentence. The CCKS-2017 Task 2 dataset was applied in the experimentation. The outcomes exhibited that the precision obtained from the designed algorithm was counted 90.63%, recall was 92.02% and F1-score 91.32% in comparison with the existing techniques.
- Li et al., projected the fine-tuned pre-trained contextual language models to recognize the named entity (NE) on the clinical trial eligibility criteria [8]. For this purpose, BioBERT, BlueBERT, PubMedBERT and SciBERT were implemented, and two systems were deployed for the open domain with regard to three clinical trial eligibility criteria corpora. The outcomes revealed that the PubMedBERT system performed more effectively as compared to others. Moreover, domain-specific transformer-based language models were proved more applicable in recognizing the NE in clinical criteria.
- Li et al., presented an approach in which conditional random field (CRF) was integrated with BiLSTM for recognizing and extracting the NEs in unstructured medical texts [9]. The asthma was cured by deriving 804 drug specifications from the Internet. Thereafter, a vector was utilized as the input of the neural network (NN) to quantize the normalized field of drug specification word. The presented approach led to enhance the accuracy by 6.18%, recall by 5.2% and F1 value by 4.87% in contrast to the existing algorithms. Additionally, the adaptability of this approach was proved for extracting the information regarding NE from drug specification.
- Yu et al., focused on implementing the BioBERT model on the basis of Google BERT model to define the clinical problems, treatments and tests automatically in EMR [10]. At first, the text was transformed into a numerical vector by pre-training the BioBERT model on the corpus of medical fields. After that, the BiLSTM-CRF was adopted for training the processed vectors and accomplishing the task to tag the entity. The experiments were conducted on I2B2 2010 dataset. The experimental results proved that the presented approach was assisted in enhancing efficiency to recognize the named entity for EMR and provided the F1 score up to 87.10%.
- Qiu et al., intended a RD-CNN-CRF system to recognize the Chinese clinical named entity (CNE) [11]. Initially, dense vector representations were employed to project the Chinese characters and dictionary attributes whose implementation was done later in RDCNN for capturing the contextual features. Eventually, the dependencies among neighboring tags were captured, and the optimal tag sequence was acquired for the whole sequence using CRF. The CCKS-2017 Task 2 dataset was utilized to evaluate the intended system. The outcomes confirmed the effectiveness concerning training time.

- Liu et al., constructed a medical dictionary to recognize the CNE [12]. Afterward, the CRF system was utilized to determine the impact of diverse kinds of attributes in task to recognize CNE. The experiments were conducted on 220 clinical texts whose selection was done from Peking Anzhen Hospital at random. The experimental results depicted the adaptability of these attributes in varying degrees to recognize the CNE.
- Luu et al., investigated a new technique to recognize the CNE on the basis of deep learning (DL) technique [13]. This technique was relied on two models, namely feed forward network (FFN) and recurrent neural network (RNN) which assisted in increasing the efficiency with respect to diverse parameters such as precision, recall and F-score. CLEF 2016 Challenge task 1 dataset was applied to quantify the investigated technique. The results demonstrated that the investigated technique offered the F-score of 66% using RNN as compared to other algorithm.

3 Research Methodology

This section of the research work is divided into two phases which include dataset description, and second part includes the methodology used for the CER.

3.1 Dataset Description

Annotated Corpus for NER using Groningen Meaning Bank (GMB) corpus for entity classification with enhanced and popular features by NLP applied to the dataset. Essential info about entities:

- geo = Geographical entity
- org = Organization
- per = Person
- gpe = Geopolitical entity
- tim = Time indicator
- art = Artifact
- eve = Event
- nat = Natural phenomenon

Total Words Count = 1,354,149.Target Data Column: "tag". Inspiration: This dataset is getting more interested because of more features added to the recent version of this dataset. Also, it helps to create a broad view of feature engineering with respect to this dataset. Why this dataset is helpful or playful? It might not sound so interested for earlier versions, but when you are able to pick intent and custom named entities from your own sentence with more features, then it is getting interested and helps you solve real business problems (Fig. 2).

Fig. 2 Distribution of length
of each sentence in dataset

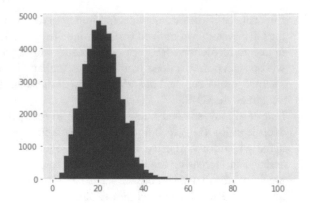

3.2   Proposed Methodology

This research work is based on the CER. The CER techniques have three steps which include feature extraction, boundary detection and classification.

– Feature extraction: In the phase of feature extraction, the whole input data will be converted into the certain set of patterns. The patterns will help us to classify data into certain set of categories. To apply step of feature extraction, the input data needs to be tokenized and needs to apply part of speech tags, and in the last step, we need to extract the context words.
– Boundary detection: In the boundary detection technique, the noun phrases will be removed which are not required. To apply step of boundary detection techniques, inverse document frequency will be applied.
– Entity classification: In the phase of classification, the frequency of the words will be checked, and U-Net approach will be applied for the classification. It is represented by initiating a layer-hopping connection, fusing feature maps in the coding phase to the decoding phase, which is helpful to obtain the details of the segmentation process.

The U-Net method is able to segment the lesion part into a complicated background and still yields high accuracy. The network design has many similarities to that of a convolutional auto encoder, which consists of a contracting trail ("encoder") and an expansive trail ("decoder"). The skip connection between the encoder and the decoder is the basic characteristic of U-Net that differentiates it from a typical auto encoder. Skip connections are responsible for recovering the spatial information lost in down-sampling, which is important for segmentation operations (Fig. 3).

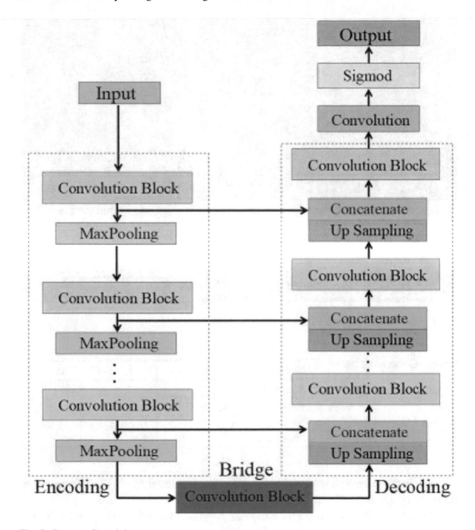

Fig. 3 Proposed model

4 Result and Discussion

This research work is based on the CER. The entity recognition has various phases which include pre-processing, boundary detection and classification. The dataset is collected from the Kaggle for the entity recognition. The phase of pre-processing will tokenize the input data in the next of boundary detection the nouns will be removed from the data. In the last phase of classification, U-net architecture will be applied which will recognize the clinical entity. The performance of the proposed model is analyzed in terms of accuracy, precision and recall.

```
1   # def FUNCTION TO MAKE PREDICTIONS
2   def predictmodeln(model):
3       #y_pred = model.predict_classes(x_test)
4       y_pred = model.predict(x_test)
5       y_pred = np.round(y_pred).astype(int)
6       f,t,thresholds = medtrics.roc_curve(y_test,y_pred)
7       cn = metrics.confusion_matrix(y_test,y_pred)
8       print("Score:", metrics.auc(f,t))
9       print("Classification report:")
10      print(metrics.classification_report(y_test,y_pred))
11      print("Confusion Matrix")
12      print(cn)
```

```
1   # DEFINING THE NEURAL NETWORK
2   model = Sequential()
3   model.add(Dense(256,activation='sigmoid',input_din=29))
4   model.add(Dense(128,activation='sigmoid'))
5   model.add(Dense(64,activation='sigmoid'))
6   model.add(Dense(1,activation='sigmoid'))
```

Fig. 4 Execution of proposed model

Table 1 Comparative results

Models	Accuracy (%)	Precision (%)	Recall (%)
LSTM model	87.78	88.12	88
LSTM + CNN model	92.67	91	91
U-Net model	98	99	99

As shown in Fig. 4, the proposed model is implemented for CER. The proposed model is the U-Net model for the entity recognition.

As shown in Fig. 5, the performance of proposed model is compared with the LSTM and LSTM + CNN models. The proposed model achieves accuracy up to 98% which shows approx which is shown in Table 1. 5% improvement in the results.

5 Conclusion

The biomedical entity recognition models are generally derived for the entity recognition. The entity recognition model is used to generate new entities based on the input data. When analyzed from the previous research work, the deep learning and pattern recognition algorithms are commonly used for the entity recognition. It is analyzed that latest efficient designed algorithm is combination of two deep learning models which has high complexity. In this research work, classification method

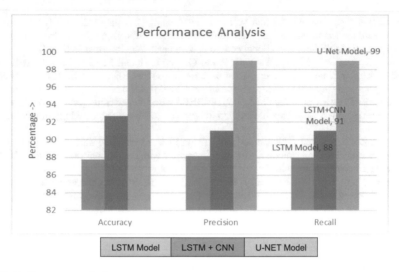

Fig. 5 Performance analysis

will be designed which is combination of three steps which are feature extraction, boundary detection and classification. The model of U-Net is implemented in Python. It analyzed the results in terms of accuracy, precision and recall. The LSTM + CNN model shows accuracy: 92.67%, precision: 91% and recall: 91%, and the proposed model shows accuracy: 98%, precision: 99% and recall: 99% which is approx. 5% improvements in the results for the entity recognition.

References

1. Li L, Zhao J, Hou L, Zhai Y, Shi J, Cui F (2019) An attention-based deep learning model for clinical named entity recognition of Chinese electronic medical records. In: 2015, The second international workshop on health natural language processing (HealthNLP 2019), pp 1–11
2. Bose P, Srinivasan S, Sleeman WC, Palta J, Kapoor R, Ghosh P (2021) A survey on recent named entity recognition and relationship extraction techniques on clinical texts. Appl Sci 2021:1–30
3. Gong L, Zhang Z, Chen S (2020) Clinical named entity recognition from Chinese electronic medical records based on deep learning pretraining. J Healthcare Eng 1–8
4. Ghiasvand O, Kate RJ (2015) Biomedical named entity recognition with less supervision. Int Conf Healthcare Inf 495–495
5. Xing Z, Sun P, Xiaoqun L (2021) Chinese clinical entity recognition based on pre-trained models. In: 2021, international conference on big data analysis and computer science (BDACS)
6. Li L, Hou L (2019) Combined attention mechanism for named entity recognition in chinese electronic medical records. In: 2019, ieee international conference on healthcare informatics (ICHI)
7. Qiu J, Wang Q, Zhou Y, Ruan T, Gao J (2018) Fast and accurate recognition of chinese clinical named entities with residual dilated convolutions. In: 2018, IEEE international conference on bioinformatics and biomedicine (BIBM)

8. Li J, Wei Q, Ghiasvand O, Chen M, Lobanov V, Weng C, Xu H (2021) Study of pre-trained language models for named entity recognition in clinical trial eligibility criteria from multiple corpora. In: IEEE 9th international conference on healthcare informatics (ICHI)
9. Li WY, Song WA, Jia XH, Yang JJ, Wang Q, Lei Y, Huang K, Li J, Yang T (2019) Drug specification named entity recognition base on BiLSTM-CRF model. In: IEEE 43rd annual computer software and applications conference (COMPSAC)
10. Yu X, Hu W, Lu S, Sun X, Yuan Z (2019) BioBERT based named entity recognition in electronic medical record. In: 10th international conference on information technology in medicine and education (ITME)
11. Qiu J, Zhou Y, Wang Q, Ruan T, Gao J (2019) Chinese clinical named entity recognition using residual dilated convolutional neural network with conditional random field. In: IEEE transactions on nanobioscience
12. Liu K, Hu Q, Liu J, Xing C (2017) Named entity recognition in Chinese electronic medical records based on CRF. In: 14th Web information systems and applications conference (WISA)
13. Luu TM, Phan R, Davey R, Chetty G (2018) Clinical name entity recognition based on recurrent neural networks. In: 18th international conference on computational science and applications (ICCSA)

Neural Machine Translation from English to Marathi Using Various Techniques

Yash Jhaveri, Akshath Mahajan, Aditya Thaker, Tanay Gandhi, and Chetashri Bhadane

Abstract Machine translation (MT) is a term used to describe computerized systems that generate translations from one linguistic communication to another, either with or without the need of humans. Text can be used to evaluate knowledge and converting that information to visuals can help in communication and information acquisition. There have been limited attempts to analyze the performance of state-of-the-art NMT algorithms on Indian languages, with a significant number of attempts in translating English to Hindi, Tamil and Bangla. The paper explores alternative strategies for dealing with low-resource hassle in neural machine translation (NMT), with a particular focus on the English-Marathi NMT pair. To provide high-quality translations, NMT algorithms involve a large number of parallel corpora. In order to tackle the low-resource dilemma, NMT models have been trained, along with transformers and attention models, as well as try hands-on sequence-to-sequence models. The data has been trained for sentence limit of 50 words and then fine-tune the default parameters of these NMT models to obtain the most optimum results and translations.

Keywords English · Marathi · Natural language processing · Neural machine translation · Rule-based machine translation · Transformers · Attention models · Seq2Seq models

1 Introduction

Communication has been an integral part of human life ever since the beginning of time. As per Census of India, 2011, Marathi is the third most frequently spoken language in India and ranks 15th in the world in terms of combined primary

---

Yash Jhaveri, Akshath Mahajan, Aditya Thaker, Tanay Gandhi have contributed equally.

Y. Jhaveri (✉) · A. Mahajan · A. Thaker · T. Gandhi · C. Bhadane
DJ Sanghvi College of Engineering, Mumbai, India
e-mail: yashj65@gmail.com

© The Author(s), under exclusive license to Springer Nature Singapore Pte Ltd. 2023
Y. Singh et al. (eds.), *Proceedings of International Conference on Recent Innovations in Computing*, Lecture Notes in Electrical Engineering 1011,
https://doi.org/10.1007/978-981-99-0601-7_3

and secondary speakers [1]. It is spoken by roughly 83 million of the world's 7 billion people [2]. In today's modern world, majority research and other materials are written in English, which is ubiquitously recognized and valued. Existing Marathi documents must be translated into English for them to be universally used. However, manual translation is time consuming and expensive, necessitating the development of an automated translation system capable of performing the task efficiently. Furthermore, there hasn't been much advancement in translating Indian languages. English is a Subject-Verb-Object language, whereas Marathi is Subject-Object-Verb with relatively free word order. Consequently, translating it is a difficult task.

MT refers to computerized systems that generate translations from one linguistic communication to another, either with or while not human involvement. It is a subset of natural language processing (NLP) wherein translation from the source language to the target language is undertaken while conserving the same meaning of the phrase. Furthermore, neural machine translation (NMT) has achieved tremendous progress in recent years in terms of enhancing machine translation quality (Cheng et al. [3]; Hieber et al. [4]). The encoder and decoder, which are commonly based on comparable neural networks of different sorts, such as recurrent neural networks (Sutskever et al. [5]; Bahdanau et al. [6]; Chen et al. [7]), and more recently on transformer networks, make up NMT as an end-to-end sequence learning framework (Vaswani et al. [8]).

The proposed work seeks to improve the English-to-Marathi translations and vice-versa and try to mitigate the low-resource problem. The paper proposes a method to enact translations using the paradigmatic NMT models accompanied by the state-of-the-art models like Sequence2Sequence models, attention and transformer models taking into consideration models like SMT along with rule-based learning as the baselines.

2 Literature Survey

In recent years, NMT has made significant progress in improving machine translation quality. Google Translate [9], Bing Translator [10] and Yandex Translator [11] are some of the most popular free online translators, with Google Translator [9] being one of the most popular locations for machine translation.

The state-of-the-art approaches for machine translation, which includes rule-based machine translation and NMT, have been widely used [12–16]. Rule-based MT primarily connects the structure of given input sentences to the structure of desired output sentences, ensuring that their distinctive meaning is preserved. Shirsath et al. [12] offer a system to translate simple Marathi phrases to English utilizing a rule-based method and a NMT approach, with a maximum BLEU score of roughly 62.3 in the testing set. Garje et al. [13] use a rule-based approach to develop a system for translating simple assertive and interrogative Marathi utterances into matching English sentences. Due to the lack of a large corpus for translation, Govilkar et al. [14] used rule-based techniques to translate only the components of speech for the sentence. The proposed system uses a morphological analyzer to locate root words and then compare the root word to the corpus to assign an appropriate tag. If a word contains more than one tag, ambiguity can be eliminated using grammatical rules. Garje et al. [15] present an online parts of speech (POS) tagger and a rule-based system for translating short Marathi utterances to English sentences. Garje et al. [16] primarily focus on the grammar structure of the target language in order to produce better and smoother translations and employ a rule-based approach to translate sentences, primarily for the English–Marathi pair, with a maximum BLEU score of 44.29. Banerjee et al. [17] specifically focus on the case of English–Marathi NMT and enhance parallel corpora with the help of transfer learning to ameliorate the low-resource challenge. Techniques such as phrase table injection (PTI) have been employed and for augmenting parallel data, pivoting and multilingual embeddings to leverage transfer learning, back-translation and mixing of language corpora are used.

Jadhav [18] has proposed a system where a range of neural machine Marathi translators were trained and compared to BERT-tokenizer-trained English translators. The sequence-to-sequence library Fairseq created by Facebook [19] has been used to train and deduce with the translation model.

In contrast with the NMT model, there has been a quite significant upscale in other models that can be used along with the state-of-the-art NMT models for MT. Vaswani et al. [8] have deduced that when compared to conventional recurrent neural network (RNN)-based techniques, the transformer model provides substantial enhancements in translation quality which was proposed by Bahdanau et al. [6], Cho et al. [20] and Sutskever et al. [5]. Self-attention and absence of recurrent layers can be used alongside state-of-the-art NMT models that enable training quicker and a better performance in the case of absence of a huge corpus for translation.

3   Research Gap

Google Translate [9] mainly uses statistical MT models, parameters of which are obtained through analysis of bilingual text corpora, i.e., sentences that have poor quality text translations. Furthermore, BLEU score of the translation received for sentences less than 15 words is 55.1, and above 15 words is 28.6.

The rule-based technique employed by [12–16] is now obsolete and is being replaced by transformers, deep learning models that employ the mechanism of self-attention. Furthermore, Shirsath et al. [12] have provided a maximum BLEU score of about 62.3 in the testing set using rule-based techniques, whereas the paper has achieved a maximum BLEU score of about 65.29 using the proposed methodology. Govilkar et al. [14] translated only the parts of speech for the sentence using rule based techniques. In order to increase the system's performance, extra meaningful rules must be added. Garje et al. [16] have also used rule-based techniques for translation but have provided a maximum BLEU score of around 49, whereas the paper has achieved a maximum BLEU score of about 65.29 using the proposed methodology. Moreover, the problem with rule-based learning lies with exploring with the incomprehensible grammar, which is on the other hand eliminated by the approach presented by the paper. Newer techniques such as phrase table injection (PTI), back-translation and mixing of language corpora have been applied by Banerjee et al. [17], yet have failed to achieve an adequate BLEU score having used a huge corpus of around 2.5 lakh sentences. From the results from the proposed system of Jadhav [18], it can be observed that the proposed transformer-based model can outperform Google Translation for sentence length up to 15 words but not more than 15 words. This paper, on the other hand, focuses on sentences more than 15 words length and tries to model accurate predictions.

4   Methodology

4.1   Data Used and Data Preprocessing

The dataset used is the parallel corpus data from "https://www.manythings.org/anki/". Processing of around 44486 samples from the dataset has been carried out. The sentences were almost clean, but some preprocessing was required. The special characters, extra spaces, quotation marks and digits in the sentences were removed, and the sentences were lowercase. The paper compares the performance of language translation by restricting the length of the sentences to 15 and 50. The target sentences were prefixed and suffixed by the START and END keywords. The authors padded the shorter sentences after the sentence using the Keras pad_sequences method. The dataset was tokenized using the TensorFlow dataset's SubWordTextEncoder (Table 1).

4.2   Model Architecture

Statistical MT [21] is one of the most widely used techniques in which conditional probabilities are calculated using a bilingual corpus, which is used to reach the

Table 1 Dataset examples

English	Marathi
Could you get me some tea?	मला थोडा चहा आणून देशील का?
I'm doing what I can	मी जे करू शकतो ते मी करतोय
Do you really live alone?	तुम्ही खरच एकटे राहता का?
Tom was also shot in the leg	टॉमला पायातसुद्धा गोळी लागली
I also like cakes	मला केकसुद्धा आवडतात

most likely translation. As a baseline model, SMT model has been employed to convert English sentences to Marathi. This was achieved through a word-based SMT model, trained by calculating the conditional probabilities of Marathi words given an English word, and using it to translate input sequences token by token. Most translation systems are based on this technique but do not achieve precise translations.

In order to tackle this, newer methods like rule-based MT and NMT had been introduced with the most accurate method being NMT. This method employees NLP concepts and includes models like sequence-to-sequence, attention and transformers.

Sequence-to-sequence. RNNs [22] are a type of artificial neural networks, which were one of the first to be used to work with sequential or time series data. RNNs require that each timestep be provided with the current input as well as the output of the previous timestep. Although it stores context from past data in the sequence, it is also prone to vanishing and exploding gradient problems. LSTMs were introduced to overcome this problem, by maintaining forget, input and output gates within each cell, that controls the amount of data which is stored and propagated through the cell.

Sequence-to-sequence (seq2seq) models [23] are a class of encoder–decoder models that are used to convert sentences in one domain to sentences in another domain. This encoder–decoder architecture comprises the encoder block, the decoder block and context vector.

1. Encoder block: This block consists of a stack RNN layer, preferably with LSTMs cells. The outputs of the encoder block are discarded, as the hidden states of the last LSTM cell are used as a context vector and sent to the decoder block.
2. Decoder block: This block consists of the same architecture as that of the encoder block. It is trained for a language modeling task, in the target language taking only the states of the encoder block as input (Fig. 1).

The image above describes the architecture of the encoder–decoder model. During the training phase of the decoder, teach forcing is used, which feeds the model

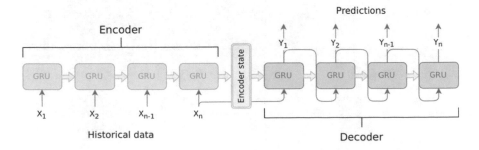

Fig. 1 Seq2Seq [24]

ground truth instead of the output of the previous states. In the testing phase, a <START> token is provided as input to the first cell of the decoder block that marks the start of a sequence, along with the hidden states of the encoder block. The outputs of this cell are used as input to the next cell to make a prediction for the next word. This procedure continues, until the <END> token is generated which marks the end of the sequence. This token is used so that the model can be assured that the sentence translation procedure has finished.

A single RNN layer has been used consisting of LSTM cells for the encoder block and a similar architecture for the decoder block. Embedding layers are used to translate the sentences from words to word vectors before it can be used by the encoder. Another embedding layer is used to convert the outputs of the decoder block into words in target language, after which a softmax function gives a probability distribution over the vocabulary.

Attention. In recent years, NMT problems have found major success using the encoder–decoder framework, which first encodes the source sentence, that is used to generate the translation by selecting tokens from the target vocabulary one at a time. [22, 23]

This paradigm, however, fails on long sentences where the context required to correctly predict the next word might be present at a different position in the sentence which might be forgotten. An attention mechanism is used to refine translation results by focusing on important parts of the source sentences [25] (Fig. 2).

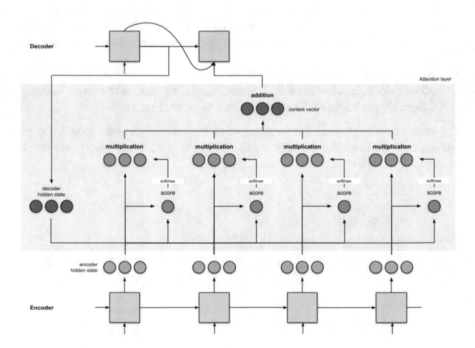

Fig. 2 Illustrated attention [26]

The proposed encoder network consists of three LSTM layers having 500 latent dimensions. On the other hand, the decoder network first has an LSTM that has its initial state set to the encoder state. The attention layer is then introduced that takes the encoder outputs and the outputs from the decoder LSTM. Finally, the outputs from the decoder LSTM and the attention layer are combined and passed through a time-distributed dense layer.

The authors have used the "Teacher Forcing" method to train the network faster. The model was set to train for 40 epochs using the RMSProp optimizer along with sparse categorical cross-entropy loss but observed early stopping after just 22 epochs.

The trained weights are then saved, and an inference model is generated using the encoder and decoder weights to predict and evaluate the translation results. This is done by adding a fully connected softmax layer after the decoder in order to generate a probability distribution over the target vocabulary.

Transformers The work by Ashish Vaswani et al. [8] proposes a novel method for avoiding recurrence and depending solely on the self-attention mechanism. This new architecture is more precise, parallelizable and faster to train (Fig. 3).

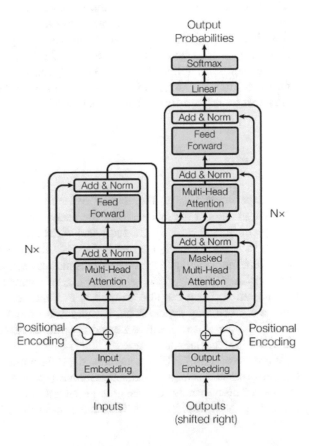

Fig. 3 Transformer architecture [8]

In the transformer model, a stack of six encoders and six decoders is used. The input data is first embedded before it is passed to the encoder or decoder stacks. Because the model lacks recurrence and convolution, the authors injected some information about the relative or absolute positions of the tokens in the sequence to allow the model to use the sequence's order. Positional encoding was added to the input embeddings to achieve this. The positional encodings and embeddings have the same dimension, therefore can be added together.

There are two levels to each encoder. The multi-head attention layer is the initial encoder in the stack through which the embeddings with their positional encoding are passed and subsequently supplied to the feed-forward neural network. The self-attention mechanism uses each input vector in three different ways: the query, the key, and the value. These are transmitted through the self-attention layer, which calculates the self-attention score by taking the dot product of the query and key vectors. To have more stable gradients, this is divided by the square root of the dimensions of key vectors and then supplied to the softmax algorithm to normalize these scores. This softmax score is multiplied by the value vectors, and then the sum of all weighted value vectors is computed. These scores indicate how much attention should be paid to other parts of the input sequence of words in relation to a certain word. Because the self-attention layer is a multi-headed attention layer, the word vectors are broken into a predefined number of chunks and transmitted through various self-attention heads to pay attention to distinct parts of the words. To generate the final matrix, the output of each of these pieces is concatenated and multiplied by the specified weight matrix. This is the final output of the self-attention layer, which is normalized and added to the embedding before being sent to the feed-forward neural network.

5   Results

After experimenting with the number of layers in the model and fine-tuning the hyperparameters of the models used, the paper compares the results of the translations produced using the BLEU score and WER score.

The sacreBLEU score is a metric for assessing the quality of machine translations from one language to another. The link between a machine's output and that of a human is characterized as quality. It was created to evaluate text generated for translation, but it can also be used to evaluate text generated for other natural language processing applications. Its output is often a score between 0 and 100, indicating how close the reference and hypothesis texts are. The higher the value, the better the translations.

Word error rate (WER) computes the minimum edit distance between the human-generated sentence and the machine-predicted sentence. It calculates the number of discrepancies between the projected output and the target transcript by comparing them word by word. The smaller the value, the better the translations.

From the Tables 2, 3, 4, 5, 6, 7 and Fig. 4, it can be observed that the best performing model with respect to SacreBLEU score and WER score metrics is the
transformer model, while the worst performing model is SMT. This is so because
the transformer model keeps track of the various word positions in the sentences
and uses the attention mechanism while the SMT depends upon the probability of
the next word which makes it less accurate and reliable.

Table 2 Comparison of various metrics for various models

Metrics	Sequence-to-sequence	Attention	Transformers	SMT
SacreBLEU score	64.49	61.8	65.29	48.22
WER	1.87	4.0	1.55	3.4

Table 3 Translation result

Input	I really didn't have time
Required	माझ्याकडे खरच वेळ नव्हती।
SMT	वेळ खरच तुमच्याकडे मी
Seq2Seq	माझ्याकडे खरच वेळ नव
Attention	माझ्याकडे खरच वेळ नव्हता
Transformer	माझ्याकडे खरच वेळ नव्हता

Table 4 Translation result

Input	Ive already finished reading this book
Required	हे पुस्तक माझं आधीच वाचून झालं आहे
SMT	वाचत आधीच पुस्तक या हे
Seq2Seq	पुस्तक माझं आधीच वाचून झालं
Attention	मी आधीच हे पुस्तक वाचून काढलं आहे
Transformer	माझं आधीच हे पाच वाजलं आहे

Table 5 Translation result

Input	I don't understand your answer
Required	मला तुमचं उत्तर समजलं नाही
SMT	समजत उत्तर तुमचं मी
Seq2Seq	मचं उत्तर समजलं
Attention	मला तुझं उत्तर समजत नाही
Transformer	मला तुझं उत्तर समजलं नाही

Table 6 Translation result

Input	Did you drink coffee yesterday
Required	काल तू कॉफी प्यायलास का
SMT	केलं तुम्ही पीत कॉफी काल
Seq2Seq	कॉफी प्यायला
Attention	काल तू कॉफी काल केलास का
Transformer	काल तू कॉफी प्यायलीस का

Table 7 Translation result

Input	Whose umbrella is this
Required	ही छत्री कोणाची आहे
SMT	आहे कोणाची या ही
Seq2Seq	छत्री कोणाची
Attention	ही छत्री कोणाची आहे
Transformer	ही छत्री कोणाची आहे

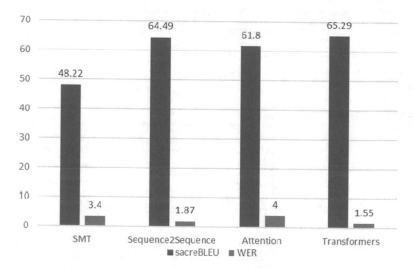

Fig. 4 SMT versus Sequence2Sequence versus attention versus transformer

6   Conclusion

After scrutinizing and implementing different models like Sequence2Sequence, attention models, transformers and SMT, the authors have arrived at the conclusion that after training all mentioned models over a low corpus, the leading fidelity has been obtained by the transformers model. The BLEU Score of about 65.29 and The WER Score of 1.55 state an upper bound on the efficiency of this model. To conclude, the authors did not only mitigate the low-resource problem but also discerned how exactly the translation works and moreover provides almost the exact translations of the given sentence.

References

1.   https://en.wikipedia.org/wiki/List_of_languages_by_total_number_of_speakers.   Last accessed 03 Sept 2022
2.   https://censusindia.gov.in/2011Census/C-16_25062018_NEW.pdf. Last accessed 03 Sept 2022

3. Cheng Y, Xu W, He Z, He W, Wu H, Sun M, Liu Y (2016) Semisupervised learning for neural machine translation. Assoc Comput Linguistics, 1(Long Papers):1965–1974
4. Hieber F, Domhan T, Denkowski M, Vilar D, Sokolov A, Clifton A, Post M (2017) Sockeye: A toolkit for neural machine translation, arXiv preprint, arXiv:1712.05690
5. Sutskever I, Vinyals O, Le QV (2014) Sequence to sequence learning with neural networks. Int Conf Neural Inf Process Syst 2:3104–3112
6. Bahdanau D, Cho K, Bengio Y (2015) Neural machine translation by jointly learning to align and translate. Int Conf Learn Represent
7. Chen MX, Firat O, Bapna A, Johnson M, Macherey W, Foster G, Jones L, Parmar N, Schuster M, Chen Z (2018) The best of both worlds: combining recent advances in neural machine translation. Assoc Comput Linguistics 76–86
8. Vaswani A, Shazeer N, Parmar N, Uszkoreit J, Jones L, Gomez AN, Kaiser Ł, Polosukhin I (2017) Attention is all you need. Adv Neural Inf Process Syst 5998–6008
9. https://translate.google.com/. Last accessed 03 Sept 2022
10. https://www.bing.com/translator. Last accessed 03 Sept 2022
11. https://translate.yandex.com/. Last accessed 03 Sept 2022
12. Shirsath N, Velankar A, Patil R, Dr. Shinde S Various approaches of machine translation for Marathi to English language. ICACC-2021 (vol 40)
13. Garje GV, Bansode A, Gandhi S, Kulkarni A (2016) Marathi to English sentence translator for simple assertive and interrogative sentences. Int J Comput Appl 138(5):0975–8887
14. Govilkar S, Bakal JW, Rathod S (2015) Part of speech tagger for Marathi language. Int J Comput Appl 119(18):0975–8887
15. Garje GV, Gupta A, Desai A, Mehta N, Ravetkar A (2014) Marathi to English machine translation for simple sentence. Int J Sci Res (IJSR) 3(11):3166–3168
16. Garje GV, Kharate GK, Eklahare, Kulkarni NH (2014) Transmuter: an approach to rule-based English to Marathi machine translation. Int J Comput Appl 98(21):0975–8887
17. Banerjee A, Jain A, Mhaskar S, Deoghare S, Sehgal A, Bhattacharyya P (2021) Neural machine translation in low-resource setting: a case study in English-Marathi pair. In: Proceedings of the 18th biennial machine translation Summit Virtual USA, August 16–20, 2021, vol 1. MT Research Track
18. Jadhav SA Marathi To English neural machine translation with near perfect corpus and transformers. Available [Online]: https://www.researchgate.net/publication/339526359_Marathi_To_English_Neural_Machine_Translation_With_Near_Perfect_Corpus_And_Transformers
19. https://fairseq.readthedocs.io/en/latest/. Last accessed 03 Sept 2022
20. Cho K, van Merrienboer B, Bahdanau D, Bengio Y (2014) On the properties of neural machine translation: encoder-decoder approaches. CoRR, arXiv preprint arXiv:1409.1259
21. Brown PF, Cocke J, Pietra Della SA, Pietra Della VJ, Jelinek F, Lafferty JD, Mercer RL, Roossin PS (1990) A statistical approach to machine translation. Comput Linguistics 16(2):79–85
22. Sherstinsky Alex (2020) Fundamentals of recurrent neural network (RNN) and long short-term memory (LSTM) network. Physica D: Nonlinear Phenomena 404:132306
23. Sutskever I, Vinyals O, Le QV (2014) Sequence to sequence learning with neural networks. Adv Neural Inf Process Syst 27
24. https://github.com/sooftware/seq2seq. Last accessed 03 Sept 2022
25. Luong MT, Pham H, Manning CD (2015) Effective approaches to attention-based neural machine translation. arXiv preprint arXiv:1508.04025
26. Karim R (2019). Attn: illustrated attention. Towards data science, on medium, January 20. Accessed 15 May 2022

Review of Cardiovascular Disease Prediction Based on Machine Learning Algorithms

Roseline Oluwaseun Ogundokun, Sanjay Misra, Dennison Umoru, and Akshat Agrawal

Abstract The most common causes of death are cardiovascular diseases (CVDs) every year due to lifestyle which include getting stressed up, unhealthy consumption of food, and many more amongst others. Several researchers have used machine learning (ML) models for the diagnosis of CVD, but it has been discovered that a lot of them still obtain low accuracy, AUC, precision, recall, F1-score, and high false positive rate (FPR). This study therefore aims to review previous research on CVD using ML techniques. The review intends to find out the lapses encounter in this state-of-the-art review and propose a solution to how to solve the challenges aforementioned earlier. This study also examines existing technologies and provides a clear vision of this area to be aware of the utilized approaches and existing limitations in this line of research. To this end, an extensive search was conducted to find articles dealing with various diseases and methods of solving them comprehensively by reviewing related applications. SpringerLink, Talyor and Francis, and Scopus databases were checked for articles on CNN and other machine learning (ML) algorithms on the diagnosis or prediction of cardiovascular diseases. A total of 2606 publications were obtained from the databases. The

R. O. Ogundokun
Department of Multimedia Engineering, Kaunas University of Technology, 44249 Kaunas, Lithuania
e-mail: rosogu@ktu.lt

S. Misra
Department of Computer Science and Communication, Østfold University College, 1757 Halden, Norway
e-mail: sanjay.misra@hiof.no

D. Umoru
Department of Computer Science, Landmark University Omu Aran, Omu Aran, Nigeria

A. Agrawal (✉)
Amity University Haryana, Gurgaon, India
e-mail: akshatag20@gmail.com

Y. Singh et al. (eds.), *Proceedings of International Conference on Recent Innovations in Computing*, Lecture Notes in Electrical Engineering 1011,
https://doi.org/10.1007/978-981-99-0601-7_4

publications were filtered according to the type of methods used as well as the disease examined by the authors. A total of 27 publications were found to meet up with the inclusion criteria. In conclusion, the review of related works hence provided an insight into the future of cardiovascular disease diagnosis. The lapses of the presently used ML models for the diagnosis of the disease were discovered, and recommendations to solve these challenges were presented in the study.

Keywords Convolutional neural network · Cardiovascular disease · Healthcare services · Logistic regression · Machine learning

1 Introduction

Human disease is a disorder of the normal state of a human being that disrupts the normal functions [1, 2]. It always describes the infections and disabilities that could affect any living organism. Diseases can affect people mentally, physically, and even socially [3]. There are four main types of diseases: diseases of illness, inherited diseases (including both genetic and non genetic hereditary diseases), infectious diseases, and physiological diseases [4]. It can be classified into communicable diseases and diseases that cannot be communicated. The deadliest human diseases are cardiovascular diseases, followed by communicable diseases in terms of nutritional deficiencies, cancer, and chronic respiratory infections [4]. There are many forms of coronary heart disease, also known as ischemic heart disease, which is cardiovascular disease, the leading cause of death globally [4]. The description and use of convolutional neural networks (CNNs) prediction are majorly used in image classification [5]. CNN is an effective algorithm for the identification of patterns and image processing [6]. The CNN can be used to classify the electrocardiogram (ECG) images of the heart into various types of heart conditions.

Logistics regression is a statistical method for evaluating a dataset that includes one or more independent variables that decide the outcome [7]. Also, like other regression analyses, logistics regression is an analysis for prediction [8]. E-health services are trends in today's world technology. This applies to health services and information provided by IT such as the internet [9]. It is being utilized by almost everyone in the whole world ranging from products from big companies that provide health services in all their new technologies like Apple, Google, and others. It is an upcoming and trending field of medical and public health facilities [10].

This study, therefore, aims to conduct a systematic literature review on cardiovascular disease prediction based on the use of machine learning algorithms. The lapses with the present ML models used for the diagnosis of CVD were also presented in this study.

The remaining part of the article is pre-structured as thus: Sect. 2 presents the methods and materials used for the SLR. Section 3 presents the results from the SLR, while in Sect. 4, the result obtained is discussed and interpreted. The study is concluded in Sect. 5.

Table 1 Relevance of the formulated RQ

S/N	Research Questions	Relevance of the RQ
1.	What are the ML algorithms used for the diagnosis of CVD?	There are lots of ML algorithms that exist, it is relevant to know the ones that have been mostly focused on and their weakness then we can make suggestions of how to solve those weaknesses discovered
2.	What are the cardiovascular diseases that are mostly examined	It is very important to know the CVD that researchers have mainly focused on in the past and the reason behind it
3.	What are the ML accuracies obtained in previous research	Knowing the accuracies previously obtained will assist researchers in how to focus their investigation and also assist in how to make recommendations to researchers on how to improve ML accuracies

2   Methodology

Research Questions

The study formulated some research questions (RQ) for this SLR. They are as follows:

RQ 1: What are the ML algorithms used for the diagnosis of CVD?

RQ 2: What are the cardiovascular diseases that are mostly examined?

RQ 3: What are the ML accuracies obtained in previous research? (Table 1).

2.1   Method

The method used in this study is the PRISMA SLR method as shown in Fig. 1. A total of 2606 publications were obtained from three databases (SpringerLink, Taylor and Francis, and Scopus). A total of 204 duplicates were eliminated from the databases. A total of 2402 articles were examined for further elimination using abstracts, keywords, and titles. A total of 2336 publications were excluded using abstracts, keywords, and titles leading to 66 articles remaining for full-text download. The 66 articles full-text were downloaded and considered for eligibility out of which 11 were excluded due to unavailability of the full text. A total of 32 of the 55 articles downloaded for inclusion in the qualitative synthesis review were excluded because the article does not focus on CVD diagnosis and ML techniques. A total of 27 articles were later included for synthesis review (23 articles plus an extra 4 papers located when other databases were manually searched).

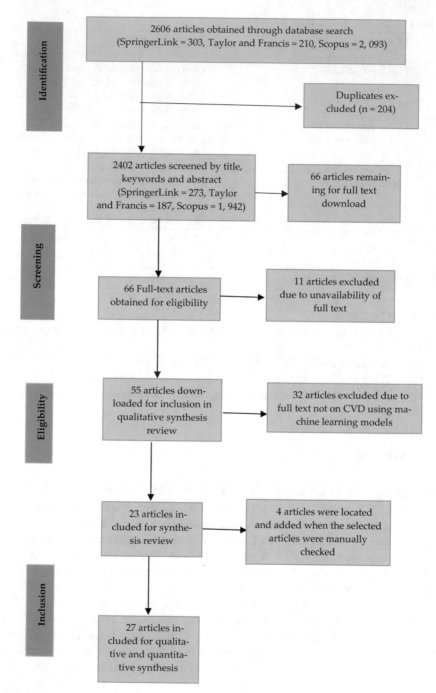

Fig. 1 PRISMA flow diagram for the SLR

Table 2 Articles obtained from each database

Databases	Publications
SpringerLink	303
Taylor and Francis	210
Scopus	2093
Total	2606

2.2 Data Sources and Searches

The study searched three databases which are SpringerLink, Taylor and Francis, and Scopus databases and obtained a total of 2606 publications as seen in Table 2. The study search string was cardiovascular, deep learning, machine learning, and diagnosis as shown in Table 3. The search included publications on CNN algorithms and other disease-related algorithms. According to the types of techniques applied between 2018 and 2022, the articles that were retrieved were filtered. Convolutional neural networks and search phrases associated with cardiac disorders were combined. Before making a choice, the complete text and, if necessary, extra materials (tables) were studied for papers lacking abstracts or without enough information in the abstract to make a choice. The databases citing references were manually searched for any additional potentially suitable research in the reference lists of the included studies.

2.3 Study Selection

Studies are required to be original to qualify for inclusion. The accuracy of the supplied indicators for evaluating predictive power may be considered. These measures are now more widely acknowledged and employed in the evaluation of incremental predictive capability. Studies only involving people were comprised.

Table 3 Databases search strings

Databases	Search string
SpringerLink	Cardiovascular AND Deep Learning OR Machine Learning AND Diagnosis
Talyor and Francis	Talyor and Francis: [All: cardiovascular] AND [All: disease] AND [All: deep] AND [[All: learning] OR [All: machine]] AND [All: learning] AND [All: diagnosis] AND [All Subjects: Medicine, Dentistry, Nursing & Allied Health] AND [Article Type: Article] AND [Publication Date: (01/01/2018 TO 12/31/2022)]
Scopus	Cardiovascular AND Deep Learning OR Machine Learning AND Diagnosis AND (LIMIT-TO (PUBYEAR, 2022) OR LIMIT-TO (PUBYEAR, 2021) OR LIMIT-TO (PUBYEAR, 2020) OR LIMIT-TO (PUBYEAR, 2019) OR LIMIT-TO (PUBYEAR, 2018)) AND (LIMIT-TO (SUBJAREA, "COMP")) AND (LIMIT-TO (DOCTYPE, "ar"))

Using the qualifying criteria provided in linked works, all of the authors assessed the citations and abstracts found through electronic literature searches. The ability of statistical models to forecast illnesses was assessed in the eligible publications, which were all published in English. Because several predicting elements that, in some way or another, are connected to neural networks have been generated using algorithms. There was no restriction on studies by diagnosis among medical populations. Excluded studies were studies based on psychiatric, surgical, and pediatric populations because datasets and results were not feasible.

2.4 Data Extraction and Quality Assessment

Data were gathered using standard risk variables as well as research populations, locations, and methodologies. Studies that were extracted and concentrated on for the examination of discrimination quality assessment showed an accuracy of over 70%. Researchers calculated the percentage of research in which they asserted an improvement in prediction when interpreting their results. Administrative, primary (such as survey, chart review), or both types of data were available. Other operational parameters including sensitivity, specificity, and predictive values are abstracted if the accuracy was not given. First and foremost, systematic information on sampling, research design, outcome criterion formulation, and statistical analysis were taken from all of the chosen studies. Then, to guarantee correct study inclusion, the aforementioned inclusion criteria were used once again.

2.5 Data Synthesis

A narrative synthesis of the available data was undertaken due to the vast variety of indicators utilized to evaluate the incremental predictive power of algorithms. Too many different types of research were included, concentrating on the populations where the model has been tested, the kinds of variables used in each model, and model discrimination.

2.6 Inclusion and Exclusion Criteria

In these reviews, we used some criteria for the elimination and inclusion of articles for this SLR. The inclusion criteria are if (i) The publication is written in English, (ii) the publication is on cardiovascular diseases, (iii) the publication used machine learning algorithms for diagnosis or detection, and (iv) the publication is a journal article.

The exclusion criteria are if (i) the publication does not focus on cardiovascular disease, (ii) the publication did not use machine learning algorithms, (iii) the publication is not written in English, (iv) the publication is a conference, preprint, chapter of books, magazine publications.

3 Results

Citations found through searches in the ScienceDirect, Research Gate, and Scopus databases were used to choose the study. A total of 128 abstracts were chosen for further screening, and 74 full-text publications were examined. After all exclusions, 54 publications covering 50 studies were included because they satisfied the qualifying requirements. The table in the connected works and its data summary includes information from research that evaluated the precision of prediction models. The following factors were taken into consideration when choosing the datasets: physical activity, dietary habits, smoking status, alcohol use, body mass index, height, weight, waist circumference, hip circumference, family history of diabetes, ethnicity, blood pressure, and/or antihypertensive medication use and steroid use. Occasionally, biological components were present (blood glucose, triglyceride, lipid variables, or uric acid levels). The RQs are answered in the sub-sections as follows:

RQ 1: What are the ML Algorithms Used for the Diagnosis of CVD?
The ML methods used by the authors of the included publications are discussed here as follows, and those methods can be seen in Table 4.

Kalaivani et al. [11] a hybridized Ant Lion Crow Search Optimization Genetic Algorithm (ALCSOGA) to accomplish operative feature selection. A combination of Ant Lion, Crow Search, and Genetic Algorithm is used in this optimization. The elite place is decided by the Ant Lion algorithm. While the Crow Search Algorithm makes use of each crow's position and memory to assess the goal function.

Singhal et al. [12] proposed a new CNN-based heart disease prediction model and compared it to existing systems, and the prediction accuracy of the prediction system is 9% relative higher than theirs. They aimed at developing a medical diagnosis system that can predict heart diseases and use the backpropagation mechanism to train datasets in CNN.

Harini and Natesh [13] gathered and examined an enormous volume of information for prediction since there were no advancements thereby using statistical knowledge, they determined the major chronic diseases and used the CNN algorithm. They also compared the prediction of risk of multimodal disease (CNN-MDRP) algorithm used with structured and unstructured hospital data, and this is 94.8% higher than the CNN-based unimodal disease risk prediction algorithm (CNN-UDRP).

Table 4 Methods used in previous research works

Authors	Methods
Kalaivani, Maheswari and Venkatesh	Ant Lion Crow Search Optimization Genetic Algorithm (ALCSOGA)
Harini and Natesh	CNN
Abdelsalam and Zahran	SVM
Cinar and Tuncer	Hybrid AlexNet-SVM
Muhammad et al.	Support vector machines, K-nearest neighbors, random trees, Naive Bayes, gradient boosting, and logistic regression
Subhadra and Vikas	Backpropagation algorithm
Kadam Vinay, Soujanya, and Singh	Artificial neural network (ANN)
Wahyunggoro, Permanasari, and Chamsudin	ANN
Sadek et al.	ANN with backpropagation
Ogundokun et al.	PCA with random forest, SVM, decision tree
Krishnan and Kumar	CNN
Thiyagarajan, Kumar, and Bharathi	Transductive extreme learning machine (TELM)
Durai, Ramesh, and Kalthireddy	J48 algorithm
Gujar et al.	Naïve Bayes
Gawande and Barhatte	CNN
Dami and Yahaghizadeh	Long short-term memory (LSTM), deep belief network (DBN)
Ogundokun et al.	Autoencoder, KNN, and DT
Bhaskaru and Devi	Hybrid differential evolution fuzzy neural network (HDEFNN)
Elsayed, Galal, and Syed	Logistic regression model

Abdelsalam and Zahran [14] offered some specific unique methods for DR early detection based on multifractal geometry to detect early non-proliferative diabetic retinopathy, and the macular optical coherence tomography angiography (OCTA) pictures are examined (NPDR), automating the diagnosing process with supervised machine learning techniques like the support vector machine (SVM) algorithm and increasing the resulting accuracy. The accuracy of the categorization method was 98.5%.

Cinar and Tuncer [15] proposed a method for the classification of normal sinus rhythm (NSR), abnormal arrhythmia (ARR), and congestive heart failure (CHF). ECG data using the highly accurate and well-liked deep learning architecture is discussed. Hybrid AlexNet-SVM is the foundation of the suggested architecture (support vector machine). There are 192 ECG signals in all, including 96 arrhythmia, 30 CHF, and 36 NSR signals. ARR, CHR, and NSR signals are initially categorized by SVM and KNN methods, reaching 68.75% and 65.63% accuracy, to show the classification performance of deep learning architectures. The signals are then categorized with an accuracy of 90.67% using long short-time memory

(LSTM). The hybrid AlexNet-SVM technique is used to apply to the photos and achieve 96.77% accuracy by collecting the spectrograms of the signals. The findings demonstrate that compared to traditional machine learning classifiers, the proposed deep learning architecture more accurately categorizes ECG data.

Muhammad et al. [16] used diagnostic CAD data gathered from the two general hospitals in Kano State, Nigeria, to construct machine learning prediction models for CAD. The dataset was used to train predictive models using machine learning techniques like support vector machines, K-nearest neighbors, random trees, Naive Bayes, gradient boosting, and logistic regression. The models were then assessed based on performance metrics like accuracy, specificity, and sensitivity using receiver operating curve (ROC) techniques.

Subhadra and Vikas [17] proposed a system implementing the concept of multilayered neural networks processed with clinical attributes thus improving the accuracy of disease diagnosis and prediction and, for effective prediction, a backpropagation algorithm was used to train the data.

Kadam Vinay et al. [18] built a new model for disease prediction with the use of predicting artificial neural network (ANN) diseases in terms of probabilistic modeling. With the concern of obtaining relevant health information data for prediction from a large hospital information database, they got the prediction accuracy and produced a confusion matrix for that data. The model gives the 95%, 98%, and 72% accuracy for predicting heart, kidney, and diabetes diseases.

Wahyunggoro et al. [19] proposed a model for prediction by using the ANN model based on datasets thereby forecasting diseases because it is a problem in terms of not covering all diseases and rare conditions. A case study of the occurrence of the number of diseases was used to compare different neural network techniques and then select the best outcome among them.

Sadek et al. [20] worked toward helping doctors know what causes Parkinson's disease in terms of earlier detection. They implemented an ANN system with a backpropagation algorithm, and then it was discovered that the symptoms come slowly and are difficult to diagnose.

Ogundokun et al. [21] proposed a PCA feature extraction technique with three classification ML methods. The methods were used to diagnose heart disease. It was discovered that the system had a low accuracy of 56.86%, but the detection rate of 98.7% was on the high side.

Krishnan and Kumar [22] proposed a big data concept of both structured and unstructured datasets using the CNN definition, multimodal disease risk prediction (CNN-MDRP) based on CNN. Due to machine learning techniques, it is not being able to handle the difficulties of little and incomplete data in disease prediction, focusing on both data prediction and big data analytics.

Thiyagarajan et al. [23] classified the output data of diabetes to improve accuracy thereby overcoming traditional methods of algorithms thereby implementing the transductive extreme learning machine (TELM) based on testing time and accuracy. The precision and execution time for the proposed TELM show high performance and thus making it a good option for the diagnosis process to classify diabetes data.

Durai et al. [24] aimed at predicting the occurrence of liver diseases based on the unhealthy lifestyles of patients. By using the J48 algorithm to make decisions on the prediction, they were able to get meaningful results from large datasets.

Gujar et al. [25] used the Naïve Bayes algorithm, which takes symptoms to predict diseases based on user input information. They aimed at analyzing clinical documents about patients' health to predict the possibility of occurrence of any diseases.

Gawande and Barhatte [26] aimed at getting reliable information from biomedical signals because of problems in consistencies and location of source or transmission of the cardiac electrical impulse and then proposed a system in which an ECG signal is given as input; then segmentation is done to analyze the signal and output is trained in one-dimensional CNN in a particular format. The obtained accuracy is nearly 99.46%.

Dami and Yahaghizadeh [27] proposed a 5 min electrocardiogram (ECG) recording and the extraction of time–frequency characteristics from ECG signals which were utilized to predict arterial events a few weeks or months in advance of the occurrence using a deep learning technique. The LSTM neural network was utilized to take into account the potential for learning long-term dependencies to swiftly recognize and avoid these situations. To choose and represent the recorded dataset's most useful and efficient characteristics, a deep belief network (DBN) was also utilized. LSTM-DBN is the short name for this method.

Ogundokun et al. [28] proposed the use of two computational intelligence techniques. They employed the use of autoencoder feature extraction algorithms with decision trees and K-nearest neighbor (KNN). The algorithms' performance was evaluated using a heart disease dataset from the Nation Health Service database. They obtained an accuracy of 56.19%.

Bhaskaru and Devi [29] proposed an accurate detection and prediction of heart diseases since medical researchers have found most results inaccurate. Hybrid differential evolution fuzzy neural network (HDEFNN) was implemented by improving the initial weight updating of the neural networks and can perform well without retraining.

Elsayed et al. [30] proposed a mobile application system that predicts coronary heart diseases based on risk factors implemented using the logistic regression model.

RQ 2: What are the Cardiovascular Diseases that are Mostly Examined?
There are several CVDs that have been examined in previous research, and a few of them were investigated in the publications included in this review. They include heart disease [12, 21, 28, 29, 30], diabetic cases [14], heart, kidney, and diabetes diseases [18], Parkinson's disease [20], liver diseases [24], and so on.

It can be deduced that many researchers worked on heart disease diagnosis. It is suggested that more investigation should be carried out on the use of ML techniques on other types of CVD to detect them early and reduce the death rate in the world.

Table 5 Authors and the accuracies obtained by them

Authors	Year	Accuracies (%)
Harini and Natesh	2018	94.8
Kadam Vinay, Soujanya, and Singh	2019	95, 98, and 72
Gawande and Barhatte	2018	99.46
Ogundokun et al.	2022	56.86
Cinar and Tuncer	2021	96.77
Ogundokun et al.	2021	56.19
Gawande and Barhatte	2018	99.46
Abdelsalam and Zahran	2021	98.5

RQ 3: What are the ML Accuracies Obtained in Previous Research?

The ML accuracies obtained from the review can be seen in Table 4 and have been precisely shown in Table 4 as well. Although some research did not present the accuracies of their proposed system, few of them that did can be seen in Table 5.

4 Discussions

CNN is mostly used, and accuracy depends on the number of layers and algorithms supporting them. Only eight related works out of the existing publication included in this review gave a straightforward accuracy. The methods consist of a series of neural networks and algorithms that support them while some are not supported. Average accuracy is being computed as per various accuracies derived in every article. Discovered in these reviews is the fact that neural networks used are being supported and then used in an intertwined manner in an approach to get high accuracy. The strengths and limitations are discussed as follows:

The removal of studies that merely provided effect estimates for the independent associations of algorithms with the outcome, such as risk ratios or relative risk, is one of this review's strengths. The justification for predictive testing is dependent less on the size of the risk ratio and more on how helpful the test findings are in enhancing illness prognosis. The restrictions stem from the fact that only English-language publications were included, making it impossible to access research that was conducted in a variety of languages also did not formally evaluate the potential for publication bias. Unpublished research, however, is more likely to have results that are discouraging; thus, they would likely reduce rather than exaggerate estimates of the predictive utility of algorithms (i.e., those reporting no improvement in the predictive accuracy).

The number of publications per year can be seen in Fig. 2, and it was discovered that 30% of the publications were published in the year 2018, 26% were published in the year 2019, 15% in the year 2020, 19% in the year 2021, and lastly 11% articles were published in the year 2022.

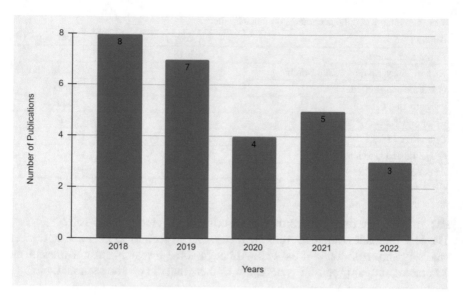

Fig. 2 Number of publication per year

5 Conclusion

More implementation studies should be carried out to improve the uptake of existing risk prediction models and, consequently, to assess their effects on the implementation and outcomes of diabetes prevention and control programs, given that
existing prediction tools based on established risk factors are already achieving
acceptable-to-good discrimination. The hybridization of algorithms such as ML
and deep learning models can be hybridized together. Various researches have
attempted to combine various techniques to boost accuracy in terms of:

a. Implementation of various characteristics of CNN in combination with other
   algorithms.
b. Some combinations appear to be individually executed and merged at a perfect
   point at the end of the implementation.
c. Tackled accuracy of results based on datasets used and therefore giving useful
   information about it.
d. Also tackled the amount of time of executions hoping that someday it can be
   improved on.
e. Most algorithms tackled the raw datasets based on patients' information.

A combination of algorithms could be the next move in the improving fastness of
result outputs in the world today. This proposed study will also tackle the accuracy of the combination of methods and would therefore provide medical professionals the aid in the prediction of various heart diseases. This study proposes the

provision of efficiency in the utility and also looks forward to reducing the rate of cardiovascular diseases in the nearest future.

References

1. Ogundokun RO, Misra S, Sadiku PO, Gupta H, Damasevicius R, Maskeliunas R (2022) Computational intelligence approaches for heart disease detection. In: Recent innovations in computing. Springer, Singapore, pp 385–395
2. Adam F (2019) Cardiovascular disease: types, symptoms, prevention, and causes. Retrieved 4 Jan 2020, from https://www.medicalnewstoday.com/articles/257484.php
3. Ahmed A, Hannan SA (2012) Data mining techniques to find out heart diseases: an overview. (Sem Qualis) Int J Innovative Technol Explor Eng (IJITEE), 1(4):18–23.
4. Ananthalakshmi AV (2017) Effective diagnosis of diabetes mellitus using neural networks and its hardware implementation on FPGA. 15(1):519–528
5. Aronson JK (2009) Medication errors: definitions and classification. Br J Clin Pharmacol 67:599–604. https://doi.org/10.1111/j.1365-2125.2009.03415.x
6. Bhaskaru O, Devi MS (2019) Accurate and fast diagnosis of heart disease using hybrid differential neural network algorithm 3:452–457.
7. Buttar HS, Li T, Ravi N (2005) Prevention of cardiovascular diseases: role of exercise, dietary interventions, obesity, and smoking cessation. Experiment Clin Cardiol
8. Chandrayan, P (2018) Logistic regression for dummies: a detailed explanation. Retrieved 9 Dec 2019, from https://towardsdatascience.com/logistic-regression-for-dummies-a-detailed-explanation-9597f76edf46
9. Christine T (2018) Structured vs. unstructured data. Retrieved 4 Jan 2020, from https://www.datamation.com/big-data/structured-vs-unstructured-data.html
10. Cochran, Q. (n.d.). Logistic regression. Retrieved January 4, 2020, from https://www.medcalc.org/manual/logistic_regression.php
11. Kalaivani K, Uma Maheswari N, Venkatesh R (2022) Heart disease diagnosis using optimized features of hybridized ALCSOGA algorithm and LSTM classifier. Network: Comput Neural Syst 1–29
12. Singhal S, Kumar H, Passricha V (2018) Prediction of heart disease using CNN. Am Int J Res Sci Technol Eng Math 23(1):257–261
13. Harini DK, Natesh M (2018) Prediction of probability of disease based on symptoms using machine learning algorithm 392–395
14. Abdelsalam MM, Zahran MA (2021) A novel approach to diabetic retinopathy early detection based on multifractal geometry analysis for OCTA macular images using a support vector machine. IEEE Access 9:22844–22858
15. Çınar A, Tuncer SA (2021) Classification of normal sinus rhythm, abnormal arrhythmia, and congestive heart failure ECG signals using LSTM and hybrid CNN-SVM deep neural networks. Comput Methods Biomech Biomed Engin 24(2):203–214
16. Muhammad LJ, Al-Shourbaji I, Haruna AA, Mohammed IA, Ahmad A, Jibrin MB (2021) Machine learning predictive models for coronary artery disease. SN Comput Sci 2(5):1–11
17. Subhadra K, Vikas B (2019) Neural network-based intelligent system for predicting heart disease. Int J Innovative Technol Explor Eng 8(5):484–487
18. Kadam Vinay R, Soujanya KLS, Singh P (2019) Disease prediction by using deep learning based on patient treatment history. Int J Recent Technol Eng 7(6):745–754
19. Wahyunggoro O, Permanasari AE, Chamsudin A (2019) Utilization of neural network for disease forecasting. گنهرف و نز, 1(4):53
20. Sadek RM, Mohammed SA, Abunbehan ARK, Ghattas AKHA, Badawi MR, Mortaja MN, Abu-Naser SS (2019) Parkinson's disease prediction using artificial neural network 3(1):1–8. Retrieved from http://dstore.alazhar.edu.ps/xmlui/handle/123456789/302.

21. Ogundokun RO, Misra S, Awotunde JB, Agrawal A, Ahuja R (2022) PCA-based feature extraction for classification of heart disease. In: Advances in electrical and computer technologies. Springer, Singapore, pp 173–183
22. Krishnan D, Kumar SB (2018) A survey on disease prediction by machine learning over big data from healthcare motivation: IOSR. J Eng 08(10):2278–8719
23. Thiyagarajan C, Kumar KA, Bharathi A (2018) Diabetes mellitus diagnosis based on transductive extreme learning machine 15(6):412–416
24. Durai V, Ramesh S, Kalthireddy D (2019) Liver disease prediction using machine learning 5(2):1584–1588
25. Gujar D, Biyani R, Bramhane T, Bhosale S, Vaidya TP (2018) Disease prediction and doctor recommendation system 3207–3209
26. Gawande N, Barhatte A (2018) Heart diseases classification using a convolutional neural network. In: Proceedings of the 2nd international conference on communication and electronics systems, ICCES 2017, 2018-Janua (June), 17–20. https://doi.org/10.1109/CESYS.2017.8321264
27. Dami S, Yahaghizadeh M (2021) Predicting cardiovascular events with a deep learning approach in the context of the internet of things. Neural Comput Appl 33(13):7979–7996
28. Ogundokun RO, Misra S, Sadiku PO, Adeniyi JK (2021) Assessment of machine learning classifiers for heart diseases discovery. In: European, mediterranean, and middle eastern conference on information systems. Springer, Cham, pp 441–452
29. Bhaskaru O, De Harini vi MS (2019) Accurate and fast diagnosis of heart disease using hybrid differential neural network algorithm 3:452–457.
30. Elsayed HAG, Galal MA, Syed L (2018) HeartCare+: a smart heart cares mobile application for Framingham-based early risk prediction of hard coronary heart diseases in the middle east. Mobile Inf Syst 2017

Exploiting Parts of Speech in Bangla-To-English Machine Translation Evaluation

Goutam Datta, Nisheeth Joshi, and Kusum Gupta

Abstract Machine translation (MT) converts one language to another automatically. One of the major challenges of MT is evaluating the performance of the system. There are many automatic evaluation metrics available these days. But the results of automatic evaluation metrics are sometimes not reliable. In this paper, we have attempted to address this issue by considering another type of evaluation strategy, i.e., syntactic evaluation in Bangla-to-English translation. We have attempted to address the problems of automatic evaluation metric BLEU and, thereby, how syntactic evaluation could be helpful in achieving higher accuracy is discussed. In our syntactic evaluation, we have exploited the use of parts of speech (POS) during computing evaluation scores. A comparative analysis is done on different types of evaluations such as syntactic, human, and automatic on a low-resourced English–Bangla language pair. A correlation indicates syntactic evaluation score correlates more with the human evaluation score compared to the normal BLEU score.

Keywords Syntactic evaluation · BLEU · Human evaluation

G. Datta (✉) · N. Joshi · K. Gupta
School of Mathematical and Computer Science, Banasthali Vidyapeeth, Banasthali,
Rajasthan, India
e-mail: gdatta1@yahoo.com

N. Joshi
e-mail: jnisheeth@banasthali.in

K. Gupta
e-mail: gupta_kusum@yahoo.com

G. Datta
Informatics, School of Computer Science and Engineering, University of Petroleum and Energy
Studies, Dehradun, India

Y. Singh et al. (eds.), *Proceedings of International Conference on Recent Innovations
in Computing*, Lecture Notes in Electrical Engineering 1011,
https://doi.org/10.1007/978-981-99-0601-7_5

1   Introduction

When developing machine translation (MT) systems that automatically convert the source to the target language, the correct evaluation of such automatic translation is a crucial task. MT has gone through different stages such as dictionary-based, rule-based, statistical, phrase-based, hybrid, and most recently neural-based [1–8]. The most popular method for MT evaluation is reference-based, which involves comparing the system output to one or more human reference translations. The majority of MT evaluation methods in use produce a computer-generated absolute quality score. The simplest scenario is comparing the similarity of translations suggested by machines (hypothesis sentence) to the human translation (reference sentence). To do this, word n-gram matches between the translation (MT output) and the reference are counted. This is the situation with BLEU [18], which has long served as the benchmark for MT evaluation. In MT evaluation, human judgment is considered to be the best, but it is time-consuming. The features used by the human evaluators are captured and applied to the automatic evaluation metrics by training with a couple of reference translations (generated by human evaluators) and hypothesis translations (translation generated by MT). How the human judgment is done when two hypotheses are given for a particular reference translation for choosing the best translation for ranking is used in automatic evaluation metrics. Such type of MT evaluation is done on the basis of ranking, i.e., without generating a direct score, we have multiple MT engines and their translation quality needs to be measured[9]. Researchers also focused on the syntactic evaluation of the MT system where they included syntactic information in the MT evaluation [10]. Apart from BLEU, some of the widely used automatic evaluation metrics are METEOR [11], National Institute of Standard and Technology (NIST) [12], Word Error Rate (WER) [13], etc. Evaluating the performance of the MT system is very important to developing a better system. Researchers are quite active in this area for a long and have explored different techniques such as precision-based, recall-based, parts-of-speech (POS)-based, neural network-based, etc. to enhance the evaluation accuracy of MT engines [14, 15]. In this paper, we attempted to evaluate the performance of two online neural-based translation systems: Google and Bing with the most popular metric BLEU in conjunction with the syntactic evaluation. In syntactic evaluation, we have taken parts of speech of the sentences (reference and hypothesis sentences) with the same approach as BLEU uses to evaluate.

The remaining part of the paper is presented as follows: Sect. 2 highlights some previous work in MT evaluation; Sect. 3 briefs methodology; Sect. 4 elaborates on our experimentation; Sect. 5 includes results and discussion. Finally, Sect. 6 has a conclusion and future direction.

2 Related Work

Researchers explored new MT evaluation metrics by exploiting the dependency parse tree. They obtained this dependency parse tree with an appropriate German parser. The reason for selecting this specific parser is such a type of parser is capable of handling ungrammatical and ambiguous data. The authors claim that their new syntactic evaluation approach correlates well with human judgment [16].

In the automatic MT evaluation process to enhance the accuracy, researchers explored a neural network-based approach. Syntactic and semantic levels of information, along with lexical information from reference and hypothesis sentences, are captured and represented in the word-embedding form before training the neural network. Researchers experimented with a benchmark dataset of WMT metrics and achieved the best result [14].

Researchers also explored the comparative analysis of MT evaluation of two frameworks such as phrase-based statistical MT (PBSMT) and neural machine translation (NMT) on English to Hindi and back translation where the translation was carried out on terminologies [17].

The most popular and widely accepted automatic evaluation metric is BLEU [18]. BLEU and another automatic metric NIST both exploit precision-based computation to generate scores. On the other hand, METEOR exploits both precision and recall.

In the NMT design, hyper-parameter plays a very vital role [19–21]. In the paper [22], researchers evaluated the performance of the transformer-based NMT model on low-resource English–Irish language pair. They performed a human evaluation to explore how the model performance varies with various hyper-parameter tunings.

There is an interesting paper on the language model's evaluation with refinement at a targeted syntax level, where the authors' proposed metric is developed with standard American English corpora [23].

3 Methodology

We collected the corpus from TDIL. Our English-to-Bangla low-resourced corpus is from the Indian Tourism domain. In our experimentation, we randomly picked a couple of Bangla sentences as the source sentence, and its corresponding English sentence is considered to be a reference sentence. We used Google and Bing as the MT engines. We computed the translation score of both the MT engines. BLEU score with $n = 1$ to $n = 4$ is computed. Further, we also attempted to compute the score syntactically, i.e., after tokenization we used the Bleus approach, but the tokens in the hypothesis sentence and reference sentence are taken as parts

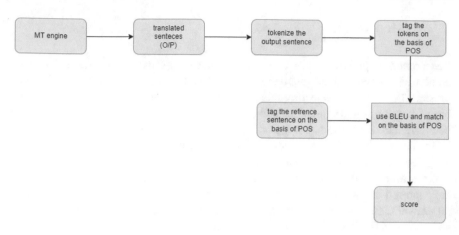

Fig. 1 Block diagram of our methodology

of speech during matching. In the paper, researchers used POS with BLEU for other language pairs [15]. BLEU score with parts of speech we named as BLEU with POS. We have tried to find the correlation of both the scores (BLEU and BLEU+POS) with the human score. Further, we compute the Pearson correlation and found BLEU with POS has a higher correlation with human scores compared to normal BLEU. Figure 1 is a diagrammatic representation of our methodology. The process of computing a human score is as follows: The source sentence is Bangla, and translated sentence is English; both are shown to five human evaluators who have expertise in both these languages: English and Bangla. On the basis of two important parameters such as adequacy and fluency, the human evaluators are asked to evaluate.

Adequacy means completeness, i.e., whether the translated sentence is complete or not that is being evaluated. Fluency ensures the correctness of the grammar of the translated text. Adequacy and fluency both are measured on a scale from 1 to 5. If all the meanings of the translated words are correct, then a full score, i.e., 5, is assigned to adequacy. If most of the meaning of the translated word is correct, then 4. In case of much meaning assigned score is 3. In the case of words that have little meaning, score is 2 and with no meaning, the score is 1. In a similar fashion, fluency scores are also decided. If the translated sentence is incomprehensible, then the score is 1, flawless translation has a score of 5, good language has 4, non-native language has 3, and if the translated sentence is disfluent, then the score is 2. On the basis of both adequacy and fluency scores, the average score is computed. And this computed average score is the human score. Next correlation is computed to figure out which automatic metric has a higher correlation with the human judgment score.

4 Experimentation

As illustrated before in Sect. 3, we have used the English-to-Bangla tourism dataset collected from TDIL. However, we have picked only a few Bangla sentences randomly from these datasets and fed them into two online MT engines Google and Bing. The same sentences and the translated sentences produced by two translators were shown to five human evaluators. In this section, we have presented our entire experimentation process with a single sentence, i.e., sentence 1. This sentence 1 is a Bangla sentence picked randomly from our dataset. Its corresponding Bangla sentence is used as a reference sentence. The entire steps of our experimentation are already mentioned in Sect. 3 (Methodology). We input sentence 1 to both the online NMT engines Google Translate and Bing one by one. Translated sentences from both engines are compared with reference sentences. Both the sentences, i.e., translated and reference sentences, are tokenized with the help of Stanford's core NLP parser's online tokenizers (Figs. 2 and 3) [24, 25].

Test sentence 1: সে আমার কাছে স্বীকার করল যে সে ঘুমিয়ে পড়েছিল।

Reference sentence: He confessed to me that he had fallen asleep.

Google Translate (Hypothesis sentence 1): He admitted to me that he fell asleep.

In Fig. 2, POS from reference sentence is as follows: PRP is a possessive preposition, VBD is verb past tense. IN is a preposition, VBN is a verb past participle, and JJ is an adjective. The corresponding parts of speech from the hypothesis sentence (Fig. 3) are compared with the help of the same approach as BLEU uses. But in n-gram matching, as a token, instead of taking words, parts of speech are taken. This is our BLEU with POS score. Along with POS with BLEU, we also computed normal BLEU with $n = 1$ to 4. The scores are also computed. The process of computing human scores is already explained in the methodology section (Sect. 3). Finally, with the Pearson correlation coefficient, we have computed the correlation of normal BLEU and BLEU with POS against the human score. With the help of the Pearson correlation coefficient, we can get the linear relationship between two different scores.

We have experimented with around five different sentences picked randomly from the dataset and observed almost the same behavior as we have observed here in this example sentence. Table 1 reports all the automatic and manual scores. As

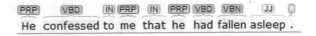

Fig. 2 Parts-of-speech tagging of reference sentence 1

Fig. 3 Parts-of-speech tagging of hypothesis sentence 1

Table 1 BLEU score with *n*-gram where *n* = 1 to 4, POS-BLEU and human score of test sentence 1 (scores are for Google Translate)

BLEU	POS-BLEU	Human score
0.66	0.88	0.98
0.37	0.75	0.95
0.29	0.73	0.95
0.17	0.70	0.98

Fig. 4 Graphical representation of automatic and human scores

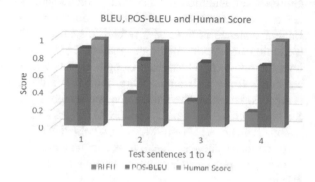

Fig. 5 Pearson coefficient has higher correlation with BLEU with POS

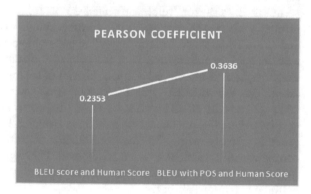

stated before, we used BLEU for multiple n-grams and accordingly also noticed variation in evaluation scores. Figure 2 is the graphical representation of all the metrics scores. Pearson's correlation score is reported graphically in Fig. 5. We can see that BLEU with POS has a higher correlation with the human score as compared to normal BLEU. Pearson's correlation coefficient values vary from −1 to +1. −1 means a negative correlation between two variables. 0 means no correlation. And +1 means positive correlation (Fig. 4).

Whereas in our second example sentence (generated by Microsoft's Bing translate), reference text exactly matches with hypothesis text and hence in this case all the scores are highest and considers to be the best translation (Figs. 6 and 7). All the metrics scores are shown in Table 4. In the automatic translation score generation process by BLEU and BLEU with POS, all tokens irrespective of the

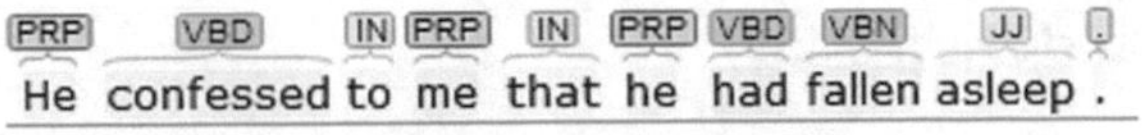

| PRP | VBD | IN | PRP | IN | PRP | VBD | VBN | JJ | . |

He confessed to me that he had fallen asleep .

Fig. 6 Parts-of-speech tagging of reference sentence 1

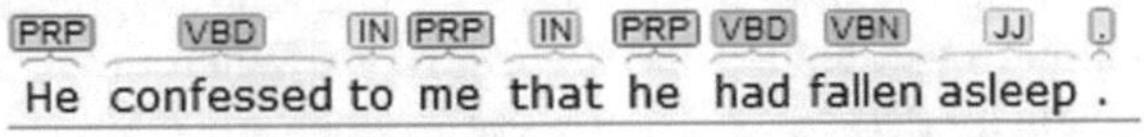

| PRP | VBD | IN | PRP | IN | PRP | VBD | VBN | JJ | . |

He confessed to me that he had fallen asleep .

Fig. 7 Parts-of-speech tagging of hypothesis sentence 2 (generated by Bing)

Table 2 Pearson's correlation between BLEU score versus human Score

BLEU	Human-score	Pearson correlation
0.66	0.98	0.2353
0.37	0.95	0.2353
0.29	0.95	0.2353
0.17	0.98	0.2353

Table 3 Pearson's correlation between POS-BLEU score versus human score

POS-BLEU	Human score	Pearson correlation
0.88	0.98	0.3636
0.75	0.95	0.3636
0.73	0.95	0.3636
0.70	0.98	0.3636

Table 4 BLEU ($n = 1$–4) scores, BLEU with POS and human score for test sentence 1 (translation generated by Bing)

BLEU (n = 1 to 4)	BLEU with POS	Human score
1	1	1
1	1	1
1	1	1
1	1	1

Hypothesis text and reference text exactly match, and hence, all metrics have the highest score

number of n-grams exactly match, and hence with BLEU's computation criteria it generates a maximum score. And in this case, the human score is also the highest because hypothesis and reference translation both are exactly the same (Tables 2 and 3).

5   Results and Discussion

From our experimentation and results as reported in tables in Sect. 4, we can see that since the BLEU metric is mainly dependent on a precision-based approach where it tries to match tokens from reference text and hypothesis text, hence n-gram with minimum n-value has the highest score. When the n value increases, the score also gradually decreases. The reason is when the n value is more than 1, then accordingly it tries to find that the many numbers of the exact token match from reference and hypothesis texts. But in reality, this is not possible because words may have multiple meanings based on context, words may be represented in a sentence in a different manner, one particular context/meaning can be represented by different words, and so on. The probability of occurrence of such type of case is higher in case of morphologically rich languages. Hence, the chances of an accurate score are also reduced in normal BLEU. On the other hand, when we use BLEU with POS, such type of exact word matching case between reference and hypothesis can be avoided and matching will be done on the basis of POS, and thereby chances of accuracy increase. This is the reason when we try to find the correlation between the human score and both the automatic metrics BLEU and BLEU with POS, the latter has a higher correlation with the human score.

6   Conclusion and Future Work

We can conclude from the above work and discussion that correct evaluation is very important in MT design. Sometimes automatic evaluation metrics fail to produce a correct score because they are primarily dependent on either n-gram-based precision or a recall-based approach. Hence, researchers also focused on machine learning-based approaches in the automatic evaluation process. We have also seen that by integrating a syntax-based approach in the evaluation process with BLEU, we are able to generate a better score, which has a higher correlation with a human score, in our Bengali-to-English translation task. Furthermore, in future work, we can exploit recently introduced ML models where we can capture the semantic, syntax, and other important information in hypothesis and reference translations, they can be embedded before training the model for our language pairs, and evaluation accuracy can be enhanced.

References

1. Hutchins J, Lovtskii E (2000) Petr Petrovich Troyanskii (1894–1950): a forgotten pioneer of mechanical translation. Mach Transl 15:187–221
2. Brown PF, Della Pietra SA, Della Pietra VJ, Mercer RL (1991) Statistical approach to sense disambiguation in machine translation 146–151. https://doi.org/10.3115/112405.112427

3. Xiong D, Meng F, Liu Q (2016) Topic-based term translation models for statistical machine translation. Artif Intell 232:54–75
4. Koehn P et al (2007) Moses: open source toolkit for statistical machine translation. In: Proceedings of the 45th annual meeting ACL interaction poster demonstration session—ACL '07 177. https://doi.org/10.3115/1557769.1557821
5. Vaswani A et al (2017) Attention is all you need. Adv Neural Inf Process Syst 2017-Decem, 5999–6009
6. Sutskever I, Vinyals O, Le QV (2014) Sequence to sequence learning with neural networks. Adv Neural Inf Process Syst 4:3104–3112
7. Stahlberg F (2020) Neural machine translation: a review. J Artif Intell Res 69:343–418
8. Vathsala MK, Holi G (2020) RNN based machine translation and transliteration for Twitter data. Int J Speech Technol 23:499–504
9. Duh K (2008) Ranking vs. regression in machine translation evaluation. In: Third workshop on statistical machine translation WMT 2008 annual meeting association on computer linguist ACL 2008 191–194. https://doi.org/10.3115/1626394.1626425
10. Liu D, Gildea D (2005) Syntactic features for evaluation of machine translation. In: Proceedings of the ACL workshop on intrinsic and extrinsic evaluation measures for machine translation and/or summarization ACL 2005 25 32
11. Banerjee S, Lavie A (2005) METEOR: an automatic metric for mt evaluation with improved correlation with human judgments. In: Proceedings of the acl workshop on intrinsic and extrinsic evaluation measures for machine translation and/or summarization 2005 65–72
12. Doddington G (2002) Automatic evaluation of machine translation quality using n-gram co-occurrence statistics 138. https://doi.org/10.3115/1289189.1289273
13. Park Y, Patwardhan S, Visweswariah K, Gates SC (2008) An empirical analysis of word error rate and keyword error rate. In: Proceedings of annual conference on international speech communication association INTERSPEECH 2070–2073. https://doi.org/10.21437/interspeech.2008-537
14. Guzmán F, Joty S, Màrquez L, Nakov P (2017) Machine translation evaluation with neural networks. Comput Speech Lang 45:180–200
15. Popović M, Ney H (2009) Syntax-oriented evaluation measures for machine translation output. In: EACL 2009—Proceedings of the Fourth Workshop on Statistical Machine Translation 29–32. https://doi.org/10.3115/1626431.1626435
16. Duma M, Vertan C, Park VM, Menzel W (2013) A new syntactic metric for evaluation of machine translation. ACL Student Res Work 130–135
17. Haque R, Hasanuzzaman M, Way A (2020) Analysing terminology translation errors in statistical and neural machine translation. Mach Transl 34:149–195
18. Papineni K, Roukos S, Ward T, Zhu WJ (2002) {B}leu: a method for automatic evaluation of machine translation. In: Proceedings of the 40th annual meeting of the association for computational linguistics 311–318 (Association for Computational Linguistics, 2002). https://doi.org/10.3115/1073083.1073135
19. Agnihotri S (2019) Hyperparameter optimization on neural machine translation. Creat Components 124
20. Lim R, Heafield K, Hoang H, Briers M, Malony A (2018) Exploring hyper-parameter optimization for neural machine translation on GPU architectures 1–8
21. Tran N, Schneider J-G, Weber I, Qin AK (2020) Hyper-parameter optimization in classification: to-do or not-to-do. Pattern Recognit 103:107245
22. Lankford S, Afli H, Way A (2022) Human evaluation of English–Irish transformer-Based NMT 1–19
23. Newman B, Ang KS, Gong J, Hewitt J (2021) Refining targeted syntactic evaluation of language models 3710–3723. https://doi.org/10.18653/v1/2021.naacl-main.290
24. Manning C et al (2015) The Stanford CoreNLP natural language processing toolkit 55–60. https://doi.org/10.3115/v1/p14-5010
25. Marcus MP, Santorini B, Marcinkiewicz MA (1993) Building a large annotated corpus of English: the Penn Treebank. Comput Linguist 19:313–330

Artificial Intelligence and Blockchain Technology in Insurance Business

Shakil Ahmad and Charu Saxena

Abstract With the use of cutting-edge technology like artificial intelligence, the Internet of Things, and blockchain, the insurance industry is about to enter a new era. Financial innovations will have a wide range of effects on economic activity. The future of the AI and blockchain-based insurance industry appears bright. The expansion of the insurance industry will be considerably aided by AI and block chain technology. The corona situation has recently brought to light some inefficiencies of the conventional system, including fewer client–insurer interaction capabilities, challenges in targeting consumers, difficulties working from home, a lack of IT support, a lack of transparency, and sparse management. This article examines the impact of cutting-edge financial technology on the insurance industry, including AI technology and blockchain, and describes the phenomena from several angles. Insightful and scientific, ICT-based insurance makes considerable use of electrical and IT equipment for technology resolution without the consumer providing the insurance company with any natural resources. Digital insurer seems to be growing more popular at the moment, but it's essential to understand the problems and challenges it faces, which result in poor penetration. Although there are many advantages for the insured and insurer, internet insurance providers are nevertheless subject to operational, regulatory, and company reputation, and their customers worry about the safety of their transactions and identity theft. This article aims to illustrate the fundamentals and significance of AI and blockchain technology in the insurance business, looking at benefits that accrue to both the provider and the customer. Although there are many studies already published, the focus of this research is the insurance industry. By fusing cutting-edge technical demands with established company competencies, this paper assists the management in expanding their understanding of competing for advantage.

S. Ahmad (✉) · C. Saxena
University School of Business, Chandigarh University, Mohali, Punjab 140413, India
e-mail: shakil.r17@cumail.in

C. Saxena
e-mail: charu.e8966@cumail.in

© The Author(s), under exclusive license to Springer Nature Singapore Pte Ltd. 2023
Y. Singh et al. (eds.), *Proceedings of International Conference on Recent Innovations in Computing*, Lecture Notes in Electrical Engineering 1011,
https://doi.org/10.1007/978-981-99-0601-7_6

Keywords Artificial intelligence · Blockchain · Technology · InsurTech · Covid-19 · Insurance business

1 Introduction

As a result of the presentation of advanced and creative innovation, the perception of conventional approaches has changed. The financial administration field is changing and being helped by financial innovations, mainly the semi of the insurance industry. Insurance advances, or "InsurTech," are currently revolutionizing financial administrations around the world, such as in the commercial fields of the insurance industry [1]. In the upcoming decade of this century, the traditional methods and procedures in insured sections of the scene are expected to change quickly. In particular, the customer and other partners in this field will benefit from the security and welfare code-based novel features, like AI technology and blockchain, to protect against financial fraud and guarantee fraud. Life in insurance exchanges, in comparison to the past, is becoming more manageable because of these cutting-edge advancements and futuristic software and lessens the hassle of actually going to the branch and interacting with the insurance expert or advocate. We believe these advancements will be much more user-friendly and client-friendly [2]. The corona pandemic has had a particularly negative impact on the insurance industry: The number of health and disaster protection claims has increased significantly, while the demand for auto and travel insurance has drastically decreased. All insurance companies have a poor understanding of the occurrence of customer collaboration. Many insurance companies have realized how pandemics may both attract and retain clients [3].

A study demonstrates how the use of technologies such as AI, blockchain, chat bots, the Internet of Things, and telematics can lessen the effects of the corona epidemic on the insurance business. The goal of this review is to examine how technology is used in the insurance sector [4]. As per Jing Tian, the growth of online finance has spawned the online insurance sector. His paper thoroughly examines the development of Online insurance and shatters the fundamental problems of the insurance business. He puts forward advancement proposals based on Huize's course of action. He investigates methodologies of the insurance business in the social consciousness of others' aspirations, item plans, and associates with the plague, and joins with the public authority to strengthen the development of life and general insurance, enhances the PPP mode, and consolidates protection innovation too [5]. Several innovative developments at the nexus of information and insurance administrations are AI and blockchain technology. Before the virus, a few standby plans began the transition to computerized technology in search of more individualized and efficient services, and many technology vendors introduced innovations into the industry [6]. On the basis of above comments the following hypothesis is formulated.

H1: There is a meaningful role of artificial intelligence and blockchain technology in the insurance business.

2 Literature Review

The move from analog-to-digital processing of information is only the beginning of the insurance firm's advanced digitalization. Eling and Lehmann [7] the combination of the digital and analogue environments with innovative technologies that improve customer engagement, data availability, and corporate operations is what is referred to as digitalization. The previous ten years have seen the emergence of InsurTechs, which are also driving the digital transformation. Cloud storage, telemetry, IoT, mobile phones, blockchain technology, AI, and predictive modeling are new technologies that are having an impact on the insurance sector [8]

Along with the insurance value chain, automation has already had a significant impact and will grow to do so as emerging technologies develop. The main changes are improved underwrite and product development, new business models, improved client relations, and improved distribution tactics. Researchers' study demonstrates that the performance of an organization of insurance businesses is significantly impacted favorably by digitization activities [9]. Six cutting-edge innovation trends, including AI, algorithmic logic, mobile finance, blockchain technology, refreshed ATMs, and the Internet of Things, were identified by the New Jen Application as having the potential to transform the bank services, insurance, and money market. Chris Smith documented how technological improvements affected the insurance industry, mainly how they changed the nature of the obligation, how the settled claims, and how the value was divided. The researcher goes into detail and clarifies how cutting-edge innovations might propel and have a significant impact on monetary administrations, notably in the financial sector. As a result, they portray various perspectives on surprise in this context. FinTech and InsurTech exhibits in these areas were examined by analysts, who then adjourned [2].

Blockchain, AI, and IoT's potential to alter how we conduct our jobs in the future will help the enterprise, governments, non-profits, and individuals alike. Trust may be established in virtually any transaction using blockchain, AI technology, and Internet of Things technology, making it possible to send money, goods, or sensitive information throughout the world more quickly. Blockchain, AI, and the IoT were studied to have a better understanding of how this innovation functions. Questions concerning Bitcoin's uses in various sectors to see how far it has come in realizing its potential have arisen from an analysis of its technological design [10]. Our economy and community are changing due to new ideas and new information genesis, and the insurance industry will follow suit. InsurTechs, or new technologies, are entering the market to provide some of the services typically offered by resident backup plans and mediators. Modern businesses are seeking favorable conditions in protection, much like newly established innovation corporations. The participants usually give generous businesses with safety net suppliers flexibility. Still, they might also emerge as direct rivals, restricting net income and testing backup plans, particularly at their client application [11].

According to Mohammad Ali (2020), financial development is significantly impacted by insurance. An ambiguity, a loss of accredited living person harmonization, a lack of a branding and promotion strategy, a lack of business ethics, inadequate IT support, ineffective officials, totally inadequate ROI, a loss of transparency and honesty, a loss of acknowledgement, and banal control are among the most critical matters in this industry, according to him. He suggests using cutting-edge marketing and advertising tools, retaining and developing talent, enhancing recognition, utilizing IT, the Internet of Things (IoT), blockchain (BC), and artificial intelligence (AI), avoiding risky competition, adopting a progressive managerial style, and enforcing substantial insurance to overcome the challenges [12].

Both the insurance industry and how insurance services are provided are being transformed by digitalization. The insurance industry's service delivery process is significantly impacted by the use of mobile devices, chatbots, big data, intelligent systems (AI), IoT, and chatbots. This includes everything from product design to underwriting risks to policy pricing to their advertising and sales to claim processing to current customer interaction and management [13]. Digitalization is reassess present tasks from fresh perspectives formed possible by technological novelty rather than just changing remain processes into more improper ones. The transition might result in new ways to do activities that are more durable or reasonable. Still, it can also be disruptive to a company's ongoing operations because digitalization fundamentally alters an organization's commercial opportunities [14].

AI has the potential to change the marketing industry completely. The focus on AI by businesses in their financial statements and its gross and net operational efficiency are compared in this study. 10-K filings are an essential source of information for research in accounting and finance, but they are still primarily ignored in marketing. Researchers create a practical guideline to demonstrate how organizations' AI focus may be tied to annual gross and operating efficiency by drawing on financial and marketing theory. The connection between AI intensity and operating efficiency is then empirically tested using a set of simultaneous equations. Their findings demonstrate the coming AI change that US-listed companies are currently experiencing. They show how an emphasis on AI may increase net revenue, net operating efficiency, and return on capital while lowering advertising costs and generating jobs [15]. AI emphasis and sales, as well as AI focus and personnel count, are positively correlated. However, it is expensive to retrain staff members and reorganize marketing roles to accommodate the mechanization of selling and marketing capabilities. It is crucial to understand whether an AI-focused organization has a good or negative impact on sales per employee. Because our guiding framework is unable to anticipate the direction of the link between AI emphasis and deals, we hypothesize that AI focus is adversely connected to sales per employee. Furthermore, our guiding framework assumes a positive correlation between the workforce size and AI focus. Since hiring more staff incurs higher costs, we anticipate that there will be a negative correlation between AI engagement and sales per employee [16].

3 Methodology

3.1 AI Technology and Blockchain Technology in Insurance Business

Artificial Intelligence: AI is becoming more popular among corporations. Insurance companies can benefit from AI in ways like increased productivity, improved customer service, and reduced extortion. AI-powered chatbots allow you to continuously respond to consumer inquiries and handle everything from sim ple approach-related problems to complaints and concerns. Safety net providers can speed out the start-finish process with the use of AI in the protection industry, from gathering information, settlement development, approval, and acknowledgement, to watching installments after rescue and recovery, dealing with legal difficulties, and contacting the board. Traditional mathematical hazard identification and pricing models will evolve as more specific risk-relevant information on policyholders becomes available through archived and real-time datasets. Insurance companies can more precisely evaluate and forecast future loss chances and total loss amount on a micro-scale. Thanks to the granular content analysis, images, and video content from internally and externally databases as well as from devices connected (such as management capabilities and health wearable). AI makes it easier for insurance companies to differentiate between good and bad risks, which decrease adverse selection. It may even encourage people who pose a high risk of loss to intensify their loss prevention efforts or alter their conduct, which lowers opportunities for corruption (e.g., usage-based insurance products). As risk-relevant behavior, including preventative endeavors, is transparent and readily measured, it also enables insurance companies to create the small, homogenous risk of violence with precise and responsive premium pricing systems for each policyholder. As a result, unfavorable chances will pay higher premiums and positive risks a lower [17].

Blockchain: Blockchain enables the creation of a computerized record that is irrevocably made. This invention can support strategies by reducing the administrative costs incurred while evaluating claims and verifying third-party installments. Blockchain ensures that all of this data is shared, easily verifiable, and misrepresentation-secured. According to PWC's investigation, blockchain might help reinsurers by facilitating the movement of the protection plan, which could result in potential private equity funds of USD 5 to 10 billion. Blockchain is projected to increase the speed, efficiency, and security and reduce the costs of a more significant portion of insurance associations' operations shortly. A blockchain is a

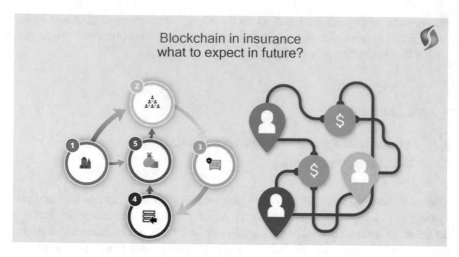

Fig. 1 How a blockchain technology works [19]

decentralized digital ledger that is shared by many peers in a network and makes it easier to record transactions and track the ownership of both tangible and intangible goods. Through the use of cryptography signatures, validated payments take the form of blocks that are sequentially added to a chain of previously confirmed blocks due to the decentralized nature of the ledger. And the fact that each new block is tagged sequentially and contains the information that references the block before it, any attempt to falsify the blockchain would necessitate falsifying every block that has already been created [18] (Fig. 1).

3.2 Impact of AI Technology and Blockchain Technology on Insurance Business

Financial year	Investment in AI and BC (US$M)	Premium volume in the world (Premium in USD Billions)			Premium volume in India (Premium in USD Billions)		
		Life	Non-life	Total	Life	Non life	Total
2020–2021	4780	2797.44 (44.50)	3489.61 (55.50)	6287.04 (100)	81.25 (75.24)	26.74 (24.76)	107.99 (100)
2019–2020	4374	2,916.27 (46.34)	3,376.33 (53.66)	6,292.60 (100)	79.67 (74.94)	26.64 (25.06)	106.31 (100)
2018–2019	5540	2820.18 (54.30)	2373.05 (45.70)	5193.23 (100)	73.74 (73.86)	26.10 (26.14)	99.84 (100)
2017–2018	2690	2657.27 (54.32)	2234.42 (45.68)	4891.69 (100)	73.24 (74.73)	24.76 (25.27)	98.00 (100)

Financial year	Investment in AI and BC (US$M)	Premium volume in the world (Premium in USD Billions)			Premium volume in India (Premium in USD Billions)		
		Life	Non-life	Total	Life	Non life	Total
2016–2017	2771	2617.02 (55.30)	2115.17 (44.70)	4732.19 (100)	61.82 (77.95)	17.49 (22.05)	79.31 (100)
2015–2016	1463	2533.82 (55.64)	2019.97 (44.36)	4553.79 (100)	56.68 (78.96)	15.10 (21.04)	71.78 (100)

Source IRDI Annual Report [20] and statista [21]

It is evident from the graph mentioned above and the chart that there is a strong correlation between technology investment and insurance performance. InsurTech investment is rising globally but slowing down around the period of the corona pandemic. The premium volume serves as a proxy for insurance performance, which is improving annually as a result of investments in technology. Due to technological investments made the year before, the premium growth rate increased during the pandemic by 2.9%. Like the rest of the world, India's insurance industry is booming. Even throughout the COVID-19 period, India's premium volume continues to rise annually. The fact that India is already a member of the technological revolution is encouraging. The majority of technology devices, or "InsurTech," are already in use by India's insurance sector. After the pandemic, we hope India will fare better than it is right now. Before the COVID-19 pandemic, the insurance companies were on a steady growth path, according to the Swiss ReInstitute's sigma research article (no. 4/2020) on global insurance. With help from the non-life industry in developed economies, the total amount of direct premium written globally climbed by about 3% in 2019 compared to the previous year. Over 60% of all insurance markets globally saw total premium growth outpace real GDP growth. The total amount of insurance premiums insured in 2019 increased from the pre-revision estimate of USD 5.4 trillion to USD 6.3 trillion, representing 7.2% of the world GDP (IRDI Annual Report) [20] (Figs. 2, 3 and 4).

4 The Core Benefits of AI and Blockchain Technology

1. *Minimization of costing*: The company's functional expenses diminished and changed in everyday activities because of the presentation of AI and blockchain technology at every single phase of business measures [22].
2. *Qualities in the dynamic interaction*: Computerization or digitalization of cycles reliably gives accurate and fast results in the insurance business. AI and blockchain technology are helping in various ways to deal with inspecting the business conditions in different business exercises and occasions in associations in a brief period. These valuable reports will help experts and key innovators in the unique cycle and stay aware of their qualities [23].

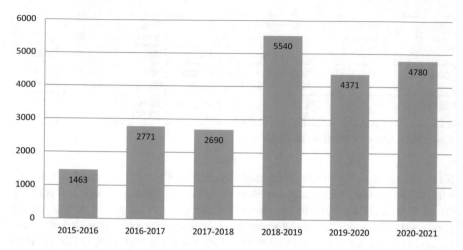

Fig. 2 Investment in AI and BC (US$M)

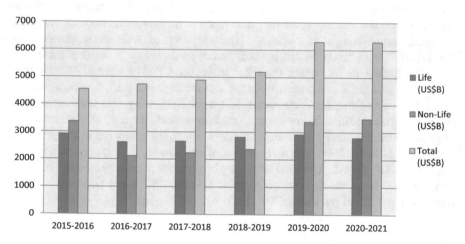

Fig. 3 Premium volume In the world (US$B)

3. *Focus on more straight forwardness*: Because of the use of AI and block-chain technology or digitalization of transaction and administrations in insurance business, protection establishments can, without much of a stretch, keep up with straightforwardness in their every single exchange, tasks of different administrations with less time at an ideal expense both the gatherings in the assistance exchange [24].

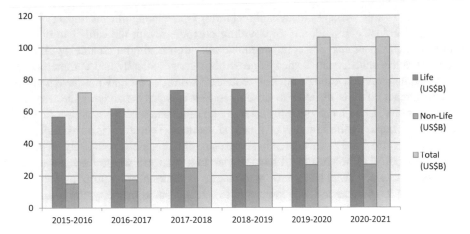

Fig. 4 Premium volume in the India (US$B)

5 Test of Hypothesis

From the above hypothetical and insightful conversation, we could gather that the new cutting-edge innovations (AI and BC) goodly affect the insurance business. In this way, our examination shows there is a meaningful role of AI and blockchain technology in the insurance business.

6 Conclusion

AI and BC play a crucial role in enhancing insurance companies, expanding the item selection, and gaining the upper hand. Underwriters and InsurTech should work together to create a setting that will boost the worth of clients by addressing their digital demands. The success of new intelligence and blockchain technology initiatives will require a combination of mechanical brilliance, marketing prowess, and knowledge of the insurance industry. After a variety of these technologies, clients will gain confidence. Plans for contingencies have become more aware of the close relationship between customer experience and computerized procedure, how to deal with change and the advancement of operations. The insurance industry will continue to make extensive use of automation and robotics, chatbots, telematics, information analysis from social networks, blockchains, AI, artificial consciousness, and anticipatory research. As a result of the discussions mentioned above, the analysts tested their hypothesis. They illustrated the significant impact that AI and BC improvements have had on the insurance industry after examining various innovations and advancements that have been used and tested in the insurance sector.

Similar to this, AI and blockchain technology have a significant impact on the insurance industry. In addition to providing customers with the caliber to browse and purchase insurance plans, the adequate adoption of digital channels will give the insurance industry the chance to grow both geographically and commercially. Because insurance is a long-term risk reduction strategy that may not have a proximate impact, if it is difficult to access, individuals may disregard the service. However, the risk reduced by doing so or by having insurance has a long-term outcome on a person's life in terms of financial security. Therefore, it must digitalize the services provided by insurance businesses to ensure quick availability of insurance services. Customers should have access to a platform where they can easily access all services, from underwriting to claims. Additionally, by utilizing the right technology to evaluate the risk connected to every insured property, the insurance business could benefit from determining the optimal premium rate.

References

1. Volosovych S, Zelenitsa I, Kondratenko D, Szymla W, Mamchur R (2021) Transformation of insurance technologies in the context of a pandemic. Retrieved from https://doi.org/10.21511/ins.12(1).2021.01
2. Chakravaram V, Ratnakaram S, Vihari NS, Tatikonda N (2021) The role of technologies on banking and insurance sectors in the digitalization and globalization era—a select study. In: Proceedings of international conference on recent trends in machine learning, IoT, smart cities and applications. Springer, pp 145–156
3. Chakravaram, Shevchuk O, Kondrat I, Stanienda J (2020) Pandemic as an accelerator of digital transformation in the insurance industry: evidence from Ukraine. Insur Mark Co 11:30
4. Ahmad S, Saxena C (2021) Reducing the covid-19 impact on the insurance industry by using technologies. EFFLATOUNIA—Multidis J 5(2). Retrieved from http://efflatounia.com/index.php/journal/article/view/343
5. Chakravara Jing T (2021) Research on the development of internet insurance in China——based on the exploration of the road of Huize insurance. In: E3S Web Conf. EDP Sciences, p 03030
6. Klapkiv L, Klapkiv J (2017) Technological innovations in the insurance industry
7. Eling M, Lehmann M (2018) The impact of digitalization on the insurance value chain and the insurability of risks. Geneva Pap Risk Insurance—Issues Practice 43:359–396. https://doi.org/10.1057/s41288-017-0073-0
8. Cappiello A (2020) The technological disruption of insurance industry: a review. Int J Bus Soc Sci 11:1
9. Bohnert A, Fritzsche A, Gregor S (2019) Digital agendas in the insurance industry: the importance of comprehensive approaches. Geneva Pap Risk Insurance—Issues Practice 44(1):1–19. https://doi.org/10.1057/s41288-018-0109-0
10. Raj R, Dixit A, Saravanakumar DA, Dornadula HR, Ahmad S (2021) Comprehensive review of functions of blockchain and crypto currency in finance and banking. Des Eng 7
11. Radwan SM (2019) The impact of digital technologies on insurance industry in light of digital transformation. Blom Egypt Invest Insur Brok Consult
12. Ali M (2020) Challenges, prospects and role of insurance on economic growth in Bangladesh. IIUM J Case Stud Manag 11:20–27
13. IAIS (2018) Issues paper on increasing digitalisation in insurance and its potential impact on consumer outcomes. Int Assoc Insuracne Supervisor

14. Parviainen P, Kaarianinen J, Tihinen M, Teppola S (2017) Tackling the digitalization challenge: how to benefit from digitalization in practice. J Acad Market Sci 5(1):63–77
15. Sagarika M, Michael T, Holly B (2022) Artificial intelligence focus and firm performance. Int J Inf Syst Project Manage 5(1)
16. Barro S, Davenport TH (2019) People and machines: partners in innovation. MIT Sloan Manage Rev 60(4):22–28
17. Martin E, Davide N, Julian S (2019) The impact of artificial intelligence along the insurance value chain and on the insurability of risks. Geneva Papers Risk Insurance—Issues Practice 2022(47):205–241. https://doi.org/10.1057/s41288-020-00201-7
18. Bonson E, Bednarova M (2019) Blockchain and its implications for accounting and auditing. Meditari Account Res 27(5):725–740
19. Blockchain in the Insurance Industry: What to Expect in the Future? https://www.dataversity.net/blockchain-in-the-insurance-industry-what-to-expect-in-the-future/
20. IRDAI Annual Report. https://www.irdai.gov.in/ADMINCMS/cms/frmGeneral_NoYearList.aspx?DF=AR&mid=11.1
21. Investment in InsurTech. https://www2.deloitte.com/us/en/pages/financialservices/articles/fintech-insurtech-investment-trends.html
22. Rai R, Dixit A, Saravanakumar A, Fathima A, Dornadula R, Ahmad S (2021) Comprehensive review of functions of blockchain and crypto currency in finance and banking. Des Eng 9:3649–3655
23. Mhlanga D (2021) Financial inclusion in emerging economies: the application of machine learning and artificial intelligence in credit risk assessment. Int. J. Financ. Stud. 9:39
24. Mhlanga D (2022) Human-centered artificial intelligence: the superlative approach to achieve sustainable development goals in the fourth industrial revolution. Sustainability 14:7804. https://doi.org/10.3390/su14137804

Classification and Detection of Acoustic Scene and Event Using Deep Neural Network

Sandeep Rathor

Abstract Sound is the basis medium to understand the world. On the basis of sound, we can analyze or recognize the events, environment or scene. Recently, artificial intelligence techniques have been prominently applied to handle speech signals, and remarkable achievement has been exhibited by utilizing the speech signal with deep neural network. Therefore, a multi-model approach to recognize acoustic event and scene is proposed by using deep neural network. In the proposed work, temporal features are captured by using LSTM, and dense layer (DL) is utilized for detention the nonlinear combination of those features. Experimental results are obtained using TUT 2017 datasets with the acceptable accuracy, i.e., 85.36%.

Keywords Acoustic scene · Acoustic event · LSTM · DNN

1 Introduction

Speech is the most convenient and fast medium of communication. On the basis of speech, communicators can understand the domain, situation, mood, and purpose of each other. The domain of communication reflects the context like political, medical, advertisement, research, games and sports, etc. [1]. If persons are discussing on a pandemic, then its domain will be "Medical"; similarly, if two persons are discussing about IPL match, then its domain will be "Games and Sports." The emotions and sentiments can also be recognized through the communication. The emotions can be classified as: happy, sad, joy, surprise, excited, etc. while sentiments can be classified as: favorable, unfavorable, and neutral. Similarly, there are different types of sound in real life like cooking, travelling, playing, watching television, passing the vehicle, singing, crying, etc. Nowadays, there has been increasing interest during analyzing numerous sounds in real-life environments equivalent to change of state sounds in a room or vehicles passing sounds in the environment [2]. Therefore, an automatic analysis can be performed to classify the environment on the basis of the sound. The

S. Rathor (✉)
Department of Computer Engineering and Applications, GLA University, Mathura, India
e-mail: sandeep.rathor@gla.ac.in

© The Author(s), under exclusive license to Springer Nature Singapore Pte Ltd. 2023
Y. Singh et al. (eds.), *Proceedings of International Conference on Recent Innovations in Computing*, Lecture Notes in Electrical Engineering 1011,
https://doi.org/10.1007/978-981-99-0601-7_7

sound can be of a bird, a car, a musical instrument, harmonic noises, and multiple noises, etc. [3]. We can also recognize critical situations. Obviously, it is considered as an event. It can be generated by acoustic. A shooting, a scream, a glass breaking, an explosion, or an emergency siren are examples of artifacts [4]. Mainly, "environmental sound detection" can be utilized for two tasks, i.e., for event detection and for scene classification. Acoustic event detection is the process of detecting the level of event like "birds singing," "mouse clicking," "travelling," "playing," etc.; it is simply identification of "Triger words," while acoustic scene classification is the process of predicting scene from the recording such as "park," "office," "train," "cricket," etc. The neural network is the most suited for recognizing events [5]. However, for the implement multimodal, i.e., for event and scene, deep neural network is more suitable [6].

Figure 1 shows two tasks that can be recognized by the sound signals, i.e., event recognition and scene recognition. The main objective of the proposed research is to recognize it by using bi-directional LSTM and deep neural network on a standard dataset.

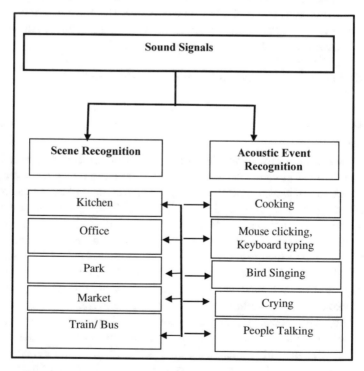

Fig. 1 Acoustic event and scene recognition from sound signals

2 Related Work

This section contains discussion of various researcher on Bi-LSTM and machine learning techniques in the context of speech signal processing to recognize the events, scene, domain, and sentiments, etc.

The acoustic scene classification is proposed by Barchiesi et al. [3]. The maximum likelihood criteria, Mel-frequency cepstral coefficients (MFCCs), and Gaussian mixture models (GMMs) were employed as the baseline tools in this paper. However, the author only found sufficient evidence to conclude that three techniques considerably surpass it. In the same context, the author also assessed the precision of human classification. The best-performing algorithm achieves a mean accuracy that is comparable to that of humans, and both computers and humans misclassify common pairs of classes. However, at least some individuals correctly classify all acoustic scenes, while all algorithms incorrectly classify some scenes.

Video event detection using LSTM is proposed by Liu et al. [7] Multiple events had been recognized through the input video. The beauty of the research is that there is no need of pre-processing for object detection. In this paper, the proposed research had been compared with SVM and found that LSTM technique is much better for events detection.

The use of deep learning technique in baby cry detection is proposed by Cohen et al. [8]. Deep learning technique with the recurrent neural network is used in this research. The result analysis represented in this paper is evident of the performance. The recital of deep neural network is better than classical machine learning approaches.

Polyphonic sound's event detection method is proposed by Hayashi et al. [9] With the support of LSTM and a hidden Markov model, a hybrid framework is used in the proposed research. The problem of non-negative matrix factorization is solved by this approach. The main focus of the proposed research was solely on polyphonic sound occurrences, which were not grouped into any categories. The speaker, on the other hand, did an excellent job of describing the need for LSTM.

An acoustic event processing by using deep learning technique is proposed by [10]. This paper contrasts a detection system-based Neural Network (NN) and an acoustic event detector-based hidden Markov model. The same database was used for both processes. Fires, glass breaks, and background noise were all part of the database. Via two hidden layers, the proposed deep neural network processes an acoustic signal. The accuracy is very good and shown by a confusion matrix. Therefore, it is evident that deep neural network works efficiently for acoustic event detection.

The recognition of events from a specific speech signal is proposed by [11]. Laughter, printer noise, keyboard and mouse clicks, phone ringing, and other occurrences in an office scene were addressed by the author. MFCC is used to extract features from the input after the author removes the noise. HMM performs the classification in order to understand the case. The scene was also classified by the author as a bus, park, store, restaurant, and so on. Furthermore, the use of CNN and spectrogram is proposed by [5] to detect acoustic scene. In the proposed research, author also used a fourfold cross-approval approach to increase the framework's accuracy. The framework's overall accuracy is calculated by averaging the four per fold accuracy. The proposed system is capable of accurately recognizing scenes such as the seashore, transportation, office, library, metro station, and so on.

A multimodal approach using LSTM is proposed by [12, 13]. To minimize the chances of overfitting, this technique employs multiple LSTM layers for each modularity, with weights shared between LSTM layers via a recurrent pattern. The results shown in the proposed research are acceptable, and it also proves that LSTM technique is better option to recognize emotions, events, and scenes.

Deep learning methods can be used in speech signal processing and pattern identification domains [14]. The main concern of this paper is to provide a complete overview of neural network-based deep learning algorithms for detecting auditory events. Different deep learning-based acoustic event detection algorithms are examined, with a focus on systems that are both strongly labelled and weakly labeled. This research also explores how deep learning approaches enhance the task of detecting acoustic events, as well as the potential challenges that need to be handled in future real-world settings.

3   Proposed Methodology

A proposed framework for acoustic scene and event recognition is shown in the Fig. 2. Acoustic features are extracted from the input speech signals by using Mel-frequency cepstral coefficient (MFCC). The input feature map is first convolved with two-dimensional filters in the convolution layer, and then its dimension is decreased via maxpooling.

The input acoustic signal is sent to the feature extraction phase, which calculates effective parameters. These parameters are specified to improve the audio signal's feature extraction. They aid in the characterization of sound segment by separating one from the other. Adapted signals are split into classes xi, for I = 1; 2;:::; n, according to classification. During the training process, the acoustic model learns from characteristics and parameters. Unsupervised and supervised learning are the two types of learning. Labels in supervised learning represent marked vectors that indicate class membership. According to unsupervised learning, train data will be divided into classes based on hidden variables [15]. The output of the maxpooling layer is then concatenated and input to a fully connected layer. Fully connected layer is used to learn relationship between the features.

In the proposed model, softmax activation function is used. The output of this layer is passed to the dropout layer to overcome overfitting problem. On the other hand, one more activation function rectified linear unit (ReLU) is used in event layers to recognized the events. The proposed network is optimized because of softmax cross-entropy objective function (Fig. 3).

The LSTM is used to discover patterns in sequential features. When learning long-term temporal dependencies, LSTM is effective. The input gate, forget gate, and

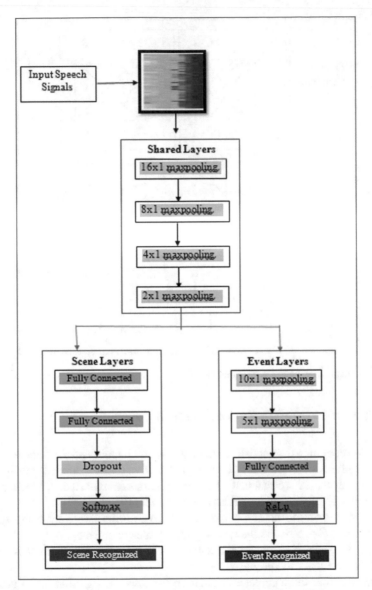

Fig. 2 Proposed framework for acoustic scene and event recognition

output gate are the three gates. The interwork dependencies are effectively learned using a bi-directional LSTM layer. This layer is in charge of teaching the sentence's word order.

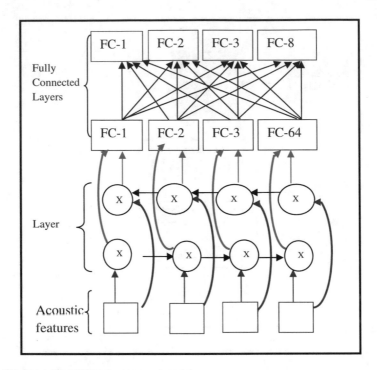

Fig. 3 Working of LSTM in the proposed model

4 Experimental Results and Discussion

In order to recognize the acoustic scene and events, I evaluated the performance of the proposed model on TUT2017 dataset [16]. In this dataset, some events were not labeled therefore, tagged manually. Measure parameters like Recall, Precision, and F-score are also mentioned to validate the research [17].

Figure 4 shows the feature extracted from an audio signal by using spectral centroid, while Fig. 5 shows the spectrogram of speech signals.

Figure 6 shows the parameter calculation for scene recognition. It is used to validate the results of the proposed model.

Figure 7 shows the parameter calculation for event recognition in the proposed model. The values show that proposed model is capable to recognize the acoustic scene and event with good accuracy.

Table 1 shows the comparison between the proposed model and other state-of-the-art methods proposed by various researchers. The results shown in Table 1 indicates that the accuracy of proposed model is better than any of the other model.

Fig. 4 Feature extraction from speech signals

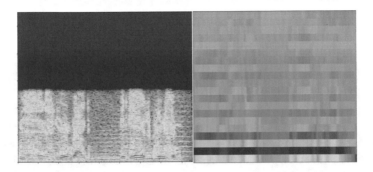

Fig. 5 Spectrogram of speech signals

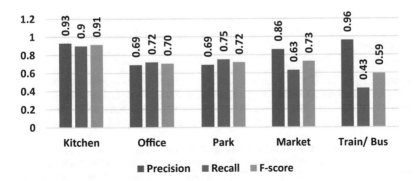

Fig. 6 Parameter calculations for scene

5   Conclusion and Future Scope

The paper explored the idea of event and scene recognition. In this paper, LSTM model is proposed to classify the acoustic events and scenes by using different fully connected layers and Bi-LSTM. The proposed model is executed on TUT2017 standard dataset to recognize events and scenes. The received accuracy is close to 85%.

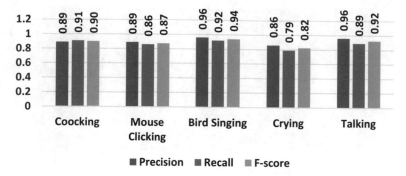

Fig. 7 Parameter calculations for event

Table 1 Comparison of proposed model with other methods

Method	Event (*F*-score)	Scene (*F*-score)
Tonami et al.,	0.776	0.66
RNN	0.683	0.59
Proposed	0.89	0.73

To implement the proposed model, Python programming language is used. İn context to parametric evaluation, precision and recall are calculated for each event and scene. İn future, the proposed research can be extended to recognize the whole environment on video data.

References

1. Rathor S, Agrawal S (2021) A robust model for domain recognition of acoustic communication using Bidirectional LSTM and deep neural network. Neural Comput Appl 33(17):11223–11232
2. Mesaros A, Diment A, Elizalde B, Heittola T, Vincent E, Raj B, Virtanen T (2019) Sound event detection in the DCASE challenge. IEEE/ACM Trans Audio Speech Lang Process 27(6):992–1006
3. Barchiesi D, Giannoulis D, Stowell D, Plumbley MD (2015) Acoustic scene classification: classifying environments from the sounds they produce. IEEE Signal Process Mag 32(3):16–34
4. Dosbayev Z et al (2021) Audio surveillance: detection of audio-based emergency situations. In: Wojtkiewicz K, Treur J, Pimenidis E, Maleszka M (eds) Advances in Computational Collective Intelligence. ICCCI 2021. Communications in Computer and Information Science, vol 1463. Springer, Cham. https://doi.org/10.1007/978-3-030-88113-9_33
5. Valenti M, Diment A, Parascandolo G, Squartini S, Virtanen T (2016) Acoustic scene classification using convolutional neural networks. In: Proceedings of Detection and classification of acoustic scenes and events 2016 Work., no. September, pp 95–99
6. Huang CW, Narayanan SS (2017) Characterizing types of convolution in deep convolutional recurrent neural networks for robust speech emotion recognition, 1–19
7. Liu AA, Shao Z, Wong Y et al (2019) Multimed Tools Appl 78:677. https://doi.org/10.1007/s11042-017-5532-x

8. Cohen R, Ruinskiy D, Zickfeld J, IJzerman H, Lavner Y (2020) Baby cry detection: deep learning and classical approaches. In: Pedrycz W, Chen SM (eds) Development and analysis of deep learning architectures. Studies in computational intelligence, vol 867. Springer, Cham

9. Hayashi T, Watanabe S, Toda T, Hori T, Le Roux J, Takeda K (2016) Bidirectional LSTM-HMM hybrid system for polyphonic sound event detection

10. Conka D, Cizmar A (2019) Acoustic events processing with deep neural network. In: 2019 29th international conference radioelektronika (RADIOELEKTRONIKA), pp 1–4. https://doi. org/10.1109/RADIOELEK.2019.8733502

11. Ford L, Tang H, Grondin F, Glass J (2019) A deep residual network for large-scale acoustic scene analysis. In: Proceedings of annual conference international speech communication association INTERSPEECH, vol 2019-Septe, pp 2568–2572

12. Ma J, Tang H, Zheng WL, Lu BL (2019) Emotion recognition using multimodal residual LSTM network. In: Proceedings of the 27th ACM international conference on multimedia, pp 176–183

13. Zhang S, Zhao X, Tian Q (2019) Spontaneous speech emotion recognition using multiscale deep convolutional LSTM. IEEE Trans Affect Comput c:1

14. Xia X, Togneri R, Sohel F, Zhao Y, Huang D (2019) A survey: neural network-based deep learning for acoustic event detection. Circuits Syst Signal Process 38(8):3433–3453

15. Tonami N, Imoto K, Niitsuma M, Yamanishi R, Yamashita Y (2019) Joint analysis of acoustic events and scenes based on multitask learning. In: 2019 IEEE workshop on applications of signal processing to audio and acoustics (WASPAA). IEEE, pp 338–342

16. Mesaros A, Heittola T, Diment A, Elizalde B, Shah A, Vincent E, Raj B, Virtanen T (2017) DCASE 2017 challenge setup: tasks, datasets and baseline system. In: Proceedings workshop on detection and classification of acoustic scenes and events (DCASE), pp 85–92

17. Tripathi R, Jalal AS, Agrawal S (2019) Abandoned or removed object detection from visual surveillance: a review. Int J Multimed Tools Appl 78(6):7585–7620

Analysis of a Novel Integrated Machine Learning Model for Yield and Weather Prediction: Punjab Versus Maharashtra

Sarika Chaudhary, Shalini Bhaskar Bajaj, and Shweta Mongia

Abstract Correct prediction is a deciding factor in modern agricultural techniques to confirm food safety and sustainability faced in crop production. As radical climatic conditions affect plants growth greatly, appropriate valuation of rainfall and yield prediction can deliver a sufficient information in advance which can be utilized to maintain the crop production quality. For accurate prediction, machine learning (ML) plays a prominent role. The collaboration of agriculture system with machine learning will lead to intelligent agriculture system that helps the farmer community in their decision-making of farm management and agribusiness activities such crop management including applications on yield prediction, disease detection, weed detection, crop quality, and growth prediction. This paper implemented and analyzed a prediction model using high potential ML algorithms, viz., random forest, linear regression, Lasso regression, and support vector machine for crop yield prediction whereas KNN, decision tree, logistic regression, Gaussian Naive Bayes, SVM, and linear discriminant analysis to find the best line method for rainfall prediction. To enhance the effectiveness of pre-processing, "SSIS-an ETL" tool is utilized. The experiments laid out for the state of Maharashtra and Punjab, and dataset collected from "ICAR-Indian Agriculture Research Institute" and www.imd.gov.in. R2-score, mean squared error, precision, recall, and F-score measure were used to ascertain the accuracy. As a result, random forest found significantly the finest algorithm for crop yield prediction by recording higher accuracy of 96.33% with maximum R2-score of 0.96 and least mean squared error (0.317), while lasso regression was found the worst with an accuracy of 32.9%. In case of rainfall prediction, Gaussian Naive Bayes secured top rank by recording considerably the highest accuracy of 91.89% as well as the maximum precision, recall, and F1-score of 0.93, 0.91, and 0.91, respectively.

Keywords Crop yield prediction · Machine learning · Rainfall prediction · SSIS · Weather prediction

S. Chaudhary (✉) · S. B. Bajaj
Amity University, Haryana, India
e-mail: sarikacse23@gmail.com

S. Mongia
MRIIRS, Faridabad, India

© The Author(s), under exclusive license to Springer Nature Singapore Pte Ltd. 2023
Y. Singh et al. (eds.), *Proceedings of International Conference on Recent Innovations in Computing*, Lecture Notes in Electrical Engineering 1011,
https://doi.org/10.1007/978-981-99-0601-7_8

1   Introduction

Agriculture is the cardinal source of livelihood for about 59% of India's population. Indian agriculture in the 21st millennium is structurally deviating and robust than the one prevalent during the green revolution era which commenced in the 1970's. Tremendous advancements and the emergence of new technologies can be witnessed in this sector [1]. Agriculture has gone through drastic alterations throughout the decades; machines are highly automated, and unmanned aerial vehicles (UAVs) and orbital satellites are becoming essential. Agriculture stumps up about 17% of the gross domestic product (GDP) of Indian economy and provides employment to over 60% of the population [2]. The Government of India ratified a number of measures to ameliorate the system of agricultural marketing, standardization of weights and measures, establishment of warehouses, open regulated markets, and commencing policies like MSP and PDS.

India is a multiproduct agricultural powerhouse and produces an enormous range of food and non-food crops, and the yield depends on weather, soil fertility, season, water and nutrients absorbed by crops, and dosage of fertilizers and pesticides [3]. Predicting crop yield with a rimmed area of land is an arduous task in an agro-based country like India. Yield rate can be accelerated by monitoring crop growth, accurate weather predictions, field productivity zoning, crop disease prevention and management, and forecasting crop yield. Each kind of crop has its optimum growth requirements [4–6]. In India, agricultural yield predominantly recons upon weather conditions. For instance, rice cultivation primarily relies upon rainfall, and bean cultivation demands a major amount of sunlight. Weather prediction is a challenging task due to the dynamic nature of the atmosphere [7]. This research also helps the farmer in weather prediction more precisely than the previous studies so that farmers can be alerted well in advance. Prediction is done on the basis of variables like rainfall, sunlight, and pH value of soil.

Advances in machine learning have created new opportunities to revamp prediction in agriculture. Crop production rates depend on the topography and geographic conditions of the region (e.g., hilly areas, river ground, mountainous regions, and depth regions), weather conditions (e.g., rainfall, sunlight, groundwater level, temperature, and pH value of soil), soil type (e.g., sandy, clay, saline, and loam soil), and soil composition and irrigation methods [8–10]. Different prediction models are used for different types of crops grown all over India [11, 12]. This research chronicles different machine learning prediction models, concepts, and algorithms. We also resort to unearth the least traversed areas concerning the integration of machine learning techniques in the agriculture sector. This research will help prospective researchers get a better understanding of what all prediction models have been used till date, and what all still need to be focused upon. The study is segregated into various sections, each section depicting a particular aspect of agriculture.

The main objective of this research is to propose an intelligent and interactive prediction system that aims at predicting the crop yield before harvesting by learning the past 20 years' data of the farming land and helps farmers through timely weather forecasts and identification of crop condition by using machine learning techniques. Factors which are significant to crop production such as farm area, production of crop in the previous years and the seasons of farming for different Indian states, temperature of area, humidity, crop yield, growing season, and water requirements of crop were considered for experimental analysis. To anticipate continuous values, various machine learning techniques are used, and the data is pre-processed by using SSIS.

This paper is organized into following sections. Section 1 conferred with the prefatory phase. Section 2 explores the related research work done in the related field. Section 3 presents the proposed framework to deal with the limitations of the existing systems. Section 4 reveals the result and discussions, followed by conclusion in Sect. 5.

2   Literature Review

Priya et al. [13] proposed a system for predicting the yield of the crop on the basis of existing data by using random forest algorithm. Real data of Tamil Nadu, State of India, was used in this research for building the models, and the same models were tested with samples. Random forest was the only algorithm used for crop yield prediction. Jeong et al. [14], and the team proposed spawned outputs which proved that random forest is a compelling and versatile machine learning algorithm for predicting the crop yield of wheat, maize, and potato, in comparison of multiple linear regression, at both territorial and worldwide scales. The dataset consists of data of yield from US counties and northeastern seaborn regions. Manjula [15] proposed a system to predict crop yield from preceding data. This is accomplished by implementing association rule mining on agriculture data and predicting the crop yield. The paper proposed an analysis of crop yield prediction using data mining techniques for the selected region, i.e., district of Tamil Nadu in India. Raju Prasad Paswan [16] proposed extensive review of literature analyzing feedforward neural networks and traditional statistical methods to predict agricultural crop production. Traditional statistical methods included in this study are linear regression. By results, they concluded that if there is better communication among the fields of statistics and neural networks, then it would benefit both. Veenadhari [17] proposed forecasting crop yield based on parameters of climate. In this Research Crop Advisor, a software tool "Crop Advisor" has been developed as a Web page for forecasting the influence of climatic parameters on the crop yields. Main algorithm used in this study was C4.5. It produces the influencing climatic parameter on the crop yields of selected crops in selected districts of Madhya Pradesh. D Ramesh [18] proposed analysis of crop yield prediction using multiple linear regression (MLR) technique and density-based clustering technique for the district of Andhra Pradesh-East Godavari in India; in this study, results of two methods were compared according to the specific region. Shahane [19] proposed reduction on crop cultivation is basically an aggregation of sustainability, soil analysis, crop and fertilizer recommendation, and crop yield calculations based on present market conditions. Prediction on crop cultivation outperforms the existing system by revising and correcting the failures of the soil analysis processes which were manual-based. A soil analysis provides the agricultural producer with an estimate of the amount of fertilizer nutrients needed to supplement those in the soil. Ferentinos [20] proposed research on a novel artificial neural network (ANN) system which detects and classifies pesticide residues. The novel ANN is customized in a way, to a cellular biosensor operation supported by the bioelectric recognition assay (BERA) and able to, at the same time, assay eight samples in 3 min. Table1 illustrates various major contributions in this field.

Based on the detailed literature survey, the following limitations were identified:

Table1 Summarized literature review

Author	Objective	Techniques used	Accuracy achieved
Raju et al.	Regression and neural networks models for prediction of crop production	Artificial neural network (ANN)	NA
Konstantinos et al.	Pesticide residue screening using a novel artificial neural network combined with a bioelectric cellular biosensor	ANN design (steepest-descent algorithm, quasi-Newton algorithm)	ANN-P1 85% ANN-P2 85% ANN-P3 86.7% Overall 81/100 (81.0%)
Veenadhari et al.	Machine learning approach for forecasting crop yield based on climatic parameters	C4.5 algorithm	Prediction accuracy above 75 percent in all the crops
D. Ramesh et al. /2017	Analysis of crop yield prediction using data mining techniques	Multiple linear regression and density-based clustering technique	Density-based clustering between -13% and $+8\%$ Multiple linear regression between -14% and $+13\%$
E. Manjula et al.	A model for prediction of crop yield	Data clustering, data conversion, and association rule mining	Minimum support of 0.3 and minimum confidence of 0.7
Jig et al.	Predicting yield of the crop using machine learning algorithm	RF algorithm and random forest classifier	EF of -0.41 and d of 0.75.RF > RFC
Alexandros et al.	A recommended system for crop disease detection and yield prediction using machine learning approach	ANN, support vector machine, and clustering	SVM 87.9% ANN: 99.63% C RECALL:0.6066 Precision:0.9191

1. Data pre-processing is the first and crucial step while creating a machine learning model. In the previous studies, data pre-processing is done with the help of Python language which consumes a lot of time, and large codes are generated.
2. Insufficient availability of data (too few data) is a major problem in existing approaches. The studies stated that their systems worked for the limited data that they had at hand, and indicated data with more variety should be used for further testing.
3. Moreover, algorithms like "lasso regression" and "linear discriminant analysis" have never been used in crop prediction and weather forecasting.

3   Proposed Model

Modern and contemporary technologies are gaining more attention with respect to prediction and predictive analysis approaches. Predictive techniques are being favored in recent times due to their immense scope in knowing the agricultural yield in advance. This framework provides the farmer/user an approximation on how much crop yield will be produced depending upon the season, crop, area, and production. Weather has a profound influence on crop growth, total yield, amount of pesticides, and fertilizers needed by crop, and all other activities carried out throughout the growing season.

Features of the proposed model are as follows:

- The proposed model showed how beneficial the amalgam of machine learning and predicting crop yield and weather forecasting could be in agriculture.
- The proposed framework uses algorithms like lasso regression, Gaussian Naïve Bayes, and linear discriminant analysis which have never been used in crop prediction and weather forecasting and hence gives us better accuracy than the existing system.
- The framework utilizes SSIS platform by using visual studio, BIDS, for pre-processing the data which results in less time and simple codes. Python is a very powerful programming language. Combined with SSIS, it can provide robust and flexible solutions to several problems.
- The framework is well supported by measure-based evaluation which ultimately validates the performance (Fig. 1).

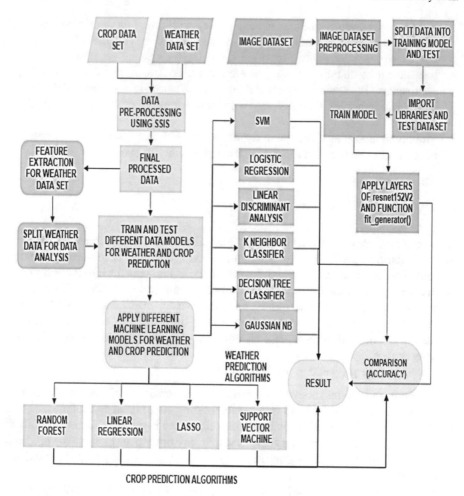

Fig. 1 Integrated ML prediction model

4 Result and Discussions

4.1 Data Pre-Processing for Yield Prediction and Weather Forecasting

Technology used for data pre-processing in this research is SSIS, SQL. Server integration services are a platform for building planned data integration and data transformations solutions. SSIS used to remove the duplicate values by sort transformation editor which worked efficiently in removing all null values, and conditional split was done as per algorithm needed, i.e., raw data was combined from various sources through SSIS.

4.2   Dataset for Crop Yield Prediction and Weather Prediction

Dataset for crop yield prediction is collected from "ICAR-Indian Agriculture Research Institute", India's national institute for agriculture research and education. The dataset collected consists of data from every district of India, but research focused on the dataset of Maharashtra and Punjab state in India as crop yield prediction for Maharashtra has yet not been carried out. The entire dataset is segregated into two parts: 70% of the dataset is used for training, and 30% dataset is used for testing. The dataset is trained on 20 years (1995–2014) of data from districts in Punjab and Maharashtra (it is noted that it starts in 1995 and ends in 2014. Merging the data frames together, it is expected that the year range will start from 1997 and ends in 2014). Dataset for weather prediction is collected from government Websites, i.e., www.imd.gov.in. Attributes of data set are state name, district_name, crop_year, season, area, and production. The features analyzed for crop yield prediction and weather prediction: season, pH value, temperature, humidity, rainfall, yield, water, and crop.

4.3   Performance Evaluation

Different machine learning algorithms such as random forest, linear regression, lasso regression, and support vector machine are applied to the Maharashtra and Punjab State of India from the dataset to predict the crop yield in advance before harvesting and were compared using the R2-score and MSE measures.

- R2-score: R-squared is a statistical measure that defines the goodness of fit of the model. The ideal R2-score value is 1. The closer the value of r-square is to 1, the better the model is. R2 value can also be negative if the model is worse than the average fitted model.

    R-square $= 1-(SS_{res}/ SS_{total})$
    SS_{res}: Residual sum of squares
    SS_{total}: Total sum of squares
- Mean Squared Error: MSE is a statistical measure that is defined as the mean or average of the square of the difference between actual and estimated values. The lesser the value of MSE, the better the model is.

Punjab, from Table 1, we can say that random forest is the best algorithm for crop yield prediction with the accuracy of 96.72% and the highest R2-score with least error SVM performs worst. Table 2 shows the accuracy representation for finding the best-fit algorithm.

For Maharashtra, from Table 3, it is concluded that random forest is giving the highest accuracy of 96.33% with the best R2-score with minimum error, and linear regression turned out to be the worst algorithm for crop yield prediction.

Table 2 Accuracy evaluation of the model (Punjab)

Algorithms	R2-score	MSE	Accuracy (%)
Random forest	0.96	0.046	96.72
Lasso regression	0.90	0.136	90.36
Linear regression	0.90	0.136	90.36
SVM	0.81	0.269	81.02

Table 3 Accuracy evaluation of the model (Maharashtra)

Algorithms	R2-score	MSE	Accuracy (%)
Random forest	0.96	0.317	96.33
Lasso regression	0.32	5.368	32.09
Linear regression	0.32	5.368	32.15
SVM	0.83	1.785	83.54

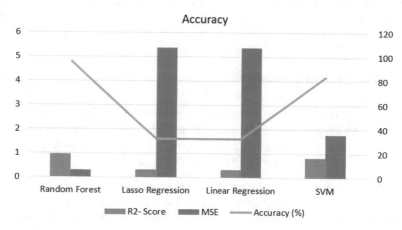

Fig. 2 Accuracy comparison of algorithms for Maharashtra State

Graph, in Fig. 2, shows the higher the R2-score and lower the MSE, better the accuracy of an algorithm is. The above graph depicts that random forest is showing the best R2-score with least MSE value. The closer the value of R2-score is to 1, the better the algorithm is.

4.4 Measures for Estimating the Accuracy of Weather Prediction

Various algorithms such as KNN, Gaussian Naïve Bayes, logistic regression, decision tree, SVM, and linear discriminant analysis applied for weather forecasting, to help the farmers to adapt to the situation and take preventive measures before harvesting

Table 4 Weather prediction accuracy

Algorithms	Precision	Recall	F1-score	Accuracy (%)
KNN	0.74	0.67	0.67	67.56
Logistic regression	0.63	0.62	0.59	62.16
Decision tree	0.90	0.83	0.85	83.78
Gaussian Naïve Bayes	0.93	0.91	0.91	91.89
SVM	0.83	0.78	0.78	78.37
Linear discriminant analysis	0.73	0.72	0.69	72.97

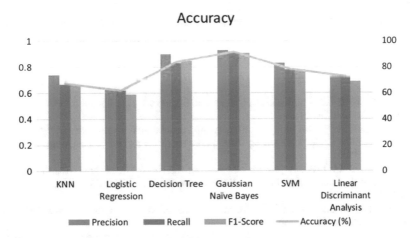

Fig. 3 Accuracy trend of algorithms against cross validation score

the crop. Different parameters such as precision, recall, and F1-score are used to find the algorithm with the best accuracy in weather prediction. From Table 4, it is concluded that Gaussian Naïve Bayes is the best algorithm for weather prediction with the accuracy of 91.89%.

The graph in Fig. 3 is plotted against cross validation score, the higher the cross validation score, the better the algorithm is. As, Gaussian Naïve Bayes (NB) is showing the highest score and is the best-fit algorithm for weather prediction.

5 Conclusion

For proposed model, random forest provides maximum accuracy of 96.33% and is highly efficient in prediction of crop yield. Gaussian Naïve Bayes is the proficient algorithm with highest accuracy 91.89% in forecasting weather. Performance of the model is found to be relatively sensitive to the quality of weather prediction, which in turn recommend the significance of weather forecasting techniques. This model lessen the troubles confronted through farmers and will serve as a delegate to offer farmers with the information they want to benefit high and maximize the profits. In future, starting from the training dataset used for the model, it can be further incorporated with crop-imaging data. Additionally, newer algorithms and learning methods could be used for prediction, enhancing the accuracy of the learning model along with better measures for calculating the accuracy of the employed classifier.

References

1. Pantazi XE, Moshou D, Oberti R, West J, Mouazen AM, Bochtis D (2017) Detection of biotic and abiotic stresses in crops by using hierarchical self-organizing classifiers. Precis Agric 18:383–393
2. Moshou D, Bravo C, West J, Wahlen S, McCartney A (2004) Automatic detection of "yellow rust" in wheat using reflectance measurements and neural networks. Comput Electron Agric 44:173–188
3. Sangeeta SG (2020) Design and implementation of crop yield prediction model in agriculture. Int J Sci Technol Res 8(01)
4. Shin HC, Roth HR, Gao M, Lu L (2016) Deep convolutional neural networks for computer-aided detection: CNN architectures, dataset characteristics and transfer learning. IEEE Trans Med Imag 35(5)
5. Torrey L, Shavlik J (2010) Transfer learning. In: Handbook of research on machine learning applications and trends: algorithms, methods, and techniques, Olivas ES et al (eds). IGI Global, pp 242–264
6. Champaneri M, Chachpara D, Chandwadkar C, Rathod M (2020) Crop yield prediction using machine learning. Int J Sci Res (IJSR)
7. Alexandros O, Cagata C, Ayalew K (2022) Deep learning for crop prediction: a systematic literature review. New Zealand J Crop Horticult Sci
8. Folnovic T (2021) Importance of weather monitoring in farm production, Agrivi. All rights Reserved
9. Pisner DA, Schnyer DM (2020) Support vector machine, in Machine Learning
10. Alonso J, Castañón ÁR, Bahamonde A, Support vector regression to predict carcass weight in beef cattle in advance of the slaughter. Comput Electron Agric
11. Suganya M, Dayana R, Revathi R (2020) Crop yield prediction using supervised learning techniques. Int J Comput Eng Technol 11(2)
12. Amit S, Nima S, Saeed K (2022) Winter wheat yield prediction using convolutional neural networks from environmental and phonological data. Sci Rep Nat
13. Priya P, Muthaiah U, Balamurugan M (2018) Predicting yield of the crop using machine learning algorithm. Int J Eng Sci Res Technol. 7(4). ISSN: 2277–9655
14. Jeong J, Resop J, Mueller N t al. Random forests for global and regional crop yield prediction, PLoS ONE J
15. Manjula E, S (2017) A model for prediction of crop yield. Int J Comput Intell Inf 6(4)
16. Paswan RP, Begum SA (2013) Regression and neural networks models for prediction of crop production. Int J Sci Eng Res 4(9)
17. Veenadhari S, Dr. Misra B, Dr. Singh CD Machine learning approach for forecasting crop yield based on climatic parameters. In: International conference on computer communication and informatics (ICCCI)
18. Ramesh D, Vardhan B (2015) Analysis of crop yield prediction using data mining techniques. Int J Res Eng Technol 4(1):47–473

19. Shahane SK, Tawale PV (2016) Prediction on crop cultivation. Int J Adv Res Comput Sci Electron Eng (IJARCSEE) 5(10)
20. Ferentinos KP, Yialouris CP, Blouchos P, Moschopoulou G, Kintzios S (2013) Pesticide residue screening using a novel artificial neural network combined with a bioelectric cellular biosensor. Hindawi publishing corporation BioMed research international

Machine Learning Techniques for Result Prediction of One Day International (ODI) Cricket Match

Inam ul Haq, Neerendra Kumar, Neha Koul, Chaman Verma, Florentina Magda Eneacu, and Maria Simona Raboaca

Abstract Cricket is the most popular sport, and most watched now a day. Test matches, One Day Internationals (ODI), and Twenty20 Internationals are the three forms in which it is played. Until the last ball of the last over, no one can predict who would win the match. Machine learning is a new field that uses existing data to predict future results. The goal of this study is to build a model that will predict the winner of a One Day International Match before it begins. Machine learning techniques will be used on testing and training datasets to predict the winner of ODI match that will be based on the specified features. The data for model will be collected from Kaggle, and some will be collected from the different cricket Web sites because the data obtained from Kaggle have only matches up until July 2021. After that prediction will be done, and the model will provide advantages to team management in terms of improving team performance and increasing the chance of winning the game. This model will be used to predict the outcomes of the next Cricket World Cup 2023, which will be, the 13th, edition of the men's ODI Cricket World Cup and will be hosted by India in 2023. Also, this work will serve as a guidance work, as there is much to be done in the field of sports.

I. Haq · N. Kumar (✉) · N. Koul
Department of Computer Science and IT, Central University of Jammu, Jammu and Kashmir, India
e-mail: neerendra.csit@cujammu.ac.in

C. Verma
Department of Media and Educational Informatics, Faculty of Informatics, Eötvös Loránd University, Budapest, Hungary
e-mail: chaman@inf.elte.hu

F. M. Eneacu
Department of Electronics, Communications and Computers, University of Pitesti, Pitesti, Romania

M. S. Raboaca
National Research and Development Institute for Cryogenic and Isotopic Technologies-ICSI, Rm. Valcea, Romania
e-mail: simona.raboaca@icsi.ro

© The Author(s), under exclusive license to Springer Nature Singapore Pte Ltd. 2023
Y. Singh et al. (eds.), *Proceedings of International Conference on Recent Innovations in Computing*, Lecture Notes in Electrical Engineering 1011,
https://doi.org/10.1007/978-981-99-0601-7_9

Keywords Machine learning · ODI · World cup · Logistic regression · Random forest · SVM · Naïve Bayes

1 Introduction

Cricket was initially brought to North America in the seventeenth century through English colonies, and it went to other areas of the world in the eighteenth century. Colonists introduced it to the West Indies, while British East India Company sailors introduced it to India. It came in Australia nearly immediately after colonization began in 1788, and in the middle of the nineteenth century, it came to New Zealand and South Africa. The International Cricket Council is in control of cricket (ICC), which is the sport's international governing body. A cricket match might end in a "win" for one of two teams or a "tie." In a limited-over's game, if the game cannot be completed on time, the game may conclude in a "draw;" in other forms of cricket, a "no result" may be feasible. When one side scores more runs than another, and all of the opponent team's innings have been completed, the game is called "won." The team with the most runs "wins," while the side with the fewest runs "loses." If the match is called off before all of the innings have been finished, the result could be a tie or no result. When the scores are tied at the end of play, the game is declared a "tie," but only if the team batting last has finished its innings. Only, two tests have ever ended in a tie, which is remarkable in cricket. In several one-day cricket formats, such as Twenty20, a Super Over or a bowl-out is commonly used as a tiebreaker to settle a result that would otherwise be a tie. If a match is ended without win or tie, the result is "draw," as described in Law 16. When one or both sides do not finish their innings before the planned end of play, the game is called a draw. No matter how many runs either side has scored, the match is ultimately drawn. If a limited-over match that has already started cannot be completed due to weather or minor disruptions, a "no result" is declared.

When rain is a factor, the match is frequently referred to be "washed out." The match can be "abandoned" or "cancelled" if weather or other conditions stop it from happening. Cricket is unique; it has 3 different forms. Test Cricket, Twenty20 International Cricket, and One Day International Cricket. One Day International (ODI) is a sort of limited-over's cricket match in which two sides compete against each other. Each team is given a set number of over's, currently 50, and the game can take up to 9 h. In which each side bats only once. Each side's innings finishes either when their allotted number of over's has been completed, or all ten wickets have been lost.

Cricket World Cup, a four-year international cricket championship that is the top one-day cricket competition and one of the supreme-viewed sporting action in the world (Table 1).

As technology progresses and applications like as fantasy 11 and betting sites grow more popular, people will trust on the predictions offered by the ML model.

Table 1 Results of the ODI Cricket World Cup 1975–2019

Year	1975	1979	1983	1987	1992	1996	1999	2003	2007	2011	2015	2019
Winner	West-Indies	West-Indies	India	Australia	Pakistan	Sri-Lanka	Australia	Australia	Australia	India	Australia	England

In a number of ways, machine learning makes life so easier. Machine learning is categorized into three groups based on the methodologies and methods of learning.

Unsupervised learning models are trained without being supervised. As the name suggests, models, on the other hand, use the data to uncover hidden patterns and insights. It is similar to how the human brain learns new information. Unsupervised learning is a type of learning in which model is trained on unlabeled data and then left to operate on it on their own. Supervised learning is a machine learning method in which systems are trained with a very well training data and used to predict output. A few of the input data already has been classified with the appropriate output, as indicated by the labeled data. The training data supplied to the machines functions as a supervisor in supervised learning, teaching the machines on how to accurately predict the output. It works on the same principle as when a pupil is instructed by a teacher. There are two types of supervised learning: regression and classification. The term "classification" is used in differentiating between categories such as blue and red. When the output is a real value, such as dollars or height, regression is applied. Reinforcement learning is a feedback-based method in which an AI agent explores its environment naturally by striking and trailing, taking action, learning from experiences, and improving its performance.

2 Literature Review

- The author of the paper [1] used data from 12 seasons of IPL matches. In this paper, the dataset of the first 11 seasons is used as a training dataset, having 580 matches. The last season is considered a testing dataset, with 60 matches. The numbers are assigned to the names of the teams. The author has used various ML modules and has assembled the modules. After the winning team is determined by computing the total percentage of all model outcomes, decision tree. regression, random forest regression, support vector machine, Naive Bayes, multiple. linear regression, and logistic regression are some of the algorithms that were utilized. The accuracy is achieved at nearly 90%.
- Chowdhury et al. [2] projected the winner of the ODI cricket match played between India and Pakistan. They manually obtained info on all ODI matches between India and Pakistan between 1978 and 2019 from www.espncricinfo.com. The data are preprocessed for model creation, with tied matches being removed, and so on. When using logistic regression, the chance of an India team winning on home court was 70.6% higher and 2.28 times higher for data when related to day night period contests. This research allows for the discovery of hindsight of a winning match in favor of Team India.
- Jalaz et al. [3] investigated the impact of two machine learning models, decision trees and multilayer perception networks, on the outcome of a cricket match. Founded on these findings, the Cricket Outcome Prediction System was created for estimating the ultimate result of a particular match; the developed method considers pregame variables such as the ground, venue, and innings. ESPN

cricinfo is used to obtain the data. All ODI matches from January 5, 1971, through October 29, 2017, are included in this dataset. There were 3933 ODI match outcomes. Some of the matches in the dataset were removed from the analysis during the cleaning phase. After comparison of accuracy, the multilayer perceptron has a score of 0.574, whereas the decision tree classifier has a score of 0.551.

- Mago et al. [4] predict the winner of an IPL match before it begins. To determine the winner of the IPL, machine learning algorithms are trained on key features. The SEMMA approach was used for the study of the IPL T20 match winner dataset. The dataset was preprocessed to ensure consistency by eliminating missing values and encoding variables into numerical format. First, a decision tree was used, which accurately predicted the winner with an accuracy of 76.9%. The parameters for the decision tree model are fine-tuned to increase model performance and get satisfactory results. The model's performance improved by 76.9–94.0%. The random forest model was then applied and predicted the winner with an accuracy of 71%. That is not enough, therefore, the random forest model was tweaked using tuning of parameters, and the results improved to 80%. The XGBoost model was used last. The outcome was 94.23% without any parameter adjustment.
- Baasit et al. [5] have examined the popular machine learning algorithms to declare the winner of the 7th edition of the 20–20 World Cup 2020, which was hosted in Australia. The ESPN cricinfo dataset was used in this analysis. This research employs four different learning methods (C4.5, random forest, extra trees, ID3, and random forest). Random forest was determined as the best algorithm using proprietary efficiency criteria. It achieved a standard efficiency of 80.86%. Australia was expected to win the 20–20 Men's Cricket World Cup for the next two years.
- Aggrwal et al. [6] have predicted results collaborative and the ability of each player to contribute to the match's result. The data were gathered from techgig.com. The database comprises data from the previous 500 IPL matches, which has been preprocessed. Support vector machine, CTree, and Naive Bayes are three machine learning approaches that achieve accuracy of 95.96%, 97.98%, and 98.99%, respectively.
- Barot et al. [7] have given a model which is used to improve a bowler or batsman's rating and performance in various match aspects are used to investigate what determines outcome of a cricket match, and outcome of cricket match is. also predicted using a variety of features. The data for analysis and prediction were collected from www.espncricinfo.com, which includes data from previous IPL editions, and retrieved from www.kaggle.com. For match predictions, machine learning methods such as SVM, logistic regression, decision, tree, random forest, and Naive Bayes were used. The best accuracy was over 87% and 95% for the decision tree and logistic regression methods, respectively.
- Islam et al. [8] presented a way of estimating a cricket player's performance in a future match. The suggested model is based on statistical data acquired from reliable sports sources on the Bangladesh national cricket team's players. As for selection methods, recursive feature elimination and univariate selection are used, and

as for machine learning algorithms, linear regression and support vector machines with linear and polynomial kernels are used. In the forthcoming match, machine learning algorithms are employed to anticipate how many runs the batsman will score and how many runs the bowler will concede. The model correctly forecasts batsman Tamim and bowler Mahmudullah with up to 91.5% accuracy, with other players' predictions being similarly accurate.

- Rameshwarie et al. [9] created a model that can forecast outcomes while the game is still being played, i.e., live prediction. The amount of wickets lost, the match's venue, the team's ranking, the pitch report, and the home team's advantages were all considered in this study. The main goal of this research is to ripen a model for predicting the final score of the first innings and the outcome of the second innings in a limited-overs cricket match. Two separate models have been provided based on prior matches, one for the first innings and the other for the second innings, utilizing the linear regression classifier and the Naive Bayes classifier, respectively. A reinforcement algorithm is also employed.

- Rudrapal et al. [10] used a deep neural network model to predict the outcome of a football match automatically. There are various obstacles and instances where the suggested method fails to predict the outcome of a match. The info was gathered from a variety of online sources. MLP, SVM, Gaussian Naive Bayes, and random forest are among the algorithms used, which show an accuracy of 73.57%, 58.77%, 65.84%, and 72.92%, respectively.

- Gagana et al. [11] predicted the number of runs for each ball by using the batsman's previous runs as observed data. Data from all past IPL matches are collected for the analysis. The Naive Bayes classifier, decision tree, and random forest were employed, with an accuracy of 42.95%, 79.02%, and 90.27%, respectively, when 70% of the data was used for training and 30% for testing.

- Kapadiya et al. [12] performed a thorough study and review of the literature in order to provide an effective method for predicting player performance in the game of cricket. This model will aid in the selection of the best team and, as a result, increase overall team performance. For player performance prediction, a meteorological dataset is used with cricket match statistics. The accuracy rates for Naive Bayes, decision tree, random forest SVM, and weighted random forest were 58.12%, 86.50%, 92.25%, 68.78%, and 93.73%, respectively.

- Passi et al. [13] developed a model that considers both teams' player performances, such as the number of runs a batsman will. score, and the no. of wickets a bowler will take. Prediction models are developed utilizing Naive Bayes, random forest, multiclass SVM, and decision. tree classifiers for both aims. Random forest was found the most accurate classifier in both experiments. Scraping tools were used to collect data from cricinfo.com. The best accurate classifier for both datasets was random forest, which correctly predicted batter runs with 90.74% accuracy and bowler wickets with 92.25% accuracy.

- Lamsal et al. [14] offered a multifactorial regression-based approach to calculating points for each player in the league, and the load of a team is determined based on the historical performance of the players who played the most for the squad. Six machine learning models have been proposed and utilized to know the result

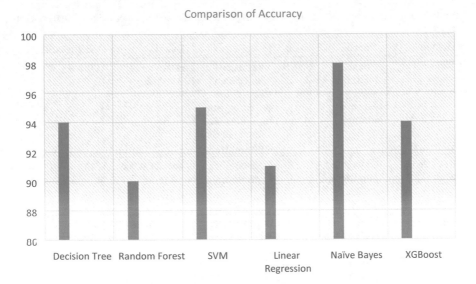

Fig. 1 Comparison of algorithms on basis of accuracy rates

of each 2018 IPL match 15 min prior to the start of the game, just after the toss. The proposed model accurately predicted over 40 matches, with the multilayer perceptron model surpassing the others by 71.6% (Fig. 1).

3   Comparative Analysis

See Table 2.

Table 2 Machine learning techniques for result prediction of One Day International (ODI) cricket match

Authors	Domain	Year	Model	Dataset	Accuracy/Result
Pallavi et al	Cricket prediction using machine learning	2020	SVM, decision tree, random forest, Naïve Bayes, logistic regression, and multiple linear regression	IPL from year 2008 to 2019	90% Aggregate
Chowdhury et al	ODI cricket forecast in logistic analysis	2020	Logistic regression	Espncriciinfo.com	NA
Jalaz et al	Using decision tree and MLP networks, predict the outcome of an ODI match	2018	Decision tree, MLP networks	Espncrickinfo com	0.574 and 0.551, resp.
Daniel Mago et al	The Cricket Winner Prediction with Application Of Machine Learning and Data Analytics	2019	Decision tree, random forest classifier, XGBoost	IPL from season 2008 to 2017	94.87%, 80.76%, 94.23%, resp.
Ab. Baasit et al	Predicting the winner of the ICC T20 Cricket World Cup 2020 using machine learning techniques	2020	Random forest, C4.5, ID3, and extra trees	espncricinfo. com	80.86%, 79.73%, 74.69%, 79.67%, resp.
Shilpi Aggrwal et al	Using machine learning to predict the outcomes of IPL T20 matches	2018	Support vector, CTree, and Naive Bayes	techgig.com	95.96%, 97.98%, 98.99%, resp.

(continued)

Table 2 (continued)

Authors	Domain	Year	Model	Dataset	Accuracy/Result
Harshiit Barot et al	Study and prediction for the IPL	2020	SVM, logistic regression, decision. tree, random forest, and Naïve Bayes	Kaggle.com espncricinfo.com	83.67%, 95.91%, 87.95%, 83.67%, 81.63% Resp.
Aminul Islam et al	Machine learning algorithms for predicting player performance in ODI Cricket	2018	Linear regression, support vector machine	Sports Web sites	Batsman Taamim has a 91.5% accuracy rate, while bowler Mahmudullah has a 75.3% accuracy rate
Rameshwarie et al	Winning prediction and live cricket score	2018	Linear regression, Naïve Bayes, reinforcement algorithm	Na	Na
Dwijen Rudrapal, et al	Predicting the outcome of a football match using deep learning	2021	MLP, SVM, Gaussian.Naive .Bayes, and random forest	Sports websites	73.57% 58.77% 65.84 and 72.92%, resp.
Gagana et al	A view on using machine learning to analyze IPL match results	2019	Naïve Bayes classifier, decision tree, random forest	dataworld.com	42.95%, 79.02%, and 90.27%, resp.

4   Open Research Challenges

Various challenges involved in the existing models of ODI cricket matches are given as follows:

- **Data shortage**: Dataset of ODI cricket match available is currently incomplete, as it only comprises matches up to July 2021. An advanced dataset is required to overcome the deficit of data.
- **ODI Model:** A very little work has been done in building a model for the ODI format of cricket. As per the latest research, main focus is given on building prediction model for Cricket World Cup 2023.
- **ODI Algorithms:** Fewer algorithms have been trained in previous researches. Trained algorithms are required in this active field of study. The algorithms that produce the best results are considered for the future use.

- **Prediction Models:** Models for outcome prediction are rapidly evolving, with various new strategies being developed and existing techniques being changed to improve performance. As per latest research, new and more advanced ways for outcome prediction are required in this active field of research.

5 Proposed Model

5.1 System Flow

The following methodology will be used in this work which consists of different phases shown in Fig. 2.

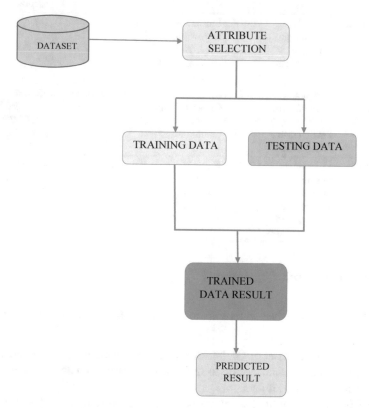

Fig. 2 Framework for prediction model

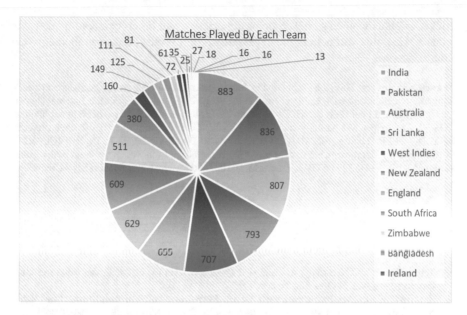

Fig. 3 Chart shows how many ODI matches each team have played

5.2 Dataset

The dataset is collected from Kaggle, and some data were added manually into the dataset. The dataset consists of 7734 ODI matches. The dataset consists of irrelevant information like date which are discarded and team 1, inn, rpo, runs, team2, overs, result, and ground are included. Machine learning model takes data only in numeric format, so the feature team 1, Ground, and team 2 are converted into numbers (Fig. 3).

6 Conclusion

Our major goal in this research is to use machine learning methods to construct a. model. That can predict the outcome of an ODI match before it starts. In this work, data of ODI matches played after July 2019 will be included in dataset. We selected 8 key features that will give the best possible prediction accuracy. If we see in Table 2, the highest accuracy is of [6], and lowest accuracy is of [11]. So, after analyzing each paper, we found all the key factors that increased the prediction accuracy. The model is based on data from previous matches between the teams. This work includes efficiency and accuracy checks. This approach can be applied to various forms of cricket, such as women's cricket, domestic cricket, and other sports, to predict the winner. Also, this will help in the development of a robust prediction model in the future.

Acknowledgements This paper supported by Subprogram 1.1. Institutional performance-Projects to finance excellence in RDI, Contract No. 19PFE/30.12.2021 and a grant of the National Center for Hydrogen and Fuel Cells (CNHPC)—Installations and Special Objectives of National Interest (IOSIN). This paper was partially supported by UEFISCDI Romania and MCI through projects I-DELTA, DEFRAUDIFY, EREMI, SMARDY, STACK, ENTA, PREVENTION, RECICLARM, DAFCC and by European Union's Horizon 2020 research and innovation program under grant agreements No. 872172 (TESTBED2) and No. 101037866 (ADMA TranS4MErs), SmarTravel.

References

1. Tekade P, Markad K, Amage A, Natekar B (2020) Cricket match prediction using machine learning. Int J Adv Sci Res Eng Trends 5(7)
2. Zayed M (2020) One day international (ODI) cricket match prediction in logistic analysis: India VS. Pakistan. Int J Hum Movement Sports Sci. https.//doi.org/10.13189/saj.2020.080629
3. Kumar J, Kumar R, Kumar P (2018) Outcome prediction of ODI cricket matches using decision trees and MLP networks. In: IEEE (ICSCCC)
4. Vistro DM, Rasheed F, David LG (2019) The cricket winner prediction with application of machine learning and data analytics. Int J Sci Technol Res 8(09)
5. Basit A, Alvi MB, Fawwad, Alvi M, Memon KH, Shah RA (2020) ICC T20 cricket world cup 2020 winner prediction using machine learning technique. IEEE https://doi.org/10.13140/RG.2.2.31021.72163
6. Agrawal S, Singh SP, Sharma JK (2018) Predicting results of indian premier league T-20 matches using machine learning. IN: IEEE international conference on communication system and network tech
7. Barot H, Kothari A, Bide P, Ahir B, Kankaria R Analysis and prediction for the Indian premier league. In: IEEE 2020 international conference for emerging technology (INCET)
8. Anik AI, Yeaser S, Hossain AGMI, Chakrabarty A (2018) Player's performance prediction in ODI cricket using machine learning algorithms. In: IEEE 2018 4th international conference on electrical engineering and information and communication technology (iCEEiCT)
9. Lokhande R, Chawan PM (2018) Live cricket score and winning prediction. Int J Trend Res Develop 5(1), ISSN: 2394–9333
10. Rudrapal D, Boro S, Srivastava J, Singh S (2020) A deep learning approach to predict football match result. https://doi.org/10.1007/978-981-13-8676-3_9, ResearchGate
11. Gagana, Paramesha K (2019) A perspective on analyzing IPL match results using machine learning. Int J Sci Res Develop 7(03)
12. Kapadiya C, Shah A, Adhvaryu K, Barot P (2020) Intelligent cricket team selection by predicting individual players' performance using efficient machine learning technique. Int J Eng Adv Technol (IJEAT), 9(3). ISSN: 2249–8958 (Online)

13. Passi K, Pandey N (2018) Increased prediction accuracy in the game of cricket using machine learning. Int J Data Mining Knowl Manage Process (IJDKP) 8(2)
14. Lamsal R, Choudhary A (2020) Predicting outcome of Indian premier league (IPL) matches using machine learning. ResearchGate arXiv:1809.09813

Recommendation System Using Neural Collaborative Filtering and Deep Learning

Vaibhav Shah, Anunay, and Praveen Kumar

Abstract Recommender systems have transformed the nature of the online service experience due to their quick growth and widespread use. In today's world, the recommendation system plays a very vital role. At every point of our life, we use a recommendation system from shopping on Amazon to watching a movie on Netflix. A recommender system bases its predictions, like many machine learning algorithms, on past user behavior. The goal is to specifically forecast user preference for a group of items based on prior usage. The two most well-liked methods for developing recommender systems are collaborative filtering and content-based filtering. Somehow, we were using the traditional methods, named content-based filtering (CB) and collaborative-based filtering (CF), which are lacking behind because of some issues or problems like a cold start and scalability. The approach of this paper is to overcome the problems of CF as well as CB. We built an advanced recommendation system that is built with neural collaborative filtering which uses implicit feedback and finds the accuracy with the help of hit ratio which will be more accurate and efficient than the traditional recommendation system.

Keywords Recommendation system · Neural collaborative filtering · Explicit feedback · Implicit feedback

1 Introduction

Search engines and recommendation systems have become an effective approach to generating relevant information in a short amount of time, thanks to the exponential increase of digital resources from the Internet. A recommendation system is a crucial tool for reducing information overload [1]. Intelligent tools for screening and

V. Shah (✉) · Anunay
Department of Computer Science and Engineering, Parul University, Vadodara, India
e-mail: shahvaibhav348@gmail.com

P. Kumar
Department of Computer Science and Engineering, Indian Institute of Technology, Chennai, India
e-mail: praveen2221cs11@iitp.ac.in

© The Author(s), under exclusive license to Springer Nature Singapore Pte Ltd. 2023   109
Y. Singh et al. (eds.), *Proceedings of International Conference on Recent Innovations in Computing*, Lecture Notes in Electrical Engineering 1011,
https://doi.org/10.1007/978-981-99-0601-7_10

selecting Websites, news articles, TV listings, and other information are among the latest developments in the field of recommendation systems. Such systems' users frequently have a variety of competing needs. There are many variations in people's personal tastes, socioeconomic and educational backgrounds, and personal and professional interests. Therefore, it is desirable to have customized intelligent systems that process, filter, and present information in a way that is appropriate for each user of them. Recommendation System Using Neural Collaborative filtering, Traditionally, relied on clustering, KNN, and matrix factorization techniques.

Deep learning has outstanding success in recent years in a variety of fields, from picture identification to natural language processing. The traditional approach for recommendation systems is content filtering and collaborative filtering. Content filtering is used broadly for creating recommendation systems that use the content of items to create features that contest the user profile. Items are compared to the previous item which is liked by the user, and then, it recommends which is the best match to the user profile [2]. Collaborative-based filtering (CF) is the most popular method for recommendation systems which exploits the data which is gathered from user behavior in the past (likes and dislikes) and then recommends the item to the user.

Collaborative filtering suffers from a cold start, sparsity, and scalability [3]. CF algorithms are often divided into two categories, such as memory-based methods (also known as nearest neighbor's methods) and model-based approaches. Memory-based approaches attempt to forecast a user's choice based on the evaluations of other users or products who have similar preferences. Locality-sensitive hashing, which implements the closest neighbor's method in linear time, is a common memory-based method methodology. On the other hand, modeling methods are developed with the help of data mining and ML techniques to reveal patterns or designs based on a training set [4]. However, an advanced recommendation system uses deep learning as it is more powerful than your traditional methods. Deep learning's ability has also improved recommendation systems. Deep learning capability to grasp nonlinear and nontrivial connections between consumers and items, as well as include extensive data, makes it practically infinite, and consequential in levels of recommendation that many industries have so far achieved. Complex deep learning systems, rather than traditional methods, power today's state-of-the-art recommender systems like Netflix and YouTube.

2 Related Work

2.1 Explicit Feedback

In recommendation systems, explicit feedback is in the form of unswerving, qualitative, and measurable responses from users. Amazon, as an illustration, permits customers to rate their purchases on a scale of 1 to 10. These ratings come straight from the customers, allowing Amazon to quantify their preferences. The thumbs-up button on YouTube is yet another example of explicit feedback from users [5]. However, the problem with this feedback is that it is seldom. Remember when you hit the like button on YouTube or contributed a response (in the form of a rating) to your online purchases? Probably not. The count of videocassettes you specifically rate is lesser than the number of videos you fob watch on YouTube.

2.2 Implicit Feedback

Implicit feedback is collected tortuously through user communications and works as a substitution for user decisions. For example, even if one does not rate the videos explicitly, the videos one watches on YouTube are utilized in the form of implicit feedback to customize recommendations for that user. Let us look at another example of implicit feedback: The products you have window-shopped on Amazon or Myntra are utilized to propose additional items that are similar to them. Implicit feedback is so common that many people believe it is sufficient.

Implicit feedback recommenders allow us to modify recommendations in here and now in short in real time, with every single hit and communication or interaction. Today, implicit feedback is used in online recommender systems, allowing the model to align its recommendations in real time with each user interaction. Though, implicit feedback has its deficiencies as well. Unlike explicit feedback, every interaction is assumed to be positive, and we are unable to capture negative preferences from users. How do we capture negative feedback? One technique that can be applied is negative sampling, which we will go through in a later section.

2.3 Collaborative Filtering

It is a technique of filtering items that a user might enjoy based on the response from other users. The cornerstone of a personalized recommender system is collaborative filtering, which involves modeling users' preferences on products grounded on their prior interaction (ex, ratings, and hits). The collaborative filtering (CF) task with implicit feedback is a common term for the recommendation problem, with the goal of recommending a selection of items to users [6].

Tapestry was one of the first collaborative filtering-based recommender systems to be implemented. The explicit opinions of members from a close-knit community, such as an office workgroup [7], were used in this method. For Usenet news and videos, the GroupLens research system [8, 9] provides a pseudonymous collaborative filtering approach. Ringo [10] and video recommender [11] are emails and Web-based systems for making music and movie suggestions, respectively.

Transforming Dataset into an Implicit Feedback Data

As the previously stated, we will be using implicit feedback to train a recommender system. The MovieLens dataset, on the other hand, is based on explicit feedback. To achieve this, we will just binarize the ratings to make them '1' (positive class) or '0' (negative class). A value of '1' indicates that the user has engaged with the piece, while a value of '0' indicates that the user has not.

It is crucial to note that providing implicit feedback changes the way our recommender thinks about the problem. Instead of attempting to forecast movie ratings, we are attempting to forecast whether the user will interact with each film, to present users with the films that offer the highest probability of interaction. After finalizing our dataset, we now have a problem because every example dataset falls under the positive class. We would also need negative examples to train our model because we anticipate users are not interested in such films.

3   System Model Approach (NCF)

Although there are other deep learning architectures for recommendation systems, we believe that the structure that he-et-al. have proposed is the most manageable to implement and is also the most straightforward one. Most recommendation systems are based on content-based, collaborative-based filtering, or hybrid filtering which are nice models but not as much as they have their disadvantages which let them down somewhere.

To overcome the disadvantages of these models, we come up with a powerful recommendation system using neural collaborative filtering which uses implicit feedback and finds the accuracy with the help of hit ratio which will be more accurate and efficient than the traditional recommendation system. The model we built (recommendation system) gives the best result and accuracy for the search of movies and is similar to it for the recommendation.

4   User Embeddings

Before we get into the model's architecture, let us get acquainted with the concept of embeddings. The similarity of vectors from a higher-dimensional space is captured by embedding a low-dimensional space. Let us take a closer look at user embeddings to better understand this notion. Let us say, we aim to serve our visitors based on their preferences for two genres of movies: action and fictional films. Assume the first dimension to represent the user's preference for action films and the second dimension as their preference for fictional films (Fig. 1).

Fig. 1 a User embedding
(2-dimensions). **b**
Representation of Bob in the
embedding Joe will be our
next user. Joe enjoys both
action and romance films. **c**
Representation of Bob and
Joe in the embedding Joe,
like Bob, is represented by a
two-dimension vector above

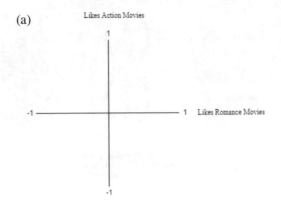

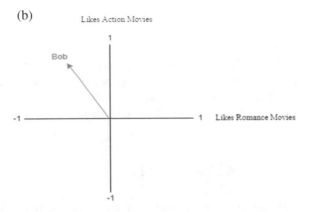

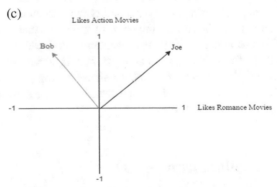

Assume Bob to be our very first user. He enjoys action films and nonetheless dislikes romance films. We place Bob in a two-dimensional vector according to his preferences.

An embedding is a name for this two-dimensional space. Embedding reduces the size of our users in a way that they could be denoted in a significant way in a

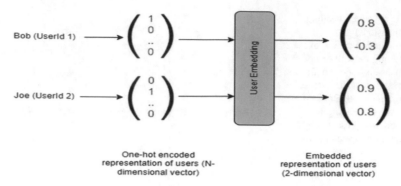

Fig. 2 Representation of users (N-dimension vector) in two-dimensional space

low-dimensional space. Users who have identical movie choices are grouped in this embedding (Fig. 2).

Of course, we are not inadequate to serve our users in only two dimensions. To represent our consumers, we employ any number or a count of dimensions. At the expense of model complexity, a larger amount of dimensions would help us capture the attributes of each user more precisely. We will use eight dimensions in our code.

5   Learned Embeddings

Furthermore, we will define the features of the objects (movies) in a lower-dimensional space using a second object embedding layer. You may wonder how we might get the embedding layer's weights so that it can present an accurate picture of users and items. We yourself generated embedding in the prior example using Bob and Joe for action and romantic movies. Is it possible to learn such decisions automatically? The answer is collaborative filtering (cf): We can recognize related users and movies by using the rating dataset and generating item and user embeddings based on existing ratings.

6   Architecture of Model

Now, since we have a good understanding of embedding, we can specify the architecture of a model. As you can see clearly, the item and user embeddings are crucial to our model. Let us have a look at the model architecture using the training sample below:

UserId: 3 movieId: 1 interacted: 1 (Fig. 3).

The embedding layer, which is a fully connected layer that converts the sparse representation into a dense vector, is located above the input layer. Then, the user and

Fig. 3 a Visual representation of the model architecture. **b** Neural collaborative filtering framework

(a)

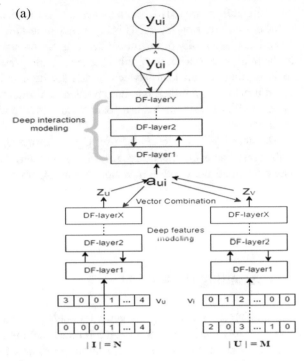

(b)

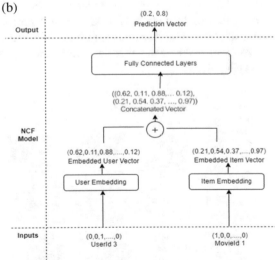

item embeddings are fed to the multi-layered neural network which we call the neural collaborative filtering layers. This layer maps the latent vectors to their prediction scores. The capability of the model depends on the size of the final hidden layer X. In the final output layer, the prediction score $\hat{y}_{ui}$ is present, and the model is trained so as to minimize the loss between y_{ui} and $\hat{y}_{ui}$. Item and user vectors for movieId = 1 and userId = 3 are then one-hot encoded as inputs to the model. The true mark (interacted) is 1 because this is a positive sample (the video was genuinely rated by the user).

The item input vector and the user input vector are, respectively, served to item embedding and user embedding, resulting in shorter, denser item, and user vectors. The embedded item and user vectors are amalgamated, here traversing through a sequence of totally connected layers that yield a prediction vector. Finally, we use a sigmoid function to arrive at the best possible class. Because $0.8 > 0.2$, the most likely class is 1 (positive class) in the case above.

7 Evaluating the Model

Our model is currently being trained and is ready to be estimated/evaluated via the test dataset. We evaluate our models in traditional machine learning (ML) projects using measures like accuracy (for classification tasks) and RMSE (for regression tasks) or MAE. For estimating recommender systems, such metrics are far too simplistic. We must first understand how advanced recommender systems are used to define suitable and meaningful metrics for evaluating them.

Take a look at Amazon's Website, which also has a list of recommendations (given below) (Fig. 4).

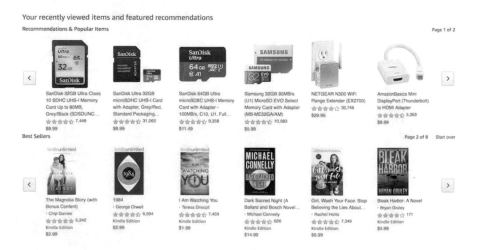

Fig. 4 Snapshot of Amazon's Website providing recommendations

The idea here is that the user is not required to interact with each and every item on the suggestion list. Rather, we simply want the user to communicate/interact with at minimum a single element from the list of recommendations; if the user does that the recommendations will work perfectly. To replicate this, use the assessment guidelines below to get a list of ten recommended items per user.

- Choose or pick 99 items arbitrarily per user that they have not associated (interact) with up till now.
- Add the over 99 items alongside with the test item (the item that the user actually interacted with). The absolute is presently 100 items.
- Apply algorithm to the above 100 objects and rank them based on their forecast prospects.
- Select top ten items from the above rundown of 100 items of the list. If the test data item is already in the topmost ten, we call it a 'hit.'
- Repeat the procedure for all users. The average hits are then used to calculate the hit ratio.

This hit ratio is generally used for evaluating recommender systems (Table 1).

$$\frac{\text{\#Of Cache Hits}}{(\text{\#Of Cache Hits} + \text{\#Of Cache Misses})} = \text{Hit Ratio}$$

OR

$$\text{Hit Ratio} = 1 - \text{Miss Ratio}$$

8 Result and Discussion

Alongside the rating, a timestamp column is present that displays the date and time when the review was submitted. Via this timestamp column, we will use the leave-one-out methodology to complete our train test split strategy. The most recent review is used as the test set for each user, while the remaining is used as training data (refer to Fig. 5).

A total of 38,700 movies have been reviewed by people. The user's most recent film review was for the 2013 blockbuster, Black Panther. For this user, we will utilize this movie as the testing data and the remaining rated movies as training data. When training and grading recommender systems, this train–test split strategy is widely utilized. We could not make a random split.

Because we could be using a user's current review for training and older ones for testing. With a look-ahead bias, it will induce data leakage, and the trained model's performance will not be generalizable to real-world performance (Graph 1).

Table 1 Number of recommendations versus calculated hit ratio

Number of recommendations (N)	Hit ratio
1	0.19
2	0.30
3	0.40
4	0.56
5	0.58
6	0.60
7	0.63
8	0.80
9	0.80
10	0.80
11	0.89
12	0.90
13	0.90
14	0.95
15	0.95
16	0.97
17	0.99
18	1.00
19	1.00
20	1.00

Fig. 5 Splitting pattern of training and test data

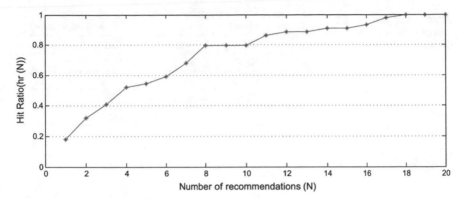

Graph 1 Number of recommendations versus calculated hit ratio

9 Conclusion

Collaborative filtering makes recommendations by simultaneously comparing similarities between people and items, addressing some of the drawbacks of content-based filtering. Serendipitous suggestions are now possible; in other words, collaborative filtering algorithms can suggest an item to user A based on the preferences of a user B who shares those interests. Furthermore, instead of relying on manually designing features, the embeddings can be learned automatically. In this paper, we demonstrated the modeling of a movie recommendation system using the principle of neural collaborative filtering (NCF) with a hit ratio of 86% which is more accurate than the other recommendation models/systems. Our model uses implicit feedback and finds the accuracy with the help of hit ratio which will be more accurate and efficient than the traditional recommendation system. The model complexity has also turned out to be less compared to the other state-of-the-art models.

The model can help users discover new interests. In an isolation, the ML system might not be aware that the user is interested in a certain item, but the model might still suggest it since other users who share that interest might be. Recommendation systems have grown to develop the most indispensable source of authentic and relevant information source in the ever-expanding world of the Internet. The relatively simple models consider a couple or a low range of parameters while the more complex ones make use of a much wider range of parameters to filter results, making them more user-friendly. With the knowledge of advanced techniques like the firefly algorithm and the restricted Boltzmann machine, robust recommendation systems can also be developed. This could be an influential measure toward the further enhancement of the current model to make it more efficient to use, increasing its business value even more (Table 2).

Table 2 Comparison of various state-of-the-art models

S. No.	Model	Model accuracy (%)
1	NCF	86
2	Firefly	84
3	Hybrid	76
4	Collaborative-based filtering	65
5	Content-based filtering	63

References

1. Gao T, Jiang L, Wang X (2020) Recommendation system based on deep learning. In: Barolli L, Hellinckx P, Enokido T (eds) Advances on broad-band wireless computing, communication and applications. BWCCA 2019
2. Wei J, He J, Chen K, Zhou Y, Tang Z Collaborative filtering and deep learning based recommendation system for cold start items. https://doi.org/10.1016/j.eswa.2016
3. Shah V, Kumar P (2021) Movie recommendation system using cosine similarity and Naive Bayes, s (IJARESM) 9(5). ISSN: 2455–6211
4. Lee H, Lee J (2019) Scalable deep learning-based recommendation systems. Article in ICT Express, published June 2019
5. Dooms S, Pessemier TD, Martens L 2011) An online evaluation of explicit feedback mechanisms for recommender systems, WEBIST 2011. In: Proceedings of the 7th international conference on web information systems and technologies, Noordwijkerhout, The Netherlands
6. He X, Liao L, Zhang H, Nie L, Hu X, Chua TS (2017) Neural collaborative filtering. In: Proceedings of the 26th international conference on world wide web (WWW '17). International World Wide Web Conferences Steering Committee, Republic and Canton of Geneva, CHE, pp 173–182
7. Goldberg D, Nichols D, Oki BM, Terry D (1992) Using collaborative filtering to weave an information tapestry. Communications of the ACM. December
8. Konstan J, Miller B, Maltz D, Herlocker J, Gordon L, Riedl J (1997) GroupLens: applying collaborative filtering to usenet news
9. Resnick P, Iacovou N, Suchak M, Bergstrom P, Riedl J (1994) GroupLens: an open architecture for collaborative filtering of netnews
10. Shardanand U, Maes P (1995) Social information filtering: algorithms for automating 'Word of Mouth'
11. Hill W, Stead L, Rosenstein M, Furnas G (1995) Recommending and evaluating choices in a virtual community of use
12. Ma H, King I, Lyu MR (2011) Learning to recommend with explicit and implicit social relations" Facing the cold start problem in recommendation systems. Expert Syst Appl 41(4):2065–2073
13. Jeong CS, Ryu KH, Lee JY, Jung KD Deep learning-based tourism recommendation system using social network analysis. Received: 2020.04.02 Accepted: 2020.04.13 Published: 2020.05.31

The Importance of Selected LMS Logs Pre-processing Tasks on the Performance Metrics of Classification Models

Janka Pecuchova⬮ and Martin Drlik⬮

Abstract Learning analytics and educational data mining are current research disciplines that can provide interesting and hidden insights into the effectiveness of different learning styles, the complexity of courses, educational content difficulties, and instructional design issues. Simultaneously, they can help to understand the concepts and reasons and estimate future performance or possible drop out of the students. However, even though the contribution of these research areas is promising, the availability of the end-to-end ML tools caused, that many scholarly papers underestimated the importance of the data understanding and data pre-processing phases of the knowledge discovery process. Subsequently, it leads to the incorrect or imprecise interpretation of the results. Therefore, this paper aims to emphasize the importance and impact of individual steps of the pre-processing phase on the quality of the ML models. The paper introduces a case study in which different data pre-processing tasks are applied to an open dataset of LMS logs before using an SVM classifier. As a result, the paper confirms the importance and significant impact of a suitably chosen set of pre-processing tasks on selected performance metrics of two ML classification models.

Keywords Data pre-processing · Learning analytics · Classification · LMS log analysis

1 Introduction

In recent years, the educational landscape has evolved dramatically. Due to the COVID-19 pandemic, it dramatically growths a number of learning management systems (LMSs) at universities, schools, and educational organizations. An LMS is a

J. Pecuchova · M. Drlik (✉)
Constantine the Philosopher University in Nitra, Nitra 949 01, Slovakia
e-mail: mdrlik@ukf.sk

J. Pecuchova
e-mail: janka.pecuchova@ukf.sk

Y. Singh et al. (eds.), *Proceedings of International Conference on Recent Innovations in Computing*, Lecture Notes in Electrical Engineering 1011,
https://doi.org/10.1007/978-981-99-0601-7_11

software application used to manage, document, track, and provide electronic educational technologies, courses, and training programs [1]. In addition, LMSs strive to use online technology in their courses to enhance the effectiveness of traditional face-to-face education [2].

In this context, educational data mining (EDM) and learning analytics (LA) have emerged as new fields of research that examine educational data to address a variety of instructive research issues, such as identifying successful students in a given course, identifying students who may drop out or require additional attention during the learning process. While EDM is a fast-emerging discipline that focuses on uncovering knowledge and extracting relevant patterns from educational information systems, LA is more general and considers the environment in which the learning process occurs. Both disciplines aim to assist students at various phases of their academic careers using digital traces they leave in the systems [3]. As a result, insights can uncover different learning styles, determine the complexity of courses, identify specific areas of the content that cause difficulties in understanding the concepts, and receive insights about the future performance or possible dropout of the students.

The visualization and analysis of this data are often carried out using a variety of machine learning included in EDM and LA methodologies to identify interesting and useful hidden patterns for predicting students' performance [4, 5]. The study of this data may yield significant information that is beneficial to both teachers and students if all phases of the knowledge discovery process are done thoroughly. Moreover, it is expected that the research team understand not only the experiment design but also the background of the learning process, resources of the educational data and the importance of the correct application of individual pre-processing techniques [6].

Even though these requirements seem ordinary, the availability of the end-to-end ML tools caused that many scholarly papers underestimated the importance of the data understanding and data pre-processing phases of the knowledge discovery process, which often led to the incorrect or imprecise interpretation of the obtained results. The paper attempts to find out the most common data pre-processing methods that may be used to improve the performance of classification models in terms of their efficiency. Therefore, the following are a summary of the aims of the work given in this paper:

1. The impact evaluation of integrating multiple pre-processing approaches on the performance of classification algorithms.
2. Determination of pre-processing techniques which leads to a more precise classification.

Main aim of this paper is to emphasize the importance and impact of individual steps of the pre-processing phase on the quality of the ML model using an open dataset of LMS logs.

The paper is structured as follows. The related works section summarizes the importance of individual data pre-processing tasks and their impact on model performance. Simultaneously, it provides an overview of the papers which deal with the educational data pre-processing, especially LMS logs. The next sections introduce a case study in which particular data pre-processing steps are presented using the open dataset of LMS logs. Finally, the results and discussion sections compare the results of two ML models, where a different set of pre-processing tasks had been done. In addition, their impact on selected performance metrics is discussed in detail.

2 Related Works

Data pre-processing has the largest influence on the possible model generalization based on machine learning algorithms. An estimate indicates that pre-processing can take up to 50 to 80 percent of the whole classification process, highlighting the significance of pre-processing phase in model development [7]. Furthermore, enhancing data quality is essential for improving ML model performance metrics.

As was mentioned before, the paper is focused on the pre-processing phase of machine learning research in education, which is frequently overlooked and unclear because the researchers do not consider it important to explain and often do not disclose what and why the data pre-processing techniques they implemented and employed.

A dataset is composed of data items known as points, patterns, events, instances, samples, observations, or entities [8]. Consequently, these data objects are usually characterized by many attributes/features that offer the key characteristics of an entity, such as the size of an object and the time at which an event occurred. A feature may be a single, measurable quality, or a component of an event. The amount of training data grows exponentially with the number of input space features. Features can be broadly classified as either categorical or numerical.

Data pre-processing is the initial step in ML techniques, in which data is transformed/encoded so that the computer can quickly study or understand it. An unprocessed dataset cannot be used to train a machine learning model. Incomplete, noisy, and inconsistent data, inherent to data obtained from original resources, continue to cause issues with data analysis. Therefore, it is necessary to address the concerns upfront and consider if the dataset is large enough, too small or fractured for further analysis. At the same time, it is necessary to identify corrupted and missing data that can decrease the prediction potential of the model. Pre-processing data with respect to correct future interpretation are a component of the form that conditions the input data to make subsequent feature extraction and resolution easier. Reducing or adding the data dimensions may increase the overall performance of the model. Additionally, if the accessible educational data does not contain a significant quantity of all types of data, which provides a complete picture of the learning process, then the information derived from the data may be unreliable because the missing or redundant attributes may reduce the model's precision [9].

In many issues, the dataset contains noisy data, and making the elimination of noisy instances is one of the most challenging tasks in machine learning [10]. Another challenging issue is distinguishing outliers from genuine data values [11]. These inliers are incorrect data values located within the center of a statistical distribution, and their localization and correction are extremely difficult [12].

The management of missing data is a frequent issue that must be addressed through data preparation [13]. Frequently, numerical, and categorical datasets must be meticulously controlled. On the other hand, it is well-known that several algorithms can handle categorical cases more effectively or exhibit greater performance primarily with such examples. Whenever this occurs, discretization of numerical data is of paramount importance [14]. The grouping or discretization of categorical data is a highly effective solution to the abovementioned difficulty. As a result, the initial dataset is converted to the numerical format.

A large amount of categorical data is difficult to manage if the frequencies of many categories vary greatly [15]. This frequently raises questions such as which subset of categories gives the most useful information and, thus, which quantity should be chosen for training the ML model [16]. In addition, a dataset with an excessive amount of characteristics or features with correlations should not be included in the learning process because these sorts of data do not provide relevant information [17–20]. Therefore, selecting features that limit unneeded information and are irrelevant to the research is required. As a result, feature selection techniques are useful in this case [21, 22].

As can be seen, partial pre-processing tasks are mentioned in learning analytics papers. However, a systematic review of the role of data pre-processing in learning analytics or educational data mining is rare. However, as stated in [21] or [22], it is evitable to actively improve researchers and educational policymakers' awareness of the importance of the pre-processing phase as an essential phase for creating a reliable database prior to any analysis.

3   Methodology

The importance and impact of individual steps of the pre-processing phase on the quality of the ML model will be demonstrated on an open dataset of LMS logs, which is available for LA research [23]. Figure 1 visualizes the sequence of individual data pre-processing steps, which should aim at the preparation of the dataset on which different ML classification tasks can be realized. They are described in the following subsections in detail.

3.1   Data Description

The dataset used within the research comes from the August 2016 session of the course "Teaching with Moodle". Moodle Pty Ltd employees teach this course twice a year in the form of online sessions.

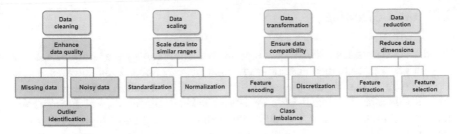

Fig. 1 Data pre-processing steps [2]

The session was offered entirely online through an LMS Moodle from August 7 to September 4, 2016. There were 6119 students registered in this course, of which 2167 gave permission to utilize their data, resulting in 32,000 entries. 735 students successfully completed the course, which consisted of completing all exercises within the course. There were a total of 1566 students who obtained a course badge, and 735 students who completed the course.

The presented dataset consists of six selected tables extracted from the Moodle database pages using SQL to CSV (comma-separated by values) files. Each dataset has a pseudonymous username to facilitate the reconstruction of linkages between individuals and their accomplishments, including activities, log entries, badges, and grades. Other columns of data were omitted due to privacy concerns or because they did not offer relevant information for this dataset [23].

Table 1 shows descriptions of six input CSV files with raw data. Each of these files has numerous features that can be used into ML techniques to construct an adequate predictive model for predicting student performance.

Table 1 Description of.csv files used to create a pre-processed dataset

Filename	Description	Records
mdl_badge_issued.csv	There are records of all badges granted to users	1844
mdl_course_modules.csv	This file contains records for each activity	60
mdl_course_modules_completion.csv	This file contains records of each user's completion of each activity in the course	30,986
mdl_grade_grades_history.csv	This table maintains a historical record of individual grades for each user and each item as they were imported or submitted by the modules themselves	70,037
mdl_logstore_standard_log.csv	This table contains entries for each "event" logged by Moodle and serves as the source for all of the Moodle "log" reports generated by the logging system	52,684
mdl_user.csv	This is the table that contains information about users	2167

3.2   Data Cleaning

The dataset consists of 2167 student records, but 213 records were removed because files, in general, contained academic information only about 1954 students. There were no records that did not provide comprehensive information but the problem of missing values had to be dealt. Missing data are acknowledged as one of the important concerns that must be carefully addressed during the pre-processing phase, prior to the use of machine learning algorithms, in order to develop effective machine learning models. The historical records were considered, which provided information about each student's grades earned through various historical activities.

As part of this step, 6 of the 16 activities had to be removed. The most prevalent strategies for handling missing values were used such as manual filling, replacement with mean or zero/null values, adaptation an imputation procedure which distorted the analysis results. Simultaneously, it was necessary to remove 752 records because there was no information about students who participated in these obsolete activities. The final dataset of 1202 records after cleaning.

3.3   Data Scaling

Data scaling is step, which is required to ensure the validity of predictive modeling; mainly in the situation, the input attributes have different scales. The ML algorithm is efficiently trained when dataset is scaled or normalized. As a result, a better prediction can be obtained and speeds up processing or training. Normalization requires extra effort. The dataset must not have a few instances with fewer features. In addition, if desirable care is no longer taken, the dataset may lose the internal structure, which leads to lower accuracy [24].

The max–min normalization and z-score standardization are two of the most widely used methods. Normalization is the process of scaling attribute values within a specific range so that all attributes have approximately similar magnitudes. While min–max normalization is sensitive to data outliers because their presence can significantly alter the data range, z-score standardization is less impacted by outliers. Typically, it is employed to transform the variable into a normal distribution with a mean of zero and a standard deviation of one [25]. Due to fact that, dataset did not contain significant outliers, and the data were normalized using min–max normalization at intervals [-1, 1].

3.4   Data Transformation

In the educational field, data transformation is mostly used to transform numerical data into discrete data categories to enable interoperability with machine learning

Table 2 Numeric to nominal value conversion

Nominal value	Attempt count	Activity history	Other modules
High	3	3	$6 \leq$
Medium	2	2	3, 4
Low	≤ 1	≤ 1	≤ 2

algorithms. Due to their usability, the equal-frequency and equal-width approaches have been widely adopted [25].

Several aspects of the performance of the students were converted from numeric to nominal values through class labeling to represent the class labels of the classification issue. As indicated in Table 2, the values of these characteristics were separated into three nominal categories (high, medium, and low). Thirds are the criteria used to transform numeric values to nominal intervals. This transformation was performed to transfer the result to the target variable *Result* more precisely.

3.5 Data Reduction

Data reduction typically consists of three main methods. The first is to directly pick variables of interest using domain expertise. The second step is to choose important variables for further study using statistical feature selection methods. The final step is to implement feature extraction techniques to generate usable features for data analysis. Sadly, the majority of datasets contain useless features, which can negatively impact the performance of learning algorithms.

Feature selection (FS) methods can be roughly categorized into three groups: filter, wrapper, and embedding approaches. The filter technique is a basic and quick way of selecting features in which variables are ranked and selected based on specified univariate parameters.

This research identified a filter-based strategy utilizing a selection algorithm based on information gain. The filter approach was based on two criteria for feature selection: correlation (correlation matrix) and information-gain attribute assessment (in this case, information-gain attribute evaluation).

A correlation matrix is suitable for checking the linear relationship between features, as shown in Fig. 2. The primary objective was to mitigate the difficulty of high-dimensional data by reducing the number of attributes without compromising classification accuracy. In this case study, the correlation matrix showed that the final outcome of the students was substantially correlated with their participation in each course activity.

Figure 3 illustrated that the highest value was received by features *badgesNo, forum,* and *feedback,* followed by categories relating to academic involvement such as *lesson, quiz, other modules,* and so on. As illustrated in Fig. 3, a significant subset of traits was picked while others were removed. Thus, the features examined in this

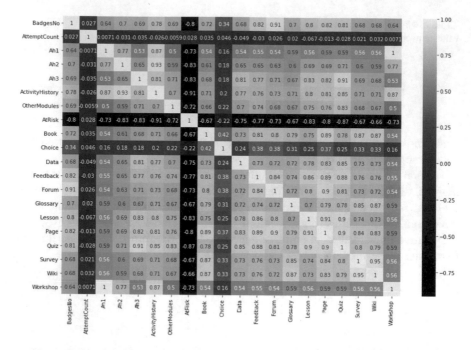

Fig. 2 Correlation matrix for selected features

study received the highest ranking, indicating that students and their involvement throughout the educational process significantly impact their academic achievements.

In contrast to feature selection, feature extraction seeks to create new features based on linear or nonlinear combinations of existing variables. Principal component analysis (PCA) and statistical methods are two representative linear feature extraction techniques. The number of extracted principal components or features is decided by the proportion of total data variance explained, e.g., the principal components should be capable of explaining at least 80 or 90 percent of the total data variation.

The following 22 attributes shown in Table 3 were selected as a result of feature selection: *"BadgesNo"*, *"AttemptCount"*, *"At1"*, *"At2"*, *"At3"*, *"ActivityHistory"*, *"OtherModules"*, *"AtRisk"*, *"M1"*, *"M2"*, *"M3"*, *"M4"*, *"M5"*, *"M6"*, *"M7"*, *"M8"*, *"M9"*, *"M10"*, *"M11"*, *"M12"*, and *"Result"*.

The *"Result"* attribute was linked to whether the student passed or failed the course based on the completion of each module. Initially, the records from *"mdl course modules.csv"* regarding each action within the course module were extracted. Therefore, *"mdl course modules completion.csv"* was parsed to extract the results of those students' participation in those activities.

In general, the dataset exhibited issues typical of this type of educational data. A large number of features caused the dataset to be multidimensional. Some of the qualities were not meaningful for classification, and others were not connected. In

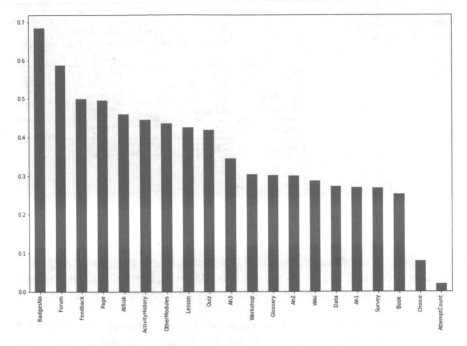

Fig. 3 Highly ranked features after applying filter-based evaluation using gain ratio

Table 3 Description of selected features

Feature	Description
BadgesNo	The number of achieved badges by a registered student
AttemptCount	The number of attempts in the course
At1–At3	Historical achieved points of each student's completion of selected three activities
ActivityHistory	A total number of completed historical activities
OtherModules	The number of any other completed activities by a registered student
AtRisk	Previously failed two or more historical activities
M1–M12	The number of completed modules within the course
Result	The final outcome of the registered student is based on the completion of each module

our situation, however, the data were not skewed, as the majority of students passed while a minority did not.

Typically, the unbalanced data problem arises as a result of learning algorithms ignoring less frequent classes in favor of more frequent ones. As a result, the resulting classifier cannot correctly identify data instances that belong to classes that are inadequately represented [5].

Table 4 Evaluation result without pre-processing

	Results of successful student test data	Results of unsuccessful student test data
Total	446	922
Classified	334	719
Precision	74.88%	77.98%
False positive	N/A	22.02%
False negative	25.12%	N/A

4 Classification Model Evaluation

Support vector machines (SVMs) classification algorithm was used to train the datasets as it was suitable for this kind of dataset composed of numerical features. SVM works for classification and prediction problems, and the idea behind it is to find a line that best isolates multi-group labels. Moreover, it is developed to deal with numeric attributes, as it deals with nominal ones after converting them to numeric data types.

The final pre-processed dataset for building the predictive model consisted of 1203 student information records stored in 20 features, 19 numerical, and one categorical. The original aim was to predict whether a student's outcome would *pass* or *fail*. The categorical feature was encoded into numerical so that the *sklearn* functions could properly work.

The final pre-processed dataset was balanced, having unsuccessful students (617), and it has successful ones (586). The data were divided into the training and testing dataset to increase the efficiency and stability of ML models.

The confusion matrix shows the efficiency of the final model along with the overall accuracy. Attention was paid to false positive and false negative values within the confusion matrix. A false positive identifies successful students as failures. A false negative identifies the unsuccessful students as successful. A false positive can have more impact than a false negative.

The evaluation of the result without the pre-processing of the dataset is shown in Table 4. The results with pre-processing of the dataset are shown in Table 5. An evaluation was carried out in the test dataset, which consisted of 394 successful students and 398 dropouts. The results showed that false positive values were greatly reduced for the pre-processed dataset, and the overall accuracy was increased by 16%.

5 Conclusion

Data preparation is an essential stage in finding knowledge from educational data. Unfortunately, educational data tend to be incomplete, inconsistent, noisy, and

Table 5 Evaluation result with pre-processing

	Results for successful student test data	Results for unsuccessful student test data
Total	394	398
Classified	367	360
Precision	93.14%	90.45%
False positive	N/A	9.55%
False negative	6.86%	N/A

CourseID	StudentNo	BadgesNo	AttCount	AttCount	At1...At3	ActivityH	ActivityH	OtherMod	OtherMod	AtRisk	M3...M22	ModSum	Result
10464	Student 1	0	1	Low	0	1	High	0	High	Yes	1	5	Fail
10464	Student 2	1	1	Low	6	3	Low	0	High	No	3	20	Fail
10464	Student 3	0	1	Low	0	0	High	0	High	Yes	0	1	Fail
10464	Student 4	3	1	Low	6	3	Low	3	Medium	No	4	33	Pass
10464	Student 5	2	1	Low	6	3	Low	4	Low	No	4	33	Pass
10464	Student 6	1	2	Medium	0	2	Medium	0	High	Yes	3	19	Fail
10464	Student 7	2	1	Low	6	3	Low	3	Medium	No	4	33	Pass
10464	Student 8	0	1	Low	0	0	High	0	High	Yes	0	1	Fail
10464	Student 9	2	1	Low	6	3	Low	3	Medium	No	4	33	Pass

Fig. 4 Final dataset after applying pre-processing techniques

missing. Thus, data pre-processing is an important matter involving converting the obtained data to a more manageable format. In this paper, the impact of various data pre-processing approaches was examined.

Conventional data pre-processing tasks, such as imputation of missing value, outlier identification, data scaling, data reduction, and data transformation, were applied to the open dataset. Data scaling improved to a prediction and speeds up processing or training through normalization. Consequently, the data transformation method modified data types. Afterward, the feature selection method was utilized to determine the best set of features with the highest scores. Figure 4 represents the final dataset created from the processed data.

The results shown in Tables 4 and 5 confirmed that data pre-processing techniques have an efficient, effective, and important role in the knowledge discovery process of educational data. Furthermore, the comparison of both outputs shows improved precision with simultaneous degrees of false positives in the case of the pre-processed dataset. Thus, the test results show that SVM classification with thoroughly pre-processed data can improve the prediction accuracy and confirmed that suitably selected pre-processing tasks have a significant impact on the performance of the SVM classifier, especially on the number of false positives.

Limitations of this paper are that the data pre-processing cannot be entirely automated due to the wide range of decisions which must be made about the features, which can influence the learning process or identification of students' success or failures. Therefore, this process is still iterative and often has a character of a trial-and-error procedure that strongly relies on subject expertise and immediate practical tasks. It can be assumed that the impact of the adopted pre-processing techniques would vary from one classification algorithm to another.

Future research will investigate the impact of various pre-processing strategies on other classification and clustering algorithms. In addition, the appropriate pre-processing procedures for such datasets must be determined in order to address problems that are less frequent.

Acknowledgements This work was supported by the Scientific Grant Agency of the Ministry of Education of the Slovak Republic and Slovak Academy of Sciences under Contract VEGA-1/0490/22, and by the European Commission ERASMUS+ Program 2021, under Grant 2021-1-SK01-KA220-HED-000032095.

References

1. Skalka J, Švec P, Drlík M (2012) E-learning and quality: the quality evaluation model for e-learning courses. In: Divai 2012 - 9th International scientific conference on distance learning in applied informatics
2. Amrieh EA, Hamtini T, Aljarah I (2015) Preprocessing and analyzing educational data set using X-API for improving student's performance. In: 2015 IEEE Jordan conference on applied electrical engineering and computing technologies (AEECT). IEEE, pp 1–5
3. Alcalá-Fdez J, Sanchez L, Garcia S, del Jesus MJ, Ventura S, Garrell JM, Herrera F (2009) KEEL: a software tool to assess evolutionary algorithms for data mining problems. Soft Comput 13(3):307–318
4. Chouldechova A (2017) Fair prediction with disparate impact: a study of bias in recidivism prediction instruments. Big Data 5(2):153–163
5. Kabathova J, Drlik M (2021) Towards predicting student's dropout in university courses using different machine learning techniques. Appl Sci 11(7):3130
6. Chawla NV, Bowyer KW, Hall LO, Kegelmeyer WP (2002) SMOTE: synthetic minority over-sampling technique. J Artif Intell Res 16:321–357
7. Kadhim AI (2018) An evaluation of preprocessing techniques for text classification. Int J Comput Sci Inf Secur (IJCSIS) 16(6):22–32
8. Cui ZG, Cao Y, Wu GF, Liu H, Qiu ZF, Chen CW (2018) Research on preprocessing technology of building energy consumption monitoring data based on a machine learning algorithm. Build Sci 34(2):94–99
9. Davis JF, Piovoso MJ, Hoo KA, Bakshi BR (1999) Process data analysis and interpretation. Adv Chem Eng 25:1–103. Academic Press
10. Zhu X, Wu X (2004) Class noise vs. attribute noise: a quantitative study. Artif Intell Rev 22(3):177–210
11. Khoshgoftaar TM, Van Hulse J, Napolitano A (2010) Comparing boosting and bagging techniques with noisy and imbalanced data. IEEE Trans Syst Man Cybern Part A Syst Hum 41(3):552–568
12. van Hulse J, Khoshgoftaar TM, Napolitano A (2007) Experimental perspectives on learning from imbalanced data. In: Proceedings of the 24th international conference on Machine learning, pp 935–942
13. Farhangfar A, Kurgan LA, Pedrycz W (2007) A novel framework for imputation of missing values in databases. IEEE Trans Syst Man Cybern Part A Sys Hum 37(5):692–709
14. Elomaa T, Rousu J (2004) Efficient multisplitting revisited: optima-preserving elimination of partition candidates. Data Min Knowl Disc 8(2):97–126

15. Guyon I, Elisseeff A (2003) An introduction to variable and feature selection. J Mach Learn Res 3(2003):1157–1182
16. Skillicorn DB, McConnell SM (2008) Distributed prediction from vertically partitioned data. J Parallel Distrib Comput 68(1):16–36
17. Czarnowski I (2010) Prototype selection algorithms for distributed learning. Pattern Recogn 43(6):2292–2300
18. Xiao W, Ji P, Hu J (2022) A survey on educational data mining methods used for predicting students' performance. Eng Rep 4(5):e12482
19. Mingyu Z, Sutong W, Yanzhang W, Dujuan W (2022) An interpretable prediction method for university student academic crisis warning. Complex Intell Syst 8(1):323–336
20. Ismael MN (2022) Students performance prediction by using data mining algorithm techniques. Eurasian J Eng Technol 6:11–25
21. Feldman-Maggor Y, Barhoom S, Blonder R, Tuvi-Arad I (2021) Behind the scenes of educational data mining. Educ Inf Technol 26(2):1455–1470
22. Luna JM, Castro C, Romero C (2017) MDM tool: a data mining framework integrated into Moodle. Comput Appl Eng Educ 25(1):90–102
23. Dalton E (2017) Learn Moodle August 2016 anonymized data set. [Dataset]
24. Munk M, Drlík M (2011) Impact of different pre-processing tasks on effective identification of users' behavioral patterns in web-based educational system. Procedia Comput Sci 4:1640–1649
25. Munk M, Kapusta J, Švec P (2010) Data preprocessing evaluation for web log mining: reconstruction of activities of a web visitor. Procedia Comput Sci 1(1):2273–2280

Analysis of Deep Pre-trained Models for Computer Vision Applications: Dog Breed Classification

Meith Navlakha, Neil Mankodi, Nishant Aridaman Kumar, Rahul Raheja, and Sindhu S. Nair

Abstract Machine perception is one of the most lucrative domains in the modern landscape and one of the most challenging tasks embodied by this domain is analyzing and interpreting images. Image recognition has seen several advancements over the years such as the introduction of pre-trained models that have expunged the complexity associated with developing high performance deep neural networks. In this paper, we have proposed several eminent pre-trained models that have the ability to categorize dogs as per their breeds. The main objective of this paper is to present a fair comparison of the proposed models and establish the nonpareil model on the basis of comparison metrics such as accuracy, validation accuracy, time requirements, precision and recall scores of the model. It was observed that ResNet 152V2 performed the best with respect to the accuracy, precision and recall scores, Inception-ResNet gave the best validation accuracy and NASNet-Mobile had the highest efficiency albeit with inferior performance in accordance with the other evaluation metrics.

Keywords Transfer learning · Image recognition · Deep learning · Convolutional neural networks · Dog breed classification · Inception-v3 · ResNet · Xception · DenseNet · NASNet-Mobile · Inception-ResNet · VGG

M. Navlakha (✉) · N. Mankodi · N. A. Kumar · R. Raheja · S. S. Nair
Department of Computer Engineering, Dwarkadas J. Sanghvi College of Engineering, Mumbai, India
e-mail: meithnavlakha@gmail.com

N. Mankodi
e-mail: neilmankodi13@gmail.com

N. A. Kumar
e-mail: nishu.ak99@gmail.com

R. Raheja
e-mail: rraheja49@gmail.com

S. S. Nair
e-mail: sindhu.nair@djsce.ac.in

1 Introduction

There has been severe debate among the prominent governing bodies of canine registries regarding the official number of recognized dog breeds. For example, the American Kennel Club (AKC) only recognizes 195 breeds, while the Federation Cynologique International (FCI) recognizes 360 breeds officially. There isn't an exact, worldwide, internationally agreed-upon number, but it would be safe to say that there are several hundreds of canine breeds and having these many breeds poses a serious problem. Many of these breeds have easily perceptible differences in their characteristics such as height, weight and color that make them easy to distinguish, but at the same time, there are several dog breeds that are nearly identical with small, imperceptible differences that make it very difficult for us as human beings to distinguish them from each other. The task of classifying dogs on the basis of their breeds is a suitable candidate for being included as a machine perception problem that can be managed with the help of convolutional neural networks (CNNs). CNNs are a type of artificial neural networks that are primarily used in image recognition and processing that are explicitly designed to process pixel data.

One of the most impactful advancements in the usage of CNNs has been the introduction of transfer learning and pre-trained models. Transfer learning is a subset of artificial intelligence (AI) and machine learning (ML) which aims to apply the knowledge obtained from one task to a different but similar task. Pre-trained models are deep neural networks which have been trained on an extensive dataset, typically to solve some substantial image classification task. One can either use this pre-trained model as it is or you can personalize it to fit your given task. This paper puts the spotlight on the following pre-trained models:

- DenseNet 201

Fig. 1 Examples of dog breeds from Stanford Dogs dataset [1]

- InceptionV3
- Inception-ResNetV2
- MobileNet
- NASNet-Mobile
- ResNet 152V2
- Xception

Above mentioned are just a few of the several pre-trained models available in the Keras framework and deciding which model to use is a headache that this paper aims to solve. The core objective of our research is to conduct a fair comparison of all the seven proposed pre-trained models and establish the nonpareil model. For the purpose of comparison, we will be making use of standard metrics such as the accuracy, validation accuracy, time requirements, precision and recall scores of the model.

For the purpose of classifying canine images on the basis of their breeds, we will be using the Stanford Dogs dataset that is publicly available on the Web site known as Kaggle. The Stanford Dataset is an image dataset containing 20,580 images corresponding to 120 different canine breeds, some of which are visible in Fig. 1. This dataset has been curated by using ImageNet for the sole objective of developing machine learning models to solve the challenging problem of classifying dog breeds that have nearly indistinguishable features.

2 Related Work

The majority of the papers published on dog breed classification have chosen a segment of the entire population of breeds that the dataset consists of (10, 13, 15, etc.). Even though this is computationally less resource intensive as it takes less training time, it certainly lacks practical utility. We, for the purpose of this paper, have chosen 120 breeds of the dogs present in the Stanford Dataset on Dogs [1].

Wang et al. [2] have proposed the use of two fine-tuned pre-trained models for the task of image classification of dogs and cats. VGG16 and VGG19 are the two models being considered for the research, both of which operate on only 21 separate varieties of dogs and cats. The proposed models have an accuracy of 98.47% and 98.59%, respectively. Dąbrowski et al. [3] use the Inception-v3 pre-trained model to classify among the dog breeds. It has an accuracy of 78.8%. While these studies achieved excellent levels of accuracy, it falls short in certain aspects. The study only focuses on a segment of breeds which severely limits its usability in real-world applications. These papers also have a very limited scope of comparison as it chooses to target only two of the several pre-trained models available for image classification.

Bouaafia et al. [4] have proposed the use of four pre-trained models for the task of traffic sign identification in the domain of computer vision. The four selected models are VGG16, VGG19, AlexNet and ResNet-50. The authors of the paper have managed to obtain exemplary levels of accuracy for all of the aforementioned pre-trained models, but there are certain aspects that the research fails to take into account. The study has a very confined scope of comparison as it chooses to only focus on four pre-trained models. This makes it difficult for the readers to get the complete overview of the pre-trained model landscape.

Varshney et al. [5] have proposed the use of a transfer learning approach on two neural networks which are VGG16 and Inception-v3. The scope of the study is however restricted to only two models; this restricts the comparison as the study chooses to target only two of the several pre-trained models available for image categorization. Furthermore, the accuracy (validation accuracy of 0.545 and training accuracy of 0.69281) achieved by the study is not competitive in nature rendering the model practically unusable. Also, the overhead of deploying two pre-trained models for the purpose of breed prediction given the accuracy achieved does not classify as a reasonable trade-off.

Junaidi et al. [6] have proposed a transfer learning approach for the task of image classification of an egg incubator. The proposed approach consists of two pre-trained models which are VGG16 and VGG19. The scope of research is severely limited due to the fact that it only takes two models into account and both of them are versions of VGG. Comparing only VGG models results in the paper not being able to provide the reader with more significant results and insights. In addition to this drawback, the paper achieves a maximum accuracy of 92% which has been adequately surpassed by some of the models considered in our research paper.

Nemati et al. [7] have proposed the use of transfer learning in order to classify the patients as COVID-19 positive or negative using chest X-ray images. This paper inspired us to extensively use pre-trained models in our research. They have compared the accuracies of 27 different models some of them being Inception-v3, ResNet, Xception, NASNet, VGG16, VGG19, etc. This paper became the source of inspiration as the accuracy which they managed to attain was extraordinary. Moreover, the baseline problem statement of our as well as their paper was similar, i.e., related to image classification using CNNs.

Akhand et al. [8] have proposed the use of transfer learning to perform facial emotion recognition (FER) using images of an individual. This paper makes use of the following pre-trained models VGG16, VGG19, ResNet-18, ResNet-34, ResNet-50, ResNet-152, Inception v3 and DenseNet 161 to carry out the above task. This paper uses the base pre-trained models for feature extraction and an additional dense layer for the classification task which led to accuracies of 96.51 and 99.52% in the FER task. This was a major inspiration for the methodologies used in our paper. In addition to this, the paper also has a similar framework to the methodology proposed in this paper.

All the papers related to dog breed prediction had a restricted scope. The scope of all the related papers was restricted to at the most three models. We have considered eight models in total making it more elaborate and comprehensive.

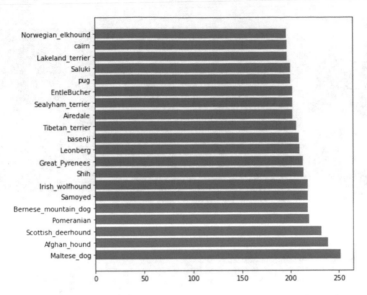

Fig. 2 Top 20 classes in the ImageNet dataset

3 Dataset

Images of 120 different dog breeds are included in the Stanford Dogs dataset. The purpose of this dataset was to do fine-grained image categorization utilizing images and annotations from ImageNet [9]. It was initially gathered to help with fine-grained image categorization, a difficult task, given that several dog breeds have almost similar traits or differ in color and age. It comprises 20,580 images of 120 dog breeds. Figure 2 contains the top 20 classes. It contains images with class labels, bounding box dimensions of the main subject of that image and the dimensions of the image itself. The bounding box consists of the dimensions of the subject of the image. The dataset is made up of 2 folders, i.e., the images folder and the annotations folder. The names of the objects in each folder are the same thus linking them together.

4 Data Augmentation

The data augmentation technique is used to enhance the diverseness in our training data by applying stochastic, but rational transformations, like image rotation, skewness, scaling, horizontal, or vertical shifting on the existing dataset. This helps to reduce the chances of overfitting. Figure 3 displays the augmented images. Data augmentation is implemented using the ImagedataGenerator class available in TensorFlow's Keras library.

Fig. 3 Images after data augmentation

5 Proposed Work

5.1 Models

Inception-v3 Inception-v3 [3] is a convolutional neural network used in object detection and image analysis. It originated as a component for GoogleNet. Inceptionv3 was created to permit more layers of the deep network all the while preventing the number of parameters from increasing too much. The network accepts 299 × 299 image input.

ResNet 152V2 ResNet-152 V2 [10] was released by Microsoft Research Asia in 2015. ResNet's architecture was able to obtain impressive results in the MS-COCO and ImageNet competition. The main idea behind this model, residual connections, is found to largely enhance gradient flow, hence permitting training of considerably deeper models with many more layers. The network accepts 299 × 299 image input.

MobileNet The MobileNet [11] model, which is TensorFlow's earliest mobile computer vision model, was developed for use in mobile apps, as suggested by its name. MobileNet [11] makes use of depth-wise separable convolutions and thus considerably brings down the number of parameters in comparison with a network with regular convolutions with similar depth in the network hence resulting in light deep neural networks. The network accepts 224 × 224 image input.

NASNet-Mobile The mobile version of NASNet (i.e., NASNet-Mobile [12]) has 564 million multiply accumulates and 5.3 million parameters in 12 cells. More than a million photos from the ImageNet dataset were utilized to train the CNN. The model can classify images in 1000 categories, like animals, pencil, mouse, keyboard,

Table 1 Properties of pre-trained models

Model	Depth	Size(MB)	Trainable Parameters	Input Image Size
Inception-v3	189	92	15,728,760	299 × 299
ResNet 152v2	307	232	245,880	299 × 299
MobileNet	55	16	9,953,400	224 × 224
NASNet-Mobile	389	23	6,209,400	224 × 224
Xception	81	88	24,576,120	299 × 299
DenseNet 201	402	80	120,120	224 × 224
Inception-ResNet	449	215	11,796,600	299 × 299

etc. The network was able to learn significant feature patterns for a wide variety of pictures as a result. The network accepts 224 × 224 image input.

Xception Xception [13] is a CNN with 71 deep layers that derives upon depthwise separable convolutions. It is trained on ImageNet, allowing the network to categorize images among thousand distinct subcategories. The network accepts 299 × 299 image input.

DenseNet 201 DenseNet 201 [14] is a CNN with 201 deep layers. It is trained on ImageNet, allowing the network to categorize images among thousand distinct subcategories. The network accepts 224 × 224 image input.

Inception-Resnet Inception-ResNet [15] is a CNN that incorporates residual connections and gleans from the Inception architecture family, having 164 deep layers. ResNet reduces network complexity by making it more uniform. It is trained on ImageNet, allowing the network to categorize images among thousand distinct subcategories. The network accepts 299 × 299 image input.

Table 1 depicts all the key characteristics of all the pre-trained models being considered in our research on dog breed classification.

5.2 Proposed Framework

The dataset was split into training, testing and validation segments. Each segment was preprocessed and augmented as mentioned in Sect. 4. The resultant images of training and testing segments were fed to the pre-trained model. While training the model, accuracy, precision, recall were observed for every epoch. Finally, the validation accuracy was computed by evaluating the complete model. The entire workflow is demonstrated in Fig. 4.

Fig. 4 Flowchart

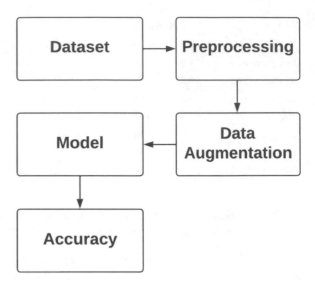

6   Experimental Performance

6.1   *Experimental Setup*

The experiment was performed using Python and Keras package with Tensorflow-gpu on Kaggle having 13 GB RAM and 2-core Intel Xeon CPU, along with Tesla P100 16 GB VRAM GPU. The parameters used in the learning process for all pre-trained models were as follows:

- Batch-Size: 32
- Epochs: 25
- Loss function: Categorical Crossentropy
- Optimizer: Adam, RMSProp and Adamax
- Learning Rate: 0.001 and 0.01
- Metrics: Accuracy, Precision and Recall
- Activation: Softmax

7   Results and Analysis

7.1   *Evaluation Metrics*

The ratio of true positives to actual positives is used to calculate precision. Precision is a metric used in classification and pattern recognition. It helps us to check how well the model can classify positive samples.

$$Precision = \frac{TP}{TP + FP}$$

The ratio of correctly categorized positive (+ve) records to all the positive (+ve) records is termed as recall. Higher recall score denotes the fraction of positive samples correctly detected.

$$Recall = \frac{TP}{TP + FN}$$

TP   *True Positives.*
FN   *False Negatives.*
FP   *False Positives.*

The F1-score is computed by taking the harmonic mean of a model's recall and precision, combining the two into an effective statistical measure. It is generally used to compare among the models.

$$F1\ score = 2 \times \frac{Precision \times Recall}{Precision + Recall}$$

7.2 Analysis

We have implemented the above pre-trained models using the proposed framework as shown in Fig. 4; the comparison between them is displayed in Table 2, and the optimal versions for each pre-trained model are presented in Table 3. We have emboldened the best results for the chosen hyper-parameters across all the pre-trained models. We have plotted the accuracy versus epochs, validation accuracy versus epochs and F1-score versus epochs graphs as visible in Figs. 5, 6 and 7, respectively. As witnessed in Table 3, the accuracy, precision and recall scores for ResNet 152V2 appear the best although it may have overfit as the validation accuracy is not comparable to the training accuracy.

The validation accuracy is highest for Inception-ResNet at approximately 82.5% as witnessed in Table 3, which demonstrates that it has generalized the best with the pet dataset, along with an overall decent accuracy, precision and recall scores. It can also be observed from Table 3 that NASNet-Mobile is the most efficient model as it requires the least amount of time to train, but overall the model falls short due to inferior values achieved for metrics such as accuracy, validation accuracy, precision and recall.

Table 2 Model accuracy for hyper-parameter tuning

Model	LR[a]	Adam	RMSProps	Adamax
Inception V3	0.01	0.890	0.891	0.902
	0.001	0.901	0.896	**0.910**
	0.01	0.909	0.880	**0.948**
ResNet 152v2	0.001	0.89	0.911	0.909
	0.01	0.876	0.883	0.888
	0.001	0.84	0.896	**0.899**
MobileNet	0.01	**0.823**	0.807	0.817
NASNet-Mobile	0.001	0.82	0.816	0.816
	0.01	0.923	0.914	0.925
Xception	0.001	0.92	0.921	**0.933**
	0.01	0.894	0.882	0.744
	0.001	0.9	**0.912**	0.893
Inception-ResNet	0.01	0.891	0.885	0.887
	0.001	0.889	0.886	**0.903**

a Learning Rate Parameter

Table 3 Comparison of optimally performing models from Table 1

Model	A^a	V^b	T^c	P^d	R^e
Inception V3	0.910	0.801	11,451	0.911	0.910
ResNet 152v2	**0.948**	0.764	12,060	**0.956**	**0.942**
MobileNet	0.899	0.676	11,708	0.901	0.899
NASNet-Mobile	0.823	0.676	**7,761**	0.823	0.823
Xception	0.933	0.808	12,283	0.935	0.933
DenseNet 201	0.912	0.799	11,632	0.926	0.902
Inception-ResNet	0.903	**0.825**	12,401	0.905	0.903

a Accuracy
b Validation Accuracy
c Time (sec)
d Precision
e Recall

8 Conclusion

This paper successfully provides a fair comparison of the aforementioned pre-trained models by using the target task of classifying images of dogs on the basis of their breeds. All of the pre-trained models operated on the Stanford Dogs dataset that comprises 20,580 images corresponding to 120 different breeds of dog. The comparison was conducted by using the technique known as hyper-parameter tuning, which

Fig. 5 Accuracy versus epochs

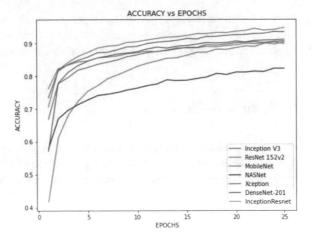

Fig. 6 Validation accuracy versus epochs

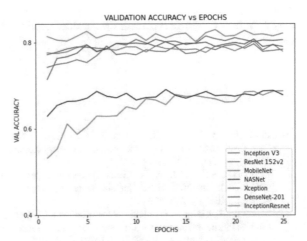

Fig. 7 F1-scores versus epochs

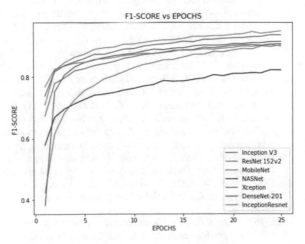

involved training models by varying hyper-parameters which included learning rate and optimizer. The results represented in Figs. 5, 6 and 7 will facilitate readers in making an informed decision regarding the model to be used.

9 Future Work

As we propose convolutional neural networks which efficiently extract features of a dog which in turn help in determining the breed of the dog, the pre-trained CNN model used thereof can be used to predict the breeds of other animals like cats, cows, etc., with minimal adjustments in later layers of the neural network. Furthermore, the scope of the research may further be expanded to other pre-trained models in future. Hypertuning of parameters may also be undertaken in future with a view to improve the existing accuracies. Further the later layers of the neural networks may be adjusted to improve accuracy for dog breed prediction in order to deploy the model.

References

1. Khosla A, Jayadevaprakash N, Yao B, Li F-F (2011) Novel dataset for fine-grained image categorization: Stanford dogs. In: Proc. CVPR workshop on fine-grained visual categorization (FGVC), vol 2, no 1. Citeseer
2. Wang I-H, Lee K-C, Chang S-L (2020) Images classification of dogs and cats using fine-tuned VGG models. In: 2020 IEEE Eurasia conference on IOT, communication and engineering (ECICE). IEEE, pp 230–233
3. Dabrowski A, Lichy K, Lipiński P, Morawska B (2021) Dog breed library with picture-based search using neural networks. In: IEEE 16th International conference on computer sciences and information technologies (CSIT). IEEE, pp 17–20
4. Bouaafia S, Messaoud S, Maraoui A, Ammari AC, Khriji L, Machhout M (2021) Deep pre-trained models for computer vision applications: traffic sign recognition. In: 2021 18th International multi-conference on systems, signals & devices (SSD). IEEE, pp 23–28
5. Varshney A, Katiyar A, Singh AK, Chauhan SS (2021) Dog breed classification using deep learning. In: 2021 International conference on intelligent technologies (CONIT). IEEE,pp 1–5
6. Junaidi A, Lasama J, Adhinata FD, Iskandar AR (2021) Image classification for egg incubator using transfer learning of VGG16 and VGG19. In: 2021 IEEE International conference on communication, networks and satellite (COMNETSAT). IEEE, pp 324–328
7. Nemati MA, BabaAhmadi A (2022) An investigation on transfer learning for classification of COVID-19 chest x-ray images with pre-trained convolutional-based architectures. In: 2022 30th International conference on electrical engineering (ICEE). IEEE, pp 880–884
8. Akhand MAH, Roy S, Siddique N, Kamal MAS, Shimamura T (2021) Facial emotion recognition using transfer learning in the deep CNN. Electronics 10(9):1036
9. Deng J, Dong W, Socher R, Li L-J, Li K, Fei-Fei L (2009) Imagenet: a large-scale hierarchical image database. In: 2009 IEEE Conference on computer vision and pattern recognitio. IEEE, pp 248–255
10. Ojha N, Kumar A (2020) A comparison based breast cancer high microscopy image classification using pre-trained models. In: 2020 IEEE Students conference on engineering & systems (SCES). IEEE, pp 1–6

11. Howard AG, Zhu M, Chen B, Kalenichenko D, Wang W, Weyand T, Andreetto M, Adam H (2017) Mobilenets: efficient convolutional neural networks for mobile vision applications. arXiv preprint arXiv:1704.04861
12. Zoph B, Vasudevan V, Shlens J, Le QV (2018) Learning transferable architectures for scalable image recognition. In: Proceedings of the IEEE conference on computer vision and pattern recognition, pp 8697–8710
13. Chollet F (2017) Xception: deep learning with depthwise separable convolutions. In: Proceedings of the IEEE conference on computer vision and pattern recognition, pp 1251–1258
14. Ovreiu S, Paraschiv E-A, Ovreiu E (2021) Deep learning & digital fundus images: glaucoma detection using DenseNet. In: 2021 13th International conference on electronics, computers and artificial intelligence (ECAI). IEEE, pp 1–4
15. Szegedy C, Ioffe S, Vanhoucke V, Alemi A (2017) Inception-v4, inception-ResNet and the impact of residual connections on learning. In: Thirty-first AAAI conference on artificial intelligence

An Efficient Recognition and Classification System for Paddy Leaf Disease Using Naïve Bayes with Optimization Algorithm

Vandana Prashar, Deepika Sood, and Santosh Kumar

Abstract In today's world, paddy is the main crop used by most people. Several diseases occur in paddy crops. Due to numerous diseases, only a limited quantity of paddy crop is yielded. Due to lack of technical and scientific knowledge, it is not easy for farmers to predict the diseases. Physically recognize and categorize paddy diseases has required more time. The research topic is classification, and recognition of the diseases from the leaf is one of the recent research topics in artificial intelligence. Consequently, an automatic and exact recognition scheme has turned into vital to reducing this issue. A novel method has been developed to find the paddy diseases brown spot, bacterial blight, blast diseases, and sheath rot through the application of machine learning (ML) classification methods. In this research article, a robust methodology is proposed that is categorized into different phases. In the preprocessing phase, removal of the background using the RGB images converted into HSV images and the segmentation phase of the normal fraction, diseased fraction, and the background a k-means clustering method is used. The machine learning method has been implemented to enhance the accuracy rate and the planned method. The firefly optimization has been implemented to fuse with an ML named Naïve Bayes classification method. Afterward, 494 leave images of our dataset of five dissimilar paddy leave diseases are used to instruct the NB with firefly proposed model. NB classification is then trained with the characteristics, which are extracted from NB with firefly method. The proposed method capably recognized and classified paddy leave diseases of five different categories and achieved 98.64% accuracy rate.

Keywords Machine learning · Paddy leaf diseases · Naïve Bayes · Firefly optimization algorithm

V. Prashar (✉) · D. Sood · S. Kumar
Department of Computer Science, Chandigarh University, Gharuan, India
e-mail: vandana.e12751@cumail.in

1 Introduction

Department of Agriculture is one of the most essential earning sources of human beings in many countries [1]. The environmental situation of soil and different food plants are then harvested by the plant. Apart from it, the farmer is covering many problems which are created by nature such as plant diseases, inevitable accident, shortage of water, and earthquake [2]. The challenging task is diseases management by the human being. Several diseases are seen on the leaf or stem or root of a plant. Some advance technical facilitates solving, most of the problems. Prevention is a beneficial step from leaf diseases that may improve the productiveness of agricultural food. Moreover, the ambitious step to recognize the leaf diseases is challenging.

Comparable social networking and some other research fields are equally useful in the field of agriculture to classify the diseases using different machine learning approaches. There are various categories and symptoms of various plant leaf diseases. Categorize on the behave of disease leaffeature such as size, color, and shape. The geographical factors and characteristics of the plant are changing due to environmental activities day by day. Therefore, manually identifying leaf conditions based on symptoms is challenging. For forecasting diseases, the statically classification algorithm is ineffective. As a result, in order to identify the disorders, make the appropriate choice, and choose the best course of therapy, an expertise computational intelligence system is required [3]. In addition to this, paddy is one of the main staple food and crops to seventy percent of the whole population in India. In the world, paddy is a 2nd central crop of South India [4]. Paddy establishes about 52 percent of the total food gain and cereal manufacture. The quantity and quality of the paddy plant can be enhanced by its precise and timely analysis, which can enhance a financial growth of the county.

Certain normal pests searched in NORTH MAHARASHTRA are blast paddy leaf, brown spot paddy leaf, bacterial blight paddy leaf, and sheath rot paddy leaf [5, 6]. The proposed system defines is classified within the four most normal diseases in the North-east India (NEI), namely sheath rot, bacteria blight, and blast leaf which are shown in Fig. 1. Several scholars and analysts who can recognize the diseases according to their signs are not simply accessible in remote edges. The computer vision (CV) is a non-destructive and possible solution to deal with this issue and gives consistent outcomes.

This research work main motives at implementing an automatic scheme to categorize the paddy diseases by executing the subsequent phases: (i) Paddy leave image acquisition (ii) paddy leave the picture preprocessing (iii) image segmentation in Paddy leave (iv) feature extraction and selection (v) classification in paddy leave diseases [7, 8].

Paddy is playing a significant role in equally the peoples of West Bengal and Asian countries. However, the paddy production excellence, as well as the amount, maybe reduced because of paddy leave diseases. Having a disease in paddy plants is moderately normal. If the completion stage is not acquired in this view, then its origins severe effects on paddy plants that affect likely; product amount, productivity, and

Fig. 1 **i** Bacterial blight paddy leaf **ii** Blast paddy leaf **iii** Brown spot paddy leaf **iv** Sheath rot paddy leaf

quality. Generally, plant infections attack by several pathogens like bacteria, viruses, and fungus, etc. Normally, these viruses destruct the plant photosynthetic procedure and interrupt the growth of plants. Thus, it is the main task to categorize the paddy plant disease in the premature phase [9].

At this time, farmers used their personal knowledge and knowledge to classify and identify infections. Without identifying the plant infections, farmers use insecticides in extreme quantity, which cannot assist in the inhabitation of disease but may enclose an effect on plants. However, some paddy leaf diseases may generate the same spot area. It also different lesions could be formed from similar diseases because of various paddy leaf varieties and local situations or climate [10]. Therefore, at times, their misclassification creates a bad effect on paddy cultivation. Consequently, it requires a guidance form paddy infection professionally. In rural areas, paddy disease professionals cannot able to offer rapid remedies to the farmers at the correct time, and they need the costly devices and a large amount of time for manual recognition and classification of paddy diseases. The second phase, for physical image processing, needs additional eyes to verify and crosscheck for its accuracy rate. However, an automatic scheme recognition and classification infections affected pictures more accurately than the manual classifying procedures.

The research approach defined a system to recognize and categorize the infection of paddy leaves. It guides the farmers to take the precise evaluation and encourage them to augment construction. In the study work, for the categorization of paddy leave disease images, it has developed an ML system by implementing NB with an FF optimization algorithm. The extraction of features is assumed to train a NB algorithm. The research system is then, calculated on a test, analyzes the database of 200 paddy leaf images, and achieved a calculation AUC of 98.64 percent.

In this research paper, explained several recognition and classification methods of diseases in paddy crops are analyzed in Sect. 1. In the Sect. 2, an approach is implemented for classifying of paddy leave diseases. Detailed feature extraction that defines the ROI for classification is elaborated in Sect. 4. Experimental analysis is given in Sect. 5. The conclusion is given in Sect. 6.

2 Related Work

In the advancement technology the recent research model is related to recognize and classification of plant leaf diseases. It is playing the important role in the machine learning to recognize the diseases.

The classification and recognition of the paddy leaf diseases was based on DNN Jaya approach. Using acquisition, the image of paddy plant leaves was acquired from yield for normal paddy leave image, bacterial blight paddy leaf, brown spots images, and blast infections. During the image preprocessing, the background elimination of the RGB images was transformed into HSV pictures, and it was dependent on the H (Hue) and S (Saturation) by Ramesh et al. [11] based binary pictures that were extracted to segment the unhealthy and non-infected parts. During the segmentation of the infected parts, the standard parts, and contextual process, the clustering algorithm was used. In addition, the classification of the diseases was considered utilizing optimized DNN along with Jaya algorithm. Experimental analysis was done and comparable to ANN, DAE, and DNN. The planned model has acquired the accuracy for the blast diseased up to 98.9%; bacterial blight was 95.34%, and sheath rot was 92.1%; brown spot was 94.56, and the standard leaf picture was 90.54% by Nidhis et al. [12].

On the foundation of techniques, there is a deep convolutional neural network; maize leaf disease detection was developed by Sun et al. [13]. They used a dataset divided into three sets like a validation set, testing set, and training set of maize leaf, and the total images 8152 include terrible diseases; these images were calibrated by human–plant pathologists. The detection performance achieved by single-shot multibox detector (SSD), and the observational outcomes depicted the feasibility and efficiency of their proposed model.

The latest method caffeNet, a deep learning framework formulated by the learning center and Barkley vision, which was used to perform the deep convolutional neural network introduced by Sladojevic et al. [14]. Apart from it, the dataset of the total number of images in 4483 that were downloaded from the Internet included 15 categorizes of disease images and non-disease images. The classifier named softmax was trained from scratch, and they used the backpropagation algorithm, and the overall accuracy of 96.3 percent was achieved by a deep learning model.

Using machine learning approaches, Liu et al. [15] developed a significant model for detecting and classification diseases names Mosaic Rust, Brown spot, Alternaria leaf caused by the apple leaves. The proposed AlexNet model and GoogleNet were efficient and reliable, and they achieved the 97.62 percent accuracy and convex optimization algorithm used for faster convergence rate.

To automatically detect the plant, leave diseases using the new approaches designed called as support vector machine (SVM) by Zang et al. [16]. RGB model was converted to HSI (Hue, saturation, and intensity), gray models, and YUV model for identify the color from the images and the segmented image by used the region growing algorithm. The most important features were selected by genetic and correlation-based feature selection algorithm. At the end of the feasible and effective model, the classification phase done by SVM and achieved the correctness was 90 percent.

Sengupta et al. [17] used the supervised incremental classifier named as a particle swarm optimization algorithm to perform the identify rice diseases with 84.02 percent. The proposed method became a model which had time complexity of the classification system constantly increment, and the system became more reliable and efficient for finding rice plant diseases. Moreover, the increment classifier model is suitable for applying on the dataset of postulation rice plant diseases such as the aspects and the change behavior of rice leaf due to nature, environmental condition, geological, and biological component.

In [18], Jiang et al. proposed an apple diseases identification model that is based on image processing techniques and deep learning methods, namely INAR-SSD and the deigned method by introducing the GoogleNet. In the experimental terms, the database of diseases image is divided between training and testing phases by the ratio 4:1. This approach accomplished the accuracy level which is 96.52 percent. This model is a feasible solution for finding the real-time detection of apple diseases.

Ramcharan et al. [19] used the proposed method to find virus diseases named as cassava (Manihot esculenta Crantz) disease considered as food security crops. It is a source of carbohydrates for the human food chain. The dataset was collected of 11,670 images of cassava leaves from an International Institute of Tropical Agriculture (IITA). Apart from it, researchers used the deep learning method which had the dataset divided into training and testing phases. During the model, 10 percent was used for training phase, and the other 90 percent was used for testing phase. The transfer learning method used for training cassava images and the deep learning convolutional neural method to detect the disease with the accuracy of 96 percent.

Image processing and the neural network method proposed to identify cotton leave diseases by Batmavady et al. [20]. Researchers collected the dataset of cotton leaves from the plant village. Firstly, the process started from there using processing techniques named as filter the image and then removes the noise and converted the input image into grayscale conversion. A set of features extracted from the processing image was used by the radial basis function neural network classifier. Further, SVM classifier divides the dataset into the testing and training phase and then used to recognize cotton disease with the better accuracy.

Ma et al. [21] proposed the method of deep learning convolutional network conducting the symptom-wise identification of four type of cucumber diseases, i.e., downy mildew, anthracnose, target leaf spots, and powdery mildew. The DCNN achieved a good accuracy of 93.4% on the results of 14,208 symptom images, and the comparative scientific research used the classifiers support vector machine and the random forest algorithm along with AlexNet.

Bai et al. [22] investigated cucumber leaf spot diseases implemented the segmentation method named morphological operations and also the watershed algorithm for working against the complex background of the disease images. The algorithm is demonstrated the neighborhood greyscale information which improves the capacity of noise filtering algorithm. The proposed method is an impressive and robust segmentation method for recognizing the cucumber diseases. The success rate of the segmentation method is 86.32%.

Spares representation classifier (SR Space) proposed by Zhang et al. [23] for recognizing cucumber diseases and the segmentation done by k-means algorithm. The advantages of the SR space classifier are effectively reducing the computational cost, and this method was improving the performance of the recognition rate that is 85.7%. Moreover, the lesion feature extraction method was used to extract the important feature the cucumber leaf dataset.

The survey on various IP and ML methods used for recognition of the rice plant (RP) viruses depends on the pictures of diseased RP by Shah et al. [24]. This research presented different methods for the recognition and categorization of the diseases. It carried out different surveyed papers on rice plant disease. The methods were carried out with the picture database, quantity of classes, preprocessing segmentation methods, classifiers, and so forth. They surveyed and studied the detection and arrangement of the RP diseases. Proposed research by Singh et al. [25] on the blast infection of the paddy leaf observes the picture of the plant leaf by specialists with required actions. The disease recognition approach was a color slicing method that perceived the infected spots and destructed proportion of the complete leaf that provides the advice, if the disease took place and removes in a required period of time to prevent the losses. In this research, the recognition of the blast disease of the paddy leaf was presented. In the planned model, the accuracy was acquired up to 96.6%. The new recognition provides better outcomes as compared to edge recognition models.

Sharma et al. [26] examined numerous machine learning (ML) algorithms and evolutionary computation with deep learning methods for identifying paddy diseases. This research took into consideration three main paddy ailments including, paddy blasting, bacterial leaf blight, and brown spot. The findings of a comprehensive comparison investigation concluded that transfer learning approaches were better than the traditional ML algorithms. The findings could be used to farming practices recognize the paddy disease early as then immediate strategy could be done moving forward. To extrapolate the research's results, emphasis needs to be placed on working with large-scale datasetsin subsequent investigations shown in(Table 1).

3   Problem Definition

The human-based perceptions are conservative methods to monitor the leaf diseases in the previous year. In these cases, it is very complicated, time-consuming, and expensive to try the expert advice. Many farmers who are not educated and have lack of knowledge about the type of diseases hence the eye-catching method suffer from many downsides. Moreover, the machine learning methods for recognize the leaf diseases and to make the right decision for selecting the proper treatment is also challenging. To overcome the downside of the conservative methods, a new classification method is required for a new machine learning. Machine learning approaches are very few documented specially in the plant leaf diseases detection. The disease

Table 1 Summary of different classification techniques detect different diseases

Author name, Year	Type of leaves	Classification techniques	Algorithm	Type of diseases detect	Accuracy
Ramesh et al. [11]	Rice/paddy leaf	DNN	Jaya optimization algorithm	Blast, bacteria blight, brown spot, sheath rot	97%
Sun et al. [13]	Maize leaf	Convolutional neural network	Single-shot multibox detector (SSD)	Northern maize leaf blight	91.83%
Sladojevic et al. [14]	Infected	CaffeNet	Backpropagation algorithm	13 different diseases	96.3%
Liu et al. [15]	Apple leaf	GoogleNet and CaffeNet	Convex optimization algorithm	Mosaic rust, brown spot, alternaria leaf	97.62%
Zhang et al. [16]	Apple leaf	SVM	Genetic algorithm and region growing	Powdery mildew, mosaic and rust	90.00%
Sengupta et al. [17]	Rice	Particle swarm optimization	IPSO (Incremental PSO)	Rice diseases	84.02%
Jiang et al. [18]	Apple leaf	Convolutional neural networks—INAR-SSD model	Oriented object detection algorithm	5 type of apple diseases	96.52%
Ramcharan et al. [19]	Cassava	Convolutional neural network and machine learning method—SVM	Transfer learning algorithm	Mosaic disease, red mite camage, healthy and brown leaf spot, brown streak disease	96%
Batmavady et al. [20]	Cotton leaf	RBF neural network, SVM	–	Cotton leaf diseases	Sensitivity-92.00% Accuracy-90.00% Specificity-96.00%

(continued)

Table 1 (continued)

Author name, Year	Type of leaves	Classification techniques	Algorithm	Type of diseases detect	Accuracy
Ma et al. [21]	Cucumber leaf	AlexNet	Random forest and support vector machine	Downy mildew, anthracnose, target leaf spots, and powdery mildew	93.4%
Bai et al. [22]	Cucumber leaf	Morphological operations	Water-shed algorithm	Leaf spot	86.32%
Zhang et al. [23]	Cucumber leaf	Sparse representation	k-means cluster algorithm	Downy mildew, bacterial angular, corynespora cassiicola, scab, gray mold, anthracnose	85.7%

can be identified, and a solution for disease can be found by the classifier then using the classification method as well as detection methods.

4 Proposed Methodology

With the five phases that include the image acquisition, image preprocessing, segmentation, feature extraction, and image classification, the proposed structure is developed. The compute vision is an automatic system that performs disease and non-disease paddy leave recognition and categorization is established in implemented work. Various stages include in the investigation are given in Fig. 2.

4.1 Image Acquisition

The basic method of gathering the dataset of various paddy leave disease photos from the https://archive.ics.uci.edu/ml/datasets/Paddy+Leaf+Diseases site, that are used for this presented design, is known as image acquisition. The research system is using three disease categories of paddy images such as brown spot, bacteria light, and sheath rot. In this proposed system the paddy leaves are collected by using higher-resolution digitalized camera (HRDC). After that the detection or classification of disease, complete capture paddy pictures are located to processor, where the establishment process may take place.

Fig. 2 Proposed flowchart

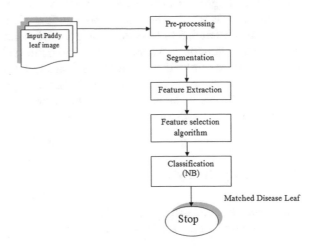

4.2   Image Preprocessing

In the second phase, preprocessing for reducing the images in the database is re-sized and cropped into 256*256 image pixels. The significant things to remove the image environment are done by considering the hue values. Originally, the paddy leave image in red, green, and blue model is transformed into HSV model [27]. Using the HSV model, the normal S value is measured in the model, though it completes the whiteness. It depends on the threshold rate of about 90 degrees, and leaf picture is transformed into a binary picture, and the binary image is associated with the actual RGB picture to create the label. The leaf region has the available disease segment where the contextual part hides the picture.

4.3   Image Segmentation

In this phase shows the segmentation of a paddy leave picture, k-mean clustering model is analyzed in this research. Normally, clustering is a technique to cluster the picture into groups. The infected region is eliminated from the leaf picture through grouping. In the paddy leaf picture, clusters are developed that estimate the infected region and non-disease part.

k-means Clustering
Images are selected exploitation of the picture reading operated and displayed. After that, the transformation of the colors is completed from the actual picture for selecting the picture. The grouped pictures are based on the component. The bunch methods are segmented that is based on the segmented algorithms [20]. It is a simple unsupervised learning algorithm that follows the simple model to classify the essential index by a specified amount of the clusters. It focused on the route of different region. The following phase considers the centroid as center of clusters. k-centroid recomputed where k-centroid of the cluster where similar data points and closest center are arranged. With the creation of the new loop, the k-center changes with the new position unless more alterations occur. Figure 3 shows the diseased portion.

4.4   Feature Extraction

This research work extracts both the text and color feature. The color characteristic comprises the extraction of the mean and standard deviation. The textual

Fig. 3 Highlighted diseased portion

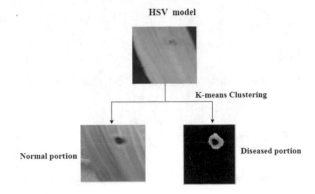

characteristics contain the histogram-oriented gradient (HoG) features like as local features.

Feature Extraction Using HoG Method

Histogram-oriented gradient (HoG) is the characteristic descriptor (FD) used in computerized vision and digital IP for recognizing the data objects. This technique selects the existence of the gradient orientation in the local parts of the picture. It is same as the boundary orientation histogram that used the local contrast normalization for improving the accuracy.

Pseudo-code with Proposed System (HoG) Method

Input Segmented picture classes are leaf blast, brown spot, sheath blight

Output Color Features

For complete segmented picture of every class

Step 1: Read RGB pictures as Img;

Step 2: Distinguish R, G, B element of RGB pictures,

$$R = img(a, b);$$
$$G = img(a, c);$$
$$B = img(d, b);$$

Step 3: Normalization of the RGB elements ranges from 0 and 1;

$$r = \frac{R}{254.5}, \; g = \frac{G}{254.5}, \; \frac{B}{254.5}$$

Step 4: Searching the standard deviation of r, g, b and measure the correlated factor;

Step 5: Combine the extracted features as;

$$RGB_{factor} = [R_{sd}, \ G_{sd}, B_{sd}, \text{RGcorrelted}]$$

End

Here, Img = Image, R = Red, G = Green and B = Blue.

4.5 Feature Selection

The research work selects the valuable feature using a firefly optimization algorithm. This feature selection is a met heuristic method that is nature inspired by the irregular behavior of fireflies and the phenomenon of bio-luminescent communication.

Feature Selection Using FFOA

Feature selection is also used for optimizing the dimension of the databases, for reducing the calculation time, price and classification fault. Various researchers used the feature selection for numerical methods [28]. Firefly algorithm is used for searching the association of the features that improved the desired fitness function. FFA is used for searching the new output that depends on the lighting of the fireflies. The modification in the output location given the direction vector linking the better output, the present situation is updated in direction vector linking the best output, and present output is updated with diverse intensity. The random values of the parameters are given by β, and motion of the firefly is another firefly k. The individual size is represented by the feature space for single feature and duration of every size range from zero to one for searching the optimum value that improved the fitness function. The fitness function for firefly algorithm is given for improving the classification rate for trained values.

Pseudo Code (Firefly Optimization Methods)

Step 1: Selective function of $f(y)$ where $y = (y_1, \ y_2, \ \ldots \ y_n)$;

Step 2: Create initial population of firefly;

Step 3: Define the absorbed coefficient β;

While ($s <$ max_generation);

For $j = 1$ to n;
   For $j = 1$ to n;
     If $(l_j > 1)$

Moving firefly j in the direction of k;

Change data with the distance r through β value.

End if;

Step 4: Compute the new outcome and updating the light intensity;

End for j;

End for k;

Step 5: Rank firefly and search the current best value;

End while.

Step 6: Process the resulted values and get the output;

End.

4.6 Classification

The classification process is the most important factor of the paddy leave disease prediction. The research system has implemented the Naïve Bayes (NB) method to classify the disease leave. It is calculating the feature probability or depends only a particular feature in a group, and it does not depend on any other group. When the research system is developed for detection method, then implement a Naïve Bayes method, which is a supervised learning method. This method is separated into 2 phases like (i) training and (ii) testing.

Classification Using Naïve Bayes
The classification is a technique for the classification of the plant species in accordance to the accurate set based the major features. The classification of features is depends on the extraction of features. Naïve Bayes is utilized for the classifier procedure [28]. The textual and shape variables are the identified features. Naïve Bayes is the theorem with the statistical method for the classification of the prediction classes that depends on the possibilities. It is measured as a more effective method for comparing the decision tree and other applications. It is based on the assumption of the autonomy between the predicted values. Naïve Bayes (NB) classifier depends on the specific feature in the class, and it may not base on the other features. For instance, the image color and shape of the paddy leaf are determined; then, the features are considered based on the detection of the plant. Hence, NB is simple in construction of the large database.

Performance Parameters

Accuracy Rate
The research model has proposed in the paddy leave detection system to enhance the accuracy parameter as compared with the existing algorithm. The research model accuracy has increased, according to the increase the number of images. The proposed accuracy rate value is 98.64%.

Cross-Entropy Loss
The research model has proposed in the paddy leave detection system to decrease the entropy loss parameter as compared with the existing algorithm. The research model

entropy loss has decreased, according to the increase the probability. The proposed entropy loss value is 0.0064%.

True Positive Rate

In deep learning concept, the TPR is also denoted to recall, is utilized to consider the percentage of real positives which are appropriately verified. Thus, the true positive rate (TPR) is 0.9964.

True Negative Rate

In deep learning concept, the TNR is also denoted to specificity, considers the proportion of real incorrectly values that are correctly verified. Thus, the true negative rate value is 0.990.

False Positive Rate

The FPR is called false alarm ratio. FPR is the probability of wrongly discarding the null hypothesis for a certain test analysis. The FPR is evaluated as the ratio among the no. of negative actions incorrectly considered as FPs and the complete number of real wrong actions. Thus, the FPR value is 0.0100.

False Discovery Rate

The FDR is the predictable proportion of kind of expectations. The main kind of exception is where incorrectly discarding the null hypothesis. Thus, FDR value is 0.0036.

5 Result Analysis

In this result explanation, comparative analysis with various parameters and mathematical expression described. In this, research work is performed in MATLAB simulation tool used, and GUIDE has designed a project desktop application. In this experimental result analysis, totally with 494 images which include 125 brown spot, 102 bacteria blight, 61 sheath rot, 120 normal, and 86 blasts are given.

It is almost 494 images of paddy leaf images were reserved for the training module. The testing module presentation of the research model Naïve Bayes classification method, long-term database images of paddy leaf diseases were selected for testing module. The research flowchart shows the working of the proposed model. All images were stored in .png and .jpg format. The characteristics like segment area, edges, HSV model, etc are very essential.

The complete proposal is commonly concerned with different regions, the edges with segmentation, HSV color model, and filtration is performing an important role in the research system. Different types of performance metrics are calculated in DIP methods such as segmented area. This research phase has improved the performance and compared with the existing classification method (DNN-Jaya), which can be seen in several kinds of global features. It calculates the system performance with accuracy rate, loss, TPR, TNR, FDR, FPR, etc.

This window creates by MATLAB GUIDE environment toolbox used. The design information is displayed, and a code is also written in the script window. The first step is to separate two groups into classification and recognition systems. The training_module and testing module are defined in this conceptual framework. The training_module demonstrates how to upload many photos at once. Preprocessing techniques should be used to spot the distortions in the submitted image. Implementing the filtering approach to eliminate the unwanted noise if distortion is present. The HSV model is definite color space. After that, the KMC algorithm developed in this system to divide the data into different number of clusters or groups. The segmentation process involves locating the filtered image's region. Then, it developed the FE algorithm using HoG method. This method is used to eliminate the global-based features and calculate the segmented image features such as energy, mean, SD, and image contrast. After that, to train the Naïve Bayes network, given data and labels or targets are used. In testing, module shows the upload of the test picture. It transforms the color picture to grayscale picture. Segment the regional area, extract unique features, and calculate the performance metrics.

The firefly optimization algorithm method is based on a performance metric modification, varied search strategy and change the solution space to create the search simply using various probability distributions. The detection process is done by Naïve Bayes classification. Figures 4 and 5 demonstrate the two kinds of graph plot with compare among planned and previous parameters like as accuracy, cross-entropy rate. The planned approach has improved the accuracy compared with various kinds of approaches like as DNN, DNN_JAO, and OF-Naïve Bayes.

Table 2 displays the results of the research effort, including characteristics like accuracy rate (98.64%), cross-entropy loss (0.0064), TNR, TPR, and FPR values (0.990, 0.9964, and 0.0036 respectively). The accuracy rate and loss in the paddy leave disease detection technique have been increased through the methodology.

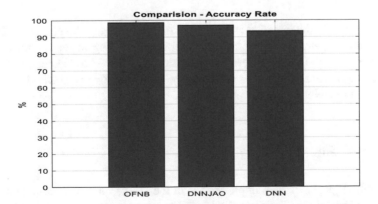

Fig. 4 Comparative analysis with the proposed and existing method using accuracy rate comparison

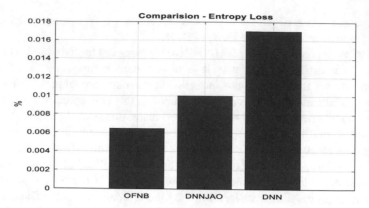

Fig. 5 Comparative analysis with the proposed and previous method using entropy loss

Table 2 Proposed parameter with research work

Parameters	Values
Cross-Entropy Loss (%)	0.0064
Accuracy Rate (%)	98.64
TNR (%)	0.9900
TPR (%)	0.9964
FPR (%)	0.0100
FDR (%)	0.0036

Table 3 Comparison: Proposed and conventional models (OFNB, DNN + JAO, DNN)

Parameters	OF-Naïve Bayes	DNN_JAO	DNN
Accuracy Rate	98.64	97	93.50
Cross-Entropy Loss	0.0064	0.0100	0.01700

Table 3 compares the intended and prior techniques utilizing the classifiers OFNB, DNN, and DNN + JOA. The accuracy performance rating of the methodology is 98.64 percent with OFNB, 97 percent with accuracy, and 93.50 percent with DNN. Cross-entropy loss study process performance value is 0.0064 with OFNB; accuracy is 0.0100, and DNN is 0.01700 percent.

6 Conclusion

This research paper defines improvement in the image quality, a large number of operations carried out. The normal phase adds image preprocessing of paddy leave images. It involves RGB to grayscale conversion and RGB to HSV model conversion. It applied the filtration methods such as Gaussian and three-dimensional box filtration methods to enhance the picture quality and eliminate the unwanted noise. After that, implement a k-means clustering approach for calculating the region of the image. From these images, inner and outer quality characteristics like color, global, and texture are removed using HoG method. The color image feature is eliminated through mean and standard deviation. Feature metrics extract through HoG. The classification is done by optimized firefly Naïve Bayes (OFNB) or OFNB method. After that compute the performance metrics such as TP, TN, FN, and FP and improve the accuracy rate as compared with the existing methods (ODNN) or OFNB. The automatic paddy leaf disease detection system using different categories of feature like image contrast, entropy, homogeneity, energy, and color is implemented. For the extraction of features, histogram-oriented gradient (HoG) is used resulting in unique feature calculation. The firefly optimization approach uses the results in low dimension and complication. k-mean clustering segments the paddy leaf picture, which gives maximum accuracy rate with less error rate and entropy loss. This proposed system also compared with OFNB, DNN_JAYA and DNN classification, where OFNB classifier gives higher disease detection accuracy rate (98.64%), DNN_JAYA accuracy rate (97%), and then DNN (93.50%). Future work, it will develop image processing method and deep learning method using F-CNN algorithm to improve the system performance. In addition, other options such as hybrid approaches like as clustering model or RNN can be used for improving the PSNR and precise values through classification procedure. In addition, the number of image processing algorithms can be developed for the identification of leaf disease.

References

1. Islam MA, Hossain T (2021) An automated convolutional neural network based approach for paddy leaf disease detection. Int J Adv Comput Sci Appl 12(1):280–288
2. Pantazi XE, Moshou D, Tamouridou AA (2019) Automated leaf disease detection in different crop species through image features analysis and one class classifiers. Comput Electron Agric 156:96–104. https://doi.org/10.1016/j.compag.2018.11.005
3. Sun G, Jia X, Geng T (2018) Plant diseases recognition based on image processing technology. J Electr Comput Eng 2018:1–7. https://doi.org/10.1155/2018/6070129
4. International Rice Research Institute (2009). Crop health: diagnostic of common diseases of rice. Retrieved from http://www.knowledgebank.irri.org/ipm/terms-and-definitions.htm. Accessed on Aug 2013
5. Phadikar S, Sil J, Das AK (2013) Rice diseases classification using feature selection and rule generation techniques. Comput Electron Agric 90:76–85. https://doi.org/10.1016/j.compag.2012.11.001
6. Retrieved from http://www.rkmp.co.in/research-domain
7. Pratheba R, Sivasangari A, Saraswady D (2014) Performance analysis of pest detection for agricultural field using clustering techniques. In: 2014 International conference on circuits, power and computing technologies [ICCPCT-2014]. IEEE, 1426–1431. https://doi.org/10.1109/ICCPCT.2014.7054833

8. Han L, Haleem MS, Taylor M (2015) A novel computer vision-based approach to automatic detection and severity assessment of crop diseases. In: 2015 Science and information conference (SAI). IEEE, pp 638–644. https://doi.org/10.1109/SAI.2015.7237209

9. Lichtenthaler HK (1996) Vegetation stress: an introduction to the stress concept in plants. J Plant Physiol 148(1–2):4–14. https://doi.org/10.1016/s0176-1617(96)80287-2

10. Yao Q, Guan Z, Zhou Y, Tang J, Hu Y, Yang B (2009) Application of support vector machine for detecting rice diseases using shape and color texture features. In: 2009 International conference on engineering computation. IEEE, pp 79–83. https://doi.org/10.1109/ICEC.2009.73

11. Ramesh S, Vydeki D (2019) Recognition and classification of paddy leaf diseases using optimized deep neural network with jaya algorithm. Inf Process Agric 7(2):249–260. https://doi.org/10.1016/j.inpa.2019.09.002

12. Nidhis AD, Pardhu CNV, Reddy KC, Deepa K (2019) Cluster based paddy leaf disease detection, classification and diagnosis in crop health monitoring unit. In: Computer aided intervention and diagnostics in clinical and medical images. Lecture notes in computational vision and biomechanics, vol 31, pp 281–291. Springer, Cham. https://doi.org/10.1007/978-3-030-04061-1_29

13. Sun J, Yang Y, He X, Wu X (2020) Northern maize leaf blight detection under complex field environment based on deep learning. IEEE Access 8:33679–33688. https://doi.org/10.1109/ACCESS.2020.2973658

14. Sladojevic S, Arsenovic M, Anderla A, Culibrk D, Stefanovic D (2016) Deep neural networks based recognition of plant diseases by leaf image classification. Comput Intell Neurosci 2016:1–11. https://doi.org/10.1155/2016/3289801

15. Liu B, Zhang Y, He D, Li Y (2017) Identification of apple leaf diseases based on deep convolutional neural networks. Symmetry 10(1):11. https://doi.org/10.3390/sym10010011

16. Zhang C, Zhang S, Yang J, Shi Y, Chen J (2017) Apple leaf disease identification using genetic algorithm and correlation based feature selection method. Int J Agric Biol Eng 10(2):74–83. https://doi.org/10.3965/j.ijabe.20171002.2166

17. Sengupta S, Das AK (2017) Particle swarm optimization based incremental classifier design for rice disease prediction. Comput Electron Agric 140:443–451. https://doi.org/10.1016/j.compag.2017.06.024

18. Jiang P, Chen Y, Liu B, He D, Liang C (2019) Real-time detection of apple leaf diseases using deep learning approach based on improved convolutional neural networks. IEEE Access 7:59069–59080. https://doi.org/10.1109/access.2019.2914929

19. Ramcharan A, Baranowski K, McCloskey P, Ahmed B, Legg J, Hughes DP (2017) Deep learning for image-based cassava disease detection. Front Plant Sci 8:1852. https://doi.org/10.3389/fpls.2017.01852

20. Batmavady S, Samundeeswari S (2019) Detection of cotton leaf diseases using image processing. Int J Recent Technol Eng 8(2S4):169–173. https://doi.org/10.35940/ijrte.b1031.0782s419

21. Ma J, Du K, Zheng F, Zhang L, Gong Z, Sun Z (2018) A recognition method for cucumber diseases using leaf symptom images based on deep convolutional neural network. Comput Electron Agric 154:18–24. https://doi.org/10.1016/j.compag.2018.08.048

22. Bai X, Li X, Fu Z, Lv X, Zhang L (2017) A fuzzy clustering segmentation method based on neighborhood grayscale information for defining cucumber leaf spot disease images. Comput Electron Agric 136:157–165. https://doi.org/10.1016/j.compag.2017.03.004

23. Zhang S, Wu X, You Z, Zhang L (2017) Leaf image based cucumber disease recognition using sparse representation classification. Comput Electron Agric 134:135–141. https://doi.org/10.1016/j.compag.2017.01.014

24. Shah JP, Prajapati HB, Dabhi VK (2016) A survey on detection and classification of rice plant diseases. In: 2016 IEEE International conference on current trends in advanced computing (ICCTAC). IEEE, pp 1–8. https://doi.org/10.1109/ICCTAC.2016.7567333

25. Singh A, Singh ML (2015) Automated color prediction of paddy crop leaf using image processing. In: 2015 IEEE Technological innovation in ICT for agriculture and rural development (TIAR). IEEE, pp 24–32. https://doi.org/10.1109/TIAR.2015.7358526

26. Sharma M, Kumar CJ, Deka A (2022) Early diagnosis of rice plant disease using machine learning techniques. Arch Phytopathol Plant Prot 55(3):259–283
27. Junhua C, Jing L (2012) Research on color image classification based on HSV color space. In: 2012 Second international conference on instrumentation, measurement, computer, communication and control. IEEE, pp 944–947. https://doi.org/10.1109/IMCCC.2012.226
28. Salfikar I, Sulistijono IA, Basuki A (2018) Automatic samples selection using histogram of oriented gradients (HOG) feature distance. EMITTER Int J Eng Technol 5(2):234–254. https://doi.org/10.24003/emitter.v5i2.182

Development of Decision-Making Prediction Model for Loan Eligibility Using Supervised Machine Learning

Raj Gaurav, Khushboo Tripathi, and Ankit Garg

Abstract Nowadays, banking sectors/financial institutions are facing acute problems of loan default. Their loan assets are converting into non-performing assets rapidly. This problem makes these institutions running out of capital money and making their profit into heavy debt. For the last few years, there has been news of merging of banks & other financial institutions due to heavy loss, and the main reason for their loss is defaulting of loans. The fundamental objective of this paper is to forecast the risk of loan default of applying persons, institutions, or any other organization sooner approving them a loan. Various parameters like educational background, age, dependents member size, certain income, nature of income, these are some basic basis, would be seen into noticeable before accepting the NPA-free loan. This paper also has an objective to automate the process that will reduce the process time and human energy and deliver the service more efficiently. In this paper, multiple ML models are compared with the several loan forecasting models, and accuracy has reached more than 90 percent which is much better than existing models where accuracy is up to 80 percent. This accuracy would be helpful for financial institutions in the practical world to forecast if the loan should be accepted or not, making them NPA while returning back their loans. That will be helpful in fastening the loan lending process with less required resources to accept the loan applications and filter the creditworthy applicants for lending NPA-free loans.

Keywords Supervised machine learning · Loan prediction · Data science · KNN · SVM · Random forest · Preprocessing

R. Gaurav (✉) · K. Tripathi · A. Garg
Amity School of Engineering & Technology, Amity University, Gurugram, India
e-mail: emailtorajgaurav@gmail.com

K. Tripathi
e-mail: ktripathi@ggn.amity.edu

A. Garg
e-mail: agarg1@ggn.amity.edu

© The Author(s), under exclusive license to Springer Nature Singapore Pte Ltd. 2023   169
Y. Singh et al. (eds.), *Proceedings of International Conference on Recent Innovations in Computing*, Lecture Notes in Electrical Engineering 1011,
https://doi.org/10.1007/978-981-99-0601-7_14

1 Introduction

Machine learning and artificial intelligence are the most using technology nowadays to enhanced automation in day-to-day business. This paper also comes under the machine learning technology to enhance and automate the day-to-day business of loan lending process. Nowadays, financial institutions are suffering from huge pressure of non-performing asset. To reduce non-performing asset problem, it is important to predict before lending the loan for the customer which is capable to repay of loan or not. The prime objective of this models to lower the fear of defaulting of loan repayment and also filters out the creditworthy applicants to lending fear-free loans. It is also noticed that only good credit score is not a guarantee of loan defaulter. Huge number of cases shown that also good credit scorer person also defaulting in loan repayment. So, financial experts believe that there should be other mechanism could also apply to cope with loan defaulting problem. Because loan defaulting problem has different factors also. So, the objective of this paper is to analyze existing loan eligibility system using supervised machine learning, to develop decision-making prediction model for loan eligibility through supervised machine learning algorithm, and to compare the results with existing model and enhanced the accuracy results with developed model for loan eligibility system.

2 Related Works

I have studied so many research paper to understand the different aspects of machine learning model related to loan prediction model and find some useful information to propose some more robust model.

The authors, Shinde et al. [1] in their research paper used two machine learning model logistic regression and random forest with k-folds cross-validation to predict the eligibility of loan applicants. After using this method, their ML model reached the accuracy of 79 percent with 600 cases in dataset.

A work by Ramya et al. [2] were used dataset having the attributes of the loan customer applicants is accumulated with six hundred cases. This dataset is classed among two sets: Train dataset that is exercised for training the machine learning algorithm model. This carries each of the individualistic variables also with objective variable, and test dataset carries each of the individualistic variables but not the objective variable. Applying the machine learning model to forecast the objective variable to the test dataset. The logistic regression machine learning algorithm model is exercised to forecast the dual result.

Another work by Sarkar et al. [3] describes the "A Research Paper on Loan Delinquency Prediction", to design and develop a system that assists in detecting and formulating the loan delinquency in a detailed way that can be easily understood by borrowers. Since it involves prediction work, so, they used number of machine learning algorithm, i.e., neural networks, linear regression.

Sheikh et al. [4] used logistic regression model for their proposed machine learning model on "prediction of loan approval". They used 1500 cases of dataset with different number of features and apply various data preprocessing technique to reach optimal accuracy of the machine learning model.

Vaidya [5] introduces regrading to applied region moving swiftly to automating the process, to importance of automating the process, and the introduction of artificial intelligence as well as machine learning algorithm in it. Decision-making machines are the further enhancement shifting of this trend after machine learning model. Explained models are implemented with several machine learning model algorithms. With more distant ample of co-related affairs, the proposed paper exercised logistic regression as machine learning model technique for likelihood and forecasted approach and achieved the accuracy between 70 to 80%.

After analyzing above related works, it can be seen that their ML prediction accuracy is not reached more than 80 percent. In the financial sector, loan lending is the main source of profit. To increase the profit, it is important for these financial sectors that their lending loans must be returned on specified time without defaulting of their applicants. This proposed paper will help to reduce their defaulting applicants and filter the creditworthy loan applicants with the accuracy more than 90 percent. This accuracy will be reached with the help of a higher number of generated cases in the dataset to train the model. The main challenge in the proposed model is to generate the dataset near the actual dataset because due to the sensitivity nature of information, higher cases of dataset are not freely available. It can be seen in the above related works model where the maximum number of cases used in the dataset is 1500.

3   Problem Statement

Financial institutions, financial organizations, and big and small non-banking finance company transaction in with several variety of loans like house loan, vehicle loan, organization loan, personal loan, etc., in every region of state/countries. Those financial companies are providing services in cities, towns and village region. Through file an application for loan by applicant that financial companies validate creditworthiness of applicants to sanction the loan rejecting the application. This model paper is providing resolution for automating loan eligibility processes by deploying several machine learning model algorithms. Therefore, those applicants will apply on company's portal with simple application form and get the status of their application. That application form consists fields like Gender, Married Status, Academic Qualifications, Dependents Member Details, Yearly Income, Loan Amount, Credit History of Customers, and other related members. Using machine learning algorithm to automate this process, start the process that model will pick out that facts of the applicants whom is deserving to sanctioning the loan amounts; therefore, financial companies could attention on those applicants. Loan eligibility forecast is a real-life matter means almost each financial/NBFC organization address their loan sanctioning process. Whether the loan qualified procedure is unmanned, this could cut back too many hours of resources and upgrade the capacity services to their customers in other operations. The enhancement in customer assistant with satisfactory service and benefits with processing costs are always important. Nevertheless, the financial gain can only be dissecting whether the financial companies will have a sturdy and trustworthy models to error-free forecast which customer's loan that should admit and which customer's loan to reject, so that decrease the fear of loan defaulting problem or NPA.

4 Proposed Model

Decision-making predication to find eligible customer to borrow the loan from banking or financial institutions is the proposed model. This model is based on classification target, so supervised machine learning classification model, KNN algorithm model, random forest algorithm model, and support vector machine algorithm model are used to develop the model and compare the accuracy and execution time between them to select best model for prediction.

Preprocessing of dataset is the major area which consumes lot of time. Augmentation the figure of cases in dataset is also a complex factor because in available dataset has no more than 1500 cases due to sensitivity of data. So, increase the number of cases in dataset, dataset generation tools are used.

Exploratory data analysis is used to understand the features of dataset, and then, feature engineering is also explored to enhance the prediction of proposed model.

Python language and their robust libraries is applied to develop the supervised ML model.

4.1 Collection of Dataset

Dataset is downloaded from Kaggle which provides the huge source of dataset for learning purpose. But, this dataset is very sensitive in nature so financial institution whom published the data for learning purpose very less number of cases. Dataset downloaded from have two sets of dataset one for training and one for testing. Training dataset is used to train the model with divided into 70:30 ratios. Bigger part of dataset is for train the model, and smaller part is test for model, so accuracy is calculated for this developed model.

Structure of Dataset in this section, describing attributes of dataset to be consider in to develop machine learning model.

In dataset, there is 13 attributes with different data types. Loan_ID is an object data type with hold the data of loan identification. Gender is an object data type with hold the data of applicant's gender, i.e., male/female. Married is an object data type with hold the data of that applicant is married or not in yes/no options. Dependents are an object data type with hold the data that how many dependents of applicant in numbers. Education is an object data type with hold the data that applicant's qualification. Self_Employed is an object data type with hold the data of applicant's self-employment. ApplicantIncome is an Int64 data type with hold the data of income of applicant. CoapplicantIncome is a float64 data type with hold the income data of co-applicant. LoanAmount is a float64 data type with hold the data of loan amount in USD. Loan_Amount_Term is float64 data type with hold the data of loan term in month. Credit_History is a float64 data type with hold the data of meet the criteria or not. Property_Area is an object data type with hold the data that property lies in which area means city, town, and village. Loan_Status is an object data type to hold the data that whether loan is approved or rejected.

Cases Generation of Dataset have seen that in all previous related works their cases in dataset is small. Low number of cases in dataset impact the performance of machine learning model. So, in this model generated more number of cases using technique of data generation with the help of following online dataset case generation tool:

- Faker
- Mockaroo
- Mock Turtle.

Now, after using these tools, have increased number of cases from 614 to 4298. Increased number of cases in this dataset impacted positively to enhance the accuracy of developed machine learning model.

Figure 1 showing sample cases of dataset after case generation, i.e., 4298 rows × 13 columns.

	Loan_ID	Gender	Married	Dependents	Education	Self_Employed	ApplicantIncome
0	LP1001	Male	No	0	Graduate	No	5849
1	LP1002	Male	Yes	1	Graduate	No	4583
2	LP1003	Male	Yes	0	Graduate	Yes	3000
3	LP1004	Male	Yes	0	Not Graduate	No	2583
4	LP1005	Male	No	0	Graduate	No	6000
...	...	...	...	...	...	...	...
4293	LP5294	Female	No	0	Graduate	No	2900
4294	LP5295	Male	Yes	3+	Graduate	No	4106
4295	LP5296	Male	Yes	1	Graduate	No	8072
4296	LP5297	Male	Yes	2	Graduate	No	7583
4297	LP5298	Female	No	0	Graduate	Yes	4583

4298 rows × 13 columns

Fig.1 Sample data-1

4.2 Exploratory Data Analysis of Dataset

After data are collected, then next step is exploring the data for better understanding. Understanding of data is very important to filter out those value which decreases the execution and precision of the deployed model. Because all the values are fitted into the model, then outliers or extreme value will impact the execution and precision of deployed model. Missing values are also important to be filed because it also impacts the performance of ML model. Data analysis of dataset is also important to understand the relation of their attributes.

Figure 2 showing statistical analysis of data.

Exploratory data analysis (EDA) of gender feature in dataset is showing 80 percent of applicants is male, and 20 percent isfemale applicants (see Fig. 3).

Figure 4 is showing credit history with loan status of paid and unpaid debts applicants where 1.0 is indicating paid debts applicants and 0.0 is unpaid debts applicants. It can be seen that loan approved for 85 percent of paid debts paid applicants (1.0) and loan disapproved for 85 percent of unpaid debts applicants (0.0) as per dataset after EDA.

	ApplicantIncome	CoapplicantIncome	LoanAmount	Loan_Amount_Term	Credit_History
count	6754.000000	6754.000000	6512.000000	6600.00000	6204.000000
mean	5403.459283	1621.245798	146.412162	342.00000	0.842199
std	6104.516792	2924.080938	85.521575	65.07105	0.364584
min	150.000000	0.000000	9.000000	12.00000	0.000000
25%	2876.000000	0.000000	100.000000	360.00000	1.000000
50%	3812.500000	1188.500000	128.000000	360.00000	1.000000
75%	5800.000000	2302.000000	168.000000	360.00000	1.000000
max	81000.000000	41667.000000	700.000000	480.00000	1.000000

Fig. 2 Statistics of data

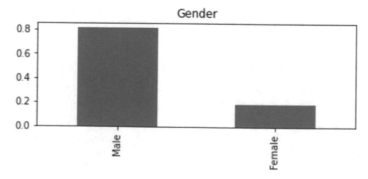

Fig. 3 Gender of the applicants

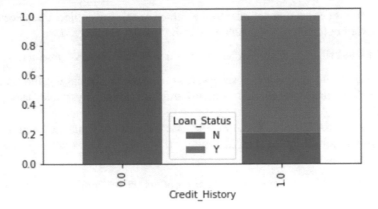

Fig. 4 Credit history—paid and unpaid applicants with loan status

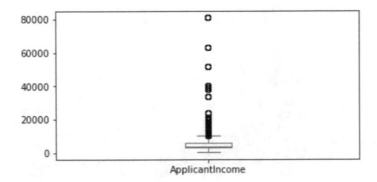

Fig. 5 Boxplot of applicant income

Figure 5 which is a box plot is describing dataset has lot of outliers values after exploration of dataset. These outliers will be treated in data preprocessing section to reduce outliers from the dataset.

4.3 Preprocessing of Data

Data preprocessing is the most important part to develop machine learning model. Accuracy and performance are much dependent on data preprocessing. To development in machine learning steps, data preprocessing takes maximum amount of time. Exploratory data analysis shows that the following outcome:

- Dataset has some missing values. Before applying machine learning model, that is important to fill that missing values.

- Dataset has certain number of extreme values or can say outliers that necessarily addressed before applying machine learning model (see Fig. 5).

  Above described problem can be addressed with the following manner:

- Addressing missing values: Missing values are loaded with applying mode method for each attributes for classed variables and for continued variables using mean method.
- Addressing outliers or extreme values: Extreme values are handled with normalized/standardized by log transformation method to attributes which has outliers or extreme values.

  Figures 6 and 7 are showing the loan amounts values prior and after standardized/normalization of data systematically.

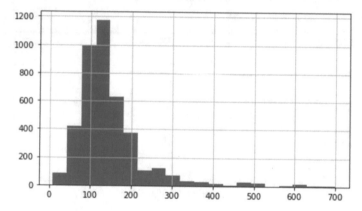

Fig. 6 Before standardization

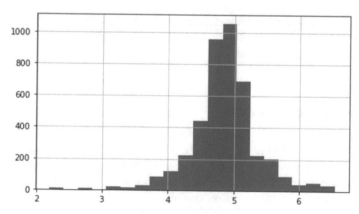

Fig. 7 After standardization

4.4 Feature Engineering with Dataset

Understanding of domain knowledge, have applied some new features that impacted positively on model accuracy and performance. Have created new three features in dataset:

- Total Income: Combined the applicants income and co-applicant's income because chances of loan approval are higher whether the total income of applicant is higher.
- EMI: Higher EMI's for sanctioned loan might difficult for the applicant to payback.
- Balance Income: The idea behind balance income feature is that if balance income is higher after paying EMI's then it is better chance that applicant can repay the loan amount.

4.5 Design of System

See Fig. 8.

- First dataset is collected, and then, exploratory data analysis (EDA) process is performed.
- On the basis of EDA, dataset is preprocessed, and both train dataset, test dataset ready for deploying in machine learning model process.
- Machine learning model is built, and model validation is performed.

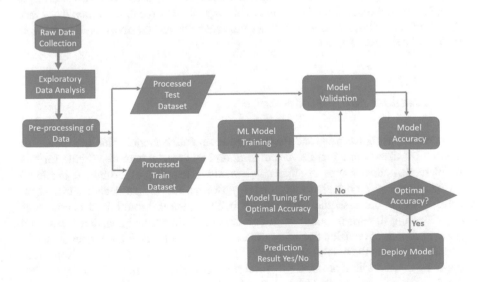

Fig. 8 Architecture of machine learning model system

After reaching the optimal accuracy of machine learning developed model, model forecasts if loan applicant is eligible to indorsation the loan or not.

5 Machine Learning Model Selection

This section of paper describes the selection of machine learning model. Here, working on the classification problem where prediction of loan applicant considered that loan applicant is eligible or not for sanctioning the loan. There are so many classification machine learning model available, and each different model has different merits and limitations. For this proposed model, are using k-nearest neighbor (KNN) algorithm model, random forest algorithm model, and support vector machine algorithm model and compare their optimal accuracy for this loan eligibility prediction model.

5.1 k-Nearest Neighbor Model

This algorithm model is used for supervised machine learning algorithms that is best suited for classed and regressive problems of prediction. This is type of lazy learning algorithm because it does not follow the special training phase and only stores the dataset and performs action on dataset for the time of classification also known for non-parametric learning algorithm because it does not make any assumption about dataset. After applying k-nearest neighbor model (k-NN) algorithm with *value of k* = 5, ML model reaches the *Maximum Accuracy* 99.94 percent, *Minimum Accuracy* 99.77 percent, *Overall Accuracy* 98.84 percent with *Recall Score* 0.96, *F1 Score* 0.97, and *Precision Score* 0.98.

5.2 Random Forest Model

Next model using for proposed model is random forest model. This model is also applied for classes and regressive problem, and this model is more reliable for this problem. It is also a type of supervised machine learning algorithm. It is rooted on decision tree model algorithm, but it performs better than decision tree algorithm because the essential idea behind, it combines the multiple decision trees in processing the last term output in comparison to individual decision trees. And higher number of decision tree in random forest model leads to the higher accuracy and also prevents the overfitting problem of machine learning algorithm. Now, after applying this algorithm on dataset with *maximum depth of* 10, this ML model reaches the *Maximum Accuracy* 95.09 percent, *Minimum Accuracy* 93.63 percent, *Overall*

Accuracy 93.7 percent with *Recall Score* 1.0, *F1 Score* 0.95, and *Precision Score* 0.91.

5.3   Support Vector Machine Model

After random forest model, now implementing support vector machine (SVM) model for proposed model. Support vector machine is also a supervised machine algorithm that is applied for the classification problem, regressive, and outlier's detection. This model is completely different from other classification machine learning algorithm. SVM for classification problem chooses decision boundary that utmost the distance from closer data points for all the classes. This decision boundary called hyperplane which helps to find best boundary or best line for the classes. After applying SVM model algorithm with *kernel* = *'linear'*, this ML model reaches the *Maximum Accuracy* 76.93 percent, *Minimum Accuracy* 74.03 percent, *Overall Accuracy* 75.52 percent with *Recall Score* 1.0, *F1 Score* 0.84, and *Precision Score* 0.73.

5.4   Stratified k-Fold Cross-Validation Algorithm

Cross-validation algorithm is much stronger tool which supports to preferential utilization of data. This provides so many facts about model execution. It also helps in tuning the machine learning model. Here, used this model to increase performance of this machine learning model. Because this algorithm is very useful when data are imbalanced and in very small number. It splits data into in k-number of fold and helps the machine learning model to enhance accuracy and performance. In this model, have used k-fold with n_splits value 5 which means data will be stratified 5 number of folds and tune the best performance to the model.

6   Conclusion and Future Scope

In this research paper, supervised machine learning model is used to forecast the loan eligibility of the applicants on the basis of their credit history, total income means sum of applicant's income and co-applicant's income, number of dependent, EMI, employment status to approve the eligible loan applicant for the further process. After developing the ML model, it is found that the KNN model has better accuracy than random forest model and SVM model, and also, SVM model has the lowest accuracy. Cross-validation stratified k-fold algorithm also applied on machine learning models to reach optimal accuracy with accuracy validation. All three models are tuned with optimal and satisfactory levels of accuracy. It will definitely help to reduce the non-performing asset problem of financial institutions.

This model could enhance further with more knowledge of domain and their features. In this model, three features are addressed, i.e., EMI, total income, and balance income; in future, more features could be addressed like age-related illness, progress of startup businesses, etc., with experience of loan handling and explore the issue about applicant means if more understanding about applicant's failure to repay the loan. As much as diverse scenario of loan payable or failure collected, then it is always a future scope to enhance the system.

References

1. Shinde A, Patil Y, Kotian I, Shinde A, Gulwani R (2022) Loan prediction system using machine learning. In: ICACC, vol 44, article no. 03019, pp 1–4. https://doi.org/10.1051/itmconf/202244 03019
2. Ramya S, Jha PS, Vasishtha IR, Shashank H, Zafar N (2021) Monetary loan eligibility prediction using machine learning. IJESC 11(7):28403–28406
3. Sarkar A, Sai KK, Prakash A, Sai GVV, Kaur M (2021) A research paper on loan delinquency prediction. IRJET 08(4):715–722
4. Sheikh MA, Goel AK, Kumar T (2020) An approach for prediction of loan approval using machine learning algorithm. In: 2020 International conference on electronics and sustainable communication systems (ICESC). IEEE, pp 490–494
5. Vaidya A (2017) Predictive and probabilistic approach using logistic regression: application to prediction of loan approval. In: 2017 8th International conference on computing, communication and networking technologies (ICCCNT). IEEE, pp 1–6. https://doi.org/10.1109/ICC CNT.2017.8203946
6. Madan M, Kumar A, Keshri C, Jain R, Nagrath P (2020) Loan default prediction using decision trees and random forest: a comparative study. IOP Conf Ser Mater Sci Eng 1022(012042):1–12
7. Supriya P, Pavani M, Saisushma N, Kumari NV, Vikash K (2019) Loan prediction by using machine learning models. Int J Eng Tech (IJET) 5(2):144–148
8. Raj JS, Ananthi JV (2019) Recurrent neural networks and nonlinear prediction in support vector machine. JSCP 1(1):33–40
9. Jency XF, Sumathi VP, Sri JS (2018) An exploratory data analysis for loan prediction based on nature of clients. Int J Recent Technol Eng (IJRTE) 7(4S):176–179
10. Turkson RE, Baagyere EY, Wenya GE (2016) A machine learning approach for predicting bank credit worthiness. In: 2016 Third international conference on artificial intelligence and pattern recognition (AIPR). IEEE, pp 1–7. https://doi.org/10.1109/ICAIPR.2016.7585216
11. Kim H, Cho H, Ryu D (2018) An empirical study on credit card loan delinquency. Econ Syst (Elsevier) 42(3):437–449
12. Tariq HI, Sohail A, Aslam U, Batcha NK (2019) Loan default prediction model using sample, explore, modify, model and assess (SEMMA). J Comput Theor Nanosci 16(8):3489–3503
13. Prasad P, Tripathi K (2021) Natural scene text localization and removal: deep learning and navier-stokes inpainting approach. In: 2021 IEEE International conference on electronics, computing and communication technologies (CONECCT). IEEE, pp 1–5. https://doi.org/10. 1109/CONECCT52877.2021.9622609
14. Malhotra R, Kaur K, Singh P (2021) Wavelet based image fusion techniques: a comparison based review. In: 2021 6th International conference on communication and electronics systems (ICCES). IEEE, pp 1148–1152
15. Gomathy CK, Charulata, Aakash, Sowjanya (2021) The loan prediction using machine learning. Int Res J Eng Technol 08(10):1322–1329
16. Dosalwar S, Kinkar K, Sannat R, Pise N (2021) Analysis of loan availability using machine learning techniques. IJARSCT 9(1):15–20

Prediction of Osteoporosis Using Artificial Intelligence Techniques: A Review

Sachin Kumar Chawla and Deepti Malhotra

Abstract Osteoporosis is a condition that goes unnoticed until it causes fragility fractures. Reduced bone mass density (BMD) raises the risk of osteoporotic fractures. Currently, osteoporosis is detected using traditional procedures such as DXA scans or FEA testing, which need a lot of computer resources. However, early diagnosis of osteoporosis, on the other hand, allows for the detection and prevention of fractures. The radiologist was able to successfully segment the region of interest in medical imaging to improve disease diagnosis using automated segmentation, which is not possible with traditional methods that rely on manual segmentation. With more advancement in technology, various AI learning techniques have been introduced which tends to generate results efficiently in less time. Keeping this essence, some researchers have developed AI-based diagnostic models based on biomarkers for effective osteoporosis prediction. This article seeks to thoroughly investigate efforts on automated diagnostic systems using artificial intelligence techniques based on quantitative parameters (biomarkers) for early osteoporosis prediction considering the data from 2016–2021. Based on the existing studies, this article highlights the open issues existing in the literature that needs to be addressed and also presented an outline of the proposed work based on knee X-ray and AI techniques which will be implemented in the future and could be an aid to the clinicians.

Keywords PR · Panoramic radiography · DXA · Dual X-ray absorptiometry · ANN · Artificial neural network · RF · Random forest · LR · Logistic regression · FEA · Finite element analysis · SVM · Support vector machine · KNN · k-nearest neighbor · CV · Cross-validation · ML · Machine learning · MLR · Multivariate linear regression · DT · Decision tree · CNN · Convolutional neural network · CT · Computed tomography · BMD · Bone mineral density · LBP · Local binary pattern · PPV · Positive predictive value · FRAX · Fracture risk assessment tool ·

S. K. Chawla
Computer Science and IT, Central University of Jammu, Jammu & Kashmir, India
e-mail: scsachin110@gmail.com

D. Malhotra (✉)
Department of Computer Science and IT, Central University of Jammu, Jammu & Kashmir, India
e-mail: deepti.csit@cujammu.ac.in

DB · Database · AUROC · Area under the receiver operator characteristic · FEDI · Fuzzy edge detection algorithm · OA · Osteoarthritis

1 Introduction

Over 200 million people worldwide, mostly the elderly, suffer from osteoporosis, a metabolic bone disease [1–5]. Reduced bone density increases bone fragility and fractures susceptibility; spinal compression fractures are the most common types. According to the World Health Organization, osteoporosis is defined as having a bone mineral density that is 2.5 standard deviations or more below the mean for a young, healthy adult T-score (DXA measure 2.5) [6–9].

Because not all physicians have access to this technology, it may not be feasible to routinely test the general population using DXA, even if early detection can reduce the risk of future morbidity and mortality from fracture-related consequences. To accurately screen patients and decrease over-diagnosis and misdiagnosis, extensive study has been done on when and how to utilize DXA and how to avoid giving patients a false sense of security [10–14]. Osteoporosis is commonly diagnosed in clinical settings using a battery of clinical tests, while some of the treatments can be pricey and a few of the tests might generate erroneous results as a consequence of anthropogenic or other chemical flaws [15–20].

As a result, we need an expert system that can scan medical records in a variety of formats and deliver accurate, dependable results without fatigue or error [21]. With the growing trend of technology, AI techniques have evolved in medical diagnosis and found to be a promising solution to improving health care [22]. Many CAD systems based on demographic parameters (Age, gender, eating habits, and life style) and vision-based modalities (MRIs, DXA, and CT scans) have recently been developed in the field of osteoporosis with better accuracy [23–25]. However, such systems are not deployed in clinical practices. This may be due to high-cost involvement or difficult data acquisition procedure [13]. Keeping this essence in mind, the researchers are attempting to develop a cost-effective system using AI techniques and X-ray images which could be convenient and can be used in real time and yield better results with greater accuracy (Fig. 1).

2 Literature Review

This section provides a brief insight into the existing osteoporosis detection techniques introduced by the various researchers covering the period 2016–2021.

- **Gao et al.** [1] provided a systematic study to show how using medical images and AI techniques to diagnose osteoporosis has been developed. Using the Quality Assessment of Diagnostic Accuracy Studies (QUADAS-2) methodology, the

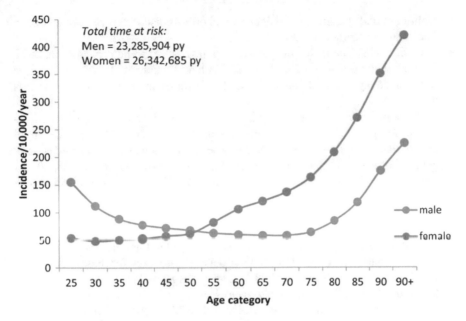

Fig. 1 Annual prevalence rate of osteoporosis worldwide

included studies' bias and quality were evaluated. They had an accuracy of 95 percent.

- **Ho et al.** [2] Dual-energy X-ray absorptiometry is the old standard in this discipline and has been recommended as the primary method for determining bone mineral density (BMD), which is an indicator of osteoporosis (DXA). It is recommended to utilize standard radiography for osteoporosis screening as opposed to diagnosis. They had an accuracy of 88 percent.
- **Fang et al.** [3] developed a fully automatic method for bone mineral density (BMD) computation in CT images using a deep convolutional neural network and outlined a plan for using deep learning for individuals with primary osteoporosis (DCNN). They were 82 percent accurate on average.
- **Shim et al.** [4] proposed Ml algorithms are highly accurate at predicting the risk of osteoporosis. It could assist primary care professionals in identifying which female patients should get a bone densitometry examination if they are at high risk of osteoporosis. They had an accuracy of ANN 0.742%, SVM 0.724%, and KNN 0.712%.
- **Yamamoto et al.** [5] built a model using convolutional neural networks (CNNs) based on hip radiographs to diagnose osteoporosis at an early stage. They had an accuracy of 93 percent.
- **Yasaka et al.** [6] propounded a method that is based upon unenhanced abdominal computed tomography (CT) images. A deep learning method can calculate the lumbar vertebrae's bone mineral density (BMD). They had an accuracy of 97 percent.

- **Kalmet et al.** [7] Compared to doctors and orthopedists, CNN performed better. They had an accuracy of 99%.
- **Park et al.** [8] The major purpose of this research is to help doctors make an accurate diagnosis of colorectal cancer. With an accuracy of 94.39 percent, one fully linked layer and 43 convolutional layers make up our recommended CNN design.
- **La Rosa** [9] developed a robotized computer-aided diagnostic strategy that incorporates a variety of previously offered standardizations, like MLR and ICA highlight extraction, for the detection of OA. The recommended CAD has higher categorization rates than those found in past investigations. The multivariate linear regression (MLR) technique was used to analyze 1024 X-ray pictures of people with osteoarthritis.
- **Ferizi et al.** [10] Deep learning algorithms were applied to analyze MRI images of knee bones using the 3D MRI datasets SK10, dataset OAI Imorphics, and dataset OAI ZIP. The 3D CNN and SSM models were used in this analysis. The SK110, OAI Imporphics, and OAI ZIb 3D MRI dataset's accuracy for these models was 75.73 percent, 90.4 percent, and 98.5 percent, respectively.
- **Rehman et al.** [11] In this study, deep learning methods were used to X-ray pictures of osteoarthritis patients. KNN and SVM algorithms were used SVM. KNN was found to be 100% accurate and SVM to be 79% accurate for normal images, whereas KNN was found to be 100% accurate and SVM was found to be 100% accurate for aberrant images.
- **Jaiswal et al.** [12] X-ray images are used to analyze osteoarthritis in the knee using deep learning techniques. The authors described a novel method for noninterventional detection and analysis of knee OA using X-rays. It might make the process of finding a treatment for knee pain more efficient. With a separate testing set, their approach yields the best multi-class grouping results: a normal multi-class accuracy of 66.71 percent, a radiographical OA accuracy of 0.93, a quadratic weighted Kappa of 0.83, and an MSE of 0.48. This could be compared to how people typically think. The study included 3000 patients from the osteoarthritis initiative dataset.
- **Reshmalakshmi and Sasikumar** [13] The purpose of this study is to use MRI imaging of knee bones to diagnose knee bone disorders. This technique made use of online MRI datasets. This work employs the scale space local binary pattern feature extraction method. The accuracy of the research was 96.1 percent.
- **Antony et al.** [14] In this work, MR images of the knee joint were used to identify cartilage lesions, such as cartilage softening, fibrillation, fissuring, focal defect broad thinning due to cartilage degeneration, and acute cartilage damage. They had achieved 87.9% accuracy.
- **Liu et al.** [15] The results of total knee arthroplasty (TKA) using patient-specific instruments (PSIs), which are described in this study, are comparable to those of TKA using conventional equipment in terms of post-operative radiography results. They had a 95% accuracy rate.
- **Ebsim et al.** [16] This research describes a strategy for integrating different detection techniques to find femur and radius fractures. To detect fractures,

these algorithms extract several types of characteristics. They were 82.6 percent accurate.

- **Deniz et al** [17] This study's objective was to assess conventional radiography and MRI for early diagnosis and classification of OVFs and to measure the rate of misdiagnosis of OVFs. They were 81 percent accurate.
- **Reshmalakshmi and Sasikumar** [18] This study's goal is to demonstrate a deep convolutional neural network-based automatic proximal femur segmentation technique CNN. They had a 95% accuracy rate.
- **Gornale et al.** [19] To determine if quantitative susceptibility mapping (QSM) accurately measures postmenopausal women's osteoporosis. They had a 86 percent accuracy.
- **Schotanus et al.** [20] By integrating the results of the traditional X-ray image processing method with the fuzzy expert system used to calculate the degree of osteoporosis, a conclusion is made. To indicate the likelihood of osteopenia, the authors in this study display the percentage reduction in bone density. Furthermore, it is determined that the ratio of trabecular to total bone energy is 0.7985, indicating a loss of bone density of 7.985 percent.
- **Chen et al.** [21] This work uses DCCN to automatically assess the degree of knee osteoarthritis from radiographs. When adjusted for regression loss, the network's multi-class grade 0–4 classification accuracy in this study is 59.55 percent.
- **Hordri et al.** [22] In this study, authors have segmented a knee X-ray picture using the active contour algorithm before using several feature extraction methods. Using the random forest classifier, the retrieved features revealed an accuracy rate of 87.92 percent.

3 Comparative Analysis

The study on various osteoporosis detection methods based on a variety of deep learning and machine learning technologies offered by various researchers is summarized in this section.

Table 1 summarizes a comparative analysis of various existing AI prediction techniques for osteoporosis detection. From this table, it was observed that researchers have employed various datasets depending on various modalities (MRI, CT scans, and X-ray), but these models cannot be generalized as it employs a small sample size. In the case of MRI and CT scans, the data acquisition is tedious and more expensive, and it sometimes requires patients to be still which causes more difficulty for the affected subjects. On contrary, an X-ray is a more convenient method for detecting osteoporosis in clinical settings. Keeping this essence, most of the researchers have worked on X-ray imaging data for osteoporosis prediction.

From Fig. 2, it has been observed that out of all ML or DL techniques, the DCCN classifier yields the best accuracy independent of the modalities. The rationale behind the adoption of such a model is that it can learn to perform actions from text, auditory,

Table 1 Comparative analysis of existing AI prediction techniques for osteoporosis detection

Author	Objectives	Data	Input data amount	Methods	Trained data	Main results
Gao et al. [1] (2021)	Employing artificial intelligence to analyze medical photos to detect osteoporosis	X-ray CT	3186 patients metanalysis	DCNN, CNN	Training samples set 940 In which 610 adverse cases and 330 decent ones	95%
Ho et al. [2] (2021)	Using a deep learning neural network, plain X-ray radiography is used to predict bone mineral density	X-ray	3472 pairs of pelvis X-ray images	CNN	2800 (male = 609, female = 2191) pelvis X-ray 5027 unilateral femur images (male = 1115, female = 3912)	0.88%
Fang et al. [3] (2020)	Screening for osteoporosis in multi-detector CT imaging making use of deep convolutional neural networks	CT	(244 images, 16.8%), osteopenia (605 images, 41.8%), and normal (600 images, 41.4%)	DCNN model	Set 1 463 images, set 2 200 images, and set 3 200 images	0.823, 0.786, 0.782

(continued)

Table 1 (continued)

Author	Objectives	Data	Input data amount	Methods	Trained data	Main results
Shim et al. [4] (2020)	Use of machine learning techniques to predict osteoporosis risk in postmenopausal women	X-ray MRI CT	1792 34% Osteoporosis	ANN, SVM, KNN	76% fivefold CV_24%	ANN 0.742%, SVM 0.724%, KNN 0.712%
Yamamoto et al. [5] (2020)	Hip radiographs and patient clinical covariates are used in deep learning for osteoporosis classification	X-ray	1131 (53% Osteoporosis)	ResNet-18, ResNet-34, Google-Net, Effective-Net b3, Effective-Net b4	80% 10% 10%	AUROC 0.937%
Yasaka et al. [6] (2020)	Deep learning is used to forecast bone mineral density from CT tomography using a convolutional neural network	CT	2045 (% not stated)	CNN(4 layers)	81% 9% 10% (external validation)	0.96%
Kalmet et al. [7] (2020)	A narrative overview of deep learning in fracture detection	X-ray	1,891 69 percent fracture images	ResNet-152	90 percent trained 10 percent testing	1%, compassion 0.99%, specificity 0.97%
Park et al. [8] (2020)	Using optimized CNN, adenocarcinoma recognition in endoscopy images	Colonoscopy images dataset	49,458 images	CNN	49,048 images	94.3%

(continued)

Table 1 (continued)

Author	Objectives	Data	Input data amount	Methods	Trained data	Main results
La Rosa [9] (2019)	A decision support tool for the early identification of knee OA utilizing X-ray imaging and machine learning was developed using data from the OA initiative	X-ray	1024 images	MLR	Train data 80% Tested data 20%	AUROC 82.9%
Ferizi et al. [10] (2019)	Automated knee bone and cartilage segmentation for osteoarthritis initiative using statistical shape knowledge and convolutional neural networks	MRI	3D MRI datasets SKI10 dataset-OAI Imorphics dataset OAI ZIB	3D CNNs, SSMs	3D MRI datasets SKI10 Dataset OAI Imorphics, dataset OAI ZIB	75.8%, 90.4%, 98.4%
Rehman et al. [11] (2019)	Identification of knee osteoarthritis using texture analysis	X-ray	X-ray images of OA patients	KNN SVM	Train data 70% Tested data 30%	Normal: KNN = 99%, SVM = 79% Abnormal: KNN = 98%, SVM = 97%
Jaiswal et al. [12] (2018)	A deep learning-based automatic diagnosis of knee osteoarthritis from plain radiographs	X-ray	The osteoarthritis dataset contains 3000 participants	CNN	Trained data 70% Tested data 30%	Auroc = 66.71%, ROC curve = 0.93%, Auroc = 0.93%, MSE = 0.48%

(continued)

Table 1 (continued)

Author	Objectives	Data	Input data amount	Methods	Trained data	Main results
Reshmalakshmi and Sasikumar [13] (2018)	A framework based on SSLBP for feature extraction in knee MRI scans to identify bones	MRI	Online MRI dataset	LBP SVM	Trained data 90% Tested data 10%	96.1% 88.26%
Antony et al. [14] (2018)	Deep learning approach for knee MR image evaluation	MRI	17,395 images	CNN	16,075 images trained 1320 images testing	87.9%
Liu et al. [15] (2018)	Total knee arthroplasty with positive alignment using MRI-based patient-specific tools	MRI	841 knee images	PSI	70% trained 30% testing	89.1%
Ebsim et al. [16] (2018)	Using Optimized CNN, adenocarcinoma recognition in endoscopy images	Colonoscopy images dataset	49,458 images	CNN	49,048 images	94.39%
Deniz et al. [17] (2018)	The use of MRI in determining the kind of osteoporotic vertebral fractures and their impact on diagnosis	MRI	173 patients images dataset	AO spine classifier	70% trained 30% testing	81%
Reshmalakshmi and Sasikumar [18] (2018)	DCCN for proximal femur segmentation from MR images	MRI	44 patients images dataset	DCCN	70% trained 30% testing	95%

(continued)

Table 1 (continued)

Author	Objectives	Data	Input data amount	Methods	Trained data	Main results
Gornale et al. [19] (2018)	An additional and accurate indicator of osteoporosis in postmenopausal women is bone susceptibility mapping with MRI	MRI	QCT images dataset	CNN	70% trained 30% testing	87%
Schotanus et al. [20] (2016)	Osteoporosis detection using fuzzy inference	X-ray	20 patients images	FEDI	20 patients images	79%
Chen et al. [21] (2016)	DCCN for determining the severity of radiographic osteoarthritis in the knee	X-ray	8892 images of knee joints	DCCN	7030 images	95%
Hordri et al. [22] (2016)	Image from a knee X-ray was used to identify osteoarthritis	X-ray	200 knee images	Random forest classifier	40% trained 60% testing	87.92%

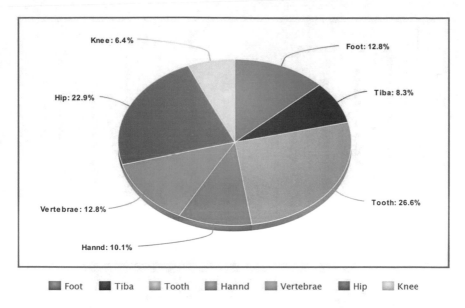

Foot   Tiba   Tooth   Hannd   Vertebrae   Hip   Knee

Fig. 2 Performance % accuracy of existing AI techniques for early osteoporosis detection

or visual input and can do so with astounding accuracy—in certain cases, even better than human performance.

Figure 3 depicts the % usage rate of various medical descriptions for analysis of osteoporosis. From this figure, it has been observed that there is extensive research on X-ray image data for early osteoporosis prediction due to its ease of availability and less cost in comparison to other modalities (MRI and CT scans).

4   Open Gaps and Challenges

This section proffers the existing gaps and challenges in the field of early osteoporosis prediction.

1. **Insufficient Standardized Dataset** Specifically for images of the knee joint, there aren't many available standardized sample datasets. The generalizability of the model may be constrained by the fact that researchers have created their datasets under controlled conditions. Hence, there is a dire need to create open-source datasets that will be beneficial for the research community in building a robust model for early osteoporosis prediction [17–20, 22].

2. **Sample Size in Osteoporosis Prediction Models** In the case of X-ray images, an open-source dataset is available on multiple sites, but there is a huge amount of heterogeneity in data which in turn affects the performance of the model. To overcome heterogeneity, transfer learning techniques could be employed.

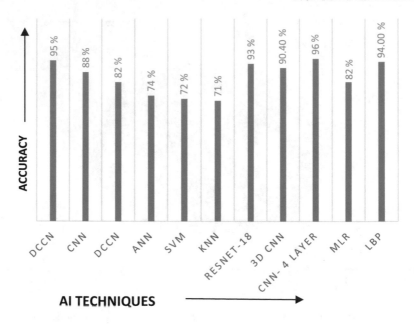

Fig. 3 %age utilization ratio of data modalities in early osteoporosis prediction

3. **High Computational Cost** Diagnosis of osteoporosis is a challenging task as the current traditional medical diagnostic systems, such as (DXA, FEA, and FRAX), are not widely available and, among other limitations, are expensive for low economics[1, 3, 4].
4. **Delayed Diagnostic Rate** Conventional methods for diagnosing osteoporosis involves manual segmentation which is time-consuming and more prone to human errors. With the advent of artificial intelligence, the radiologist was able to successfully segment the region of interest in medical imaging to improve disease diagnosis, which could not be possible with traditional methods that rely on manual segmentation. [4, 9, 23, 24].
5. **Lack of Interpretability** The long-term goal of the computational model is to present a methodology that is more transparent and reliable, specifically in the context of appropriately classifying osteoporosis patients. In the real world, health practitioners are apprehensive to utilize AI techniques. Hence, a model should be more transparent and robust to ease osteoporosis diagnosis in clinical settings.
6. **Severity Estimation** Most of the studies have only focused on classifying osteoporosis and did not consider the severity of the disorder or its subtypes. [6, 9, 13].

5   Proposed Intelligent Osteoporosis Classifier

From the exhaustive literature, it has been observed that various researchers have developed X-ray imaging data and intelligent systems used for osteoporosis detection, but no study has been conducted to classify osteoporosis subclasses. Hence, there is a need to develop an efficient framework that has the potential capability to classify osteoporosis and its subclasses. Figure 4 shows the percentage usage of various body regions in detecting early osteoporosis. From this figure; we observed that knee regions have very less work down in this field due to the less dataset availability and heterogeneity in the osteoporosis dataset. Our proposed framework is based on knee X-ray images. Keeping this essence in mind, this study proposes an intelligent osteoporosis classifier based on knee X-ray imaging data. We proposed a deep learning model that can have the potential capability to classify osteoporosis and its subclasses. Data collection, data preprocessing, prediction model, and results are among the main elements (Fig. 5).

5.1   Dataset Description

Data collection is the most important aspect of the diagnosis system, and selecting an appropriate sample for machine learning trials is crucial. This dataset will be taken from the Kaggle repository and contains knee X-ray data on knee joints (1656 images). The severity of the offense includes images from a variety of categories. The images descriptions are as follows:

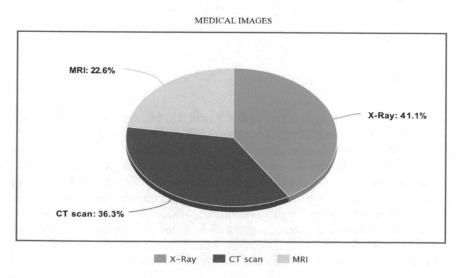

MEDICAL IMAGES

MRI: 22.6%

X-Ray: 41.1%

CT scan: 36.3%

■ X–Ray   ■ CT scan   ▨ MRI

Fig. 4 %age usage of various body regions in detecting early osteoporosis

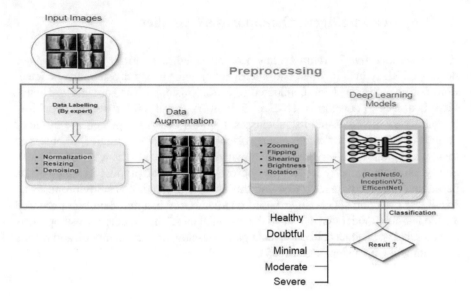

Fig. 5 Intelligent osteoporosis classifier

CLASSES	NO. OF IMAGES
Class (0) Healthy	639
Class (1) Doubtful	296
Class (2) Minimal	447
Class (3) Moderate	223
Class (4) Severe	51

5.2 Data Preprocessing

After data collection, preprocessing of the dataset will be employed to improve the images. The obtained images may be improved so that the data might aid in the early detection of osteoporosis cases in various stages. We will resize each input image while maintaining the aspect ratio in the initial preprocessing stage to lower the training expenses. Additionally, to balance the dataset, we will perform up-sampling and down-sampling. During the up-sampling procedure, areas are randomly cropped to increase minority classes. Flipping and 90o rotation are commonly used to balance the samples of the various classes, improve the dataset, and prevent overfilling. To satisfy the cardinality of the smallest class, more instances of majority classes can be removed throughout the down-sampling process. Each image in the generated distributions is mean normalized to remove feature bias and minimize training before being flipped and rotated.

5.2.1 Data Labeling

In a machine learning model, data labeling is an essential part of the data preprocessing stage. The quality of a dataset can be significantly impacted by any error or inaccuracy made in this procedure. Additionally, a predictive model's overall effectiveness could be ruined and result in misunderstanding. The proposal uses the dataset acquired from the Kaggle repository which further divided the images into 5 classes: Class 0 is Healthy; Class 1 is Doubtful; Class 2 is Minimal; Class 3 is Moderate; Class 4 is Severe.

5.2.2 Normalization

Image normalization is a popular method of modifying the pixel intensity range in image processing. One option is to utilize a function that creates a normalization of the input images.

5.2.3 Resizing

Image resizing is essential to make sure that all of the input images are the same size because their sizes vary. Deep learning models must scale all input photos to the same size before feeding them into the model because they often learn quickly on smaller images and accept inputs of the same size.

5.2.4 Denoising

The procedure of eliminating noise or distortions from an image is known as image denoising. Different types of noise, such as Poisson, Speckle, Gaussian, and Salt & Pepper, may be present in an image. Many vision denoising filters, such as fuzzy-based filters and classical filters, can be used to achieve the purpose of denoising.

5.2.5 Data Augmentation

Data augmentation techniques create several duplicates of a dataset to artificially increase its size. To collect additional data, we only need to make a few minor changes to our old dataset. Minor alterations such as flips or translations, rotations, cropping, scaling, zooming, adding noise, and shearing can be made to increase the size of the dataset.

5.2.6 Results and Prediction

DCNN will be used in the proposed model to identify and categorize images. It makes use of a hierarchical model to construct a network that resembles a funnel before producing a fully connected layer where the output is processed and all neurons are connected. The main advantage of DCNN over the existing techniques is that it automatically extracts key elements without the need for human participation. DCNN would therefore be a great option for spotting osteoporosis in its early phases.

6   Conclusion

After reviewing the existing literature relating to the automated prediction models for early osteoporosis detection, the findings suggest that machine learning and deep learning techniques have been frequently used in osteoporosis diagnostic field. From the exhaustive study, it has been observed that nearly 70% of the studies employed machine learning techniques, and 30% of the studies employed deep learning techniques. However, the work in the field of osteoporosis prediction based on deep learning techniques is very less. The prime focus of this study is to critically access and analyze AI-based models for early osteoporosis prediction using several modalities (X-ray, MRI, CT scans) and AI techniques over the years 2016–2021. After carefully examining the existing gaps and challenges, this paper elucidates some future directions that need to be addressed and proposed an intelligent osteoporosis classifier using X-ray imaging data and a deep convolutional neural network that will be implemented in the future and will pose as a potential aid to research scholars and health practitioners by providing a more precise, effective, and timely diagnosis of osteoporosis.

References

1. Gao L, Jiao T, Feng Q, Wang W (2021) Application of artificial intelligence in diagnosis of osteoporosis using medical images: a systematic review and metaanalysis. Osteoporos Int 32(7):1279–1286. https://doi.org/10.1007/s00198-02105887-6
2. Ho C-S, Chen Y-P, Fan T-Y et al (2021) Application of deep learning neural network in predicting bone mineral density from plain x-ray radiography. Arch Osteoporos 16:153. https://doi.org/10.1007/s11657-021-00985-8
3. Fang Y, Li W, Chen X et al (2021) Opportunistic osteoporosis screening in multi-detector CT images using deep convolutional neural networks. Eur Radiol 31:1831–1842. https://doi.org/10.1007/s00330-020-07312-8
4. Shim J-G, Kim DW, Ryu K-H et al (2020) Application of machine learning approaches for osteoporosis risk prediction in postmenopausal women. Arch Osteoporos 15:169. https://doi.org/10.1007/s11657-020-00802-8

5. Yamamoto N, Sukegawa S, Kitamura A, Goto R, Noda T, Nakano K, Takabatake K, Kawai H, Nagatsuka H, Kawasaki K, Furuki Y, Ozaki T (2020) Deep learning for osteoporosis classification using hip radiographs and patient clinical covariates. Biomolecules 10(11):1534. https://doi.org/10.3390/biom10111534

6. Yasaka K, Akai H, Kunimatsu A et al (2020) Prediction of bone mineral density from computed tomography: application of deep learning with a convolutional neural network. Eur Radiol 30:3549–3557. https://doi.org/10.1007/s00330-02006677-0

7. Kalmet PHS, Sanduleanu S, Primakov S, Wu G, Jochems A, Refaee T, Ibrahim A, Hulst LV, Lambin P, Poeze M (2020) Deep learning in fracture detection: a narrative review. Acta Orthop 91(2):1–6. https://doi.org/10.1080/17453674.2019.1711323

8. Park H-C, Kim Y-J, Lee S-W (2020) Adenocarcinoma recognition in endoscopy images using optimized convolutional neural networks. Appl Sci 10(5):1650. https://doi.org/10.3390/app10051650

9. La Rosa F (2017) A deep learning approach to bone segmentation in CT scans. PhD thesis, The University of Bologna, Bologna, Italy

10. Ferizi U, Besser H, Hysi P, Jacobs J, Rajapakse CS, Chen C, Saha PK, Honig S, Chang G (2018) Artificial intelligence applied to osteoporosis: a performance comparison of machine learning algorithms in predicting fragility fractures from MRI data. J Magn Reson Imaging 49(4):1029–1038. https://doi.org/10.1002/jmri.26280

11. Rehman F, Shah SIA, Riaz M, Naveed; Gilani SO, Faiza R (2019) A region-based deep level set formulation for vertebral bone segmentation of osteoporotic fractures. J Digit Imaging 33:191–203. https://doi.org/10.1007/s10278-019-00216-0

12. Jaiswal AK, Tiwari P, Kumar S, Gupta D, Khanna A, Rodrigues JJPC (2019) Identifying pneumonia in chest x-rays: a deep learning approach. Measurement 145:511–518. https://doi.org/10.1016/j.measurement.2019.05.076

13. Reshmalakshmi C, Sasikumar M (2017) Trabecular bone quality metric from x-ray images for osteoporosis detection. In: 2017 International conference on intelligent computing, instrumentation and control technologies (ICICICT), pp 1694–1697. https://doi.org/10.1109/ICICICT1.2017.8342826

14. Antony J, McGuinness K, O'Connor NE, Moran K (2016) Quantifying radiographic knee osteoarthritis severity using deep convolutional neural networks. In: 2016 23rd International conference on pattern recognition (ICPR), pp 1195–1200. https://doi.org/10.1109/ICPR.2016.7899799

15. Liu F, Zhou Z, Samsonov A, Blankenbaker D, Larison W, Kanarek A, Kijowski R et al (2018) Deep learning approach for evaluating knee MR images: achieving high diagnostic performance for cartilage lesion detection. Radiology 289(1):160–169. https://doi.org/10.1148/radiol.2018172986

16. Ebsim R, Naqvi J, Cootes T (2016) Detection of wrist fractures in x-ray images. In: Clinical image-based procedures. Translational research in medical imaging. CLIP 2016. Lecture notes in computer science, vol 9958. Springer, Cham. https://doi.org/10.1007/978-3-319-46472-5_1

17. Deniz CM, Xiang S, Hallyburton RS et al (2018) Segmentation of the proximal femur from MR images using deep convolutional neural networks. Sci Rep 8:16485

18. Reshmalakshmi C, Sasikumar M (2016) Fuzzy inference system for osteoporosis detection. In: 2016 IEEE Global humanitarian technology conference (GHTC), pp 675–681. https://doi.org/10.1109/GHTC.2016.7857351

19. Gornale SS, Patravali PU, Manza RR (2016) Detection of osteoarthritis using knee x-ray image analyses: a machine vision-based approach. Int J Comput Appl 145(1):0975–8887

20. Schotanus MGM, Thijs E, Heijmans M et al (2018) Favourable alignment outcomes with MRI-based patient-specific instruments in total knee arthroplasty. Knee Surg Sports Traumatol Arthrosc 26:2659–2668. https://doi.org/10.1007/s00167-017-4637-0

21. Chen Y, Guo Y, Zhang X et al (2018) Bone susceptibility mapping with MRI is an alternative and reliable biomarker of osteoporosis in postmenopausal women. Eur Radiol 28:5027–5034. https://doi.org/10.1007/s00330-018-5419-x

22. Hordri NF, Samar A, Yuhaniz SS, Shamsuddin SM (2017) A systematic literature review on features of deep learning in big data analytics. Int J Adv Soft Comput Appl 9(1):32–49

23. Giornalernale SS, Patravali PU, Manza RR (2016) Detection of osteoarthritis using knee X-ray image analyses: a machine vision-based approach. Int J Comput Appl 145:20–26

24. Madani A, Moradi M, Karargyris A, Syeda-Mahmood T (2018) Semi-supervised learning with generative adversarial networks for chest X-ray classification with the ability of data domain adaptation. In: 2018 IEEE 15th International symposium on biomedical imaging (ISBI 2018), pp 1038–1042. https://doi.org/10.1109/ISBI.2018.8363749

25. Sharma AK, Toussaint ND, Elder GJ, Masterson R, Holt SG, Robertson PL, Ebeling PR, Baldock P, Miller RC, Rajapakse CS (2018) Magnetic resonance imaging-based assessment of bone microstructure as a non-invasive alternative to histomorphometry in patients with chronic kidney disease. Bone 114:14–21. https://doi.org/10.1016/j.bone.2018.05.029

26. Marongiu G, Congia S, Verona M, Lombardo M, Podda D, Capone A (2018) The impact of magnetic resonance imaging in the diagnostic and classification process of osteoporotic vertebral fractures. Injury 49(Suppl 3):S26–S31. https://doi.org/10.1016/j.injury.2018.10.006

Analyzing Stock Market with Machine Learning Techniques

Kirti Sharma and Rajni Bhalla

Abstract The financial market is extremely volatile, and this unstable nature of the stock market is not easy to understand. But technological advancements have given a ray of hope that it might be possible that one can make the machines understand this level of volatility and can make accurate predictions about the future market prices. This paper emphasizes various techniques by which machines can learn the financial markets and their future trends/movements. This paper has made use of four such techniques along with sentiment analysis on the news related to the undertaken tickers. This study shows that classification techniques give a good estimate of unusual highs and lows of the market, which in turn can prove helpful for the traders in taking timely and accurate decisions, i.e., bullish or bearish trends. This study is focused on determining the trends of the market while considering not only the stock trends but also the sentiments of the news headlines, using the polarity scores. The ensembled technique has given better results than other techniques in terms of R2 score and mean absolute error.

Keywords Sentiment analysis · Gradient boosting regressor · Decision tree regressor · Ensembled technique · Machine learning

1 Introduction

A number of investors in the stock market are growing enormously every day. From the very early days of the stock market, investors are using traditional ways of trading, i.e., using statistical analysis, fundamental analysis, the past episodes of any happenings in the markets, and technical indicators. These techniques need continuous involvement of the investor and are very time-consuming. Increasing numbers of

K. Sharma (✉) · R. Bhalla
School of Computer Application, Lovely Professional University, Phagwara, Punjab, India
e-mail: kirtisharmaa.11@gmail.com

R. Bhalla
e-mail: rajni.12936@lpu.co.in

traders are making the investors always look for better, accurate, trustworthy, and faster ways to do the predictions.

Technology advancements are making it possible to make all these things automated. But matching the accuracy of the predictions with the predictions made by experienced and professional traders who are in this field of stocks for generations is not easy. With the availability of big data and machine learning techniques, developing new algorithms or methodology for stock market predictions is very much possible. Mehta et al. [1] show in their paper that news related to finances has a significant effect on next-day stock prices. News headlines that were scraped from reputed web portal, i.e., moneycontrol.com [2] using the customized Python-based script, then were classified as per the tickers and the corresponding sentiments which were derived from the news dataset.

Qualitative analysis shows that news data affects the stock price movements [3], and these two are interdependent. This study was done considering only news headlines instead of the entire news article, presuming that the headlines alone have a significant effect on investors, and these headlines are trustworthy and a rapid way of influencing the stock market. A quantitative study is done with the data which was gathered using yahoo-finance-API.

2   Review of Literature

The majority of stock market forecasting revolves around statistical and technical analysis, though today various researches have shown clear shreds of evidence of great interdependency of news data and the stock movements. Mohan et al. [4] in their paper, they have used variations of RNN-LSTM in terms of input, i.e., prices, text polarity, text, and multivariate model along with Facebook Prophet and ARIMA. The results show a good relationship between stock prices and news data. They have got promising results with RNN, and their model failed when it came to lower stock prices and high volatility, whereas Mehta et al. [1] have done an implementation of SM prediction tool taking in the sentiments, news data, and past data for forecasting the SM prices. Authors have used SVM, LR, NB, and long short-term-memory.

Akhtar et al. [5] have preprocessed the historical dataset and then machine learning techniques, i.e., Random Forest and SVM classifier were applied. Authors have achieved the accuracy of 78.7% with SVM and 80.8% with RF classifier. As per them, additional parameters such as financial ratio and sentiments of masses to figure out the interrelationship of clients with employees can be used as input. Emioma and Edeki [6] have used the least-square LR model with close price as the dependent variable and everyday stock value as an independent variable. Their model made good predictions with 1.367% MAPE and 0.512 RMSE. Their model is reusable with other tickers of other stock markets also. Sharma et al. [7] have discussed technical analysis, fundamental analysis, basic technical indicators, and some of the most popular machine learning techniques to make the forecasting. In their further study have proposed a hybrid model for forecasting of stock market on the basis of news sentiments and historic dossier on Nifty50 data [8].

Thormann et al. [9] long short-term memory with lagged closing price served as the basis of their study. Their study acts as a guide to use and preprocess Twitter data, which in turn is combined with technical indicators to make the forecasting of APPL. Their model can do better than the undertaken base model. Kedar [10] has used fresh Twitter data and integrated the results of sentiment analysis of these tweets along with the results of the ARIMA model's application over the historical dataset. Their study shows that change in the company leads to a change in accuracy level due to COVID-19.

Chen et al. [11] have suggested a hybrid model based on XGBoost integrated with firefly algorithm to do forecasting. Firefly helps in the optimization of hyperparameters of XGBoost. Then, stocks with high potency are undertaken, and MV is applied over them. Their hybrid model is performing way better than the other state of the-art models. They listed the reasons behind the success of the model as they considered characteristics of the SM that may affect the upcoming prices, and they have also increased the accuracy using their own IFAXGBoost model and also used the MV model to better utilize allocation of the assets.

Sarkar et al. [12] have proposed a model which uses the ways, investors, and traders use for making predictions in combination with technical and financial analysis using, news headings and made the predictions about Google prices. They have used sentiment analysis for headings of collected news along with a long short-term-memory model for historical data. Their approach is showing prominent improvement in the forecasting results.

Gondaliya et al. [13] have worked with the dataset which is influenced by COVID-19. They have selected and applied the top six algorithms, i.e., DT, KNN, LR, NB, RF, and SVM to the data. LR and SVM have shown better forecasting solutions. They have suggested that these top algorithms can be used as input for building a robust forecasting model.

Gupta and Chen [14] have analyzed the effect of sentiments derived through StockTwits. An analysis is done via extraction of features from the tweets and application of ML techniques. The relation between the average SM prices on an everyday basis along with daily headlines is studied. The sentiment feature extracted is then used with past data to improvise the SM price predictions.

Li et al. [15] have proposed a new approach with the integration of both technical indicators along with sentiments. Two-layered LSTM is proposed to understand time-series data of five years. Base models MKL and SVM are used for comparison with their proposed approach. LSTM gave better results, i.e., accuracy and f1-score in comparison with MKL and SVM. Yadav and Vishwakarma [16] have examined various important architectures related to sentimental analysis. They have presented DL models such as various types of neural networks including bidirectional recurrent neural networks. They have derived the conclusion that LSTM is giving more appropriate predictions than other DL techniques. They have also highlighted various languages being used for performing SA. Authors have concluded that researchers find it difficult to do SA, because of constrained textual data's topic-wise categorization.

Reddy et al. [17] have proposed a ML model based on RF which is optimized with PSO. Derivations were depicted on plots. Comparison with other state-of-the-art techniques was done with Twitter tweet sentiments as well. They have concluded that Random Forest is the least expensive among all. In this study, Suhail et al. [18] provide a method for trading stocks that utilizes a Reinforcement Learning (RL) network and market sentiments. To identify patterns in stock prices, the system applies sentiment analysis to the daily market news. Their findings demonstrate the impact of market mood on stock price forecasts.

Subasi et al. [19] have employed prices with both per-day and per-minute frequencies and have utilized SVM to forecast stock prices for both large and small cap stocks in the three separate marketplaces. In comparison with the chosen benchmarks, the model delivers better profits.

(Rouf et al. [20]) In this article, studies using a general framework for stock market prediction (for 2011–2021) were examined (SMP). Based on inputs and methodologies, various performance measures that various studies utilized to quantify performance were examined. Additionally, a thorough comparison analysis was conducted, and it was shown that SVM is the most widely utilized method for SMP. However, methods like ANN and DNN are frequently employed because they offer quicker and more precise forecasts. Raubitzek at el. [21] reviewed the subject of stock market as a random walk and seek to ascertain whether there are patterns in forecasting and variability for the data under study, whether there is any correlation, and what impact inflation has on the data and its certainty.

3   Proposed System

3.1   Overview

This study aimed to categorize the stocks as per the respective industries and derive the sentiments from their corresponding news data. Figure 1 shows the architecture of the proposed system. There are three major parts (1) data collection using Yahoo Finance API and preprocessing of collected news headlines. (2) Categorization of tickers using various models. (3) Analyzing the categorized (derived) labels.

3.2   Data Collection

Yahoo finance API with Python is used to collect the historical dataset of two years, whereas a custom Python script is used for scraping news headlines data. This way made it possible to get specific words out of the news headlines, which is the basic

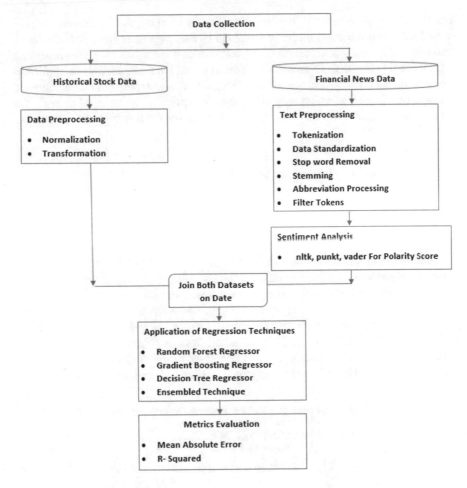

Fig. 1 Overall flow diagram of system

requirement of our model. Specific rules were defined for generating a set of rules for the dictionary building. Nifty fifty is one of the topmost indices of India. Features included were open, close, volume, and only headline data. Detailed news articles were not taken into consideration. The news was gathered on the daily basis to get up to the date data about the latest tendency of the stock market.

3.3 Data Preprocessing

News headlines from the reputed and trustworthy financial website, i.e., moneycontrol.com [2] were used. While collecting the news, some factors like consideration of news headlines only specific to undertaken tickers and elimination of detailed news

articles were kept in mind. After finalizing the list of news, the same were preprocessed with the usage of regular expressions [11]. Hashtags and other references were
also rejected, so that classification of the text can be done unbiasedly. Preprocessing
steps including lowercase conversion, stemming, and lemmatization were applied to
the news data, and preprocessing of historical data was also performed. Features are
selected on the level at which they affect the close price of the stocks. For both, the
datasets were checked for any null values and were replaced with the mean values.

3.4 Application of Various Models

This study made use of the news headlines which were scraped and filtered with
a specific ticker, the company name, i.e., INFY. Polarity scores of all the text data
combined based on the date were calculated. Python libraries nltk, punkt, and vader
for sentiment analysis were used. After the calculation of polarity scores, datasets
were combined based on the date column (intersection of dates). As there were days
when the stock market was closed due to weekends or national holidays, and there
were days when there was no major or meaningful news available. The news dataset
was then used as input for different machine learning techniques. Our model has
used Random Forest Regressor [17], Gradient Boosting Regressor [22], Decision
Tree Regressor [23], and finally a voting regressor technique that integrates all the
three listed above. The split ratio used was 0.2 (Figs. 2, 3 and 4).

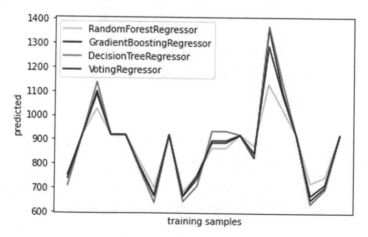

Fig. 2 Polarity score as input for close price predictions

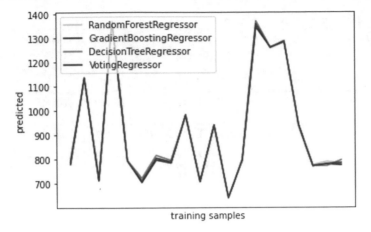

Fig. 3 Polarity Score and open price as inputs for close price predictions

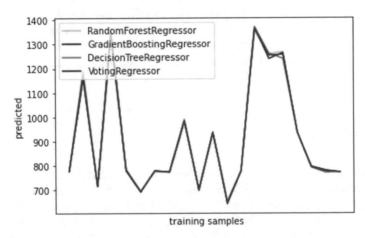

Fig. 4 Polarity score, open price, no. of trades, low price, high price as inputs for close price predictions

4   Accuracy Test of Various Models

Accuracies of RF Regressor, GB Regressor, DT Regressor, and ensembled technique which was made with the all three previously mentioned models were derived and are presented in the summary Table 1. The model was built and checked for accuracy with three different feature input sets. Firstly, only the polarity score was used as input to the model, which did not give the desired results. Then, one more input feature, i.e., the open price was included, which resulted in better results as compared to the previous one. Then, the input feature set was used which consisted of polarity score, no. of trades, open, low, and high prices.

Table 1 Table for accuracy comparison

Model name	Polarity score		Polarity score, open price		Polarity score, open price, no. of trades, low price, high price	
	R2	MAE	R2	MAE	R2	MAE
Random forest regressor	0.32792	169.77811	0.9984	16.1495	0.9991	9.1502
Gradient boosting regressor	0.40803	178.15511	0.99948	17.1172	0.9999	9.4666
Decision tree regressor	0.412627	183.0247	0.99996	19.54499	1.0	11.377
Ensembled technique	0.399825	172.72410	0.99961	16.92102	0.9998	8.67906

As depicted in the summary Table 1, increasing the number of features as input to the model gave very good accuracy with our ensembled model over others. Although the R2 score with the decision tree regressor achieved the best results but at the same time, its mean absolute error was also very high. Comparatively, the R2 score was a little lower in the case of our ensembled model but the mean absolute error achieved was also the lowest, which is the best of all.

5   Conclusion

This paper shows the analysis and comparison of four regression techniques named RF Regressor, GB Regressor, DT Regressor, and ensembled regression technique to do stock market predictions. Python's nltk, punkt, and vader APIs were implemented for sentiment analysis. The model can be used for forecasting of next three to five days for any NIFTY indexed stock. This can be concluded that using polarity as the only feature comes out as a non-dominant feature. But when combined with open price, it gives excellent results. In the future, better-improved methods of polarity score decisions for sentiment analysis of news can be used. To derive sentiments of stake holders, financial annual reports or the companies, tweets can also be used, which may enhance the prediction results. Moreover, hyperparameters tuning of the models can be used further to improve the model.

References

1. Mehta P, Pandya S, Kotecha K (2021) Harvesting social media sentiment analysis to enhance stock market prediction using deep learning. PeerJ Comput Sci 7:1–21. https://doi.org/10.7717/peerj-cs.476

2. Business News | Stock and Share Market News | Finance News | Sensex Nifty, NSE, BSE Live IPO News. Retrieved from https://www.moneycontrol.com/. Accessed on 10 Feb 2022

3. Zhao W et al (2018) Weakly-supervised deep embedding for product review sentiment analysis. IEEE Trans Knowl Data Eng 30(1):185–197. https://doi.org/10.1109/TKDE.2017.2756658

4. Mohan S, Mullapudi S, Sammeta S, Vijayvergia P, Anastasiu DC (2019) Stock price prediction using news sentiment analysis. In: 2019 IEEE Fifth international conference on big data computing service and applications (BigDataService), pp 205–208. https://doi.org/10.1109/BigDataService.2019.00035

5. Akhtar MM, Zamani AS, Khan S, Shatat ASA, Dilshad S, Samdani F (2022) Stock market prediction based on statistical data using machine learning algorithms. J King Saud Univ Sci 34(4):101940. https://doi.org/10.1016/j.jksus.2022.101940

6. Emioma CC, Edeki SO (2021) Stock price prediction using machine learning on least-squares linear regression basis. J Phys Conf Ser 1734:012058. https://doi.org/10.1088/1742-6596/1734/1/012058

7. Sharma K, Bhalla R (2022) Stock market prediction techniques: a review paper. In: Second international conference on sustainable technologies for computational intelligence. Advances in intelligent systems and computing, vol 1235. Springer, Singapore, pp 175–188. https://doi.org/10.1007/978-981-16-4641-6_15

8. Sharma K, Bhalla R (2022) "Decision Support Machine- A hybrid model for sentiment analysis of news headlines of stock market." Int J Electr Comput Eng Syst 13(9):791–798. https://doi.org/10.32985/ijeces.13.9.7

9. Thormann ML, Farchmin J, Weisser C, Kruse RM, Safken B, Silbersdorff A (2021) Stock price predictions with LSTM neural networks and twitter sentiment. Stat Optim Inf Comput 9(2):268–287. https://doi.org/10.19139/soic-2310-5070-1202

10. Kedar SV (2021) Stock market increase and decrease using twitter sentiment analysis and ARIMA model. Turk J Comput Math Educ 12(1S):146–161. https://doi.org/10.17762/turcomat.v12i1s.1596

11. Chen W, Zhang H, Mehlawat MK, Jia L (2021) Mean–variance portfolio optimization using machine learning-based stock price prediction. Appl Soft Comput 100:106943. https://doi.org/10.1016/j.asoc.2020.106943

12. Sarkar A, Sahoo AK, Sah S, Pradhan C (2020) LSTMSA: A novel approach for stock market prediction using LSTM and sentiment analysis. In: 2020 Int Conf Comput Sci Eng Appl (ICCSEA), pp 4–9. https://doi.org/10.1109/ICCSEA49143.2020.9132928

13. Gondaliya C, Patel A, Shah T (2021) Sentiment analysis and prediction of Indian stock market amid Covid-19 pandemic. IOP Conf Ser Mater Sci Eng 1020(1):012023. https://doi.org/10.1088/1757-899X/1020/1/012023

14. Gupta R, Chen M (2020) Sentiment analysis for stock price prediction. In: Proc 3rd Int Conf Multimed Inf Process Retrieval (MIPR), pp 213–218. https://doi.org/10.1109/MIPR49039.2020.00051

15. Li X, Wu P, Wang W (2020) Incorporating stock prices and news sentiments for stock market prediction: a case of Hong Kong. Inf Process Manag 57(5):102212. https://doi.org/10.1016/j.ipm.2020.102212

16. Yadav A, Vishwakarma DK (2020) Sentiment analysis using deep learning architectures: a review. Artif Intell Rev 53(6):4335–4385. https://doi.org/10.1007/s10462-019-09794-5

17. Reddy NN, Naresh E, Kumar VBP (2020) Predicting stock price using sentimental analysis through twitter data. In: Proc (CONECCT) 6th IEEE Int Conf Electron Comput Commun Technol, pp 1–5. https://doi.org/10.1109/CONECCT50063.2020.9198494

18. Suhail KMA et al (2021) Stock market trading based on market sentiments and reinforcement learning. Comput Mater Contin 70(1):935–950. https://doi.org/10.32604/cmc.2022.017069

19. Subasi A, Amir F, Bagedo K, Shams A, Sarirete A (2021) Stock market prediction using machine learning. Procedia Comput Sci 194(November):173–179. https://doi.org/10.1016/j.procs.2021.10.071

20. Rouf N et al (2021) Stock market prediction using machine learning techniques: a decade survey on methodologies, recent developments, and future directions. Electronics 10(21):2717. https://doi.org/10.3390/electronics10212717

21. Raubitzek S, Neubauer T (2022) An exploratory study on the complexity and machine learning predictability of stock market data. Entropy 24(3):332. https://doi.org/10.3390/e24030332
22. Polamuri SR, Srinivas K, Mohan AK (2019) Stock market prices prediction using random forest and extra tree regression. Int J Recent Technol Eng 8(3):1224–1228. https://doi.org/10.35940/ijrte.C4314.098319
23. Yang JS, Zhao CY, Yu HT, Chen HY (2020) Use GBDT to predict the stock market. Procedia Comput Sci 174(2019):161–171. https://doi.org/10.1016/j.procs.2020.06.071

Machine Learning Techniques for Image Manipulation Detection: A Review and Analysis

Suhaib Wajahat Iqbal and Bhavna Arora

Abstract Low-cost modified or tampered image enhancement processes and advanced multimedia technologies are becoming easily obtainable as image editing methods, different editing software, and image altering tools are becoming advanced. These manipulated multimedia images or videos can be used to fool, attract or mislead the public or readers, malign a person's personality, business, political opinions and can affect in criminal inquiry. Most of the research have been undertaken this research work to find the solution to the image manipulation using the deep learning (DL) methodologies that deal with the problem of determining and distinguishing the tempered regions in real and fake images. In this work, we present a study of existing machine as well deep learning-based image manipulation detection approaches. The survey found that the researchers have paid more attention to image content while paying less attention to tempering artifacts and other image features. The primary focus of this research is on ML-based solutions of image manipulation detection. The presented study examines various techniques for determining whether image is genuine or tempered. Besides, the review paper presents a comparative study of various effective approaches in image manipulation field. A brief summary of various datasets used for image manipulation detection is also presented. Finally, the research challenges that open for further research work are listed. Although, lot of work has been done to find the manipulations in the images and videos but these approaches still fall short in some areas.

Keywords Image manipulation · Machine learning · Deep learning · Tempering artifacts · Image features · Tempered regions

S. W. Iqbal · B. Arora (✉)
Department of Computer Science & IT, Central University of Jammu, Jammu, Jammu Kashmir 181143, India
e-mail: bhavna.aroramakin@gmail.com

S. W. Iqbal
e-mail: Suhaibwajahat@gmail.com

1 Introduction

Image manipulation is the use of image editing techniques to create an illusion or deception on images using analog or digital means. Splicing, copy-move, and removal are the popular tampering techniques [1]. As photo manipulation technology advances and is more widely used, it is necessary to evaluate the efficacy of manipulation detection methods against images produced by a variety of tools, manipulations, and manipulation comprehension. Moreover, not even all tampering alters the image's semantics [2]. So, these tempered images are used to mislead the public personality, and defame person images can be used to fool or mislead the public, defame a person's character, business, and political opinions as well and affect criminal inquiry. After these tampering techniques, although post-processing such as Gaussian smoothing can be used with care, people have had trouble recognizing altered areas. As a result, telling the difference between genuine and tampered images has become extremely difficult. Image forensic investigation study is important in order to prevent hackers from accessing fiddled images for unethical commercial or political gain. Unlike existing image recognition systems, that further attempt to identify all objects in various types of images, image manipulation networks attempt to detect only the tampered regions. The significant proportion of these methodologies is focused on a specific tinkering method. A newly enacted architecture-based LSTM segment altered patches by attempting to learn to identify altered corners, demonstrating dependability against various tampering techniques [1]. We have seen that visual feature on both the physical and conceptual stages can help us identify altered images. We have seen that visual features both physical and semantic levels are useful for identifying altered images. Bringing together visual data from the frequency and pixel domains may thus improve image detection performance. Inherently, not all components contribute similarly to the purpose of detecting manipulated images, implying that few visual features are more crucial than many in determining whether or not a given image has been tempered [3].

The primary findings of this work are as follows (a–d):

(a) The paper presents a taxonomy for the image manipulation detection technique.
(b) The paper discusses an in-depth study and analysis of several image manipulation literature approaches.
(c) A brief illustration for the comparison of various datasets available.
(d) We highlight few open research gaps and challenges in this field of study.

The organization of the remaining part of this paper is depicted in Fig. 1. The remainder of the paper is organized as follows: The Sect. 2 provides an overview of image processing in ML, feature extraction, and image processing techniques in ML. Section 3 gives a brief about the types of manipulations and Sect. 4 discusses about the manipulation detection technique. Section 5 provides a summary of the recent literature survey, whereas Sect. 6 discusses about the datasets available for image tempering detection. Section 7 discuss about datasets and Sect.8 highlight open research challenges and Sect. 9 conclude the paper, respectively.

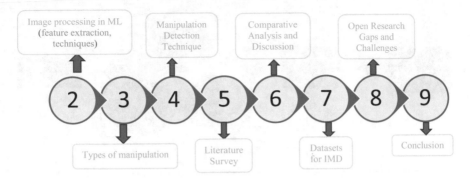

Fig. 1 Organization of the paper

2 Image Processing in ML

ML algorithms learn from data and follow a predefined pipeline or set of steps. To begin, ML algorithms require a large amount of high-quality data in order to learn and predict highly accurate results. As the scale of data in machine learning increases, so does the performance. The images should be well processed, interpret, and generic for ML image processing. Computer Vision (CV) comes into play in this which is a field concerned with machines' ability to understand image data.CV can be used to process, load, transform, and manipulate images in order to create an ideal dataset for the machine learning model [4].

Preprocessing steps used are as:

1. Changing all of the images to the same format.
2. Removing extraneous areas from images.
3. Converting them to numbers so that algorithms can learn from them (array of numbers).

These features (processed data) are then used in further steps involved in selecting and developing a machine learning models to classify unknown feature vectors by classifying the large database of feature vectors. After this, we need to select a suitable algorithm like Bayesian Nets, Decision Trees, Genetic Algorithms, Nearest Neighbors, and Artificial Neural Networks, etc., can be used for this.

The Fig. 2 diagram explains the working of a traditional machine learning image processing method for image data [4].

2.1 Feature Extraction

The dimensionality reduction divides and reduces the set of raw data to more manageable groups. So that processing becomes simpler. The important feature is the large number of variables in these large datasets. These variables help significantly in

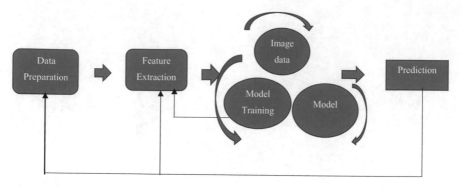

Fig. 2 Machine learning workflow of image processing for image data [4]

the amount of computing power to process. It helps in attaining the best features from large datasets by merging the selected and combined variables into feature sets which effectively reduces the amount of data. Processing these features is simple which helps in identifying the accurate and unique narrating the actual data [5].

Feature Extraction Techniques

- **Bag of words** It is the most commonly used NLP technique. This process involves extracting words or attributes from a sentence, manuscript, internet site, or other source and categorizing them in terms of frequency of use. As a result, one of the most important aspects of this entire process is feature extraction.
- **Image Processing** Image processing is a fantastic and fascinating field. In this domain, you will primarily start playing with your images to understand them. We use a variety of methodologies, as well as feature extraction and algorithms, to identify and process features such as shapes, edges, or motion in a digital image or video.
- **Auto-encoders** The primary goal auto-encoders is data efficiency which is unsupervised in nature.so the feature extraction procedure is used to identify the key features from the data to code by learning from the original data to derives the new data [5]

The last step after the feature extraction is training and modeling the image data using various algorithms to learn from the patterns with specific parameters so that we can use the trained model to predict the previously unknown data.

3   Types of Manipulations

Any operation performed on an image or video that causes the visual content to differ from its original version is referred to as a digital manipulation. Moreover, many images processing approaches, such as rotation, downsizing, and the application of global filters on images, partially manipulates the information represented by visual content. As a result, image forensic approaches are becoming increasingly effective at detecting maliciously manipulated visual content. Image manipulation techniques are classified into two types: content preservation and content altering. According to the authors, steganography is also a type of image manipulation technique because it alters the image content invisibly. The sections of the paper that follow explain various types of image tempering and manipulation detection techniques. The manipulations like copy-paste, splicing, and retouching are possible by employing both basic image processing techniques and advanced methods based, for example, on GAN [6] (Fig. 3).

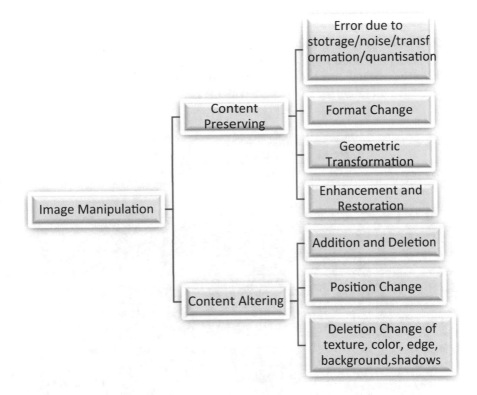

Fig. 3 Types of manipulations [7]

4 Manipulation Technique Detection

It can lead to distinct manipulative tactics thanks to its rich feature illustration. Then analyzing the manipulation detection technique and comparing the performance. NIST16 includes splicing, removal, and copy-move tempering techniques for detecting multi-class image manipulation. To become familiar with unique visual tinkering artifacts and acoustic aspects for each class, we change the manipulation classification classes. Below is shown performance of each temper class (Table 1).

Splicing is the simplest and most likely to produce RGB artifacts such as artificial corners, contrast discrepancies, and noise artifacts. Because the inpainting which occurs after the removal process has a major impact on the noise features, removal detection outperforms copy-move. The most difficult tamper technique is copy-move which results in same noise distribution perplexes the noise stream resulting in same contrast between the two regions [1].

The Fig. 4 shows for splicing, copy-move, and removal manipulation detection, and the RGB and noise maps focus on providing conflicting information. RGB-N produces the recognition accuracy for various tampering techniques by incorporating the characteristics from the RGB image with the noise features.

Table 1 Performance comparison of various detection techniques on NIST16 dataset [1]

	Splicing	Removal	Copy-Move	Mean
Performance	0.960	0.939	0.903	0.934

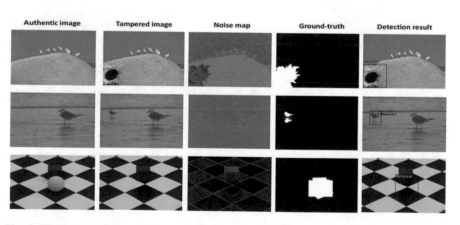

Fig. 4 Multi-class image manipulation detection [1]

5   Literature Survey

This section gives us a brief summary of existing image manipulation detection approaches.

Qi et al. [3] used a novel framework for MVNN was suggested for detecting fake news by fusing visual and pixel domains. They used a CNN-RNN model for extracting visual features in the pixel domain. They used Seino Weibo dataset. The evaluation metrics of the paper got 84.6% accuracy outperformed the existing approaches by 9.2%. Jin et al. [8] proposed the visual and statistical features for fake detection. They also used the Seino Weibo dataset and have applied the Random Forest model for this paper. The dataset used was Seino Weibo dataset. They have got 83.3% of accuracy for image features than existing image features and have got 7% more accuracy for non-image features. Zhou et al. [1] in their paper proposed a Faster R-CNN network and trained it to detect tampered regions in an image that had been manipulated. They used RGB and noise streams to detect manipulation artifacts such as intensity variations and unnaturally tampered regions. They also identified genuine images derived from tampered images which has been difficult to detect. They used the coco dataset and Nist16 dataset for training and testing purposes. The training (90%) and testing (10%) sets are separated to make sure that the same background and tampered object do not appear. They generated a 42 K image pairs that were both tampered with and authentic. They also tested their method against standard methods such as NIST Nimble 2016, CASIA, COVER, and the Columbia dataset. Tolosana et al. [9] proposed a survey for the detection of face manipulation and fake detection. The authors used a technique of deep learning GAN for the detection of face manipulation. The DFFD database and Celeb-DF database were used are only available for public research. The Deepfake database (DFFD) acquired an AUC result of 100% while the Celeb database acquired only 60% AUC result. Heller et al. [10] proposed a PS-Battles collection of images dataset for the detection of image manipulation. The visual domain features were taken from Photoshop battles subreddit. The dataset consists of 102,028 original as well as varied images grouped in 11,142 subsets. Shi et al. [11] proposed that uses a dual-domain convolution neural networks for image manipulation detection and localization. They proposed two sub-networks SCNN and FCNN to localize tempered regions. In this paper, the model is evaluated on CASIAv2.0, Columbia, and Carvalho datasets. They used pixel-wise region localization features. Their proposed model performed efficiently to locate the tempered regions and outperforms existing methods of image manipulation. Dong et al .[12] proposed MVSS network for detecting image manipulation. They took features from both pixel and image levels. The dataset in the paper was CASIAv2 and DEFACTO dataset. The model used was MVSS-Net and ConvGeM. It outperformed existing methods in both within and across dataset scenarios. Bayar et al. [2] also proposed a novel approach using CNN neural network. They took a large-scale dataset for this paper. The model used was CNN and ET classifier (Extra Tree). The feature taken was from the image content. They achieved an accuracy of 99%. Nataraj et al. [13] suggested a holistic image alteration by using pixel matrices. They used the Media Forensics (Medi For) dataset for the proposed work. The model used was CNN. They achieved an accuracy of 81%. Kwon et al. [14] presented localization and detection of JPEG artifacts for the manipulation of image. The feature used is DCT coefficients and JPEG artifacts. They used the CASIA dataset, fantastic Reality dataset, and IMD2020 dataset in this paper. The accuracy achieved was 81%. (Horvath et al. [15]). In this paper, they used vision transformer model for detecting manipulation in satellite taken images. They used the xview2 and WorldView3 dataset consisting of satellite manipulated images. Their model performed better than previously unsupervised detection techniques. Dang et al. [16] gave face image manipulation detection based on CNN. The face region features were used. They used framework HF-MANFA and MANFA on the DSI-1 dataset and Imbalanced dataset. It outperformed existing expert and intelligent systems. Shi et al. [17] presented a paper on Image Manipulation Detection Using a Global Semantic Uniformity System. The features used to exploit were texture and semantic information. The datasets that were used in this paper are NIST2016 and CASIA. They proposed GSCNet for this paper. It significantly outperforms the previous methods in terms of performance. Zhou et al. [18] proposed image manipulation detection using a neural network architecture based on geometric rectification. The features extracted were at the pixel level. They gave RNN model for the detection of manipulation. The datasets used were Pascal VOC07, Imbalanced dataset, and CASIA. It achieved the desirable performance with common tempering artifacts. Wei et al. [19] developed an algorithm based on Edge Detection and Faster R-CNN. The tamper feature and edge were used. They used the model using Faster R-CNN and RPN. The datasets used were NIST2016, Columbia, and CASIA. The proposed model was more effective than other traditional algorithms. In 2021, Bekci et al. [20] presented Cross-Dataset Face Manipulation Detection utilized Deepfake. They used metric learning and steganalysis-rich model. The datasets on which experiments were done are FaceForensics++, DeepfakeTIMIT, CelebDF. They also gave a Deepfakes detection framework for the face manipulation detection. Under unseen manipulations, the framework improved accuracy by 5% to 15% and showed high degree of generalization.

6   Comparative Analysis and Discussion

A comparison of various manipulating detection approaches: Table 2 summarizes the findings of a study of several machine learning and deep learning-based manipulation detection and tempered regions proposed by various researchers.

The above survey shows an insight of common image fudging forms, publicly available image tampered datasets, and cutting-edge tamper methodologies. It provides a distinct viewpoint on rethinking varied opinions of manipulation indications beside various detection methods. People have an incredibly limited potential to perceive and locate photo manipulation in real pictures even though no original images are provided for comparing. The method proposed in [1] experimented on

Table 2 Literature survey of various image manipulation detection

Year	Authors	Objective	Features used	Modal	Dataset	Matrices	Results
2017	Jin et al. [8]	Novel visual and statistical image features for microblog news verification	Visual content and statistical features	Random Forest	Seino Weibo Dataset	ACCURACY	83.6%
2018	Zhou et al. [1]	Image manipulation detection With learning-rich features	RGB stream, noise stream	Faster R-CNN	CoCO dataset, Nist16 dataset	ACCURACY	It performed better than individual RGB stream
2018	Heller et al. [10]	The PS-Battles dataset is a set of images used to detect image manipulation	Visual domain	Photoshop battles subreddit	PS-Battles dataset	ACCURACY	Does not achieve higher accuracy
2018	Shi et al. [11]	Detection and localization of image manipulation using dual-domain convolutional neural networks	Pixel-wise region localization	CNN	CASIAv2.0, Columbia, Carvalho datasets	ACCURACY	It efficiently detects the tempered areas and outperforms current image manipulation methods

(continued)

Table 2 (continued)

Year	Authors	Objective	Features used	Modal	Dataset	Matrices	Results
2018	Bayar et al. [2]	Constrained convolutional neural networks: A novel approach to detect image manipulation	Image content	CNN, ET classifier	Large-scale dataset	ACCURACY	99%
2019	Qi et al. [3]	Detecting fake news using multi-domain visual information	Visual feature, pixel domain	CNN and CNN-RNN	Seino Weibo dataset	ACCURACY	84.6%
2020	Toloson et al. [9]	Deepfakes and beyond: Facial manipulation and fake detection	Image content	GAN	DFFD database, Celeb-DF database	ACCURACY	100% but Celeb-DF database showed only 60% AUC Accuracy
2021	Dong et al. [12]	MVSS-Net: Image manipulation detection using multi-view multi-scale supervised networks	Pixel level, image level	MVSS-Net, ConvGeM	CASIAv2, DEFACTO dataset	ACCURACY	It outperformed existing methods for both inside of and throughout dataset situations

(continued)

Table 2 (continued)

Year	Authors	Objective	Features used	Modal	Dataset	Matrices	Results
2021	Nataraj et al. [13]	Detection of holistic image manipulation using pixel co-occurrence matrices	Pixel domain	CNN	Media Forensics (MediFor)	ACCURACY	81%
2021	Kwon et al. [14]	Learning to detect and localize JPEG compression artifacts of image manipulation	DCT coefficients JPEG compression artifacts	CNN, DCT stream	Park et al. proviced JPEG images, CASIA v2, Fantastic Realit-, IMD2020	ACCURACY	93%
2021	Horvath et al. [15]	Detecting manipulation in satellite images using vision transformer	Image splicing detection	Vision transformer	xView2 dataset, WorldView3 dataset	ACCURACY	Superior performace to previously unsupervised splicing detection techniques
2020	Bassi et al. [21]	A review of fake news identification methods and techniques	Content and social context	MVAE,SAFE,CARMN,FANG,FAKEDETECTOR	BuzzFace, CredBank, Emergent, Fever, LIAR	ACCURACY	Does not achieve higher accuracy
2019	Dang et al. [16]	A CNN is used to detect face image manipulation	Face regions	MANFA & HF-MANFA	DSI-1, Imbalanced dataset	ACCURACY	It outperformed existing expert and intelligent systems

(continued)

Table 2 (continued)

Year	Authors	Objective	Features used	Modal	Dataset	Matrices	Results
2020	Shi et al. [17]	Image Manipulation Detection Using a Global Semantic Uniformity System	Texture and semantic	GSCNet	NIST2016 and CASIA	ACCURACY	It outperformed previously compared methods with improved performance
2021	Zhou et al. [18]	Geometric rectification-based neural network architecture for image manipulation detection	Pixel level	RNN	Pascal VOC07, Imbalanced dataset, CASIA	ACCURACY	It achieved the desirable performance with common tempering artifacts
2019	Wei et al. [19]	Creating a Faster R-CNN image manipulation detection algorithm using edge detection	Tamper feature, edge detection	Faster R-CNN, RPN	NIST2016, Columbia, and CASIA	ACCURACY	It was more effective than other traditional algorithms
2021	Bekci et al. [20]	Cross-Dataset face manipulation detection	Deepfakes	Deepfakes detection framework	FaceForensics++, DeepfakeTIMIT, Celeb-DF	ACCURACY	It improved the accuracy by 5% to 15%

four standard image manipulation datasets using two stream Faster R-CNN framework showed that method used not only detects tempering artifacts but also differentiates various tempering techniques with improved performance. The Random Forest approach in [10] achieves an accuracy of 83.6%; accuracy on Seino Weibo dataset means it does not achieve higher accuracy, while researchers build a MVSS-Net approach [14] which outperformed recent studies that have attempted in both intra- and inter-database instances. The CNN ET classifier approach [2] uses large-scale dataset and achieves an accuracy of 99%. Furthermore, the vision transformer approach in [16] uses image splicing detection on xview2 dataset which achieves superior performance to previously unsupervised splicing techniques, while the deep flakes and beyond approach proposed in [11] achieves an accuracy of 100% but Celeb-DF database showed only 60% AUC accuracy on GAN model.

7 Datasets for Image Tempering Detection

This section outlines the datasets that are accessible for image manipulation detection (IMD). Table 3 compares and contrasts several commonly used datasets. These datasets were gathered from different online platforms and mainstream media portals.

When comparing datasets, few trends emerge out. CASIA provides spliced and copy-move images of various objects. The tampered portions are specifically selected, and post-processing techniques such as filtering and obscuring are

Table 3 Comparison of various available datasets [22]

Dataset	Release Year	Tampering Types	Authentic / Tempered	Image Size	Format	Mask	Post-processing
CASIA v1.0	2009	Cut-paste	800/921	384 × 256	JPEG	NO	NO
CASIA v2.0	2009	Cut-paste Copy-move	7491/5123	240 × 160/ 900 × 600	TIFF/JPEG	NO	YES
IMD	2012	Copy-move	48/48**	3000 × 2300	JPEG/PNG	YES	Optional
CoMoFoD	2013	Copy-Move	5200/5200	512 × 512	JPEG/PNG	YES	YES
Wild Web	2015	Cut-paste Copy-move Erase-fill	0/10646	Various	Various	YES	YES
COVERAGE	2016	Copy-move	100/100	Various	TIFF	YES	NO

employed. The differential among both manipulated and authentic images is fuzzified to create ground truth masks. COVER is a tiny dataset that focuses on copy-move operations. It conceals tampering artifacts by covering common items as spliced areas, and it also includes ground truth masks. The Columbia dataset concentrates on lossless copy-move. Masks with underlying data are presented [1]. The CoMoFoD dataset was created to detect counterfeit of a copy-move. It includes 260 crafted pictures categorized into two: tiny (512 × 512 pixels) and big (512 × 512 pixels) (3000 × 2000 pixels). Each set comes with a forgery, a mask of the tampered area, and the authentic image. The manipulation of images is classified into five categories: translation, rotation, scaling, combination, and distortion [6]. Datasets that are provided are outdated and out of date. Such datasets are inadequate for tackling the issue of image manipulation for recent image data because manipulating image data producers' techniques change over time. The altered area in the CASIA v2.0 dataset has linguistic features besides a living creature or a vehicle, and certain relevant data could be derived from the uniformity in between two datasets. The pixel value in the feature space is close together. Furthermore, the image and consistency are divided using correlation among nearby pixels. The term region is used by some clustering algorithms. As a result, 3 tests are carried out depending on multiple traditional features extraction in order to split the pictures into various portions.

8   Research Gaps and Challenges

There are several challenges associated with detecting image manipulation in order to identify and locate the manipulated regions on the image, which are as follows:

(a) **Real-time Data Collection** The manual task of identifying image manipulation is extremely subjective. Typically, it needs to detect the manipulated image from various sources using manual detection techniques. Detecting these manipulated images can impose a considerable challenge.

(b) **Less Distinct Data** The disparity in the number of datasets available has less distinct images presents a challenge in image detection.

(c) **Exploring More Features** For this problem, researchers in the field of detecting image manipulation are trying to construct a larger multimodal dataset by fusing current datasets available to explore more generalization on different datasets is still a challenging task.

(d) **Other Challenges** The existing integration of the machine learning preprocessing and post-processing operation to search for higher level cues information for effective detection and identifying tempered regions in manipulated is a challenging task.

9 Conclusion

In this modern world, the manipulation of digital images has become very popular. The easily available editing software, tools, and many image altering tools have become widely used to alter the image. These manipulations on an image can mislead the public and can malign person's character, business, criminal inquiry, and political opinions. Hence, it has become very important for us to detect if the image is tempered or not. The present survey shows that various techniques are available to detect image manipulation using machine and deep learning technique but because of the high complexity of deep learning models, machine learning approaches are more preferred. Also, they do not perform well when the data is very small so they require a vast data, training them requires a lot of computational power, which makes them more time-consuming and resource intensive. The CNN, RNN, and LSTM models are among the deep learning techniques used for this purpose of manipulation detection. Faster R-CNN outperforms the MFCN and LSTM. So, to address these challenges, machine learning techniques will be better to use as these techniques use large amount of data to learn from the patterns and creates self-learning algorithms so that machines can learn fast by themselves and make decisions based on that learned data. This study investigates various datasets available for detecting tempered regions. Even though several methods and techniques for combating manipulation have advanced over the last decade, there are still several ongoing research challenges such as Jpeg compression, low accuracy, less data available, lack of rapid, and real-time discovery.

References

1. Zhou P, Han X, Morariu VI, Davis LS (2018) Learning rich features for image manipulation detection. In: Proceedings of the IEEE conference on computer vision and pattern recognition (CVPR), pp 1053–1061
2. Bayar B, Stamm MC (2018) Constrained convolutional neural networks : a new approach towards general purpose image manipulation detection. IEEE Trans Inf Forensics Secur 13(11):2691–2706
3. Qi P, Cao J, Yang T, Guo J, Li J (2019) Exploiting multi-domain visual information for fake news detection. In: Proc IEEE IntConf Data Mining (ICDM), pp 518–527. https://doi.org/10.1109/ICDM.2019.00062
4. Machine learning image processing. Retrieved from https://nanonets.com/blog/machine-learning-image-processing/. Accessed on 23 Jan 2022
5. What is feature extraction? Feature extraction in image processing". Retrieved from https://www.mygreatlearning.com/blog/feature-extraction-in-image-processing/. Accessed on 23 Jan 2022
6. Novozámský A, Mahdian B, Saic S (2021) Extended IMD2020: a large-scale annotated dataset tailored for detecting manipulated images. IET Biometrics 10(4):392–407. https://doi.org/10.1049/bme2.12025
7. Retrieved from https://www.researchgate.net/figure/Types-of-Image-Manipulation_fig1_320703095

8. Jin Z, Cao J, Zhang Y, Zhou J, Tian Q (2017) Novel visual and statistical image features for microblogs news verification. IEEE Trans Multimed 19(3):598–608. https://doi.org/10.1109/TMM.2016.2617078

9. Tolosana R, Vera-Rodriguez R, Fierrez J, Morales A, Ortega-Garcia J (2020) Deepfakes and beyond: a survey of face manipulation and fake detection. Inf Fusion 64(June):131–148. https://doi.org/10.1016/j.inffus.2020.06.014

10. Heller S, Rossetto L, Schuldt H (2018) The PS-battles dataset - an image collection for image manipulation detection. arXiv:1804.04866v1 pp 1–5. Retrieved from https://arxiv.org/pdf/1804.04866.pdf

11. Shi Z, Shen X, Kang H, Lv Y (2018) Image manipulation detection and localization based on the dual-domain convolutional neural networks. IEEE Access 6:76437–76453. https://doi.org/10.1109/ACCESS.2018.2883588

12. Dong C, Chen X, Hu R, Cao J, Li X (2022) MVSS-Net : Multi-view multi-scale supervised networks for image manipulation detection. IEEE Trans Pattern Anal Mach Intell 45(3):3539–3553

13. Nataraj L, Goebel M, Mohammed TM, Chandrasekaran S, Manjunath BS (2021) Holistic image manipulation detection using pixel co- occurrence matrices. arXiv:2104.05693v1, pp 1–6. Retrieved from https://arxiv.org/pdf/2104.05693.pdf

14. Kwon MJ, Nam SH, Yu IJ et al (2022) Learning JPEG compression artifacts for image manipulation detection and localization. Int J Comput Vis130:1875–1895. https://doi.org/10.1007/s11263-022-01617-5

15. Horvath J, Baireddy S, Hao H, Montserrat DM, Delp EJ (2021) Manipulation detection in satellite images using vision transformer. In: Proceedings of the IEEE/CVF conference on computer vision and pattern recognition (CVPR) workshops, pp 1032–1041. https://doi.org/10.1109/CVPRW53098.2021.00114

16. Dang LM, Hassan SI, Im S, Moon H (2019) Face image manipulation detection based on a convolutional neural network. Expert Syst Appl 129:156–168. https://doi.org/10.1016/j.eswa.2019.04.005

17. Shi Z, Shen X, Chen H, Lyu Y (2020) Global semantic consistency network for image manipulation detection. IEEE Signal Process Lett 27:1755–1759. https://doi.org/10.1109/LSP.2020.3026954

18. Zhou Z, Pan W, Wu QMJ, Yang C-N, Lv Z (2021) Geometric rectification-based neural network architecture for image manipulation detection. Int J IntellSyst 36(12):6993–7016. https://doi.org/10.1002/int.22577

19. Wei X, Wu Y, Dong F, Zhang J, Sun S (2019) Developing an image manipulation detection algorithm based on edge detection and faster R-CNN. Symmetry (Basel) 11(10):1–14. https://doi.org/10.3390/sym11101223

20. Bekci B, Akhtar Z, Ekenel HK (2020) Cross-dataset face manipulation detection. In: 2020 28th Signal Process Commun Appl Conf (SIU), pp 34–37. https://doi.org/10.1109/SIU49456.2020.9302157

21. Bassi MA, Lopez MA, Confalone L, Gaudio RM, Lombardo L, Lauritano D (2020) Enhanced Reader.pdf. Nature 388:539–547

22. Zheng L, Zhang Y, Thing VLL (2019) A survey on image tampering and its detection in real-world photos. J Visual Commun Image Representation 58:380–399

Improving Real-Time Intelligent Transportation Systems in Predicting Road Accident

Omolola Faith Ademola, Sanjay Misra, and Akshat Agrawal

Abstract Advances in the use of an intelligent transportation system (ITS) have been deployed in most of the world, which presents new opportunities for developing sustainable transportation system. This paper focuses on improving intelligent transportation systems using Big Data tools in predicting road accidents in Nigeria, based on real-time data gotten from Twitter. The work gives a review of common problems associated with the intelligent transportation system, and how this can be improved by utilizing Apache Spark Big Data. The revolution in intelligent transportation systems can be impacted by the availability of large data that can be used to generate new functions and services in intelligent transportation systems. The framework for utilizing Big Data Apache Spark will be discussed. The Big Data Apache Spark applications will be used to collect a large amount of data from various sources in intelligent transportation system; in return, this will help in predicting road accidents before it happens, and also, a feedback system for alerting can be projected. The use of machine learning algorithms is being used to make necessary predictions for the intelligent transportation system. The result obtained shows that for the classified data relating to road accidents, KNN gave a 94% accuracy when compared to other classification algorithms such as the Naïve Bayes, support vector machine, and the decision tree.

Keywords Road accident · Machine learning · Big data · Intelligent transportation system

O. F. Ademola
Covenant University, Ogun State, Ota, Nigeria
e-mail: omolola.ademola@covenantuniversity.edu.ng

S. Misra
Ostfold University College, Halden, Norway
e-mail: sanjay.misra@hiof.no

A. Agrawal (✉)
Amity University, Haryana, India
e-mail: akshatag20@gmail.com

1 Introduction

Almost 15 million people die every year as a result of traffic collision which indicates that more than 3000 deaths are recorded every day; also, 2 to 5 million people get injured due to road accident [1].

We can say that one of the worst affected countries of road accident is Nigeria, despite integrated efforts to reduce fatal road accidents, yet the country still falls victim of such mishap. With a human population of approximately 207 million, a high vehicle population estimated at more than 30 million, a total road length of about 194,000 km (34,120 km of federal, 30,500 km of state and 129,580 km of local roads), and the country have suffered severe losses due to fatal car accidents [1]. Survey results have shown that the death rate in road accidents among young adults is very high, which is a major part of the global economy [2]. The problem of road accident can worsen in the future due to high increase of population growth and migration to urban areas around the country. Various road safety strategies, methods, and countermeasures have been proposed and applied in resolving this problem. Such as training and retraining of drivers in adherence to safety tips Federal Road Safety Corps (FRSC) and Vehicle Inspection Officers are responsible for ensuring compliance with speed limit regulations, ensuring that vehicles are in perfect condition, building sustainable roads, and repairing damaged roads. Hence, it is vital to develop more technologically driven and practical solutions to reduce road accidents.

Advances in the use of an intelligent transportation system (ITS) have been deployed in most of the world, which presents new opportunities for developing sustainable transportation system [3]. In several cities around the world, especially the developed countries incorporate the use of ITS. One might claim that ITS stands out to be one of the oldest technology that constitutes the Internet of Things, but still presently it can be seen as a leading-edge in sustainable cities. In Madrid, for example, every public transport network and part, including trains, trams, busses, and bus stops, is linked to a central control center, which collects and processes data in real time to deliver smart and efficient services and applications to end users [4]. A large amount of data can be generated using intelligent transport systems. The technology advancement in ITS, such as smart card, GPS, sensors, video detectors, social media, and so on, has increased the complexity, the variety, and quantity of information generated and collected from vehicles and the movements of persons [5]. Massive volume of data is being recorded from different device that makes up the ITS; however, traditional data management systems are inefficient and cannot fully analyze the data being produced for deployment of an effective transportation system. This is because the data volume and complexity are not compactible. To combat this problem, a candidate solution is the use of Big Data analytics tools such as Apache Spark, has been found to process vast volumes of data, and has been used extensively in academies, stock markets, organizations, and industries [6]. An efficient structure is necessary to design, implement, and manage the transportation system in order to meet the computational necessity of the massive data analysis. In this context, Apache Spark has become a centralized engine for large-scale analysis of data across a variety of ITS services. The deployment of the Apache Spark in intelligent transportation system can be developed for applications, in the area of real-time traffic control, and estimating the average speed and the congested sections of a highway. Apache Spark is much quicker and simpler to use with this advanced model [7]. Apache Spark uses the memory of a computer cluster to minimize reliance on the distributed network underlying it, which results in significant improvements of Hadoop's map reduction.

The contribution this paper offers will be the use of Apache Spark in filtering out data gotten from Twitter that is relating to road accident. This data will be used to predict the likelihood of road accident occurring with the use of machine learning algorithms, and based on those predictions, the algorithm that predicts best can be deployed for use in the intelligent transportation system.

The rest of this paper is divided into four sections. In Sect. 2, we discuss related works and identified research gaps which has led to this study. In Sect. 3, the methodology of the research which deploys the use of the Apache Spark in organizing and cleaning of the raw data, this processed data are then analyzed using the machine earning algorithm to make predictions. Furthermore, the results and discussion is given in Sects. 4, and 5 concludes on the research study.

2 Literature Review

The digitization of trackers, nomadic sensors, and smart meters, summarized as Internet of Things (IoT), has been incorporated into the world at large [8]. This linked system contributes to advanced transport management techniques, for example, the possibility of gathering the quantity of multi-source traffic data required to make precise forecasts of traffic [9].

Some researchers have installed equipment used on vehicles such as RADAR sensor, GPS, infrared sensor, GLONASS, and cameras. These devices are used in order to gather data about road conditions, environmental settings, as well as technical details of vehicles involved in road accidents, information about the behavior of the driver like drowsiness or stress level, geographical position, and other relevant information, were gathered from the installed equipment, which was used in analyzing inevitable road accident occurrence [10].

In similar research, data sources about road accidents were gotten from Waze, Google Maps, Twitter, and Inrix. Social media were used in traffic and road collisions as a data source [11]. Although the use of social media can be seen to be efficient in gathering all information about road accident, yet it can be termed unreliable because of the inability to authenticate its source or publisher. Also, the social media information can be challenging to interpret because users use local terminology to publish their contents and text may include ideology and grammar errors. As a result, social media data may be classified as inaccurate, incomplete, and hard to read. However, there is a need to work on the data by filtering it and make the information gotten online to be efficient for use in improving the intelligent transportation system.

ITS devices such as microwave vehicle detection system (MVDS) are also considered as means of collecting road accident data, data relating to vehicle speed, distance, occupancy, and vehicle type; these road devices can generate thousands of records weekly of the road activities [12].

The use of decision tree classifiers, PART induction rules, Naive Bayes, and multi-layer perceptron was used by [13] to establish essential variables for the prevention of accident seriousness. Through comparing the various models obtained, the authors concluded that with a value of 0.08218, the tree classification and the rule induction are the most accurate. Age, gender, nationality, year of the accident, and accident form were the most significant variables in accident fatality.

A proposed software structure by [14], draw a significant relationship between the variables linked to the road accident, applied to the dataset on road accidents in Morocco, the proposed work selected the appropriate rules employing multiple criteria analysis. Ultimately, the system will forecast death and injury based on time series analysis using the selected regulations. The consistency of the regulations is calculated by the assist value or the occurrence frequency of rules and compliance.

A study examined the use of Twitter as a data source and natural language processing to improve the efficacy of road incident detection by [15]. The results showed that only 5 percent of the information received was useful after analysis, suggesting that the tweets were connected with traffic accidents and were able to geocode information on a map. For the complete classification of the dataset as traffic accidents, the researchers registered an accuracy of 0.9500. On the other hand, the precision value for the geocoding phase from tweets was 0.5200. Public data sources, such as the road monitoring network and the police recording of incidents of traffic, have been checked. The authors affirmed that the frequency of the postings, which culminated at weekends, was steady.

The results of a proposed ANN were evaluated, and a correlation of 0.991, an R-squared value of 0.9824, and an average 4,115 square error, a root mean square (RMSE) error of 2.0274 were reported by [16]. The model was proposed for predicting road accidents based on an artificial neural network and taking into account not only the accident details, such as the behavior of drivers, cars, time and hour, and road structure but also rather certain information on the geometry of road transport and road volume statistics. The authors consider the variable vertical degree of road curvature to be the critical parameter influencing the number of road accidents. Artificial neural networks have been developed by [17] to predict the severity of road accidents by preprocessing road accident data using a K-means cluster to sort the data and improve the prediction. To verify the results obtained, the authors applied the ordered test model, finding that the ANN yields a higher precision, with a value of 0.7460 above the 0.5990, which was obtained from the other models.

Predictive analyzes were performed using supervised learning perspective incorporated with autoregressive integrated moving average (SARIMA), and a Kalman filter was developed by [18]. The work done in [19], suggested multi-task learning (MTL) can be integrated into a deep learning model in order to learn the efficiency of unattended flow prediction features. This profound learning model allowed for the automatic prediction process while guaranteeing a high level of precise learning.

Another work considers the use of deep learning approach in finding the range in object detection in this case car which aid in improving the safety in self-service vehicles [20]. For highway scenarios, the model may reduce errors in the range estimation to an appropriate amount. The behavior of drivers in delivering the required support to ensure safety is often taken into account in independent vehicles.

The research reported by [21], as used the Bayesian network, the J48 decision treaties, and the neural network artificially to identify the most significant variables in order to predict the frequency of road accidents. This research shows that the Bayesian network has produced the most accurate 0.8159, 0.7239, 0.7239, and an F-measure of 0.723. Findings showed that the lighting, road condition, and weather condition could result to accidents on the road. Bad roads were identified by the system and were marked as a likelihood of causing road accident.

In [22], the authors proposed a method for automated detection by machine learning and Big Data technologies of road traffic events from tweets in the Saudi Dialect. First, they create a classifier and train it with four machine learning algorithms for the filtering of tweets in a relevant and irrelevant way, support vector machine, decision tree algorithms, k-nearest neighbor (KNN), and Naïve Bayes algorithm. Then, they train other rank classifiers for the identification of different types of accidents, bridges, road closures, traffic damage, fire, weather and social occurrences. Analyzes of one million tweets have shown that their method automatically detects road traffic events, their location and time without having been aware of the events beforehand. To the best of their knowledge, the Apache Spark Big Data Platform was the first task in detecting traffic events from Arabic tweets. The research gap in this paper couldn't extract the exact location of events that occurred in the location detection approach, and the variety of data gotten from this work was well filtered enough to get the relevant information about road accidents.

If we are to consider real-time processing data, Apache Spark is of best choice. It has a function called Spark Streaming which has the advantage manage lots and stream workloads by a single execution engine, thus overcoming the constraints of conventional streaming systems. Spark Streaming allows Spark to improve its core planning capacity to display data in real time.

3   Methodology

The intelligent transportation system (ITS) incorporates evolutionary technologies such as smart control technologies, Transmission of data technologies and also different sensors are being used in the intelligent transportation system. The advances in these technologies produce a lot of data got from diverse sources incorporated into the system. This large amount of data must be run, tracked, and controlled by data-driven models like Apache Spark. With the use of a Big Data analytics, the problem of data storage, data analysis, and data management is resolved. Sometimes handling this data to make accurate and precise decision about the intelligent transportation system can sometimes be challenging and difficult to handle. However, it is becoming highly difficult to process huge data in reasonable time and make decisions in real time. A number of important problems, namely the right preprocessing, real-time analytics, and a model of communication, are posed from literature. Therefore, we are exploring the criteria for a resourceful communication model based on big data analysis and proposing a standardized architecture in an intelligent transport network for processing data in real time. We will make use of data gotten from twitter based on the transportation system, which will help us improve the intelligent transportation system.

The Spark Streaming is used for real-time data processing. With the Spark Streaming, live data are processed, scalable and fault-tolerant, high throughput, and real-time data support of about 0.5 s. Spark uses RDD to arrange data and recover from failures.

To illustrate the proposed architecture for predicting road accident using Apache Spark. It consists of four main section: Big Data gathering and regulation, data preprocessing, data processing, data prediction.

3.1 Big Data Gathering and Regulation

Data are collected using Twitter streaming API. A social media accounts will be created, and keyword sets will be defined making use of the hashtag. This accounts will be used to trend about the keyword relating to road accidents tweets consisting different road accidents occurring in different locations will be posted using the hashtag. This data can be logged and kept for record purpose in order to be used in making a decision.

The raw data are gotten and stored as Javascript Object notation (JSON) file extension. This file extension is stored in MongoDB. For the proposed work, attributes are used for the event prediction. The attribute is defined by; the timestamp, a user who has made a post using the hashtag, the location the accident occurred, the text context of the user, and the road name detected location of event. Each attribute will be separated by a delimiter "\\" character.

3.2 Data Preprocessing

The next action to take after collecting the data is preprocessing. The data gotten from social media are vulnerable to incompatible data from various foundations, misplaced text, unnecessary words, illegal characters, and noise. The preprocessing technique helps to filter out unnecessary information and reduce the noise. The preprocessing will be applied before the actual processing. This preprocessing helps to clean out the inaccuracy of the collected information. Spark SQL is utilized to preprocess the data. The words or text gotten from tweets can be transformed into Token. The tweets gotten from social media cannot be analyzed directly upon because of the noise. A supervised machine learning is then feed with this preprocessed data. The preprocessed data are stored back to the MongoDB as cleaned and parsing data. Social media comprises of different kind of noise; this noise can be reduced by making use of an optimal estimator known as Kalman filter. The Kalman filter is utilized for fast response of the data processing in filtering out the noisy data. The expression for the Kalman filter is given below:

$$f_k = M f_{k-1} + A v_{k-1} + w_{k-1} \tag{1}$$

M is the state estimation matrix applied to the former state $f_{(k-1)}$,

A is the input control matrix applied to the controller vectorvg $(k-1)$,

$w_{(k-1)}$ is the noise vector.

The actual processing is then conducted which is based on the idea of parallel processing in which multiple processes are conducted simultaneously in order to save time. The formulations of a program to run faster since it runs multiple engines (CPUs) are parallel processing. Parallel processing is the principle of load balancing. Load management increases workload distribution over multiple computer processors. The implementation is performed using the MapReduce programming method in order to be really accurate in the data processing scenario. In this step, MapReduce and HDFS are used for the same structure. The MapReduce removes all numbers and punctuations such as commas, (,), period (.), semi-colons (;), and question marks (?). The removal of these punctuation marks reduces the size of the text we want to analyze and make it easy for us to identify our input. However, we can say, it is possible for a punctuation mark not to be used properly, since the tweets on social media are informal and not formal, and sometimes the words used with the hashtag (#) may not add up.

In addition, other than HDFS, we may also use the alleged HBASE and HIVE SQL to store historical information for the administration of a database (offline or in-memory) in regulating the data.

Stream processing and memory retaining competencies, Spark ecosystem operates alongside a high-speed resilience distributed dataset (RDD), are also carried out. Spark-based large data ecosystem is considered when processing data streaming on a large scale in the memory phase is needed for nearly real time.

Additionally, the preprocessed words are checked before going ahead to the processing stage, and any words that cannot be transformed to a tokenized form is removed as this word may be written in error. Figure 1 shows the applied step to the sample tweet.

Fig. 1 Preprocessing steps
applied to a sample tweet

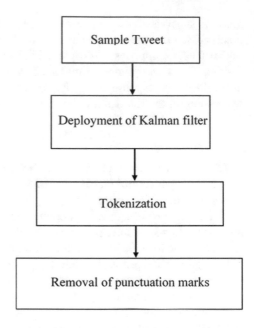

3.3   Data Processing

Text Feature Extraction

We will make use of frequency–inverse document frequency method (TF–IDF) for
the text feature extraction. The TF–IDF can perform good result for learning and
used widely in the area of text mining. This is provided in the Spark ML package.
The TF–IDF measures important words that are used in the tweets. The product of
term frequency (TF) and inverse document frequency (IDF) will give TF–IDF. The
TF (j, n) term is a frequency, which is said to be a function of t and d, is how many
appearance the term t makes in a tweets d while IDF measures the number of tweets
that term j provides. The IDF is a numerical approach and calculated by:

$$IDF(t, \ D) \ = \ \log \frac{|D|}{DF(j, \ N)} \tag{2}$$

|N| is denoted as the total number of documents in the collection N.
DF (j, N) is denoted as the number of documents where the term j appears.
The resultant logarithm used in the above equation will return 0 for terms
appearing in all documents. Therefore, the product of TF and IDF is written as

$$TFIDF \ (j, \ n, \ D) \ = \ TF \ (j, \ n) \ * \ IDF \ (j, \ N) \tag{3}$$

The algorithm converts the input which will be the lists of tokens into vector
matrix of tokens. The results of the term frequency will be transferred to the IDF

algorithm. After which, the IDF will sort content vector producing an output that will be stored in the content set. The content set is passed as an input to the classification algorithm.

Tweet Classification

The collected tweet, as we know, is not all relating to traffic. A sort of binary classification will be applied that will categorize the tweets to two classes: traffic related or non-traffic related.

After that a model is build which can be evaluated using evaluation metrics such as accuracy, recall, specificity, and precision. The four classification models used are support vector machine, decision tree algorithms, k-nearest neighbor (KNN), and Naïve Bayes. The model of best fit with the traffic event detection will be used in classifying real-time data relating to areas in Nigeria.

Parallel computing is employed for building and training a model using MLlib in Apache Spark. Label 1 and label 0 are given which signifies traffic related and non-traffic, respectively. The models are built and trained with the default input parameters. Training data are incorporated which the models learns from; the pattern is found from label of each tweet text with the training data. Next, the model's accuracy is evaluated using evaluation metrics and cross-validation approach. Furthermore, with the best selected model, new tweets are predicted and categorized as 0 and 1 iteratively. For further processing, we filter out tweets that are not associated with traffic. We also summarize the data for interpretation and gain insight through the application of various data, such as hourly counting of number of tweets and the show of the traffic events distribution place.

3.4 Data Prediction

For training set and test set, we use a ratio 70/30 test split. We need to know the output of our model, in particular with unnoticed data. The first is to split the dataset into training and tests using the cross-validation approach. Finally, through measurement metrics, we analyze the prediction results from test data. The output of four classification models (Support vector machine, decision tree algorithms, k-nearest neighbor (KNN), and Naïve Bayes) was compared. Moreover, to forecast actual twitter data in the world, we pick the best model.

In our experiment, we use different training/test split data ratios, includes 50/50, 60/40, 70/30, and 80/20. We use a train test break, one of the methods of cross-validation assessment. We divides our dataset into two un-overlapping sections (workout set and test set). It is easy but efficient for validation purposes. The training package is used to train our model while a test set and/or stop set are used to evaluate our models' output using the evaluation calculation for handling invisible data.

In the context of a geo-mapping, we need to locate spatial positions for the distribution of the traffic status in the context of a Cartesian coordinate (latitude, longitude) in order to define the geographical distribution of traffic status. This helps to analyze tweets and to extract valuable information. Google Maps Geocoding API is used for this. Geocode is a transformation of certain addresses into geographical coordinates (latitude, longitude) to identify a position of the input given on the map. The opposite is facilitated by reverse geocoding. It transforms geographical coordinates into a readable address for humans. Geocoding reverse provides information for the location of the particular place, which can be read and understood by people like postal code, the name of the road, the town, the road number and area.

4 Results and Discussion

This section shows the result of the application of four models using support vector machine, decision tree algorithms, k-nearest neighbor (KNN), and Naïve Bayes models in predicting the certainty of the causes of road accidents and what might actually make this road accident to occur.

Before we look into the models used in predicting road accidents, we have observed that a number of attribute such as relating to a person being involved in a crash, or due to roadway reconstruction and environmental conditions results to factors that are used to study the causes which are fire, road closure, road damage, social events, weather condition such as heavy rainfall can also be said to be a cause of road accidents. These attributes that are likely to occur can be calculated using statistics as demonstrated below. The formula for the statistic is expressed below and as well as descriptive statistics of the explanatory variables is presented in Eq. (4).

$$\overline{Z} = \sum_{k=1}^{n} \frac{\hat{z}_{kt1}}{n} \tag{4}$$

where $\hat{z}_{kt1}$ is the random variable individually drawn from the sample data collected from various event relating to road accident.

n is the finite sample size of the various events relating to road accident.

From Table 1, the result obtained for the standard deviation shows that, with the different events that have occurred, the higher the value of the standard deviation, the more likely that variable will occur more often.

4.1 Result for the Filtered Tweets

Figure 2, support vector machine, decision tree algorithms, k-nearest neighbor (KNN), and Naïve Bayes algorithm are the four classification algorithm considered in this paper. The performance of this four algorithm was measured using the accuracy, recall, specificity, and precision evaluation metrics. Figure 3 shows that KNN performs better than support vector machine (SVM), decision tree, and Naïve Bayes

Table 1 Descriptive statistics of road traffic crashes data

Variable	Number of months	Minimum	Maximum	Mean	Standard Deviation
Accident	108	0	32	16.89	9.66954
Fire	108	0	10	5.80	3.316625
Road shutdown	108	0	17	8.50	5.338539
Road damage	108	0	6	3.00	2.160247
Social events	108	0	24	12.00	7.359801
Traffic congestion	108	0	19	9.472	5.91608
Weather condition	108	0	9	4.62	3.02765
Road work	108	0	14	6.667	4.472136

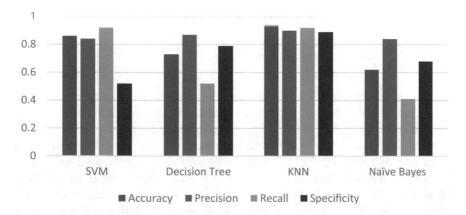

Fig. 2 Result for the evaluation of filtered tweets

in terms of accuracy, specificity, and precision. However, 92% recall was achieved for both KNN and support vector machine.

Different events have also been measured using the classification models of support vector machine, decision tree algorithms, k-nearest neighbor (KNN) algorithm, and Naïve Bayes algorithm when measured with the evaluation metrics. These events, which are also referred to as the variable as seen in Table 1, are likely causes of road accident, and they have been evaluated using the four classification models considered in this paper. The result obtained shows that the one with a better performance is KNN. The chart in Fig. 3 shows that damaged roads, accidents, traffic, weather condition, road shutdown, social events, road work, and fire using the KNN give a better yield.

Figure 4 represents the number of vehicles in one of the roads in Port Harcourt city at a particular time, which amounts to causing traffic. As at 7:00am–9:00am (during the breaking of the day) and 11:25am–12:30 pm (when the sun is up), the roads tends to be extremely busy with high traffic when many vehicles are on the road. This is due to the fact that from 7:00am–9:00am, office time resumes, school children on

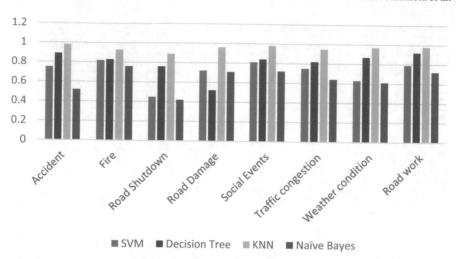

Fig. 3 Machine learning results for the analyzed data

their way to school, and much people tend to be on the road at that time. Therefore, this system tells us when and where the traffic is at the highest at a particular time. The system has the ability to recognize obstacles on the road; information from the tweet is being extracted about the cause of the current traffic, and information about the number of cars at a particular road can be extracted. Large amount of cars on the road can result to road accident as this vehicles struggle to move forward; this can give a possible red flag as care must be taken in ensuring angry drivers don't create a scene of causing accident on that path, due to hesitation of one getting to their destination on time (Fig. 5).

The tweets gotten are measured in millions. The words extracted from these tweets are road, congestion, highway; after gathering the tweets, it was observed that the major cause of the traffic congestion was due to blockage of one side of the main roads making it difficult for drivers to get to their stipulated destination on time; at another time, there was a heavy tanker that fell on the road and caught fire, making it difficult for cars to pass as it may also make those cars passing to catch fire as well.

The more the number of vehicles on the road the lesser the average speed of the cars, which will as well result to longer time in getting to one's destination. We can as well get real-time traffic information to estimate the best path between two points. The objective of this paper is to process data gotten from tweets concerning road accidents, road traffic, road events, and other various road activities that occur on the road using Apache Spark to yield intelligent transportation system. Based on the results given, we can conclude that the system operates well in real time when implementing it on Apache Spark system. The Apache Spark system has helped in classification of the Big Data gotten from tweets and categorized in a format where we can make better prediction of when road accident is likely to occur.

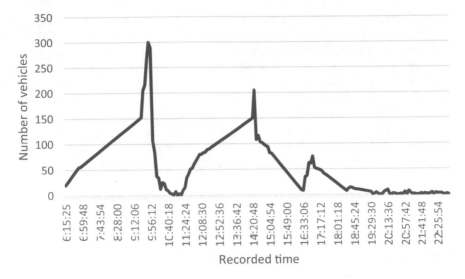

Fig. 4 Number of vehicle on the road at a particular point in time

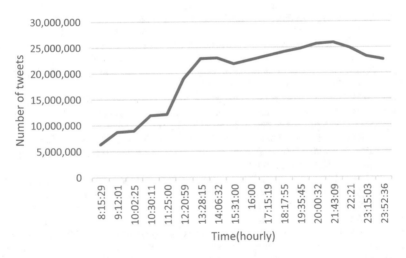

Fig. 5 Number of tweets per hour for the traffic condition

5 Conclusion

The data collected from the tweets are used to make predictions for the intelligent transport system. The Apache Spark was used to filter out irrelevant tweets from the raw data. The models support vector machine, decision tree algorithms, k-nearest neighbors (KNNs), and Naïve Bayes algorithm were used to make predictions based on the relevant tweets of occurrence of road accidents happening around Nigeria.

The location of happenings was extracted. The models showed it was able to detect the events of road accidents and make further predictions of how the intelligence transportation systems is effective. Future research area can be on incorporating an alarming system that sends information to the intelligent transportation system to avoid road accidents from occurring. Given huge amount of data size, this system is efficient enough for processing and offers the solution for real-time processing in an intelligent transportation system. The incorporation of the Apache Spark into intelligent transportation system is of great advantage in reducing the number of road accident. Additionally, it will be helpful to equate prediction output with other sophisticated simulation methods, such as the artificial neural network.

Acknowledgements This paper was sponsored by Covenant University, Ota, Ogun State Nigeria.

References

1. Favour OI, et al (2016) Statistical analysis of pattern on monthly reported road accidents in Nigeria. Sci J Appl Math Stat 4(4):119–128. https://doi.org/10.11648/j.sjams.20160404.11
2. Igho OE, Isaac OA, Eronimeh OO (2015) Road traffic accidents and bone fractures in Ughelli, Nigeria. IOSR J Dent Med Sci 149(4):2279–861. https://doi.org/10.9790/0853-14452125
3. Rezaei M, Klette R (2014) Look at the driver, look at the road: No distraction! No accident!. In: Proceedings of the IEEE conference on computer vision and pattern recognition (CVPR), pp 129–136. Retrieved from https://openaccess.thecvf.com/content_cvpr_2014/html/Rezaei_Look_at_the_2014_CVPR_paper.html
4. Menouar H, Guvenc I, Akkaya K, Uluagac AS, Kadri A, Tuncer A (2017) UAV-enabled intelligent transportation systems for the smart city: applications and challenges. IEEE Commun Mag 55(3):22–28. https://doi.org/10.1109/MCOM.2017.1600238CM
5. Buele J, Salazar LF, Altamirano S, Aldás RA, Urrutia-Urrutia P (2019) Platform and mobile application to provide information on public transport using a low-cost embedded device. RISTI-Rev Iber Sist e Tecnol Inf 476–489
6. Babar M, Arif F, Jan MA, Tan Z, Khan F (2019) Urban data management system: towards big data analytics for internet of things based smart urban environment using customized Hadoop. Futur Gener Comput Syst 96:398–409. https://doi.org/10.1016/j.future.2019.02.035
7. D'Silva GM, Khan A, Gaurav, Bari S (2017) Real-time processing of IoT events with historic data using Apache Kafka and Apache Spark with dashing framework. In: 2017 2nd IEEE International conference on recent trends in electronics, information & communication technology (RTEICT), pp 1804–1809. https://doi.org/10.1109/RTEICT.2017.8256910
8. Contreras-Castillo J, Zeadally S, Guerrero-Ibañez JA (2017) Internet of vehicles: architecture, protocols, and security. IEEE Internet Things J 5(5):3701–3709. https://doi.org/10.1109/JIOT.2017.2690902
9. Beil C, Kolbe TH (2017) CiytyGML and the streets of New York — a proposal for detailed street space modelling. In: Proceedings of the 12th International 3D GeoInfo Conference. ISPRS Ann Photogramm Remote Sens Spat Inf Sci vol IV-4/W5, pp 26–27
10. Cao G, Michelini J, Grigoriadis K et al (2015) Cluster-based correlation of severe braking events with time and location. In: 2015 10th System of systems engineering conference (SoSE), pp 187–192. https://doi.org/10.1109/SYSOSE.2015.7151986
11. Salas A, Georgakis P, Petalas Y (2017) Incident detection using data from social media. In: 2017 IEEE 20th International conference on intelligent transportation systems (ITSC), pp 751–755

12. Shi Q, Abdel-Aty M (2015) Big data applications in real-time traffic operation and safety monitoring and improvement on urban expressways. Transp Res Part C Emerg Technol 58(Part B):380–394

13. Alkheder S, Taamneh M, Taamneh S (2017) Severity prediction of traffic accident using an artificial neural network. J Forecast. 36(1):100–108

14. Ait-Mlouk A, Agouti T (2019) DM-MCDA: a web-based platform for data mining and multiple criteria decision analysis: a case study on road accident. SoftwareX. https://doi.org/10.1016/j.softx.2019.100323

15. Gu Y, Qian Z, Chen F (2016) From twitter to detector: real-time traffic incident detection using social media data. Transp Res Part C Emerg Technol 67:321–342. https://doi.org/10.1016/j.trc.2016.02.011

16. Çodur MY, Tortum A (2015) An artificial neural network model for highway accident prediction: a case study of Erzurum, Turkey. Promet Traffic Transp 27(3):217–225

17. Taamneh M, Alkheder S, Taamneh S (2017) Data-mining techniques for traffic accident modeling and prediction in the United Arab Emirates. J Transp Saf Secur 9(2):146–166

18. Lippi M, Bertini M, Frasconi P (2013) Short-term traffic flow forecasting: an experimental comparison of time-series analysis and supervised learning. IEEE Trans Intell Transp Syst 14(2):871–882

19. Deng S, Jia S, Chen J (2019) Exploring spatial–temporal relations via deep convolutional neural networks for traffic flow prediction with incomplete data. Appl Soft Comput 78:712–721. https://doi.org/10.1016/j.asoc.2018.09.040

20. Parmar Y, Natarajan S, Sobha G (2019) DeepRange: deep-learning-based object detection and ranging in autonomous driving. IET Intell Transp Syst 13(8):1256–1264

21. Castro Y, Kim YJ (2016) Data mining on road safety: factor assessment on vehicle accidents using classification models. Int J Crashworthiness 21(2):104–111

22. Alomari E, Mehmood R, Katib I (2019) Road traffic event detection using twitter data, machine learning, and apache spark. In: 2019 IEEE SmartWorld, ubiquitous intelligence & computing, advanced & trusted computing, scalable computing & communications, cloud & big data computing, internet of people and smart city innovation (SmartWorld/SCALCOM/UIC/ATC/CBDCom/IOP/SCI), pp 1888–1895. https://doi.org/10.1109/SmartWorld-UIC-ATC-SCALCOM-IOP-SCI.2019.00332

Consumer Buying Behavior Analysis During COVID-19 Phase Using Statistical Data Mining and Multi-verse Stochastic Algorithm

Anurag Sinha⑩, Mopuru Bhargavi, N. K. Singh⑩, Devendra Narayan, Namit Garg⑩, and Siddhartha Pal

Abstract COVID-19 has changed particularly the marketing and retail sectors, the epidemic has changed peoples' habits. Analyzing client purchase patterns and flows is a common use for consumer behavior analysis. For the purpose of analyzing consumer behavior and purchase trends across the COVID-19 era, this work combined a statistical strategy with a data mining approach. Furthermore, a survey-based (online) data analysis was conducted for the evaluation by retailers and customers, filling out a questionnaire that comprised demographic information and product associations obtained during the epidemic. In order to test the association rule of data mining, the information for this study was gathered from a nearby grocery. Additionally, main data was converted to secondary data using the meta-verse algorithm (MOA) (balanced). One of the newest meta-heuristic optimization algorithms, multi-verse optimization (MVO), imitates the multiversity hypothesis of physics and simulates the interaction of many universes. This MOA is based on natural phenomena that employed a stochastic method to accomplish its objective. Finally, statistical analysis has been carried out to look into the purchasing and selling trends of retailers within the dataset. In this method, association rules are generated via pincher search. It counts the supports of the candidate in each run

A. Sinha (✉)
Department of Computer Science, IGNOU, New Delhi, India
e-mail: anuragsinha257@gmail.com

M. Bhargavi
Department of Computer Science and Engineering, Koneru Lakshmaiah Education Foundation, Vaddeswaram, Guntur, India

N. K. Singh
Department of Computer Science, BIT Mesra, Ranchi, Jharkhand, India

D. Narayan
Department of Biotechnology, Amity University, Ranchi, Jharkhand, India

N. Garg
Department of Mathematics and Statistics, IIT Kanpur, Kanpur, Uttar Pradesh, India

S. Pal
Department of Physics, IIT Kanpur, Kanpur, Uttar Pradesh, India

© The Author(s), under exclusive license to Springer Nature Singapore Pte Ltd. 2023   241
Y. Singh et al. (eds.), *Proceedings of International Conference on Recent Innovations in Computing*, Lecture Notes in Electrical Engineering 1011,
https://doi.org/10.1007/978-981-99-0601-7_19

using a bottom-up approach in addition to the supports of chosen item sets using a top-down approach. The Maximum Frequent Candidate Set refers to these. The proposed binary versions transfer a continuous version of the MVO algorithm to its binary counterparts using the idea of transformation functions.

Keywords Consumer behavior · Buying pattern · Data mining · Statistics · Association rule mining · Multi-verse algorithm · Swarm intelligence · Data optimization

1 Introduction

A sizable amount of consumer transactional information, such as credit card transactions, is recorded every day. This information describes the trajectories of consumer activities. It is a well-known fact that past behaviors can anticipate future ones. In order to anticipate future risk based on past behavior, credit risk models are among the most frequently employed to rely on this observation. In reality, a variety of models, including those for acquiring new customers, maintaining accounts, and queuing collections, are frequently used to help decision-makers in the card industry make daily decisions throughout the credit cycle [1, 2]. The models' complexity must be at least as advanced as the related behaviors they are forecasting. Consumer behavior analysis is the activity of examining the customer's buying pattern and his interest in a particular set of products. In some cases, customer behavior is far more complex than a straightforward model can capture, and this means that crucial information may be concealed in the subtle interactions and correlations between the variables [3]. An example of this is the credit card fraud detection model, where the interactions between the transaction variables offer crucial hints about the nature of the transactions, such as whether they are fraudulent or not. A model's improvement due to multivariate interactions can be discovered by contrasting non-interactive models against models constructed taking into account all potential interactions [3–6].

It is clear that the apriori algorithm uses a bottom-up, breadth-first search strategy. The computation progresses upward from the lowest set of frequent item sets to the largest frequent item set. The maximal size of the frequent item set and the number of database passes are equivalent. The method must go through numerous rounds as a result of the lengthening of any one of the frequent item sets, which lowers performance [4]. Customers have different kinds of needs and inclinations toward different sets of products. Thus, the buying patterns of the customers cannot be equivalent to each other due to different parameters such as financial condition, psychological inclination, or other kinds of factors associated with the customer. Ultimately, this factor affects the shopping behavior of a particular customer. In today's era, with the rise of e-commerce, the buying patterns of the maximum customer segment are inclined toward online purchasing habits. Due to the limited physical availability of markets, many customers have turned to e-commerce websites to purchase groceries and other everyday items. In recent years, transaction data has been vitally used for this kind of research, where the associations between the product and the buying investigation have to be measured [7, 8]. The major problem with this data is the clustering and establishing the correlation between each other. Thus, with the introduction of meta-heuristic algorithms, the more complex kinds of optimization problems have been solved. A multi-verse algorithm has been recently proposed based on a meta-heuristic swarm intelligence algorithm, which is inspired by multi-verse astrophysics. In this paper, we have used this MVO algorithm for the data [9].

To overcome the complexity of the distance measures between datasets, the Manhattan and Euclidean distance clustering optimizations are implemented. Data mining is one of the most widely used methods for consumer behavior analysis by implementing association rule mining. In this paper, we have used an association rule mining-based statistical mining approach for examining the frequent set of items that have been purchased during COVID-19 by the same set of customers. In the part of statistical data mining in which we have examined the maximum threshold associations of products with the data mining approach, business decision-making is one of the critical factors that can be improved when a company comes into contact with a customer segment and their preferences for a product, whether through an offline or online marketing framework. As a result, implementing consumer behavior analysis using a data mining approach increases business efficiency and productivity [4]. The major contribution of this paper is that in feature selection, a collection of M features is selected features from a collection of N features in the data, MN, so that the value of a certain evaluation criterion or function is optimized over the set of every conceivable feature subset. In this article, we suggest, test, and discuss an effective strategy based on the most recent multi-verse optimizer (MVO) for feature selection and parameter optimization to increase the apriori algorithm's accuracy of variable. To the best of our knowledge, this is the first time using data optimization using MVO. I suggested an enhanced architecture to increase the data resilience and generalization ability. This paper is arranged in a chronological fashion where Sect. 1 contains an introduction, Sect. 2 literature review, Sect. 3 materials and method, Sect. 4 proposed method, Sect. 5 result and discussion, Sect. 6 conclusion, and future work, respectively.

2 Literature Review

In Ref. [1], author has used consumer behavior analysis for neuro-marketing application in which the maximum threshold of 90% was obtained from the features that have been collected from that 20 people. It was used for obtaining the maximum efficiency in the application area of the marketing using the EEG signal. In Ref. [2], author has used association rule mining for consumer behavior analysis on the supermarket data for examining the buying pattern of the consumer toward the products of the supermarket using conventional data mining approach. In Ref. [10], Author has used web-based association rule mining for examining the customer factor toward the particular association of the product on e-commerce website using conventional data mining approach. In Ref. [4], author has used association rule-based market basket analysis for examining the engagement of the customer in one supermarket in Thailand country based on the different threshold metrics on the product like support and gain that is converted on the feature vector after calculation. In Ref. [5], author has employed data mining-based association rule mining approach for increasing the sales and predicting the consumer behavior with the accuracy of the confidence level. In Ref. [3], to create voltage amplitude and discrete phase distributions in the dipole elements for the creation of flat-top beam/pencil beam patterns, the multi-verse optimization algorithm is used. While the phase distributions of these two patterns are different, they both have similar amplitude distributions. Simulation findings show that this algorithm completed its mission successfully and that it is also better than other algorithms like Particle Swarm Optimization, Gray Wolf Optimization, and Imperialist Competitive Optimization algorithms. In Ref. [6], author has used different statistical approach for consumer behavior analysis, such as chi-square and ANOVA test, which gives the maximum entropy on the validation data that have been performed on the secondary dataset. In Ref. [11], author has used different machine learning algorithms which have been implemented on Python for consumer behavior analysis based on the supermarket data with the optimized algorithm, which gives the maximum threshold confidence level of the model. In few work, the high exploration and local optima avoidance of the MVO algorithm are the source of the MVO-based SVM's dependability and robustness. The rapid shifts in the solutions produced by the use of white/black holes highlight the exploration process and aid in the removal of the local optima stagnation. The WEP and TDR parameters also help MVO to accurately utilize the promising regions during the course of iterations in order to increase the generalization power and resilience of the SVM after initially performing a thorough broad search of the search space [9, 12]. Anxiety, COVID-related dread, and sadness all predicted consumer behavior toward necessities, whereas necessities-only behavior was predicted by anxiety. Furthermore, personality characteristics, perceived economic stability, and self-justifications for buying were all found to predict consumer behavior toward needs and non-necessities. We now know more about how consumer behavior changed during the COVID-19 pandemic thanks to the current study. The findings may be used to create marketing plans that take psychological elements into account in order to cater to the requirements and feelings of genuine consumers [13]. Due to the development of new optimization approaches that were effectively used to address such stochastic mining challenges, data mining optimization has drawn a lot of interest in recent years. In order to build evolutionary optimization algorithms (EOAs) for mining two well-known machine learning data sets, this research applies four alternative optimization strategies. Iris dataset and Breast Cancer dataset are the chosen datasets utilized to assess the proposed optimization strategies [9]. [14] discusses how actions like increasing home cold storage capacity could undermine system resilience by exacerbating bullwhip effects, or amplifying consumer demand shocks that are propagated to upstream food supply chain actors, whereas responses like improving food skills can reduce the propagation of shocks through the supply chain by allowing greater flexibility and less waste.

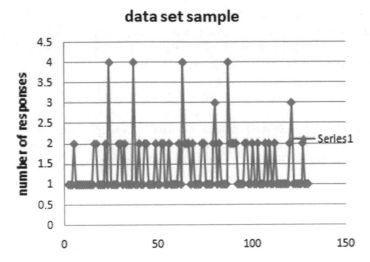

Fig. 1 Sample of responses

3 Materials and Methods

3.1 Dataset Description

For this particular research, we have conducted a survey in which we gathered data in the form of categorical values where the several sections like demographic data, psychological buying pattern, and products that they have bought during COVID is segmented [2]. The data is then further combined to close values using secondary form of the data, which is later converted into secondary form in which we have implemented statistical analysis for getting entropy-level, variance, and for hypothesis testing, which is discussed in the later part of the paper. The data questionnaire is represented in the figure below. We have also gathered associations of products which they have brought by collecting payment receipt data from various local markets, which has been later preprocessed using data discretization and a multi-verse stochastic algorithm as shown in Fig. 1. [4, 5, 10]

3.2 Multi-verse Algorithm

The multi-verse algorithm is one of the recently developed areas in the field of metaheuristics and nature-inspired optimizers. As per the concept of the multi-vision algorithm, it shows that the universe has an infinite number of universes within itself which exist, and the theories which underline the different warm holes and white holes. So in this algorithm, the wormholes represent the total exploration and exploitation parts combined with white and black holes and the total variables are referred to as the object which is used as the inflation rate for the solution finiteness problem and finally refers to the iterations [6, 11]. The core mathematical model of the MVO method is Eq. (1), which are as follows:

$$X_j - 1 = X_{jk}, r1 < NI(Ui)X_{ji}, r1 \geq NI(Ui)^2 \tag{1}$$

The multi-verse algorithm's mathematical modeling depicts the object having the interchange particles between the universes, and this is done using the roulette wheel selection. And in every iteration, this universe is depicted as being the best one where the D is used as the variable. And N is used as the number of universes, and U is the total solution formulated for the total set of these universes, where the universe is categorized as a normalized inflammation rate of the M universe [10].

3.3 Data Preprocessing

Data preprocessing is the total scheduling and cleaning of the data that has been collected and transformed from the various sources of data by employing the different noising techniques. The data cleaning is done, and the segregated part of the data is transformed to the normalized form. In this process, several anomalies and irregularities are eliminated from the data. As an outcome, the information that is being mined in the use of the data is shown using the abstracted form of the information and the portrayal of this total in the collection of the data delivers total equivalent logical outcomes of the total information being processed [15]. In dimensionality decrease, information encoding plans are applied in order to get a decreased or "packed" portrayal of the first information. Models also include information pressure procedures (such as wavelet changes and head part examination). As for trait subset choice (e.g., eliminating unessential characteristics) and quality development (e.g., where a little arrangement of more helpful properties is gotten from the first set) [4]. You have chosen, say, that you might want to utilize a distance-based digging calculation for your investigation, like brain organization, closest neighbor classifiers, or clustering. 1 Such strategies produce better results if the information to be dissected is standardized, or at the very least scaled to a more modest reach, such as [0.0, 1.0] [5]. The client information, for instance, contains the credit age and yearly compensation. The annual compensation trait generally prioritizes quality over age. Hence, assuming that the qualities are left normalized, the distance estimations taken on yearly compensation will, for the most part, offset the distance estimations taken on age. Discretization and idea order age can also be beneficial in cases where raw information values for credits are replaced by ranges or higher applied levels [3]. For instance, crude qualities for age might be supported by more elevated levels of ideas, like youth, grown-up, or senior [6]. Furthermore, idea ordered progression age is an integral asset for information mining in that it permits the mining of information at various degrees of deliberation. For instance, the expulsion of excess information might be viewed as a type of information cleaning, as well as an information decrease [4, 16].

3.4 Association Rule Mining

An everyday example of a many-to-many interaction between two different types of things is described by the market basket model of data. We have things, or what are sometimes referred to as "transactions", on the one hand, and baskets, on the each basket contains an assortment of objects, or an item set, and often we suppose that small compared to the overall number of objects, a basket contains few items of things. Typically, a much higher, larger number of baskets is assumed beyond what can fit in the primary memory [17]. We define the interest of an association rule I_j as the difference between its confidence and the percentage of baskets containing j. In other words, if I has no bearing on j, then we would anticipate that the proportion of baskets that include I and contain j would match identically. Such a rule has no interest. However, it is intriguing if a rule has either high interest, which denotes that the existence of I in a basket somehow encourages the presence of j, or significantly negative interest, which denotes that the presence of I prevents the appearance of j [11, 18]. Let us say that we have identified all item sets that have support levels that are above a certain threshold and that we have the precise support figures generated for each of these item sets. All of the association rules that have broad acceptance and high confidence can be found in them. To put it another way, if J is a set of n items that is discovered to be frequent, then there are only n potential association rules involving this set of items, namely J_j for each j in J. J must be at least as frequent as J if J is. Therefore, it too is a common item set, and we have previously determined the support of both J and J_j. Their ratio is the rule J_{jj}'s confidence quotient.

4 Proposed Method

4.1 Data Optimization Using Multi-verse Algorithm

The multi-verse-based data optimization methodology that has been suggested. It is composed of four distinct stages, the first of which is the tenderization stage, the second of which is the wellness evaluation, and the third of which is the updating of the total centroid variables. In this complete process, the total connection with the universe $U1$ and $U2$ is introduced. The overall randomization value, which is provided at each node of this space where UI exists in the universe, displays the overall standardization vectors of these syndrome aids. And this centralizes since the range from C_{ij-1} to C_{ij+1} is the predetermined group centroid value, which is equal to the total number of universe variables. Each centroid in this paragraph emphasizes

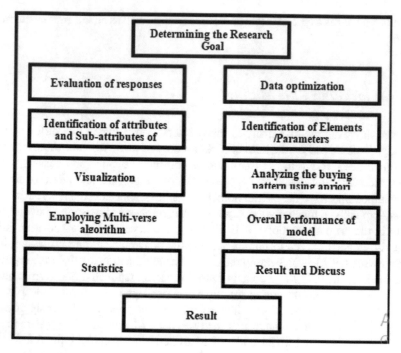

Fig. 2 Proposed method

this subvector, which contains the M components of the dataset's quantities [15, 16]. In Fig. 2, the model is proposed which shows the hierarchy of work.

$$\text{sse} = \sum_{x}^{1x} gc\sigma(cj, ri) \tag{2}$$

The equation gives the Euclidean distance between the centroid c_j and the Ith data point r_i, which is represented with m-dimensions as r_i $(r_{i1}, r_{i2},..., r_{im})$.

$$\sigma(c_j, r_i) = \sum 1 \times gc\ \sigma(c_j, r_i) \tag{3}$$

4.2 Statistical Analysis

We have conducted a statistical analysis using an analytical hierarchical process on the data that we have collected. It shows that the total segments of the responses are categorized into the nine different items in the pair-wise comparison of the total association of the product, which has the highest frequency rate in the transactional data. This data indicates how many characteristics are preferred. Total support is preferred on the point scale. Then, the respondents are asked to show the pair-wise comparisons in the meantime, the range of attributes, which depicts the relative importance of each of them [11, 19]. The two-way steps show that the pair-wise comparison with the metrics N attributes is taken to show the Kth individual variable where the pair-wise comparison variable shows. The greater the importance of the reciprocal, the more the data is reformatted into the pair-wise comparison metrics format. This particular case shows the conducting AHP as an integrated part of the data. It denotes the importance of the total variable, where the support of the variables is reciprocal to the pair-wise matrix and this coded information. Either shows a true or false positive in nature and a false negative in nature. Association rule mining is a well-researched approach for finding the interrelation between the variables of the items within a large, scalable transactional database. It is done to identify these strong rules within the data using the different measures of the variables based on this concept.

The total items in the database have N attributes, which is called items. And the total transactions within those items are what is called in the database. And each transaction has a unique transaction ID and attributes ID where it shows the rule can be defined as a support, confidence, and antecedent between the data [4].

$$s = \sum t_a + t_c \frac{1}{\sum t} \tag{4}$$

$$\text{support} = \frac{\text{the number of transaction on } a}{\text{total transaction}} \tag{5}$$

where S is support; $t_a + t_{c1}/(\sum t)$ is the number of transaction that contains antecedent and consequent; and is the number of transaction.

$$\text{Confidence} = \frac{\text{number of transaction on } A \text{ and } b}{\text{total number of transaction } a \text{ contains}} \tag{6}$$

where C is the confidence; is the number of transaction that contains antecedent and consequent; and is the number of transaction that contains antecedent [11] of given series $[t_1 + t_2...t_n]$. However, there is a chance that this point estimate will be off by a little bit. Therefore, creating an interval that includes the value would be a better estimate a population-level variable. The range known as the confidence interval consists of the point estimate and conceivable error margin. How likely is it that the selected confidence interval typically contains the population parameter as 0.95. The confidence level refers to this probability.

Here, y predicted is the predicted value of the response variable (y), x is the explanatory variable, a is the intercept with regard to y, and b is referred to as the slope of the regression line. There will typically be some disparity between the expected and observed values of the data when you fit a linear regression line to a collection of data (say y observed). The residual is the difference between the observed value and the expected value, expressed as (y observed $-$ y predicted). The least squares method, which minimizes the sum of squares of these residuals, is one of the most used techniques for locating the regression line. For computing residual, use the following equations:

$$\text{Residual of item shingles} = \frac{y \text{ observed set}}{\text{number of otliers}} \tag{7}$$

5 Result and Discussion

- Accuracy: The number of real positive tests compared to the total number of predicted benefits.

$$\text{Accuracy} = TP/TP + FP \tag{8}$$

- Review: The number of authentic positive tests among the true positive instances is taken into account.

$$TP/TP + FN \tag{9}$$

- F1-measure: The weighted normal of accuracy and review represents the two measures. It may give precedence to pieces of information over accuracy due to the lopsidedness of the classifications.

In Fig. 3, the frequent item purchased during lockdown that has been retrieved from transaction data is shown metrics [7, 18, 19]. In Fig. 3, the maximum confidence metrics of product association is shown. In retail and e-commerce settings, its any particular item is likely to account for a small share of transactions. Here, we have aggregated up to the product category level and very popular categories are still only present in 5% of transactions. Consequently, item sets with 2 or more item categories will account for a vanishingly small share of total transactions (e.g., 0.01%). Only 0.014% of transactions contain a product from both the sports and leisure and health and beauty categories. These are typically the type of numbers we will work with when we set pruning thresholds in the following section [4].In Fig. 4, the frequency range of data in cluster is given based on internal data coverage distance and its skewness is measured in Fig. 4, the result of customer segment as per their income and purchasing habit shown using clustering. Figure 5 shown confusion matrix of classified accurate data. The frequency level of data in e-commerce can vary widely depending on the specific type of data being analyzed. The convergence plot MVO is given below which shows the correction in data optimization also a box plot graph is given to depict the accuracy of classification based on SSE aftereffects of MVO,

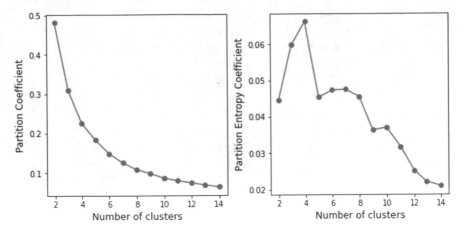

Fig. 3 Frequent item set cluster 1

PSO, DE, and GA in treating all experiments. According to results on Figs. 3 and 4, it is seen that the MVO can recognize the moderately best groups with the base SSE results for all datasets. The same data is classified and test in data mining testing environment using j48 classifier [4].

6 Conclusions and Future Scope

In this paper, we have shown that how data processing can optimize and enhance the result of model. The result of the paper discusses the classification and segmentation of customer purchasing patterns during COVID-19, which submit based on the response which is collected from the custom on their transaction data where the co-relation of their response is tested using AHP method. We have also used multiple optimizations for data clustering and Manhattan distance analysis after the data preprocessing. We have implemented apriori algorithm and customers buying pattern based on association run by which gives the 87% of the accuracy which have been tested on the weaker data mining environment and machine learning algorithm. We have amalgamated fuzzy-c clustering method with association mining to segment the customer type based on data collected in the survey. The major limitation of proposed system is that it is tested on smaller dataset, more imbalanced and unstructured data lies on web to be tested for consumer behavior analysis and usability prediction.

Out[41]:

	consequent	antecedent	support	confidence	lift
4	root vegetables	yogurt, tropical fruit	228	0.463636	2.230611
5	sausage	shopping bags, rolls/buns	59	0.393162	2.201037
8	tropical fruit	yogurt, root vegetables	92	0.429907	2.156588
1	citrus fruit	whole milk, other vegetables, tropical fruit	66	0.333333	2.125637
10	yogurt	whole milk, tropical fruit	199	0.484211	1.891061
2	other vegetables	yogurt, whole milk, tropical fruit	228	0.643836	1.826724
6	shopping bags	soda, sausage	50	0.304878	1.782992
0	bottled water	yogurt, soda	59	0.333333	1.707635
9	whole milk	yogurt, tropical fruit	228	0.754098	1.703222
3	rolls/buns	yogurt, tropical fruit	97	0.522222	1.679095
7	soda	yogurt, sausage	95	0.390625	1.398139

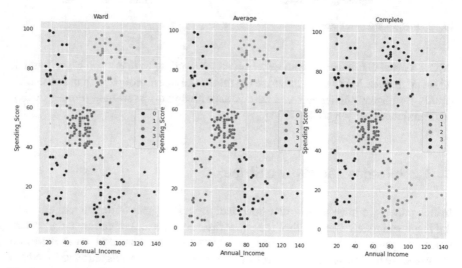

Fig. 4 Predicted association

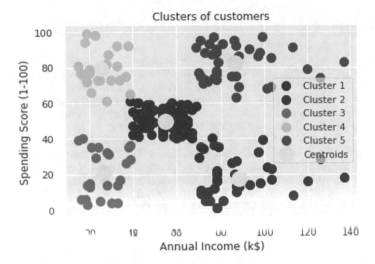

Fig. 5 Types of purchasing habit of customer as per their income

References

1. Watada J, Yamashiro K (2006) A data mining approach to consumer behavior. In: First international conference on innovative computing, information and control - volume I (ICICIC'06), Beijing, China, vol 2, pp 652–655. https://doi.org/10.1109/ICICIC.2006.191
2. Gol, M, Abur A (2015) A modified Chi-Squares test for improved bad data detection. In: 2015 IEEE Eindhoven PowerTech, Eindhoven, Netherlands, pp 1–5. https://doi.org/10.1109/PTC. 2015.7232283
3. Mujianto AH, Mashuri C, Andriani A, Jayanti FD (2019) Consumer customs analysis using the association rule and apriori algorithm for determining sales strategies in retail central. E3S Web Conf 125:23003. https://doi.org/10.1051/e3sconf/201912523003
4. Yingzhuo X, Xuewen W (2021) Research on community consumer behavior based on association rules analysis. In: 2021 6th International conference on intelligent computing and signal processing (ICSP), Xi'an, China, pp 1213–1216. https://doi.org/10.1109/ICSP51882.2021.940 8917
5. Amin CR et al (2020) Consumer behavior analysis using EEG signals for neuromarketing application. In: 2020 IEEE Symposium series on computational intelligence (SSCI), Canberra, ACT, Australia, pp 2061–2066. https://doi.org/10.1109/SSCI47803.2020.9308358
6. Singh SP, Kumar A, Yadav N, Awasthi R (2018) Data mining: consumer behavior analysis. In: 2018 3rd IEEE International conference on recent trends in electronics, information & communication technology (RTEICT), Bangalore, India, pp 1917–1921. https://doi.org/10. 1109/RTEICT42901.2018.9012300
7. Bender KE, Badiger A, Roe BE, Shu Y, Qi D (2022) Consumer behavior during the COVID-19 pandemic: an analysis of food purchasing and management behaviors in U.S. households through the lens of food system resilience. Socio Econ Plann Sci 82:101107. https://doi.org/ 10.1016/j.seps.2021.101107
8. Peighambari K, Sattari S, Kordestani A, Oghazi P (2016) Consumer behavior research: a synthesis of the recent literature. SAGE Open 6(2):215824401664563. https://doi.org/10.1177/ 2158244016645638

9. Gourisaria MK et al (2022) Semantic analysis and topic modelling of web-scrapped COVID-19 tweet corpora through data mining methodologies. Healthcare 10(5):881. https://doi.org/10.3390/healthcare10050881

10. Habeeb E, Ghazal N, Majzoub S (2019) Behavior analysis tool for autistic children using EEG Signals. In: 2019 Advances in science and engineering technology international conferences (ASET), Dubai, United Arab Emirates, pp 1–5. https://doi.org/10.1109/ICASET.2019.8714498

11. Jia H, Peng X, Song W, Lang C, Xing Z, Sun K (2019) Hybrid multiverse optimization algorithm with gravitational search algorithm for multithreshold color image segmentation. IEEE Access 7:44903–44927. https://doi.org/10.1109/ACCESS.2019.2908653

12. Di Crosta A et al (2021) Psychological factors and consumer behavior during the COVID-19 pandemic. PLoS ONE 16(8):e0256095. https://doi.org/10.1371/journal.pone.0256095

13. Zimand-Sheiner D, Levy S, Eckhaus E (2021) Exploring negative spillover effects on stakeholders: a case study on social media talk about crisis in the food industry using data mining. Sustainability 13(19):10845. https://doi.org/10.3390/su131910845

14. Zhang C, Jiang J, Jin H, Chen T (2021) The impact of COVID-19 on consumers' psychological behavior based on data mining for online user comments in the catering industry in China. IJERPH 18(8):4178. https://doi.org/10.3390/ijerph18084178

15. Fatimah B, Javali A, Ansar H, Harshitha BG, Kumar H (2020) Mental arithmetic task classification using Fourier decomposition method. In: 2020 International conference on communication and signal processing (ICCSP), Chennai, India, pp 46–50. https://doi.org/10.1109/ICCSP48568.2020.9182149

16. Huang C, Xiao Y, Xu G (2021) Predicting human intention-behavior through EEG signal analysis using multi-scale CNN. IEEE/ACM Trans Comput Biol Bioinf 18(5):1722–1729. https://doi.org/10.1109/TCBB.2020.3039834

17. Faris H, Hassonah MA, Al- AM, Mirjalili S, Aljarah I (2018) A multi-verse optimizer approach for feature selection and optimizing SVM parameters based on a robust system architecture. Neural Comput Applic 30(8):2355–2369. https://doi.org/10.1007/s00521-016-2818-2

18. Song R, Zeng X, Han R (2020) An improved multi-verse optimizer algorithm for multi-source allocation problem. Int J Innovative Comput Inf Control 16(6):1845–1862. https://doi.org/10.24507/ijicic.16.06.1845

19. Hans R, Kaur H (2020) Binary multi-verse optimization (BMVO) approaches for feature selection. IJIMAI 6(1):91. https://doi.org/10.9781/ijimai.2019.07.004

IoT and Networking

Addressing Role of Ambient Intelligence and Pervasive Computing in Today's World

Jaspreet Singh, Chander Prabha, Gurpreet Singh, Muskan, and Amit Verma

Abstract Intelligence refers to the ability to learn and apply knowledge in new innovative situations. Recent computational and electronic advances have increased the level of autonomous semi intelligent behavior exhibited by systems, so the new terms like ambient intelligence and pervasive computing started to emerge. Artificial intelligence (AI) and ambient intelligence (AmI) are inextricably technology cooperation agreements with each other. Artificial is something made by human beings and ambience is something that surrounds us while ambient intelligence assumed to be something artificial. Pervasive computing is the growing trend of embedding computational capability into everyday objects to make them effectively communicate and perform useful tasks to satisfy end user's resource needs. The paper provides a bibliometric information using VosViewer software about the present trends of ambient intelligence and pervasive computing technologies using Scopus and Web of Science databases, which will help researchers getting ideas toward research development in these domains. It also highlights a comprehensive study on ambient intelligence and pervasive computing covering the application areas that are dramatically affected by them in today's world. Undoubtedly, both clean technologies have strongly influenced the recent developments in the past few years. Furthermore, we can expect that the scope will continue to multiply in upcoming years.

J. Singh (✉) · G. Singh
Department of Computer Science and Engineering, Chandigarh University, Mohali, Punjab, India
e-mail: cec.jaspreet@gmail.com

C. Prabha
Chitkara University Institute of Engineering and Technology, Chitkara University, Mohali, Punjab, India

Muskan
Chandigarh University, Mohali, Punjab, India

A. Verma
Department of Research, Chandigarh University, Mohali, Punjab, India

© The Author(s), under exclusive license to Springer Nature Singapore Pte Ltd. 2023
Y. Singh et al. (eds.), *Proceedings of International Conference on Recent Innovations in Computing*, Lecture Notes in Electrical Engineering 1011,
https://doi.org/10.1007/978-981-99-0601-7_20

Keywords Ambient intelligence · Pervasive computing · Application of AmI · Application of pervasive computing · Ambient intelligence versus pervasive computing

1 Introduction

In coming future, explicit source and destination devices would not be an issue since sensors as well as processors will be built throughout ordinary objects, and the system would smoothly adjust to that same user's needs and preferences. Here is what ambient intelligence (AmI) envisions regarding intelligent computing. A complex AI system called ambient intelligence can respond immediately to human presence. The response users receive from Siri or Alexa when it recognizes that your voice is one specific example of ambient intelligence. Using ambient intelligence that is concealed in the network connecting multiple devices, ambient intelligence would enable gadgets to work together to support persons in carrying out routine activities, chores, and traditions inside of an intuitive way [1]. Pervasive computing is the capacity of implanting knowledge in regular objects such that the individual who interacts with this object diminishes the degree of interaction with the electronic gadget [2]. Pervasive as well as ubiquitous are phrases that often are frequently misused and confused with one another. Pervasive is related with the term penetrate, which means to distribute throughout, whereas pervasive is derived from either the word ubiquity, which means to be something else. In other words, it refers to that same transparency between technology-embedded different objects since there is man-to-machine conversation. In contrast, pervasive computing corresponds to the public accountability in a level where certain technology-embedded objects are completely invisible or are embraced mostly by environment, i.e., beyond man-to-machine interaction. In contradiction to pervasive computing, which is seen as an effective implications, usage of Internet is viewed as a paradigm.

Furthermore, the paper is organized in such a way that Section II represents a bibliometric analysis on the ambient intelligence and pervasive computing fields. In Section III, lights have been thrown on the technology of ambient intelligence covering its properties and applications in today's world. The Section IV unfolds the prospect related to pervasive computing covering its properties and applications in today's world. At last, in Section V, paper summarizes current challenges for ambient intelligence and pervasive computing.

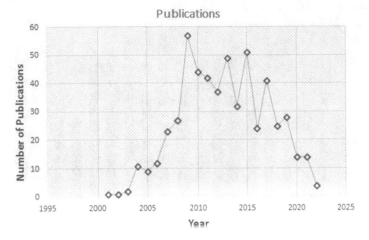

Fig. 1 Web of science core collection

2 Bibliometric Analysis for Ambient Intelligence and Pervasive Computing

2.1 Bibliometric Analysis Web of Science Data

Web of Science Data Core Collection
The keywords ambient intelligence "AND pervasive computing" are used to search the documents from the Web of Science database and total of 548 documents are found using the search; Fig. 1 graph shows the year wise distribution of the documents found using the keywords. The documents or the research starts from the year 2001, and there is increase in the numbers from 2004 to 2015. Then, from 2015 to 2022, there is decline in the number of publications.

Citations and Publications Report
548 results from Web of Science Core Collection. Keywords Used: pervasive computing AND ambient intelligence (All Fields).

The figure has been retrieved from Web of Science database. The graph shows the number of publications and number of citations trend year wise. Number of publications is highest in the year 2009 as shown in Fig. 2.

2.2 Bibliometric Analysis SCOPUS Data

SCOPUS Core Collection
For the country-wise research and number of publication published, the keywords ambient intelligence "AND pervasive computing" are used to search from the Scopus

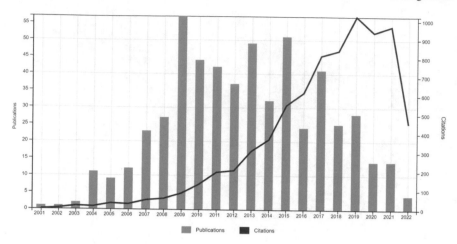

Fig. 2 Citations' and publications' year-wise report

database. The total of 405 documents was found from the database. The UK leads the list with highest number of publications 53. India is at number 9 with 13 publications. The result given in Fig. 3 graph shows the top 14 countries of the world with highest number of publications. We use the filter of minimum of nine number of publications, and the graph shows the countries which matched the said condition. About 405 document results are retrieved from the Scopus database. Keywords Used: "pervasive computing" AND "ambient intelligence".

Fig. 3 SCOPUS core collection

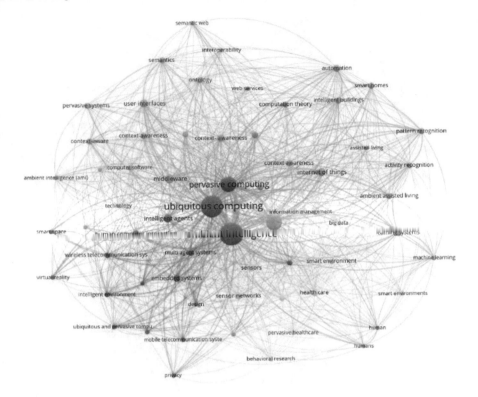

Fig. 4 Co-occurrence keywords with ambient intelligence and pervasive computing

Co-occurrence Keywords

Out of 2944 keywords, 63 meet the threshold (minimum number of occurrences of a keyword is 10). Highest occurrences are of ambient intelligence with 254 number of occurrences. Occurrences of pervasive computing arc 119 as shown in Figs. 4, 5, and 6.

3 Ambient Intelligence

A new field called ambient intelligence adds intelligence to the surroundings we live in. Research addressing ambient intelligence (AmI) draws on advancements made in pervasive computing, artificial intelligence, including sensor networks. Intelligent, pervasive, highly unobtrusive electronic systems that are integrated often in human made systems and therefore are adapted to the demands of the individuals include known as ambient intelligence as well as pervasive computing. Such interactions of different contemporary information and communication technology (ICT) aid people

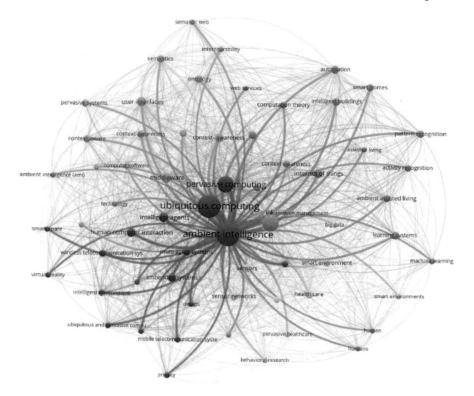

Fig. 5 Ambient intelligence with occurrence keywords

by providing knowledge as well as direction across a range of application domains. The properties of AmI systems [3] are given in Table 1.

3.1 Applications of Ambient Intelligence in Today's World

Ambient intelligence can always have a significant impact on life in a variety of contexts. AmI experts have already researched several of these applications. Current AmI implementations are highlighted below.

1. Smart homes: A smart home "would be an illustration of a surroundings enhanced by ambient intelligence". The above-mentioned intelligence looks into home automation technology. A home can always be enhanced with sensors to learn more about how they are used and, inside some situations, to operate without personal communication [4].

2. Ambient-assisted living and e-health care: E-health care with ambient group homes could provide intelligent services that help people with their real activities. A living person smart environment is created by an ambient-assisted living

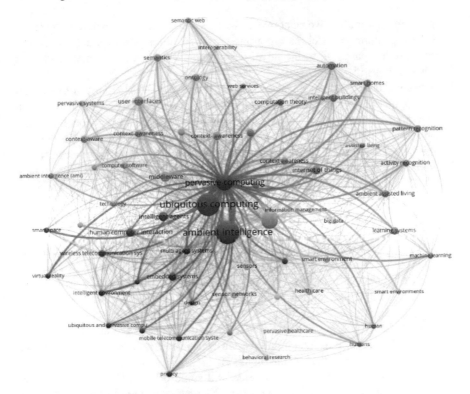

Fig. 6 Occurrence keywords with pervasive computing

Table 1 Properties of AmI systems

Properties	Description
Embedded	Instruments that are embedded become "invisible" to that same environment
Context-aware	They identify a consumer, maybe along with all the person's current condition and the context in which it occurs
Personalized	They frequently take the participant's wants into account
Adaptive	As the recipient's physically or psychologically state varies, Aml systems adapt their functionality
Anticipatory	It are able to foresee user wants despite active cognitive involvement
Unobtrusive	Discreet only giving additional gadgets and humans everything information they need to know more about individual
Non-invasive	It act independently and thus do not enable the company to take any action

platform that is attentive, adaptive, as well as responsive to human requirements, habits, gestures, as well as emotions. Ambient intelligence is a good choice for geriatric care due to the novel communication between humans and their surroundings [5].

3. Transportation: Humans travel extensively during our lives in a variety of ways. It is possible to install technology in automobiles, buses, as well as train stations that can provide users a basic understanding of how well the system is operating right now. Based on this knowledge, preventive measures can be taken, and by using the system more efficiently, the pleasure of those who use that conveyance can be improved. AmI technology, which includes GPS-based spatial location, vehicle identification, as well as image processing, can help public transportation become more fluid, efficient, and safe [6]. By offering route planners, Microsoft also uses AmI technology for driver assistance. Additionally, they produce inferences regarding potential preferred routes and offer drivers personalized route ideas [7].

4. Education: There is a difficulty in learning concerning technology like ambient intelligence (AmI) in the classroom, but support of AmI will enhance the educational experience. Institutions involved in education can utilize software to ensure students' progress on assignments as well as how frequently they attend important events. AmI techniques have many opportunities to carefully analyze and classify the learners' qualities as well as learning demands as a result of the novel pedagogical as well as interactional approaches used by learners and indeed the e-learning platforms that enable them [3].

5. Emergency services: Fire departments and some other safety-related services should enhance the response to a concern by many more accurately locating the scene of an accident and by planning any path to get there in conjunction with street services. The e-road program's image processing and traffic monitoring can be used to achieve this. Additionally, this tool may immediately identify a location where such a risk has been present or imminent and set up improved access and security staff to a certain location [8].

4   Pervasive Computing

Pervasive computing, furthermore referred to as interconnected devices, is the growing trend of integrating computing power in terms of microchips into common objects because even though that they can effectively communicate with one another and carry out useful tasks while minimizing the need for end consumers to engage with machines as computers [9]. Technology for ubiquitous computing is continually active and connected to the network. In contrast to conventional computing, ubiquitous computing can occur with almost any device, at anytime, anywhere, and in any standard format across such a networking. In the near future, ubiquitous computing the next stage of personal computing will surely improve our workplace communication and interaction practices [10]. The pervasive computing, compact, portable personal secretarial devices including high speed, wireless connectivity, decreased real-time energy rate, cloud computing with everlasting memory, coin-sized disk device, small color display video, as well as voice processing technologies, will be available. Users would be able to communicate and access the data at anytime from anywhere else in the world thanks to most of these capabilities [11]. Three technologies have converged together form ubiquitous computing [12]:

- Compact, powerful, as well as energy-efficient devices as well as displays are produced by microelectronic technology.
- Universal roaming and enhanced bandwidth as well as data transfer rates are also benefits of modern telecommunication technology.
- In order to include a foundation for connecting various components into an inter-connected power system comprising security, service, and billing systems, the Internet has been regulated by numerous management system standards as well as industry.

Homogeneous environment featuring complete Internet connectivity is delivered through pervasive computing. A variety of techniques, including Internet connec-tivity, speech recognition, networking, artificial intelligence, and wireless computing, provide pervasive computing. Daily computer operations are incredibly approach-able because to pervasive computing devices. Numerous possible implementations for pervasive computing exist ranging from intelligent transport system but also geographic tracking to home care and other services. The properties of pervasive computing are enlisted below in Table 2.

4.1 Applications of Pervasive Computing in Today's World

Pervasive computing has the potential to significantly alter a variety of contexts in our life. Current pervasive computing implementations are highlighted below [14]:

1. Retail: There has been a need in the retail industry toward quicker and less expen-sive method of delivering goods from stores to customers online. Computers are now used by individuals to make purchases. The list of items to be purchased can then be completed on these devices, registered to the store, and indeed, the transaction can be mailed to that same customer.
2. Airlines booking and check-in: Carriers are interested in targeting bags of clothing from verification to baggage claim thanks to tagging that are attached toward each travel case and recognized at various points all across the journey.
3. Sales force automation: Portable computers are necessary for workforces to access and analyze data while traveling. They can now access business information while on the go thanks to the availability of portable modems.
4. Health care: In healthcare provision, a sizable emergence of digital devices, sensors, and agents are already in use. Access to diagnostic results, surgical reports, procurement procedures, as well as medical directory searches can all be

Table 2 Pervasive computing in today's world [13]

Accessibility	Members can view information with ubiquitous computing systems from any location, at whichever time
Immediacy	A ubiquitous computing system provides quick, useful information at all times
Interactivity	A ubiquitous computing system links among peers, analysts, instructors, students, and coaches even during process of learning
Dynamic connection	That whenever a sensor is momentarily unreliable, it constitutes a lost or unavailable link that needs to somehow be established as soon as it is time after time easily accessible
Adaptability and personalization	A context-aware system should be able and customizable One of the most crucial elements of learning purposes is this omnipresent computing system that gathers the right thing at the right moment everywhere
Privacy and security	Technologies that are context-aware are particularly private and secure. Proactiveness: the context-aware method detects behavior and anticipates what the user intends to really do
Presentation delivering	The precise services as well as information offered to both the end user
Programmed delivery	A introduction of improved independently
Marking	Wording showing vendors with outside data collection along with data

expanded [6]. It would be necessary for administrative essential nursing statistics to be exchangeable and accessible when required, but mostly in specific circumstances.

5. Tracking: Numerous industries have been converted through use of barcodes. Only a small number of participants can be encoded by a one-dimensional barcode.

5 Current Challenges for Ambient Intelligence and Pervasive Computing

- New distributed gadgets and services would have to be incredibly easy to use and install because professional administrators are really not typically found in home situations. They would not need to be configured, updated, or retired by professional programmers. Users must therefore be given the ability to control the settings and activities of respective home environments, while somehow allowing for some degree of "autonomy" throughout the form of self-configuration and self-adaptation of those ecosystems [15].
- It is necessary that AmI systems are informed of the preferences, intentions, and needs of the user. AmI technologies should be aware of when it is more practical

to withhold from providing a recommendation, and that it is more practical to interrupting a user, and when it is both. At times, taking action may be necessary to save a life or stop an accident. If the user becomes weary of the system's assistance and are decided to stop paying attention, the systems may be insufficient or even rendered worthless. The social skills that people acquire across their lives are not easy to master [16, 17].

- The major problem of ubiquitous computing is to provide a framework capable of responding to everyday life situations while still providing privacy to each and every single user in an extremely dynamic pervasive environment. Pervasive computing delivers applications of ambient administrations that enable people and gadgets in many physical places to communicate organically [18, 19].
- The fact that pervasive computing is seriously unsafe is one of its main problems. Traditional data security is not very well matched to the hardware and software used in ubiquitous computing. This is because to the random way in which they join the omnipresent network [20]. As a result, trust mechanisms need to be built with improved security in mind. Numerous other disadvantages of ubiquitous computing included frequently downed lines, slow connections, severe financial requirements, host efficient utilization of the few resources available, and destination data. All of these conditions expose various system weaknesses, putting the security of pervasive computing at risk [21].

6 Conclusion

The current scenario and research potential in said fields using bibliometric analysis, numerous applications in real world for said technologies, and current challenges in underlying fields are envisaged in this paper. Bibliometric analysis conducted in this paper which unfolds the trend, i.e., user interest in the field of ambient intelligence and pervasive computing technologies according to various parameters considered such as country-wise and year-wise publications. Around the globe, these fields are gaining attentions. The scope of the keywords with their respective keywords and citations of the respective documents is analyzed and results that are achieved are presented with figures, which are showing the continuous increase in the publications and research trends in the utilizations of respective fields. In recent years, the scope of ambient intelligence and pervasive computing has expanded tremendously, with almost all branches of software research and practice strongly feeling its impact. In the context of this paper, increasing use of ambient intelligence as well as pervasive computing environments and applications has been determined, and study presented will give the most important suggestions for the future researchers in developing and employing enhanced applications. We are aware that the goals put forth for AmI and pervasive systems are not easily achievable, but the field is gaining momentum at fast pace. So, looking at current growth in future, we need the researchers who can build effective systems utilizing the properties of above-said technologies for implementation in numerous sectors which will be the need of time in nearby future.

References

1. Pantoja CE, Soares HD, Viterbo J, Seghrouchni A E-F (2018) An architecture for the development of ambient intelligence systems managed by embedded agents. In: SEKE, pp 214–215
2. AbdulSattar K, Al-Omary A (2020) Pervasive computing paradigm: a survey. In: 2020 International conference on data analytics for business and industry: way towards a sustainable economy (ICDABI). IEEE, pp 1–5
3. Gams M, Gu I Y-H, Härmä A, Muñoz A, Tam V (2019) Artificial intelligence and ambient intelligence. J Ambient Intell Smart Env 11(1):71–86. https://doi.org/10.3233/AIS-180508
4. Muskan, Singh G, Singh J, Prabha C (2022) Data visualization and its key funda- mentals: a comprehensive survey. In: 2022 7th International conference on communication and electronics systems (ICCES). IEEE, pp 1710–1714
5. Dobre C, Mavromoustakis CX, Garcia NM, Goleva RI, Mastorakis G (2016) Ambient assisted living and enhanced living environments: principles, technologies and control. Butterworth Heinemann
6. Rakotonirainy A, Tay R (2004) In-vehicle ambient intelligent transport systems (I-VAITS): towards an integrated research. In: Proceedings of the 7th international IEEE conference on intelligent transportation systems (IEEE Cat. No.04TH8749). IEEE, pp 648–651
7. Letchner J, Krumm J, Horvitz E (2006) Trip router with individualized preferences (TRIP): incorporating personalization into route planning. In: AAAI, pp 1795–1800
8. Dashtinezhad S, Nadeem T, Dorohonceanu B, Borcea C, Kang P, Iftode L (2004) TrafficView: a driver assistant device for traffic monitoring based on car-to-car communication. In: 2004 IEEE 59th vehicular technology conference. VTC 2004-Spring (IEEE Cat. No.04CH37514), vol 5. IEEE, pp 2946–2950
9. Devi A, Rathee G, Saini H (2022) Secure information transmission in intelligent transportation systems using blockchain technique. In: Intelligent cyber-physical systems for autonomous transportation. Internet of Things. Springer, pp 257–266. https://doi.org/10.1007/978-3-030-92054-8_15
10. Papageorgiou N, Apostolou D, Verginadis Y, Tsagkaropoulos A, Mentzas G (2018) A situation detection mechanism for pervasive computing infrastructures. In: 2018 9th International conference on information, intelligence, systems and applications (IISA). IEEE, pp 1–8
11. Xu W, Xin Y, Lu G (2007) A system architecture for pervasive computing. In: Third international conference on natural computation (ICNC 2007), vol 5. IEEE, pp 772–776
12. Henricksen K, Indulska J, Rakotonirainy A (2002) Modeling context information in pervasive computing systems. In: International conference on pervasive computing. Lecture Notes in Computer Science, vol 2414. Springer, pp 167–180
13. Satyanarayanan M (2001) Pervasive computing: vision and challenges. IEEE Pers Commun 8(4):10–17
14. Hansmann U, Merk L, Nicklous MS, Stober T (2013) Pervasive computing handbook. Springer Science & Business Media, Springer-Verlag Berlin Heidelberg
15. Becker C, Julien C, Lalanda P, Zambonelli F (2019) Pervasive computing middleware: current trends and emerging challenges. CCF Trans Pervasive Comp Interact 1:10–23. https://doi.org/10.1007/s42486-019-00005-2
16. Silva-Rodríguez, Nava-Muñoz SE, Castro LA, Martínez-Pérez FE, Pérez-González HG, Torres-Reyes F (2021) Predicting interaction design patterns for designing explicit interactions in ambient intelligence systems: a case study. Pers Ubiquit Comput 26:1–12
17. Singh J, Singh G, Verma A (2022) The anatomy of big data: concepts, principles and challenges. In: 2022 8th International conference on advanced computing and communication systems (ICACCS). IEEE, pp 986–990
18. Shaheed SM, Abbas J, Shabbir A, Khalid F (2015) Solving the challenges of pervasive computing. J Comput Commun 3:41–50

19. Singh J, Singh G, Bhati BS (2022) The implication of data lake in enterprises: a deeper analytics. In: 2022 8th International conference on advanced computing and communication systems (ICACCS). IEEE, pp 530–534
20. Singh J, Bajaj R, Kumar A (2021) Scaling down power utilization with optimal virtual machine placement scheme for cloud data center resources: a performance evaluation. In: 2021 2nd Global conference for advancement in technology (GCAT). IEEE, pp 1–6
21. Singh J, Duhan B, Gupta D, Sharma N (2020) Cloud resource management optimization: taxonomy and research challenges. In: 2020 8th International conference on reliability, infocom technologies and optimization (trends and future directions) (ICRITO). IEEE, pp 1133–1138

Improving the AODV Routing Protocol Against Network Layer Attacks Using AntNet Algorithm in VANETs

Rand S. Majeed, Mohammed A. Abdala, and Dulfiqar A. Alwahab

Abstract Vehicular Ad Hoc Network (VANET) is the ad hoc wireless technology which uses vehicle nodes connected with each other's to form a network. The basic function of this network is providing a connection between vehicles to enable the transmission of data. Security is an important issue that must be provided by VANETs. Moreover, one of the most challenges of this network is their vulnerability of security since it has a large number with high speed of vehicles which may lead it to lay under different types of security attacks. Attacks of the network layer can be classified based on routing operation phase. One of the most widely used routing protocols is Ad Hoc on Demands Distance Vector (AODV). AODV is a reactive routing protocol, which is established to make connection between nodes only as required. This research aims to eliminate the effect of attacks and improve the performance of AODV in VANET. Three types of network layer attacks were used to examine their effect on VANET (blackhole, flooding, and rushing attacks). In addition to proposing the combination of more than one type of routing attacks at the same time to show their effect on AODV in VANET, we also proposed the use of AntNet algorithm to eliminate the effect of attacks. The simulation results show that the blackhole attack has a higher effect on AODV compared with the flooding and rushing attacks. The proposed combination of attacks on network showed a high damage on its performance compared with using individual network attacks on AODV in VANET. Simulation results also show various AntNet algorithm efficiencies in the improvement of VANET performance under network attacks. They show in this research that the AntNet has higher efficiency in reducing blackhole attack effect with 29.83%, then, with rushing attack,

R. S. Majeed
Faculty of Computer Engineering, Al-Esraa University College, Baghdad, Iraq
e-mail: rand@esraa.edu.iq

M. A. Abdala
Department of Medical Instrumentation Techniques Engineering, AL-Hussain University College, Karbala, Iraq
e-mail: dr.m.ahmed@huciraq.edu.iq

D. A. Alwahab (✉)
Faculty of Informatics, Eötvös Loránd University, Budapest, Hungary
e-mail: aalwahab@inf.elte.hu

© The Author(s), under exclusive license to Springer Nature Singapore Pte Ltd. 2023   271
Y. Singh et al. (eds.), *Proceedings of International Conference on Recent Innovations in Computing*, Lecture Notes in Electrical Engineering 1011,
https://doi.org/10.1007/978-981-99-0601-7_21

the percentage of enhancement is 16.82%, and with flooding attack, AntNet has the lowest enhancement efficiency with value of 9.04%.

Keywords VANET · AODV · Blackhole attack · Rushing attack · Flooding attack · Hybrid attacks · AntNet algorithm

1 Introduction

Because of the latest controlling packages of software's and automatic electronic innovation, communication networks become one of the wide domain services in all fields used by human. The communication journey started from telephone landline to communication entity like Internet of Things (IoT). As communication of real-world objects is started, it is motivated to find solutions for traffic managements using vehicular networks. In these types of networks, vehicles are using wireless connection type to share information with each other's [1].

VANET is a technology that creates a mobile network using moving cars. It converts every car into a wireless router or node, allowing them to be connected approximately 100 up to 300 m with each other to create a network with a wide range [2]. The vehicular network architecture categories can be divided into (as shown in Fig. 1):

- **Vehicle-to-vehicle communication (V2V)**: This type of communication allows vehicles to communicate effectively with each other's [3].
- **Vehicle-to-infrastructure communication (V2I)**: It allows vehicles to communicate with infrastructures laid on the road [3].
- **Roadside units (RSUs) to roadside unites (R2R)**: It allows the RSUs to connect with the other RSUs with wired or wireless medium [1].

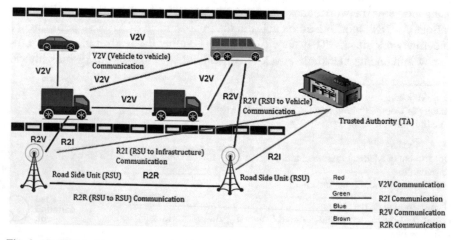

Fig. 1 VANET architecture categories [1]

- **RSUs to trusted authority (TA)**: TA also transfers some traffic-related policies or emergency messages to all RSUs [1].
- **TA to TA**: This communication type between TAs is used for sharing common information between them [1].

1.1 VANET Security Challenges

Security is the most serious issue that impacts the performance of VANET. Due to scale of network and mobility of node, it is vulnerable to different types of attacks and also may lead VANET to be vulnerable to jamming [4]. Due to sensitivity nature of data being transmitted by VANET, there is need for designing applications to be protected from malicious nodes' manipulation and losing. In network security, attacks can be divided into three main classifications of threats; threats deal with Authenticity, Confidentiality, and Availability of resources [5].

Attacks may affect the vehicular networks in different situations, and some of them may raise the network delay. Others may cause congestion of network and create routing loops. Other types of attacks may prevent the sending vehicle node from finding correct route to destination [6]. Three types of attacks at the network layer are tested in this research, these are:

- **Blackhole attack**: This type of attack shows itself as having a shortest path to the received node that wants to attack its packets. The intermediate nodes in the vehicular network advertise its availability of fresh routes and always have the ability to reply the route requests and take the data packets [7]. Figure 2 shows blackhole attacker node function on ad hoc network. Source node S wants to send packets to the receiver node D. Firstly, it will send a Route Request (RREQ) message to all neighbor nodes (nodes A, B, and C). Assuming node C has the best route to node S, it replies with Route Reply (RREP) message firstly to the node S. Before this, the blackhole node M sends a false RREP to node A with form of having a high sequence number than the destination node, so node S sends the packets among node A, assuming that it has the shortest path to the receiver node. After that, the attacker node M will receive the packets and dropped them [8].
- **Flooding attack**: The main goal of this type of attack is to spend the resources of network like bandwidth by broadcasting fake packets to non-existence nodes in the network and make it go down, so that users become unable to access the network [9]. The flooding attack can be classified into: data flooding, RREQ flooding, and synchronization flooding. The RREQ flooding type is used in this research. The attacker vehicle node floods RREQ packets to exist or not exist nodes in the network. It will fill the routing table of neighbor nodes with RREQs. So, only few numbers of data packets will be able to reach to destination. It will spend the bandwidth and resources of the network. The high mobility of ad hoc network makes it difficult to recognize this attack in network. Figure 3 shows the RREQ flooding attack, which was used in this research [10].

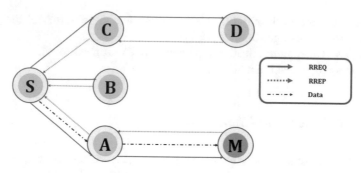

Fig. 2 Blackhole node function in ad hoc network [8]

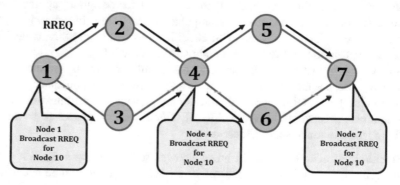

Fig. 3 RREQ flooding attack [8]

- **Rushing attack**: This type of attack (also known as denial of service or novel attack) behaves as denial-of-service attack with all currently proposed ad hoc routing protocols in network (e.g., AODV). Usually, the source node finds a suitable route by looking for route cache of previously learned routes. In this attack, the attacker node exploits property of the operation of route discovery [11] as shown in Fig. 4.

1.2 AODV Routing Protocol

AODV is a reactive routing protocol which establishes to make connection between nodes only as required by sending nodes and it helps them to maintain the routes if needed [13]. AODV protocol algorithm contains the following steps [14]:

- Step 1: Before sending the data, the node firstly checks the route availability by sending Route Request (RREQ) message to find destination. Otherwise, it will broadcast RREQ to find another path by another node.

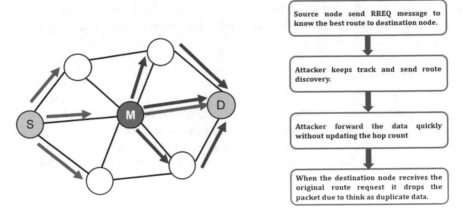

The flowchart contains:

Source node send RREQ message to know the best route to destination node.

Attacker keeps track and send route discovery.

Attacker forward the data quickly without updating the hop count

When the destination node receives the original route request it drops the packet due to think as duplicate data.

Fig. 4 Rushing attack [12]

- Step 2: If the received node is a ready destination node, it will reply by sending an Route Reply (RREP) message to the source node in order to start transmission of data. If not, it will start a search of a local route over neighboring node to find available route to destination.
- Step 3: Finally, the checking of transmission will be done. If it is not successful, the route discovery procedure will be started again by activating route maintenance to find the available transmission path.

1.3 AntNet Algorithm

An AntNet routing algorithm is one of optimization algorithms inspired by the real behavior of ants. It is the first intelligence routing protocol that developed and used for finding the optimal route between the source and target (destination) nodes. There are two types of the agents in the AntNet algorithm: backward ants and forward ants. The forward ant's behavior uses the data collection of topology status from a source node to a group of destination nodes. The backward ants handle the information and updates of the routing table [15]. In an ant colony algorithm, the effect of malicious nodes is eliminated by using an anomaly detection to eliminate the effect of attacks by using a colony of asynchronous ants that move among the problem situation and provide a partial solution of the problems. Attacks' elimination and network performance enhancement with this solution can be achieved in several steps. Firstly, the AODV features are used to discover the initial path. Secondly, if exist, activating the anomaly detection for attacks using AntNet Colony algorithm. Finally, rediscovering path again using the same algorithm [16].

The information and updates of AODV routing table are handled by backward ant. The following are a repetitive process to give better solutions to a problem [15]:

1. Using artificial agents called ants to generate a solution to the given problem.
2. Using the collected information during the past to create better future solutions.

2   Proposed Method

In this research, three types of network layer attacks were used to examine their effect on VANET (blackhole, flooding, and rushing attacks). In addition to proposing the combination of more than one type of routing attacks at the same time to show their effect on AODV. The solution used in this research for eliminating the effect of network attacks was the AntNet Colony Optimization (ACO) algorithm.

The process of VANET network using AntNet solution to blackhole attack includes the following steps as shown in Figs. 5 and 6:

- First, to find destination, the node who wants to transmit data will send RREQ message to all nodes in the network.
- If the blackhole node received the RREQ message, it advertises its ability of fresh route. Otherwise, the procedure to find destination will continue and start data transmission if it is successful.
- If the ACO algorithm is used, it will detect the attacker node and find suitable route to destination by rediscovering suitable path again. Otherwise, the network performance will be bad.
- Finally, the data transmission will be started successfully with attack effect elimination.

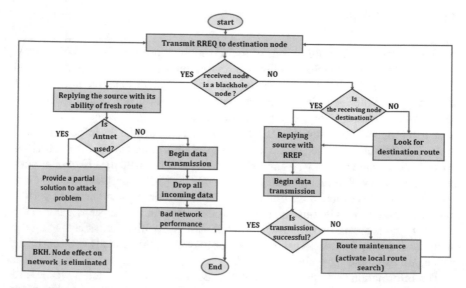

Fig. 5 Flowchart of blackhole with AntNet algorithm

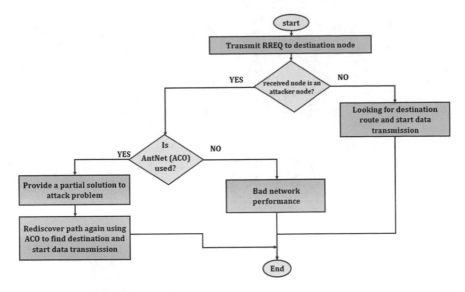

Fig. 6 AntNet algorithm flowchart

For the rushing, flooding attacks, and using hybrid attacks including combination of (blackhole + rushing) and (blackhole + flooding) on the same time on VANET, the same process of eliminating using ACO algorithm is also used to eliminate their effect on AODV as shown in Fig. 7.

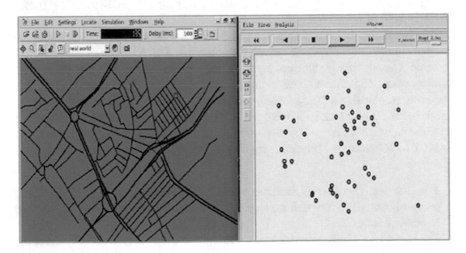

Fig. 7 Integration between SUMO and NS2

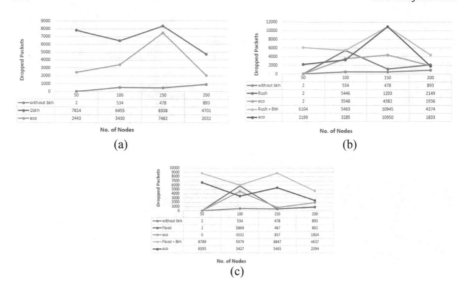

Fig. 8 Total number of dropped packets of VANET for situations: without, with attacks and after implementing the ACO algorithm with speed of cars equals to 30 kmph for **a** the blackhole (Bkh), **b** rushing and using hybrid of (Bkh + rushing at the same time), and **c** flooding and using hybrid of (Bkh + flooding at the same time)

2.1 Simulation Design

Two simulators were jointly used to build the VANET simulation: SUMO [17] 0.25.0 which is used as a traffic simulator to simulate vehicular real-time simulation and NS2 (network simulation) [18] with version 2.35 as a network simulation to configure suitable parameters for VANET network as shown in Fig. 8.

Table 1 shows the parameters used for the simulation. In this simulation, for wireless connection, the wireless channel was chosen; the AODV routing protocol was used; for the signal propagation, omni antenna was used; the number of nodes equals to 50, 100, 150, and 200 were positioned randomly over (2700 * 2700) m^2 of area; the simulation time was 600 s, and the IEEE 802.11p protocol is used as the VANET MAC protocol. The following three sets of scenarios were simulated and analyzed:

- The first set of scenarios includes running the simulation of AODV routing protocol to analyze the performance of the network while changing speeds and number of nodes.
- The second set of scenarios includes running the simulation of AODV under the effect of attacks to see how it affects the performance of VANET.
- The third set of scenarios includes running simulation after implementing the ACO algorithm and then compares the results of all scenarios.

Table 1 Network simulation parameters

Parameters	Value
Type of channel	Wireless channel
Type of antenna	Omnidirectional
Propagation model	Two-ray ground
Packet length	512 bytes
Traffic type	UDP
Simulation time	600 s
Routing protocols	AODV
Type of channel	Wireless channel
No. of nodes	50, 100, 150, 200
Speeds (kmph)	30, 80
Simulation area (m^2)	2700 * 2700

3   Results

In this research, the total dropped packets and throughput metrics were measured to study and analyze the overall performance of VANET network under the effect of flooding, blackhole, and rushing attacks with combination between them and how network performance was improved after implementing the proposed ACO algorithm solutions.

The following performance metrics are chosen:

- Total number of dropped packets is the difference between sent and received packets.
- Average throughput [19] is the rate of successful packet delivery over a communication channel. In Eq. (1), the number 8 indicates the packets received converted to bits. It is usually measured in bps, Kbps, or Mbps.

$$Avg\ Thr = (Pr * 8)/(Tr - Ts), \tag{1}$$

where Avg Thr is average throughput; Pr is number of packets received; Tr is the time of last received packet; and Ts is the time of the first packet sent.

Figures 8 and 9 show the results of total dropped packets and average throughput performance of VANET with vehicles' speed equal to 30 kmph.

Figures 10 and 11 show the results of total dropped packets and average throughput performance of VANET with vehicles speed equal to 80 kmph.

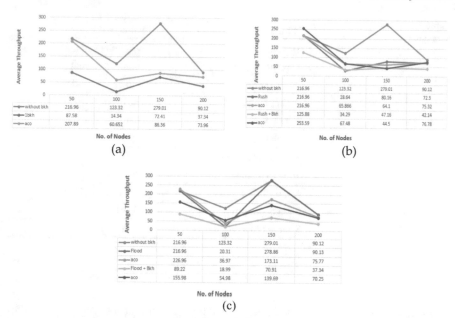

Fig. 9 Average throughput of VANET for situations: without, with attacks and after implementing the ACO algorithm with speed of cars equals to 30 kmph for **a** the blackhole (Bkh), **b** rushing and using hybrid of (Bkh + rushing at the same time), and **c** flooding and using hybrid of (Bkh + flooding at the same time)

4 Discussion

After computing and analyzing the results of the total number of dropped packets and average throughput for 50–150 nodes with all situations implemented, a comparison between results is done and discussed to show how the network security attacks affect the performance of AODV in VANET and how the implementation of AntNet algorithm eliminated the effect of these attacks. Table 2 shows the percentage of effected values of total number of dropped packets without and with attacks and after implementing ACO algorithm nodes as follows:

- The percentage of effected values for total dropped packets with blackhole attack is highly compared with flooding and rushing attacks.
- Increasing speed of nodes from 30 to 80 kmph leads to increase the percentage of effected values of total dropped packets.
- The hybrid of rushing with blackhole attack at the same time values is few compared with using only blackhole attack.
- ACO percentage values with blackhole are few compared with other attacks; for example, with the number of nodes 50 and speed of 30 kmph, the effect of blackhole attack is decreased from 53.86% into 16.84%.

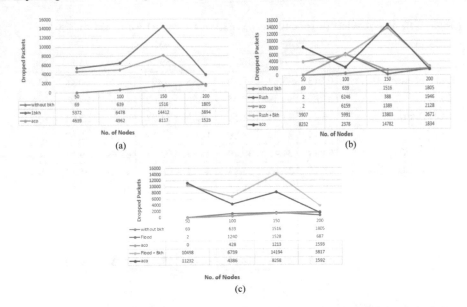

Fig. 10 Total number of dropped packets of VANET for situations: without, with attacks and after implementing the ACO algorithm with speed of cars equals to 80 kmph for **a** the blackhole (Bkh), **b** rushing and using hybrid of (Bkh + rushing at the same time), and **c** flooding and using hybrid of (Bkh + flooding at the same time)

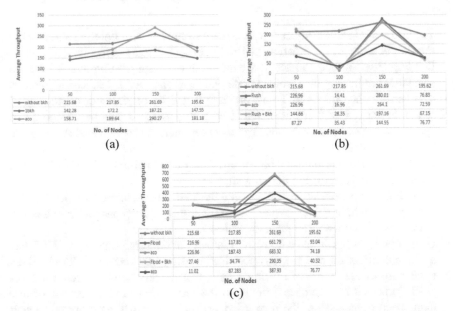

Fig. 11 Average throughput of VANET for situations: without, with attacks and after implementing the ACO algorithm with speed of cars equals to 80 kmph for **a** the blackhole (Bkh), **b** rushing and using hybrid of (Bkh + rushing at the same time), and **c** flooding and using hybrid of (Bkh + flooding at the same time)

Table 2 Percentage values (%) of total dropped packets for VANET with attacks and ACO

Network status	30 kmph				80 kmph			
	50	100	150	200	50	100	150	200
Without attack	0.013	3.67	3.29	6.15	0.47	4.4	10.44	12.44
Bkh	53.85	44.48	57.45	32.39	37.02	44.6	99.3	26.83
ACO (Bkh)	16.84	23.64	51.56	14.00	31.97	34.2	55.94	10.5
Rushing	0.014	37.53	8.29	14.80	0.014	43.04	2.67	13.41
ACO (rushing)	0.014	24.45	30.19	13.34	0.014	42.44	9.57	14.66
Bkh + rushing	42.06	37.65	75.43	30.14	26.92	41.28	95.12	18.41
ACO (Bkh + rushing)	15.15	22.64	75.46	12.63	56.73	16.38	98.42	12.64
Flooding	0.014	40.45	3.35	6.15	0.014	8.55	10.53	4.73
ACO (flooding)	0	31.23	5.76	13.12	0	2.95	8.36	10.97
Bkh + flooding	60.57	41.20	60.96	31.95	72.35	46.44	97.82	26.30
ACO (Bkh + flooding)	45.43	23.61	37.66	16.49	77.40	30.22	56.90	10.97

5 Conclusions

After implementing the AntNet algorithm on AODV in VANET, computing and comparing the results of all situations used in this research, the following were concluded:

- The effect of blackhole attack is very highly compared with flooding and rushing attacks, because of its function that it hacks and deletes along its path, while flooding attack affects the network bandwidth only and rushing attack function specializes in finding the received node and starts packets' transmission quickly.
- The performance of VANET network is affected by the number of nodes and speed. In this research, it was seen that with increasing the number of nodes and speed, the number of totals dropped packets was increased which leads to decrease throughput values.
- ACO algorithm solution has a higher efficiency in reducing the blackhole attack compared with rushing and flooding attacks.
- There is an opposite relationship between the total number of dropped packets and average throughput values. When the total number of drop packets is increased, the values of average throughput are reduced.
- Using hybrid of blackhole with rushing attacks at the same time on VANET has a less bad effect on network compared with using blackhole attack only. This is because of the nature of rushing attack, that quickly forwards route request (RREQ) packet to gain access firstly to the forwarding group. Because of this high-speed transmission, the forwarded packet by the rushing node will reach firstly to the destination node. The destination will accept rushing RREQ and discard it from the other nodes. In this way, rushing node will successfully access in the communication between sender and receiver nodes faster than blackhole

attacker node. This will lead to decrease its effect on VANET and hence decrease the number of total dropped packets which is high with blackhole attack.

- Implementing the hybrid of (Bkh + flooding) has a higher bad effect on VANET than (Bkh + rushing), because both of blackhole and flooding attacks aim to increase the values of drop packets while the goal of rushing attack is to gain access the communication between source and destination nodes.
- The speed and number of nodes in the network affects ACO algorithm solution. The efficiency of algorithm is reduced with increasing those parameters.

References

1. Botkar SP, Godse SP, Mahalle PN, Shinde GR (2021) VANET challenges and opportunities, 1st edn. CRC Press, U.S
2. Kaur M, Kaur S, Singh G (2012) Vehicular Ad-hoc networks. J Global Res Comput Sci 3(3):61–64
3. Qays R (2015) Simulation and performance enhancement of VANET routing protocols. Al-Nahrain University, Baghdad, Iraq, College of Information Engineering
4. Xiao L, Zhuang W, Zhou S, Chen C (2019) Learning-based VANET communication and security techniques, 2nd edn. Springer Nature, Switzerland AG
5. Rehman S, Arif Khan M, Zia TA, Zheng L (2013) Vehicular Ad-hoc networks (VANETs)—an overview and challenges. J Wirel Networking Commun 3(3):29–38
6. Qureshi K, Abdullah H, Mirza A, Anwar R (2015) Geographical forwarding methods in vehicular Ad-hoc networks. Int J Electr Comput Eng (IJECE) 5(6):1407–1416
7. Salih R, Abdala M (2017) Blackhole attack effect elimination in VANET networks using IDS-AODV, RAODV and AntNet algorithm. J Telecommun 36(1):1–5
8. Salih R (2017) Development of security schemes for VANETs. Al-Nahrain University, Baghdad, Iraq, College of Information Engineering
9. Jindal S, Maini R (2014) An efficient technique for detection of flooding and jamming attacks in wireless sensor networks. Int J Comput Appl (IJCA) 98(10):25–33
10. Rani S, Narwal B, Mohapatra AK (2018) RREQ flood attack and its mitigation in Ad-hoc network. Springer Nature Singapore Pte Ltd, pp 599–607
11. Kumar S, Sahoo B. Effect of rushing attack on DSR in wireless mobile Ad-hoc network. Department of Computer Science & Engineering, NIT Rourkela, Orissa
12. Mishra P, Sharma S, Bairwa A (2019) Mitigating the effect of rushing attack on AODV routing protocol in mobile Ad-hoc network. Int J Res Anal Rev (IJRAR) 6(1):1154–1170
13. Kaur R, Rana S (2015) Overview on routing protocols in VANET. Int Res J Eng Technol (IRJET) 2(3):1333–1337
14. Abdullah A, Aziz R (2014) The impact of reactive routing protocols for transferring multimedia data over MANETs. J Zankoy Sulaimani 16(4):9–24
15. Sudhakar T, Inbarani H (2019) Improving the performance of AntNet protocol using performance measures modeling in mobile Ad-hoc network. Int J Recent Technol Eng (IJRTE) 8(4):733–737
16. Pal S, Ramachandran K, Dhanasekaran S, Paul ID, Amritha (2014) A review on anomaly detection in manet using AntNet algorithm. Middle-East J Sci Res 22(5):690–697
17. Dlr: http://www.dlr.de/ts/en/desktopdefault.aspx/tabid-9883/16931_read-41000/
18. ISI: http://www.isi.edu/nsnam/ns/
19. Hota L, Nayak B, Kumar A, Sahoo B, Ali GN (2022) A performance analysis of VANETs propagation models and routing protocols. Sustainability 1–20

A Comparative Study of IoT-Based Automated Hydroponic Smart Farms: An Urban Farming Perspective

Swati Jain and Mandeep Kaur

Abstract With the increase in world population and the simultaneous decline in the resources available for farming, meeting the food demands of all the inhabitants is a substantial challenge. Adopting advanced technologies to maximize agricultural yield with the judicious use of available resources is the means to tackle these challenges. Internet of Things (IoT), one such budding technology, has made automation possible in various facets of human lives and changed the way human–machine interaction takes place. IoT is a network of inter-related things which have the ability to transfer data within each other and the surrounding environment. Integration of IoT technology with modern farming facilitates various cost-effective intelligent farming methods such as indoor farming and precision agriculture which, subsequently, leads to a reduction in resource wastage and optimal utilization of resources such as land, water, labor and fertilizers. This paper aims to review the most recent papers in the IoT-assisted smart farming domain. The intention is to provide a comparative analysis of all the recent developments in the area of IoT-based hydroponic smart farms to future researchers.

Keywords Internet of things (IoT) · Smart farming · Precision agriculture · Indoor farms · Hydroponics

1 Introduction

Farming is the eldest means of support in the history of mankind. With global population explosion, high urbanization rate, climate change and resource crisis, the gap between demand and supply of food is widening at a significant and alarming rate. Total arable land area used for agriculture has reduced from 39.47% in 1991 to 37. 73% in 2015 [1]. This has put the agriculture industry under immense pressure to produce more food even when the available resources are declining. Luckily,

S. Jain (✉) · M. Kaur
School of Engineering and Technology, Department of Computer Science and Engineering,
Sharda University, Greater Noida, India
e-mail: jainswati3107@gmail.com

© The Author(s), under exclusive license to Springer Nature Singapore Pte Ltd. 2023   285
Y. Singh et al. (eds.), *Proceedings of International Conference on Recent Innovations in Computing*, Lecture Notes in Electrical Engineering 1011,
https://doi.org/10.1007/978-981-99-0601-7_22

modern-age technologies have introduced new ways to meet challenges of tradi-tional farming. IoT-based smart farms adapt the use of sensors, precision agriculture and take predictive measures against the changes in farm [2]. With the automation of multiple aspects of farming, production of higher quantities of food with less use of resources is possible even in the areas with climate conditions not suitable for farming. To achieve these automations, researches have integrated IoT with the farming methods to create different monitoring systems for the farm with varied levels of automation [3, 4].

In this paper, systematic literature review (SLR) technique is applied to review the recent developments in the field of IoT-based smart hydroponic systems from multiple perspectives—IoT components (i.e., sensors and actuators), type of automation, level of automation, farm type (i.e., indoor or outdoor). The rest of the work is organized as follows. Related work is presented in Sect. 2 followed by the review article selection methodology in Sect. 3. In Sect. 4, existing smart farms are categorized and studied based on the kind of automations implemented, and in Sect. 5, the comparative study and analysis of the recent work on the above-said parameters are discussed followed by the summarization in a tabular form. Study conclusion, additional developments and future scope are discussed in Sects. 6 and 7 of the paper.

2 Related Work

Related surveys by Srilakshmi et al. [5], Elijah et al. [6], Farooq et al. [7], Raneesha et al. [8] and Farooq et al. [9] have been published in the past few years in the use of IoT in the agriculture domain.

Srilakshmi et al. [5] analyzed the three major implementations of IoT in smart agriculture as irrigation system automation using Smart Irrigation Decision Support System (SIDSS), use of IoT in detecting Nitrate levels in the soil through the use of planar-type interdigital sensors and the use of IoT in precision agriculture to sense all the environmental parameters remotely and maintain them remotely. Elijah et al. [6] identified five broad categories of the applications of IoT in the agriculture sector as: (a) monitoring and maintenance of farm and livestock, (b) tracking and tracing of parameters, (c) IoT-assisted agricultural machinery, (d) IoT-based precision agricul-ture and (e) greenhouse production. The study also discussed already existing IoT solutions for agriculture in the market and compared them based on their features. Farooq et al. [7] discussed three main IoT applications in agriculture as follows: (a) precision farming which includes climate and irrigation monitoring, disease moni-toring, farm management, etc.; (b) livestock management which includes animal health monitoring, GPS-based monitoring, heat stress level, etc.; and (c) greenhouse monitoring which includes weather monitoring, water management, plant monitoring and agricultural drones. In this article, the authors also identified over 60 smartphone applications for agricultural practices along with their features. Raneesha et al. [8] reviewed 60 scientific publications of recent IoT applications in the farming sector based on their sensors, used technology and IoT sub-vertical. The study identified that water and crop management are the sub-verticals with the highest ratios and livestock and irrigation management with the lowest ratios for making the use of IoT. Environment temperature was the most commonly used sensor, and Wi-Fi is being the most used communication technology. Farooq et al. [7] conducted a review of research work published during the years 2015–2020 on the IoT-based farming domain. The study reviewed existing literature of IoT frameworks for water management, intelligent irrigation systems, crop monitoring, pest control, etc., by highlighting their scope and methodology along with a comparison of their architecture and pros and cons.

3   Review Selection Methodology

The aim of this study is to properly recognize current advancements of IoT for the automation of smart farms, and the study has been done carefully by examining, comparing and analyzing the published work in the domain. This comprehensive study acts as a resource to support the understanding of the basics of IoT in smart agriculture in a very precise way and helps future researchers to conduct further research in the domain. A thorough and systematic review of the existing literature is conducted to meet the aim of this study.

To identify vital literature, following keywords were searched in the scholarly databases: "Smart Farming" OR "IoT Farming" OR "Precision Agriculture" OR "IoT in Hydroponics" OR "Indoor Farming" OR "Automated Indoor Farms" OR "Automated Irrigation" OR "Automated Nutrient Control".

In total, 64 papers were identified through the keyword search. After identifying papers, the first level of paper selection was on the basis of title, abstract and text, and then, the selected papers were shortlisted again based on the type of indoor farm, technology used, type of automation implemented. With this two-step shortlisting procedure, 38 were selected for this study.

4   IoT-Based Smart Hydroponic Farms

Based on the existing research work, three broad categories of IoT-based smart hydroponic farms have been formulated and discussed in this paper, namely, IoT-based farm climate monitoring and control systems, IoT-based automated irrigation systems and IoT-based nutrient control systems.

Jaiswal et al. [10] designed a prototype for a fully automated greenhouse for hydroponics and vertical farming with security provisions using machine learning. The greenhouse farm is equipped with sensors to measure the parameters which are sent to the cloud for decision-making. When the temperature, humidity and light intensity reach their threshold values, automatic actions are taken by the system to control the parameters. It is equipped with the security feature through facial recognition, and livestream surveillance is available to users at all times.

Dholu et al. [11] proposed to implement precision agriculture using cloud-based IoT application by sensing all the required parameters and controlling the actuations. The proposed system contains sensors which collect data of soil moisture level, temperature and humidity of the farm and send it to the microcontroller unit which forwards the data to the ThingSpeak platform for data analysis. Users can see the current sensor reading through a mobile application which fetches data from the cloud.

Bhojwani et al. [12] proposed an IoT-based model to monitor and analyze different parameters affecting the growth and production of crops in an agricultural field. Sensors are used to send live weather and soil condition data to the cloud server where they are analyzed and presented to the user in the form of graphs. Proposed model helps the farmers in deciding the ideal crop for farming by analyzing the weather and soil conditions of the field.

Doshi et al. [13] developed an IoT-based plug and sense, portable prototype powered by a power bank notification system to send farm updates to remote farmers. Greenhouse was equipped with the prototype, and ESP32s takes readings from the sensor, sends it to the cloud for notification and then goes to sleep for 18 min. LEDs are also used to notify the farmers.

Herman et al. [14] proposed a monitoring and controlling system for hydroponics based on IoT and fuzzy logic to control the parameters for precision agriculture. Data from sensors will be read by the Arduino Uno, and with the application of fuzzy logic, a decision will be made based on the range of values determined for the parameter. Output of the fuzzy logic will determine the time duration to turn on the tap valve for pH tube, nutrition water and open/close para net curtain.

Palande et al. [15] built a cost-effective and completely automated hydroponic system which requires no human interaction for the control of parameters of plant growth. The system consists of two Arduino nodes and Raspberry Pi as the main controller. Nodes collect the data from the sensors and send it to a gateway connected to Raspberry Pi. It makes all the data available on the web interface which can also take user inputs. The system controls the growth parameters automatically and informs the user of any abnormal data from sensors.

Aliac et al. [16] designed an integrated IoT system for surveillance and maintenance of the hydroponic garden with the aim to provide the perfect habitat for the plants to grow. Raspberry Pi controls the data received from the sensors in real time and uploads into the firebase database and takes necessary actions based on the commands received in return to maintain the ideal growth conditions and control the irrigation and nutrient solution intake for the crops. Fully functional automated system was built where a web application monitored and controlled the drainage system, fan, sprinkler and water pump of the farm and a warning system was built to notify users for the farm conditions, and the recommended conditions for crop types are displayed on the application.

Nurhasan et al. [17] implemented an automated water level monitor and control using fuzzy Sugeno and website application. The system obtains data from the sensors and stores it in a database contained in MCU. Ignition of water pumps is controlled by the output of fuzzy based on current data received from sensors. The system automatically monitors and controls the water level of DFT at all times using fuzzy to control the relay switch of the water pump. Trial tests were conducted on automated and manual systems, and automated systems performed better in terms of height and number of leaves that grow.

5 Comparative Study and Analysis

This section presents a comparative study of 38 IoT-based smart hydroponic farms identified from existing research work (Table 1). These recognized farms are analyzed on the basis of following four criteria (Fig. 1):

(i)   Type of Sensing is implemented which is further categorized into Weather Monitoring achieved through temperature, humidity, light and CO_2 sensors, Irrigation and Nutrient Monitoring achieved through pH, EC, water temperature, water level and soil moisture sensors.

(ii) Type of Automation control is implemented and then further categorized into Weather Control, Irrigation Control and Nutrient Control.

(iii) Level of Automation implemented is whether full automation or semi-automation.

(iv) Type of user interface available is whether mobile application or web application.

From the comparison of sub-vertical automations implemented in the recent 38 studies (Table 1), it is identified that irrigation automation, weather monitoring and control and nutrient feeding were achieved in 53%, 47% and 21% of the IoT smart systems, respectively (Fig. 2a), while there was no system which implemented all three automations in a single system.

Second observation from the study was that for all the automations implemented, only 45% of them were full automation, while the remaining 55% were semi-automation needing human intervention in some form (Fig. 2b). Lastly, only 50% of the systems were equipped with either a mobile or a web application to monitor and control the system remotely without the need of physical presence.

6 Conclusion

IoT empowers hydroponics and makes urban farming and precision agriculture attainable. With the implementation of IoT, the precision agriculture aspect of hydroponics can be automated. This will lead to intelligent use of available resources and minimize the resource wastage associated with the human intervention in the farm. The

Table 1 Comparative study of functionality of 38 hydroponic and IT smart systems

References	Sensing									Automation			Automation level	Mobile/web application
	Weather (air)				Irrigation and nutrients (water)					Weather	Irrigation	Nutrient		
	Temperature	Humidity	Light	CO$_2$	pH	EC	Water temperature	Water level	Soil moisture					
Aliac and Maravillas [16]	✓	✓	✗	✗	✓	✗	✗	✓	✗	✓	✓	✗	Full	✓
Ani and Gopalakirishnan [18]	✓	✓	✗	✗	✓	✓	✓	✗	✗	✗	✓	✗	Full	✓
Bharti et al. [19]	✓	✗	✗	✗	✓	✓	✗	✓	✗	✗	✓	✗	Semi	✗
Bhattacharya et al. [20]	✗	✗	✗	✗	✗	✗	✗	✓	✓	✗	✓	✗	Full	✓
Bhojwani et al. [12]	✓	✗	✓	✓	✗	✗	✗	✓	✓	✗	✓	✗	Semi	✗
Changmai et al. [21]	✓	✓	✓	✗	✓	✓	✗	✗	✗	✓	✗	✓	Full	✓
Cho et al. [22]	✗	✗	✗	✗	✓	✓	✗	✓	✗	✗	✓	✓	Full	✓
Chowdhury et al. [23]	✓	✗	✗	✗	✓	✓	✗	✓	✗	✓	✗	✓	Full	✓
Sen et al. [24]	✗	✗	✗	✗	✗	✗	✗	✗	✓	✗	✓	✗	Full	✓
Dholu and Ghodinde [11]	✓	✓	✗	✗	✗	✗	✗	✗	✓	✓	✗	✗	Semi	✗
Domingues et al. [25]	✗	✗	✗	✗	✓	✓	✓	✗	✗	✗	✗	✓	Full	✓
Doshi et al. [13]	✓	✓	✓	✗	✗	✗	✗	✓	✓	✗	✗	✗	Semi	✗

(continued)

Table 1 (continued)

References	Sensing									Automation			Automation level	Mobile/web application
	Weather (air)				Irrigation and nutrients (water)					Weather	Irrigation	Nutrient		
	Temperature	Humidity	Light	CO$_2$	pH	EC	Water temperature	Water level	Soil moisture					
Dudwadkar, and Asawari [26]	✓	✓	✓	✗	✓	✓	✗	✓	✗	✓	✗	✗	Full	✓
Gori et al. [27]	✗	✗	✗	✗	✓	✗	✗	✗	✓	✗	✓	✗	Semi	✗
Herman and Surantha [14]	✗	✓	✗	✗	✓	✓	✗	✓	✗	✗	✓	✗	Semi	✗
Jaiswal et al. [10]	✓	✓	✗	✗	✗	✗	✗	✗	✓	✓	✓	✗	Semi	✗
Kaburuan et al. [28]	✓	✓	✓	✓	✗	✗	✓	✗	✓	✓	✓	✗	Semi	✗
Lakshmanan et al. [29]	✓	✓	✗	✗	✓	✗	✓	✗	✗	✓	✗	✗	Full	✓
Lakshmiprabha and Govindaraju [30]	✓	✓	✗	✗	✗	✗	✗	✓	✗	✗	✓	✗	Semi	✓
Mahale and Sonavane [31]	✓	✓	✗	✓	✗	✗	✗	✗	✗	✓	✗	✗	Semi	✗
Mahendran et al. [32]	✓	✓	✓	✓	✗	✗	✗	✗	✓	✓	✗	✗	Full	✓

(continued)

Table 1 (continued)

References	Sensing										Automation			Automation level	Mobile/web application
	Weather (air)				Irrigation and nutrients (water)						Weather	Irrigation	Nutrient		
	Temperature	Humidity	Light	CO₂	pH	EC	Water temperature	Water level	Soil moisture						
Mehboob et al. [33]	×	×	×	×	✓	✓	×	×	×	×	×	✓	Semi	×
Mohanraj et al. [34]	✓	✓	✓	×	×	×	×	✓	✓	✓	✓	×	Full	✓
Montoya et al. [35]	×	×	×	×	✓	✓	✓	✓	✓	×	✓	×	Semi	×
Munandar et al. [36]	✓	✓	✓	×	✓	✓	✓	×	×	×	✓	×	Full	✓
Nurhasan et al. [17]	✓	✓	×	×	✓	×	×	✓	×	×	✓	×	Semi	×
Palande et al. [15]	✓	✓	✓	×	✓	✓	×	×	×	✓	×	×	Full	✓
Pallavi et al. [37]	✓	×	✓	✓	×	×	×	×	✓	✓	×	×	Semi	×
Perwiratama et al. [38]	✓	✓	✓	×	✓	✓	×	✓	×	✓	×	✓	Semi	×
Ramachandra et al. [39]	✓	✓	×	×	✓	×	×	✓	✓	×	✓	×	Full	✓

(continued)

Table 1 (continued)

References	Sensing									Automation			Automation level	Mobile/web application
	Weather (air)				Irrigation and nutrients (water)					Weather	Irrigation	Nutrient		
	Temperature	Humidity	Light	CO_2	pH	EC	Water temperature	Water level	Soil moisture					
Deepika et al. [40]	✓	✓	×	×	✓	×	×	✓	×	✓	×	×	Full	✓
Saraswathi et al. [5]	✓	✓	×	×	✓	✓	×	×	×	✓	×	×	Semi	✓
Shekhar et al. [41]	✓	×	×	×	×	×	×	×	✓	×	✓	×	Semi	×
Siddiq et al. [42]	✓	✓	×	×	×	×	×	×	×	✓	×	×	Semi	×
Valiente et al. [43]	✓	×	✓	×	✓	×	×	×	×	✓	×	✓	Semi	×
Van et al. [44]	✓	✓	×	✓	✓	×	×	✓	×	×	✓	×	Full	✓
Velmurugan et al. [45]	✓	✓	×	×	×	×	×	×	✓	×	✓	×	Semi	×
Yolanda et al. [46]	×	×	×	×	✓	✓	×	✓	×	×	×	✓	Semi	×

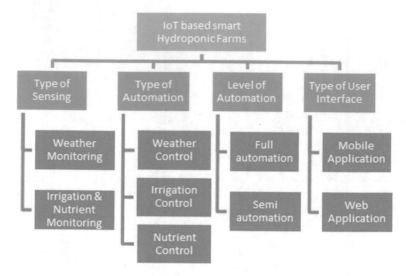

Fig. 1 Classification criteria for existing IoT-based smart hydroponic farms

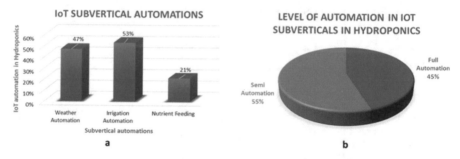

Fig. 2 **a** Percentage of each sub-vertical automation implemented in the 38 smart hydroponic systems studied; **b** level of automation achieved category wise in 38 IoT systems

conclusion drawn from the study is that IoT is an evolutionary technology and when implemented with hydroponics can open a new level of urban farming. This research study brings forth a thorough literature survey on the existing IoT which enables hydroponic farms and highlights that the need for an IoT-based hydroponic farm with automation for climate monitoring and control, irrigation control and nutrient manipulation and control. A system with all the before-stated automations will be fully automated farm in truest sense and will have the capabilities of revolutionizing urban farming.

7 Future Scope

From the analysis of farms presented in Sect. 5, following research gaps were identified:

- Less than 50 percent farms were automated with weather monitoring and control.
- Less than 25% systems were equipped with automated nutrient control.
- Further, less than 50% of the systems were able to implement full automation of their respective verticals, and yet no systems had provisions for all three categories of automation, i.e., weather automation, irrigation automation and nutrient automation identified in Sect. 4.

Thus, it is self-evident that an indoor hydroponic farm with complete automations for all levels has not been designed yet. Thus, to be capable of promoting urban farming using indoor hydroponic plants and make the use of available resources judiciously, the research gap in the existing work needs to be bridged and a hydroponic system which is automated on all levels of its day-to-day activity is the pressing priority.

References

1. Khongdet P et al (2022) Applicability of internet of things in smart farming. J Food Qual, Hindawi. https://doi.org/10.1155/2022/7692922
2. Boursianis AD et al (2022) Internet of things (IoT) and agricultural unmanned aerial vehicles (UAVs) in smart farming: a comprehensive review. Internet Things 18:100187. ISSN 2542-6605. https://doi.org/10.1016/j.iot.2020.100187. https://www.sciencedirect.com/science/article/pii/S2542660520300238
3. Thakur D, Kumar Y, Kumar A et al (2019) Applicability of wireless sensor networks in precision agriculture: a review. Wireless Pers Commun 107:471–512. https://doi.org/10.1007/s11277-019-06285-2
4. Malhotra P, Singh Y, Anand P, Bangotra DK, Singh PK, Hong W-C (1809) Internet of things: evolution concerns and security challenges. Sensors 2021:21. https://doi.org/10.3390/s21051809
5. Srilakshmi A, Rakkini J, Sekar KR, Manikandan R (2018) A comparative study on internet of things (IoT) and its applications in smart agriculture. Pharmacogn J 10(2):260–264
6. Elijah O, Rahman TA, Orikumhi I, Leow CY, Hindia MN (2018) An overview of internet of things (IoT) and data analytics in agriculture: benefits and challenges. IEEE Internet Things J 5(5):3758–3773. https://doi.org/10.1109/JIOT.2018.2844296
7. Farooq MS, Riaz S, Abid A, Abid K, Naeem MA (2019) A survey on the role of IoT in agriculture for the implementation of smart farming. IEEE Access 7(2019):156237–156271. https://doi.org/10.1109/ACCESS.2019.2949703
8. Raneesha AA, Halgamuge MN, Wirasagoda WH, Ali S (2019) Adoption of the internet of things (IoT) in agriculture and smart farming towards urban greening: a review. Int J Advan Comput Sci Appl 10:11–28. https://doi.org/10.14569/IJACSA.2019.0100402
9. Farooq H, Rehman H, Javed A, Shoukat M, Dudely S (2020) A review on smart IoT based farming. Ann Emerg Technol Comput 4:17–28. https://doi.org/10.33166/AETiC.2020.03.003
10. Jaiswal H et al (2019) IoT and machine learning based approach for fully automated greenhouse. In: 2019 IEEE Bombay section signature conference (IBSSC). https://doi.org/10.1109/ibssc47189.2019.8973086

11. Dholu M, Ghodinde KA (2018) Internet of things (IoT) for precision agriculture application. In: 2018 2nd international conference on trends in electronics and informatics (ICOEI). https://doi.org/10.1109/icoei.2018.8553720

12. Bhojwani Y et al (2020) Crop selection and IoT based monitoring system for precision agriculture. In: 2020 international conference on emerging trends in information technology and engineering (Ic-ETITE). https://doi.org/10.1109/ic-etite47903.2020.123

13. Doshi J et al (2019) Smart farming using IoT, a solution for optimally monitoring farming conditions. Procedia Comput Sci 160:746–751. https://doi.org/10.1016/j.procs.2019.11.016

14. Herman, Surantha N (2019) Intelligent monitoring and controlling system for hydroponics precision agriculture. In: 2019 7th international conference on information and communication technology (ICoICT). https://doi.org/10.1109/icoict.2019.8835377

15. Palande V et al (2018) Fully automated hydroponic system for indoor plant growth. Procedia Comput Sci 129:482–488. https://doi.org/10.1016/j.procs.2018.03.028

16. Aliac CJG, Maravillas E (2018) IOT hydroponics management system. In: 2018 IEEE 10th international conference on humanoid, nanotechnology, information technology, communication and control, environment and management (HNICEM). https://doi.org/10.1109/hnicem.2018.8666372

17. Nurhasan U et al (2018) Implementation IoT in system monitoring hydroponic plant water circulation and control. Int J Eng Technol 7(4.44):122. https://doi.org/10.14419/ijet.v7i4.44.26965

18. Ani A, Gopalakirishnan P (2020) Automated hydroponic drip irrigation using big data. In: 2020 second international conference on inventive research in computing applications (ICIRCA). https://doi.org/10.1109/icirca48905.2020.9182908

19. Bharti NK et al (2019) Hydroponics system for soilless farming integrated with android application by internet of things and MQTT broker. In: 2019 IEEE Pune section international conference (PuneCon). https://doi.org/10.1109/punecon46936.2019.9105847

20. Bhattacharya M et al (2020) Smart irrigation system using internet of things. In: Applications of internet of things lecture notes in networks and systems, pp 119–129. https://doi.org/10.1007/978-981-15-6198-6_11

21. Changmai T et al (2018) Smart hydroponic lettuce farm using internet of things. In: 2018 10th international conference on knowledge and smart technology (KST). https://doi.org/10.1109/kst.2018.8426141

22. Cho WJ et al (2017) An embedded system for automated hydroponic nutrient solution management. Trans ASABE 60(4):1083–1096. https://doi.org/10.13031/trans.12163

23. Chowdhury MEH et al (2020) Design, construction and testing of IoT based automated indoor vertical hydroponics farming test-bed in Qatar. Sensors 20(19):5637. https://doi.org/10.3390/s20195637

24. Sen D, Dey M, Kumar S, Boopathi CS (2020, June) Smart irrigation using IoT. Int J Advan Sci Technol 29(4s):3080–9. http://sersc.org/journals/index.php/IJAST/article/view/22479

25. Domingues DS et al (2012) Automated system developed to control PH and concentration of nutrient solution evaluated in hydroponic lettuce production. Comput Electron Agric 84:53–61. https://doi.org/10.1016/j.compag.2012.02.006

26. Dudwadkar A (2020) Automated hydroponics with remote monitoring and control using IoT. Int J Eng Res Technol 9(6). https://doi.org/10.17577/ijertv9is060677

27. Gori A et al (2017, Sept) Smart irrigation system using IOT. Int J Advan Res Comput Commun Eng 6(9):213–216

28. Kaburuan ER et al (2019) A design of IoT-based monitoring system for intelligence indoor micro-climate horticulture farming in Indonesia. Procedia Comput Sci 157:459–464. https://doi.org/10.1016/j.procs.2019.09.001

29. Lakshmanan R et al (2020) Automated smart hydroponics system using internet of things. Int J Electr Comput Eng (IJECE) 10(6):6389. https://doi.org/10.11591/ijece.v10i6.pp6389-6398

30. Lakshmiprabha KE, Govindaraju C (2019) Hydroponic-based smart irrigation system using internet of things. Int J Commun Syst. https://doi.org/10.1002/dac.4071

31. Mahale RB, Sonavane SS (2016) Smart poultry farm monitoring using IoT and wireless sensor networks. Int J Advan Res Comput Sci 7(3):187–190. http://www.ijarcs.info/index.php/Ijarcs/article/view/2665
32. Mahendran M et al (2017) Implementation of smart farm monitoring using IoT. Int J Current Eng Sci Res (IJCESR)
33. Mehboob A et al (2019) Automation and control system of EC and PH for indoor hydroponics system. IEEC
34. Mohanraj I et al (2016) Field monitoring and automation using IOT in agriculture domain. Procedia Comput Sci 93:931–939. https://doi.org/10.1016/j.procs.2016.07.275
35. Montoya AP et al (2017) Automatic aeroponic irrigation system based on arduino's platform. J Phys: Conf Ser 850:1–11. https://doi.org/10.1088/1742-6596/850/1/012003
36. Munandar A et al (2018) Design and development of an IoT-based smart hydroponic system. In: 2018 International seminar on research of information technology and intelligent systems (ISRITI). https://doi.org/10.1109/isriti.2018.8864340
37. Pallavi S et al (2017) Remote sensing and controlling of greenhouse agriculture parameters based on IoT. In: 2017 international conference on big data, IoT and data science (BID). https://doi.org/10.1109/bid.2017.8336571
38. Perwiratama R et al (2019) Smart hydroponic farming with IoT-based climate and nutrient manipulation system. In: 2019 international conference of artificial intelligence and information technology (ICAIIT). https://doi.org/10.1109/icaiit.2019.8834533
39. Ramachandran V et al (2018) An automated irrigation system for smart agriculture using the internet of things. In: 2018 15th international conference on control, automation, robotics and vision (ICARCV). https://doi.org/10.1109/icarcv.2018.8581221
40. Deepika S et al (Jan 2020) Enhanced plant monitoring system for hydroponics farming ecosystem using IOT. GRD J Eng 5(2):12–20
41. Shekhar Y et al (2017) Intelligent IoT-based automated irrigation system. Int J Appl Eng Res 12(8):7306–7320. https://www.ripublication.com/ijaer17/ijaerv12n18_33.pdf
42. Siddiq A et al (2020) ACHPA: A sensor based system for automatic environmental control in hydroponics. Food Sci Technol 40(3):671–680. https://doi.org/10.1590/fst.13319
43. Valiente FL et al (2018) Internet of things (IOT)-based mobile application for monitoring of automated aquaponics system. In: 2018 IEEE 10th international conference on humanoid, nanotechnology, information technology, communication and control, environment and management (HNICEM). https://doi.org/10.1109/hnicem.2018.8666439
44. Van L-D et al (2019) PlantTalk: a smartphone-based intelligent hydroponic plant box. Sensors 19(8):1763. https://doi.org/10.3390/s19081763
45. Velmurugan S et al (2020, 2 May) An IOT-based smart irrigation system using soil moisture and weather prediction. Int J Eng Res Technol (IJERT) Electr–2020 8(7)
46. Yolanda D et al (2016) Implementation of real-time fuzzy logic control for NFT-based hydroponic system on internet of things environment. In: 2016 6th international conference on system engineering and technology (ICSET). https://doi.org/10.1109/icsengt.2016.7849641

Smart Healthcare System Based on AIoT Emerging Technologies: A Brief Review

Chander Prabha⬤, Priya Mittal⬤, Kshitiz Gahlot⬤, and Vasudha Phul⬤

Abstract The artificial intelligence Internet of Things (AIoT), one of the most rising topics, has become a hot topic of conversation all the times. In recent years, it has been successful in attracting the attention of many people. Many researchers have focused now their research on AIoT; however, in health care, the research is still in the Stone Age. Smart Healthcare System is nothing new but introducing an innovative and more helpful version of traditional medical facilities, artificial Intelligence, IoT, and cloud computing which will surely take our healthcare system to new heights. This new technology is getting more important with increasing days as it is more convenient and more personalized for both doctors and of course for patients too. Role in the healthcare domain is also mandatory, as it focuses on intrabody monitoring services as well as maintains the healthcare records of patients. In the introductory part of this paper, the discussion is made on the great use of Smart Healthcare Systems by humans. In a brief literature survey, ideas about various techniques proposed in smart health care are presented. The later section expounds on various sensors used in smart health care and is of great benefit for old-age people. Sensors like nap monitors or breath monitors also help people to take care of their health and reduce the risk of any type of health issues in the future. The existing challenges with the use of AIoT in smart health care are discussed along with their applications favoring real-life examples used in the healthcare industry which are also part of this research paper.

C. Prabha (✉) · P. Mittal · K. Gahlot · V. Phul
Chitkara University Institute of Engineering and Technology, Chitkara University, Rajpura, Punjab, India
e-mail: chander.prabha@chitkara.edu.in

P. Mittal
e-mail: priya1669.be20@chitkara.edu.in

K. Gahlot
e-mail: kshitiz1599.be20@chitkara.edu.in

V. Phul
e-mail: vasudha1342@chitkara.edu.in

Keywords Artificial intelligence (AI) · Internet of things (IoT) · Health care · Sensors · Machine learning

1 Introduction

Can machines think? A question raised by Alan Turing in 1950 during the Second World War to break German's Computer Code gave a great idea of how AI improves the living standards of an individual. In the journey of answering this question, giving thought that how the Smart Healthcare System can be considered a milestone turned in favor of human evolution. The innovative idea of smart health care was born out of the concept of "Smart Planet" proposed by IBM (Armonk, NY, USA) in 2009 [1]. Smart health care is a health service system that uses technology such as wearable devices, IoT, and mobile internet to dynamically access information [1]. The idea of AIoT came into existence when artificial intelligence was tried to implement in combination with the Internet of Things (IoT). AIoT establishes itself in various aspects of everyday life by providing the users an ambient living, working, and domestics [2]. AIoT is flourishing its wings in e-health. In health care, AI along with IoT is used to improve the methods of treatment, for doing predictive analysis of disease, to monitor patients in real-time, and for patient care and medication. Smart Healthcare System is a relatively recent concept, whose diffusion has been rapidly increasing in the last years [3]. There are still many nations in the world that still need to go far to satisfy the condition of the WHO to have at least one doctor over thousand patients. So, this new technology can resolve today's one of the biggest problems in a very efficient manner. For the improvement of the quality of healthcare services provided by government or private hospitals and in reducing the burden on health professionals, smart health applications will act as a catalyst [4]. In the healthcare field, it can promote interaction between all parties, help the parties make informed decisions, ensure that participants get the services they need, and facilitate the rational allocation of resources [1].

Various studies by other prominent authors have been discussed in Sect. 2 as a part of the literature survey. Section 3 of the paper discusses various sensors used in different types of devices to achieve a Smart Healthcare System. Section 4 discussed the challenges faced in smart health care that we need to overcome so that no problems could be created in the future due to this technological development. Some of the applications of AIoT are discussed in Sect. 5. Finally, Sect. 6 concludes the research paper and discussed the doors for future scope.

2 Literature Survey

Many research papers, journals, and magazines have already discussed the concept of a Smart Healthcare System using AIoT. Some of the studies done by various authors have been discussed below.

Tian et al. [1] give the great view that smart health care is a boon to today's scenario. This research paper makes us aware of the born-out idea of smart health care from the smart planet. This paper mainly focuses on applications of the Smart Healthcare System. It also introduces the concept of Smart Hospitals and virtual assistance. Then, it discusses various problems and solutions associated with the innovative new idea of medical facilities.

Priyadarshini et al. [2] explained various technologies for IoT. The work in this paper focuses on devices that gather and share information directly with every other connected cloud. This paper is also focusing on Internet of Nano-Things (IoNT) and Internet of Bio-Nano Things (IoBNT). This study does not comprise any profound concept of overcoming the challenges talked about in the research paper.

Ivan-Garcia et al. [5] emphasize the fact that wearable technology and applications improve the quality of life. It gave an idea that the new technology implemented in daily life can evaluate the risk of disease in the future by keeping a regular record of our physical activity and symptoms and warning us to take preventive measures. Mobile applications can also keep track of our emotional evaluation.

Shweta et al. [6] explained the IoT in big healthcare data. This research mainly prioritizes the usage of IoT and machine learning for disease prediction and diagnosis systems for health care. In the world, death is caused by various diseases and some of those diseases can be rehabilitated if the problem is identified earlier. The chance of death can be reduced. This work gathers an analysis based on data collection. Based on the survey, the work in this paper discusses various diseases like early detection of dementia, brain tumor, heart diseases, and the IoT model for predicting lung cancer. The work in this paper also focuses on healthcare applications of ML and IoT and briefly describes and also talks about their advantages. The study in this paper does not describe the disadvantages of devices used for health care.

Chander et al. [7] provides details about case studies and success stories on machine learning and data mining for cancer prediction. Among the copious diseases, cancer still stays on the top of the list for the deadliest disease to cure, and an early prediction can be helpful for rehabilitation. To analyze that, this paper proposed machine learning and data mining, and the techniques can work with unusual optimization procedures to classify the patterns on the other hand data mining, in brief, which involve finding patterns in data. Discussed various kinds of cancer in the body for both males and females and their detection. Also, includes a comparative analysis of work done by various researchers on various types of cancers. Issues and challenges are also a part of discussion but no substantial concept to overcome those challenges is presented.

Ganapathy et al. [8] show confidence in drones used in health care and to provide various facilities. This research paper mainly focuses on how drones ease the process of providing medical facilities in a limited time as in Africa drones speed up the process of HIV testing. The paper also gives a great example of Rawandi, where 7000 units of blood were supplied without the wastage of a single drop of blood. It also gives an idea of the design and working of drones. Some challenges that need to be overcome are also discussed, leaving behind the idea and initial of using drones in health care for better outcomes.

Gagandeep et al. [9] considered AIoT in health care as a novel concept and can create an intelligent environment. This research paper divides the work into different layers such as the observation layer, middleware layer, application layer, and business layer. It introduces different wearable gadgets which can give new wings to traditional methods of medical care along with various applications in health care. Smartphone applications that can provide virtual assistance can detect disease and can advise precautionary measures to patients are also very well described. In short, this writing considered the Smart Healthcare System as a great boon to technological development.

3   Sensors Used in Health Care

Figure 1 shows how the data is collected from the sensors on the devices and then further processed and stored in the cloud for analytic processing for a future prescription from past reports which are also helpful for doctors for diagnosis. The data which was stored on the cloud also sends real-time alerts and healthcare reports onto the devices.

Table 1 shows some of the sensors used in health care. AIoT is nothing new but a traditional way of problem-solving with innovative ideas. The applications of AIoT are using already invented ideas fitted with new invented and intelligent sensors which give new wings to problem-solving. Different sensors are designed for different purposes. Some sensors can detect changes in the environment and give the required amazing results.

4   Challenges

The challenges in front of the materialization of the Smart Healthcare System equipped with IoT are copious which are technical as well as social. The challenges can be sub-classified into Security, Privacy, Legal/Accountability, and General [13]. The challenges faced in health care are depicted in Fig. 2.

High cost: Good and proper medical assistance can be counted as the fourth basic need for the survival of mankind. It must be cheap and accessible to every single human being. According to the census, a big ratio of people is beneath the poverty line [14]. They can hardly manage two–three meals a day. The software and hardware requirements of AI are very expensive because it requires plenty of

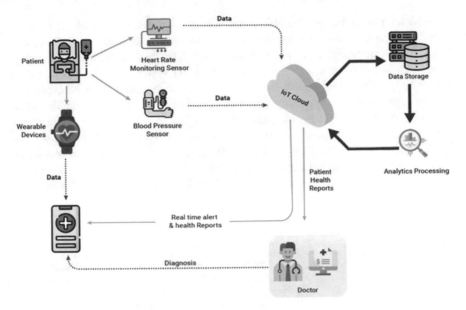

Fig. 1 Data tracking and processing using sensors

renovation to satisfy current global necessities. Dealing with the cost issues of the Smart Healthcare System is the biggest challenge.

Limited thinking: Many times, doctors opt for a treatment that is not acceptable to laws and rules of medical science and pop out with a declaration that claims "It's unbelievable, it's a miracle!!!" But, the machines cannot think out of the field and treat every different patient in the same way they are trained for. AI machines cannot be innovative and ingenious as they cannot beat the power of human intelligence. Without creating machines that can sincerely think like humans, smart health care is barely viable.

No feelings and emotions: For suitable treatment and medical help, a healthy doctor–patient relationship is necessary. A doctor cannot treat his patient unless he cannot feel his pain. The doctor has to introduce his different forms while treating the patient. Sometimes hard as a stone and sometimes soft as a flower. AI machines can be excellent performers, but still, it does not have the feeling. This results in non-emotional attachment with humans and may sometimes be noxious to users if the right care is not taken.

Increase dependency on machines: Human beings need to apprehend and take delivery of the reality that extra of the entirety is awful. If this practice of dependency on machines for even small tasks exists, then in the future the glorious solar will set no matter how high we touch the height of technology and development. Technology is there to assist us, not to make us lazy. So, creating stability between healthful lifestyles and smart destiny is vital. Technological development is amazing, and it handiest goes wrong when machines remind you of how powerful you are.

Security and privacy: The ubiquitous connected sensors [15] collect massive user data. Through IoT networks, this data is transmitted and stored in the cloud. The biometric information is contained within this data which adds a level of security. Data leakage may occur due to ciphertext on the AIoT system as in the universe, data doubles every two years, and quintillions of bytes of data are generated every day [16]. So, data security and privacy are crucial areas of concern in AIoT applications.

Multi-model data: In health care, AIoT consists of a large number of heterogeneous sensors, generating a huge stream of data of numerous formats and sizes (multi-model data), thereby challenging further their processing, storage, and transmission [17]. This results in transmission delay at the cloud/edge/fog node of the AIoT system, and managing and increasing the efficiency are still a challenge.

Table 1 Features and functions of sensors used in health care

Sensors	Functions and features
Infrared (IR) temperature sensors	• Enable specific non-contact temperature quantification in medical devices. Calculating ear temperature, forehead temperature, or pores and skin temperature can be considered one of the finest applications of the temperature sensor. The object's infrared energy is calculated by multiple thermos couples on a silicon chip fitted in the sensing element • IR sensors sense electromagnetic waves in the 700 nm–14,000 nm range. The sensors used a very thin plastic that acts as a transparent medium for electromagnetic waves and protects the sensor from dust, dirt, and other foreign objects. The IR thermometer boasts a wide temperature range of 0–710 °C and can be used at temperatures of up to 85 °C without additional cooling
Skin vision sensor	• Skin vision is an awareness tracking mobile application that supports one in detecting suspicious moles and skin situations in the comfort of their houses with the aid of using their smartphones • Application's advanced and finest camera quality provides the opportunity of clicking high-quality pictures which can easily detect skin cancer. Risk within the short duration of just 30 s
Pedometer sensor	• Pedometer readings conventionally assist as a motivational aspect for physical activity behavior change. Pedometers are designed to stumble on the vertical motion on the hip and so measure the wide variety of steps and offer an estimate of the gap walked • Pedometer sensors are very fast and accurate. It is power-saving device as continues to work while the rest of the system can be turned off. They cannot provide information on the temporal pattern of physical activity or the time spent in different activities at different intensities

(continued)

Table 1 (continued)

Sensors	Functions and features
Heart rate monitor	• These monitors perform by calculating electrical signals from your heart. They are transmitted to a wristwatch or data center • Many models analyze data via a computer and are having that data allows you to interpret your workout and better understand the benefits of exercise done • Hard estimate can be gotten with the aid of monitoring your pulse the old-fashioned way, feeling it in your wrist or neck, but that can be disruptive to your actual workout • A digital heart rate monitor can give you precise, genuine-time records • Heart rate monitors used in hospitals are wired and contain multiple sensors. Consumer heart rate monitors are for everyday use and do not contain wires • Modern heart rate monitors use either optical or electrical signals to record heart signals • They are usually 90% accurate in their readings, but sometimes more than 30% error can persist for several minutes
Calorie counter	• Counting calories can assist you to lose weight via bringing awareness to what you orally consume each day • This can also act as a helping hand that can guide you about the ingesting patterns that you may need to alter, retaining you on course to lead a healthy life
Pulse oximeter	• It is an affordable sensor used to quantify the oxygen saturation degree of hemoglobin in the blood • The sensor is very easy to use as it is simply positioned on the fingertip and the photodetector in the sensor measures the intensity of transmitted light at each wavelength; thus, oxygen quantity is calculated within the blood • It is a very quick device as it gives results within 5 s. It has a bright screen for easy readability. The device can give an incorrect reading when indicates a low battery sign on the screen [10]
CT/MRI/ultrasound scanner	• Helpful examining the X-ray images of body components using X-rays, electromagnetic waves, or magnetic waves • CT scan is not suggested during pregnancy. It can also cause a small dose of radiation, but it is not harmful. Ct scanner takes 5 min while MRI takes approximately 30 min [11]
Breathe monitor	• The respiratory steps, i.e., inhaling and exhaling flow of air in the trachea can be examined by breathe monitor • It is placed on the throat to detect the breathing frequency and alert the doctors for sudden clinical response. It is helpful in cases where death occurs due to late detection of respiratory deterioration • Spire (the breathing monitor device) has a battery life of around 7 days and is charged wirelessly, using a special pad that the device can simply be placed upon, negating the need for extra wires and plugs [12]

(continued)

Table 1 (continued)

Sensors	Functions and features
Nap monitor	• It uses accelerometers to track the quantity and quality of sleep • It consists of small motion detectors. It measures the movement a person makes while he is asleep

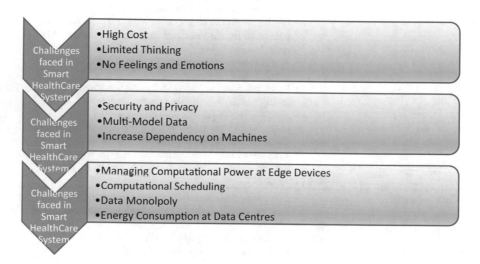

Fig. 2 Challenges faced in the smart healthcare system

Managing computational power at edge devices: The edge devices are having the limitation of limited computational and storage power. The real-time deployment of models on these devices is crucial. Further, on data, network pruning, quantization, and compression need to be done that still a challenge.

Computational scheduling: A computational scheduling challenge may occur in real-world AIoT systems, as from the edge devices to the cloud center or fog node, some deep computation might be needed to offload. Further, to deal with user requirements over time and with unbalanced data, a more robust and dynamic scheduling strategy would be needed [18].

Data monopoly: Most data collected in AIoT is unlabeled. It would be expensive and time-consuming to label unlabeled data [19]. A lot of advancements have been done in unsupervised learning in terms of self-supervised learning for leveraging AIoT multi-model data and providing solutions to challenges of rare cases and new classes of healthcare problems. Data monopoly might be another challenge in the AI era, as AIoT companies providing AIoT support in smart health care, restrict market entry to new competitors. This challenge faced by new entities is due to the non-accessibility and non-availability of vast data collected/protected by already established parties.

Energy consumption at data centers: According to [20], another important challenge is the increase in energy consumption at data centers. Out of the total global usage, 21% of the electricity is consumed by various communication technologies, and data centers' contribution is more than one-third to this. Therefore, for a sustainable future, techniques are required to enhance energy efficiency at data centers. The increasing growth in the quantity of cloud data centers depicts the growth of AIoT applications. Therefore, continuous efforts are needed to address the challenge of energy consumption at data centers.

5   Applications of AIoT in Smart Health Care

The applications of AIoT are uncountable. In every field, AIoT has spread its roots. It plays a momentous role in a wide range of healthcare applications as it allows patients and the elderly to live self-sufficiently [4]. The healthcare industry remains reckless to adopt AIoT, and the reason to adopt this drift is due to the integration of AIoT features into medical devices. This greatly improves the quality of life and effectiveness of the service, especially in cases of elderly patients with chronic conditions and who require regular supervision. Moreover, in case minor problems persist or are identified, the IoT system itself can advise patients accordingly [21]. AI has proven to be a boon for the healthcare industry.

Some of the applications of AIoT are presented in this section, and Fig. 3 gives the pictorial idea of the applications of AIoT in the healthcare system. It shows that the health data is collected from the person using wearable devices or devices used in hospitals, and then, the cloud gateway streams the collected data. The unstructured data is processed in the data lake. Then, the data warehouse stores the processed data. Further, processing of data is done using machine learning (ML) algorithms and ML models, analysis is done, and the cloud server processes all the information in the form of Electronic Health Records (EHRs), which is further analyzed and inspected by medical staff, patient, and admin interface showing the transparency in the system. AI comprises machine learning (ML), neural networks (NNs), and deep learning (DL). The ML is further divided into supervised, (a) unsupervised, (b) semi-supervised, and (c) reinforcement learning. The artificial neural networks (ANNs) are trained in deep learning and further categorized into convolution neural networks (CNNs) and recurrent neural networks (RNNs). All the above approaches are used in health care to make systems more robust and as per patients' needs to develop smart healthcare instruments to be managed either remotely or at a specific doctor's/hospital location. AI along with IoT is used in the following areas of health care as shown in Fig. 4. The further section discusses various applications, where AIoT is used as a technology.

Wearable technology: Wearable technology has boosted several solutions to improve the health and quality of people's lives. Wearable belts are of great use. It helps to monitor physical activity by counting steps during the day and ensure that

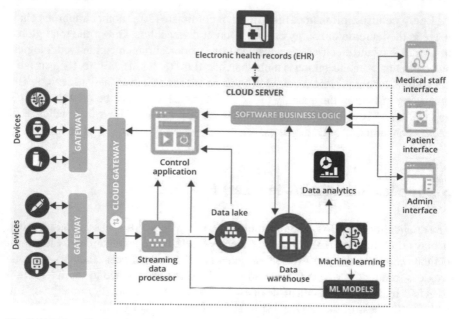

Fig. 3 AIoT applications in health care

Fig. 4 AIoT areas of health care

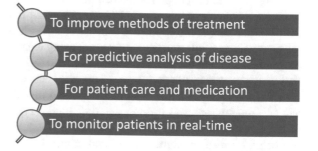

the wearer's activity keeps the heart rate in a non-risky and healthy range by tracking the heart's activity [5].

Blood pressure monitoring system: Irregular downfall or sudden rise in blood pressure is one of the biggest problems people are facing these days. There are many popular blood pressure devices that are safe and simple to use [22]. To collect the patient's BP in real time, an electronic BP monitor is attached to an IoT sensor. It is of great advantage to both doctors and patients as it provides quick service in the condition of a medical emergency.

Wheelchair management: The AIoT comes out with a very innovative solution, i.e., Smart Wheelchairs [23]. They are genuinely considered to be of great help for elderly and incapacitated individuals who cannot move their chairs genuinely, rather they will be given the benefit of moving them through hand gestures.

Remote diagnostics and telemedicine: In such a large population of 7.9 billion, it is very difficult to reach every patient and provide medical facilities as we hardly have one doctor for over a thousand patients. In this condition, remote monitoring allows the user to access the medical facilities wirelessly via text, voice, audio, or video [6].

Drones: Some tasks demand a lot of manpower and time. In that case, drones provide us the opportunity to complete the tasks in minimum time with utmost efficiency. During the COVID pandemic period, drones helped a lot in sanitizing hospital areas, remote monitoring, spraying disinfectants, and many others [6]. Drones are also being used by drowning victims at beaches as a public health surveillance modality and in the identification of mosquito habitats [8].

Smart Trash Bin: It is popularly said that clean surroundings lead to a satisfied and healthful existence [24]; however, usage of the common bin in public locations can lead to the spread of communicable sickness as a healthy person indirectly comes in contact with an unhealthy person. Smart Trash Bin is the approach to this hassle. Smart Trash Bin automatically senses the person or object using an ultrasonic sensor and sends the message to the servo motor using Arduino Uno. As a result, the instance a person comes in the direction of the bin, the bin cap opens and closes after some time on its very own. This innovation is of great use in public areas specifically in hospital areas.

6 Conclusion and Future Scope

The AIoT holds the promise of ameliorating life quality of people. If technology is at the next level today, then it is only because of remarkable work done in the area of AI along with the IoT. Now, we are aiming to achieve much more like this as technology is considered a great growling engine of change. This paper surveys how AIoT when implemented in traditional medical facilities became a major benefactor to our medical distribution system, and no doubt, we came up with the idea of a Smart Healthcare System. To provide health care to everybody at all times and anywhere, smart health care plays an active role and simultaneously increases its attention in society by keeping quality high. The main aim of this review paper is to give an idea about the Smart Healthcare System, the challenges faced, the solutions to the problems which are yet to be discovered, and many more. In short, it is considered a stepping stone for those researchers who are interested in the field of AIoT, and particularly, a smart and intelligent way of dealing with medical health issues of generations.

References

1. Tian S, Yang W, Le Grange JM, Wang P, Huang W, Ye Z (2019) Smart healthcare: making medical care more intelligent. 3(3):62–65
2. Priyadarshini S, Sahebzathi S, Narmatha M, R Delshi Howsalvya Devi (2016, July) The internet of things in healthcare: a survey. GRD J|Global Res Dev J Engl|Int Conf Innovations Eng Technol (ICIET). 559–564. e-ISSN: 2455-5703
3. Marr B (2018, May) How much data do we create every day? The mind-blowing stats everyone should read. Forbes
4. Sanghavi J (2019) Review of smart healthcare systems and applications for smart cities. In: ICCCE 2019. Springer Singapore
5. Magarino I-G, Sarkar D, Lacuesta R. Wearable technology and mobile applications for healthcare. 2019:6247094. published 21 May (2019)
6. Agarwal S, Prabha C (2021) Diseases prediction and diagnosis system for healthcare using IoT and machine learning, smart healthcare monitoring using IoT with 5G: challenges, directions, and future predictions. CRC Press, pp 197–228
7. Prabha C, Sharma G (2021, Dec) Applications of machine learning in cancer prediction and prognosis, cancer prediction for industrial IoT 4.0: a machine learning perspective. CRC Press (Taylor & Francis). eBook ISBN 9781003185604
8. Ganapathy K (2022) Drones in healthcare, issue 47. https://www.asianhhm.com/healthcare-management/drones-in-healthcare
9. Kaur G, Gupta M, Kumar R (2021) IOT-based smart healthcare monitoring system: a systematic review. 25(1):3721–3728
10. Kapoor A. Pulse oximeter buying guide: Price, how to buy the right oximeter types and more. https://timesofindia.indiatimes.com/most-searched-products/health-and-fitness/buying-guide/pulse-oximeter-buying-guide-features-price-how-to-buy-the-right-one-more/articleshow/79992362.cms. Accessed 13 Jan 2022
11. Fletcher J (2019, Oct) What is the difference between CT scans and MRI scans? https://www.medicalnewstoday.com/articles/326839
12. McIntosh J (2014) https://www.medicalnewstoday.com/articles/278847, June 26
13. Zhang K, Ni J, Yang K, Liang X, Ren J, Shen X (2017) Security and privacy in smart city applications: challenges and solutions. IEEE Commun Mag 122–129
14. World Bank. Poverty and inequality platform. https://data.worldbank.org/indicator/SI.POV.NAHC
15. Yang Q, Liu Y, Chen T, Tong Y (2019) Federated machine learning: concept and applications. ACM Trans Intell Syst Technol 10(2):1–19
16. Nazir S, Ali Y, Ullah N, García-Magariño I (2019) Internet of things for healthcare using effects of mobile computing: a systematic literature review. Wirel Commun Mobile Comput. https://doi.org/10.1155/2019/5931315
17. Duan L-Y, Liu J, Yang W, Huang T, Gao W (2020) Video coding for machines: a paradigm of collaborative compression and intelligent analytics, arXiv preprint arXiv:2001.03569
18. Jing L, Tian Y (2020) Self-supervised visual feature learning with deep neural networks: a survey. IEEE Trans Pattern Anal Mach Intell
19. He K, Fan H, Wu Y, Xie S, Girshick R (2020) Momentum contrast for unsupervised visual representation learning. In: Proceedings of the IEEE/CVF conference on computer vision and pattern recognition (CVPR), pp 9729–9738
20. Yuan H, Bi J, Zhou M, Liu Q, Ammari AC (2021) Biobjective task scheduling for distributed green data centers. IEEE Trans Autom Sci Eng 18(2):731–742. Art no 8951255
21. Zhang J, Tao D (2020, Nov) Empowering things with intelligence: a survey of the progress, challenges, and opportunities in artificial intelligence of things. IEEE Internet Things J 20(10)

22. Islam SMR, Kwak D, Kabir H, Hossain M, Kwak K-S (2015) The internet of things for health care: a comprehensive survey. IEEE Access 3:678–708
23. Yash KS, Sharma T, Tiwari P, Venugopal P, Ravikumar CV (2020, June) A smart wheelchair for health care systems. IJEET 11(4):22–29
24. Huh J-H, Choi J-H, Seo K (2021) Smart trash bin model design and future for smart city. 11(11):4810. https://doi.org/10.3390/app11114810

Intelligent Communication for Internet of Things (IoRT)

Kaleem Ullah Bhat, Neerendra Kumar, Neha Koul, Chaman Verma, Florentina Magda Enescu, and Maria Simona Raboaca

Abstract To enable Machine-to-Machine (M2M) communication and data transmission using fundamental network protocols like TCP/IP, the Internet of Things (IoT) provides a sturdy framework for attaching things to the internet. Its cutting-edge technology is having an impact on numerous fields of study and development of smart solutions. IoRT is the integration of diverse technologies which encompasses cloud computing, machine learning, artificial intelligence, and Internet of Things (IoT). With so many devices and so few resources, security issues like device authorization and authentication and maintaining the integrity and privacy of data arise. IoT was forced to apply conventional security techniques because of a variety of new circumstances, including restricted computational power, RAM size, ROM size, register width, different operating environments, and others. The consequences of the several cryptanalyses performed on these ciphers by various cryptanalysts are also examined in this study. This paper provides background information on lightweight block ciphers in order to protect transmitted data within the network. For more secure communication, we have also implemented the PRESENT block cipher due to its lightweight behavior. A future dimension is also proposed to develop a good lightweight cipher.

K. U. Bhat · N. Kumar (✉) · N. Koul
Department of Computer Science and IT, Central University of Jammu, Jammu, Jammu & Kashmir, India
e-mail: neerendra.csit@cujammu.ac.in

C. Verma
Department of Media and Educational Informatics, Faculty of Informatics, Eötvös Loránd University, Budapest, Hungary
e-mail: chaman@inf.elte.hu

F. M. Enescu
Department of Electronics, Communications and Computers, University of Pitesti, Pitești, Romania

M. S. Raboaca
National Research and Development Institute for Cryogenic and Isotopic Technologies-ICSI, Rm. Valcea, Romania
e-mail: simona.raboaca@icsi.ro

© The Author(s), under exclusive license to Springer Nature Singapore Pte Ltd. 2023
Y. Singh et al. (eds.), *Proceedings of International Conference on Recent Innovations in Computing*, Lecture Notes in Electrical Engineering 1011,
https://doi.org/10.1007/978-981-99-0601-7_24

Keywords Internet of Things (IoT) · Robotics · Internet of Robotic Things
(IoRT) · Intelligent communication · Encryption · Lightweight · Block ciphers

1 Introduction

With the use of Internet of Things (IoT) technology, consumers may make their
current equipment smarter and connect it to the Internet so that it can communicate
with other devices. The Internet of Things is growing immensely, and large number
of devices are getting associated with IoT making the industry enter $300 billion
business. IoT is the idea of connecting any contraption to the Internet and other associated gadgets (as long as it has an on/off switch). IoT is a vast network of connected
humans and objects that all gather and share information about how they are used
and their environment. The usage of an IP address as a distinctive identification is
implied by this incorporation with the Internet by the devices. The idea is to manage
objects through the Internet, and Arduino is the ideal model for these applications
[1]. Technologies like sensors and actuators are widely used to integrate people,
process device in Internet of Things. In terms of communications, cooperation, and
technical analytics, this total integration of IoT with humans allows for real-time
decision-making. Robotics is one area where the technology of Internet of Things is
gaining traction. Robotics is a cutting-edge, rapidly expanding technology that has
radically altered various elements of human society in recent decades.

Robotics generally perform tasks which are tedious and repetitive in nature; also,
dangerous and critical tasks which are unaffordable by a human being are performed
by robots. Initially, the robots utilized in these applications were single machines
with hardware and computational capabilities that limited them. To address these
concerns, the robots were first associated in a correspondence organization [2], either
wired a or remote association, resulting in the creation of a Networked Robotic
System. Nowadays, the concept of "Cloud Robotics" is evolving which is another
type of proficient mechanical frameworks that depends of "Distributed computing"
foundation to get to a huge volume of registering power and information to empower
its tasks. Contrary to networked robots, no one independent system is used for all
sensing, processing, and memory. There are many open issues and challenges faced.
Maintaining a balance between the real-time requirements of diverse situations and
performance precision is a difficult task [3], even for robot memory, because the
notion of cloud robotics is founded on real-time requirements. Because cloud storage
refers to data that is stored remotely, there is a requirement for better cloud security.
Other challenges that tend to stand in the way of the efficient performance of cloud
robots are Rapidity, Remoteness, Network, etc.

1.1 Internet of Robotic Things

The Internet of Robotic Things is an idea in which events can be tracked by intelligent devices, coordinate sensor information from a number of sources, utilize nearby and circulated knowledge to sort out the fitting methodologies, and afterward act to modify or control objects in the real world, sometimes while physically moving through it. IoRT has many technologies such as sensors and actuators, communication technologies, processing, data fusion, modelling and dynamic mapping, virtual and augmented reality, voice recognition, voice control, decentralized cloud [4], adaptation, machine learning, end-to-end operation, and internet technologies safety and security framework [5]. Robotic devices act intelligently in the sense that they have monitoring (and sensing) capabilities while also being able to obtain sensor data from other sources that are fused for the device's "acting" purpose. The machine can also use local and distributed "intelligence", which is another "intelligent" feature. In other words, it may analyze data from the events it watches (which, in many cases, entails the use of edge computing or fog computing) and has access to analyzed data. IoRT takes use of the IP protocol and its IPv6 version to achieve the benefits of current communication and cloud-based interoperable technologies. By connecting external resources to robots through the Internet, IoRT aims to increase Internet usage [6]. IoRT delivers sophisticated robotic capabilities, allowing multidisciplinary solutions to emerge for a variety of fields. The Robotic Internet Things is envisioned as being built on top of the cloud*robotics*paradigm, utilizing characteristics of cloud*computing such as cloud storage. Three service models and virtualization technology, i.e., software, platform, and infrastructure, while taking advantage of The Internet of Things (IoT) and its enabling technologies have the potential to enable enormous amounts of data. To meet the provisional goal for networked robotics new application design and implementation flexibility are suggested. As a core utility, distributed computer resources are used. Other technologies, such as multi radio access to link smart devices, artificial intelligence to provide optimized solutions for difficult challenges, and intelligent systems to provide superior service, support IoRT solutions in addition to IoT capabilities. The Internet of Robotic Objects (IoRT) is positioned at the pinnacle of cloud robotics, combining IoT technology with the independent and self-instructed behavior of associated automated things to foster refined and creative arrangements that utilize appropriated assets.

2 Architecture

The architecture of an Internet of Robotic Things system is mainly composed of three layers, Physical, Network and Control, Service and Application. The robotic part of IoRT is integrated into the OSI model with a different perspective.

2.1 Physical Layer

The physical layer comprises the bottom layer of the IoRT architecture. It comprises the hardware part which is related to components, such as vehicles, robots, mobile phones [7], healthcare equipment's, home appliances, sensors, and actuators. Robots work as perceptive agents, collaborative with each other to perform specific tasks and achieve desired objectives in a distributed fashion. A robot can be described, for example, a healthcare assistant which helps a person to reinstate the walking ability and a "TUG" robot that is intended to independently move around a hospital to provide supplies like medication and clean linens. The sensors installed in them are used for the perception of data from the external environment, sense the movement of objects, actions happening around while actuators perform the required actions like turn on/off smart lights, air conditioning, etc. Sensors and actuators can both be incorporated into a single system to improve and optimize performance and accomplish a shared objective through dispersed operations.

2.2 Network Layer

Network layer is the second layer arriving after physical layer. It provides for the connectivity of the robotic systems such as cellular network connectivity including 3G and LTE/4G. Internet of Things (IoT) development will speed up thanks to 5G technology [8]. To provide constant connectivity, some short-range technologies are employed between the robotic objects, including WiFi, Bluetooth Low Energy (BLE) [7], 6LoWPAN Broad Band Global Area Network (BGAN) [9], and Near-Field Communication. Common long-range protocols are Sigfox, LoRa WAN, and NB0IoT, while NB-IoT operates on a licensed spectrum with less interference than Sigfox and LoRaWAN [5], which operate on unlicensed frequency bands. Zigbee and 6LowPAN are based on IEEE 802.15.4. The Routing Protocol for Low Power and Lossy Networks and 6LowPAN is popular IoT protocols used at the network layer (RPL) [10]. For efficient information transmission among the robotic network infrastructure located at a greater distance [10], LoRA has been implemented [1].

2.3 Service Layer and Application Layer

The IoRT architecture's top layer, which handles end-user actions, is entirely dependent on software implementations. Standard and user programs for monitoring, processing, and managing environmental elements and agents are necessary to achieve the objectives of the integrated Internet of Robotic Things. Sensors, actuators, and robotics are installed and used in the smart environment at the service and application layer. A variety of settings, such as commercial organizations, R&D

facilities, data centers, can use cloud computing. IoRT can be thought of as a field with endless potential. Modern robotics must include machine learning techniques because they may be applied to tasks like mapping, localization (knowing where a robot is), and learning the environment.

3 Application Domains

The socioeconomic circumstances of people have undergone a significant transformation as a result of the Internet of Robotic Things. The development of a digital, networked society where each "thing" can sense its environment, share information, solicit feedback, and act is facilitated by the Internet of Things. By fusing IoT/IIoT technology with robotics, the Internet of Robotics Things improves robotic capabilities. Robots are becoming more autonomous as a logical by-product of having connected IoT deployment systems with robotic systems. If they are a component of ubiquitous intelligence solutions, they are just more effective and efficient overall.

3.1 Manufacturing

With the advancement of the Fourth Industrial Revolution, the manufacturing technology has developed responding to new inventive technologies creating high-quality goods and services. It has resulted in smart manufacturing, which incorporates adaptability, monitoring, and change. The Fourth Industrial Revolution's increased digital technology breakthroughs which must be expanded with improvements in production methods and raw materials. Task scheduling in manufacturing has become more adaptable and expandable with the arrival of cloud robotics. Robotic fleets are used for planning, delivery, and the moving of various items like equipment and cartons. Robots communicate with the cloud to obtain information about the industrial architecture and other robots already installed in the system. Robotics has adapted to manufacturing uncertainties.

3.2 Agriculture

Agriculture has benefited greatly from advanced technology as part of Industry 4.0. Due to the integration of IoT-based systems in the agriculture sector, effective changes are seen with heavy labor-intensive tasks being performed by robots such as harvesting of fruit, remote controlling of tractors by hosting agricultural machinery with smartphones. Rough patches' codes are memorized and gears are altered accordingly while smooth navigation for the protection of crops and equipment is being employed [12, 13]. Automatic harvesting of crops is done through robots preventing from untimely harvests. There are numerous factors that affect agricultural productivity, including the weather, poorly trained staff, and inadequate farm management. Through the application of internet of robotic things technologies, farmers may raise high-value crops. Self-guiding robotic drones use ground control systems, GPS, image processing, and infrared cameras. Farmers employ drones as a service (DaaS) to monitor fields, forecast crop yields, and identify insect infestations. These robots come with a variety of sensors, including ones that measure temperature, humidity, and crop hazards.

3.3   Health Care

IoRT has several uses and offers health, societal, and economic benefits, especially for patients who require specialized treatment, such as those with mental disabilities, patients at higher risk for stroke. The future of edge computing includes applications for virtual reality and wireless health monitoring [14] and robotics where a quick response to sensor input is required. Robotics combined with sensors and Internet of Things devices offers various benefits [4], by providing real-time health information, identifying patient problems, lowering the likelihood of an incorrect diagnosis, adjusting prescription dosages, etc. [15].

3.4   Education

Adaptable and intelligent tactics are applied by robotics in education domain in order to build and maintain social interactions with humans. Robotics also provide support services such as homework and learning. Electrodermal activity (EDA) is a shift in skin reflectivity [16] that furthermore, because minors are unable to respond to these impulses in the same way that adults do, the EDA feedback in children may differ significantly from the average adult response.

3.5   Surveillance

To safeguard a specific region, IoRT systems are created and deployed providing rapid and reliable information. Monitoring of sites and individuals is essential to reduce the security risks in sensitive regions, military areas, public places, and common houses. In order to reduce the limitation of blind areas, a number of CCTV cameras can be installed outdoors as well indoors which cover most of the areas. A versatile field monitoring robot was used to identify mines and hazardous chemicals. NodeMCU stands for NodeMCU WiFi, which is used to connect controllers and develop robots that can navigate and collect data on any terrain. The data assembled by the robot's sensors is shipped off cloud servers [17]. An embedded web server makes it simple to monitor and control any device that is located in a remote location [18].

3.6 Disaster Response

Every year, a number of catastrophes like typhoons, earthquakes, and tsunamis occur. When a catastrophe of this size occurs, time is extremely valuable. The current top priority is to save as many lives as possible, minimize casualties, and quickly restore vital services. After receiving deep learning training, robots can be quite helpful in these kinds of circumstances. The first step in achieving this is to utilize robots to collect data about the immediate area. An external network expands on these inputs to produce an AI model, in which the inside system then evaluates. For extra performance testing, the prototype was also transferred from the cloud to the local computer. Once the model reaches a certain level of reliability, it is eventually implemented into the robots for the following stage of the learning process (Fig. 1).

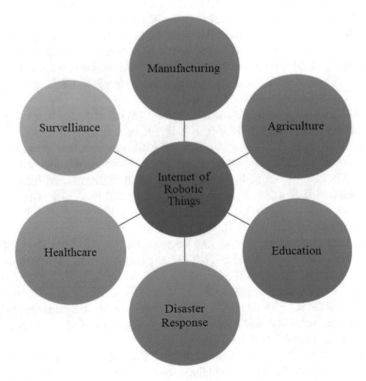

Fig. 1 Application areas

4 Literature Survey

Craft is a flexible block cipher designed to defend against differentiable fault analysis attacks. The 64-bit plaintext, 128-bit key, and 64-bit public tweak make up the algorithm's structure, which is based on involuntary building blocks. The round function implements S-box, mixing columns, permutations based on data nibble positions, constant addition, and tweaked addition processes [19].

Block cipher CHAM was only invented in 2017 by Korean experts. The algorithm's generalized four-branch FN structure is designed for IoT devices with limited resources. The key schedule is quite easy to utilize [19].

Lilliput with Extended Generalized Feistel Network is a lightweight block cipher that Mumthaz and Geethu optimized (EGFN). The suggested strategy has put in place the PRESENT S-Box [7] and key schedule corresponding to the DES key schedule. LILLIPUT uses an 80-bit key, a 64-bit block size, and 30 rounds, where a nibble-level round function is acting [20].

By enhancing PRESENT, Subhadeep Baniket et al. proposed GIFT, a tiny, quick, energy-efficient, and more secure SPN block cipher. Bit permutation is employed for diffusion since the S-box of PRESENT is expensive. By combining permutation with the Difference Distribution Table (DDT)/Linear Approximation Table (LAT) of the S-Box, GIFT enhanced the PRESENT cipher. GIFT-64, with 28 rounds, and GIFT-128, with 40 rounds, are the two variations of GIFT that have been proposed. The key size for both variants is 128 bits. To improve resistance against linear and differential cryptanalysts, linear hulls, and clustering effect, GIFT uses a smaller, less expensive S-box than PRESENT [21].

Proposed SIT (Secure IoT) is a compact 64-bit symmetric key block cipher with five rounds and a key size of 64 bits. SIT is a hybrid strategy that combines SPN with Feistel structure. The suggested method incorporates a few logical processes as well as some switching and replacement. Energy efficiency is increased by using five distinct keys for five rounds of encryption. SIT algorithm uses a Feistel network of substitution diffusion functions for confusion and diffusion. Key expansion block makes use of a 64-bit cipher key supplied by the user. Key generation F-function is based on modified KHAZAD [22].

By utilizing several guidelines and suggestions provided by the authors of, Sufyan Salim Mahmood AlDabbagh [23] suggested a new lightweight Feistel block cipher algorithm called Design 32-bit Lightweight Block Cipher Algorithm (DLBCA). With 15 rounds and an 80-bit key size, DLBCA is built with 32-bit block size. On the four levels, DLBCA performs the following operations: S-box, bit-permutation, Exclusive-OR rotation, and key operations. It offers strong defense against boomerang and differential strikes.

SKINNY and MANTIS were presented by Christof Beierle et al. in 2016. SPN is used by the SKINNY family of adaptable block ciphers. By deleting the final NOT gate at the end, it uses a small S-box S4 that is extremely similar to the S-Box of PICCOLO. Block, key, and tweak sizes are all movable in SKINNY, and data is loaded in rows like in AES. A new sparse diffusion layer and a new, light-colored key schedule were both introduced by SKINNY [24].

In order to provide greater search space, SKINNY makes advantage of a larger key size compared to block size and fewer rounds. Implementations that are leakage resilient are provided via SKINNY's tweakable capabilities.

MANTIS is a low latency, 64-bit block size, 128-bit key-size, 64-bit tweakable block cipher. By employing its S-box and linear layer for quick diffusion, it improves upon MIDORI. Mantis employs the row-wise added round constant of PRINCE. In MANTIS, tweak-scheduling is used to guarantee a large number of active S-boxes and boost security against related tweakey attacks by increasing the number of rounds. Because MANTIS is primarily a copy of its identical MIDORI counterpart, security analysis is not performed effectively [24].

A series of lightweight symmetric-key block ciphers called SPARX and LAX based on the ARX (Modular Addition/Bitwise Rotation/XOR) algorithm was proposed by Daniel Dinu and colleagues in 2016. To add nonlinearity and enough diffusion, LTS uses sparse linear layers and a huge ARX-based S-Box termed arx-box. Through the avoidance of database lookups, ARX lessens the impact of side-channel attacks. By reducing the number of actions needed, this design style enables quick software implementations. The LTS (a dual of WTS and LAX complete) long trail design technique was used to create the cipher SPARX. Sparx-64/128 has eight steps with three rounds each, and Sparx-128/256 has eight stages with four rounds, each use 10 phases with 4 cycles [26].

5   Research Challenges

5.1   Data Processing

IoRT's dual cyber-physical nature makes it necessary to gather enormous volumes of data in a wide variety of forms from a wide variety of sources, containing information for robot control or data from several sorts of sensors. To be processed remotely, these data must be sent from the nodes to the cloud. This calls for more advanced data processing capabilities than a normal IoT application and may result in the following problems.

In an IoRT network, large amounts of data processing and streaming typically call for a lot of communication capacity. For instance, when on patrol, a group of security robots would need to simultaneously disseminate a variety of visual and auditory data. To ensure that the system can react quickly to any issues, fast data processing is necessary.

In a complex environment, unstable communication can greatly impact system performance. For instance, the complicated indoor layout of a hospital and interference from other medical devices like magnetic resonance imaging may be a problem for a medical IoRT system there (MRI) [11].

5.2 Robot Ethics

The major concern for IoRT is cooperation as a result of coexistence of robotics and humans within a common space. The effective handling of those challenges is essential for the peaceful cohabitation of humans and robots in contemporary civilization. Since the beginning of the robot concept, there have been discussions on the ethics of robot systems [26]. The core concept is simple: Consumers should not be harmed by robots. IoRT, a technology with ethical implications, cannot, however, achieve this objective on its own. Instead, to stop the misuse of robots, the relevant society—including governments, users, and creators of robots—must take action on effective policies and cutting-edge practices.

5.3 Privacy

Who should have access to the data and how should it be used are two more potential ethical issues. Things become more serious, particularly for sensitive data like financial or medical information. No data collector should be given full, unrestricted access to or use of any data without the necessary authorization or authorization, according to the existing cultural agreement on privacy [27]. This problem needs to be appropriately addressed because the majority of IoRT programs have a high capability to collect data from both their working environment and interacting clients.

5.4 Security Issues

In robots, security and trust are the main concerns. Particularly, in the case of IoRT, where cloud involvement is essential, we will confront two significant security challenges. First and foremost, the IoRT-VM environment needs to be reliable. For instance, in military applications, IoRT-enabled robotic objects must be able to distinguish among a variety of IoRT-VM infrastructures that may be trusted, allowing them to connect to that infrastructure while avoiding known harmful IoRT-VM infrastructure. Three strategies can be used to address this issue: building trust, measuring trust, and reputation-based trust [16]. Second, the owners or controllers of future robotic systems must have confidence in them to start computing tasks on IoRT-based clouds where the cloud must be set up to allow for owner or controller verification. Here, we must make sure that these outsourced duties are not being carried out by malicious code. Secret data may be stored permanently on IoT-enabled cloud servers while also replicating the logical shadow of the data to private cloud servers. As a result, strict approaches are required to safeguard the integrity, trust, and secrecy of IoRT data.

6 Proposed Methodology

Our methodology is based on mixed communication models wherein the device communicates directly and through edge node. For secure communication between devices, we propose the use of lightweight encryption algorithm, namely, PRESENT [28]. We have implemented this encryption algorithm to evaluate its lightweight behavior. Moreover, propose the use of ML and DL-based intelligent algorithm at the fog layer. This is because the robotic devices and edge node have limited abilities of computational power and battery capacity that are prerequisites for executing a ML and DL model.

6.1 PRESENT Cipher Implementation

PRESENT is an ultra-lightweight cryptographic algorithm, and this is also considered as a block cipher. In order to create the encrypted data, PRESENT combines the data and key in 32 steps, known as rounds, using 64 bits for the data and 80 bits for the key. The 64 most significant bits of the key are used in each cycle to perform a bit-by-bit XOR operation on the data. The resultant data is passed through two operation blocks: sBoxLayer and pLayer, which update the information acquired. Additionally, a key-update block modifies the key before each round. This procedure is depicted in Fig. 2 diagram.

The sBoxLayer replaces each of the 16 possible combinations with the information provided in Table 1, where x is each nibble and S[x] is the replacement value, changing the input data in groups of four bits (referred to as "nibbles") at a time. In

Fig. 2 General block working diagram of the PRESENT algorithm [29]

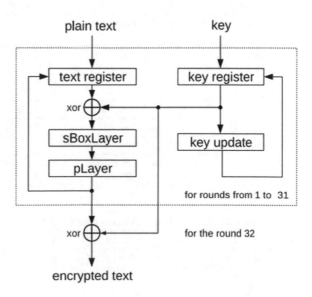

contrast, the pLayer exchanges or permutes each input bit individually in the order depicted in Table 2, where I denotes the input data bit position and P(i) denotes the output position. The key is updated by rotating the register left by 61 bit positions, passing the leftmost four bits through the sBoxLayer, and using an XOR with the current round's counter to act on the bits from 19 to 15.

The process is done 30 more times, and in the final round (round number 32), the data is only exclusive-ORed with the key without passing via the sBoxLayer and the pLayer. During each cycle, the processed data and the key are both updated. The encrypted data is the data that was obtained after round 32. Since the decipher key is same, the decipher must perform 32 rounds of cipher key updating before beginning the decryption process (Tables 3 and 4).

Table 1 Robotics-relevant protocols at various tiers of the protocol stack [11]

Layer	IoRT Protocols	Usage
Application	API, RIP, SNMP, TELNET	Data streaming
Transport	UDP, TCP	End-to-end communication
Network	IPv6, ARP, ICMP	Internet connectivity
Data link	Bluetooth, Wi-Fi, PPP	Robot-to-robot, access point, gateway communication
Physical	Robots, sensors, actuators	Data collection

Table 2 Summary of the related work

Cipher	Year	Technique	Key size	Block size	No. of rounds
CRAFT [18]	2019	SPN	128	64	32
CHAM [19]	2017	FEISTAL	128/256	64/128	80/96
IMPROVED LILLIPUT [19]	2017	EGFN	80	64	30
GIFT [20]	2017	SPN	128	64/128	28/40
SIT [21]	2017	FEISTEL + SPN	64	64	5
DLBCA [22]	2017	FEISTEL	80	32	15
LiCi [25]	2017	FEISTEL	128	64	31
SKINNY [24]	2016	SPN	64–384	64/128	32–56
MANTIS [24]	2016	SPN	128	64	10/12
SPARX	2016	SPN with ARX-based S-Boxes	128/256	64/128	20–40

Table 3 General block working diagram of the PRESENT algorithm [30]

x	0	1	2	3	4	5	6	7	8	9	A	B	C	D	E	F
S[x]	C	5	6	B	9	0	A	D	3	E	F	8	4	7	1	2

Table 4 Permutation bit order performed by the pLayer [29]

i	0	1	2	3	4	5	6	7	8	9	10	11	12	13	14	15
$P(i)$	0	16	32	48	1	17	33	49	2	18	34	50	3	19	35	51
i	16	17	18	19	20	21	22	23	24	25	26	27	28	29	30	31
$P(i)$	4	20	36	52	5	21	37	53	6	22	38	54	7	23	39	55
i	32	33	34	35	36	37	38	39	40	41	42	43	44	45	46	47
$P(i)$	8	8	40	56	9	25	41	57	10	26	42	58	11	27	43	59
i	48	48	50	51	52	53	54	55	56	57	58	59	60	61	62	63
$P(i)$	12	12	44	60	13	29	45	61	14	30	46	62	15	31	47	63

Table 5 System configuration

RAM	4 GB
Processor	i5
HDD	512 GB
Python	3.9.7

6.2 Results

See Table 5.

Algorithm name: **PRESENT**

Input word length: 8 bits

Average execution time for encryption: 0.00100088119506835941 ms

Average execution time for decryption: 0.002008676528930664 ms

See Fig. 3.

Fig. 3 Results

7 Conclusion

IoT and robotics are two phrases that each encompasses a wide range of technology and ideas. An emerging field called the Internet of Robotic Things aims to incorporate robotic technologies into IoT contexts. Robotic systems can share information and communicate with each other over the Internet of Robotic Things in order to carry out complex tasks. IoRT principles, technologies, applications, and current obstacles have been discussed in this paper. Energy-constrained devices and sensors will constantly communicate with one another, and the security of these communications cannot be compromised. In this paper, a lightweight security method called PRESENT is introduced for this purpose. PRESENT is one of the most lightweight encryption techniques. As a result, many researchers have been fascinated by the creation of lightweight block ciphers, notably over the past seven years.

Acknowledgements This paper supported by Subprogram 1.1. Institutional performance-Projects to finance excellence in RDI, Contract No. 19PFE/30.12.2021 and a grant of the National Center for Hydrogen and Fuel Cells (CNHPC)—Installations and Special Objectives of National Interest (IOSIN). This paper was partially supported by UEFISCDI Romania and MCI through projects AISTOR, FinSESco, CREATE, Hydro3D, EREMI, STACK, ENTA, PREVENTION, RECICLARM, UPSIM, SmartDelta, USWA, SWAM, DISAVIT, MEWS, iPREMAS and by European Union's Horizon 2020 research and innovation program under grant agreements No. 872172 (TESTBED2), No. 883522 (S4ALLCITIES).

References

1. Simoens P, Dragone M, Saffiotti A (2018) The internet of robotic things: a review of concept, added value and applications. Int J Adv Robot Syst 15(1):1–11. https://doi.org/10.1177/172 9881418759424
2. Yfantis EA, Fayed A (2014) Authentication and secure robot communication. Int J Adv Robot Syst 11(1):1–6. https://doi.org/10.5772/57433
3. Schwan KS, Bihari TE, Taulbee GM, Weide BW (1987) High-performance operating system primitives for robotics and real-time control systems. ACM Trans Comput Syst 5(3):189–231. https://doi.org/10.1145/24068.24070
4. Dixon C, Frew EW (2009) Maintaining optimal communication chains in robotic sensor networks using mobility control. Mobility Netw Appl 14:281–291. https://doi.org/10.1007/s11036-008-0102-0
5. Sabry SS, Qarabash NA, Obaid HS (2019) The road to the internet of things: a survey. In: 2019 9th Annual information technology, electromechanical engineering and microelectronics conference (IEMECON). IEEE, pp 290–296. https://doi.org/10.1109/IEMECONX.2019.8876989
6. Yoshino D, Watanobe Y, Naruse K (2021) A highly reliable communication system for internet of robotic things and implementation in RT-middleware with AMQP communication interfaces. IEEE Access 9:167229–167241. https://doi.org/10.1109/ACCESS.2021.3136855
7. Liao B, Ali Y, Nazir S, He L, Khan HU (2020) Security analysis of IoT devices by using mobile computing: a systematic literature review. IEEE Access 8:120331–120350. https://doi.org/10.1109/ACCESS.2020.3006358

8. Tsai W-C, Tsai T-H, Xiao G-H, Wang T-J, Lian Y-R, Huang S-H (2020) An automatic key-update mechanism for M2M communication and IoT security enhancement. In: 2020 IEEE International conference on smart internet of things (SmartIoT). IEEE, pp 354–355. https://doi.org/10.1109/SmartIoT49966.2020.00067

9. Ray PP (2016) Internet of robotic things: concept, technologies, and challenges. IEEE Access 4:9489–9500. https://doi.org/10.1109/ACCESS.2017.2647747

10. Gulzar M, Abbas G (2019) Internet of things security: a survey and taxonomy. In: 2019 International conference on engineering and emerging technologies (ICEET). IEEE, pp 1–6. https://doi.org/10.1109/CEET1.2019.8711834

11. Villa D, Song X, Heim M, Li L (2021) Internet of robotic things: current technologies, applications, challenges and future directions. arXiv:2101.06256v1, pp 1–8. Retrieved from https://arxiv.org/pdf/2101.06256.pdf

12. Park J-H, Baeg S-H, Ryu H-S, Baeg M-H (2008) An intelligent navigation method for service robots in the smart environment. IFAC Proc Volumes 41(2):1691–1696

13. Kumar N, Takács M, Vámossy Z (2017) Robot navigation in unknown environment using fuzzy logic. In: 2017 IEEE 15th International symposium on applied machine intelligence and informatics (SAMI). IEEE, pp 279–284. https://doi.org/10.1109/SAMI.2017.7880317

14. Zhu Y, Zhang-Shen R, Rangarajan S, Rexford J (2014) Cabernet: connectivity architecture for better network services. In: CoNEXT '08: proceedings of the 2008 ACM CoNEXT conference, pp 1–6. https://doi.org/10.1145/1544012.1544076

15. Alsulaimawi Z (2020) A privacy filter framework for internet of robotic things applications. In: 2020 IEEE security and privacy workshops (SPW). IEEE, pp 262–267. https://doi.org/10.1109/SPW50608.2020.00059

16. Khalid S (2021) Internet of robotic things: a review. J Appl Sci Technol. Trends 2(3):78–90. https://doi.org/10.38094/jastt203104

17. Romeo L, Petitti A, Marani R, Milella A (2020) Internet of robotic things in smart domains: applications and challenges. Sensors (Switzerland) 20(12):1–23. https://doi.org/10.3390/s20123355

18. Rajkumar K, Kumar CS, Yuvashree C, Murugan S (2019) Portable surveillance robot using IoT. Int Res J Eng Technol 6(3):94–97

19. Hadipour H, Sadeghi S, Niknam MM, Song L, Bagheri N (2019) Comprehensive security analysis of CRAFT. IACR Trans Symmetric Cryptol 2019(4):290–317. https://doi.org/10.13154/tosc.v2019.i4.290-317

20. Adomnicai A, Berger TP, Clavier C, Francq J, Huynh P, Lallemand V, Le Gouguec K, Minier M, Reynaud L, Thomas G (2019) Lilliput-AE: a new lightweight tweakable block cipher for authenticated encryption with associated data. Submitted to NIST Lightweight Project, pp 1–58

21. Rani DJ, Roslin SE (2021) Optimized implementation of gift cipher. Wireless Pers Commun 119(3):2185–2195. https://doi.org/10.1007/s11277-021-08325-2

22. Mishra Z, Mishra S, Acharya B (2021) High throughput novel architecture of SIT cipher for IoT application. In: Nanoelectronics, circuits and communication systems. Lecture notes in electrical engineering, vol 692. Springer, pp 267–276. https://doi.org/10.1007/978-981-15-7486-3_26

23. AlDabbagh SSM (2017) Design 32-bit lightweight block cipher algorithm (DLBCA). Int J Comput Appl 166(8):17–20. https://doi.org/10.5120/ijca2017914088

24. Beierle C et al (2016) The SKINNY family of block ciphers and its low-latency variant MANTIS. In: Advances in cryptology – CRYPTO 2016. CRYPTO 2016. Lecture notes in computer science, vol 9815. Springer, Berlin, Heidelberg, pp 123–153. https://doi.org/10.1007/978-3-662-53008-5_5

25. Patil J, Bansod GV, Kant KS (2017) LiCi: a new ultra-lightweight block cipher. In: 2017 International conference on emerging trends & innovation in ICT (ICEI). https://doi.org/10.1109/ETIICT.2017.7977007

26. Seok B, Lee C (2019) Fast implementations of ARX-based lightweight block ciphers (SPARX, CHAM) on 32-bit processor. Int J Distrib Sens Networks 15(9):1–9. https://doi.org/10.1177/1550147719874180

27. Sengupta J Ruj S, Bit SD (2019) End to end secure anonymous communication for secure directed diffusion in IoT. In: ICDCN '19: Proceedings of the 20th international conference on distributed computing and networking, pp 445–450. https://doi.org/10.1145/3288599.3295577
28. Kumari AS, Mandi MV (2019) Implementation of present cipher on FPGA for IoT applications. IJERT 8(8):26–29
29. Santa FM, Jacinto E, Montiel H (2019) PRESENT cipher implemented on an ARM-based system on chip. In: Data mining and big data. DMBD 2019. Communications in computer and information science, vol 1071. Springer, pp 300–306. https://doi.org/10.1007/978-981-32-9563-6_31
30. Jacinto EG, Montiel HA, Martínez FS (2017)Implementation of the cryptographic algorithm 'present' in different microcontroller type embedded software platforms. Int J Appl Eng Res 12(19):8092–8096

Energy-Efficient Priority-Based Routing Model for Smart Health Care

Umer Gull, Yashwant Singh⑩, and Bharti Rana⑩

Abstract The Internet of Things is a network that connects a huge number of plans to collect data over the internet. The internet of possessions has a wide variety of applications including smart health care. Using IoT, the patient can monitor medical data at any time and from any location in this circumstance. Smart health care is one of the key areas where IoT infrastructures and solutions are widely used to facilitate the best possible patient surveillance, accurate diagnosis, and timely operation of patients with existing diseases. Smart healthcare systems, on the other hand, face various challenges including energy consumption, data integrity, data privacy, and data transmission speed. Keeping in view the challenges, we have suggested a system model based on priority routing. The proposed model uses RPL along with priority approach for prioritizing the healthcare data. The healthcare data is classified into different classes: medium, critical, and normal to deliver the most critical data to the healthcare centers first. As a result, patients are not limited to a certain selection of health centers and specialists at particular instance. RPL works for small power and lossy networks, and the QoS provides a certain bandwidth to the data to reach the destination. The congestion overhead of the channels is controlled by the time division multiple access. The time division multiple access time slot is used for synchronization between the source and destination to reduce energy use. The data in a TDMA slot is used to check the traffic. If the traffic is high, normal data is sent; otherwise, priority data is sent. Our proposed model promises the delivery of healthcare data on time by using RPL with TDMA. The DODAG's topology ensures healthcare data to be traveled by the shortest possible path to improve performance metrics.

Keywords Internet of Things (IoT) · Smart health care · Sensors · Priority-based routing

U. Gull · Y. Singh · B. Rana (✉)
Department of Computer Science and Information Technology, Central University of Jammu, Jammu, Jammu and Kashmir 181143, India
e-mail: 0250519.csit@cujammu.ac.in; rana9bharti@gmail.com

Y. Singh
e-mail: Yashwant.csit@cujammu.ac.in

© The Author(s), under exclusive license to Springer Nature Singapore Pte Ltd. 2023 329
Y. Singh et al. (eds.), *Proceedings of International Conference on Recent Innovations in Computing*, Lecture Notes in Electrical Engineering 1011,
https://doi.org/10.1007/978-981-99-0601-7_25

1 Introduction

The proliferation of the Internet of Things and related computers have resulted in the development of smart systems such as smart cities, smart transportation systems, smart energy, and smart home. The IoT interconnects all objects (living and non-living) to form a physical network, and all processes such as sensing, processing, and communication are impulsively controlled and supervise without human intervention [1]. The healthcare business has developed immensely by adopting some level of automation since there is so much room for improvement. By bringing together medical devices, healthcare professionals, and patients, the present healthcare business has progressed beyond hospitals [2]. By transforming traditional healthcare systems into current smart healthcare systems, the Internet of Things has great change in the field of health care [3]. SHS was created to deal directly with patient health-related information [4]. SHS gives in-depth insights on illness symptoms and minimizes the need for frequent health checks, which can aid the elderly, diabetics, and others with secure management. The process of selecting a pathway for data to go from its source to its destination is known as routing. Routing is done by a router, which is unique equipment. Under the OSI paradigm, a router operates at the network layer, whereas in the TCP/IP model, it operates at the internet layer.

Smart health care uses a replacement generation of knowledge like the Web of Things, big data, and cloud computing to remodel the usual medical system in an all-around result. Nowadays, health care becomes more efficient, convenient, and personalized. With an aim of introducing the concept of smart health care, we first review the literature concerning the prioritized routing protocols for better quality of services. The prevailing problems in smart healthcare system are addressed to formulate solutions. Lastly, we attempt to propose a priority-based model to reduce delay during the transmission of data. Priority routing is the priority assigned to data for accessing a destination when capacity becomes available [5]. The priority is allocated to two or more entities or classes of data that are waiting for a routing destination to become available [6]. The priority-based routing breaks a pull by deciding which entity has access to the destination location first when it becomes available. Routing could be a process of choosing a path along which the information transits starting from the source to the destination. Routing is done by a router, which is unique equipment inside this OSI model; the router operates at the network topology using the internet protocol inside the TCP/IP model [7]. The router usually forwards packets using information from packet header and sending table. The packets are routed using routing techniques. The routing protocol is nothing more than code that determines the best path for the packets to travel.

Routing protocols use metrics to work out the simplest path for the packet delivery from source to the destination such as hop count, bandwidth, latency, and current network load. The routing algorithm generates and maintains the routing table for the path determination process. Routing performance is determined by performance measures such as hop count. The hop count indicates how many times a packet must pass through internetworking procedure such as a router on its way from source to destination. Using hop as a major criterion, the path with the fewest hops will be deemed as the optimum way for getting from source to destination. Similarly, the time it takes the router to process queue and transfer a datagram to a workflow is referred to as delay. This measure is used by the protocols to calculate the delay morals for all acquaintances along the path from beginning to end. The optimum path will be the one with the smallest delay.

Following is the rest of the analysis: The review of literature is covered in Sect. 2. The problems of smart health care are discussed in Sect. 3. Section 4 describes the proposed smart healthcare model; in this model, we discuss RPL and its working and QoS with algorithm and various benefits in smart health care. Conclusion and future work are presented in Sect. 5.

2 Literature Survey

The [8] used RPL-based solution for reducing IoT device energy consumption. Time division multiple access time slot is employed to synchronize the correspondent and recipient and decrease energy usage. The proposed technique was tested using NS-2 to evaluate the proposed and traditional methods' routing overhead, energy consumption, and end-to-end delay.

In [9], the author presented priority-based energy-efficient routing protocol (PEERP) for reliable data transmission in smart health Care. The PEERP divides health data into two categories: emergency circumstances and important health data, based on their importance. Based on the division and importance provided by PEERP to healthcare data packet, the efficiency of the model is tested, using simulation software. After the simulation experiment, the PEERP technique extended the lifetime of the networks.

In [10], RPL protocol is used for minimizing carbon emissions and energy consumption with significant improvement in QoS metrics. The RPL protocol was applied on low-power healthcare gadgets including smart processes, which use a hybrid paradigm to restore power efficiency and transmitter status of RPL, and procedures are used to calculate the ETX value for a connection after sending a data packet.

The [11] devised a routing and transference power control algorithm to outline an energy-efficient, reliable, and cost-effective RPL system for IoT approach. The restriction of this suggested model is that all of the nodes that are investigated are homogenous and have an RF transceiver CC2420.

The [12] used a routing strategy that will improve energy efficiency and, as a result, sensor longevity. The efficient protocol for dual statistical method supports the suggested idea for IoT in health care (EERP-DPM). MATLAB software and the MySignals HW V2 hardware platform were used to design and test the proposed system.

The [4] formulated a system in which healthcare information identification was included in the IP data packet at the sensor level, QoS was modified at the router level, and the uppermost priority was given to the healthcare data package routing. This system was tested by using TI LaunchPad.

The [13] developed an energy-efficient solution for IoT applications in which nodes consume less power and preserve the network's lifespan. For IoT applications, a network and communication power quality algorithm is built for stable, energy-efficient operation. For a network containing nodes, using different types of RF transceivers, the suggested solution might not perform well. The comparative analysis of various priority based routing mechanisms are depicted in Table 1.

3   Smart Health Care: Challenges

With data in this crowded environment coming from different IoT devices, the basic challenge is to manipulate the data and give high priority to health-related data so that patients will be taken care of in a smarter way in this modern world. There are different challenges faced during transmission of data so that it may reach to its destination in proper time. These challenges include energy consumption, volume of data, data integrity and privacy, data transmission speed, throughput, and delay that are discussed subsequently.

- **Energy Consumption**

  Energy consumption is the major concern that various researchers are dealing within the IoT-driven environment [18]. The energy is consumed due to the frequent occurrence of collisions, repetitive congestion due to limited bandwidth of the channel, and data transmission to longer distances. Another factor that contributes to energy consumption is the limited battery capacity, computing speed, and memory of IoT devices.
- **Volume of Data**

  IoT devices generate the big amount of data known as big data that is characterized by volume, velocity, and variety. As the bandwidth of the channel is limited, it becomes difficult to route the data. Therefore, employing priority to the transmitted data solves the problem to some extent [19]. The massive amount of data generation results into data loss and delay and minimizes the throughput. Therefore, efficient and reliable mechanisms must be developed to tackle the big data generated from smart motes.
- **Data Integrity and Privacy**

  In today's healthcare industry, data integrity is a chronic issue. Data integrity assures that the information is accurate and is not tampered with in any way [20]. The data integrity is approached by the fabricated hackers that have continuous access to the data that serves the purpose of money making and gets useful insights from data. Therefore, data integrity must be ensured in smart healthcare system. For the healthcare business to preserve patient's personal information and comply with laws, data privacy is critical. Due to the influence of various attacks like DDoS

Table 1 Comparative analysis of priority-based routing techniques in the state-of-the-art IoT

Author/year	Model	Techniques	Simulation tools	Analysis	Demerits	Performance metrics
El Zouka and Hosni [4]/2017	RPL-based system for IoT	Self-rejoining algorithm	RPL(QU-RPL)	A RPL-based IoT system that is reliable and low-cost	Nodes using multiple types of RF transceiver do not perform well	Throughput, latency
Ambarkar and Shekokar [10]/2020	RPL model	Ant colony-inspired algorithm Decentralized ant-based algorithm	TDMA	Growing network efficiency in provisos of best packet	Node's maximum speed was 25 m/s, only	Bandwidth, delay
Hathaliya et al. [3]/2020	For reliable data transmission in PEERP	Reliable routing algorithm	MATLAB platform	Enlarge the lifetime of the system	ATTEMPT protocol not perform well after 2500 rounds	Network lifetime, throughput and path loss
Choudhary et al. [14]/2020	In a congested IoT environment, a sensor-driven approach to priorities healthcare data	Priority decision algorithm	TICC1310 launchpads	Proposed technology could greatly enhance remote medical operations	Routing issue of healthcare data packets	Delay, latency
Debroy et al. [15]/2021	Dual prediction model (EERP-DPM)	Decision-making algorithm LMS algorithm	EERP-DPM	Increasing the network lifetime of sensors in IoT	Network duration is less	Throughput, end-to-end delay
Proposed model/2022	Priority-based smart healthcare model	Priority algorithnm using QoS	TI LaunchPad	Saving time, cost, and more importantly, human lives		Bandwidth, delay

(continued)

Table 1 (continued)

Author/year	Model	Techniques	Simulation tools	Analysis	Demerits	Performance metrics
Safara et al. [8]/2020	Three-tier clustering technique	Routing algorithm	NS-3	Extend network lifetime	Many problems from general to definite problems	Delay, latency
McGhin et al. [16]/2020	BHEEM	ANT colony-inspired algorithm	NS-3	To cesign and analyze a smart health monitoring system	Implementation of smart contract is not defined formally	Bandwidth, throughput
Ray et al. [17]/2019	S2SH framework	Priority decision algorithm	TICC1310 Launchpads	A detailed framework for a smart healthcare system is developed	The integration of various subsystem must be validate before approval	Pass loss, end-to-end delay

attack, man in the middle, and battery drainage attacks, maintaining data privacy becomes crucial. Advanced machine learning mechanisms and blockchain-based solutions must be imposed to ensure data privacy of data.

- **Data Transmission Speed**

  The primary source of energy consumption in IoT devices is data transmission [21]. Processing and delivering redundant data consume part of the energy used by these devices. Therefore, reducing the number of redundant transmissions preserves the energy of the network. There must be a trade-off between the data transmission speed and the required bandwidth to ensure the reliable transmission of data.

- **Throughput**

  The amount of packets that preserve be delivered to the end medical server is decided by this performance indicator, with more throughput indicating better network quality. Routing services by means of high performance along with low packet loss are required for the patient monitoring system [22]. The normal network life, which corresponds to the amount of active sensor nodes, determines the amount of packets received at the medical server.

- **End-to-End Delay**

  End-to-end delay is the amount of time it takes for a packet of data to get from the origin to the destination node [23]. The IoT is implemented in healthcare applications to communicate sensitive data from the sensor nodes to a health server. The discovered data is not always ordinary; in certain circumstances, it is critical; as a result, it must be transmitted to the target system fast.

4   Proposed Smart Healthcare Model

The proposed smart healthcare model entails sensing layer followed by assigning priorities by adding a label M, prioritized routing layer, and healthcare services as illustrated in Fig. 1. The functionality of each layer is discussed in detail in the following section.

(A) **Sensing Layer**

  IoT systems are being developed for a range of applications, urban services, including health care and smart city transportation. On the opposite side, collecting massive volumes of enormous data for multimedia content from these networks usually causes traffic jam within the principal network. We suppose that a sensor is installed on somebody's body to capture certain healthcare data like diabetes sugar levels, A1C, and so on. The sensors capture this information and classify data into three classes. The classified data is sent as input to priority algorithm which prioritizes the information and provides critical data with the highest priority. The priority will be provided to the headers of critical data as M/11, normal data headers will be assigned as M/01, and medium data will be

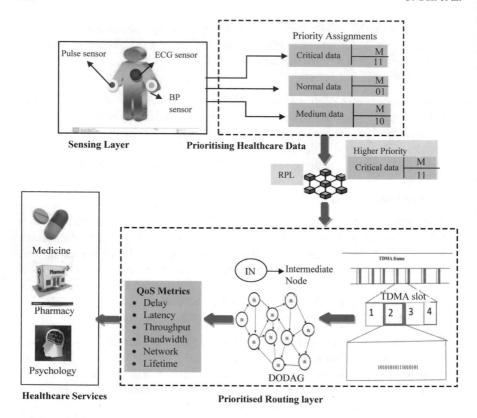

Fig. 1 Prioritized routing model in smart healthcare system

provided as M/10. Critical data is then fed into RPL, which then goes through
DODAG's topology [24] to locate the best data delivery method.

(B) **Prioritized Routing Layer**

RPL uses a vigorous process to enlarge and maintain a directed, non-circular
graph, destination-oriented routing architecture. The information is focused on
the DODAG root during this graph [25]. From each node to the DAG's root,
the sides form a route. If the system is in an extremely constant condition, RPL
employs a limited DODAG information object (DIO) beacon technique to keep
the structure of the DODAG router. The DIO beacon is controlled by using
trickle timer. The trickling timer is employed to come up with RPL messages.
Trickle timers permit nodes to limit the amount of messages they send while
maintaining network strength. As long as a node gets communication that is
companionable with its own data, it will exponentially increase the amount of
direct packets it sends until it reaches its highest point. The suggested DODAG
is described by the routing input with DODAG $= (V, N)$, where V is the
collection of nodes and N represents number of bits to communicate in an
extremely packet in accordance with Eq. (1):

$$V = (V_0(N_s), \ V_1(N_{10}), \ V_2(N_{11}), \ V_3(N_{20}), \ V_4(N_{21}),$$
$$nV_i(N_{i0}), \ V_i + 1(N_i), \ V_i + 2(N_0(i + 1), \ V_i + 3(N_1(i + 1)))) \tag{1}$$

Here, V_0 is indeed the DODAG root, N_s denotes the first bit, and $N_0 i$ and N_1 i denote the value of 0 and 1 in binary coding, respectively. The time division multiple access slot is used to match between the source and destination and reduce energy utilization. From the information during a TDMA slot [26], first, the traffic is examined. Regular data is sent while traffic is heavy; otherwise, priority data is sent. Node priorities and data rates are used to allocate space for sending packets. It is the coordinator's fault for allocating time slots. If a number of nodes have the identical priorities, choosing TDMA would be supported during the transmission rate, Frame values can be changed to use the data unit to put a frame in an idle mode.

To synchronize the transmitter and recipient, preliminary bits are used in the first section of the frame [27, 28]. While no data can be found on the channel, the receiver detects that acceptable data is placed on the channel by acquiring a sample of *01,010,101*. Once the receiver gets *10,101,011*, which contains two consecutive bits of one, it indicates that genuine data begins after two consecutive ones and is ready to receive. The subsystem modifies its preliminary bits such that the rest of the bits are received. The number of repeats is determined by the initial settings for the module established by the information center inside the preliminary bits. The following bits are delivered as the destination when the data on the channel has been validated, i.e., after getting the consecutive bits *11* of the instance *10,101,011* inside the preliminary bits. The sender's location of the frame is represented by the source node. The address of the equipment to which the message is transmitted is saved in the receiver's location field to determine which is to be sent. The size of the payload component of the packet is defined in the control part of the packet. If a notification must be transmitted, the acknowledgment bit inside the packet test area is set to true. The burden also refers to the space used by the signal processor to receive data from the sensor via the serial connection between both the signal processor and the sensor. The payload size can be specified anywhere between *0* and *32* bytes in the device settings page. The cyclic redundancy check at the frame's top is responsible for the frame's integrity. The validity of the frame is tested when the error checking code is enabled if the CRC does not equal the frame which is incorrect.

(C) **Quality of Service (QoS)**

The QoS receives prioritized data via DODAG [29]. Since all the data is health-care data, QoS will further prioritize the health care into critical and normal automatically by using priority mechanism [30]. The data headers will further be modified to give priorityto critical data. QoS will halt all other data and transmit the critical data by providing a particular bandwidth and hence reduce delay. QoS returns to standard processing when critical healthcare data has been sent.

Algorithm 1: Modified QoS

Data: Data with a header M = higher priority
Result: Prioritizing the data
1 if if received data indication with header M = higher priority **then**
2   **while** Data with a header M! = NULL **do**
3   Grant channel for specific data and suspend
             other ongoing processing;
4   Data with a header M+ + ;
5   **end**
6 else
7   Execute a normal transaction;
8 end

Thus, the healthcare data passing via the proposed model will reach to healthcare centers, and hence, the health of patients will be monitored, and patients will receive appropriate instructions or treatment in time.

5 Conclusion

In this study, we present insights to the researchers regarding routing challenges in smart healthcare system (SHS). In smart healthcare system, most critical data is often delayed to the destination location due to the inappropriate routing models incorporated in SHS. This delay in delivery of data to the health centers poses the lives of patients at risk. Therefore, our proposed model has a potential to mitigate several challenges along with maintaining QoS metrics. The devised model is a hybrid model that includes priority mechanism, RPL layer, and TDMA. The priority mechanism assigns priority as per the criticality of data. This mechanism provides due importance to the critical data by attaching the label to the classes of data. RPL protocol is used for less power and lossy networks as IoT networks are constrained in resources. Also, DODAG topology is used for routing so that the data will be transmitted through the shortest path. The congestion overhead of the channels is controlled by TDMA. If the traffic is high, normal data will be sent first; otherwise, prioritized data will be sent. Our proposed model promises the delivery of healthcare data on time to avail the services in an energy-efficient manner. We use the suggested approach in automobile networks in the future. Furthermore, we will utilize meta-heuristic approach for managing moving packets in IoT during routing.

References

1. Kumar A, Krishnamurthi R, Nayyar A, Sharma K, Grover V, Hossain E (2020) A novel smart healthcare design, simulation, and implementation using healthcare 4.0 processes. IEEE Access 8:118433–118471. https://doi.org/10.1109/ACCESS.2020.3004790

2. Akkaş MA, Sokullu R, Çetin HE (2020) Healthcare and patient monitoring using IoT. Internet Things 11:100173. https://doi.org/10.1016/j.iot.2020.100173

3. Hathaliya JJ, Tanwar S, Tyagi S, Kumar N (2019) Securing electronics healthcare records in Healthcare 4. 0: a biometric-based approach R. Comput Electr Eng 76:398–410. https://doi.org/10.1016/j.compeleceng.2019.04.017

4. El Zouka HA, Hosni MM (2019) Secure IoT communications for smart healthcare monitoring system. Internet Things 13:100036. https://doi.org/10.1016/j.iot.2019.01.003

5. Rana B, Singh Y, Singh PK (2021) A systematic survey on internet of things: energy efficiency and interoperability perspective. Trans Emerg Telecommun Technol 32(8):1–41. https://doi.org/10.1002/ett.4166

6. Rana B, Singh Y (2021) Internet of things and UAV: an interoperability perspective. In: Unmanned aerial vehicles for internet of things (IoT): concepts, techniques, and applications, pp 105–127. https://doi.org/10.1002/9781119769170.ch6

7. Rana B, Singh Y, Singh H (2021) Metaheuristic routing: a taxonomy and energy-efficient framework for internet of things. IEEE Access 9:155673–155698. https://doi.org/10.1109/ACCESS.2021.3128814

8. Safara F, Souri A, Baker T, Ridhawi IA, Aloqaily M (2020) PriNergy: a priority-based energy-efficient routing method for IoT systems. J Supercomput 76:8609–8626. https://doi.org/10.1007/s11227-020-03147-8

9. Soufiene BO, Bahattab AA, Trad A, Youssef H (2020) PEERP: An priority-based energy-efficient routing protocol for reliable data transmission in healthcare using the IoT. Procedia Comput Sci 175(2020):373–378. https://doi.org/10.1016/j.procs.2020.07.053

10. Ambarkar SS, Shekokar N (2020) Toward smart and secure IoT based healthcare system. In: Internet of things, smart computing and technology: a roadmap ahead. Studies in systems, decision and control, vol 266. Springer, Cham, pp 283–303. https://doi.org/10.1007/978-3-030-39047-1_13

11. Newaz AI, Sikder AK, Rahman MA, Uluagac AS (2020) A survey on security and privacy issues in modern healthcare systems : attacks and defenses. ACM Trans Comput Healthc 2(3):1–44. https://doi.org/10.1145/3453176

12. Mawgoud AA, Karadawy AI, Tawfik BS (2019) A secure authentication technique in internet of medical things through machine learning. arXiv:1912.12143, pp 1–24

13. Reena JK, Parameswari R (2019) A smart health care monitor system in IoT based human activities of daily living: a review. In: 2019 International conference on machine learning, big data, cloud and parallel computing (COMITCon). IEEE, pp 446–448. https://doi.org/10.1109/COMITCon.2019.8862439

14. Choudhary A, Nizamuddin M, Singh MK, Sachan VK (2019) Energy budget based multiple attribute decision making (EB- MADM) algorithm for cooperative clustering in wireless body area networks. J Electr Eng Technol 14(1):421–433. https://doi.org/10.1007/s42835-018-000 06-8

15. Debroy S, Samanta P, Bashir A, Chatterjee M (2019) SpEED-IoT: Spectrum aware energy efficient routing for device-to-device IoT communication. Futur Gener Comput Syst 93:833–848. https://doi.org/10.1016/j.future.2018.01.002

16. McGhin T, Choo K-KR, Zhechao CL, He D (2019) Blockchain in healthcare applications : research challenges and opportunities. J Netw Comput Appl 135:62–75. https://doi.org/10.1016/j.jnca.2019.02.027

17. Ray PP, Dash D, Salah K, Kumar N (2021) Blockchain for IoT-based healthcare: background, consensus, platforms, and use cases. IEEE Syst J 15(1):85–94. https://doi.org/10.1109/JSYST.2020.2963840

18. Hossain MS, Muhammad G (2018) Emotion-aware connected healthcare big data towards 5G. IEEE Internet Things J 5(4):2399–2406. https://doi.org/10.1109/JIOT.2017.2772959
19. Chen SK et al (2012) A reliable transmission protocol for zigbee-based wireless patient monitoring. IEEE Trans Inf Technol Biomed 16(1):6–16. https://doi.org/10.1109/TITB.2011.2171704
20. Shakeel PM, Baskar S, Dhulipala VRS, Mishra S, Jaber MM (2018) RETRACTED ARTICLE: Maintaining security and privacy in health care system using learning based deep-Q-networks. J Med Syst42:186. https://doi.org/10.1007/s10916-018-1045-z
21. Luo E, Bhuiyan MZA, Wang G, Rahman MA, Wu J, Atiquzzaman M (2018) PrivacyProtector: Privacy-protected patient data collection in IoT-based healthcare systems. IEEE Commun Mag 56(2):163–168. https://doi.org/10.1109/MCOM.2018.1700364
22. Elhoseny M, Ramírez-González G, Abu-Elnasr OM, Shawkat SA, Arunkumar N, Farouk A (2018) Secure medical data transmission model for IoT-based healthcare systems. IEEE Access 6:20596–20608. https://doi.org/10.1109/ACCESS.2018.2817615
23. Moosavi SR et al (2015) SEA: A secure and efficient authentication and authorization architecture for IoT-based healthcare using smart gateways. Procedia Comput Sci 52(1):452–459. https://doi.org/10.1016/j.procs.2015.05.013
24. Mahmud MA, Abdelgawad A, Yelamarthi K (2017) Energy efficient routing for Internet of Things (IoT) applications. In: 2017 IEEE International conference on electro information technology (EIT). IEEE, pp 442–446. https://doi.org/10.1109/EIT.2017.8053402
25. Besher KM, Beitelspacher S, Nieto-Hipolito JI, Ali MZ (2021) Sensor Initiated healthcare packet priority in congested IoT networks. IEEE Sens J 21(10):11704–11711. https://doi.org/10.1109/JSEN.2020.3012519
26. Koutras D, Stergiopoulos G, Dasaklis T, Kotzanikolaou P, Glynos D, Douligeris C (2020) Security in IoMT communications: a survey. Sensors 20(17):4828. https://doi.org/10.3390/s20174828
27. Redondi A, Chirico M, Borsani L, Cesana M, Tagliasacchi M (2013) An integrated system based on wireless sensor networks for patient monitoring, localization and tracking. Ad Hoc Netw 11(1):39–53. https://doi.org/10.1016/j.adhoc.2012.04.006
28. Balasubramanian V, Otoum S, Aloqaily M, Ridhawi IA, Jararweh Y (2020) Low-latency vehicular edge: a vehicular infrastructure model for 5G. Simul Modell Pract Theory 98:101968. https://doi.org/10.1016/j.simpat.2019.101968
29. Saha HN et al (2017) Health monitoring using Internet of Things (IoT). In: 2017 8th Annual industrial automation and electromechanical engineering conference (IEMECON). IEEE, pp 69–73. https://doi.org/10.1109/IEMECON.2017.8079564
30. Sciences H (2016) 済無 4(1):1–23

An Internet of Things-Based Mining Worker Safety Helmet Using ESP32 and Blynk App

N. Umapathi⬤, Poladi Harshitha, and Ireddy Bhavani

Abstract In underground mining industries, there are many factors that may risk the lives of workers. We have seen many people who lost their livelihood due to emergency situations in the mines. These emergency situations occur due to the leakage of harmful gases or sudden explosions, and also workers may undergo some kind of ventilation problems. As we know, coal contributes more in the production of energy, so mining of a coal plays a major role. By considering all these factors, we have introduced a smart helmet system which helps the workers to get warned under any emergency circumstances. The only objective of this prototype is to ensure the safety of the workers. And, the technique we use to do this is to provide a proper communication between the worker and the monitoring system. The communication can be provided using various methods such as Bluetooth, ZigBee, Wi-Fi modules. Here, we have chosen Wi-Fi module, i.e., ESP32 as it provides higher transmission rate along with covering longest distance. The data can be received by a monitoring system through the Blynk app which is an open-source IoT platform. The gases exist in the environment are detected by our smart helmet using the related sensors and are sent to the monitoring system. By implementing this model, we have created a communication so that the worker can feel safe to enter the mines and can concentrate on his/her work.

Keywords ESP32 · Blynk app · Wireless communication · Sensors

1 Introduction

The main problem we face in the underground mining is toxic atmosphere, ventilation-related problems, and unexpected explosions, and these may result in developing diseases like tuberculosis, silicosis, etc., and breathing problems that affect the worker's complete life. Due to the high digging levels in the mines, the

N. Umapathi (✉) · P. Harshitha · I. Bhavani
Department of Electronic and Communication Engineering, Jyothishmathi Institute of Technology & Science, Karimnagar, Telangana, India
e-mail: nrumapathi@gmail.com

© The Author(s), under exclusive license to Springer Nature Singapore Pte Ltd. 2023   341
Y. Singh et al. (eds.), *Proceedings of International Conference on Recent Innovations in Computing*, Lecture Notes in Electrical Engineering 1011,
https://doi.org/10.1007/978-981-99-0601-7_26

temperature also gets raised which is abnormal to the normal conditions. "Deeper the mines, greater the risk". In the world, India holds the fifth largest and considered as a home for vast coal reserves. In the year 2020, environment ministry has cleared the way for 14 new coal mine projects. And, it is estimated that by 2024, one billion tons of coal production can be raised. Along with the increase in the production and development, the risk is also being increased for the mine workers. As per the records, coal mines have recorded for highest number of accidents in mines. According to the ministry data, averages of 549 deaths were reported every year.

At present, the workers in mining industry only use a normal helmet for protection against the probable dangerous bumps. So, by adding some technologies to that helmet, which helps in communication between worker and the monitor, helps to a larger extent to identify the hazardous atmosphere inside the mine. As of now, the prototype helmets made for the safety of mining workers were done using a different wireless technologies like Bluetooth, ZigBee, etc., and with different kinds of sensors. Our prototype is developed in the advancement for the existing systems. We have used Wi-Fi technology with the combination of four different sensors to cover the maximum possibilities for the indication of hazardous occurrences in the mines. For overcoming and maintaining the friendly atmosphere in underground mines, our design helps in implementing a better communication technology for sensing and warning. This monitoring system allows the mine to intellectually make changes based on the data monitored in the Blynk app which is an open source of IoT. This kind of progression using IoT technology plays a prominent role in ensuring safety of mining workers.

2 Literature Survey

In 2017, Sravani and Rambabu implemented ZigBee technology in order to ensure the safety of the people who works in a mining industry and also to control and adapt the changes in the mining environment [1]. In 2016, Behr et al. designed a system using a various sensors for determine hazardous gases anda also which helps in warning about the condition which causes human to suffocate [2]. In 2018, Kumar and Reddy used sensors like micro electromechanical system (MEMS) and infrared (IR) sensors which are proficient enough in divulging the conditions and to keep a track of the conditions where there is a precarious situations. For this Prevailing, they have used ZigBee transmitter so as to monitor frequently and efficiently [3]. In 2017, Dhanalakshmi et al. modeled a module that monitors the continuous tracking using the GPS location and also to update the data which will be received through the sensors [4].

In 2017, Deokar et al. displayed parameters when harmful gases like CO2, methane, carbon monoxide, etc., out scripts the general limit and helps to get aware of the surroundings which helps the worker out there to get noticed and be cautious [5]. In 2018, Borkar and Baru mapped data that is composed regarding the circumstances and also displaying that data on light-emitting diode and further that is upgraded on the monitoring system using ThingSpeak which is an IoT open-source platform [6, 7]. In 2020, Tajane et al. made a module that imparts a well-founded communication using ZigBee module between the worker and the monitoring [8]. In 2013, Pradeepkumar and Usha recognized the high-risk cases that exist in the mines like accretion of unhealthy gases and also determine the collisions etc. [9]. In 2015, Pradeepkumar et al. provided a workable and the full fledge outcome for establishing private LoRa network using both the hardware and software algorithms [10].

In 2020, Kumar et al. aimed and researched regarding the analysis of providing a secure environment using a technology named LoRa [11]. In 2017, Jagadeesh and Nagaraj implemented a smart helmet for the detection of unhealthy situations in the mining areas using Internet of Things [12]. In 2014, Shabina created a helmet using technologies like wireless sensor network and radio frequency for ensuring the safety of the underground mining workers [13]. In 2013, Ussdek et al. designed a system detecting high sensitivity gas and monitoring system [14]. In 2010, Sheng and Yunlong Accident analyzed and counter measured the coal and gas for mining safety and environmental protection [15, 16]. In 2009, Zhou et al. made a module in order to monitor the gases exist using a chain-type WSN [17].

3 Proposed Method

In this proposed method, we use four input devices, namely, alcohol sensor, gas sensor, temperature sensor, and smoke sensor. In these, the alcohol sensor senses the ethanol presents in the air, gas sensor detects the gases present in atmosphere, temperature sensor measures the temperature which is of air, liquid, or solid temperatures, and smoke sensor module senses the smoke which is an indicator of fire. And, the output devices are liquid crystal display for self-monitoring and buzzer for warning in any case of dangers detected. All these are connected to the ESP32 Wi-Fi module, and it acts as the main device which helps in the communication in between input and output devices. Power supply is given for ESP32 Wi-Fi module (Fig. 1).

4 Circuit Diagram of Proposed Method

ESP32 is the main component in this project. The components used in this project are alcohol sensor, smoke sensor, gas sensor, DHT-11 temperature sensors, LCD (16 * 2), and buzzer. The negative terminal of MQ-6, MQ-3, and DHT-11 is linked to the ground pin of ESP32.The positive terminal of gas sensor is coupled to the G18 pin of ESP32, whereas alcohol sensor is tie up with the G19 pin of ESP32, DHT-11 sensor is linked to G15 pin of ESP32, and the another smoke sensor is bridged to G21 pin of ESP32 Wi-Fi module. Buzzer positive terminal is connected to G13 pin (Fig. 2).

Fig. 1 Block diagram of proposed method

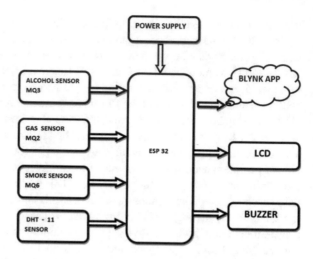

Fig. 2 Circuit diagram of proposed method

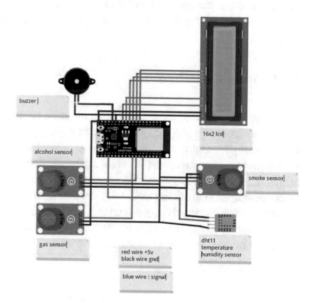

5 Components

1. ESP32

    ESP32 is a general-purpose microcontroller which is a System on Chip (SoC) along with the capabilities of handling wireless connections like Wi-Fi and Bluetooth. It is a low-power system which can be handled on a dual mode. And, it is low in cost compared to other microcontroller systems with all capabilities. ESP32 has 26 general-purpose input and output pins along with Vin, gnd, and

Fig. 3 ESP32

enable pins, out of which 15 pins can be used as ADC pins and 2 pins as DAC pins (Fig. 3).

2. Gas Sensors

We use three kinds of gas sensors for three different sensing operations. One is MQ-3 sensor for detecting alcohol, i.e., for sensing harmful gases like isopropyl. And, the other sensors are MQ-2 and MQ-6 for sensing combustible gases and smoke, respectively. The operation involved in all these MQ series of sensors is they contain a coil or a heater inside them, which gets heated when it comes in contact with reactants (Fig. 4).

3. DHT-11 Sensor:

DHT-11 is a digital humidity and temperature sensor which is generally of low cost. It contains a capacitive sensor for measuring humidity and a thermorcsistor for measuring the temperature. It does not need any analog input pins. And, it can be easily interfaced with any kind of microcontrollers (Fig. 5).

Fig.4 Gas sensors

Fig.5 DHT-11

6 Result Analysis

Blynk App Monitoring System Output
See Fig. 6.

The change in relative humidity affects the temperature. As the cool air contains less vapor than the hot air, we observed that the temperature gets raised as there is no moisture is presented in the air. That is, humidity is inversely proportional to the temperature, and here, we have observed it practically (Table 1).

Blynk App Monitoring System Output

Fig. 6 Temperature and humidity monitored values

Table. 1 Temperature vs. humidity values

Trial	Temperature	Humidity
Trial 1	27.1	59
Trial 2	32.3	57
Trial 3	33.3	55

Fig.7 Blynk app monitoring output for alcohol sensor

Fig. 8 Blynk app monitoring output for smoke sensor

Fig. 9 Blynk app monitoring output for gas sensor

When presence of harmful gases like isopropyl phenol is detected, alcohol sensor senses it; we get an alert message like shown in Fig. 7.

If there is any fire around or nearer to mining worker, MQ-6 sensor senses the smoke, then we receive an alert message as shown in Fig. 8.

Combustible gases like methane, carbon monoxide, LPG, etc., are detected by MQ-2 sensor, and then, alert message we receive is shown in Fig. 9.

Figure 10 shown is prototype of our proposed system—"Mining worker safety helmet using IOT".

7 Conclusion

Finally, we conclude that a smart helmet which helps mining workers to get rid of various risk factors has been developed. It is cost-effective and very efficient, result is now that workers can have safe place to work. Being operated using IoT makes its operation easily tracked and also helps in better transfer of information. The components being used in it have integrated features which also help in ease of operation. And, monitoring of the output can be accessed or monitored by any individual using any open-source IoT platform; here, we make use of Blynk app, which is easy to understand and monitor. The only aim or objective of this proposed

Fig. 10 Prototype model of
proposed method

system is to create a safe environment for mining workers and to provide better
communication, which was achieved finally.

8 Future Scope

In future, this prototype can be implemented in the advancement of existing model for
the recognition of various kinds of dangerous conditions such as the accumulation of
carbon monoxide gases, exhalation in the mine areas. Additionally, this model can be
implemented by adding the sensors like pressure sensor, IR sensor, etc. This system
can also be enhanced by the implementation of few attributes such as the strength
and ranges of the signal, and also, the worker's heartbeat and the blood pressure can
be checked and monitored by adding the required equipment.

References

1. Sravani B, Rambabu K (2017) A smart and secured helmet for mining workers. Int J Adv Res
   Trends Eng Technol (IJARTET) 4(3):112–118
2. Behr CJ, Kumar A, Hancke GP (2016) A smart helmet for air quality and hazardous event detection for the mining industry. In: 2016 IEEE International conference on industrial technology
   (ICIT). IEEE, pp 2026–2031. https://doi.org/10.1109/ICIT.2016.7475079
3. Kumar GR, Reddy BK (2018) Internet of things based an intelligent helmet for wireless sensor
   network. Int J Eng Sci Res Technol (IJESRT) 7(6):88–92
4. Dhanalakshmi A, Lathapriya P, Divya K (2017) A smart helmet for improving safety in mining
   industry. Int J Innov Sci Res Technol (IJISRT) 2(3):58–64
5. Deokar SR, Kulkarni VM, Wakode JS (2017) Smart helmet for coal mines safety monitoring
   and alerting. Int J Adv Res Comput Commun Eng (IJARCCE), ISO 3297:2007 Certified.
   6(7):1–7

6. Umapathi N, Teja S, Roshini, Kiran S (2020) Design and implementation of prevent gas poisoning from sewage workers using Arduino. In: 2020 IEEE International symposium on sustainable energy, signal processing and cyber security (iSSSC). IEEE, pp 1–4. https://doi.org/10.1109/iSSSC50941.2020.9358841
7. Borkar SP, Baru VB (2018) IoT based smart helmet for underground mines. Int J Res Eng Sci Manage (IJESM) 1(9):52–56
8. Tajane PS, Shelke SB, Sadgir SB, Shelke AN (2020) IoT mining tracking & worker safety helmet. Int Res J Eng Technol (IRJET) 7(4):5587–5590
9. Pradeepkumar G, Usha S (2013) Effective watermarking algorithm to protect electronic patient record using image transform. In: 2013 International conference on information communication and embedded systems (ICICES). IEEE, pp 1030–1034. https://doi.org/10.1109/ICICES.2013.6508251
10. Pradeepkumar G, Prasad CV, Rathanasabhapathy G (2015) Effective watermarking algorithm to protect electronic patient record using DCT. Int J Softw Hardware Res Eng 3(11):16–19
11. Kumar GP, Saranya MD, Tamilselvan KS, SU Jhanani, Iqbal MJL, Kavitha S (2020) Investigation on watermarking algorithm for secure transaction of electronic patient record by hybrid transform. In: 2020 Fourth international conference on I-SMAC (IoT in social, mobile, analytics and cloud) (I-SMAC). IEEE, pp 379–383. https://doi.org/10.1109/I-SMAC49090.2020.9243411
12. Jagadeesh R, Nagaraj R (2017) IoT based smart helmet for unsafe event detection for mining industry. Int Res J Eng Technol 4(1):1487–1491
13. Shabina S (2014) Smart helmet using RF and WSN technology for underground mines safety. In: 2014 International conference on intelligent computing applications. IEEE, pp 305–309. https://doi.org/10.1109/ICICA.2014.105
14. Ussdek MEM, Junid SAMA, Majid ZA, Osman FN, Othman Z (2013) High-sensitivity gas detection and monitoring system for high-risk welding activity. In: 2013 IEEE Conference on systems, process & control (ICSPC). IEEE, pp 256–261. https://doi.org/10.1109/SPC.2013.6735143
15. Sheng XZ, YunlongZ (2010) Accident cause analysis and counter measure of coal and gas outburst nearly two years of our country. Min Saf Environ Prot 37(1):84–87
16. Umapathi N, Sabbani S (2022) An Internet of Things (IoT)-based approach for real-time kitchen monitoring using NodeMCU 1.0. In: Futuristic communication and network technologies. VICFCNT 2020. Lecture notes in electrical engineering, vol 792. Springer, Singapore, pp 35–43. https://doi.org/10.1007/978-981-16-4625-6_4
17. Zhou G, Zhu Z, Chen G Hu N (2009) Energy-efficient chain-type wireless sensor network for gas monitoring. In: 2009 Second international conference on information and computing science. IEEE, pp 125–128. https://doi.org/10.1109/ICIC.2009.140

Time Series-Based IDS for Detecting Botnet Attacks in IoT and Embedded Devices

Sonal Sharma, Yashwant Singh, and Pooja Anand

Abstract The existing intrusion detection systems (IDS) have found it very demanding to detect growing cyber-threats due to the voluminous network traffic data with increasing Internet of Things (IoT) devices. Moreover, security attacks, on the other hand, tend to be unpredictable. There are significant challenges in developing adaptive and strong IDS for IoT to avoid false warnings and assure high detection efficiency against attacks, especially as botnet attacks become more ubiquitous. Motivated by these facts, in this paper, different types of botnet attacks have been studied and how they are more conveniently launched with open and vulnerable IoT devices. Then, the growing trend of deep learning (DL) techniques is being studied extensively for their ability to detect botnet attacks by learning from time series data specifically in the IoT environment. Hackers are exploiting the Internet of Things (IoT), creating millions of new vulnerability points in critical infrastructure. We must build greater consensus on IoT security standards and trust in security across critical infrastructure.

Keywords IoT · Intrusion detection · Deep learning · RNN · LSTM · GRU

1 Introduction

The Internet of Things (IoT) has gotten enormous attention in the recent past as a result of its unique applications and support for a variety of fields, considering industrial applications, health care, automation, environmental sensing, and so on. The Internet of Things needs an infrastructure upon which programs, devices, and

S. Sharma (✉) · Y. Singh · P. Anand
Central University of Jammu, Jammu, India
e-mail: isonal0000@gmail.com

Y. Singh
e-mail: yashwant.csit@cujammu.ac.in

P. Anand
e-mail: poojaanand892@gmail.com

© The Author(s), under exclusive license to Springer Nature Singapore Pte Ltd. 2023   351
Y. Singh et al. (eds.), *Proceedings of International Conference on Recent Innovations in Computing*, Lecture Notes in Electrical Engineering 1011,
https://doi.org/10.1007/978-981-99-0601-7_27

services may be interconnected and used to receive, interact, store, analyze, and exchange information from the physical world [1]. The Internet of Things (IoT) is an expanding media concept in which things embedded with sensors and actuators can detect their surroundings, interact with one another, and exchange information over the Internet. To provide useful information and make timely decisions, a large number of sensors and actuators are required for real-time monitoring and the environment of various industrial sectors [2]. There are approximately 50 billion IoT devices connected to the Internet, and this amount is likely to expand in the next few years. These devices provide a large amount of data that can be used by various applications. Regardless of the fact that it provides a vast variety of services and applications, it is vulnerable to cyberattacks [3]. Cyber attackers can harm systems in several different ways, such as restricting access, theft of information, or eliminating a specific target [4]. Because of its diversity, immense scale, limited hardware resources, and universal availability of IoT systems, IoT security has become a challenge [5]. However, several aspects of IoT security, such as data validation, transparency, and access controls, have been strengthened. These security mechanisms are designed to work in conjunction between the user and the IoT, yet they still have security flaws. These security flaws in IoT can lead to a wide variety of issues, and research is needed to address them.

As a result, a different module is required to provide IoT network security []. Network security can be handled by the use of a software application known as an intrusion detection system (IDS), which examines the system's or networks' harmful things [6]. The basic goal of any IDS is to distinguish between abnormal and normal traffic patterns [7]. The presence of anomalies often wastes a lot of resources and leads to a serious situation. The detection of anomalies may have a significant impact on the overall efficiency of monitoring systems. Several techniques to implement anomaly based intrusion detection systems have been presented, but machine learning (ML) is currently a popular approach among researchers [8]. Machine learning-based IDSs for security against IoT networks have been described in many surveys. Machine learning advances have brought a new era for artificial intelligence and opened the path for the creation of intelligent intrusion detection systems [9]. Machine learning (ML) is widely regarded as one of the best computational paradigms for providing embedded intelligence in IoT devices. In tasks including classification, regression, and density estimation, machine learning is applied. Various applications such as fraud detection, virus detection, authentication, speech recognition, and bioinformatics use machine learning techniques. Machine learning algorithms are complex; they require a huge amount of domain knowledge as well as humanitarian assistance and are only willing of doing what they are designed for.

Deep learning models are frequently used to tackle problems involving sequential input data, such as time series. Time series forecasting is part of predictive analytics [10]. In the time series analysis, it is possible to do regression analysis against a set of past values of the variables. In most anomaly intrusion detection approaches, time series patterns are used to train the models. In the IoT environment, numerous methods have been developed to expose the abnormality in historical data using real-time analytics and the prediction of abnormal actions [11]. However, the use of DL algorithms in anomaly detection techniques has made it easier to distinguish most of the features that humans may not be able to notice [12]. Deep learning is a subset of machine learning. It follows unsupervised learning [13]. Deep learning solves complex machine learning problems; its architecture is based on the human brain [14]. The network learns to map the input features to the output due to numerous levels of abstraction. Deep learning network intrusion detection systems are categorized according to their architecture and methodologies [15]. Various deep learning models used for the analysis of intrusion detection are deep neural net classifier (similar to logistic regressor), autoencoder-based self-taught learning model, and RNN [16]. RNN is a type of artificial neural network. RNN takes into account not only the present input instance but also what they have learned in past. RNN works on the time series data and is used to solve problems using sequential input data. RNN has the limitation of processing input in a specific order. RNN suffers from the problem of vanishing gradient to overcome this LSTM is used.

This research aims to probe deeper into the development of deep learning approaches for detecting intrusions in IoT environments. In this paper, we will investigate and analyze the existing deep learning-based techniques used to detect intrusions in the IoT environment with the goal of precise understanding of existing methods and introducing a deep learning approach that would help us to detect intrusions in the IoT environment.

2 Different Types of Botnet Attacks

Botnet is a network of infected nodes turning out to be army of different network zones, aiding hackers to control multiple smart systems to launch malicious activities [17]. Botnet identification has become a subject of the growing interest in cybersecurity. Botnets are a form of Internet attack assault that aims to simultaneously hijack several computers and put them into "zombie" systems. Botnets are a web of infected or hijacked computers that are used for things like sending spam emails, distributing malware, and framing DDoS assaults. The device owner's authorization is not required for botnet activation. Botnets are quickly becoming the most dangerous to online environments and computation assets [18]. DDoS assaults, spams, malwares, and phishing emails, pirated movies including softwares, click fraud, prompt action, stealing information and computing resources, corporate extortion, and identity theft are all common uses for malevolent botnets [19].

The processes of creating a botnet can be divided into the following categories: activate by exposing, infecting, and growing. To infect the victim with ransomware, the hacker will first find a weakness in either a website, program, or user activity. A bot herder's role is to fool victims via hiding the exposure and subsequent malware infestation. They make use of websites/application programs vulnerabilities to transmit malware through emails, drive/Trojan horse downloads. In the next step, malware infects the victim devices, allowing it to take control of them. Using techniques such as web downloads, exploitation tools, popup advertisements, and email attachments, hackers can produce zombie devices after the initial malware infection. For the centralized botnet, the herder works differently and leads the network of bots to a command and control server (C&C). In peer propagation, there is a peer-to-peer botnet, and thus infected devices work on connecting with additional zombies. When the bot herder has corrupted a large number of bots, they can organize their attacks in stage 3. To receive their order, the zombie devices will download the most frequent response from the command and control channel. The bot then executes its commands and participates in hostile behavior. This way its herder continues to maintain and expand its bot network from afar, allowing them to carry out a variety of horrible crimes. Botnets do not pinpoint individuals because the purpose of the bot herder is to infect as many devices as possible so that malicious assaults can be carried out.

Figure 1[21] shows the working of a botnet. This procedure is divided into several steps. Botnets, when fully operational, may carry out large-scale attacks. To improve a botnet's ability, hackers must provide it with additional machinery or gadgets. Bot herder is necessary to guide the network's connected infected devices. It works with remote commands and directs the gadgets to complete specific tasks. Mirai botnet attacks and Zeus botnets attacks are an example of botnets [21]. Botnet identification is difficult due to the diversity of botnet protocols and architecture. Botnet attacks in specific are extremely challenging. After capturing the botnet, the intruder can use the C&C server to manage the devices and perform attacks against the target hosts. Any attacker will be attracted as it becomes easy to infect and the produced bot population has become stable.

3   IoT-Based Botnet Attacks

IoT is widely accepted and used in a variety of fields, including healthcare, smart homes, and agriculture [22]. On the other hand, the Internet of Things faces resource constraints and varied surroundings, such as low processing power and storage. These limitations make it difficult to provide and deploy a secure system in IoT devices. These limits worsen the IoT environment's already-existing issues. As a result of the vulnerability of IoT devices, numerous types of attacks are feasible [23]. The IoT-based botnet threat is among the most common; it grows quickly and has a larger effect than other types of attacks. A botnet attack is a substantial cyberattack launched by fully automated ransomware machines. For a botnet controller, it turns infected devices into "zombie bots" [24]. The smart bot is generally referred as a software robot which will hunt for vulnerable smart nodes and will infect them to make them the part of a larger botnet as conventional bots do. It is similar to malware propagation procedure running in the background. The management of the smart botnet is handled by a malicious node that deploys these bots to complete tasks cohesively. Distributed attacks, spams, phishing, clicking frauds, spambots, brute force attacks, and spyware are all examples of coordinated action [25]. The dearth

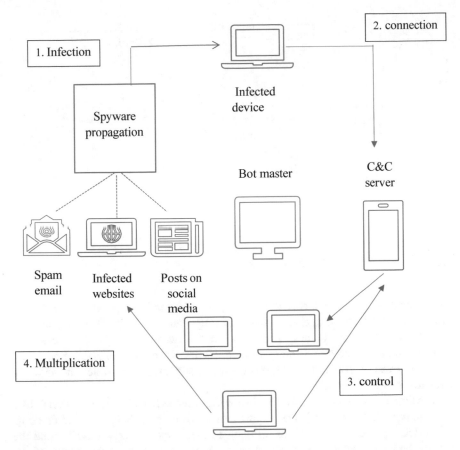

Fig. 1 Working of botnet

of fundamental protection leads to malware infestation, in which victim opens the attachment in a malicious email, turning IoT devices into bots. The lifecycle of an IoT botnet is comparable to that of a typical botnet. It is divided into four stages. The initial infection is the first step, followed by C&C, then attack and recovery phases the final levels.

Botnets are posing a substantial risk to the growing IoT networks. According to Hamid et al. [26], there are no security standards in place for IoT devices, such as authentication, encryption, or access control techniques [27]. Figure 2 shows different types of malware in IoT environments [28]. In safeguarding against security assaults such as botnets, a lack of oversight standards poses a significant issue for the IoT sustainability. A botnet being the network of maliciously infected computers, with infected devices known under the name of bots, and they may be managed remotely using a device known as the bot master. With the predicted rise in botnet attacks, research and industries have proposed a number of botnet detection/mitigation solutions. The controversial Mirai botnet being the most damaging DDoS assaults in

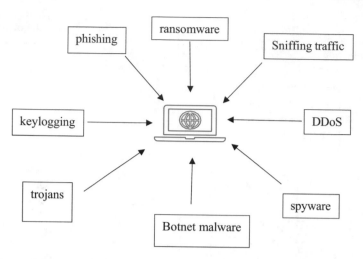

Fig. 2 Types of malware in IoT environment

recent years needs to be addressed in priority. By constantly propagating to weavers, Mirai conducts a DDoS against a group of target servers. With the rising severity of DoS assaults, smart botnet target both small offices/home offices devices with large organizations covering enterprises. In the first six months of 2019, Kaspersky discovered around 100 million assaults on smart devices, among which the majority were IoT-based botnet attacks.

The first IoT botnet reported by Malware Must Die in late 2016 was written in Lua programming language [29]. The majority of its army is made up of cable modems using ARM processors as well as utilizing Linux. Rapidity Networks17 found the Hajime botnet in October 2016, and it uses a similar infection mechanism to Mirai. Hajime uses the BitTorrent distributed hashtag (DHT) protocol for peer finding and the Torrent Transport Protocol for data transfer. Each message is encrypted with RC4 and signed with public and private keys [30]. Similar to Mirai, Bricker-Bot, a Busy-Box-based IoT botnet was discovered by Radware researchers in April 2017. This malware launches a permanent denial-of-service (PDoS) assault against IoT systems by exploiting SSH server default username and password, misconfigurations, or known vulnerabilities to deface a device's firmware, erase all files from its memory, and reconfigure network parameters.

Persirai15 is an IoT botnet that uses Mirai's software and has been active since April 2017 [31]. TCP port 81 is used to access the interface of specific vendors' webcams. If successful, it then uses a universal plug and play (UPnP) vulnerability to gain access to the client's router, downloads the malicious binaries, and then deletes them after execution [32]. Rather than brute-forcing camera credentials, the virus spreads by making use of a known zero-day vulnerability that allows attackers to acquire the password file directly [33]. User Datagram Protocol flooding attacks are part of the DDoS attack arsenal. Persirai is a threat to an estimated 120,000 devices in the ecosystem. Mirai mutants are still being generated daily, and the notion that they might spread and inflict real harm using the same infiltration techniques as the original threat demonstrates IoT device manufacturers' persistent disdain for even basic security procedures [34]. If the security industry does not respond more quickly and build new countermeasures, increasingly sophisticated attacks may become the norm, threatening the Internet's infrastructure [35].

4 Time Series-Based Models Used for Detecting Botnet Attacks

With the rapid development of threats and the variability of attack methods, Internet of Things (IoT) devices faces significant challenges in detecting security flaws and attacks [36]. Many detection technologies and methodologies that use full time series data during malware operation and are based on machine learning/deep learning are becoming more popular. Almost all current intrusion detection and prevention systems do not harness the power of time series modeling [37]. Software developers will be able to better manage resource allocation and system readiness to fight against malicious activity by using time series models. Time series analysis is a technique for forecasting future events based on the premise that future predictions would be similar to previous patterns. Forecasting is the practice of fitting models to historical data to forecast future values [38]. Time series analysis is a dynamic study area that has gained a lot of attention because of its potential applications in a wide variety of fields [39]. Much effort was put into developing and refining the development and refinement of time series analysis models during the last several decades. One of the most critical concerns among the areas of study ranging from dimensionality reduction to data segmentation is time series prediction for acquiring future trends and tendencies [40]. The results can be used for a variety of applications, such as production planning, control, optimization, and so on. As a result, different models for handling this problem have been presented, such as Auto-Regressive Integrated Moving Average (ARIMA) filtering-based approaches support vector machines, and so on.

Machine learning and deep learning are important in every sector making machines intelligent. Examples of machine learning are everywhere. Machine learning algorithms are complex and require a lot of domain knowledge and human intervention, whereas deep learning holds more promise for AI creators and the rest of the world in this aspect. Researchers have made considerable progress in detecting botnets using classical machine learning methods such as Naive Bayes, SVM, random forests, and networking algorithms such as DBSCAN and X-means based on a variety of aspects to create a model that can distinguish malicious network traffic over the last few decades [39]. These detection models have low false-positive and false-negative rates. Artificial neural networks (ANNs) are versatile computational frameworks that may be used to accurately solve a wide range of time series problems. However, using artificial neural networks to model linear problems has shown varied results, so it's not a good idea to use them indiscriminately on any data.

Through network traffic, the convolutional neural network (CNN) primarily learns spatial characteristics from the spatial dimension. Malicious behavior affects IoT devices; therefore, a CNN-based deep learning algorithm was proposed to detect small variations in power consumption data. CNN is most suitable for spatial data and classification such as images, it takes fixed-size input and generates fixed-size output [41]. From the time dimension, a recurrent neural network (RNN) learns the properties of data traffic in time series. In 1, author proposed a model based on the DL method, i.e., CNN+RNN that extracts features from two dimensions time and space automatically [39]. RNN detects botnet by characterizing communication patterns in a network as a series of time-varying states. Bi-LSTM—RNN model was built to identify botnet activities in consumer IoT devices and networks [42]. The benefit of using the Bi-LSTM neural network to obtain data for detection was that it could extricate features for classification by learning the perspective correlation of vectors in the sequence more effectively. LSTM model is used to detect network anomaly detection. LSTM networks are a sort of RNN that includes a hybrid of special and standard units. RNN suffers from the problem of exploiting and vanishing gradient to solve this problem LSTM and GRU were introduced [43]. LSTM addresses these issues by introducing new gates, such as input and forget gates that allow for greater control of the gradient flow and improved preservation of "long-range dependencies". GRU was introduced as a modification for LSTM to do machine translation for easy structure and convenience to solve problems.

5   Conclusion

Botnets have been seen as a severe threat to cybersecurity as they serve as a platform for a variety of cybercrimes, including DDoS attacks against critical targets, virus distribution, phishing, and click fraud. Despite the fact that malevolent botnets have been around for a long time, still very difficult to detect them at the growing phase, and botnet research is still in its infancy. In this paper, we focused on IoT intrusion detection systems, and we studied different machine learning and deep learning models to detect various forms of IoT network attacks on time series data, which are typically carried out by botnets. We also analyzed the detrimental consequences of Mirai, its variations, and other similar botnets on the Internet depicting the reality that IoT devices reflect. We find that machine learning models are complex and are commonly utilized in projects that require predicting outcomes or identifying trends. Machine learning is used when there is limited, structured data available. Deep learning models work well with time series data and perform well for predictive analysis. Because of its greater accuracy when trained with massive volumes of data, deep learning is becoming increasingly popular with RNN as the best intrusion detection method when compared to other ML algorithms. Further, the implementation of an RNN-based technique to detect botnet attacks with less false alarm rate will be carried out in the future.

References

1. Al-Taleb N, Saqib NA (2022) Towards a hybrid machine learning model for intelligent cyber threat identification in smart city environments. Appl Sci (Switzerland) 12(4):1863. https://doi.org/10.3390/app12041863
2. Malhotra P, Singh Y, Anand P, Bangotra DK, Singh PK, Hong W-C (2021) Internet of things: evolution, concerns and security challenges. Sensors 21(5):1–35. https://doi.org/10.3390/s21051809
3. Anand P, Singh Y, Selwal A, Alazab M, Tanwar S, Kumar N (2020) IoT vulnerability assessment for sustainable computing: threats, current solutions, and open challenges. IEEE Access 8:168825–168853. https://doi.org/10.1109/ACCESS.2020.3022842
4. Anand P, Singh Y, Selwal A, Singh PK, Felseghi RA, Raboaca MS (2020) IoT: internet of vulnerable things? Threat architecture, attack surfaces, and vulnerabilities in the internet of things and its applications towards smart grids. Energies (Basel) 13(18):4813. https://doi.org/10.3390/en13184813
5. Smys S, Basar A, Wang H (2020) Hybrid intrusion detection system for internet of things (IoT). J ISMAC 2(4):190–199. https://doi.org/10.36548/jismac.2020.4.002
6. Putchala MK (2017) Deep learning approach for intrusion detection system (IDS) in the internet of things (IoT) network using gated recurrent neural networks (GRU). Retrieved from https://etd.ohiolink.edu/apexprod/rws_etd/send_file/send?accession=wright1503680452498351&disposition=inline
7. Lazarevic A, Ertoz L, Kumar V, Ozgur A, Srivastava J (2003) A comparative study of anomaly detection schemes in network intrusion detection. In: Proceedings of the 2003 SIAM international conference on data mining, pp 1–12. Retrieved from https://epubs.siam.org/doi/epdf/https://doi.org/10.1137/1.9781611972733.3
8. Sinclair C, Pierce L, Matzner S (1999) An application of machine learning to network intrusion detection. In: Proceedings 15th annual computer security applications conference (ACSAC'99). IEEE, pp 371–377. https://doi.org/10.1109/CSAC.1999.816048
9. 2018 10th Computer Science and Electronic Engineering (CEEC). IEEE, 2018.
10. Wu Y, Liu Y, Ahmed SH, Peng J, El-Latif AAA (2020) Dominant data set selection algorithms for electricity consumption time-series data analysis based on affine transformation. IEEE Internet Things J 7(5):4347–4360. https://doi.org/10.1109/JIOT.2019.2946753
11. Saufi SR, Ahmad ZAB, Leong MS, Lim MH (2019) Challenges and opportunities of deep learning models for machinery fault detection and diagnosis: a review. IEEE Access 7:122644–122662. https://doi.org/10.1109/ACCESS.2019.2938227
12. Ahmad Z, Khan AS, Shiang CW, Abdullah J, Ahmad F (2021) Network intrusion detection system: a systematic study of machine learning and deep learning approaches. Trans Emerg Telecommun Technol 32(1):e4150. https://doi.org/10.1002/ett.4150
13. Shone N, Ngoc TN, Phai VD, Shi Q (2018) A deep learning approach to network intrusion detection. IEEE Trans Emerg Top Comput Intell 2(1):41–50. https://doi.org/10.1109/TETCI.2017.2772792
14. Jiang K, Wang W, Wang A, Wu H (2020) Network intrusion detection combined hybrid sampling with deep hierarchical network. IEEE Access 8:32464–32476. https://doi.org/10.1109/ACCESS.2020.2973730
15. Otoum S, Kantarci B, Mouftah HT (2019) On the feasibility of deep learning in sensor network intrusion detection. IEEE Networking Lett 1(2):68–71. https://doi.org/10.1109/LNET.2019.2901792

16. Saharkhizan M, Azmoodeh A, Dehghantanha A, Choo K-KR, Parizi RM (2020) An ensemble of deep recurrent neural networks for detecting IoT cyber attacks using network traffic. IEEE Internet Things J 7(9):8852–8859. https://doi.org/10.1109/JIOT.2020.2996425

17. Ahmed AA, Jabbar WA, Sadiq AS, Patel H (2020) Deep learning-based classification model for botnet attack detection. J Ambient Intell Human Comput 13:3457–3466. https://doi.org/10.1007/s12652-020-01848-9

18. Malik R, Alankar B (2019) Botnet and botnet detection techniques. Int J Comput Appl 178(17):8–11

19. Stahlbock R, Weiss GM (2016) In: DMIN'16: the 12th International conference on data mining (Proceedings of the International Conference on Data Mining DMIN'16). Retrieved from http://www.dmin-2016.com/

20. Zhu Z, Lu G, Chen Y, Fu ZJ, Roberts P, Han K (2008) Botnet research survey. In: 2008 32nd Annual IEEE international computer software and applications conference. IEEE, pp 967–972. https://doi.org/10.1109/COMPSAC.2008.205

21. Kolias C, Kambourakis G, Stavrou A, Voas J (2017) DDoS in the IoT: mirai and other botnets. Computer (Long Beach Calif) 50(7):80–84. https://doi.org/10.1109/MC.2017.201

22. Ali I et al (2020) Systematic literature review on IoT-based botnet attack. IEEE Access 8:212220–212232. https://doi.org/10.1109/ACCESS.2020.3039985

23. Hussain F et al (2021) A two-fold machine learning approach to prevent and detect IoT botnet attacks. IEEE Access 9:163412–163430. https://doi.org/10.1109/ACCESS.2021.3131014

24. Jain LC, Tsihrintzis GA, Balas VE, Sharma DK (eds) (2019) Data communication and networks, vol 1049. Singapore: Springer Nature Singapore Pte Ltd. https://doi.org/10.1007/978-981-15-0132-6

25. Anton SD, Ahrens L, Fraunholz D, Schotten HD (2018) Time is of the essence: machine learning-based intrusion detection in industrial time series data. In: 2018 IEEE International conference on data mining workshops (ICDMW). IEEE, pp 1–6. https://doi.org/10.1109/ICDMW.2018.00008

26. Hamid H et al (2021) IoT-based botnet attacks systematic mapping study of literature. Scientometrics 126(4):2759–2800. https://doi.org/10.1007/s11192-020-03819-5

27. Institute of Electrical and Electronics Engineers (2016) In: 2016 3rd International conference on electronic design (ICED)

28. Malik R, Singh Y, Sheikh ZA, Anand P, Singh PK, Workneh TC (2022) An improved deep belief network IDS on IoT-based network for traffic systems. J Adv Transp 2022:1–17. https://doi.org/10.1155/2022/7892130

29. Mishra N, Pandya S (2021) Internet of things applications, security challenges, attacks, intrusion detection, and future visions: a systematic review. IEEE Access 9:59353–59377. https://doi.org/10.1109/ACCESS.2021.3073408

30. Cook AA, Mısırlı G, Fan Z (2020) Anomaly detection for IoT time-series data: a survey. IEEE Internet Things J 7(7):6481–6494. https://doi.org/10.1109/JIOT.2019.2958185

31. Samy A, Yu H, Zhang H (2020) Fog-based attack detection framework for internet of things using deep learning. IEEE Access 8:74571–74585. https://doi.org/10.1109/ACCESS.2020.2988854

32. International Joint Conference on Neural Networks, IEEE Computational Intelligence Society, International Neural Network Society, and Institute of Electrical and Electronics Engineers (2018) In: 2018 International joint conference on neural networks (IJCNN): 2018 proceedings. IEEE

33. Popoola SI, Ande R, Adebisi B, Gui G, Hammoudeh M, Jogunola O (2022) Federated deep learning for zero-day botnet attack detection in IoT-edge devices. IEEE Internet Things J 9(5):3930–3944. https://doi.org/10.1109/JIOT.2021.3100755

34. Bekerman D (2017) New-mirai-variant-launches-54-hour-DDoS-attack-against-US-college. In: Imperva. Retrieved from https://www.imperva.com/blog/new-mirai-variant-ddos-us-college/

35. Popoola SI, Adebisi B, Hammoudeh M, Gui G, Gacanin H (2021) Hybrid deep learning for botnet attack detection in the internet-of-things networks. IEEE Internet Things J 8(6):4944–4956. https://doi.org/10.1109/JIOT.2020.3034156

36. Denning DE (1987) An intrusion-detection model. IEEE Trans Software Eng SE-13(2):222–232. https://doi.org/10.1109/TSE.1987.232894
37. Gao X, Shan C, Hu C, Niu Z, Liu Z (2019) An adaptive ensemble machine learning model for intrusion detection. IEEE Access 7:82512–82521. https://doi.org/10.1109/ACCESS.2019.2923640
38. Henderson T (2020) TIME Series Analysis for Botnet Detection. Master's thesis, George Mason University
39. Han Z, Zhao J, Leung H, Ma KF, Wang W (2019) A review of deep learning models for time series prediction. IEEE Sens J 21(6):7833–7848. https://doi.org/10.1109/JSEN.2019.2923982
40. Zhang J, Pan L, Han Q-L, Chen C, Wen S, Xiang Y (2022) Deep learning based attack detection for cyber-physical system cybersecurity: a survey. IEEE/CAA J Automatica Sinica 9(3):377–391. https://doi.org/10.1109/JAS.2021.1004261
41. Wurzinger P, Bilge L, Holz T, Goebel J, Kruegel C, Kirda E (2009) Automatically generating models for botnet detection. In: Computer security – ESORICS 2009. ESORICS 2009. Lecture notes in computer science, vol 5789. Springer, Berlin, Heidelberg, pp 232–249. https://doi.org/10.1007/978-3-642-04444-1_15
42. Viinikka H, Debar L, Mé L, Séguier R (2006) Time series modeling for IDS alert management. In: ASIACCS '06: Proceedings of the 2006 ACM symposium on information, computer, and communications security, pp 102–113. https://doi.org/10.1145/1128817.1128835
43. Fu R, Zhang Z, Li L (2016) Using LSTM and GRU neural network methods for traffic flow prediction. In: 2016 31st Youth academic annual conference of Chinese association of automation (YAC), pp 324–328. https://doi.org/10.1109/YAC.2016.7804912

Swarm Intelligence-Based Energy-Efficient Framework in IoT

Simran, Yashwant Singh⑩, and Bharti Rana⑩

Abstract Internet of things (IoT) has been developed for use in a variety of fields in recent years. The IoT network is embedded with numerous sensors that can sense data directly from the environment. The network's sensing components function as sources, observing environmental occurrences and sending important data to the appropriate data center. When the sensors detect the stated development, they send this world data to a central station. Sensors, on the other hand, have limited processing, energy, transmission, and memory capacities, which might have a detrimental influence on the system. We have concentrated our current research on lowering sensor energy consumption in IoT network. This study chooses the most appropriate potential node in the IoT network to optimize energy usage. Throughout this paper, we suggest a fusion of techniques that combines PSO's exploitation capabilities with the GWO's exploration capabilities. The fundamental concept is to combine the strengths of the PSO's capability to exploit with Grey Wolf Optimizer's ability for efficient potential node selection. The proposed method is compared to the traditional PSO, GWO, Hybrid WSO-SA, and HABC-MBOA algorithms on the basis of several performance metrics.

Keywords Internet of Things (IoT) · Grey Wolf Optimizer (GWO) · Particle Swarm Optimization (PSO) · Swarm Intelligence (SI)

1 Introduction

The Internet of things (IoT) [1] has rapidly grown technologically in recent years, and there is an abundance of enabling gadgets that use this technology. The relations "Internet" and "Things" talk about a global network that is interconnected and based

Simran (✉) · Y. Singh · B. Rana
Department of Computer Science and Information Technology, Central University of Jammu, Jammu, Jammu and Kashmir 181143, India
e-mail: simrangupta665@gmail.com

Y. Singh
e-mail: Yashwant.csit@cujammu.ac.in

© The Author(s), under exclusive license to Springer Nature Singapore Pte Ltd. 2023
Y. Singh et al. (eds.), *Proceedings of International Conference on Recent Innovations in Computing*, Lecture Notes in Electrical Engineering 1011,
https://doi.org/10.1007/978-981-99-0601-7_28

on sensory, communication, networking, and information processing technologies. Every piece of equipment, including kitchen appliances, automobiles, thermostats, and air conditioners, can access the Internet. These devices include everything from domestic appliances to industrial machinery [2]. The new wireless sensory technologies have considerably increased the sensory capabilities of devices, extending the basic concept of IoT to ambient intelligence and autonomous control [3].

The term 'Internet of Everything has been used by both CISCO and Qualcomm (IoE) [4]. The term CISCO has a broader definition. People, data, processes, and objects are the "four pillars" of IoE. The IoE also enhances people's lives by extending corporate and industrial operations. IoE can collect and analyze factual data from billions of sensors that are linked to it, then utilize that data to enhance "automated and human-based activities". Other advantages include the application of IoE to aid in the achievement of national priorities, ecological responsibility, and socioeconomic objectives [5].

Embedded applications find a home on the Internet of Things. The majority of these applications rely on deeply embedded systems that must function on restricted energy sources like batteries or energy harvesters [6]. Meeting the application's energy requirements is a huge difficulty [7]. Low-cost wireless sensor nodes make up the Internet of Things network. Due to the enormous number of sensor nodes in WSN, routine maintenance such as battery replacement is difficult [8]. The node's energy source, which is often a battery, depletes its energy faster. The power usage increases as the range within the nodes grow. This study uses Swarm Intelligence (SI) [9, 10] based Particle Swarm Optimization (PSO) and Grey Wolf Optimization (GWO) approach to overcome the issue. By substituting a particle of the PSO with a modest chance with a particle partially enhanced using the GWO, we combine two methodologies. This has a significant impact on the remaining energy of nodes, as longer distances cost more energy. A hybrid model of PSO and GWO is developed with energy consumption and distance in mind, and it iteratively selects the optimal next node. In contrast to the greedy method, it seeks to minimize node distance and hence extend the network lifespan.

Mirjalili et al. [11] proposed the GWO method, which is focused on wolf poaching technique. GWO can be a better alternative for deployment in actual applications than Particle Swarm Optimization (PSO) and other optimization algorithms since it has less configurable parameters and is simpler. When applied with high-dimensional nonlinear objective functions, the GWO, like other optimization methods, has several drawbacks, such as being easily caught in the local optima [12]. Furthermore, the faster convergence speed of GWO makes managing the balance between exploitation and exploration more challenging. To overcome these limitations, a hybrid model of PSO and GWO is deployed in this paper. The stability of the PSO method is preserved with our hybrid approach, while exploration is aided by the GWO algorithm.

Table 1 Acronyms and their meaning

Acronym	Meaning
SI	Swarm Intelligence
IoT	Internet of Things
WSN	Wireless Sensor Network
PSO	Particle Swarm Optimization
ABC	Artificial Bee Colony
GWO	Grey Wolf Optimizer
CH	Cluster Head
pbest	Particle Best
gbest	Global Best

Table 1 lists all of the acronyms used in the text. The remainder of the paper is organized as follows: The literature work is discussed in Sect. 2. In Sect. 3, the GWO–PSO is examined. The described energy-efficient model is described in Sect. 4. The conclusion is presented in Sect. 5.

The research contributions of the study are as follows:

- To conduct comparative analysis of energy-efficient PSO-based technique in IoT.
- To define and design the energy-efficient framework for IoT.
- To hybridize the working of PSO and GWO for efficient potential node selection in IoT.

2 Literature Survey

Researchers are currently facing major issues in building an energy-efficient model. Various academics have worked on energy-efficient ways to create an optimized model, which will extend the network lifespan in WSNs as a result. The application of metaheuristic approaches in the energy sector is discussed subsequently.

Devika et al. [13] discussed a wireless sensor network (WSN) energy-efficient clustering approach. The author explains why, how, and where Swarm Intelligence (SI) is a technique for reducing energy use and making networks more energy efficient in this article. Among the numerous SI methods, the author discovered that the PSO technique is more efficient. SI algorithm is broken down into groups depending on the social behavior of insects, bacteria, birds, fish, and other creatures, with insect-based SI accounting for over half of the effort. The author used a variety of SI-based WSN clustering techniques, including ACO, PSO, and ABC, to reduce redundant data inside clusters.

Alqattan and Abdullah [14] found that PSO is more reliable than the ABC. Author used the ABC technique along with PSO algorithm to evaluate the Protein Structure Prediction. Different metrics are utilized to evaluate the performance of two algorithms, including Colony size (S) [total number of working and watching bees], Swarm population size (N), S1 stands for Self-confidence, and Swarm-confidence stands for (S2). Using these various parameters, the author demonstrates that the PSO methodology outperforms the Artificial Bee Colony in terms of Time, Average Number of Function Evaluation, and accuracy figures by 70%, 73 percent, and 3.6 percent, respectively.

Cluster heads (CHs) in a wireless sensor network (WSN) consume more energy due to increased overload for receiving and gathering data, according to Rao et al. [15] The author of this paper revealed a cluster head selection approach using Particle Swarm Optimization for energy efficiency. The CH selection is influenced by the fitness function of remaining energy, load, temperature, and aliveness of nodes. As a result, the cluster head is chosen to optimize network speed and network lifetime. Based on a high-energy node with low load, latency, range, and power heat, CH is chosen to enhance the fitness function should be maximized to increase the net-stability works and efficiency. In addition, Iwendi et al. [16] describe the fitness function to determine cluster head (CH) using α, β, Φ, ω, and θ as weighted parameters, and the computational parameters used in calculating fitness function are FFenergy (Energy computation), FFdistance (Distance computation), FFdelay (Delay computation), FFtemperature (Temperature computation), and FFload (Load computation). And the total of these values is the fitness function [8].

Vijayalakshmi and Anandan [17] discussed the Tabu-PSO model, a hybrid PSO and Tabu method to select the cluster head with the least power utilization rate in the cluster and to increase the flexibility to pick the CH in a IoT network by utilizing a hybrid heuristic approach, in [5]. By expanding the number of clusters and enhancing the node survival rate, Tabu research was used to increase the ethnic variety of PSO in order to prevent local optimal issues. In comparison with the low-energy adaptive clustering hierarchy algorithm and Particle Swarm Optimization, their suggested methodology effectively reduces the overall packet loss rate by 27.32 percent and the average end-to-end delay by an average of 1.2 s.

Further, in [11], the author introduces the GWO, a novel SI-based algorithm influenced with grey wolves. Twenty-nine test functions were used to evaluate the proposed algorithm's performance in terms of search, attack, avoidance of local optima, and convergence. The author discovered that GWO gave exceptionally competitive outcomes when contrasted to well-known heuristics such as PSO [18], GSA [19], DE, EP, and ES.

Furthermore, on the basis of network longevity and efficiency, a comparison is made by author in [20] between the GWO, ABC, and AFSA. The author concluded that GWO has a longer network lifespan than ABC and AFSA. In an IoT setting, the GWO used less energy than the ABC and AFSA. Also According to the author's results, GWO has some extent higher throughput than the ABC and AFSA.

Moreover, the author suggests a unique hybrid technique in [20] that combines Particle Swarm Optimization's (PSO) exploitative ability with the GWO's searching ability. By substituting a particle of the PSO with a tiny chance with a particle partially enhanced using the GWO, the author integrates two approaches. The findings show that the hybrid strategy successfully combines the two algorithms and outperforms all other approaches tested.

Table 2 summarizes the relative studies of several frameworks/models in IoT by various writers. The authors predicted the model's analysis, strategy, strengths, drawbacks, and quality metrics. In the model, numerous attributes were used to characterize metaheuristic methods and strategies.

3   Particle Swarm Optimization (PSO)

In 1995, John Kennedy and Eberhart introduced the PSO concept [18]. PSO is a methodology for population-based stochastic optimization. It is made up of a swarm of particles (fishes, birds, etc.) wandering across a search area for probable solutions to complex problems. Each individual has a velocity vector and a position vector that represents a possible solution to the issue. The velocity here refers to processing time or coverage, and the position here to the rank of a test case in a testing process. In addition, each particle has a little memory that remembers its own best position so far as well as a global best position achieved through interaction with its neighbors. Particle Swarm Optimization took what it had acquired from the situation and applied it to the optimization challenges. In Particle Swarm Optimization, every solution is a "bird" in a solution area. It's referred to as a "particle or individual". Every individual possesses fitness values that the fitness function evaluates to optimize them, along with velocities that guide their flight. Through the solution area, the individual follows the current optimal individual. PSO starts with a set of random solutions and then iterates over generations in search of an optimal solution. Each particle is restructured every cycle by comparing two "best" values. Currently, the first option (fitness) is the most effective. (In addition, the fitness value is kept.) pbest is the name given to this integer. Another "finest" value recorded by the particle swarm optimizer is the best value reached yet by each individual in the swarm and gbest, which refers as "global best", is the highest value. Figure 1 depicts the overall concept of Particle Swarm Optimization.

Due to its advantage over other algorithms like GA [25], PSO pays greater emphasis to maximizing WSN lifespan. It has a number of advantages, including ease of use, the ability to avoid local optima, and quicker convergence. Its fitness function takes into account the leftover energy of nodes as well as the distance between them,

Table 2 Comparative study of PSO-based frameworks in cutting-edge IoT industries

Author	Year	Contribution	Techniques used	Merits	Future Scope	Performance Metrics
Rao et al. [15]	2016	IoT network CH selection configuration	PSO-ECHS,	Energy consumption rate is reduced	Work for heterogeneous networks	Energy consumption, network lifetime, packets receiving
Vijayalakshmi and Anandan [17]	2021	WSN's effective cluster head selection	Tabu search (TS), PSO	Reduce the average packet loss	Improved by integrating firefly and TS methods	Alive nodes, Average packet loss, Network lifespan
Şenel et al. [21]	2018	Optimization approach using a hybridized methodology	PSO, GWO	Hybrid technique converges to more optimum solution with least iterations	The hybridization process between PSO and GWO will improve the search performance	Energy, throughput
Kaur and Mahajan [22]	2018	Energy-efficient hybrid metaheuristic approach for sensor nodes	Ant Colony Optimization (ACO), PSO	Enhances network lifetime by conserving the energy	Improve the node's performance	Network lifetime, throughput residual energy
Rastogi et al. [23]	2021	Optimization approach in wireless sensor network	PSO, ACO	In the context of network longevity, this gives the best performance	To work better in virtually all of the performance factors mentioned, use a variety of alternative computational intelligence approaches	Alive node, dead node, throughput, remaining energy
Sundaramurthy and Jayavel [24]	2020	Optimization approach for prediction of arthritis	PSO, GWO	Enhances energy efficiency of nodes	Propose a hybrid methodology to enhance node reliability	Network's lifetime, throughput, average energy consumption

giving the WSN an optimum path thanks to PSO's ability to avoid local optima. PSO is also utilized for node location, CH selection, and cluster creation, among other things. The goal of PSO implementations is to aid energy management by lowering energy costs per process and thereby increasing node lifespan.

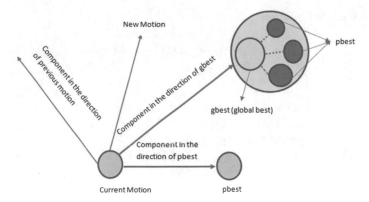

Fig. 1 Diagrammatical representation of PSO

Pseudocode of PSO

Step 1: **Begin**
Step 2: **Initialization**
For each particle
    (a) Initialize particles position with uniform distribution
    (b) Initialize particle velocity
End For
Step 3: **Do**
For each individual
    Evaluate the fitness function
    If the fitval (fitness value) in the history is superior than pbest
        Set the current value to the new pbest value
    End If
End For
Select the individual with best fitval (fitness value) among all the individuals as gbest
For each individual
    Update the $vel_i(t + 1)$ as determined by Eq-1
    Update the $x_i(t + 1)$ as determined by Eq-2
End For
Until stopping criteria is fulfilled
End Begin

The individual or particle adjusts its velocity and locations using equations (Velocity Update equation) and (Position Update equation) after selecting the two optimal values (Position Update equation).

Velocity Update Equation

$$vel_i(t + 1) = w \cdot vel_i(t) + l_1 r_1 [pbest_i(t) - x_i(t)] + l_2 r_2 [gbest_i(t) - x_i(t)]$$
$$(1)$$

Position Update Equation

$$x_i(t + 1) = x_i(t) + vel_i(t) \tag{2}$$

where

i	Particle index
w	Inertial coefficient
l_1, l_2	Learning elements $(0 \leq l_1, l_2 \leq 2)$
$r_1 \cdot r_2$	Random variables $(0 \leq r1, r2 \leq 1)$
$vel_i(t)$	i^{th} Particle's velocity having time t
$x_i(t)$	Current position of particle having time t
$pbest_i(t)$	Particle's best solution having time t
$gbest_i(t)$	Global best solution at time t

3.1 Grey Wolf Optimizer

Mirjalili et al. [11] introduced the GWO algorithm. Grey wolves' social structure and hunting behavior have influenced the GWO. The testing findings revealed its capability and great performance in handling a variety of traditional engineering design challenges, including spring tension, welded beams, and so on. Grey wolf leadership is a source of inspiration for the GWO algorithm. Grey wolves are the most powerful predators on the planet. Within the leadership structure, there are four different sorts of grey wolves, i.e., α, β, δ, and $\hat{\omega}$.

The optimal answer in the GWO algorithm is represented by alpha (α) wolves. Beta (β) and delta (δ) wolves are the population's second and third best solutions. Omega $(\hat{\omega})$ wolves are the finest prospects for a solution. The GWO method assumes that alpha, beta, and delta wolves hunt, with omega wolves trailing after them. The three primary aspects of grey wolf hunting are as follows: (1) following the prey, chasing it down, and approaching it. (2) Pursuing, surrounding, and torturing the prey until it comes to a complete stop. (3) Attacking the prey by surprise.

The formula is as follows:

$$
\begin{aligned}
&D = |C \cdot Xp - X(t)| \quad D\alpha = |C_1 \cdot X\alpha - X| \quad X_1 = X\alpha - a_1 \cdot (D\alpha) \\
&X(t+1) = X_p(t) \qquad D\beta = |C_2 \cdot X\beta - X| \quad X_2 = X\beta - a_2 \cdot (D\beta) \\
&a = 2l \cdot .r_1; C = 2 \cdot r_2 \quad D\delta = |C_3 \cdot X\delta - X| \quad X_3 = X\delta - a_3 \cdot (D\alpha\delta)
\end{aligned}
$$

The number of iterations is t, the position of prey is Xp, and the position of a grey wolf is X. While a and C are vector coefficients, r_1 and r_2 specify random numbers. $D\alpha$, $D\beta$ and $D\delta$ are the fitness functions for alpha, beta, and gamma groups.

Pseudocode of Grey Wolf Optimization

Step 1: Begin
Step 2: Initialize a, C, and $t = 1$
Step 3: Calculate each individual's fitness in the population
      (a) $X\alpha$ = individual having best fitness value
      (b) $X\beta$ = individual having second best fitness value
      (c) $X\delta$ = individual having third best fitness value
While (i < Maximum_itr)
For each individual
      Position of current individual is updated using equation
$X(t + 1) = (X_1 + X_2 + X_3)/3$ where X_1, X_2 and X_3 are position vector of α, β and δ wolves
End For
      Update t, a, C
      Calculate the fitness of all individual
      Update $X\alpha$, $X\beta$, $X\delta$
      $i = i + 1$
End While
Return $X\alpha$
Step 4: Return best solution
Step 5: End

---

4 Proposed Framework

On terms of the suggested structure as depicted in Fig. 2, an energy-efficient sensor network must be designed in the bottom layer, where the energy usage of sensor nodes may be lowered before it is transferred to the middle layer.

Our hybridized proposed approach (PSO–GWO) is created without altering the main functionality of the Particle Swarm Optimization and Grey Wolf Optimization algorithms. The PSO approach may be used to address almost any real-world problem. There must, however, be a mechanism to reduce the likelihood of the Particle Swarm Optimization algorithm duping in a local minimum. In our recommended technique to reduce the likelihood of collapsing into a local minimum, the Grey Wolf Optimization algorithm is employed to support the Particle Swarm Optimization algorithm. The PSO algorithm, as previously noted, guides certain particles to random sites with a slight chance of avoiding local minimums. As a nutshell, these paths might lead to a departure from the global minimum. To circumvent these problems, the GWO algorithm's exploration capability is used to send particular particles to GWO-enhanced locations rather than random locations.

Hence by deploying the above hybrid methodology, an energy-efficient framework could be attained as it will improve energy efficiency by minimizing energy consumption. The hybrid model aids in determining the best way to higher throughput and lower energy consumption. The proposed model is based on three-layer IoT architecture.

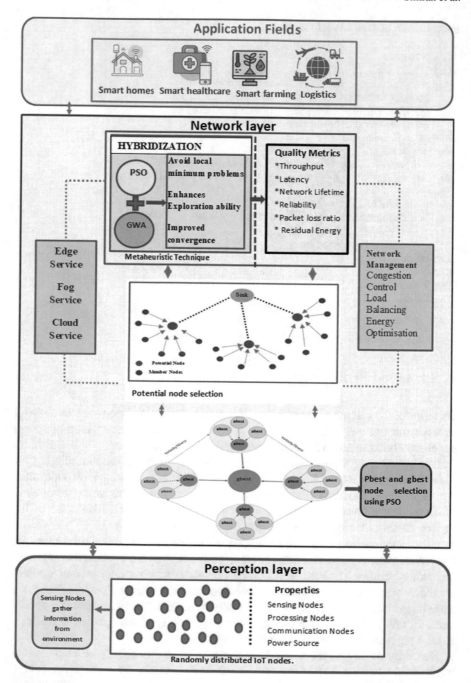

Fig. 2 Swarm intelligence-based energy-efficient framework using PSO–GWO in IoT

- *Physical Layer*: Diverse nodes are generally dispersed across the geographical region in the perception layer. The intelligent gadgets that operate on the bottom layer are each given a unique identifier. A detector, a processor unit, a transceiver unit, and a legitimate power supply describe intelligent devices on the bottom layer. Manufacturers are creating a variety of smart devices with varying specifications, standards, and technologies.
- *Network Layer*: Middle layer is the core layer where transmission of data takes place. The data collected by sensor nodes in the perception layer is sent to the network layer for processing. The network layer is broken into three portions:

    1. *pbest and gbest node selection*: pbest and gbest is the particle best and global best solution respectively. Pbest and gbest node selection is done via PSO approach. The particle adjusts its velocity and locations using the particle velocity update equation as Eq. 1 and particle position update position equation as Eq. 2 after obtaining the two best values.
    2. *Potential node selection in each region for energy efficiency*: This stage entails aggregates sensor nodes and choosing Potential nodes (PNs) for all of the clusters in each region. The information acquired from all of the nodes will be compiled by the potential node and transferred to the IoT base node. To determine the potential node, a PSGWO methodology is used.
    3. *Attain optimal solution and performance metrics*: To obtain the best solution, a PSO model is used with GWO. By employing the GWO technique's exploration capabilities, the PSO method will be prevented from trapping into local minimums and hence an optimal solution can be attained. The performance metrics utilized to pick the potential node include the amount of load, living nodes, energy, network lifetime and throughput

- *Application Layer*: Mobile consumers, businesses, and huge organizations all benefit from the services provided by the applications layer. It is the top most user interactive layer. The real communication is initiated and reflected at the application layer. Cost function may be assessed using quality criteria such as latency, node lifespan, and residual energy.

5 Conclusion

Despite the fact that IoT has huge potential in a variety of applications in the modern period, there are a number of barriers to overcome. Privacy, energy optimization, networking, hardware configuration concerns, data network congestion, and other issues must be solved to improve the resilience of IoT. We choose to focus on the energy optimization problem in this study. To resolve this issue, a hybrid meta-heuristic framework based on PSO–GWO is devised to reduce the sensor's energy usage in IoT networks in this study. The key advantage of PSO is that there are fewer parameters to tweak. PSO gets the optimal result through particle interaction;

however, it converges at a relatively slow speed to the global optimum through a high-dimensional search area. To overcome this, PSO is hybridized with GWO's exploration capabilities to avoid local minima problems. In this study, several performance parameters including energy consumption, network lifetime, alive nodes, temperature, and throughput are taken into account to choose the best potential node for the IoT network. Using various simulations, we will evaluate the performance of the proposed algorithm and compare it to other metaheuristic techniques such as PSO, GWO, Hybrid WSO-SA, and HABC-MBOA algorithms.

References

1. Chopra K, Gupta K, Lambora A (2019) Future internet: the internet of things-a literature review. In: 2019 International conference on machine learning, big data, cloud and parallel computing (COMITCon). IEEE, pp 135–139. https://doi.org/10.1109/COMITCon.2019.8862269b
2. Alaa M et al (2017) A review of smart home applications based on internet of things. J Netw Comput Appl 97:48–65. https://doi.org/10.1016/j.jnca.2017.08.017
3. Rana B, Singh Y (2021) Internet of things and uav: an interoperability perspective. In: Unmanned aerial vehicles for internet of things (IoT): concepts, techniques, and applications, pp 105–127. https://doi.org/10.1002/9781119769170.ch6
4. da Costa VCF, Oliveira L, de Souza J (2021) Towards a taxonomy for ranking knowledge in internet of everything. In: 2021 IEEE 24th International conference on computer supported cooperative work in design (CSCWD). IEEE, pp 775–780. https://doi.org/10.1109/CSCWD4 9262.2021.9437857
5. Miraz MH, Ali M, Excell PS, Picking R (2015) A review on internet of things (IoT), internet of everything (IoE) and internet of nano things (IoNT). In: 2015 Internet technologies and applications (ITA). IEEE, pp 219–224. https://doi.org/10.1109/ITechA.2015.7317398
6. Rachman T (2018)济無. Angew Chem Int Ed 6(11):951–952 (pp 10–27)
7. Rana B, Singh Y, Singh PS (2021) A systematic survey on internet of things : energy efficiency and interoperability perspective. Trans Emerg Telecommun Technol 32(8):e4166 (pp 1–41). https://doi.org/10.1002/ett.4166
8. Rana B, Singh Y, Singh H (2021) Metaheuristic routing: a taxonomy and energy-efficient framework for internet of things. IEEE Access 9:155673–155698. https://doi.org/10.1109/ACCESS.2021.3128814
9. Kumoye AO, Prasad R, Fonkam M (2020) Swarm intelligence algorithm and its application:a critical review. In: 2020 International conference in mathematics, computer engineering and computer science (ICMCECS). IEEE, pp 1–7. https://doi.org/10.1109/ICMCECS47690.2020. 246996
10. Sun W, Tang M, Zhang L, Huo Z, Shu L (2020) A survey of using swarm intelligence algorithms in IoT. Sensors (Switzerland) 20(5):1420. https://doi.org/10.3390/s20051420
11. Mirjalili S, Mirjalili SM, Lewis A (2014) Grey Wolf Optimizer. Adv Eng Softw 69:46–61. https://doi.org/10.1016/j.advengsoft.2013.12.007
12. Long W, Xu S (2016) A novel grey wolf optimizer for global optimization problems. In: 2016 IEEE Advanced information management, communicates, electronic and automation control conference (IMCEC). IEEE, pp 1266–1270. https://doi.org/10.1109/IMCEC.2016.7867415
13. Devika G, Ramesh D, Karegowda AG (2020) Swarm intelligence–based energy-efficient clustering algorithms for WSN: overview of algorithms, analysis, and applications. In: Swarm intelligence optimization: algorithms and applications, pp 207–261

14. Alqattan ZNM, Abdullah R (2013) A comparison between artificial bee colony and particle swarm optimization algorithms for protein structure prediction problem. In: Neural information processing. ICONIP 2013. Lecture notes in computer science, vol 8227, no Part 2. Springer, Berlin, Heidelberg. https://doi.org/10.1007/978-3-642-42042-9_42

15. Rao PCS, Jana PK, Banka H (2016) A particle swarm optimization based energy efficient cluster head selection algorithm for wireless sensor networks. Wireless Netw 23:2005–2020. https://doi.org/10.1007/s11276-016-1270-7

16. Iwendi C, Maddikunta PKR, Gadekallu TR, Lakshmanna K, Bashir AK, Piran MJ (2021) A metaheuristic optimization approach for energy efficiency in the IoT networks. Softw Pract Exp 51(12):2558–2571. https://doi.org/10.1002/spe.2797

17. Vijayalakshmi K, Anandan P (2019) A multi objective tabu particle swarm optimization for effective cluster head selection in WSN. Cluster Comput 22:12275–12282. https://doi.org/10.1007/s10586-017-1608-7

18. Okwu MO, Tartibu LK (2021) Particle swarm optimisation. In: Metaheuristic optimization: nature-inspired algorithms swarm and computational intelligence, theory and applications. Studies in computational intelligence, vol 927. Springer, Cham, pp 5–13. https://doi.org/10.1007/978-3-030-61111-8_2

19. Duman S, Güvenç U, Sönmez Y, Yörükeren N (2012) Optimal power flow using gravitational search algorithm. Energy Convers Manag 59:86–95. https://doi.org/10.1016/j.enconman.2012.02.024

20. Manshahia MS (2019) Grey wolf algorithm based energy-efficient data transmission in internet of things. Procedia Comput Sci 160:604–609. https://doi.org/10.1016/j.procs.2019.11.040

21. Şenel FA, Gökçe F, Yüksel AS, Yiğit T (2019) A novel hybrid PSO–GWO algorithm for optimization problems. Eng Comput 35:1359–1373. https://doi.org/10.1007/s00366-018-0668-5

22. Kaur S, Mahajan R (2018) Hybrid meta-heuristic optimization based energy efficient protocol for wireless sensor networks. Egypt Inform J 19(3):145–150. https://doi.org/10.1016/j.eij.2018.01.002

23. Rastogi R, Srivastava S, Tarun, Manshahia MS, Varsha, Kumar N (2021) A hybrid optimization approach using PSO and ant colony in wireless sensor network. In: Mater today: proceedings. https://doi.org/10.1016/j.matpr.2021.01.874

24. Sundaramurthy S, Jayavel P (2020) A hybrid grey wolf optimization and particle swarm optimization with C4.5 approach for prediction of rheumatoid arthritis. Appl Soft Comput 94:106500. https://doi.org/10.1016/j.asoc.2020.106500

25. Lambora A, Gupta K, Chopra K (2019) Genetic algorithm- a literature review. In: 2019 International conference on machine learning, big data, cloud and parallel computing (COMITCon). IEEE, pp 380–384. https://doi.org/10.1109/COMITCon.2019.8862255

Performance Analysis of MADM Techniques in Cognitive Radio for Proximity Services

Mandeep Kaur and Daljit Singh

Abstract It is envisioning that Cognitive Radio (CR) networks are boundless to end up spectrum scantiness and upgrade spectrum efficiency by spectrum allocation techniques. Cognitive Radio authorizes to use the unlicensed spectrum with the ease of spectrum sensing. Energy detection is one of the prominent methodologies of spectrum sensing. Energy detection does not require any prerequisite knowledge about the primary user. The energy detector differentiates the calculated energy with the threshold value to check the channel availability. Energy detection with adaptive double threshold meliorates than Adaptive Threshold. The regulation of radio frequencies procedure is managed by spectrum handoff. Spectrum handoff is the main aspect of Spectrum Management techniques. The phenomenon of spectrum handoff starts with the appearance of Cognitive Radio. The decision of selecting the top notch network by network selection ranking in spectrum handoff using Multiple Attribute Decision Making schemes (MADM). MADM itself is a procedure used to obtain the ideal alternative from different alternative with particular attribute. The objective of this paper is to implement Grey Relational Analysis (GRA) and Cost Function method and in contrast with simple additive weighting (SAW) and technique of order preference to ideal solution (TOPSIS). The comparison of multiple attribute based on quad-play services in CR network to reduce complexity and extent quality of services. The numerical results indicate that Grey Relational Analysis is reliable for optimal network using Relative Standard Deviation having quad-play services such as voice, video, data, and mobile voice over Internet are 4.34, 16.06, 10.80, 9.9 used in this paper.

Keywords Adaptive double threshold · Probability of detection · Cognitive radio network · Spectrum handoff · Spectrum sensing

M. Kaur (✉) · D. Singh
Guru Nanak Dev Engineering College, Ludhiana, PB 141006, India
e-mail: Mandeepghattaura123@gmail.com

© The Author(s), under exclusive license to Springer Nature Singapore Pte Ltd. 2023
Y. Singh et al. (eds.), *Proceedings of International Conference on Recent Innovations in Computing*, Lecture Notes in Electrical Engineering 1011,
https://doi.org/10.1007/978-981-99-0601-7_29

1 Introduction

The overcrowding spectrum seems like an old chest nut due to the burgeoning of wireless devices and mobile devices and makes the spectrum scare. Each wireless service has its own licensed spectrum. Some of the wireless devices have wide spectrum with small number of users in consequence the spectrum remains underutilized while in other remains over utilized [1]. To overthrow the spectrum underutilized dynamic spectrum access comes into play. A scheme that uses DSA is referred as Cognitive Radio [2].

The capability of CR to vanquish the insufficiency of spectrum underutilized. It permits secondary users to access licensed spectrum of primary use while evading intervention with PU. The prospective of CR to enhance the spectrum efficiency and spectrum becomes well founded [3–5].

The key features for the Cognitive Radio for proximity services are spectrum sensing and spectrum database. CR devices follow the spectrum band in their region to discover its numerous authorized user and white spaces. These spectrum holes are generated and separated synchronously and can be utilized without an authorization. Spectrum sensing can be distinguished into two categories such as non-cooperative and cooperative. In cooperative method, the spectrum data can be divided by unlicensed user devices while in other each unauthorized user independently. Federal Communication Commission (FCC) suggested a spectrum database idea to get rid of problem of spectrum sensing approach and to utilize TV void space. To carry the increasing number of devices that uses the radio frequency spectrum, a merged approach is advantageous. It confirms that devices can rapidly detect unexploited spectrum and upgrade quality of services.

When underutilized frequency band is used by unlicensed user, then maintenance of data transmission is necessary on channel by spectrum handoff scheme [4]. Spectrum handoff implies on spectrum sensing to observe the suitable optimum channel. This handoff triggering relies on three techniques are proactive handoff, reactive handoff, and hybrid handoff scheme [6]. These schemes concede secondary users to switch to other channel without intrusion their transmission when PU is in licensed band [7, 8].

2 MADM Methodologies

Multiple Attribute Decision Making are mathematical approaches to handout the decision problems with some aspects. The selection of ideal network is based on their substitutes and attributes. To choose a network, there are having some strategies means simple additive weighting (SAW), technique for ideal preference by similarity to ideal solution (TOPSIS), Grey Relational Analysis (GRA), and cost function method.

Table 1 Multiple substitutes of various networks

Networks	Multiple attributes			
	D	DR	PLR	P
Wi MAX	60	35	25	15
Wi-Fi	50	11	30	8
Cellular	30	2	50	40
Satellite	200	2	80	8

Fig. 1 Procedure steps of multiple attribute decision making

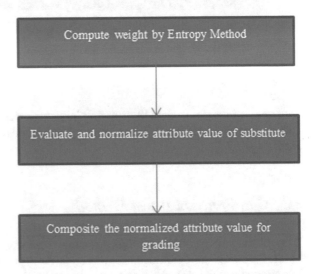

MADM coincides with quad play services having fixed voice, fixed video, and paid TV services with mobile voice and data services.

The number of attributes is Wi MAX, Wi-Fi, Cellular and Satellite with some of the alternatives is Delay, Data rate, Packet loss ratio, Price is shown in Table 1. Various attributes are considered as instrumental such as data rate while others as detrimental. The decision of MADM depends on the precedence level of alternatives. MADM methods are having best decision making capabilities (such as prioritization, selection) from the available alternatives. Therefore, this is the reason to choose MADM for spectrum handoff for selecting optimum network. To resolve multiple attributes decision making issues few mechanisms are as follows in Fig. 1.

2.1 Grey Relational Analysis Method

It is the method of calculating Grey relational degree [9]. GRA utilizes grey system theory. The strength of this procedure is simple in calculations and in priority decision, and it decreases error probability. The weakness of this method is used to handle

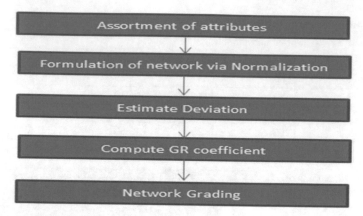

Fig. 2 Grey relational analysis flowchart

only uncertainties in the data like missing information and partial information in the data or other data set is small for processing. The methods of Grey Relational Analysis are in Fig. 2.

2.2 Cost Function Based Method

It is a method of determining the performance of networks in Cognitive Radio [10]. This procedure of multiple attributes is based on the approximation of cost of network. The strength of this method is that it is as easy as ABC for execution due to the simply usage of various attributes and it totally dependent on attribute value the limitation of this procedure is that the value of cost function should be maximum for best network selection. The algorithm is shown in Fig. 3.

Fig. 3 Flowchart of cost function

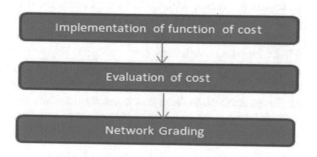

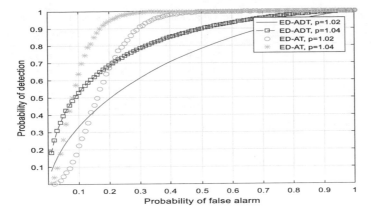

Fig. 4 ROC curve for energy detection with different noise uncertainty factor

3 Performance Analysis of Comparison Between Adaptive Threshold and Adaptive Double Threshold

Adaptive Threshold and Adaptive Double threshold is based on energy detection technique of spectrum sensing. It outperforms optimal results in the non-appearance of noise uncertainties [11]. But in practical situation, noise uncertainty can't be snubbed. In this method, the threshold takes into account the uncertainty in noise parameter. ρ represents the size of noise uncertainty ED_ADT achieves preferable results than ED_AT in the existence of noise uncertainty with $N = 1000$, SNR $= -9$ dB. ROC curve between Probability of Detection and Probability of False Alarm demonstrates that there is an effect in Adaptive Threshold and Adaptive Double Threshold due to the presence of noise uncertainty factor. Figure 4 shows that there is increase in Probability of Detection in Adaptive Double Threshold rather than Adaptive Threshold. ED_ADT performs better than Adaptive Threshold because it diminishes sensing time and enhances the throughput of the CR network.

4 Results and Discussions

In proposed scheme, we implement the entropy, GRA, cost function, SAW, and TOPSIS methodologies to estimate the performance using MATLAB. Results acquired from this scheme for approach with precedence. Networks such as WIMAX, WIFI, Cellular, Satellite were esteemed for ranking.

Table 2 Network selection ranking matrix formulation using alternatives for quad play services

Alternatives	Voice	Video	Data	Mobile VOIP
Wi Max	0.8055	0.8248	0.7154	0.6100
Wi Fi	0.8183	0.6196	0.5848	0.5176
Cellular	0.8879	0.85314	0.6804	0.6266
Satellite	0.8285	0.91966	0.7587	0.6558
Ideal for handoff	Cellular	Wi Max	Satellite	Satellite
RSD (%)	4.34	16.06	10.80	9.9

4.1 Implementation of MADM Algorithms for Quad Play Services: Grey Relational Analysis

RSD means Relative Standard Deviation. RSD helps us to evaluate the precision or accuracy of data. To achieve more précised data, the value of RSD is small. If value of RSD is large, the results are more extended from the mean of data. This will degrade the performance of data. The best network for spectrum handoff is Cellular (0.8879) and best service from quad play service is video services that is given in Table 2 and represent data in bar graph in Fig. 5.

Cost Function-Based Method

The excellent network for spectrum handoff is Wi Max (0.8530) and the best service is video service is in Table 3, and data is shown in bar graphs using MATLAB (Fig. 6).

TOPSIS Method

The optimum network for spectrum handoff is Cellular (0.8261) in Table 4. Based on this Table, data represents in Fig. 7.

SAW Method

The ideal network for spectrum handoff is Satellite (0.8707) that is given in Table 5. SAW network also represents data in bar graph in Fig. 8.

5 Performance Comparison of MADM Algorithm Using Relative Standard Deviation

The graph estimates the performance between Relative Standard Deviation (RSD) and MADM algorithms. When the value of RSD is lower, then the precision of data is more that is shown in Fig. 9. The GRA depicts the RSD for voice, video and data and mobile voice over Internet services are 4.34, 16.06, 10.80, and 9.9. These values are small when compared with other methodologies. Therefore, this method is superior for spectrum handoff.

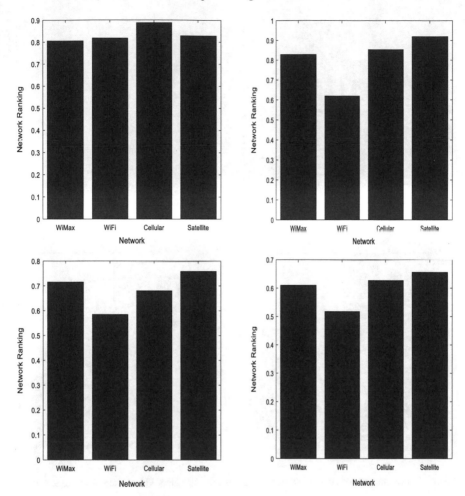

Fig. 5 Representation of GRA algorithm implement to voice, video, data, and mobile VOIP services

Table 3 Cost function network ranking for quadruple services

Networks	Voice	Video	Data	Mobile VOIP
Wi Max	0.5815	0.8530	0.7189	1.3003
Wi Fi	0.6235	0.4661	0.7964	1.4198
Cellular	0.7990	0.2883	0.4825	1.2815
Satellite	0.2310	0.1776	0.4807	0.7116
Ideal for handoff	Cellular	Wi Max	Wi Fi	Wi Fi
RSD (%)	42.6	66.3	26.19	26.8

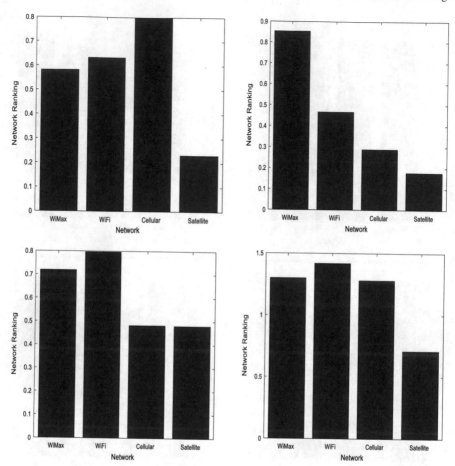

Fig. 6 Representation of cost function implementation to voice, video, data, and mobile VOIP services

Table 4 TOPSIS network ranking for quad play services

Networks	Voice	Video	Data	Mobile VOIP
Wi Max	0.8260	0.9481	0.8334	0.8289
Wi Fi	0.5000	0.5000	0.5000	0.5000
Cellular	0.8261	0.2141	0.4220	0.6623
Satellite	0.1048	0.0980	0.4676	0.2732
Ideal for handoff	Cellular	Wi Max	Wi Max	Wi Max
RSD (%)	60.62	85.99	33.52	41.80

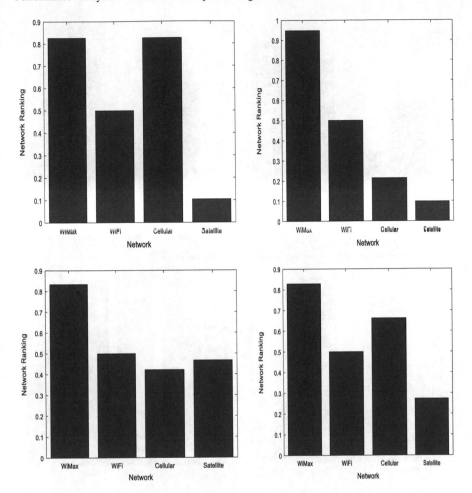

Fig. 7 Representation of TOPSIS algorithm implement to voice, video, data, and mobile VOIP services

Table 5 SAW network ranking for quad play services

Networks	Voice	Video	Data	Mobile VOIP
Wi Max	0.4748	0.5543	0.5756	1.0500
Wi Fi	0.2579	0.2352	0.2759	0.5338
Cellular	0.9012	0.8050	0.5994	1.5003
Satellite	0.3349	0.8707	0.6075	0.9425
Ideal for handoff	Cellular	Satellite	Satellite	Cellular
RSD (%)	33.08	11.67	31.03	79.56

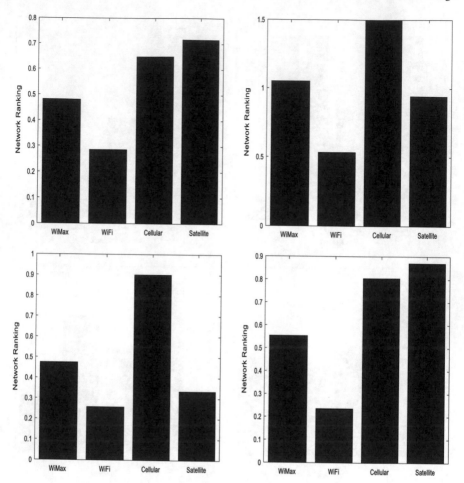

Fig. 8 Representation of SAW algorithm implement to voice, video, data, and mobile VOIP services

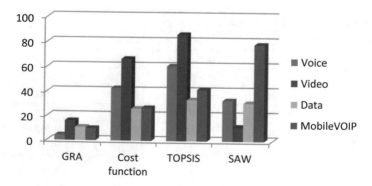

Fig. 9 Effect of RSD on MADM algorithms

6 Conclusion

In this paper, we have proposed a spectrum handoff scheme for the best network selection according to quad play services. In the CR network, MADM algorithm plays a pivot role with the combination of entropy method. This method helps us to estimate the attribute weights according to the CR preference. All MADM methods were advantageous for determining the best network for spectrum handoff process. The usage of attributes values is directly for GRA method to reduce the error probability. Computational method of GRA is straightforward. Cost function method relies on attribute values, and its implementation is uncomplicated. GRA overtops the other entire algorithm because its value is more precise which is achieved by Relative Standard Deviation. According to RSD method lower the value of RSD, the precision of data is more. In the future research, these algorithms applied to network mobility in CR vehicular network.

References

1. Yucek T, Arslan H (2009) A survey of spectrum sensing algorithms for cognitive radio applications. IEEE Commun Surv Tuts 11(1):116–130 (1st Quart.)
2. Song M, Xin C, Zhao Y, Cheng X (2012) Dynamic spectrum access: From cognitive radio to network radio. IEEE Wireless Commun 19(1):23–29
3. Ghasemi A, Sousa ES (2008) Spectrum sensing in cognitive radio networks: requirements, challenges and design trade-offs. IEEE Commun Mag 46(4):32–39
4. Mahendru G, Shukla AK, Patnaik LM (2021) An optimal and adaptive double threshold based approach to minimize error probability for spectrum sensing at low SNR regime
5. Akyildiz IF, Lee W, Vuran MC, Mohanty S (2008) A survey on spectrum management in cognitive radio networks. IEEE Commun Mag 46(4):40–48
6. Akyildiz IF, Lee WY, Chowdhury KR (2009) Spectrum management in cognitive radio ad hoc networks. IEEE Netw 23(4):6–12
7. Jaiswal M, Sharma AK, Singh V (2013) A survey on spectrum sensing techniques for cognitive radio. Proc Conf ACCS 1–14
8. Kumar K, Prakash A, Tripathi R (2016) Spectrum handoff in cognitive radio networks: a classification and comprehensive survey. J Netw Comput Appl 61:161–188
9. Bicen AO, Pehlivanoglu EB, Games S, Akan OB (2015) Dedicated radio utilization for spectrum handoff and efficiency in cognitive radio networks. IEEE Trans Wireless Commun 14(9):5251–5259
10. Verma R, Singh NP (2013) GRA based network selection in heterogeneous wireless networks. Wirel Pers Commun 72:1437–1452
11. Divya A, Nandakumar S (2019) Adaptive threshold based spectrum sensing and spectrum handoff using MADM methods for voice and video services. IEEE Commun

Image Processing and Computer Vision

Comparison of Debris Removal in Pap-Smear Images Using Enhanced Erosion and Dilation

Soumya Haridas and T. Jayamalar

Abstract The pap-smear test is considered one of the most common methods available for cervical cancer screening. Women above a particular age are supposed to undergo the cervical screening procedure at least once a year to identify whether there is the presence of cancerous cells. Since the manual screening of each cell from a pap-smear slide is tedious and time-consuming, automated pap-smear analysis is the need of the day. In the slides, during automated analysis, the presence of unwanted materials such as bacteria and blood particles may produce a false diagnosis. The cell structure is one of the main challenges faced during the process. Usually, debris removal is performed by many researchers along with segmentation. While enhancing the images debris removal can be performed. This can improve the quality of enhanced images. Therefore, debris removal plays a very important role in the automated pap-smear analysis. In this paper, debris removal is performed using enhanced erosion and dilation along with the image enhancement which makes it easy for the subsequent processes. The debris removal methods are compared by using two different binarization techniques. Initially, the images are preprocessed using diffusion stop function-based CLAHE algorithm, and then, the images are undergone kittler binarization. Mathematical morphological operations like erosion and dilation are applied to the image to remove the debris. Next, instead of kittler binarization, Otsu's method is used. The results are evaluated with the performance measures sensitivity, specificity, and accuracy. From the results obtained with a sensitivity value of 99%, specificity of 99%, and accuracy of 98%, it is clear that the first method i.e., kittler binarization with morphological operations gives better performance.

Keywords Pap-smear images · Debris removal · Morphological operation

S. Haridas (✉)
Department of Computer Science, Avinashilingam Institute for Home Science and Higher Education for Women, Coimbatore, India
e-mail: soumya.smya@gmail.com

T. Jayamalar
Department of Information Technology, Avinashilingam Institute for Home Science and Higher Education for Women, Coimbatore, India
e-mail: jayamalar_it@avinuty.ac.in

1 Introduction

Cervical cancer is one of the most dangerous diseases that can be identified and treated if diagnosed at an early stage. Different diagnosis methods are available for cervical cancer detection, out of which pap-smear analysis is the most common method available. In a pap-smear, the doctor takes cells out of the cervical portion of the patient, and it is sent to the laboratory for analysis. The laboratory technicians prepare slides and analyze them under the microscope to identify the nature of the cells. The cytotechnician analyzes the images by considering the signs of malignancy [1]. These slides prepared may contain artifacts or debris due to blood particles, air drying, or bacteria. Liquid-based cytology (LBC) slide preparation is used to avoid this. In LBC, the sample is immersed in a solution, which is subsequently processed to normalize it, remove unwanted components (such as red blood cells), and deposit an adequate mono-layer sample on a glass slide [2]. The cytologists can find the cancerous and other types of cells, such as precancerous ones, after analyzing the slides [3]. Besides manual analysis, automated pap-smear analysis is also performed in which the software checks every slide based on specific features to classify them as cancerous or non-cancerous. Usually, the automated analysis consists of preprocessing the images, segmentation, feature extraction and selection, and finally, classification. The accurate classification of cancerous as well as non-cancerous cells is possible only if the nucleus and cytoplasm are correctly identified and segmented [4]. Preprocessing is improving the quality of input images by some methods such as noise removal and image enhancement. After preprocessing, the region of interest is identified using segmentation, and the remaining processes follow this. Many researchers have suggested methods for debris removal during the segmentation phase. If the debris is removed early, it will help reduce the time required for the following processing phase. In this paper, we propose a method in which debris removal is performed after preprocessing. Here, the image is first preprocessed for contrast enhancement and noise removal. Kittler binarization is applied to this image to separate the background and foreground to identify the objects. Morphological erosion and dilation are used in this image to fill the gaps and remove protrusions. This method helps in removing the debris. Since segmentation separates the nucleus and cytoplasm from the image, if the debris is removed at an earlier stage, it will reduce the possibility of false identification of cells. Some debris resembles the nucleus in shape. Suppose these are removed, the possibility of false identification of debris as nucleus can be avoided. The time required for segmentation can also be reduced using this approach.

2 Literature Survey

Kuko et al. have performed debris removal by thresholding [5]. After creating a threshold image, contour pixels are found, and bounding rectangles are formed based on these pixel values. The bounding rectangles below a threshold value of 2% of the area of the sliding window are rejected and considered debris. The main drawback of the method is that there is a chance of missed cells due to the intensity change among some debris particles.

According to William et al., the reason for the failure of many of the existing automated pap-smear analysis methods is the presence of debris [6], which may generate some objects that may have similarities with the actual cells. A three-phase elimination scheme is considered in their paper to classify the cells and debris. The size, shape, and texture of the objects are analyzed. In size estimation, a threshold is applied, in shape analysis, P2A descriptor is considered, and in texture analysis, Zernike moments have been used.

Malm et al. considered a sequential elimination approach for debris removal [7]. During initial segmentation, the object that does not resemble a cell nuclei is rejected. This will reduce the complexity in the next stages since there will be less amount of objects left after elimination. The objects are analyzed and classified into cells or debris in a sequence of steps. First, the area is considered, then the shape (region-based and contour-based), then elliptical deviation, then texture, and finally, the average gray value. The proposed method works efficiently by reducing the classification dimensionality and saving computational power. The major drawback is that the training data is to be updated to adapt to new variations.

According to Agarwal et al. [8], adaptive thresholding is applied to the image after preprocessing, which will convert the image into a binary format. Mathematical morphological operations such as closing and erosion are applied to the image after that, and later the final nuclei objects are obtained.

Martin et al. developed an automatic rejection of non-cellular outlines [9]. The process is carried out based on some quantitative metrics. Four different filters based on the coefficient of variance of cytokeratin and Hoechst fluorescence (C&HF), the correlation coefficient of C&HF, standard deviation of C&HF, and circularity value of C&HF were considered, and the outlines were rejected accordingly. Cutoff values for the filters were determined manually. Only less no of the samples were studied, and with a large dataset, only the effectiveness of the filters can be further analyzed.

Moshavegh et al., in their paper, performed artifact rejection in such a way that the nucleus is retained, thereby removing other unwanted objects [10]. The features used here are size, shape, and nuclear granularity. These features are calculated for each segmented nuclei-like object and then checked for shape, texture, and shape. The objects that do not satisfy the rules about the features were rejected.

Kumar et al., in their paper, performed artifact rejection using SVM after feature extraction [11]. After segmenting the images using Laplacian of Gaussian, the objects in the image were classified as cervical cells or artifacts based on a set of ranked features. The sequential minimal optimization method was used to optimize the selected features. The selected features were ranked using the maximization function, and features were selected based on ranking and histogram analysis. They had achieved an excellent level of proper classification.

Oscanoa et al., in their paper, performed artifact rejection by extracting certain features that can differentiate cells from debris [12]. Area and perimeter of the objects are calculated. Using thresholding the objects with values less than a fixed value are rejected. They have achieved a nuclei detection efficacy of 92.3%. The main drawback is missing nuclei due to less edge contrast and faint staining.

Various methods for debris removal have been discussed in the literature review, and the need for a method that can remove the debris at an earlier stage of automated pap-smear analysis is identified. Most of the methods discussed above do the debris removal either after segmentation or during the segmentation process. Since the segmentation stage requires the correct identification of the nucleus and cytoplasm presence of debris may cause false results. If a method can remove debris from the image before segmentation is performed, it will reduce the effort in the segmentation phase.

3   Materials and Methods

The images were taken from the Sipakmed dataset, which contains 4049 isolated cells. It falls into five categories, namely superficial-intermediate cells, parabasal cells, koilocytotic cells, dyskerarotic cells, and metaplastic cells [12].

The images from the dataset are given to the DSF-CLAHE algorithm for enhancement [13]. The DSF-CLAHE algorithm enhances the images so that each object in the image is transparent for further processing. There is a chance of debris such as bacteria, air drying, dye concatenation, and other unwanted things in the image. Debris may hamper or reduce the performance of the cervical cancer diagnosis and generate a large number of suspicious objects. Due to this, the classification accuracy gets degraded. So the removal of debris is a necessary process. If the debris removal process can be carried out earlier, it is effective during the subsequent processing. In this paper, the debris is removed using the enhanced erosion and dilation operation. Here, two binarization techniques, namely Otsu's method and the kittler binarization are applied to the images separately. Then, the erosion and dilation operation takes place on the results separately. The kittler enhanced erosion and dilation (KED) method considerably improves the debris removal performance than Otsu enhanced erosion and dilation (OED) (Fig. 1).

The steps of the KED are given by.

Step 1: Initially, the images are binarized by using kittler binarization method. To find a threshold value, the kilter approach utilizes a mixture of Gaussian distribution functions. The image is split into two portions using this threshold value: background and foreground, both of which are modeled by Gaussian distribution. The mathematical formulation of the kittler method is given by

$$K_{\text{mix}}(g) = uK_{\text{BG}}(g) + (1 - u)K_{\text{FG}}(g), \tag{1}$$

where $K_{\text{BG}}(g)$ and $k_{\text{FG}}(g)$ are the Gaussian distributions of the background and foreground regions, and the $K_{\text{mix}}(g)$ is the mixture of these two Gaussian distribution,

Fig. 1 Steps in debris removal

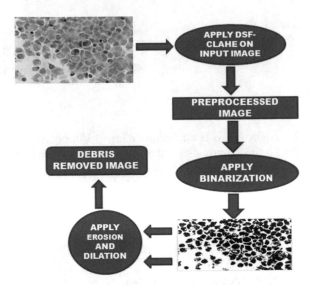

and u is determined by the position of background and foreground in the images. Thus, the binarized image is represented as $br(I)$.

Step 2: Perform the morphological operations (dilation and erosion) on the binarized images.

(a) Dilation operation: Dilation is the process of lengthening or thickening of binarized image $br(I)$. The extent of the thickening process is controlled by structural elements. The dilation of an element α by a structuring element β is mathematically defined as follows:

$$\alpha \oplus \beta = \{\Delta \big| (\hat{\beta})_\Delta \cap \alpha \neq \Omega \}. \tag{2}$$

Here, α is dilated by β, β is a structure element, Ω signifies the empty set, and $\hat{\beta}$ stands for the reflection of collection of β. After mapping and translation, β at least has one overlap with α.

(b) Erosion operation: Erosion operation shrinks or thins the objects in the preprocessed image. The structure element β controls the shrink and the extended process. The erosion of the object α by a structuring element β is given by

$$\alpha \ominus \beta = \{\Delta | (\beta)_\Delta \subseteq \alpha \}, \tag{3}$$

where α_c is the supplement of collection α, β is the structure element. Thus, the debris removed images are given by

$$\mathrm{Dr(I)} = \{di_1, di_2, di_3, ..., di_n\} \tag{4}$$

where $\{di_1, di_2, di_3, ..., di_n\}$ denotes the number of debris removed images.

Table 1 Comparison of debris removal

Method	Sensitivity (%)	Specificity (%)	Accuracy (%)
KED	99	99	98
OED	84	94	90

Algorithm for KED

1. Apply kittler binarization on image I to obtain binarized image br(I).
2. Perform morphological operations on br(I) using structuring element S to obtain the debris removed image Dr(I).

$$\left. \begin{array}{l} br(I) \oplus S \\ br(I) \ominus S \end{array} \right\} Dr(I)$$

4 Results and Discussion

The images from the database are preprocessed using the diffusion stop function-based CLAHE algorithm. After preprocessing, the enhanced images are given as input into the debris removal process. Initially, kittler binarization is applied to the images to make the objects in the image visible. Morphological operations remove the smaller objects present in the image during the process. The morphological operations help to remove the small protrusions and fill the gaps in the images. By this method, the debris in the image is removed up to a large extent. In the second method, instead of kittler binarization, Otsu's method is used along with morphological erosion and dilation. The results are evaluated using performance measures like sensitivity, specificity, and accuracy. Table 1 shows the performance values of the two methods when compared with each other.

KED performs with a sensitivity of 99%, specificity of 99%, and accuracy of 98% when compared with OED having 84%, 94%, and 90%, respectively, as the performance values. The image results are shown in Figs. 2 and 3. Before segmentation, the debris is removed, which causes less effort and time during the segmentation. The obtained output shows that the debris removal method using kittler binarization performs better than Otsu's method.

5 Conclusion

The process of segmenting pap-smear image is a challenging task due to the presence of unwanted particles in the image that resembles nucleus, and therefore, debris removal is one of the primary concerns in automated analysis. In this paper, we have proposed a debris removal method using binarization and morphological operations.

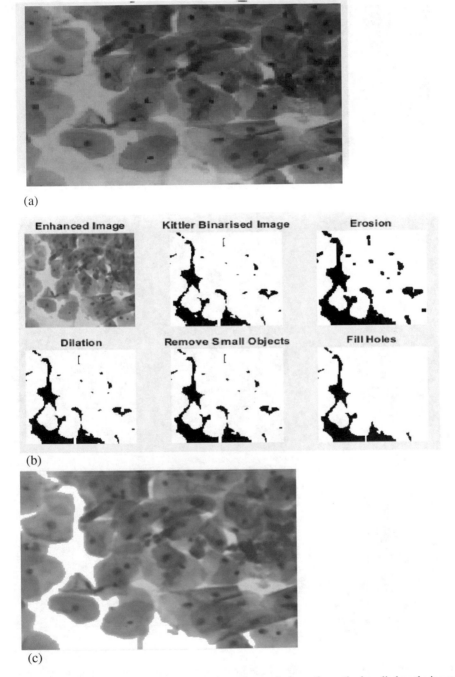

Fig. 2 Images before and after KED. **a** is the input image, **b** shows the method applied on the input image, and **c** shows the debris removed output image

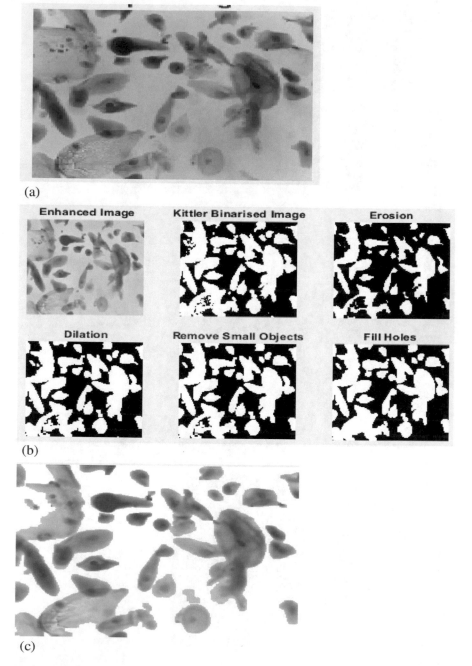

Fig. 3 Images before and after KED. **a** Input image, **b** method applied on the input image, and **c** Debris removed output image

Here, two debris removal methods have been compared. By applying mathematical morphological operations, the unwanted objects in the images can be removed up to a large extent. By applying our method on the images, objects other than cells in the image are eliminated. The results are evaluated based on performance measures like sensitivity, specificity, and accuracy. The KED method performs well with a sensitivity of 99%, specificity of 99%, and an accuracy of 98%. The image outputs clearly show that the debris removal method using KED performs well. Each cell can be seen clearly and is ready to use for further processing. Different methods are available for debris removal, but the proposed method gives a precise result. In the future, other binarization methods can be used along with morphological operations and this debris removal method can be applied on more images from different datasets.

References

1. World Health Organization (2006) Comprehensive cervical cancer control: a guide to essential practice. WHO Press
2. Grohs H, Husain O, (eds) (1994) Automated cervical cancer screening, IGAKU-SHOIN Medical Publishers, Inc. https://doi.org/10.1002/dc.2840130221
3. Saslow D, Solomon D, Lawson H, Killackey M, Kulasingam S, Cain J, Garcia F, Moriarty A, Waxman A, Wilbur D, Wentzensen N, Downs L, Spitzer M, Moscicki A, Franco E, Stoler M, Schiffman M, Castle P, Myers E (2012) American cancer society, American society for colposcopy and cervical pathology, and American society for clinical pathology screening guidelines for the prevention and early detection of cervical cancer. Am J Clin Pathol 137:516–542. https://doi.org/10.1309/AJCPTGD94EVRSJCG
4. Wasswa W, Obungoloch J, Basaza-Ejiri A, Ware A (2019) Automated segmentation of nucleus, cytoplasm and background of cervical cells from pap-smear images using a trainable pixel level classifier. In: 2019 IEEE applied imagery pattern recognition workshop (AIPR). IEEE, pp 1–9. https://doi.org/10.1109/AIPR47015.2019.9174599
5. Kuko M, Pourhomayoun M (2019) An ensemble machine learning method for single and clustered cervical cell classification. In: 2019 IEEE 20th international conference on information reuse and integration for data science (IRI), IEEE, pp 216–222. https://doi.org/10.1109/IRI.2019.00043
6. Wasswa W, Ware A, Bazaza-Ejiri A, Obongoloch J (2019) Cervical cancer classification from Pap-smears using an enhanced fuzzy C-means algorithm. Inform Med Unlocked 14:23–33
7. Malm P, Balakrishnan B, Sujathan N, Kumar VK, Bengtsson RE (2013) Debris removal in pap-smear images. Comp Methods Progr Biomed 111(1):128–138
8. Agarwal P, Sao A, Bhavsar A (2015) Mean-shift based segmentation of cell nuclei in cervical PAP-smear images. In: 2015 fifth national conference on computer vision, pattern recognition, image processing and graphics (NCVPRIPG), IEEE, pp 1–4. https://doi.org/10.1109/NCVPRIPG.2015.7490039
9. Martin D, Sandoval TS, Ta CN, Ruidiaz ME, Cortes-Marteos MJ, Messmer D, Kummel AC, Blair SL, Rodrriguez JW (2011) Quantitative automated image analysis system with automated debris filtering for the detection of breast carcinoma cells. Acta Cytologica 55:271–280. https://doi.org/10.1159/000324029
10. Moshavegh R, Bejinordi BE, Menhert A, Sujathan K, Malm P, Bengston E (2012) Automated segmentation of free-lying cell nuclei in Pap smears for malignancy-associated change analysis. In: 2012 34th annual international conference of the IEEE EMBS San Diego, IEEE, pp 5372–5375. https://doi.org/10.1109/EMBC.2012.6347208

11. Kumar RR, Kumar AV, Kumar SPN, Sudhamony S, Ravindrakumar R (2011) Detection and removal of artifacts in cervical cytology images using support vector machine, In: 2011 IEEE international symposium on IT in medicine and education, IEEE, pp 717–721. https://doi.org/10.1109/ITiME.2011.6130760

12. Oscanoa J, Mena M, Kemper G (2015) A detection method of ectocervical cell nuclei for Pap test images, based on adaptive thresholds and local derivatives. Int J Multimed Ubiquotus Eng 10(2):37–50

13. Plissiti ME, Dimitrakopoulos P, Sfikas G, Nikou C, Krikoni O, Charchanti A (2018) SIPAKMED: a new dataset for feature and image based classification of normal and pathological cervical cells in Pap smear images, In: 2018 IEEE international conference on image processing (ICIP) , IEEE, pp 7–10. https://doi.org/10.1109/ICIP.2018.8451588

14. Haridas S, Jayamalar T (2022) Pap smear image enhancement using diffusion stop function based CLAHE algorithm. In: 2022 8th international conference on advanced computing and communication systems (ICACCS), IEEE, pp 1048–1054. https://doi.org/10.1109/ICACCS 54159.2022.9785050

Image Enhancement with Deep Learning and Intensity Transformation of No-Reference Grayscale Fluorescence Microscopy Images

Aruna Kumari Kakumani and L Padma Sree

Abstract One of the essential and important preprocessing steps for raw microscopy images is to enhance the contrast between background and the foreground. In this work, we implement a deep learning-based image enhancement without using any reference microscopy images for the task. The deep neural network is combined with intensity transformation curves and trained with a loss function to obtain the enhanced results. The proposed framework is shown to have a great potential to enhance the low contrast grayscale fluorescence microscopy images. We show the superiority of this method over some traditional image processing techniques using quantitative metrics. For example, edge-based contrast measure is 13.1528, 84.8378, and 146.0890 for Fluo-N2DH-GOWT1, Fluo-C2DL-Huh7, and Fluo-C2DL-MSC datasets, respectively, which is a significant improvement when compared with some classical methods like contrast limited adaptive histogram equalization. One of the advantages of this method is that, it does not need any reference images for training. Image enhancement is an essential preprocessing step in many image analysis applications, and this preprocessing technique may be used to improve the contrast, to better differentiate the background from the cells in microscopy images. This further helps in the tasks such as cell segmentation, quantification, and tracking, which are essential for computational biology research.

Keywords Deep learning · Intensity transformation curve · Microscopy images · Contrast enhancement

A. K. Kakumani (✉) · L. P. Sree
VNR Vignana Jyothi Institute of Engineering and Technology, Hyderabad, Telangana 500090, India
e-mail: arunakumari_k@vnrvjiet.in

L. P. Sree
e-mail: padmasree_l@vnrvjiet.in

© The Author(s), under exclusive license to Springer Nature Singapore Pte Ltd. 2023
Y. Singh et al. (eds.), *Proceedings of International Conference on Recent Innovations in Computing*, Lecture Notes in Electrical Engineering 1011,
https://doi.org/10.1007/978-981-99-0601-7_31

1 Introduction

Microscopy image analysis plays a pivotal role in biomedical and computational biology research, which is evident from plethora of research publications in the areas of cell segmentation, cell classification, cell tracking, and drug discovery to mention a few. Microscopy images are captured with microscopes equipped with image acquisition systems. There are several challenges [1] involved in processing microscopy images due to limited spatial resolution, limited signal-to-noise ratio (thus poor contrast, low illumination), poor halo artifacts, irregular shapes of cells, and heterogeneous microscopy imaging modalities like fluorescence microscopy, phase-contrast microscopy, etc. Most of the times, microscopy image analysis involves accurate detection of cells as the initial step. In order to detect cells and delineate cell boundaries accurately, image enhancement techniques would be of much help. There are innumerable number of image enhancement methods discussed in the literature, which are mostly traditional image processing techniques, however, deep learning methods have become popular recently which opens new methodologies for the task and potential advantages may be explored.

Global histogram equalization (HE) is one of the earliest and famous method to enhance the image contrast automatically by distributing the values of intensities of an image uniformly over the entire intensity range. This process enhances the image globally, but when it comes to enhancing the local details, this method is little effective and moreover introduces certain artifacts. To overcome these shortcomings, many modified HE techniques were studied. One such method is contrast limited adaptive histogram equalization (CLAHE) where the contrast is restricted to certain limit and removes the defects produced by mapping the two near grayscale values to remarkably different values. Logarithmic transformation followed by histogram equalization is also used as a preprocessing step [2]. Non-uniform illumination is often found in microscopy images. In order to address this problem, top-hat filtering is used as a preprocessing step [3]. Top-hat filtering is a morphological filtering technique, which calculates the morphological opening of the image and then deducts the result from the original input image. Linear filtering methods like mean filtering [4], Gaussian filtering [5], and bandpass filtering [5] techniques were used as a preprocessing step to enhance the microscopy images. Nonlinear filtering methods like median filtering were used to remove the noise and thereby enhance the microscopy images [6]. Anisotropic diffusion algorithm aims at removing noise in the images while preserving the edges [7]. Frequency domain method in which variations in image phase is transformed into variations of magnitude is utilized to improve the structural details of the phase-contrast image [8]. In the article [9], image enhancement of scanning electron microscopy images was proposed which first regularizes the intensity values of the original image while the next step maps the image intensities to their corresponding natural dynamic range to improve the contrast.

Learning-based methods are data driven, which may be broadly classified into paired, unpaired, and no-reference contrast enhancement methods. Paired image enhancement is a process of learning a mapping function which modifies the input image into enhanced image via low contrast/high contrast image pairs. For the input images, multiple exposure images were generated and the highest quality image is considered as the reference image and then a convolutional neural network (CNN) model was trained in a supervised manner from end to end, to get the output enhanced image [10]. A neural network architecture was designed inspired by the bilateral grid processing that is capable of image enhancement which needs pairs of input–output images [11]. In general, implementing these enhancement techniques requires raw image and their ideal reference image pairs to train a deep learning network and usually obtaining ideal reference image is a tough task.

In unpaired image contrast enhancement, a set of images are used as the source domain and their improved images are used as the target domain, and the correlation between source and target domains is avoided. Unpaired learning method for image enhancement was proposed, where starting with a given set of input images with desired characteristics, this method learns an image enhancer which transforms any input image into an enhanced image with the desired characteristics. This framework utilizes a two way generative adversarial networks [12]. CURve Layers (CURL) [13] is a neural block which is augmented to an encoder/decoder backbone network. CURL uses global image adjustment curves. These curves are estimated while training the network to enhance image characteristics. These adjustments are controlled by a well-defined loss function.

Finally, in the no-reference image enhancement, there is no need of paired or unpaired images. In one such methods, a fully convolutional network is utilized to learn the weighted histograms from the original input images. This technique effectively enhances the areas with less contrast and have the areas with acceptable contrast unaltered [14]. In another method, low light image was enhanced without paired or unpaired images using deep intensity transformation curves and well formulated no-reference loss functions [15].

Inspired by the article based on Zero-DCE [15], in this work, we design a deep learning-based image enhancement method for grayscale fluorescence microscopy images. One of the attractive feature of this method is that it does not require any reference images for training the network. In this work, we train the network three times for three different types of fluorescence microscopy grayscale images. The raw microscopy images with low contrast are given to a deep convolutional neural network. The pixels in the output layer of the network are then mapped to a well-designed second order quadratic curve. This curve is applied multiple times (iterations) to achieve higher order curves which helps in achieving better image enhancement. The network is then trained with a well formulated no-reference loss functions to achieve the desired enhancement results. An example of enhancing the low contrast grayscale fluorescence microscopy image of mouse stem cells belonging to Fluo-N2DH-GOWT1 dataset [16] is shown in Fig. 1.

The contributions of this work are listed below:

1. We investigate a deep learning-based image enhancement method for no-reference grayscale fluorescence microscopy images.

Fig. 1 Sample input (left) and enhanced output (right) of an enhanced microscopy image

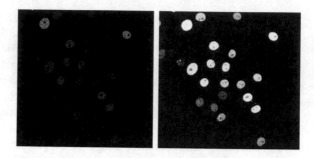

2. We adjust the deep learning architecture and total loss function of Zero-DCE to suit the grayscale image enhancement.

3. We demonstrate the performance of the deep learning framework with no-reference image quality metrics and compare our results with some of the state-of-the-art techniques.

The organization of the remaining article is as follows. Section 2 depicts the methodology used in this work. Section 3 illustrates the results obtained. Section 4 mentions the conclusion which is followed by the references.

2 Materials and Methods

2.1 Reference Data

The dataset for the current study is taken from ISBI cell tracking challenge [16]. Specifically, we study the enhancement of Fluo-N2DH-GOWT1, Fluo-C2DL-Huh7, and Fluo-C2DL-MSC 2D fluorescence microscopy image datasets. The Fluo-N2DH-GOWT1 dataset has two time-lapse image sequences of 92 frames each, thus it consists of a total of 184 frames of 2D fluorescence microscopy images of mouse stem cells. The Fluo-C2DL-Huh7 dataset has two time-lapse image sequences of 30 frames each, thus consisting of a total 60 frames of 2D fluorescence microscopy images. The Fluo-C2DL-MSC dataset has two time-lapse image sequences of 48 frames each, thus consisting of a total of 96 frames of 2D fluorescence microscopy images.

2.2 Methodology

We adopt the deep learning framework and the intensity curves as mentioned in zero-DCE [15]. However, we modify and tune the process to suit the grayscale images and we believe that this is the first deep learning-based image enhancement for

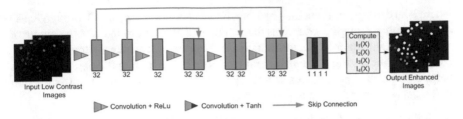

Fig. 2 Deep learning framework used in our work

no-reference grayscale microscopy images. This work utilizes combination of deep learning and intensity transformation curves with well-designed no-reference loss function to attain the desired task of image enhancement. In this section, we describe the deep learning framework, intensity transformation curves, and the loss function used for this study.

Deep Learning Framework. We modify the DCE-Net [15] to suit for the grayscale images. The framework comprises of seven convolutional layers with skip concatenation connections as shown in Fig. 2. Each convolutional layer uses 32 convolutional kernels with the size of 3 × 3 and stride 1, thus giving 32 channels output at each layer. These 32 channels in each of the first six convolutional layers are followed by ReLu activation function, whereas the seventh convolutional layer is followed by Tanh activation function. Finally, the network outputs four layers corresponding to four curve parameter maps also indicating four iterations.

Intensity Transformation Curves. In basic pixel intensity transformation, an input image pixel is mapped to an output image pixel guided by a transformation function like gamma transformation, power law transformation, etc., while doing so both input and output images are kept in the same intensity range.

In this work, we normalize the input image intensity values to lie between 0 and 1 by dividing each value by 255. The mapping curve is designed such that output image pixel values also lie between 0 and 1. We adopt the light enhancement curves as shown in [15] which is carefully designed to achieve the objective of image enhancement through training with appropriate loss function. The intensity transformation curves are applied iteratively and are given in Eq. 1 through 4.

$$IE_1(X) = I(X) + A_1(X)I(X)(1 - I(X)) \tag{1}$$

$$IE_2(X) = IE_1(X) + A_2(X)\,IE_1(X)(1 - IE_1(X)) \tag{2}$$

$$IE_3(X) = IE_2(X) + A_3(X)\,IE_2(X)(1 - IE_2(X)) \tag{3}$$

$$IE_4(X) = IE_3(X) + A_4(X)\,IE_3(X)(1 - IE_3(X)) \tag{4}$$

where $I(X)$ denotes the input image, X are the pixel coordinates, $A_1(X)$ through $A_4(X)$ are the parameter maps produced as the output of channel 1 through 4 at the output of the network. Higher order iterations increase the order of the intensity transformation curves, thus increasing the dynamic range. Moreover, applying the parameter maps helps to train the network to select the best parameter pixel-wise to produce best fitting curves according to the pixel value.

No-Reference Loss Function. The network is trained with exposure control loss. The design of this loss function is critical since the ground truth reference images are not available for the network to learn. The exposure control loss L_{exp} computes the distance between the brightness level E and the average value of intensity of a local region, Y_K. The value of E is set to 0.6. This loss helps in moderating the brightness level. The exposure control loss L_{exp} is represented as

$$L_{exp} = \frac{1}{M} \sum_{K=1}^{M} |Y_K - E| \tag{5}$$

Here, M denotes the number of 16×16 non-overlapping local regions and Y denotes the average intensity value of a local area in the enhanced image.

Deep Learning Training. We train the network three times for each of Fluo-N2DH-GOWT1, Fluo-C2DL-Huh7, and Fluo-C2DL-MSC datasets. Data augmentation is performed by horizontal and vertical flips, 90 degrees left and right rotations. For each dataset, five images were set apart as test set, ten percent of the remaining images are used for validation, and all others were used for training. The network is trained with total loss function for 30 epochs. Input image pixel values are divided by 255 so that pixels are normalized to lie between 0 and 1. The weight initialization for the filters is done with zero mean and 0.02 standard deviation Gaussian function. Batch size of 4 is applied. Fixed learning rate of 10^{-4} is used, and network is optimized with ADAM optimizer. Four channels are used at the output of the network which performs pixel-wise intensity transformation curves for four iterations. The network is trained using Google Colaboratory.

3 Results

3.1 Evaluation Metrics

Image enhancement enables to perceive image details in a better way when compared to its original image. Quantifying image enhancement is not an easy task especially when the original image/reference image pairs are not available. In this study, we use no-reference image quality metrics namely contrast enhancement-based contrast-changed image quality measure (CEIQ) [17], screen image quality evaluator (SIQE), accelerated screen image quality evaluator (ASIQE) [18], histogram spread (HS) [19], edge-based contrast measure (EBCM) [20], and discrete entropy (DE) [21].

CEIQ is a no-reference image quality assessment to measure the quality of image without reference image. CEIQ learns a regression module with the features like histogram equalization, structural-similarity index (SSIM), histogram-based entropy, and cross entropy to infer the quality score. A high value of CEIQ means better image quality.

SIQE and ASIQE are no-reference methods to evaluate the perpetual quality of screen content pictures with big data learning. This method extracts certain features from the image, and a regression method is trained on a number of training images which are labeled with visual quality prediction score. SIQE and ASIQE higher values indicate better quality. Humans can perceive edges more significantly in an image, this observation is considered for designing EBCM. Contrast $c(i, j)$ for a pixel X located at (i, j) is defined as

$$c(i, j) = \frac{|x(i, j) - e(i, j)|}{|x(i, j) + e(i, j)|} \tag{6}$$

where $e(i, j)$ is the average edge gray level and is given by

$$e(i, j) = \frac{\sum_{(k,l) \in \mathcal{N}(i,j)} g(k, l) x(k, l)}{\sum_{(k,l) \in \mathcal{N}(i,j)} g(k, l)} \tag{7}$$

where $\mathcal{N}(i, j)$ represents all the neighboring pixels of pixel (i, j) and $g(k, l)$ represents the edge value at pixel (k, l). We consider 3×3 neighborhood, and $g(k, l)$ is the magnitude of the image gradient calculated using the Sobel operators. The EBCM for an image X is calculated as the average value

$$\text{EBCM}(X) = \sum_{i=1}^{M} \sum_{j=1}^{N} c(i, j)/MN \tag{8}$$

where M and N are the number of rows and columns of the image, respectively. For enhanced image, EBCM is higher than the original image.

Histogram spread is defined as

$$\text{HS} = \frac{\text{Quartile distance of histogram}}{\text{Possible range of pixel values}} \tag{9}$$

$$\text{HS} = \frac{(\text{3rd quartile} - \text{1st quartile}) \text{ of histogram}}{(\text{max} - \text{min}) \text{ of the pixel value range}} \tag{10}$$

It is observed that low contrast images with narrow histograms have a low value of HS when compared with the high contrast images with uniform histogram. Discrete entropy of an image X measures its content. If the value of discrete entropy is higher, it suggests that the image has richer details. It is defined as

$$DE(X) = -\sum_{\forall i} P(x_i)\log(p(x_i)) \tag{11}$$

where $P(x_i)$ is the probability of the pixel intensity of the pixel x_i, which is obtained from the normalized histogram.

3.2 Quantitative Results

The quantitative results of the above metrics for the grayscale fluorescence datasets [14] are shown in Table 1. The quality evaluation metrics used are CEIQ, EBCM, DE, HS, SIQE, and ASIQE. We have compared our approach to the other two well-known methods—Autocontrast and CLAHE. It can be observed in Table 1 that the metrics are significantly superior for the proposed method. Our method has very good quality evaluation metrics for the datasets Fluo-N2DH-GOWT1 and Fluo-C2DL-Huh7. The proposed method also did very well for the image Fluo-C2DL-MSC in all the quality metrics except for DE and SIQE.

Figure 3 displays the raw images and their corresponding output images. It is apparent that the proposed method is giving better visual output. The improvement of the results is because we used a combination of deep learning and intensity transformation curves with well-defined no-reference loss function to train the network for obtaining the desired results. Moreover, this method does not require a reference image for training the deep learning network.

Table 1 Performance comparison of different image quality metrics. The best result for each image type is highlighted

Dataset	Method	Quality metrics					
		CEIQ	EBCM	DE	HS	SIQE	ASIQE
Fluo-N2DH-GOWT1	Autocontrast	1.6295	1.0611	2.3487	0.0040	0.6603	0.7132
	CLAHE	1.6526	1.0428	2.3800	0.0120	0.6622	0.7210
	Proposed	**1.9331**	**13.1528**	**3.0800**	**0.0862**	**0.6747**	**0.7338**
Fluo-C2DL-Huh7	Autocontrast	1.9857	6.4561	5.1894	0.0392	0.6528	0.6988
	CLAHE	2.2823	12.6114	5.9950	0.0826	0.6850	0.7078
	Proposed	**3.1450**	**84.8378**	**6.3552**	**0.3220**	**0.7392**	**0.7505**
Fluo-C2DL-MSC	Autocontrast	2.0078	0.4007	3.9220	0.0274	0.6380	0.7026
	CLAHE	2.0230	0.2014	**5.3151**	0.0681	**0.7315**	0.7246
	Proposed	**2.0238**	**146.0890**	3.9029	**0.1213**	0.7267	**0.7618**

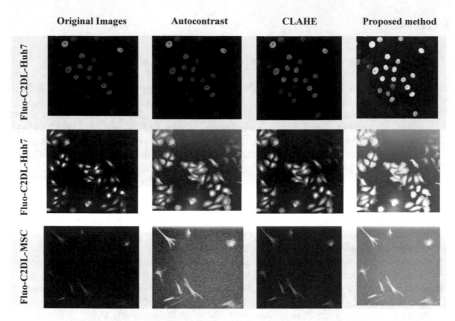

Fig. 3 Original microscopy images and their enhanced outputs for different image enhancement methods

4 Conclusion

In this article, a combination of deep learning and intensity transformation curves trained with no-reference loss functions for the image enhancement of grayscale fluorescence microscopy images without the reference data is studied. The proposed method is evaluated with no-reference image quality metrics, and comparison with state-of-the art techniques indicates the superiority of this method. For instance, EBCM metric gave 13.1528, 84.8378, and 146.0890 for Fluo-N2DH-GOWT1, Fluo-C2DL-Huh7, and Fluo-C2DL-MSC datasets, respectively, which is a significant improvement when compared to CLAHE and autocontrast method. The proposed method could be explored for image enhancement applications, in particular, to process the microscopy images. This method may be utilized as the preprocessing step for the enhancement of low contrast microscopy images, thus may contribute in improved results of cell segmentation and cell tracking. This method is particularly useful in all the applications, where finding an ideal reference image for training a deep learning network is difficult. In the future, this method could be extended to work for other microscopy imaging modality like phase-contrast microscopy, bright-field microscopy. The loss function could be redesigned, the intensity transformation curve could be modified, and the deep learning framework may be revisited so that the methodology could be used for wide varieties of image enhancement problems.

References

1. Meijering E, Dzyubachyk O, Smal I (2012) Methods for cell and particle tracking. Methods Enzymol 504(February):183–200
2. Panteli A, Gupta DK, De Bruijn N (2020) Siamese Tracking of Cell Behaviour Patterns Efstratios Gavves. Proc Mach Learn Res [Internet]. 121:570–587
3. Al-Kofahi Y, Zaltsman A, Graves R, Marshall W, Rusu M (2018) A deep learning-based algorithm for 2-D cell segmentation in microscopy images. BMC Bioinformatics 19(1):1–11
4. Das DK, Maiti AK, Chakraborty C (2015) Automated system for characterization and classification of malaria-infected stages using light microscopic images of thin blood smears. J Microsc 257(3):238–252
5. Harder N, Mora-Bermúdez F, Godinez WJ, Wünsche A, Eils R, Ellenberg J et al (2009) Automatic analysis of dividing cells in live cell movies to detect mitotic delays and correlate phenotypes in time. Genome Res 19(11):2113–2124
6. Luengo--Oroz MA, Pastor-Escuredo D, Castro-Gonzalez C, Faure E, Savy T, Lombardot B et al (2012) 3D+t morphological processing: Applications to embryogenesis image analysis. IEEE Trans Image Process 21(8):3518–3530
7. Liu H (2013) Adaptive gradient-based and anisotropic diffusion equation filtering algorithm for microscopic image preprocessing. J Signal Inf Process. 4(01):82–87
8. Cakir S, Kahraman DC, Cetin-Atalay R, Cetin AE (2018) Contrast enhancement of microscopy images using image phase information. IEEE Access. 6:3839–3850
9. Al-Ameen Z (2018) Contrast enhancement for color images using an adjustable contrast stretching technique. Int J Comput 17(2):74–80
10. Cai J, Gu S, Zhang L (2018) Learning a deep single image contrast enhancer from multi-exposure images. IEEE Trans Image Process 27(4):2049–2062
11. Gharbi M, Chen J, Barron JT, Hasinoff SW, Durand F (2017) Deep bilateral learning for real-time image enhancement. ACM Trans Graph 36(4)
12. Chen YS, Wang YC, Kao MH, Chuang YY (2018) Deep photo enhancer: unpaired learning for image enhancement from photographs with GANs. Proc IEEE Comput Soc Conf Comput Vis Pattern Recognit, 6306–6314
13. Moran S, McDonagh S, Slabaugh G (2020) CuRL: neural curve layers for global image enhancement. Proc—Int Conf Pattern Recognit, 9796–9803
14. Xiao B, Xu Y, Tang H, Bi X, Li W (2019) Histogram learning in image contrast enhancement. IEEE Comput Soc Conf Comput Vis Pattern Recognit Work, 1880–1889, June
15. Guo C, Li C, Guo J, Loy CC, Hou J, Kwong S, et al (2020) Zero-reference deep curve estimation for low-light image enhancement. Proc IEEE Comput Soc Conf Comput Vis Pattern Recognit, 1777–1786
16. Cell Tracking Challenge. http://celltrackingchallenge.net/, last accessed 2022/8/17
17. Li L, Yan Y, Lu Z, Wu J, Gu K, Wang S (2017) No-reference quality assessment of deblurred images based on natural scene statistics. IEEE Access. 5(7):2163–2171
18. Gu K, Zhou J, Qiao JF, Zhai G, Lin W, Bovik AC (2017) No-reference quality assessment of screen content pictures. IEEE Trans Image Process 26(8):4005–4018
19. Tripathi AK, Mukhopadhyay S, Dhara AK (2011) Performance metrics for image contrast. In: ICIIP 2011—Proc 2011 Int Conf Image Inf Process (Iciip), 0–3
20. Celik T, Tjahjadi T (2012) Automatic image equalization and contrast enhancement using Gaussian mixture modeling. IEEE Trans Image Process 21(1):145–156
21. CE S (1948) A mathematical theory of communication. Bell Syst Tech J XXVII(3):379–423

Real-Time Human Action Recognition with Multimodal Dataset: A Study Review

Kapil Joshi, Ritesh Rastogi, Pooja Joshi, Harishchander Anandaram,
Ashulekha Gupta, and Yasmin Makki Mohialden

Abstract Due to difficulties including a cluttered background, partial occlusion, and variations on dimensions, angle, illumination, or look, identifying human endeavors using video clips or still images is a challenging process. A numerous mechanism for recognizing activity is necessary for numerous applications, such as robotics, human–computer interaction, and video surveillance for characterizing human behavior. We outline a classification of human endeavor approaches and go through their benefits and drawbacks. In specifically, we classify categorization of human activity approaches into the two broad categories based on whether or not they make use of information from several modalities. This study covered a depth motion map-based approach to human recognizing an action. A motion map in depth created by adding up to the fullest differences with respect to the two following projections maps for each projection view over the course of the entire depth video series. The suggested approach is demonstrated to be computationally effective, enabling real-time operation. Results of the recognition using the dataset for Microsoft Research Action3D show that our method outperforms other methods.

Keywords Human action recognition · Depth motion map · Multimodal

K. Joshi (✉) · P. Joshi
Department of CSE, Uttaranchal Institute of Technology, Uttaranchal University, Dehradun, India
e-mail: kapilengg0509@gmail.com

R. Rastogi
Department of IT, Noida Institute of Engineering and Technology, Greater Noida, India

H. Anandaram
Centre for Excellence in Computational Engineering and Networking, Amrita Vishwa Vidyapeetham, Coimbatore, Tamil Nadu, India
e-mail: a_harishchander@cb.amrita.edu

A. Gupta
Department of Management Studies, Graphic Era (Deemed to Be University), Dehradun, India

Y. M. Mohialden
Computer Science Department, Mustansiriyah University, Baghdad, Iraq
e-mail: ymmiraq2009@uomustansiriyah.edu.iq

© The Author(s), under exclusive license to Springer Nature Singapore Pte Ltd. 2023   411
Y. Singh et al. (eds.), *Proceedings of International Conference on Recent Innovations in Computing*, Lecture Notes in Electrical Engineering 1011,
https://doi.org/10.1007/978-981-99-0601-7_32

1 Introduction

Computer vision research is currently focused on human activity recognition. Action recognition has already been attempted using video sequences recorded by cameras. Recognizing human behaviors frequently uses spatiotemporal characteristics, e.g. [1]. Real-time depth data gathering is now possible because of advancements in image technology. Depth maps can also deliver 3D details for distinguish behavior that is challenging to characterize utilizing standard images and are less susceptible to changes with lighting conditions than traditional photos. Figure 1 shows two illustrations of the actions of golf swing and a kick forward, each with nine depth maps. Numerous studies on human action detection utilizing depth images have been conducted but since introduction inexpensive depth sensors, in particular, ASUS Xtion with Microsoft Kinect, e.g., [2]. Observed in, additional information is provided to complete action recognition by the 3D multiple objects of skeleton of a person that are calculated using depth photographs.

There are two key queries regarding various classification techniques which action? (specifically, the issue with recognizing) and "Where in the video?" (specifically, the localization issue). The kinetic states of such a person must be known when trying to recognize human activity because then the computer can do so effectively.

Examining actions from still photos or video clips is the aim of human activity recognition. This fact serves as the driving force behind human activity identification systems' quest to accurately classify data input into the relevant activity category. Different types of human behavior are six categories, depending in terms of

a) Golf Swing

b) Forward Kick

Fig. 1 Actions of a golf swing and a kick forward are examples of depth map sequences [3]

Fig. 2 Decomposition of human activities

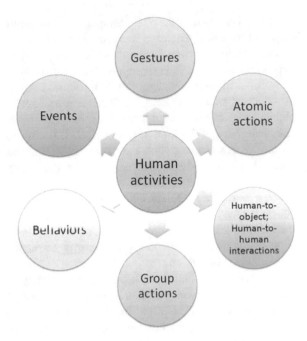

complexity: gestures, atomic acts, human-to-human or object group actions, interactions, behaviors, and things. According to their complexity, human activities are divided up in Fig. 2.

The remainder of the essay the following structure Sect. 2 gives some historical context. The specifics of the depth motion maps' characteristics are explained in Sect. 3. Human activity categorization would be described in Sect. 4. Unimodal and multimodal approaches are covered in Sects. 5 and 6. Dataset collected is considered in Sect. 7. Section 8 also includes final remarks.

2 Background

For the purpose of identifying video clips of people acting recorded by conventional space-time-based RGB cameras, techniques such spatiotemporal space-time volumes characteristics, as well as trajectory, are extensively used. In [4], to recognize human action, spatiotemporal interest points and an SVM classifier were combined. Cuboids' descriptors accustomed to express actions. Activities in a series of videos were identified using SIFT-feature trajectories that were described in an order of three degrees of abstraction. In order to accomplish action categorization, several characteristics of local motion were assembled as spatiotemporal features from a bag (BoF) [3]. As motion templates to characterize the spatial and temporal properties of human motions in movies, motion energy images (MEIs) or motion-history images (MHIs) were only launched in [5]. When computing dense motion flow using MHI is occurred then, a hierarchical extension was provided with correct accuracy. The sensitivity for recognition to variations in illumination is a significant drawback of adopting either depending on hue or intensity approaches, restricting the robustness in recognition. Research with action recognition dependent on depth data has expanded with the introduction of RGBD sensors. Skeletal joint locations are retrieved from depth pictures for skeleton-based techniques. A customized spherical coordinate system and histograms of 3D joint positions (HOJ3D) were used to create a view-invariant posture representation. With the use of LDA, reprojected HOJ3D were used and grouped around K-situation visual words. A continuous hidden Markov model was used to simulate the sequential evolutions of such visual words. Based on Eigen joints, a Naive Bayes Nearest-Neighbor (NBNN) classification was used to identify human behavior. (i.e., variations in joint position) integrating data on offset, motion, and still posture. Due to some errors in skeletal estimate, many skeleton-based techniques have limits. Additionally, many programmers do not always have access to the skeleton information.

To discriminate between various actions, several techniques require spatiotemporal data extraction information based on complete [6] collection of a depth map's point's series. The use of an action graph in a group was 3D points which was also used to describe body positions and describe the dynamics of actions. The 3D points' sample technique, however, produced a lot of data, necessitating a time-consuming training phase. To efficiently describe the body shape as well as movement information for distinguishing actions, an extent motions' histogram with a map on directional gradients (HOG) has been used. A weighted sampling strategy was used to extract random occupancy frequency (ROP) features from depth pictures. The characteristics were demonstrated to be robust to occlusion by using a sparse coding strategy to effectively encode random occupancy sequence features during action recognition. In order to preserve spatial and geographic context statement while managing intra-class conflict variability, 4D advanced patterns were being used as features. Then, for action recognition, a straightforward technical design here on cosine distance was applied. A hybrid system for action recognition method incorporating depth and the skeleton data was employed. Local occupancy patterns and 3D joint position were employed as features then, to characterize each action and account for intra-class variances; another action let accuracy of the model was learned.

3   Depth Motion Maps as Features

The 3D structure but also shape information can be recorded using a depth map. Alemayoh et al. [7] suggested to characterize the motion of an action by imposing depth pictures across three Cartesian orthogonal planes. Because it is computationally straightforward, the same strategy is used throughout the work while the method for getting DMMs is changed. In more detail, any 3D depths are frame also used like create three map v 2D mapped projections that represent the top, side, or front perspectives

$$\text{Where } v = \{f, s, t\} \tag{1}$$

To illustrate (x, y, z) with in a frame depth z, the number of pixels in three projected maps is denoted by the value of depth in such an orthogonal coordinate system, z, x, and y, respectively.

Separated from, the actual distinction between these two separate maps before thresholding is used in this calculation to determine the motion energy for each projected map. The depth gesture map DMM_v is created in-depth video series N frame's worth by stacking all motion energies throughout the full sequence as follows:

$$DMM_v - \sum_{i=a}^{b} \left| map_v^i - map_v^{i-1} \right| \tag{2}$$

where i shows the frame index.

4 Human Activity Categorization

Over the past two decades, the categorization of human activities has remained a difficult job in computer vision. There is a lot of potential in this field based on earlier studies on describing human behavior. According to the type of sensor data they use, we first divide the acknowledgement of human action techniques into the two broad categories: (i) unimodal and (ii) multimodal identification system approaches. According on how they represent human activities, every one of those is two types, then further broken into smaller divisions. As a result, we suggest alternative classification of human activities in hierarchy techniques, as shown in Figs. 3 and 4.

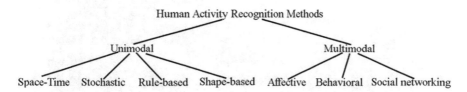

Fig. 3 Proposed hierarchical categorization of human activity recognition methods

Fig. 4 Representative frames of the main human action classes for various datasets [8]

5 Unimodal-Based Methods

Utilizing data from a single modality, single-modal identification of human action algorithms cites examples of human activity. The majority of current methods classifies the underlying activities' label using various classification models and show human activity as either a series of images elements collected from still images or video. For identifying human activities based upon motion features, unimodal techniques are appropriate. On the other hand, it can be difficult to identify the underlying class just from motion. How to maintain is the biggest challenge that the continuity of motion throughout duration of an action takes place uniformly or unevenly throughout a video sequence. Some approaches employ brief motion velocities; others track the optically flow features to employ the whole length on motion curves.

The four basic categories we use to categorize unimodal methods are (i) space-time, (ii) stochastic, (iii) rule-based, but also (iv) methods based on shapes. Depending on just the sort of representation each approach employs, every one of those sub-categories describes particular characteristics of the strategies for recognizing human activities (Fig. 5).

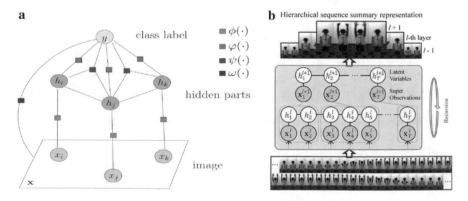

Fig. 5 Representative stochastic approaches for action recognition [9]

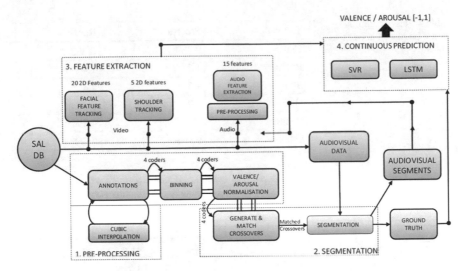

Fig. 6 Flow chart of multimodal emotion recognition [9]

6 Multimodal-Based Methods

Multimodal activity recognition techniques have received a lot of interest lately. Variety components that offer in addition to helpful information can serve as a definition an event. A number of multimodal strategies are found upon in this situation, and feature fusion could be expressed by either initial fusion or lateness fusion. Directly combining characteristics into a greater attribute vector but therefore the simplest method is to learn the underlying action benefit from numerous features. Although the resultant feature vector has a significantly bigger dimension, this feature fusion strategy may improve recognition performance.

A temporal relationship between the underlying activity and the various modalities is crucial in understanding the data since multimodal cues are typically connected in time. In that situation, audiovisual analysis serves a variety of purposes beyond just synchronizing audio and video, however, for monitoring identification of activities. Three groups of multimodal techniques are distinguished: (i) effective techniques, (ii) behavioral techniques, and (iii) social networking-based techniques. Multimodal approaches define atomic interactions or activities that may be related to the effective states of either a communicator's counterpart and depend on feelings and/or physical movements (Fig. 6).

7 Performance of Collected Dataset

Dataset is satisfied with high class variability (intra-class) and high class similarity. The following values are shown in Table 1.

Table 1 Table captions should be placed above the tables

Test data	Precision	Recall
Call	0.389	0.392
Running	1.23	1.45
Stop	1.5	1.4
Hello	0.944	0.833
Pointing	1.0	0.49
Others	0.764	0.765

Table 2 Scaling with tested data and random data

Scale	Test data	New person data
1–10	Rappel	–
10–20	Precision	Precision
20–30	Accuracy	–
30–40	–	–
40–50	–	–

In both Tables 1 and 2, we calculated the precision and recall value of tested data where some data [10] on precision, rappel and accuracy with latest relevant data. We also categorized the age scale between 1 and 10, and last range was 40–50 for monitoring the activity of human. Some results are better in age from 25 to 40, i.e., middle age. We use dataset in further study if we consider any image [11–14] pattern [15–19].

8 Conclusion

Real-time-based model can be predicted with human activity recognition, so in this paper, we conducted a thorough analysis of contemporary techniques for identifying human activity and developed a hierarchical taxonomy of grouping these techniques. According to channel of origin, many of these methods are used to identify activities of humans, and we surveyed many methodologies and divided them into two major categories (unimodal and multimodal). The motion properties of an action sequence were captured through using depth motion maps created from three projection perspectives. In future work, motion monitoring, image classification, and video classification may be useful for exascale computing with fast computing technique.

References

1. Chen C, Liu K, Kehtarnavaz N (2016) Real-time human action recognition based on depth motion maps. J Real-Time Image Proc 12(1):155–163
2. Cheng X et al (2022) Real-time human activity recognition using conditionally parametrized convolutions on mobile and wearable devices. IEEE Sensors J 22(6):5889–5901
3. Park J, Lim W-S, Kim D-W, Lee J (2022) Multi-temporal Sam pling module for real-time human activity recognition. IEEE Access
4. Mazzia V et al (2022) Action transformer: a self-attention model for short-time pose-based human action recognition. Pattern Recog 124:108487
5. Andrade-Ambriz YA, Yair A et al (2022) Human activity recognition using temporal convolutional neural network architecture. Expert Syst Appl 191:116287
6. Sun X et al (2022) Capsganet: deep neural network based on capsule and GRU for human activity recognition. IEEE Systems J
7. Alemayoh TT, Lee JH, Okamoto S (2021) New sensor data structuring for deeper feature extraction in human activity recognition. Sensors 21(8):2814
8. http://crcv.ucf.edu/data/UCF_Sports_Action.php
9. Vrigkas M, Nikou C, Kakadiaris IA (2015) A review of human activity recognition methods. Front Robot AI 2:28
10. Kumar M, Gautam P, Semwal VB (2023) Dimensionality reduction-based discriminatory classification of human activity recognition using machine learning. In: Proceedings of Third International Conference on Computing, Communications, and Cyber-Security. Springer, Singapore, pp 581–593
11. Joshi K, Diwakar M, Joshi NK, Lamba S (2021) A concise review on latest methods of image fusion. Recent Advances in Computer Science and Communications (Formerly: Recent Patents on Computer Science) 14(7):2046–2056
12. Sharma T, Diwakar M, Singh P, Lamba S, Kumar P, Joshi K (2021) Emotion analysis for predicting the emotion labels using Machine Learning approaches. In: 2021 IEEE 8th Uttar Pradesh Section International Conference on Electrical, Electronics and Computer Engineering (UPCON), pp. 1–6, November. IEEE
13. Diwakar M, Sharma K, Dhaundiyal R, Bawane S, Joshi K, Singh P (2021) A review on autonomous remote security and mobile surveillance using internet of things. J Phys: Conference Series 1854(1):012034, April. IOP Publishing
14. Tripathi A, Sharma R, Memoria M, Joshi K, Diwakar M, Singh P (2021) A review analysis on face recognition system with user interface system. J Phys: Conference Series 1854(1):012024. IOP Publishing
15. Wang Y et al (2021) m-activity: Accurate and real-time human activity recognition via millimeter wave radar. In: ICASSP 2021–2021 IEEE International Conference on Acoustics, Speech and Signal Processing (ICASSP). IEEE
16. Sun B, Wang S, Kong D, Wang L, Yin B (2021) Real-time human action recognition using locally aggregated kinematic-guided skeletonlet and supervised hashing-by-analysis model. IEEE Trans Cybernetics
17. Varshney N et al (2021) Rule-based multi-view human activity recognition system in real time using skeleton data from RGB-D sensor. Soft Comp, 1–17
18. Hossain T, Ahad M, Rahman A, Inoue S (2020) A method for sensor-based activity recognition in missing data scenario. Sensors 20(14):3811
19. AlShorman O, Alshorman B, Masadeh MS (2020) A review of physical human activity recognition chain using sensors. Indonesian J Elect Eng Inform (IJEEI) 8(3):560–573

KnowStress: A Mobile Application Prototype Detection of Stress and Stress Related Disorders

Meith Navlakha, Neil Mankodi, and Pranit Bari

Abstract In today's growing world, we often overlook the need of paying heed to the health of some of the most important constituents of our body viz., mental health. This might be due to working hours, which do not allow a person to pay due attention that this facility desires, this in turn may lead to a lot of disorders such as bipolar disorder, ADHD, PTSD and anxiety disorder remaining un-diagnosed which may cause further health complications. Hence, it is the need of the hour to make available a solution so as to enable the people to get themselves diagnosed within a short span of time at their own convenience and from the comfort of their homes. KnowStress has been developed with the aim of improving access to mental health resources for those who lack access to conventional support and to help people explore how technology can be used to improve general wellbeing. KnowStress provides diagnostic tests for screening stress and stress-related disorders by using well known tests and tools that are used by medical professionals. Our proposed model utilises the K-means clustering algorithm to categorise an individual's responses into the groups of "low stress", "bipolar disorder", "PTSD" and "ADHD / GAD" with a validation accuracy of 88.86%.

Keywords Stress · GST · ADHD · Bipolar disorder · PTSD · GAD · ASRS scaling · GAD-7 · PC-PTSD-5

M. Navlakha (✉) · N. Mankodi · P. Bari
Department of Computer Engineering, Dwarkadas J. Sanghvi College of Engineering, Mumbai, India
e-mail: meithnavlakha@gmail.com

N. Mankodi
e-mail: neilmankodi13@gmail.com

P. Bari
e-mail: pranit.bari@djsce.ac.in

1 Introduction

Everyone is affected by stress; there is no real way to totally escape it. Furthermore, depending on their personal, psychosocial, professional and biological backgrounds, some people are more significantly affected by its effects. Stress is essentially a human defence mechanism, but it is crucial to avoid letting it rule your life.

Stress can have many different types of origins, including physical, psychological, emotional and social ones. This application can be used independently, as a bridge to in-person therapy, or as an addition to existing therapy. There are many wonderful advantages to adopting mental health apps, including their affordability and portability. One does not have to worry about scheduling appointments, waiting lists or insurance when using mental health applications.

Additionally, this app allows for privacy and confidentiality and can serve as a safe haven for people who may be reluctant to disclose that they have mental health problems in person or who fear being disparaged or shunned by others. The current scope of this app extends to stress and stress-related disorders as follows:

- ADHD [1]: A long-term disorder characterised by impulsivity, hyperactivity and trouble paying attention.
- GAD [2]: Severe, ongoing anxiety that interferes with daily activities.
- Bipolar Disorder [3]: A condition characterised by cycles of mood swings that range from manic highs to depressive lows.
- PTSD [4]: A condition marked by an inability to recover after being exposed to or witnessing a horrific incident.

All the state-of-the-art stress-related applications focus on only one particular stress-related disorder. We first try to get an estimate of the overall stress of an individual and then redirect the user to the most probable disorder. Most of the available platforms were calibrated keeping in mind the western standards of living. This led us to design an app calibrated according to Indian standards of living. Majority of the platforms available were not acquainted with the provision of spreading awareness about the diagnosed disorder. Propagation of awareness would thereby, lead to expansion of the user base of our app. Most of the available platforms reviewed on the Internet did not seem to have incorporated personalisation. Our application deals with the issue by maintaining the records of the past tests and recommending blogs and tests accordingly.

2 Related Work

Several methodologies have been proposed to detect stress in an individual using various different techniques, some of which have been discussed below.

Fliege et al. [5] proposed the PSQ scale, an indicator for diagnosing stress which may tend to trigger or exacerbate other stress-related disorders. It is explicitly recommended for clinical facilities and is often prescribed by the doctor to have a cursory estimate of the severity of the stress. PSQ index is calculated based on the responses to the PSQ, where the respondent marks the options ranging from 1 ("Never") to 4 ("Always") depending upon the severity of stress associated feelings. Higher scores indicate greater levels of stress. However, the questions in the questionnaire are either too blunt or vague which makes it difficult for the user to understand. This could lead to a misdiagnosis. The paper also does not explain the meaning of each question. This could lead to misinterpretation and thus again, a misdiagnosis. The PSQ is a valid way of checking stress but almost always requires a doctor's guidance.

Li and Liu [6] introduced a technique that makes use of two deep neural networks that were created for the processing of physiological information captured by sensors in order to detect stress and emotions. To strengthen the robustness of the networks, the neural networks must be trained and tested on much larger datasets containing a variety of human populations. The justification for subjecting the neural networks to a dataset that is typical of the entire human population is because each individual's sensitivity to stress conditions (i.e. the conditions under which they experience stress) differs. Also, the method requires high quality sensors which is not economically feasible for everyone. Another disadvantage to this method is that it is invasive and can harm the human body instead of helping with the diagnosis. The neural networks function on unknown data and are thus unsupervised but this may lead to wrong diagnoses.

Albertetti et al. [7] have proposed a novel approach for the detection and binary classification of stressful events by taking into account both the features extracted from physiological signals and the information provided by participants via questionnaires. The signals mainly focused on blood volume pulse (BVP) and electrodermal activity (EDA). For the purpose of categorising stressful events, the research study considers three machine learning models which are gradient boosting decision tree, recurrent neural network (RNN) and convolutional recurrent neural network (CRNN). Out of these three models it was observed that the RNN performed the best with a macro F1-score of 71%. Even though the results obtained show that the novel approach proposed in this research was successful, the main drawback of this method is that it depends on receiving a constant influx of physiological signals from wearable devices such as the Empatica E4 bracelet which is not always accessible for all intended users. This drawback along with the fact that the study was conducted on a small scale imbalanced dataset with metrics such as accuracy being ignored, causes the research and its subsequent results to fall short in certain critical aspects.

Kumar et al. [8] have conducted an empirical study on eight existing machine learning algorithms for the main objective of predicting the occurrence of psychological disorders like anxiety, stress and depression. The research also involves the classification of these aforementioned disorders into classes such as "extremely severe", "severe", "moderate", "mild" and "normal" on the basis of their severity. The research also includes the proposal of a hybrid classification algorithm that will detect and classify the psychological disorder. Datasets created by using the

DASS42 and DASS21 tools were used to train and test the models with a train-test split of 75:25. On the DASS42 data, it was observed that the radial basis function network (RBFN) performed the best amongst the eight existing machine learning models with a maximum accuracy of 97.48% and while the hybrid model gave a comparable performance, it fell short in the aspect that it is very time consuming. On the DASS21 data gathered by the authors, it was observed that the random forest model gave the best performance with an accuracy of 100% and a score of 1 across metrics such as precision and recall. Such atypical results indicate that the model has possibly overfitted the available data and this suspicion is later confirmed by the authors themselves as they report on the imbalanced nature of the small size data obtained from the DASS21 questionnaire. Inconsistent results, poor performance of the proposed model and the lack of a better quality, more representative dataset results in the research not coming up to scratch.

Baheti et al. [9] proposed a method to detect the day-to-day stress-related expressions and emotions of the user by using their daily conversations on social media and other text-based applications based on sentiment analysis. The proposed model automatically extracted the sentiment related keywords from text and fed it to Tensi/Strength framework [10] for identifying the strength of the sentiment of the words used on the social networking sites. Tensi/Strength returned a score ranging from −5 to +5 for each word, and the final classification was performed by SVM. The proposed model provided a NLP-based sentiment analysis, but the scope was restricted to social media conversation as the model was trained on twitter sentiment dataset. A model trained on more realistic and wider scoped dataset is required that focuses on daily activities in order to gauge stress of an individual.

Despite the rapid surge in the count of smartphone-based mental health apps, the usability of these applications is still restricted. These applications were marketed as a potential solution to the corresponding stress disorder but lacked severely in its effectiveness to manage, moderate and treat the disorder. Williams and Pykett [11] conducted a study where they analysed 39 different mental health applications focusing on the social, environmental and mental angles of life, the overall user experience and whether the primary purpose was fulfilled or not. The paper clearly highlighted that the current monitoring mental health apps especially for treating anxiety and depression are still lacking in terms of efficacy. Most of the applications focused more on making the applications more engaging, rather than improvising the testing efficiency.

3 Proposed Work

3.1 System Architecture

The two-tier architecture as seen in Fig. 1 is based on the client–server model. It consists of client-application tier and database tier. The client would be running

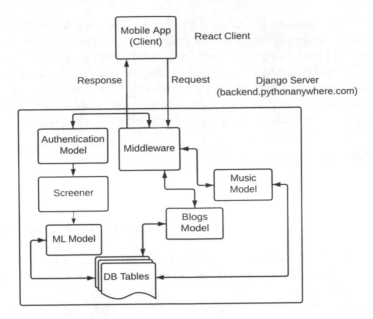

Fig. 1 Two-tier architecture

React.js, and the backend would be running Django REST framework (DRF). For our application, we have chosen SQLite database. The ML models reside on the Django server. The machine learning models are trained and ready to predict once the server has been booted successfully. Figure 1 depicts the proposed ML model which predicts the stress disorder based on the responses from the GST. The music and the blogs model illustrated in the architecture will recommend music and blogs based on the performance of the user in the GST test. An enhanced version of these models, one that also incorporates intelligence, will be completed as part of the future works of the project.

3.2 Questionnaire

Generalised stress test (GST) is a customised screening test or a set of questions (questionnaire) which we have developed by curating the questions having high correlation with stress levels in an individual. This questionnaire was adapted from the PSSQ which is a widely known stress diagnosis questionnaire. The questions included in the PSSQ are often perceived as being too blunt or too vague, making it difficult to obtain accurate responses from the user. This was considered while curating the GST questionnaire so as to improve the chances of developing a more relevant diagnostic tool.

Some of the questions (Question 2 and 5) which are included in the GST are positive in nature and hence, their influence on the stress calculation should be accordingly adjusted. Along with determining whether the user has stress or not, we have also embedded certain deterministic specialised questions in the GST. These questions help the unsupervised ML model get the probabilities of various stress-related disorders. Based on these results, the user may decide which specialised disorder test he/she needs to attempt so as to get a conclusive insight into whether the user has the disorder or not.

Generalised Stress Test:

1. People tend to feel irritable or grouchy at times. In the last month, have you felt this way?
2. Individuals tend to feel lonely or isolated. Have you felt this way in the last month?
3. Have you found yourself feeling more tired than usual in the last month?
4. People often fear that they may not be able to attain their goals. In the last month, have you felt this way?
5. In the last month, have you felt more calm than usual?
6. People often experience elevated levels of frustration. In the last month, have you felt this way?
7. In the last month, have you felt more tense than usual?
8. In the last month, have you felt as if you are always in a hurry?
9. Have you found yourself feeling more worried than usual in the last month?
10. In the last month, have you been able to enjoy and unwind?
11. Do you often feel you are doing things because you have to, and not because you want to?
12. Mental exhaustion is very common these days. In the last month, did you find yourself feeling more mentally exhausted than usual?
13. Insomnia and the inability to relax has become common in today's world. Did you have trouble relaxing in the last month?
14. In the last month, have you often felt overly active and compelled to do things?
15. Did you face difficulties in concentrating or paying attention in the last month?
16. Do you experience frequent mood swings?
17. Do you experience repeated, distressing memories or dreams about traumatic events from the past?

Using the PSQ formula and adjusting it to GST scale using,

$$GST = \frac{Raw\ Score\ -\ Minimum\ Score}{Maximum\ Score\ -\ Minimum\ Score}$$

Minimum Score $= 0$.
Maximum Score $= 60$.

Patient Name			Today's Date						
Please answer the questions below, rating yourself on each of the criteria shown using the scale on the right side of the page. As you answer each question, place an X in the box that best describes how you have felt and conducted yourself over the past 6 months. Please give this completed checklist to your healthcare professional to discuss during today's appointment.				Never	Rarely	Sometimes	Often	Very Often	
1. How often do you have trouble wrapping up the final details of a project, once the challenging parts have been done?									
2. How often do you have difficulty getting things in order when you have to do a task that requires organization?									
3. How often do you have problems remembering appointments or obligations?									
4. When you have a task that requires a lot of thought, how often do you avoid or delay getting started?									
5. How often do you fidget or squirm with your hands or feet when you have to sit down for a long time?									
6. How often do you feel overly active and compelled to do things, like you were driven by a motor?									

Fig. 2 ASRS questionnaire

Questionnaire for ADHD:

ADHD is a neuro-developmental disorder in which the patient has trouble paying attention and displays impulsive behaviours. The ADHD self-test, viz., ASRS screener was proposed by Kessler et al. [1] in collaboration with the World Health Organisation (WHO). The questions of the self-test are displayed in Fig. 2. Later the ASRS scale was validated by Brevik et al. [12].

Questionnaire for GAD:

Anxiety is an emotion that is characterised by feelings of worried thoughts, tension and physical changes like increased blood pressure. These changes are in response to some triggers. The severity or duration of an anxious feeling can sometimes be out of proportion to the original stressor or trigger. This is then known as anxiety disorder. The GAD-7 [2] questionnaire as visible in Fig. 3 is an authentic and efficient tool for screening for GAD and assessing its severity in clinical research and practice. GAD-7 was evaluated by Pranckeviciene et al. [13].

Questionnaire for Bipolar Disorder:

Bipolar disorder (BD) is a chronic condition with extremely crippling symptoms that can have a significant impact on both the victims and those who provide care for them. At least one manic or mixed-manic episode must occur during the patient's lifetime in order for there to be a diagnosis of bipolar disorder. The majority of patients also experience one or more depressive episodes at other periods. Another name for bipolar disorder is manic-depressive disease. The Rapid Mood Screener [3] is a novel tool that can be used to screen for bipolar disorder. It is a questionnaire consisting of six questions, as shown in Fig. 4, where a positive screen requires at least four

Over the last two weeks, how often have you been bothered by the following problems?	Not at all	Several days	More than half the days	Nearly every day
1. Feeling nervous, anxious, or on edge	0	1	2	3
2. Not being able to stop or control worrying	0	1	2	3
3. Worrying too much about different things	0	1	2	3
4. Trouble relaxing	0	1	2	3
5. Being so restless that it is hard to sit still	0	1	2	3
6. Becoming easily annoyed or irritable	0	1	2	3
7. Feeling afraid, as if something awful might happen	0	1	2	3

Column totals ____ + ____ + ____ + ____ =

Total score ____

Fig. 3 GAD-7 questionnaire

questions to be marked as "Yes". The accuracy of RMS has been compared against other available screening tools and validated in the study conducted by Sayyah et al. [14].

Item	Response	
1. Have there been at least 6 different periods of time (at least 2 weeks) when you felt deeply depressed?	Yes	No
2. Did you have problems with depression before the age of 18?	Yes	No
3. Have you ever had to stop or change your antidepressant because it made you highly irritable or hyper?	Yes	No
4. Have you ever had a period of at least 1 week during which you were more talkative than normal with thoughts racing in your head?	Yes	No
5. Have you ever had a period of at least 1 week during which you felt any of the following: unusually happy; unusually outgoing; or unusually energetic?	Yes	No
6. Have you ever had a period of at least 1 week during which you needed much less sleep than usual?	Yes	No

Fig. 4 Rapid Mood Screener (RMS) questionnaire

In the past month, have you...

1. had nightmares about the event(s) or thought about the event(s) when you did not want to?

    YES   NO

2. tried hard not to think about the event(s) or went out of your way to avoid situations that reminded you of the event(s)?

    YES   NO

3. been constantly on guard, watchful, or easily startled?

    YES   NO

4. felt numb or detached from people, activities, or your surroundings?

    YES   NO

5. felt guilty or unable to stop blaming yourself or others for the event(s) or any problems the event(s) may have caused?

    YES   NO

Fig. 5 PC-PTSD-5 questionnaire

Questionnaire for PTSD:

PTSD is a common disorder in which the individual fails to recover after witnessing or experiencing a terrifying event. Ranging from army veterans to people who may have lost a loved one to an accident. The PC-PTSD-5 questionnaire proposed by Sayyah et al. [4] has been shown in Fig. 5, and PTSD-5 was evaluated by Williamson et al. [15].

4 Dataset

Currently, all the available datasets on the Internet do not match the required specifications, they are either not for the decided age bracket or are not geared towards the standard of living in India. A field survey based on the GST questionnaire was conducted with the objective to develop a reliable dataset that is targeted towards the population lying in the age bracket of 15–25 residing in India. The target audience for this survey were mainly students that often suffer from various forms of stress-related disorders such as ADHD, bipolar disorder, GAD and PTSD. Finally, we procured our pertinent dataset consisting of 976 records from the survey out of which 85% (train split) records were used to train our machine learning models and the remaining 15% (test split) records were used to test the model. The data collected via the survey is independent of factors associated with time and is not affected by increased stress levels that may be observed during specific time periods such as the exam season.

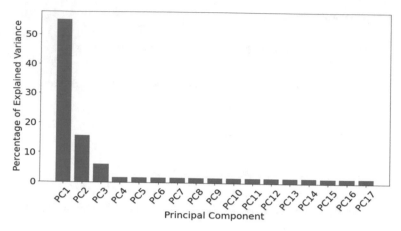

Fig. 6 Explained variance ratio of the principal axis

5 Implementation

5.1 Dimensionality Reduction

Dimensionality reduction reduces the number of independent variables in the dataset by identifying a group of most important principal components. It is a crucial aspect of the data pre-processing process since it helps to simplify the data, with minimal loss of the information in the transformed data present as compared to the raw data. Principal component analysis (PCA) is a very popular statistical technique used for performing dimensionality reduction in which a set of correlated variables is transformed into a set of uncorrelated variables using an orthogonal transformation. In our use case, on applying the technique of PCA on the original dataset containing 17 variables and observing the explained ratio bar chart seen in Fig. 6, a total of three principal components were selected and the dataset was transformed to three-dimensional data and fed to the machine learning models for stress categorization. Performing dimensionality reduction to obtain the new processed data will significantly decrease the training time and improve the model's learning capabilities.

5.2 Models

The data samples in the aforementioned dataset are clustered on the basis of their similarities by utilising clustering algorithms such as (1) density-based spatial clustering of applications with noise (DBSCAN) and (2) K-means.

DBSCAN: It clusters the samples based on the density distribution and expands clusters from them. Good for records which include clusters of comparable density.

The DBSCAN set of rules is primarily based totally in this intuitive belief of "clusters" and "noise". The crucial concept is that for every factor of a cluster, it is mandatory for the neighbourhood of a given radius to include at least a minimal range of factors and has a reminiscence complexity of O(n). The following parameters were set for the DBSCAN model:

- algorithm: auto
- eps: 0.1
- leaf_size: 30
- metric: euclidean
- min_samples: 50

K-means: The unsupervised, iterative K-means technique separates the unlabelled dataset into k distinct clusters, with each dataset belonging to only one group with similar characteristics. Here, K specifies how many preconfigured clusters must be established during the operation. Each cluster has a centroid assigned to it because the algorithm is centroid-based. This algorithm's primary goal is to reduce the total distances between each data point and each of its corresponding clusters. We have chosen to keep four clusters as per the observations made from the elbow plot visible in Fig. 7. The following parameters were set for the K-means model:

- algorithm: lloyd
- init: k-means++
- max_iter: 300
- n_clusters: 4
- tol: 0.0001

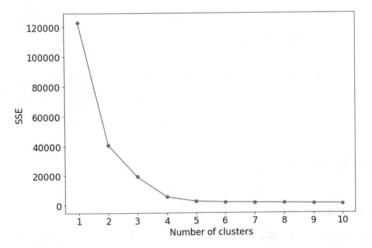

Fig. 7 K-means elbow plot

Fig. 8 Home screen

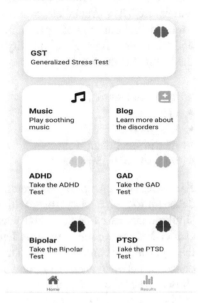

5.3 KnowStress Prototype

The screenshots of the KnowStress mobile prototype can be seen in Figs. 8, 9 and 10.

6 Results and Analysis

Both K-means and DBSCAN models were trained on the aforementioned dataset and subsequently analysed by using evaluation metrics such as the silhouette coefficient and confusion matrix.

6.1 Evaluation Metrics

Silhouette Coefficient:

Silhouette coefficient helps to determine the quality of the clusters. Higher silhouette coefficients typically have more cohesive clusters; the silhouette coefficient ranges

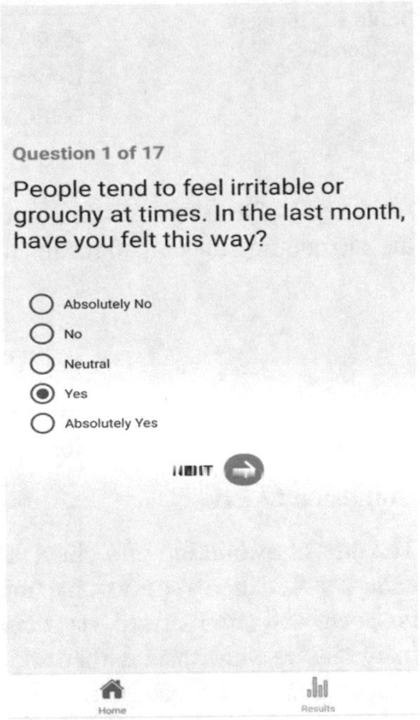

Fig. 9 GST

Fig. 10 Results screen

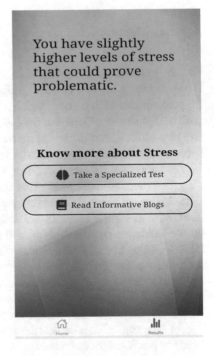

Table 1 Silhouette coefficient

Model	Silhouette coefficient	Validation accuracy (%)
K-means	0.8249	88.86
DBSCAN	0.8130	50.98

from 1 to 1. Close to + 1 silhouette coefficients suggest that the sample is far from the surrounding clusters, 0 means it is close to or on the decision boundary.

$$s = \frac{b - a}{\max(a, b)}$$

where b is cluster fragmentation and a is cluster cohesiveness.

Confusion Matrix:

The task of evaluation of a classification algorithm's performance is accomplished using a table called a confusion matrix. The effectiveness of a classification algorithm is shown and summarised via a confusion matrix, basically by summarising how many classes were misclassified and into what, to have a more intuitive understanding about the misclassifications.

6.2 Analysis

For the purpose of clustering data samples into distinct clusters, K-means and DBSCAN were the two clustering algorithms that were considered. On comparing the performance of these two models, it was observed that K-means performed better with a silhouette coefficient of 0.8249 as compared to DBSCAN that gave a silhouette of 0.8130. The comparison between DBSCAN and K-means shown in Table 1 depicts the silhouette scores and the validation accuracy of both the models.

The validation dataset consisting of 50 records was collected with the help of a reputed medical institution. This dataset was used to validate the K-means and DBSCAN models. It was observed that K-means gave a superior validation accuracy of 88.86% as visible in Table 1. DBSCAN, however, had an inferior performance on the validation dataset with just 50.98% accuracy. Moreover, another issue with DBSCAN was to decide how to handle a response which was categorised as noise. Since we focus on four disorders currently, a GST response labelled as noise can be from an extreme end of any disorder or even from a low stress score. This uncertainty on how to handle label "noise" is also a major criteria for choosing K-means over DBSCAN. The confusion matrix for the selected K-means models is displayed in Fig. 11. The vertical axis depicts the actual labels, and the horizontal axis the expected labels as predicted by the model.

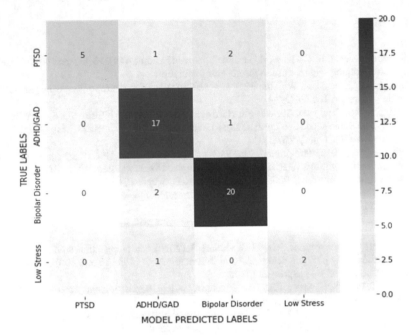

Fig. 11 Confusion matrix of K-means on validation dataset

7   Conclusion

In the current work, we have reinvented the way to classify people as stressed or not stressed by leveraging modern day machine learning techniques. The proposed application, KnowStress would help doctors as well as individuals to gauge their levels of stress in a comfortable manner thereby allowing them to prescribe tests to determine their general stress levels and possibilities of having specific stress-related disorders such as ADHD, bipolar disorder, GAD and PTSD. Our proposed GST model utilises the K-means clustering algorithm which has a validation accuracy of 88.86%. The KnowStress app also includes blogs that are aimed at spreading awareness about stress-related disorders and music that would help the users meditate and relax.

8   Future Work

The current model was trained on a small dataset consisting of only 976 records, to improve the model's performance one may increase the size of the dataset thus increasing generalisation. Furthermore, the scope of the research can be extended by incorporating additional stress-related disorders and experimenting with newer state-of-the-art models.

References

1. Kessler RC et al (2004) The World Health Organization adult ADHD self-report scale (ASRS): a short screening scale for use in the general population
2. Spitzer RL, Kroenke K, Williams JBW et al (2006) A brief measure for assessing generalized anxiety disorder. The GAD-7
3. McIntyre RS, Patel MD, Masand PS, Harrington A, Gillard P, McElroy SL, Sullivan K et al (2021) The Rapid Mood Screener (RMS): a novel and pragmatic screener for bipolar I disorder. Current Med Res Opinion 37(1)
4. Prins A et al (2016) The primary care PTSD screen for DSM-5 (PC-PTSD-5): development and evaluation within a veteran primary care sample. J General Internal Med 31(10)
5. Fliege H, Rose M, Arck P, Walter O, Kocalevent R-D, Weber C, Klapp B (2005) The Perceived Stress Questionnaire (PSQ) reconsidered: validation and reference values from different clinical and healthy adult samples
6. Li R, Liu Z (2020) Stress detection using deep neural networks. BMC Med Inform Decis Mak 20:285
7. Albertetti F, Simalastar A, Rizzotti-Kaddouri A (2020) Stress detection with deep learning approaches using physiological signals. In: International Conference on IoT Technologies for HealthCare, pp 95–111. Springer, Cham
8. Kumar P, Garg S, Garg A (2020) Assessment of anxiety, depression and stress using machine learning models. Procedia Comp Sci 171
9. Baheti RR, Kinariwala SA (2019) Detection and analysis of stress using machine learning techniques. Int J Eng Adv Tech (IJEAT). 9(1). ISSN: 2249-8958
10. Baheti RR, Kinariwala SA (2019) Survey: sentiment stress identification using tensi/strength framework. Int J Sci Res Eng Dev 2(3)
11. Williams JE, Pykett J (2022) Mental health monitoring apps for depression and anxiety in children and young people: a scoping review and critical ecological analysis. Soc Sci Med 297:114802
12. Brevik EJ, Lundervold AJ, Haavik J, Posserud M-B (2020) Validity and accuracy of the ADHD Self-Report Scale (ASRS)
13. Pranckeviciene A, Saudargiene A, Gecaite-Stonciene J, Liaugaudaite V, Griskova-Bulanova I, Simkute D, Naginiene R, Dainauskas LL, Ceidaite G, Burkauskas J (2022) Validation of the patient health questionnaire-9 and the generalized anxiety disorder-7 in Lithuanian student sample. PLoS One
14. Sayyah M, Delirrooyfard A, Rahim F (2022) Assessment of the diagnostic performance of two new tools versus routine screening instruments for bipolar disorder: a meta-analysis. Brazilian J Psychiatry 44
15. Williamson MLC et al (2022) Diagnostic accuracy of the primary care PTSD screen for DSM-5 (PC-PTSD-5) within a civilian primary care sample. J Clin Psych. https://doi.org/10.1002/jclp.23405

Handwritten Character Evaluation and Recommendation System

Rishabh Mittal, Madhulika Bhadauria, and Anchal Garg

Abstract Character recognition is a very popular machine learning application for example—OMR sheets are commonly used for grading purposes all over the world. It is a procedure of analysing human-marked data from documents like surveys and tests. The COVID-19 epidemic has had a negative influence on schooling, since it has caused students' handwriting habits to deteriorate significantly. As a result, the goal of this paper is to propose a lightweight deep learning model that uses computer vision to aid in the practice, evaluation, and analysis of nursery students' handwriting by evaluating sheets and pointing out major errors with an accuracy score that can be used to measure improvement. Recognition, evaluation, and analysis are the three primary components of the suggested paradigm. The steps are as follows: (1) Pre-processing, (2) deep learning model for feature extraction and classification, (3) predicting the letters in the supplied picture, (4) calculating the similarity index between the input image and the glyphs in the computer font, (5) matching the input image's stroke length and angles to font glyphs, and (6) displaying the character's score and analysis. The most accurate model for this process was 83.04%.

Keywords Deep learning · Computer vision · Machine learning · Keras · OpenCV · Similarity learning · Transfer learning

1 Introduction

The velocity at which handwritten data is created in the real world necessitates manual processing since computers still struggle to decipher and extract information from

R. Mittal (✉) · M. Bhadauria · A. Garg
Department of Computer Science and Engineering, Amity University, Noida, Uttar Pradesh, India
e-mail: rm.23mittal@gmail.com

M. Bhadauria
e-mail: mbhadauria@amity.edu

A. Garg
e-mail: agarg@amity.edu

© The Author(s), under exclusive license to Springer Nature Singapore Pte Ltd. 2023   437
Y. Singh et al. (eds.), *Proceedings of International Conference on Recent Innovations in Computing*, Lecture Notes in Electrical Engineering 1011,
https://doi.org/10.1007/978-981-99-0601-7_34

handwritten data. Optical character recognition (OCR) is a method that is typically [1] used to extract text from images. However, traditional OCR relies on rules and templates and is rigid in nature, in contrast to the much more recent machine learning-based OCR that uses deep learning and produces a highly accurate and adaptable system. For instance, if a student's handwriting practice sheet needs to be corrected, a traditional OCR may produce false positives if the student makes a mistake or if a new format or item is present that is not already covered by the rules. In contrast, an ML-based system may perform better on a wider variety of practice sheets.

A student's handwriting is significant since it aids in the development of their reading and spelling abilities. It also offers [2] details about the emotional, mental, and physical health. It is intimately related to phonics, psychological awareness, motor abilities, and many other characteristics. However, everything is going digital as a result of the COVID-19 pandemic, and people are more dependent on technology than ever. This is also true in schools, where students no longer write things down since everything is done online. Children in nurseries, in particular, who have just begun school, cannot practice writing as much as they might when attending school. This caused the handwriting of children to dramatically decline in quality.

As a result, a technique for character identification and analysis based on deep learning and computer vision is proposed.

- OpenCV is a programming library that focuses on computer vision. Computer vision [3] is a branch of computer science that tries to allow computers to understand images in the same way that humans do. In our model, OpenCV was utilised to provide input to the computer via a webcam or a picture uploaded to the computer. The image is next pre-processed, which includes stages such as converting RGB to grayscale, black and white conversion, smoothening, and blurring. The deep learning model that will be trained on this data will require these pre-processing steps. Finally, OpenCV was employed in the analysis to discover the extreme points of the input handwritten characters and use those to calculate angles and lengths.
- Keras is an open source software package that gives deep learning applications a Python interface. Deep learning [4] is an area of machine learning that seeks to replicate the learning process of the human brain by gradually extracting characteristics from a raw input using numerous layers. On the EMNIST dataset, we employed a deep learning model for feature extraction and classification. The pre-processed dataset is fed into a multi-layered CNN-based deep learning model, which predicts the character of the handwritten character input.

2 Literature Review

This section provides a brief of some works done in this field by various other authors.

In this study, Masoud [5] looked at the potential of deep convolutional neural networks (CNNs) for classifying multispectral remote sensing images. For wetland mapping in Canada, they examined seven well-known deep convolutional nets:

DenseNet121, InceptionV3, VGG16, VGG19, Xception, ResNet50, and Inception-ResNetV2. The top three convolutional networks offer state-of-the-art classification accuracies of 96.17%, 94.81%, and 93.57%, respectively. They are InceptionResNetV2, ResNet50, and Xception. The results of this work demonstrate that it is feasible to completely utilise current deep convolutional nets for the classification of multispectral remote sensing data, which differs significantly from sizable datasets (like ImageNet) often utilised in computer vision.

The author of this piece, Manchala et al. [6], created a way for turning a photograph of a handwritten transcription into a digital text by employing a number of techniques. Among the techniques utilised were convolutional neural network (CNN), recurrent neural network (RNN), and connectionist temporal classification (CTC). For text with the least amount of noise, the project offers the maximum level of accuracy. A conventional neural network, which has an accuracy of more than 90.3%, is used to carry out the job. In future, this research may be broadened to examine alternative embedding methods on a larger variety of datasets.

Dixit et al. [7] employed support vector machines (SVMs), multi-layer perceptron (MLP), and convolution neural network (CNN) models to show handwritten digit recognition using MNIST datasets. The main objective of this study is to determine the best digit recognition model by evaluating the accuracy and execution times of the models mentioned above. SVM (train: 99.98%, test: 93.77%), MLP (train: 99.92%, test: 98.85%), and CNN (train: 99.53%, test: 99.31%) were the results. As a consequence, among SVM, MLP, and CNN, CNN is the best contender for any prediction problem that uses image data as an input, as measured by accuracy rate.

The author, Shamim, discusses the issue of creating a reliable algorithm that can recognise handwritten numbers entered by users into a scanner, tablet, or other digital device. This article [8] discusses how to use a variety of machine learning algorithms to recognise handwritten numbers offline. This study's main objective is to make sure that techniques for reading handwritten numbers are reliable and efficient. Numerous machine learning techniques, including multi-layer perceptron, support vector machine, Nave Bayes, Bayes Net, random forest, J48, and random tree, were utilised with WEKA to recognise digits. According to the study's findings, the multi-layer perceptron has an accuracy rate that may go as high as 90.37%.

In this study, Hossain et al. [9] tried to employ deep neural networks to get around issues with handwritten recognition of Bangla numerical digits. In order to solve the handwritten numeric character recognition issue, this study suggests a fine-regulated deep neural network (FRDNN), which uses convolutional neural network (CNN) models with regularisation parameters to avoid overfitting and generalise the model. This study employed traditional deep neural network (TDNN) and fine regulated deep neural network (FRDNN) models using BanglaLekha-Isolated datasets, and the classification accuracy for the two models across 100 epochs was 96.25% for TDNN and 96.99% for FRDNN. The FRDNN model's network performance was more reliable and accurate on the BanglaLekha-Isolated digit dataset than the TDNN model, and it was reliant on experimentation.

In this study [10], Inunganbi described a novel decision forest-based strategy for transfer learning, which we utilise to recognise characters. In order to move knowledge from source tasks to a specific target task, they presented two enhancements to the decision forest architecture. The MNIST dataset shows that experiments outperform typical decision trees. They also perform well when compared to other state-of-the-art classifiers. Thus, they present a novel approach for transferring knowledge from several source tasks to a single target task in this study. As a result, a classifier that may use knowledge from related tasks to improve predicted performance on the target task has been created.

Sulaiman and Nazir, cursive text recognition [11] is considered the most difficult task in the field of machine learning and pattern recognition, but due to minor differences in character shape and a large number of characters in its character database, the Pashto language presents even more challenges to the research community. Using an MLSTM-based deep learning approach, the suggested research study demonstrates the construction of an optimal OCR system for the recognition of isolated handwritten Pashto letters. The suggested model's applicability is demonstrated utilising a decision trees classification tool based on zonal feature extraction and invariant moments-based techniques. The MLSTM-based OCR system has an overall accuracy rate of 89.03%, whereas DT-based identification rates of 72.9% are reached employing zonal feature vectors, and 74.56% is achieved for invariant moments-based feature map.

The recognition of Chinese characters with a vast character set has always been a difficult challenge that needed to be solved quickly. This study by author Dongdong and Zhang [12] focuses on the enhancement of the CRNN method and provides a character recognition algorithm based on feature fusion to address the problem of character recognition in artificial intelligence machine learning. Fine tuning's accuracy is determined to be higher than transfer learning's, which can better extract the features of the current data set and enhance accuracy, according to experimental research. A deep convolution neural network is constructed based on MNIST, which is better suited for Chinese character recognition data. Finally, on the training dataset, it achieved accuracy of 0.99 and 0.99 on the test data, indicating that the neural network model can completely suit the Chinese character recognition training set.

Syed Yasser Arafat and Muhammad Javed Iqbal address the issue that the non-Latin and cursive script of the Urdu language presents in their article [13]. They suggested utilising FasterRCNN with CNNs and regression residual neural networks to recognise Urdu writing in real situations (RRNNs). The authors made use of a collection of five unique datasets with embedded pictures in Urdu text. Using a two stream deep neural network, this technique can recognise partial portions of synthetic pictures with an accuracy of 95.20% and real-world photos with an accuracy of 76.6% (TSDNN).

Chernyshova introduced an incredibly lightweight framework that can run on an embedded or mobile device in this research [14]. The author suggested a compact ANN that operates quickly and accurately on photos of poor quality. Tesseract 4.0 OCR and ABBYY FineReader 15 are both outperformed by this effort in terms of speed and accuracy. The model was trained using MNIST and MDIV-500, and when evaluated with data from the 1961 Census Sample, the model's accuracy was 96.69% (Table 1).

3   Dataset

The EMNIST dataset [15] was used to train the model as it contains 814,255 images of handwritten characters (both capital and small mixed) and digits (zero to nine). The input is an image of size 28 × 28 which is flattened into a single dimensional array, thus the input provided to the model is a 1D array of size 1 × 784 and the final output layer gives 62 outputs as there are 62 different labels (26 for capital alphabets, 26 for small alphabets, and 10 for digits). A total of 697,923 images were used to train this model, and 116,323 were to test it (Fig. 1).

For the evaluation part of the programme, 'Times New Roman' font glyphs were used that were downloaded from a website https://graphemica.com/. This website contains various different font style glyphs that all each of size 500 × 500 × 3. So a total of 62 images were downloaded from this website as there are 62 output classes.

4   Proposed Methodology

The method proposed consists of six major steps that can be categorised in three parts (Fig. 2).

4.1   Pre-Processing

Pre-processing is a technique for preparing data for further processing processes and converting raw data into a format that may be used. The steps involved in preparing the EMNIST data for deep learning neural networks included following steps.

4.1.1   Reshaping

As the shape of the imported EMNIST data is in the form of a single dimensional array, it has to be converted into a two-dimensional single channel image as shown in Fig. 3.

Table 1 Literature review summary

S. no	Author	Dataset used	Technique used	Metrics	Performance
1	Masoud Mahdianpari, Bahram Salehi, Fariba Mohammadimanesh, and Yun Zhang	Multispectral remote sensing images	SVM, RF, DenseNet121, InceptionV3, VGG16, VGG19, Xception, ResNet50, InceptionResNetV2	Accuracy, Kappa coefficient, and F1-score	96.17% accurate
2	Sri Manchala, Jayaram Kinthali, Kowshik Kotha, and Jagilinki Kumar	IAM dataset	CNN, RNN, CTC	Accuracy rate	90.30% accurate
3	Ritik Dixit, Rishika Kushwah, and Samay Pashine	MNIST dataset	MLP, SVM, CNN	Accuracy rate	99.31% accurate
4	S. M. Shamim, Md Badrul Miah, Angona Sarker, Masud Rana, and Abdullah Jobair	Digit dataset	MLP,SVM, Naïve Bayes, Bayes Net, random forest, J48, random tree	Accuracy, time consumption, Kappa coefficient, RMS, mean absolute error	90.37% accurate
5	Md Hossain, Zainul Hossain, Manik Abadin, and Md Ahmed	BanglaLekha-Isolated multipurpose comprehensive handwritten isolated character samples dataset	FRDNN,TDNN,CNN	Accuracy rate	96.99% accurate
6	Sanasam Inunganbi and Robin Katariya	Meitei Mayek dataset	AlexNet, GoogLeNet, ResNet-18, ResNet50, VGG16	Accuracy rate	98.47% accurate
7	Sulaiman Khan and Shah Nazir	Handwritten Pashto characters database	MLSTM, decision trees	Accuracy, time consumption, recall, F-score, precision, specificity	89.03% accurate

(continued)

Table 1 (continued)

S. no	Author	Dataset used	Technique used	Metrics	Performance
8	Dongdong He and Yaping Zhang	MNIST	Inception-v3, Inception-ResNet-v2, CRNN-based feature fusion	Accuracy rate	99% accurate
9	S. Y. Arafat and M. J. Iqbal	Custom Urdu text dataset	TSDNN, FasterRCNN, RRNN, SqueezeNet, GoogLeNet, Resnet18, Resnet50	Accuracy rate	99.06% accurate
10	Y. S. Chernyshova, A. V. Sheshkus, and V. V. Arlazarov	MNIST, MIDV-500, the 1961 census for England and wales subsample	CNN, ANN	Accuracy rate	96.69% accurate

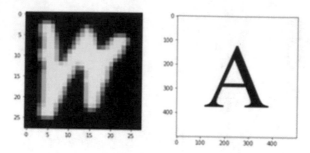

Fig. 1 Sample of EMNIST dataset and image from graphimica

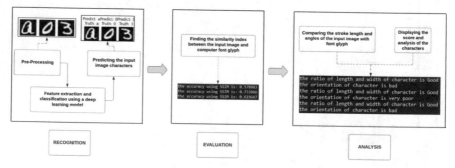

Fig. 2 Basic workflow of the proposed methodology

Fig. 3 Reshaping

4.1.2 Processing and Resizing Input Data

To identify characters in our images, we first convert the entire images in perfect black and white by converting them into grayscale, then applying a threshold. Then, we dilate the images to enhance the characters and add a Gaussian filter on top to smoothen out the image. After all the pre-processing is done, we find the contours in image and store them in a list. Finally, we get the bounding rectangle for each contour and use that to separate the characters out of the entire image (Fig. 4).

Then, finally each of the separated characters are resized to 28*28 in order to perform prediction on them.

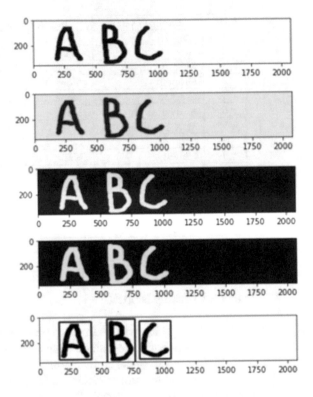

Fig. 4 Processing input image

4.2 Feature Extraction and Classification

The process of extracting information from data that may be used for categorisation purposes is known as feature extraction. Adam optimizer and the categorical-cross-entropy loss function were used to create all of the models. We utilised a variety of models for this, as seen below.

4.2.1 CNN-Based Neural Net

CNN is the most commonly applied to analyse images. In our proposed model, we used two set of convolutional layers and an output layer, which totals to 11 layers and 989,502 parameters shown in Fig. 5. The input given to the model were images with size $28 \times 28 \times 1$ as in the EMNIST without any processing.

S.No.	Layer (Type)	Output Shape	Params #
1	Conv2d	(None, 26, 26, 256)	2560
2	Activation	(None, 26, 26, 256)	0
3	Max_pooling2d	(None, 13, 13, 256)	0
4	Dropout	(None, 13, 13, 256)	0
5	Conv2d_1	(None, 11, 11, 256)	590080
6	Activation_1	(None, 11, 11, 256)	0
7	Max_poolinng2d_1	(None, 5, 5, 256)	0
8	Dropout_1	(None, 5, 5, 256)	0
9	Flatten	(None, 6400)	0
10	Dense	(None, 62)	396862
11	Activation_2	(None, 62)	0
Total params: 989,502			
Trainable params: 989,502			
Non-trainable params: 0			

Fig. 5 Structure of proposed model

4.2.2 Transfer Learning Models

It is a field where we apply knowledge from a problem into solving another problem. In our case, we used popular pre-trained deep learning models on the ImageNet dataset to extract features from the EMNIST data.

Since the dataset used to train models was ImageNet, EMNIST data had to be processed before it could be passed onto the network. Therefore, all the images that were originally of size 28×28 were scaled up to $84 \times 84 \times 3$ due to the minimum size requirement. Also the models were slightly modified at the input and output layers to accommodate the EMNIST data (Fig. 6).

The following models were tested.

VGG 16 and VGG 19

Simonyan and Zisserman [16] It contains five set of convolutional layers and an output layer for a total of 16 layers and 14,714,688 parameters. The network was trained on ImageNet with image size of $224 \times 224 \times 3$. It managed to achieve 92.7% accuracy in ILSVRC 2014, whereas VGG19 contains five set of convolutional layers and two fully connected layers for output layer for a total of 19 layers (three extra convolutional layers than VGG16) and 20,151,422 parameters.

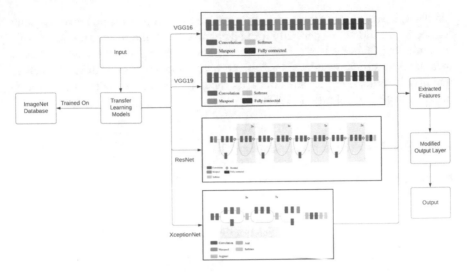

Fig. 6 Block diagram for transfer learning models

Xception

Chollet [17] It is a DNN that is an 'extreme' version of Inception network. It computes in depth by utilising 1×1 convolution across the depth to reduce the dimensions. It acts like a depth wise separable convolution network. It is a 71 layers deep network with 22,004,326 parameters that achieved an accuracy of 94.5% accuracy on ImageNet validation dataset as per Keras documentation [18].

ResNet50

Chollet et al. [19] It is a modified form of ResNet model which contains 48 convolutional layers along with one max pool and one average pool layer for a total of 23,714,750 parameters. It allows very deep neural nets due to skip connections as they remove vanishing gradient problem by giving an alternate path. It reached an accuracy of 92.5% in ILSVRC 2015.

4.3 Finding Similarity Score

After the character is recognised, the next step is to find how accurate the formation of the character is. This was done using similarity learning. For example, the cursive writing books that come have a character printed with perfect formations and we ask the child to write similar to that and then compare those to see if he/she is right/wrong.

Similarly, the idea behind this approach is that we can compare the image of character formed by a child with a perfectly formed character and find the accuracy.

Similarity, learning is a sub-field of machine learning and the goal of it is to measure how similar or related two objects are. There are a lot of different metrics that may be utilised to measure similarity index between two images, the one used here is structural similarity or SSIM.

4.3.1 SSIM or Structural Similarity [20]

It was proposed in 2004 as an evolved version to the old UQI and its performance backed its popularity as it consistently outperformed a lot of other metrics. The only downside for our use case being the fact that SSIM requires the images to be of the same size which is not practical. Thus, requiring the images to be resized which leads to a lot of distortions and errors.

4.4 Analysis of the Image

After we know what the character is and how accurately is it formed, next step is to find what is wrong in it and suggest a way to improve the character formation for the next time. This was done simply by comparing the ratio of length (as shown in Fig. 7 with red line) and width (as shown in Fig. 7 with blue line) of character under test with the length and width of Times New Roman font Glyphs and the orientation of character by calculating angle between the lines drawn from extreme points from the image contours.

The extreme points of image contours refer to the following points on the image: left most, rightmost, top, and bottom using the contours and thresholds. Then, we calculate the distances between the left–right points as shown in Fig. 8 by red colour and top–bottom by blue colour. Finally, we compare the ratio of the length of the lines and angle between them to suggest an improvement that can be made either in the ratio or orientation of the character.

Fig. 7 Calculating the ratio and angle

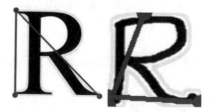

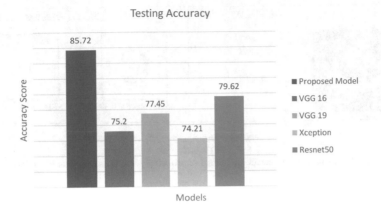

Fig. 8 Graph of summarised results

5 Results and Discussion

As we can see from the Table 2, when deep learning and transfer learning are combined, they may produce a very accurate model capable of accurately detecting characters, as seen in the image below.

Table 2 Results summary

S. no	Method	Training and testing set	Results
1	Proposed model (20 epochs)	Training images (size)—697,923 (28 × 28 × 1) Testing images (size)—116,323 (28 × 28 × 1)	Training accuracy—85.65% Testing accuracy—85.72%
2	VGG16 (10 epochs)	Training images (size)—200,000 (84 × 84 × 3) Testing images (size)—100,000 (84 × 84 × 3)	Training accuracy—78.56% Testing accuracy—75.20%
3	VGG19 (10 epochs)	Training images (size)—200,000 (84 × 84 × 3) Testing images (size)—100,000 (84 × 84 × 3)	Training accuracy—80.06% Testing accuracy—77.45%
4	Xception (10 epochs)	Training images (size)—200,000 (84 × 84 × 3) Testing images (size)—100,000 (84 × 84 × 3)	Training accuracy—76.79% Testing accuracy—74.21%
5	Resnet50 (10 epochs)	Training images (size)—200,000 (84 × 84 × 3) Testing images (size)—100,000 (84 × 84 × 3)	Training accuracy—82.68% Testing accuracy—79.62%

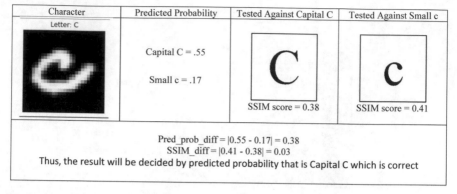

Character	Predicted Probability	Tested Against Capital C	Tested Against Small c
Letter: C	Capital C = .55 Small c = .17	C SSIM score = 0.38	c SSIM score = 0.41

Pred_prob_diff = |0.55 - 0.17| = 0.38
SSIM_diff = |0.41 - 0.38| = 0.03
Thus, the result will be decided by predicted probability that is Capital C which is correct

Fig. 9 Comparing SSIM score and predicted probability

The best accuracy was obtained with the proposed CNN-based neural net as the other transfer learning models were not properly optimised for it and were not trained on the entire dataset due to the lack of computational resources. Thus, it still leads for a huge scope of improvement.

Also, for certain characters like the capital 'C' and small 'c', since the formations are very similar in nature, most of the error comes from cases like this which leads to the problem of not being able to differentiate between the small and capital letters. This could be tackled by using similarity learning concept with the existing model prediction probability. Basically, instead of a single prediction probability, we use two prediction probabilities which give us two characters to compare the similarity of the input with times new roman glyphs and takes it into account by giving priority to the score which is more differentiated. This method provides an efficient result for cases where capital and small letter formations overlap like letter C and Z.

This has been shown via the example in Fig. 9.

6 Conclusion

Handwriting is a vital component of development, particularly in the early years, and has a significant influence on thinking and artistic abilities. As a result, it should not be overlooked in the modern age where everything is getting digitised, and proper use of technology may assist parents and teachers in guiding children with the least amount of work. The approach described in this research might assist instructors and parents in analysing their students'/kids' handwriting in a simple and painless manner.

In future, the work could be improved upon by training the transfer learning models—VGG16, VGG19, XceptionNet, and ResNet50 to extract features for the MNIST database so that it results in a lot better accuracy. As fields like deep learning and computer vision have gotten speedy as processing power has increased, better algorithms could also lead to the network becoming more resilient to cases where the formation of small and capital letters is similar. Also this same methodology could also be opted for other scripts such as Chinese and Devanagari.

References

1. Garg S, Kumar K, Prabhakar N, Ratan A, Trivedi A (2018) Optical character recognition using artificial intelligence. Int J Comp Appli 179:14–20. https://doi.org/10.5120/ijca2018916390
2. Joshi P, Agarwal A, Dhavale A, Suryavansi R, Kodolikar S (2015) Handwriting analysis for detection of personality traits using machine learning approach. Int J Comput Appl 130(15):40–45. https://doi.org/10.5120/ijca2015907189
3. Wiley V, Lucas T (2018) Computer vision and image processing: a paper review. Int J Artificial Intell Res
4. Jia X (2017) Image recognition method based on deep learning. In: 29th Chinese Control and Decision Conference (CCDC), May 28, pp 4730–4735
5. Mahdianpari M, Salehi B, Mohammadimanesh F, Zhang Y (2018) Very deep convolutional neural networks for complex land cover mapping using multispectral remote sensing imagery. Remote Sens 10:1119. https://doi.org/10.3390/rs10071119
6. Manchala S, Kinthali J, Kotha K, Kumar J (2020) Handwritten text recognition using deep learning with TensorFlow. Int J Eng Res 9. https://doi.org/10.17577/IJERTV9IS050534
7. Dixit R, Kushwah R, Pashine S (2020) Handwritten digit recognition using machine and deep learning algorithms. Inter J Comp Appl 176:27–33. https://doi.org/10.5120/ijca2020920550
8. Shamim SM, Miah B, Sarker A, Rana M, Jobair A (2018) Handwritten digit recognition using machine learning algorithms. Indonesian J Sci Tech 18. https://doi.org/10.17509/ijost.v3i1.10795
9. Hossain M, Hossain Z, Abadin M, Ahmed M (2021) Handwritten bangla numerical digit recognition using fine regulated deep neural network. Eng Int 9(2):73–84. https://doi.org/10.18034/ei.v9i2.551
10. Inunganbi S, Katariya R (2022) Transfer learning for handwritten character recognition. https://doi.org/10.1007/978-981-16-6369-7_63
11. Khan S, Nazir S (2022) Deep learning based Pashto characters recognition: LSTM-based handwritten Pashto characters recognition system. In: Proceedings of the Pakistan Academy of Sciences: A. Physical Comput Sci 58:49–58. https://doi.org/10.53560/PPASA(58-3)743
12. He D, Zhang Y (2021) Research on artificial intelligence machine learning character recognition algorithm based on feature fusion. J Phys: Conf Ser 2136:012060. https://doi.org/10.1088/1742-6596/2136/1/012060
13. Arafat SY, Iqbal MJ (2020) Urdu-text detection and recognition in natural scene images using deep learning. IEEE Access 8:96787–96803. https://doi.org/10.1109/ACCESS.2020.2994214
14. Chernyshova YS, Sheshkus AV, Arlazarov VV (2020) Two-step CNN framework for text line recognition in camera-captured images. IEEE Access 8:32587–32600. https://doi.org/10.1109/ACCESS.2020.2974051
15. Cohen G, Afshar S, Tapson J, van Schaik A (2017) EMNIST: an extension of MNIST to handwritten letters
16. Simonyan K, Zisserman A (2014) Very deep convolutional networks for large-scale image recognition. arXiv 1409.1556
17. Chollet F (2017) Xception: deep learning with depthwise separable convolutions, 1800–1807. https://doi.org/10.1109/CVPR.2017.195

18. He K, Zhang X, Ren S, Sun J (2016) Deep residual learning for image recognition, 770–778. https://doi.org/10.1109/CVPR.2016.90
19. Chollet F et al (2015) Keras. Available at: https://keras.io/api/applications/
20. Wang Z, Bovik AC, Sheikh HR, Simoncelli EP (2004) Image quality assessment: From error visibility to structural similarity. IEEE Trans Image Process

Prediction of Lung Disease from Respiratory Sounds Using Convolutional Neural Networks

R. Rajadevi, E. M. Roopa Devi, M. K. Dharani, K. Logeswaran,
S. Dineshkumar, and G. Mohan Chandru

Abstract Recently, the pollution in the world is leading into many diseases which are incurable at current situation, and also the treatments are not in the leading position. These ailments have an impact on the tubes (airways) that carry gases like oxygen into and out of the lungs. They frequently obstruct or tighten airways. Airway ailments include chronic, asthma, obstructive pulmonary disease (COPD), and bronchiectasis. People with airway illnesses frequently describe their sensation as "attempting to exhale through a straw". The main aim is to detect the lung diseases that are affected in humans and that are very vulnerable to the human bodies. Currently, a collective system for identifying and recommending lung problems is not easily available to a large number of people, and it is technically implemented by machine learning algorithms. This research makes use of deep learning techniques to learn from the respiratory audio signal of patients. The unwanted noise is removed by means of preprocessing, and it is converted to overlapping frames. The FFT is used to convert the signals into the spectrogram, and it is used for predicting the lung diseases. The convolution neural network is applied to spectrogram with eight hidden layers to predict the COPD, upper respiratory tract infection (URTI), lower respiratory tract information (LRTI), etc. This model provides 95% of accuracy, and also comparison is done with the help of different optimizers like ADAM, SGD, and RMSPROP. This system might be improved by further changing both the preprocessing techniques and fine tuning the parameters. This system detects the lungs diseases in the earlier stage itself, so that the disease may not be affect the whole body.

Keywords Audio signal analysis · Respiratory diseases · Deep learning techniques

R. Rajadevi · E. M. R. Devi (✉) · K. Logeswaran · S. Dineshkumar · G. M. Chandru
Department of Information Technology, Kongu Engineering College, Perundurai, India
e-mail: roopadevi.it@kongu.edu

M. K. Dharani
Department of Computer Science and Engineering, Kongu Engineering College, Perundurai, India

© The Author(s), under exclusive license to Springer Nature Singapore Pte Ltd. 2023   453
Y. Singh et al. (eds.), *Proceedings of International Conference on Recent Innovations in Computing*, Lecture Notes in Electrical Engineering 1011,
https://doi.org/10.1007/978-981-99-0601-7_35

1 Introduction

The main components of the human respiratory system are the lungs. On each side of the chest are two spongy, air-filled organs called the lungs (thorax). Through bronchi, tubular branches of the trachea (windpipe), air from the lungs is transferred. The bronchi were split into bronchioles, which were ultimately microscopic branches. They carry oxygen from the atmosphere into the bloodstream and release carbon dioxide from the circulation into the atmosphere, a process known as gas exchange, in the respiratory system. Lung sickness includes asthma, COPD, infections such as influenza, pneumonia, and tuberculosis, lung cancer, and a number of other breathing problems. The respiratory network, a network of tissues and organs that aids in breathing, is a network of disorders that can lead to respiratory failure. Your nasal cavity, mouth, lungs, and blood vessels are all part of your respiratory system. The respiratory system includes the muscles that operate your lungs. Together, these parts transport oxygen around the body and expel waste gases like carbon dioxide. Many disorders can have an impact on the tissues and organs that comprise the respiratory system. Some develop as a result of airborne irritants, such as viruses or bacteria that induce infection. Others arise due to illness or ageing. The following conditions can produce inflammation or otherwise impact the respiratory system: Allergies: some people develop respiratory allergies after inhaling proteins including such dust, mould, and pollen. These proteins can induce airway inflammation. Asthma: breathing becomes challenging due to the chronic (long-term) inflammation of asthma that damages the airways. Infection: infections can cause pneumonia or bronchitis. The flu (viral disease) and the common cold are both respiratory diseases. Disease: respiratory ailments include lung cancer and COPD. These conditions can compromise the respiratory system's ability to circulate oxygen through the body and filter out waste gases. An acoustic medical tool called a stethoscope is used to hear inner noises in an animal or human body. It typically has a small, skin-contact resonator in the shape of a disc, and one or two tubes that connect to two earpieces. A stethoscope can be used to listen to blood flow in arteries and veins as well as heart, lung, and gastrointestinal noises. There are different types of stethoscopes, and one of it is electronic stethoscope. The electronic stethoscopes are more costly than the normal stethoscopes, so most of the doctors use normal stethoscopes. By electronically amplifying body noises, electronic stethoscopes use modern technology to overcome these low sound levels. Acoustic sound waves collected through the chest piece must be converted into electrical signals, which are then delivered through specially developed circuitry and processed for optimal listening. This may save the sound signal in memory and use it for further processing by employing electronic stethoscopes. Convolutional neural networks (CNNs) are a hierarchical feature detector that is biologically inspired. It can learn very abstract properties and reliably recognize items. The CNN is to determine whether an individual has COPD, URTI, bronchiectasis, pneumonia, or bronchiolitis, or whether the patient is healthy. It is worth noting that the majority of the good multi-class classification results came from hybrid DL methods. Training a reliable fully convolutional architecture could be time-consuming and extremely intensive, despite the fact that it shows exceptionally promising results without the need of complicated feature engineering procedures. The training procedure is iterative as well, and it incorporates a large number of model parameters, including datasets.

2 Literature Survey

Serbes et al. [1] have developed an approach for extracting multiple feature sets from pulmonary signals using time–frequency and time–scale analysis The collected feature sets are put into three distinct machine learning methods, both individually and as a group of networks. Furthermore, up to the time analysis, bandwidths containing no-crackle data are deleted using the double tree complex wavelet transform to increase the model's effectiveness. Lin et al. [2] have announced the development of an advanced wheeze detecting system. The order cutoff average approach and a backpropagation neural network are used to create a wheeze detection system. Some characteristics from processed spectra are used to train a BPNN, which then analyzes test samples to see if they are asthmatic sounds. Chen et al. [3] have gathered a data collection of normal and pathologic heart and lung noises. An effective and automated diagnostic procedure is particularly appealing since it may help detect possible hazards at an early stage, even without the assistance of a professional doctor. Islam et al. [4] have analyzed even in the absence of wheeze, and enhanced signal processing of anterior lung sound data was used to distinguish between normal and asthmatic patients. For the multichannel signal, a spectral subband-based feature extraction strategy is developed that works with ANN and SVM classifiers. To categorize normal and asthmatic participants, a collection of statistical characteristics is calculated from each subband and used to ANN and SVM classifiers. Zulfiqar et al. [5] have analyzed that the respiratory sound (RS) features and analyzes are an important part of pneumonic pathology because they provide symptomatic information about a patient's lung. Doctors used to rely on mere listening to differentiate clinical symptoms in pulmonary sound using a standard stethoscope, which is considered a low-cost and safe approach of assessing patients. Because lung illness is indeed the third leading cause of mortality globally, properly characterizing the RS anomaly is critical in order to reduce mortality. Hafke-Dys et al. [6] have analyzed the AI analysis of sounds captured during conventional stethoscope auscultation that can be used to detect the intensities of diseased breath occurrences. The presence of aberrant noises was assessed by a panel of three physicians who were not aware of the AI predictions. The performance of each indicator in discriminating across groups was evaluated. Lang et al. [7] introduce graph moderated CNNs (GS-CNNs) for categorizing respiratory sounds into normal, crackle, and wheeze, using a low labelled sample group and a high unlabelled sample group. The results show that the suggested GS-CNNs exceed regular CNNs, and that the more graph-RS data incorporated, the better the results. Pouyani et al. [8] have presented the lung sound

(LS) analysis by computer which was investigated as a potential technique for evaluating lung function. Background noise from various sources significantly pollutes the LS signal. Traditional denoising procedures may be unsuccessful due to the LS's noisy nature and spectrum overlap with many noise sources. This study provides an adaptive strategy for filtering LS signals in a noisy environment based on wavelet transformation and ANN. The inter capacity of DWT is used with ANN as a changeable structure filter in this unique technology. You et al. [9] have proposed a method of automatic cough recognition in realistic audio recordings of patients that is crucial for diagnosing and monitoring respiratory disorders like COVID-19. To date, several detection systems have been invented, but none have reached the practical criteria. They added a convolution just before LSTM to enhance cough characteristics and maintain the sequence information in the sound source. The unique model on the final feature map contains an incorporated boundary regression for improved detection efficiency and more precise borders, which is critical for future analysis.

3   Proposed System

Lung disease is one of the symptoms of many severe diseases which may affect patient's health badly and sometimes may cause to the death. Severe pulmonary illnesses damage the bronchi or other cells of the lungs, severely affecting human lungs. Human respiratory sounds are used to assess the health from the lungs. A wheezing sound is used to identify common lung diseases like COPD, LRTI, and URTI. Respiratory sound is recorded by digital stethoscopes. These sounds can be used to find the disease of lungs. Hence in the proposed system, it employs CNN model for finding the type of lungs disease at early stage, which may save the life of patients Fig. 1 illustrates the proposed architecture model for classification lungs disease.

The following steps are involved in the processing of the model: (1) audio extraction, (2) dataset preprocessing, and (3) developing CNN model for detection of lung diseases. Every audio signal consists of many features. First step is to convert the signals from time domain to frequency domain in order to extract the features of the signal. Extracted features are converted into numerical data, and it is given as input to the training model and testing which result in the disease detection of the lungs in the specific signal.

3.1   Dataset Description

Respiratory sounds are the critical indications of respiratory health diseases. The lung sound dataset was gathered locally at Jordan University of Science and Technology's King Abdullah University Hospital in Irbid, Jordan. The freely accessible ICBHI challenge database was used to supplement this dataset. In addition to the 1176

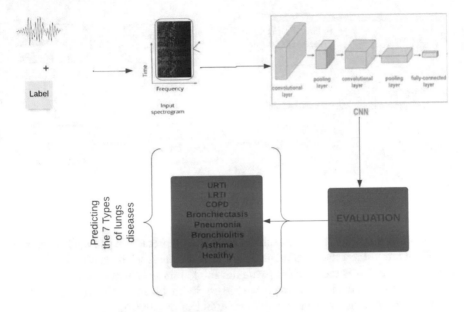

Fig. 1 Architecture of lungs diseases detection

recordings, the dataset references contained a total of 215 patients with 309 clinically obtained lung sound recordings. A total of 70 individuals with respiratory illnesses such as bronchitis, fever, heart failure, bronchiectasis, respiratory problems, and chronic obstructive pulmonary disease were included in the primary dataset (COPD). Data was also collected from 35 healthy controls. This project employs the dataset from the Kaggle; this dataset includes 920.wav soundfiles, 920 annotation.txtfiles, a file with each patient's illness, file explaining the file naming format, file listing 91 names (filename_differences.txt), and file containing demographic information for each patient. Two primary research teams from the Portuguese and Greek created the respiratory sound database. It contains 920 annotated recordings ranging in duration from 10 to 90s. These recordings were recorded from 126 patients. There are 6898 respiratory cycles in all, 1864 of which contain the crackles, 886 of which contain the wheezes, and 506 of which have both the crackles and wheezes. The data comprises both the respiratory sounds and noise recordings that mimic real-world scenarios. Patients include young people, adults, and the elderly.

3.2 Data Preprocessing

Audio signal processing is at the heart of audio recording, enhancement, storage, and transmission. Most important in audio data processing is removing the unwanted noise and sound and gaining only the wanted sound and signals that is important for

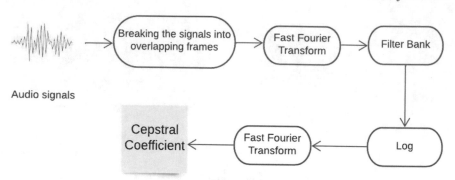

Fig. 2 Converting the audio signals into spectrogram

the further processing of the data and using it for the prediction and analysis of the audio data, and the process also involves the conversion of the audio signals into desired nature that is required for particular mode that is analogue or digital mode and it also make sure of enhancing the audio clarity that is more effective in analysis of the audio data and produces much more effective output (Fig. 2).

The audio signals consist of frames and frequency and also have unnecessary noises and unwanted signals, the first process is breaking the signals into overlapping frames, and after this, it undergoes the FFT. It deco It breaks down a signal into distinct spectral components and as a result provides frequency information about the signal. It composes a signal into discrete spectral components and offers frequency information about the signal as a result. Next, the audio signal that is processed from the FFT is passed into the many filter banks, the resultant audio signal is passed through log cube and again undergoes the FFT, and the resultant helps in producing the spectrogram of desired extraction features that is useful in the process of analysis in the audio data and predicting the lung diseases (Fig. 3).

In the frequency domain, mathematical functions and signals are expressed in terms of speed rather than duration. A bandwidth graph, for example, depicts the amount of signal present in each wavelength range, but a time-domain graph depicts changes over time. The data can, however, be translated from a time domain to a

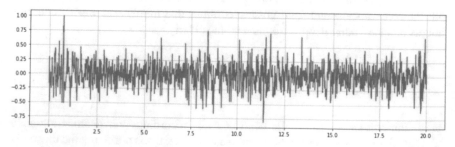

Fig. 3 Example audio wave signal

frequency-domain. A time function is transformed into a sequence of sine waves that indicate different wavelengths using the Fourier transform. The bandwidth representation of a signal is the "spectrum" of frequency content.

3.2.1 Audioframing

The FFT will induce distortions since audio is a non-stationary process. Any audio is a stationary process for a brief duration. This audio signal is divided into short frames. The FFT will be the same size as each audio frame. The frames must be overlapping. We do this to maintain some association between the frames and to avoid losing information on the frame's boundaries after applying a window function. The FFT assumes that the audio is continuous and periodic. The audio would be periodic by framing the signal. The window function is used on each frame to keep the music playing. The process will obtain high-frequency distortions if we do not do so. To solve this, first apply a window function to the framed audio before doing FFT. The window ensures that the signal terminates close to zero on both ends. Choosing the right window is difficult and time-consuming. Hanning window is used for its simplicity (Fig. 4).

Mel frequency cepstral coefficients (MFCCs) are a technique for extracting audio characteristics. The MFCC divides the frequency band into subbands using the mel scale then extracts the cepstral coefficients using the discrete cosine transform (DCT). The mel scale is based on how people discern between frequencies, making sound processing a breeze. The MFCC is a good algorithm to extract high-frequency and low-frequency information (Fig. 5).

3.3 CNN Model for Detection of Lungs Diseases

CNN is a hierarchical feature detector that is biologically inspired. It can learn very abstract properties and reliably recognize items. This CNN approach is preferred above other traditional techniques for the following reasons. CNN uses weight sharing to lower the number of parameters. This smoothes the process and stops

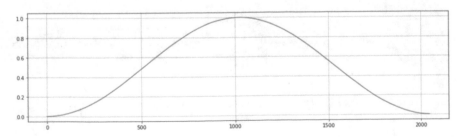

Fig. 4 Frequency domain of example audio wave signal

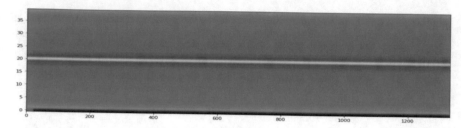

Fig. 5 Spectrogram of example audio wave signal

it from smoothing. Furthermore, the characteristic extraction and classification steps are combined. The convolutional layer, pooling layer, and fully linked layer comprise the design of a basic convolutional neural network. This model forecasts the leaf disease in the stages listed below.

3.3.1 Convolution Layer

The initial step of a CNN is the convolution layer. Convolution layers perform a convolution operation on the input and transmit the output to the following layer. A convolution is a process that turns all of the pixels or sound signals in its receptive area into a single value. For example, applying a convolution to a sound signal reduces the signal size whilst also bringing all of the information in the field together into a single signal. The convolutional layer's final output is a vector. This may employ several types of convolutions depending on the sort of issue we need to solve and the features we want to learn (Fig. 6).

The dimension of the feature extraction is lowered by pooling layers. As a result, there is a reduction in the number of parameters that must be learned and the amount of calculation carried out within the network. The preprocessing stage describes the attributes of a convolutional layer-generated feature map. Flattening is the process of combining all of the resulting pooled 2D arrays of feature maps which are combined

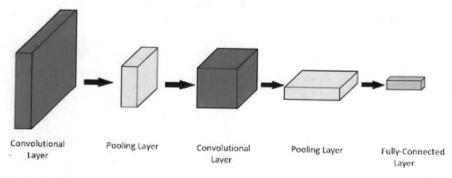

Fig. 6 Architecture of CNN

into a single sustained linear vector. To categorize image or sound data, the flattened matrix is provided as input to the fully linked layer. Softmax functions are squashing functions. Squashing functions confine the function's output from 0 to 1 range. This allows the result to be simply understood as a probability.

$$\text{Softmax}(xi) = \exp(xi)/\sum j \, \exp(xi) \qquad (1)$$

The rectified linear unit function is an extra step on top of the convolution function. The ReLU is employed in order to boost nonlinearity. This needs to enhance nonlinearity because pictures are very nonlinear and want to break up linearity. Padding is a term used in convolutional neural networks to describe the quantity of pixels added to an image during processing by the CNN kernel. Input is 1D signal having length 193, this input is applied for convolution with 64 different filters, so 64 different outputs are generated (input image [193,1]) convolution (64 filters, size [1,5]) = 64 different outputs having size (10,046,1). Convolution1D will take all 64 input channels one by one and generate 64 different output channels for each input channel; so in total, 64 × 64 = 4096 channels output should be generated. Input is 1D signal having length 193, and this input is applied for convolution with 128 different filters. Convolution1D will take all 128 input channels one by one and generate 128 different output channels for each input channel; so in total, 128 × 128 = 16,384 channels output should be generated. The resultant is send as input to the upcoming layer. Size of the filters in the pooling layer is 1 × 1, and the number of filters is 2 which are used with stride = 1. Now, the output image size is reduced to 24 × 24 × 2. The resultant values from the max pooling layer 3 are given to the next convolutional layer. Number and size of the filters in second convolution layer are 256 1 × 1. This input is applied for convolution with 256 different filters, so 256 different outputs are generated. Conv1D will take all 64 input channels one by one and generate 64 different output channels for each input channel; so in total, 256 × 256 = 65,536 channels output should be generated. With activation "ReLU". Large value sets trained on partial small datasets could overfit trained dataset. A positive drop value in a hidden layer would be in the range of 0.5–0.8. Input layers employ a higher attrition, such as 0.8. In this case, we use 0.3 as the dropout value for the 1D array value. For an example, network with $n = 100$ nodes and given dropout rate is $p = 0.3$, and this will require 200 nodes (100/0.3)—n/p whilst using dropout. After completing extracting features to flattening the result of layer 5 that is to be converted into one dimensional. The resultant from the 3rd convolutional layer is flattened, and given matrix is reduced to one vector of selective number neurons. Then, it is given to the fully connected (dense) neural network. Like other classifiers, an CNN classifier requires private and unique characteristics. This is helpful for those who require a feature vector. As a result, you must transform the output of the CNN's section into a single dimension, and feature vector can be utilized by the CNN part. Flattening is the term for this process. It takes the result of the convolutional layers and flattens all of its structure into a single dimension feature vector that the dense layer may utilize for final classification. Activation:Softmax (activation = activations.softmax)–softmax modifies a

```
Model - 'Sequential'
```

Layer (type)	Output Shape	Param #
conv1d (Conv1D)	(None, 189, 64)	384
conv1d_1 (Conv1D)	(None, 185, 128)	41088
max_pooling1d (MaxPooling1D)	(None, 92, 128)	0
conv1d_2 (Conv1D)	(None, 88, 256)	164096
dropout (Dropout)	(None, 88, 256)	0
flatten (Flatten)	(None, 22528)	0
dense (Dense)	(None, 512)	11534848
dense_1 (Dense)	(None, 6)	3078

```
Total params: 11,743,494
Trainable params: 11,743,494
Non-trainable params: 0
```

Fig. 7 Model for CNN

vector of data to a probability distribution. Amongst these, activation ReLU function is used for every convolution1D layer because this is computationally much more efficient than other activation functions (Fig. 7).

The given output from the activation function is given to the pooling layer which will be used to reduce only the specific spatial information, so that there is less chance for the overfitting, and moreover, this is down sampling process that decreases the size of datasets.

4 Experimental Analysis

Medical conditions known as respiratory illnesses, also referred to as lung diseases, cause organ and tissue damage in animals that breathe air, which results in gas production. They include issues with the trachea, bronchi, bronchioles, alveoli, pleurae, pleural cavity, nerves, and breathing muscles of the respiratory system. From minor, self-limiting conditions like the common cold, influenza, and pharyngitis to serious, life- threatening illnesses like bacterial pneumonia, pulmonary embolism, tuberculosis, acute asthma, lung cancer, and severe acute respiratory syndromes like COVID-19, there are many different types of respiratory disorders. Hence, the trained model is giving as 95% accuracy in detecting the lungs diseases with help of the audio signals recorded by the digital stethoscope. Datasets are compared with certain optimizers to test the best algorithm, and as a result, Adam and RMSprop optimizers give the best output for the system. The Adam optimizer employs a hybrid of two gradient descent methods. This approach is used to speed the gradient descent process by taking the EWA of the gradients into the data.

Confusion matrix

- A real positive result (TP) is a model which accurately predicts the positive result
- A real negative, (TN), is a result in which the model properly predicts the negative one.
- A false positive (FP) is an outcome in which the model predicts the positive class inaccurately.
- A false negative (FN) is an outcome in which the model predicts the negative class inaccurately.

5   Performance Analysis

Accuracy:

For the purposes of assessing the planned task, one of the most important aspects is accuracy. Accuracy is defined as the ratio of the total number of samples to the true positive and true negative samples.

Precision:

The proportion of correctly identified positive samples to all positive samples serves as a measure of accuracy (either correctly or incorrectly).The precision of a model is measured by its accuracy in categorizing a sample as positive.

F1-Score:

The F1-score is a single statistic that takes the harmonic mean of a classifier's accuracy and recall. It is mostly used to compare two classifiers' performance.

Recall:

Recall is determined by the true positive value, and the model's training benefits from the total of all true positive and false negative values.

The optimizer as Adam, SGD, and RMSprop helps to compare the efficiency of the system by various ways and helps in the detection of the lungs diseases and with

Table 1 Optimizer comparison for epoch 50

Optimizer	Precision (%)	Recall (%)	F1-score (%)	Accuracy (%)
Adam	95	95	95	95
SGD	80	85	88	90
RMSprop	95	95	90	95

Table 2 Optimizer comparison for epoch 70

Optimizer	Precision (%)	Recall (%)	F1-score (%)	Accuracy (%)
Adam	94	95	90	95
SGD	94	95	94	95
RMSprop	88	95	82	95

Fig. 8 Comparison of various optimizers for epoch 50 (F1-score)

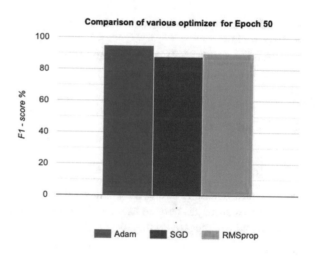

Comparison of various optimizer for Epoch 50

help of precision, recall, F1-score and accuracy helps to detect the diseases with the high accuracy and efficient manner (Tables 1 and 2; Figs. 8, 9, and 10).

6 Conclusion

This system is used to determine whether an individual has COPD, URTI, bronchiectasis, pneumonia, or bronchiolitis, or whether the patient is healthy. It is worth noting that the majority of the good multi-class classification results came from hybrid deep learning methods. Training a reliable fully convolutional architecture could be time-consuming and extremely intensive, despite the fact that it shows exceptionally promising results without the need of complicated feature engineering procedures. The training procedure is iterative as well, and it incorporates a large number of model

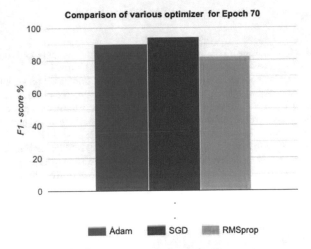

Fig. 9 Comparison of various optimizers for epoch 70 (F1-score)

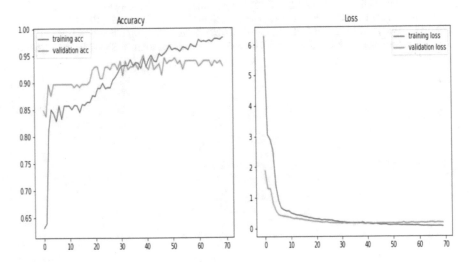

Fig. 10 Accuracy and loss of Adam optimizer

parameters, including datasets. To overcome the classification problem, this provides a simple yet successful framework that blends entropy characteristics with homogeneous ensemble classification algorithms. A deep learning model based on CNN is applied to categorize the respiratory sound collected by digital stethoscopes which were utilized. The F1-score of Adam classifier is 95 percentage, SGD classifier is 88, and with an RMSprop value of 90.26 percentage. Adam classifier performs well when compare with other classifiers. Although the suggested classification model improves performance of the system, it might be enhanced further by modifying the preprocessing procedures and trained structure of a dataset.

References

1. Serbes G, Sakar CO, Kahya YP, Aydin N (2013) Pulmonary crackle detection using time-frequency and time-scale analysis. Dig Signal Process 23:1012–1021
2. Lin BS, Wu HD, Chen S (2015) Automatic wheezing detection based on signal processing of spectrogram and backpropagation neural network. J HealthcEng 6:649–657
3. Chen Q, Zhang W, Tian X, Zhang X, Chen S, Lei W (2016) Automatic heart and lung sounds classification using convolutional neural networks. In: 2016 Asia-Pacific Signal and Information Processing Association Annual Summitand Conference (APSIPA), 1–4
4. Islam MA, Bandyopadhyaya I, Bhattacharyya P, Saha G (2018) Multichannel lung sound analysis for asthma detection. Comput Methods Programs Biomed 159:111–123
5. Zulfiqar R, Majeed F, Irfan R, Rauf HT, Benkhelifa E, Belkacem AN (2021) Abnormal respiratory sounds classification using deep CNN through artificial noise addition. Frontiers Med 8
6. Hafke-Dys H, Kuźnar-Kamińska B, Grzywalski T, Maciaszek A, Szarzyński K, Kociński J (2021) Artificial intelligence approach to the monitoring of respiratory sounds in asthmatic patients. Front Phys
7. Lang R, Fan Y, Liu G, Liu G (2021) Analysis of unlabeled lung sound samples using semi-supervised convolutional neural networks. Appl Math Comput 411:126511
8. Pouyani MF, Vali M, Ghasemi MA (2022) Lung sound signal denoising using discrete wavelet transform and artificial neural network. Biomed Signal Process Control 72:103329
9. You M, Wang W, Li Y, Liu J, Xu X, Qiu Z (2022) Automatic cough detection from realistic audio recordings using C-BiLSTM with boundary regression. Biomed Signal Process Control 72:103304

SemWIRet: A Semantically Inclined Strategy for Web Image Recommendation Using Hybrid Intelligence

M. Yash Bobde, Gerard Deepak, and A. Santhanavijayan

Abstract With a rapid increase of Websites, retrieving information from the Internet has become a very challenging and time-consuming endeavour. There is always room for improvement in terms of yielding and recommending the correct informative measures so that the user would get a very pertinent topic for which the user is searching over the Internet. To ease the work of the user, an enhanced recommendation system is best suited to recommend the optimal solution. Furthermore, relevant articles are displayed according to the user queries. It is highly likely that the relevant search article recommended will be of no use to the user if the recommendation algorithm is not effective. This paper proposes a semantic-infused hybrid intelligence approach to recommendation for user-centric queries relevant using concept and ANOVA cosine similarity and extreme gradient boosting (XGBoost) where the documents are classified. Using the flickr30k dataset, the above experiment was conducted, and 97.16% accuracy was achieved.

Keywords Concept · Cosine · XGBoost · Semantic similarity

1 Introduction

The World Wide Web as the name implies is the largest accumulation of existing information throughout the world. We are living in the era of Web 3.0, which is also known as the "semantic Web" as it has harnessed the coexisting powers of big data and

M. Y. Bobde
Department of Computer Science and Engineering, SRM Institute of Science and Technology, Kattankulathur, India

G. Deepak (✉)
Department of Computer Science and Engineering, Manipal Institute of Technology Bengaluru, Manipal Academy of Higher Education, Manipal, India
e-mail: gerard.deepak.christuni@gmail.com

A. Santhanavijayan
Department of Computer Science and Engineering, National Institute of Technology, Tiruchirappalli, India

machine learning together, where every user's data and activities are being analyzed and used to deliver more personalized web browsing experience and better richer app experiences. Information extraction from the current scenario where a tremendous amount of data is being generated on a daily basis is not just a tedious task but also involves enormous efforts to link the user queries with the precise contents throughout the World Wide Web. Even if emerging learning mechanisms are subsumed properly, most Web-based search algorithms tend to be delinquent owing to the information that looks much similar. Most of the Web-based recommendation systems are of two types query-centred and user-centred. The query-centric Web page recommendation systems pivot only on the relevance of the Web pages to the query that is being launched into the system. Although, when the Web page recommendation system is user-centric, the focus is predominantly on personalization and pleasuring the user. These systems do not ease the diversification of search results as there is a scanty of real-world knowledge that is being infused into the system. Semantic Web is constituted with a large amount of data available on the current structure of the Web by inclusion of metadata of the Web and to reason out. A strong semantic approach which can transform the existing data into knowledge or which can incorporate knowledge for the recommendation is the need of the hour. The annotation-based text recognition system is the best out of the two other options available which are context-based recommendation system and hybrid recommendation system as the data is labelled in it, and it is also a text-based classification that addresses the user-specific query efficiently as well as efficiently.

Motivation: Since the emerging technologies improve the quality and the enrichment of personalization needed by the user accordingly, the Web is also moulding itself in this and is on the Web 3.0 version, which is purely based on the semantics and keeps on improving the quality of searched queries; but still, the need for more semantic methods for the precise catering to the demanding informational needs of users is needed much, and to fulfil it, we need an annotation-based recommendation system, and add on to this, it is also the state of the art.

Contribution: Semantically inclined strategy for Web image recommendation using hybrid intelligence is proposed for Web image retrieval purposes. Several preprocessing techniques are employed in the preprocessing phase, amongst them are lemmatization, tokenization, stopword removals, and neural network extractions. Texts relevant to the topic are identified using the XGBoost classifier. The labelled categorical dataset comprising of Web images is obtained from the flickr30k dataset to enhance labels. By combining semantic similarities such as concept and ANOVA cosine, we can achieve amazing accuracy and distinguish results with the maximum level of consistency, and the accuracy is 97.17%, the F-measure of 97.16%, and an FDR value of 0.04 is attained as a result of the proposed model.

Organization: The remaining paper proceeds as follows: Section 2 discusses related works. Section 3 elaborates on the proposed system architecture. Section 4 deals with the implementation and results. Section 5 concludes the report.

2   Related Works

It is very difficult to find precise information on the Web, where we can find billions of data related to that field of search. Several people have researched this topic to arrive at a particular and effective solution to such Web image retrieval problems. Xie et al. [1] wherein authors stated that a cross-modal hashing (CMH) algorithm is an effective method of retrieving images from the Web quickly, whereas traditional CMH uses a streaming approach for the generation of the hash functions and codes that are quite ineffective for retrieval of Web images using batch learning. Chen [2] stated that with digital colour space quantization image colour distribution, the selected colour model is closely related to the colour space, and hence, we can easily retrieve the images. Li [3] asserted in their framework of isomorphic hashing, they proposed three key principles that are useful for guiding the design of hashing algorithms, with a more precise weight coefficient and a greater diversity of weights with higher precision. They proposed a novel hashing algorithm based on these guidelines called BWLH. Liu et al. [4] exploited the CNN structure by using a binary code learning framework, named deep supervised hashing (DSH), in which they learned for image data of large size, a compact similarity-preserving binary code must be created, where each image will have discrete values with approximate shapes. Gupta et al. [5] performed the study, in which they assessed existing image retrieval techniques in order to fetch information from the cyberspace by using textual annotations on the images. Liu et al. [6] analyzed the architecture of the deep convolutional neural network and concluded fusing two kinds of deep convolutional features to retrieve images which is effective. Wu et al. [7] where their framework proposes a novel local feature learning algorithm for image retrieval that only requires image-level annotations and can be trained end-to-end. Hou et al. [8] present V-RSIR as an open-access Web-based tool that requires minimum training to identify RS images. It also allows users to view RS image quantities and their spatial distributions. Huang et al. [9] proposed the MRBDL model which is multi-concept retrieval using bimodal deep learning through the use of convolutional neural networks, semantic correlations between a visual image, and its context can be effectively captured. Vijayarajan et al. [10] Google's knowledge graph is a knowledge base that the Google uses to supplement its search engine's results with semantic-search data obtained from a range of sources. Sejal et al. [11] proposed an algorithm to recommend images based on ANOVA cosine similarity, where text and visual features are compared to fill the semantic gap. This allows for visual synonyms to be computed based on semantically related visual features found in images. Kaushik et al. [12] developed a new image search algorithm for a search engine API, which filters out the relevant images from all the images retrieved from the Wikipedia API and displays them to the user. Nair et al. [13] proposed the work for a medical image retrieval where they made an effort to find the ways and means of semantic gap and inferred that the feedback got the user is a key factor. Kolahkaj [14] used the combination of wavelet transform and colour histogram which was used to extract the features, in order to bridge the semantic gap between low-level visual features of images and high-level semantics of the images. Noor et al. [15] in bandwidth-constrained situations, there is a large void in the research regarding effective retrieval and storage of progressive pictures. A novel scanning approach and a new lossy compression technique are then proposed by the authors as an orchestration methodology. In [16–21], several ontology driven semantically inclined frameworks in support of the proposed literature are depicted.

3 Proposed System Architecture

The proposed system architecture of the knowledge-centric semantic infused Web image recommendation system is depicted in Fig. 1. The user query is taken as input which is subjected to preprocessing. Preprocessing involves tokenization, lemmatization, stopword removal, and the named entity recognition (NER). The query processing is carried out using Python NLTK framework. Tokenization and lemmatization are done; wherein, tokenization is done using a blank space special character tokenizer, and the lemmatization is done using the WordNetLemmatizer. The stopword removal is done using a customized set matched stopword removal algorithm using the regular expressions, and the named entity recognition (NER) is also done using the Python NLTK library. At the end of the preprocessing steps, we have the query words which are preprocessed, and these preprocessed query words are subjected to topic modelling using LDA which stands for Latent Dirichlet Allocation. A Bayesian probabilistic model, the LDA emphasizes the parts of the data that are similar by exemplifying observed and unobserved groups. LDA offers superior disambiguation, and it assigns documents to topics with greater precision. The topic modelling is done mainly for the reason to uncover all the hidden topics and enrich the query words so that for each of the query related words, a large number of hidden and relevant topics can be discovered. At the end of the topic modelling phase, the newly discovered topics and the query words are fed into Google's knowledge graph API, in order, to yield and enrich the query words and yield enriched query words set, and parallelly, the dataset is classified using the XGBoost classification algorithm where XGBoost stands for extreme gradient boost.

In tree boosting, XGBoost is a machine learning system that can scale up is based on a tree-based ensemble machine learning algorithm; wherein, ensemble learning techniques have the potential of blending predictions coming from various machine learning models, and for determining the optimum tree model, XGBoost employs more precise approximations. Ensembles are composed of models called base learners, these techniques are applied to different statistical models, but decision trees are the most popular. Ensemble learners include both bagging and boosting and can be derived from the same learning algorithm or from different learning algorithms. Bagging is an ensemble machine learning technique used to find the best model by taking numerous weak models and aggregating them. Since weak models specialize in specific areas of the feature space, bagging can use predictions from any model to get the highest level of accuracy.

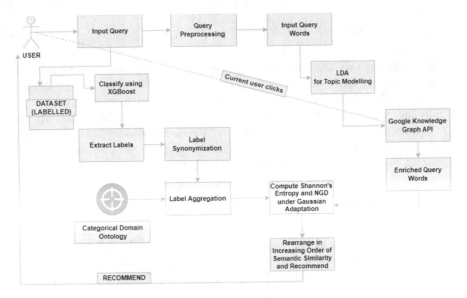

Fig. 1 Proposed system architecture

The trees are formed in a boosting order, with each successive tree striving to minimize the mistakes of the prior tree. Each tree builds on its predecessors' knowledge and corrects any faults that remain. As a consequence, the next tree in the sequence will use an updated set of residuals to learn. Scaling should be performed to numerical features as well as categorical features should be encoded for achieving the best performance over XGBoost. The classifier method uses the fit-predict method; so, after preprocessing, the data is needed to be divided into train and test sets. We preferred XGBoost because of the reasons that to handle different sorts of as the XGBoost has a sparsity-aware split finding algorithm, as well as being especially intended to make use of current technology, which is achieved by designating internal buffers in each thread for the storing of gradient information and also, for the reason of regularization as regularization helps in preventing overfitting.

The dataset is classified using the input query words; by supplying the features from the input query words, the dataset is classified using the XGBoost classifier; from the obtained classified dataset, the images are extracted from the labels because the dataset is labelled categorical dataset comprising of Web images, and so, these labels are again subjected to the synonymization; in this case, the WordNET synsets are used for the purpose of synonymization, and once, the synonyms are generated for each of the provided labels. Further, the label enrichment is done by aggregating categorical domain ontology with these labels by computing the concept similarity. Concept semantic similarity in information retrieval means that the degree of semantic matching between the text and the user's inquiry is always reflected in the similarity. The term "similarity" refers to the degree of similarity between two ideas. We set the following for two notions (A, B) and (C, D).

$$s((A, B), (C, D)) := \frac{1}{2}\left(\frac{|A \cap B|}{|A \cup B|} + \frac{|C \cap D|}{|C \cup D|}\right) \tag{1}$$

The similarity measurement in Eq. (1) is used to order the exact sibling concepts where A and B are the labels of X where they are from the label synonymization, and C and D are the labels of Y where they are from categorical domain ontology.

On the aggregation of the categorical domain ontology with the labelled synonymized classified image labels, the entity enrichment is achieved. The initial enriched query words, as well as the labelled aggregated with the ontology, are subjected to the computation of Shannon's entropy and the normalized Google distance. By comparison, the Shannon entropy computes the information measure, and the normalized Google distance determines the semantic similarity. The normalized Google distance (NGD) determines accordingly to the number of Google search hits for a set of keywords, semantic similarity is calculated. Keywords with similar or identical natural language meanings are "near" in normalized Google distance units, whereas terms with different meanings are "far" away. The normalized Google distance (NGD) between two search keywords p and q, for instance in Eq. (2), is represented by

$$NGD\ (p, q) = \frac{\max\{\log f(p), \log f(q)\} - \log f(p, q)}{\log N - \min\{\log f(p), \log f(q)\}} \tag{2}$$

where N is the total number of Web pages searched by Google times, the average number of singleton search phrases found on those pages; $f(p)$ and $f(q)$ are the number of hits for search terms p and q, respectively; and $f(p, q)$ is the number of Websites on which both p and q appear. If the NGD $(p, q) = 0$, then p and q are seen as similar as possible, whilst if NGD $(p, q) \geq 1$, then the given x and y are different.

The deviation criterion for Shannon's entropy is considered as 0.25, whilst the threshold for normalized google distance is set as 0.75, the threshold for concept similarity is set as 0.5 for the reason that we want more aggregation of categorical ontology for concept similarity. The threshold of normalized Google distance is set as 0.75 is mainly due to the fact that we want more relevant entities to be populated, Shannon's entropy is considered for computing the information measure with the step deviation of 0.25, and all this is done under the influence of Gaussian adaptation.

Signal processing systems use Gaussian adaptation, referred to as normal or natural adaptation, to maximize the yield of components due to the statistical variance of component values. A wide range of optimization problems is addressed by evolutionary algorithms and typically provides good approximation solutions using evolutionary algorithms. The stochastic adaptive process implies that an n-dimensional $x[xT = (x1, x2,\ldots, xn)]$ sample is taken from a multivariate Gaussian distribution $N(m, M)$, with mean as m and moment matrix as M.

The samples are tested to determine if they pass or fail. $M*$ and $M*$ are the Gaussian's first- and second-order moments confined to pass samples, respectively, where the $s(x)$ is a function determining the probability that x is chosen as a passing sample, $0 < s(x) < q \leq 1$. The average possibility of finding pass samples (yield) is given by the formula given in Eq. (3),

$$P(m) = \int s(x) \cdot N(x - m) \mathrm{d}x \qquad (3)$$

There is always a Gaussian probability density function (p. d. f) that is tuned for maximum dispersion for every $s(x)$ and any value of $P < q$. A local optimum requires the criteria $m = m*$ and M proportionate to $M*$.

Finally, the corresponding matching labels and the images are semantically rearranged in increasing order similarities which is the normalized google distance, and the corresponding images are recommended to the user. If the user records any further clicks, these current user clicks are again passed into the Google knowledge graph API model, or if there are no user clicks that are recorded which means the search ends, here otherwise, the search continues until there are no further user clicks that are recorded. The current user clicks are an annotation-based Web image search, so therefore, only the terms pertaining to the current user clicks is sent, and the reason we are making use of here annotation-based image search is for the reason that the World Wide Web is overpopulated with annotations labels images, and in the semantic web, it is happening that the label has to be perfectly matched concerning the images so as a result of the annotation-based semantic, search is more than sufficient compared to the traditional content-based Web search.

4 Implementation

The performance of the proposed SemWIRet which is a semantically focussed model for Web image retrieval is implemented using Python as the language of choice. The implementation was carried out using a windows 11 operating system with the Intel Core i7 9th Gen processor and also was having 16 GB RAM. Experimentation was conducted for 7742 queries and all the models that are the SemWIRet model, and the baseline models were implemented for the same environment, and also for the same numbers of queries, that is, the 7742 queries and the performance for the individual model were recorded. The queries were taken from the flickr30k dataset, although the flickr30k dataset was not used exactly as it is, however, and it was further annotated since the proposed framework is highly suited for an annotation-based image retrieval mechanism which is semantically inclined to the similar needs of the semantic Web. It is more overcrowded by crawling images of similar quality and is further annotated for much more relevant categories. Further, the dataset is modified to suit the needs of the semantic Web by crawling more likely images from the World Wide Web and also annotating the existing categories by a few more relevant categories; so, the annotation is increased as well as the size is increased by 40%; so thus, the modified flickr30k dataset is used to successfully carry out the experimentation.

The performance measurements were created and compared to baseline techniques, and the results are summarized in Table 1.

Table 1 Comparative analysis between SemWIRet and other baseline models

Model	Average precision %	Average recall %	Average accuracy %	Average F-measure %	FDR
IRACS [11]	87.12	89.44	88.28	88.26	0.13
MRDL [9]	91.16	94.41	92.78	92.75	0.09
OBIR [10]	91.37	93.62	92.49	92.48	0.09
XGBoost + KNN	90.23	91.12	90.67	90.67	0.10
Proposed SemWIRet	**96.63**	**97.71**	**97.17**	**97.16**	**0.04**

$$\text{Precision} = \frac{|(\text{relevant docs}) \cap (\text{retrieved docs})|}{|(\text{retrieved docs})|} \quad (4)$$

$$\text{Recall} = \frac{|(\text{relevant}) \cap (\text{retrieved})|}{(\text{relevant})} \quad (5)$$

$$\text{Accuracy} = \frac{\text{precision} + \text{recall}}{2} \quad (6)$$

$$F\text{-measure} = \frac{2(\text{precision x recall})}{(\text{precision} + \text{recall})} \quad (7)$$

$$\text{False discovery rate} = 1 - \text{Recall} \quad (8)$$

Here, Eqs. (4) and (5) represent the precision and recall values, respectively, whilst Eqs. (6) and (7) represent the accuracy as well as the F-measure, and Eq. (8) represents the false discovery rate (FDR).

The dataset which was used for the experimentation is flickr30k. The performance of SemWIRet was compared using IRACS MRDL and OBIR as baseline models, also XGBoost with KNN was also used as the experimental hybridization as an evaluation of the proposed SemWIRet model for receiving Web images was conducted; from Table 1 also, the efficiency measuring F-measure accuracy in precision and recall % along with the FDR value is used as potential metrics in order to evaluate the performance of the proposed SemWIRet model. Since it is a standard formulation for precision–recall accuracy F-measure % and FDR which were used, the precision–recall accuracy F-measure % indicates the relevance of the result yielded, whereas the FDR depicts the number of the false positives captured by the system. From Table 1, it is indicative that the IRACS model yields the lowest precision of 87.12%, the lowest recall of 89.44%, and the lowest accuracy of 88.28%. So, overall the lowest F-measure of 88.26% with the highest FDR value of 0.13 and also, the Table 1. Based on the MRDL model, we have the following results: 91.16% accuracy,

94.41% recall, and 92.78% accuracy, as well as 92.75% F-measure with a 0.09 FDR. Similarly, an average F-measure of 92.75% is returned by the OBIR model, plus an average recall of 93.62%, the accuracy of 92.29%, and FDR of 0.09 with an average percentage of 91.37%. By integrating XGBoost and KNN, we arrive at an average precision of 91.23%, an average recall of 91.12%, and an average accuracy of 90.67% with an FDR of 0.10. The proposed SemWIRet model yields the highest precision of 96.63%, the highest average recall of 97.71%, the highest accuracy of 97.17%, and the highest F-measure of 97.16% along with the lowest FDR value of 0.04. The reason for the proposed model to excel in terms of precision, recall, accuracy, F-measure, and FDR, for the SemWIRet model is mainly due to the fact that it incorporates first and foremost, the LDA model for topic modelling so the query words are enhanced using topic modelling by uncovering all the hidden topics relevant to the query. Furthermore, the knowledge graph API is used for incorporating the auxiliary knowledge in much more depth into the framework, it is a standard knowledge store that ensures the entity density becomes very high employing extracting content from Google's knowledge graph API, so the query words are enriched by the means of entities. Most importantly, the XGBoost model is used for the classification, and label synonymization takes place using WordNet 3.0, apart from that a standard categorical domain ontology is used for label aggregation, and in particular, the incorporation of Shannon's entropy and normalized Google distance under the Gaussian adaptation mechanism ensures that the relevancy is very high, and also, Shannon's entropy with the step-deviation value and the normalized Google distance with a threshold ensures the strength of the relevance computational mechanism, and again, there is a refinement of the initial feasible solution to a much more refined and organized solution. So, as a result, the proposed SemWIRet model performs much better, that is, it is a semantic infused model, XGBoost, as a classification model is being used along with that semantic similarity schemes, like normalized Google distance and Shannon's entropy, are being used, and most importantly, lateral auxiliary knowledge addition by means of the knowledge graph API and LDA for topic modelling ensures that a large density of global knowledge is fed into the localized framework, and as a result, the proposed model is much better as compared to baseline models.

IRACS, a proposed model yields the lowest F-measure precision–recall accuracy and the highest false discovery rate value mainly since it uses ANOVA with cosine similarity, the usage of ANOVA with cosine similarity is novel, and we trying to fill the semantic gap; however, cosine similarity is very naive and traditional scope; but still, it computes the relevance results where the relevancy factor is computed, however, the global knowledge is not fed into the algorithm rather the algorithm only uses the learning on the entities and features provided in the localized dataset alone. So as a result, the IRACS model does not perform as expected.

Similarly, the MRDL model yields only a mediocre accuracy, F-measure. precision, recall when compared to the SemWIRet model and the reason for this is that it uses deep learning techniques which is Web deep learning technique in which it uses both of the characteristics of the image, as well as the characteristics of the text and all of this, is trained using the convolutional neural network so the deep learning, however, the images do not ensure a better auxiliary knowledge for the model rather they tend to make it more primitive, however, multiple concepts by enriching the texts ensures in increasing the accuracy of the model despite this is retarded by means of the usage of the image features when compared to the text features and also, global knowledge, auxiliary knowledge, background knowledge is not supplemented into the model so as a result, the semantic correlation happens only with respect to the labelled concepts of the query rather than incorporating entity enrichment and that is the reason this model does not performs well as expected.

The reason behind the OBIR model does not perform as expected is because although it uses ontology and NLP techniques along with resource description framework (RDF), it incorporates a lot of semantic methods, and the bag-of-words is selected as the model. The relevance computation model fails, auxiliary knowledge remains the same, but the relevance computational model is not very rigid; as a result, there is a lag in the OBIR model. Ontologies increase the auxiliary knowledge, and also the RDF knowledge is quite strong but the RDF knowledge would use the subject-predicate, and object increases the complexity which again has to be handled and that is the reason there is a lag in the model.

When XGBoost and KNN are used together, the same classifier is proposed, SemWIRet is used, but nonetheless, it fails abruptly with respect to the F-measure of precision–recall accuracy, and it has a higher FDR mainly due to that there is absolutely no auxiliary knowledge fed in the framework, and relevance computation is absent; by means of strict semantic similarity model based on vary thresholds and also coming to KNN, the KNN is also a very naive classifier so the only use of the high-end classifier does not serve rather it requires relevance computation mechanism with auxiliary knowledge which is absent in this model consequences of which the proposed XGBoost with the KNN does not perform as expected.

From Fig. 2, compared to other baseline models, the proposed model SemWIRet has better accuracy vs no. of recommendations when the precision and recall curves are plotted together. It is very clear from the curve that the proposed SemWIRet is the highest precision with the increase in the number of recommendations when compared to the baseline models, named IRACS, MRDL, OBIR, and XGBoost, with KNN. The second and third position are occupied by the MRDL and OBIR, respectively; even so, they almost have a similar precision–recall accuracy F-measure, and then, the fourth-lowest position belongs to the XGBoost with KNN, and the IRACS model has the lowest precision versus the number of recommendations curve.

5   Conclusions

Using an infused hybrid intelligence model for Web image retrieval, a Web image recommendation system is developed. The proposed system first preprocesses labels based on input images, and then, XGBoost is used to weight documents based on the preprocessed labels. Concept and ANOVA cosine semantic similarities, many relevant factors are integrated to obtain the top resultant relevant queries search which

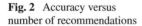

Fig. 2 Accuracy versus number of recommendations

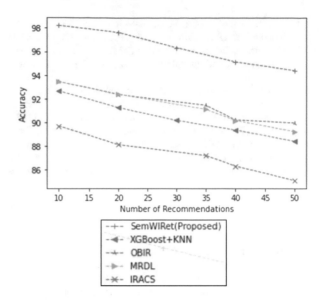

is further incorporated into Google's knowledge graph API by which the auxiliary knowledge is fed in much more depth into the framework, and also, the use of LDA in topic modelling is done to uncover all hidden topics and enrich query words.

Finally, the experiment is carried out for the modified dataset of flickr30k that yields 97.16% F-measure with the lowest FDR value of 0.04 which is achieved for the proposed SemWIRet model.

References

1. Xie et al (2016) Online cross-modal hashing for web image retrieval. In: Thirtieth AAAI Conference on Artificial Intelligence
2. Chen YY (2016) The image retrieval algorithm based on colour feature. In: 7th IEEE International Conference on Software Engineering and Service Science (ICSESS)
3. Li H (2019) A novel web image retrieval method bagging weighted hashing based on local structure information. Int J Grid Utility Comp 11(1)
4. Liu et al (2016) Deep supervised hashing for fast image retrieval. IEEE Comp Vision Pattern Recogn, 2064–2072
5. Gupta et al (2020) Comparative analysis of image retrieval techniques in cyberspace. Int J Students' Res Tech Manag
6. Liu H et al (2017) Image retrieval using fused deep convolutional features. In: ICICT
7. Wu et al (2021) Learning deep local features with multiple dynamic attentions for large-scale image retrieval. In: ICCV
8. Hou et al (2019) V-RSIR: an open access web-based image annotation tool for remote sensing image retrieval. In: IEEE
9. Huang C, Xu H, Xie L, Zhu J, Xu C, Tang Y (2018) Large-scale semantic web image retrieval using bimodal deep learning techniques. Inf Sci 430:331–348
10. Vijayarajan V, Dinakaran M, Tejaswin P, Lohani M (2016) A generic framework for ontology-based information retrieval and image retrieval in web data. HCIS 6(1):1–30

11. Sejal D, Ganeshsingh T, Venugopal KR, Iyengar SS, Patnaik LM (2016) Image recommendation based on ANOVA cosine similarity. Procedia Comp Sci 89:562–567
12. Kaushik A, Jacob B, Velavan P (2022) An exploratory study on a reinforcement learning prototype for multimodal image retrieval using a conversational search interface. Knowledge 2(1):116–138
13. Nair LR, Subramaniam K, PrasannaVenkatesan GKD, Baskar PS, Jayasankar T (2021) Essentiality for bridging the gap between low and semantic level features in image retrieval systems: an overview. J Ambient Intell Humaniz Comput 12(6):5917–5929
14. Kolahkaj M (2022) An image retrieval approach based on feature extraction and self-supervised learning. In: 2022 Second International Conference on Distributed Computing and High Performance Computing (DCHPC), pp 46–51, March. IEEE.
15. Noor J, Shanto MNH, Mondal JJ, Hossain MG, Chellappan S, Al Islam AA (2022) Orchestrating image retrieval and storage over a cloud system. IEEE Trans Cloud Comp
16. Deepak G, Priyadarshini JS (2018) Personalized and enhanced hybridized semantic algorithm for web image retrieval incorporating ontology classification, strategic query expansion, and content-based analysis. Comput Electr Eng 72:14–25
17. Gulzar Z, Leema AA, Deepak G (2018) PCRS: Personalized course recommender system based on hybrid approach. Procedia Comp Sci 125:518–524
18. Kumar A, Deepak G, Santhanavijayan A (2020) HeTOnto: a novel approach for conceptualization, modeling, visualization, and formalization of domain centric ontologies for heat transfer. In: 2020 IEEE international conference on electronics, computing and communication technologies (CONECCT), pp 1–6, July
19. Varghese L, Deepak G, Santhanavijayan A (2019) An IoT analytics approach for weather forecasting using raspberry Pi 3 model B+. In: 2019 fifteenth international conference on information processing (ICINPRO), pp 1–5, December. IEEE
20. Ojha R, Deepak G (2021) Metadata driven semantically aware medical query expansion. In: Iberoamerican Knowledge Graphs and Semantic Web Conference, pp 223–233, November. Springer, Cham
21. Rithish H, Deepak G, Santhanavijayan A (2021) Automated assessment of question quality on online community forums. In: International Conference on Digital Technologies and Applications, pp. 791–800. Springer, Cham

Optimal Drug Recommender Framework for Medical Practitioners Based on Consumer Reviews

Pooja Khanna, Sachin Kumar, Neelam Sodhi, and Apurva Tiwari

Abstract Natural language processing is a mechanized methodology by which PCs can control and perceive the regular language. That, however, is not a straightforward task. Robotized machines can easily perceive organized information in a dataset, such as accounting pages and tables, but perceiving and controlling with texts, voices, and human dialects are a time-consuming process because the information falls into an unstructured category, making it difficult for mechanized machines to remember, necessitating the use of natural language processing. Global epidemic has left the world into splits, and the entire planet came to a standstill with curbs like lockdown, social distancing, and compulsory mask as a permanent part of the day. Millions were put to edge, and everyone was suddenly placed in a state of emergency. There is inaccessibility of legitimate clinical resources like the shortage of specialists and healthcare workers, lack of proper equipment and medicines, etc. The objective of this work is to develop optimal drug recommendation system embedded with NLP algorithm supported by few machines learning models for medical practitioners to get information on the popular drugs in the market at any point of time. Work presented employed consumer reviews which were analyzed with Vader tool and language processing-assisted opinion mining techniques, and model designed was verified using random forest classifier that achieved 82.85% accuracy which is at par with existing techniques.

Keywords NLP · Recommender system · Drug · Consumer reviews · Clinical resources

P. Khanna · S. Kumar (✉) · N. Sodhi
Amity University Uttar Pradesh, Lucknow Campus, Lucknow, India
e-mail: skumar3@lko.amity.edu

P. Khanna
e-mail: pkhanna@lko.amity.edu

A. Tiwari
Chaudhary Devilal University, Sirsa, India

1 Introduction

The clinical space is quickly extending with new medications showing up consistently. Simultaneously, with increase in intricate and complex medicine routines individual often gets lost that creates a sorry state that one needs to reconsider to keep away from the undesirable aftereffects resulting from medication routine. Sometimes in certain cases, individuals often consume medications without consulting with a doctor, which consequently might be a genuine danger factor later on, particularly when certain insights about the current clinical plan or clinical records are not thought about. One of the most potential research domains with enough funding support on the net is health-based records and information. According to a survey conducted by the Pew Internet and American Life Project in 2013, 59% of Americans have looked online for health information, with 35% of those who did so focusing on identifying a medical issue. Approximately, 40% of drug errors take place as experts provide prescriptions based on their limited sample space experience. General observations that may contribute to these issues can be summarized as [1].

i. For serious health conditions, several health centers/hospitals lack either infrastructure or specialized medical experts.
ii. Efficient diagnosis is mostly dependent on the expert's expertise, particularly for inexperienced novices who are generally more prone to commit errors.

Health records about diagnosis data in hospitals have remained untouched and have not been mined, and this sometimes hides new discoveries with respect to new conclusions about data values. Patients have begun discussing their clinical reports, as well as their perspectives on the prescription, as a result of the advancement in innovation. This result is a large amount of unstructured data that must be managed and examined in order to obtain useful information. Sentiment analysis is a well-known part of NLP that focuses on analyzing the beliefs or viewpoints of people who are influenced by a substance. This domain has grown tremendously on various microblogging sites like as Twitter, Instagram, Flickr, and others, where the clients' opinions are eliminated from the text, photographs, emojis, and recordings. In any case, this field is mostly ignored in the clinical setting. The Internet's steady evolution has increased the amount of client-generated data available on the Web. Patients are increasingly uploading surveys after taking medications in order to put themselves out there and raise public awareness.

A recommendation system is a type of information filtering system that tries to predict how a user would evaluate or prefer a certain item. In layman's terms, it is an algorithm that recommends products to consumers based on their interests. For example, on Netflix, which movie to watch, on e-commerce, which product to buy, on Kindle, which book to read, and so on. There are many use-cases of it. Some are as follows:

i. Personalized content: Enhances the on-site experience by making dynamic recommendations for various audiences, similar to what Netflix does.

ii. Better product search experience: Helps to categories the product based on their features. E.g.: material, season, etc.

A drug recommendation system helps in recommending medicines to patient with a particular condition based on the previous consumer reviews and ratings. This is achieved with the help of more accurate feature selection and opinion mining from the previous experiences of consumers. Feature selection is the process of choosing, changing, and transforming raw data into features that may be used in supervised learning. Feature engineering is the process of using statistical or machine learning approaches to turn raw observations into desired attributes. While sentiment analysis on the other hand is used to determine the emotional significance of communications, analytic techniques such as statistics, natural language processing, and machine learning are used [2–5]. The paper is arranged as follows: Section 2 discusses motivation for the designing the recommender system. Section 3 discusses the methodology proposed for the work, Sect. 4 presents materials and methods employed during the work conducted, and finally, conclusion is presented in Sect. 5.

2 Related Work

In recent years, a large volume of clinical data spread over several Websites on the Internet makes it difficult for individuals to locate useful information for improving their health. Furthermore, the overabundance of medical information has made it difficult for medical professionals to make patient-centered decisions. These concerns highlight the potential requirement of recommender systems to be used in the healthcare industry to assist both end-users and medical professionals in making more efficient and accurate health-related decisions.

To deal with the growing problem of online information overload, recommender systems strive to give users with customized products and services. Since the mid-1990s, many recommender system strategies have been presented, and a range of recommender system software has lately been built for a variety of applications. The majority of recommender technologies are used in the areas of e-government, e-business, e-commerce/e-shopping, e-learning, e-tourism, and so on. However, few recommender technologies exist in the medical field, and this work focuses on the construction of a medicine recommender system as well as mining information from medical case data.

Common recommendation tactics include collaborative filtering (CF), content-based (CB), knowledge-based (KB), and hybrid recommendation systems. Each recommendation technique comes with its own set of advantages and disadvantages: Collaborative filtering (CF)-based recommendation systems aid people in making decisions based on the opinions of others who share similar interests, although CF has sparseness, scalability, and cold-start concerns. Knowledge-based (KB) recommendations promote goods to users based on information about the users, products, and/or their relationships. A functional knowledge base is often kept for KB recommendations that describes how a single item satisfies a specific user's requirement. To increase performance and overcome the disadvantages of traditional recommendation systems, a hybrid model has been designed that employs the best sections of two or more recommendation strategies into one hybrid strategy. A new recommendation system is required to solve these difficulties in a new application area [6–9].

With advent of technologies, lot of innovations have happened in diversified data mining techniques of recommender system in healthcare domain for diagnosis, treatment, or prognosis [10]. Collaborative filtering (CF) assisted and based on e-commerce to estimate risks in heart attack and outperforms existing techniques, i.e., SVM or linear regression [11]. A model titled iCARE supported by CF, and ensemble learning estimates predictions of patient disease risks based on disease history [12]. An incremental CF technique, W-InCF, employing Mahalanobis distance and fuzzy membership calculates the risk of women in pregnant condition during process of delivery [5]. Apart from these, various models were designed based on case-based reasoning (CBR). A recommendation model to estimate different state of diabetes condition according to various features shortlisted employing rough set feature reduction [13]. A CBR model can also be applied to estimate radial dose in cancer treatment, and weights of parameters are optimized by bee colony optimization [14]. Suggested work without technology support acquires lab test results in numerical form as patient feature to prepare similar case analogy in predicting the test cases. Accurate prediction for such problems is particularly quite easy with binary result as benign or infected. However, diagnostic test cases, medical inference, and treatment path in medication context can be very complex, and no direct steps for treatment may be traced. Conclusions on prescription are to be based on among diversified combination of drugs. Permissible combinations, diversified permutations, and efficient effectiveness of relationships between various drugs will further increase the complexity of the medication. Till date, no accurate and efficient model have been designed to support decisions for recommending perfect drug combination.

3   Proposed Methodology

The problem statement is straightforward, consider there is a pharmaceutical start-up that was recently acquired by one of the world's largest multi-national companies. For the knowledge acquisition process, the start-up is required to tabulate all the drugs sold till date. The start-up is also required to give an account for each drug's effectiveness. The work will likely make a complicated and common sense model by consolidating a ML and NLP-assisted techniques. This is refined by deciding the best medication for every ailment. Secret patterns and examples (if any exist) are additionally needed altogether for the association to make exact information driven decisions [15–17, 19].

Model proposed is going to span across five broad domains as depicted in Fig. 1:

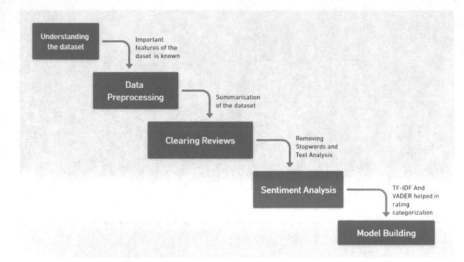

Fig. 1 Flow graph of operations

i.   Understanding the dataset
ii. Data pre-processing
iii. Clearing the reviews
iv. Sentiment analysis
v.   Model building.

4   Material and Methods

Work proposed is an effort to estimate the most popular drugs among the options available in market using NLP assisted by sentiment analysis and learning techniques, the recommender system designed will as primary step acquire data samples from the corpus and perform data cleaning before applying recommender model. Steps involved may be summarized as follows:

1. Understanding the Dataset

The design of the information is that a patient with an interesting ID buys a medication that meets his condition and composes a survey and rating for the medication he/she bought on the date. A while later, assuming the others read that audit and think that it is useful they will click useful count, which will add 1 for the variable.

The most crucial objective for the entire work will be served by the useful count column. It will confidence to the people for trusting on a drug for any specified medical condition. Figure 2 depicts dataset features to be analyzed [16, 17].

	uniqueID		drugName	condition	review	rating	date	usefulCount
11822	231989		Trazodone	Insomnia	"I changed doctors yesterday because my doctor...	1	11-Nov-09	29
32780	190158		Protonix	GERD	"I loved this drug--until my insurance company...	10	29-Oct-08	68
13110	11219		Drospirenone / ethinyl estradiol / levomefolat...	NaN	"I am on my 2nd week and I have the following ...	4	7-Aug-11	4
48290	7083		Aluminum chloride hexahydrate	Hyperhidrosis	"This thing worked wonders! I'm so glad I...	8	14-Jan-14	18
27673	58215		Acetaminophen / hydrocodone	Pain	"I have RSD a nerve disorder caused by a broke...	10	9-Aug-08	46

Fig. 2 Dataset features

2. Data Pre-processing

Putting together data in order to "change" it and observe what looks to be "normal" and what appears to be "abnormal." The allocation of a variable shows what respect the variable takes and how sometimes the variable takes these qualities.

The output shows the mean, standard deviation, min, and max value. It can be concluded that most the reviews from the consumers are positive. In addition, when compared to the other useful statistics, the max count column for the useful count column is extremely high. After the summarization of the dataset, a step further is taken. Now, the analysis of useful and useless drug will be done for the better comprehension of the task. Figure 3 summarizes rating and useful count column for the dataset.

Next important step is to unveil interesting relationships, trends, and pattern hidden inside the dataset. This can be accomplished by determining the relationship that exist between key factors in the data sample. For this step, the following will be done.

i. Check for the distribution of rating and useful count columns depicted in Fig. 4.
ii. Check for the relationship between rating and usefulness count.

Figure 5 suggests that positive and almost linear relationship exist between rating and useful count column. Useful count is increasing as the rating is increasing.

After this typical data preparation strategies were employed such as verifying and eliminating null values, duplicating rows, deleting superfluous values, and removing text from rows. After that, all null values were eliminated. Then, a visual analysis of features was done to get a better insight. Top 20 drugs per condition, bottom 20 drugs per condition, number of reviews in all the years, mean rating in the years, and number of reviews per month were some of the aspects of visualization [17–20].

Figure 6 depicts a graphic representation of the 10-star rating system's value counts. The vast majority choose four attributes; the numbers 10, 9, 1, 8, and 10 are

	rating	usefulCount
count	212053.000000	212053.000000
mean	6.989111	28.195376
std	3.276685	36.481139
min	1.000000	0.000000
25%	5.000000	6.000000
50%	8.000000	16.000000
75%	10.000000	37.000000
max	10.000000	1291.000000

Fig. 3 Summarizing rating and useful count column

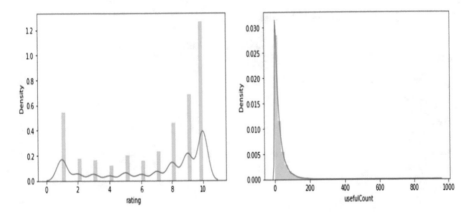

Fig. 4 Distribution of rating and useful count

nearly identical. It demonstrates that the positive side of the scale is higher than the negative, and that people's reactions are polarized.

1. Clearing the Reviews

Before vectorization, it is necessary to clean up the review text. Text preparation is another name for this procedure. Many unneeded details are included in the reviews. Punctuation marks, numbers, and stop words are useless in the next step. It is critical to get rid of these contaminants and extraneous items so that we can simply perform textual analysis. Numbers and punctuation marks have no meaning in the text. The stop words do not play any role for the analysis purpose; hence, they were removed

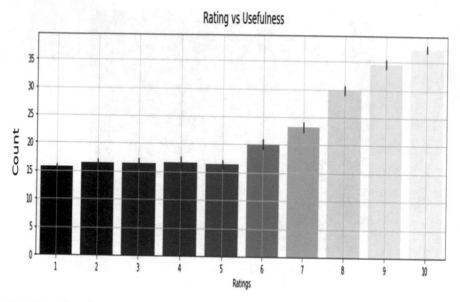

Fig. 5 Rating versus usefulness of the drugs

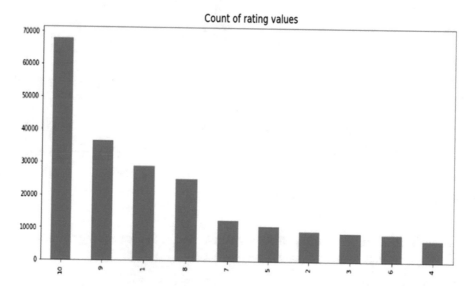

Fig. 6 Count of rating values versus 10 rating number

from the review feature. After the stop word removal process, tokenization and stemming were done. When stemming is utilized, words are reduced to their word stems. A word stem does not have to be the same root as a dictionary-based morphological root; it simply needs to be the same size as or smaller than the word. This was done by cleaning the reviews by eliminating HTML tags, punctuation, quotations, and URLs, among other things. To avoid duplication, the cleaned reviews were lowercased, then tokenization was used to break the sentences down into little chunks called tokens. Stop words such as "a, to, all, we, with, and so on" were also eliminated from the corpus. By executing lemmatization on all tokens, the tokens were returned to their foundations.

2. Sentiment Analysis

A bag of words model, which converts documents into vectors and assigns a score to each word in the document, is a popular technique for constructing sentiment analysis models. For this step, a bag of word (BOW) model was created with TF-IDF and count vectorizer. The bag of words model will help in generating a set of vectors holding the count of word occurrences in the text (reviews), and the TF-IDF model will enhance the work by including information on both the more significant and less important words. The bag of words vectors are simple to understand. TF-IDF, on the other hand, frequently outperforms TF-IDF in machine learning models. The approach was to give low weight to phrases that appeared frequently in the dataset, implying that TF-IDF measures relevance rather than recurrence. The possibility of finding a word in a document is known as term frequency (TF).

$$TF(t, d) = \log(1 + \text{freq}(t, d)) \tag{1}$$

The inverse document frequency (IDF) is the inverse of the total number of times a phrase appeared in the corpus. It detects how a particular phrase is document-specific.

$$IDF(t, d) = \log\left(\frac{N}{\text{count}(d \sum D : t \sum D)}\right) \tag{2}$$

The TF-IDF of a term is calculated by multiplying TF and IDF scores.

$$TFIDFT(t, d, D) = TF(t, d).IDFT(t, d)) \tag{3}$$

$$TF - IDF = TF * IDF \tag{4}$$

Valence aware dictionary and sentiment reasoner (VADER) is a sentiment analysis and lexicon tool built specifically for social media sentiments. VADER employs a sentiment lexicon, which is a set of lexical properties that are classified as positive or negative based on their semantic orientation. Based on its user rating, every single review was categorized as positive or negative for sentiment analysis. If the user rating ranges from 6 to 10, the review is positive; otherwise, it is negative [20–23].

3. Model Building

The data modeling for this work was done by Naïve Bayes and random forests classifier. TF-IDF vectorizer is the vectorizer used. The text vectorizer term frequency-inverse document frequency converts the text into a useable vector. Term frequency (TF) and document frequency (DF) are combined in this idea (DF). The term frequency refers to the number of times a term appears in a document. This is a typical approach for converting text into a meaningful numerical representation, which is then used to fit a machine learning algorithm for prediction.

Looking at the distribution of useful count in Fig. 7, one can see that the difference between the least and the most is roughly 1300, which is significant. The idea is that the more medications people look for, the more people read the survey, regardless of whether their review is good or negative, increasing the useful count. As a result, we normalized useful count by circumstances while developing the recommender system. Work presented simple technique for extracting opinion from the sample data, data acquired was subjected to classification for developing a recommender system, consumer reviews were analyzed using the Vader tool and NLP-based sentiment analysis algorithm, and the model was verified using random forest classifier that achieved 82.85% accuracy which is at par with existing techniques. Future work involves comparison of different oversampling techniques, using different values of n-grams, and optimization of algorithms to improve the performance of the recommender system.

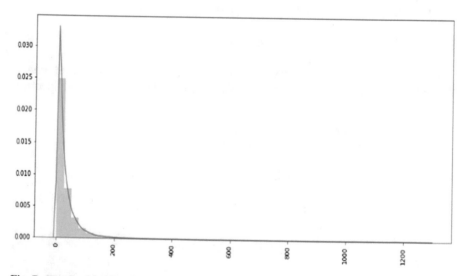

Fig. 7 Distribution of useful count

5   Conclusion

The clinical space is quickly extending with new medications showing up consistently. Simultaneously,with increase in intricate and complex medicine routines individual often gets lost that creates a sorry state that one needs to reconsider to keep away from the undesirable aftereffects resulting from medication routine. Sometimes in certain cases, individuals often consume medications without consulting with a doctor, which consequently might be a genuine danger factor later on, particularly when certain insights about the current clinical plan or clinical records are not thought about. The principle point of the drug analysis is to serve the human to make them liberated from expected sickness or anticipation of the infection. For the medication to fill its planned need, it ought to be liberated from pollutants or other impedance which may hurt people. This was simply understanding the foundation of the ventures.

The work proposed incorporated the best-predicted result of each approach to the recommendation framework. A good assembly of different expected findings is required for improved results and understanding. Work presented simple methods for extracting sentiment from data and subjected the data to classification to develop a optimal recommender system, consumer reviews were analyzed using the Vader tool and NLP-based sentiment analysis algorithm, and the model was verified using random forest classifier that achieved 82.85% accuracy which is at par with existing techniques. Future work involves comparison of different oversampling techniques, using different values of n-grams, and optimization of algorithms to improve the performance of the recommender system.

References

1. Fox S (2013) Health Online 2013. Pew Internet & American Life Project, January. https://www.pewresearch.org/internet/2013/01/15/information-triage
2. McInnes DK, Saltman DC, Kidd MR (2006) General practitioners' use of computers for prescribing and electronic health records: results from a national survey. Med J Aust 185:88
3. Ting S, Kwok SK, Tsang AH, Lee W (2011) A hybrid knowledge-based approach to supporting the medical prescription for general practitioners: Real case in a Hong Kong medical center. Knowl-Based Syst 24:444–456
4. Esfandiari A, Babavalian MR, Moghadam A-ME, Tabar VK (2014) Knowledge discovery in medicine: current issue and future trend. Expert Syst Appl 41:4434–4463
5. Lu X, Huang Z, Duan H (2012) Supporting adaptive clinical treatment processes through recommendations. Comput Methods Programs Biomed 107:413–424
6. Guo WY (2008) Reasoning with semantic web technologies in ubiquitous computing environment. J Soft 3(8):27–33
7. Hamed AA, Roose R, Branicki M, Rubin A (2012) T-Recs:Time-aware Twitter-based drug recommender system.In: 2012 IEEE/ACM International Conference on Advances in Social Networks Analysis and Mining
8. IBM (2017) IBM Watson health. http://www.ibm.com/watson/health/
9. Kitchenham B, Charters C (2007) Guidelines for performing systematic literature reviews in software engineering. 2007 Joint Report—EBSE 20 07-0 01

10. Hassan S, Syed Z (2010) From Netflix to heart attacks: collaborative filtering in medical datasets. In: Proceedings of the 1st ACM International Health Informatics Symposium, pp 128–134

11. Davis DA, Chawla NV, Christakis NA, Barabási A-L (2010) Time to CARE: a collaborative engine for practical disease prediction. Data Min Knowl Disc 20:388–415

12. Komkhao M, Lu J, Zhang L (2012) Determining pattern similarity in a medical recommender system. In: Xiang Y, Pathan M, Tao X, Wang H (eds) Data and knowledge engineering. Springer, Berlin, pp 103–114

13. Teodorovi D, Šelmi M, Mijatovi-Teodorovi L (2013) Combining case-based reasoning with Bee Colony Optimization for dose planning in well differentiated thyroid cancer treatment. Expert Syst Appl 40:2147–2155

14. Savova GK, Masanz JJ, Ogren PV, Zheng J, Sohn S, Kipper-Schuler KC et al (2010) Mayo clinical Text Analysis and Knowledge Extraction System (cTAKES): architecture, component evaluation and applications. J Am Med Inform Assoc 17:507–513

15. Kushwaha N, Goyal R, Goel P, Singla S, Vyas OP (2014) LOD cloud mining for prognosis model (case study. native app for drug recommender system), AIT 04, 03, 20–28

16. Mahmoud N, Elbeh H (2016) IRS-T2D. individualize recommendation system for type 2 diabetes medication based on ontology and SWRL. In: Proceedings of the 10th International Conference on Informatics and Systems—INFOS '16. ACM Press, New York, New York, USA, 203–209. https://doi.org/10.1145/2908446.2908495

17. Medvedeva O, Knox T, Paul J (2007) DiaTrack. Web-based application for assisted decision-making in treatment of diabetes. J Comp Sci Colleges 23(1):154–161

18. Protégé (2016) Protégé. http://protege.stanford.edu/. Accessed 15 March 2017

19. Rodríguez A, Jiménez E, Fernández J, Eccius M, Gómez JM, Alor-Hernandez G, Posada-Gomez R, Laufer C (2009) SemMed: applying semantic web to medical recommendation systems. In: 2009 First International Conference on Intensive Applications and Services

20. Sun L, Liu C, Guo C, Xiong H, Xie Y (2016) Data-driven Automatic Treatment Regimen Development and Recommendation. In: KDD'16 Proceedings of the 22nd ACM SIGKDD International Conference on Knowledge Discovery and Data Mining, 1865–1874

21. Zhang Q, Zhang G, Lu J, Wu D (2015) A framework of hybrid recommender system for personalized clinical prescription. In: 2015 International Conference on Intelligent Systems and Knowledge Engineering

22. Zhang Y, Zhang D, Hassan MM, Alamri A, Peng L (2015) CADRE. cloud-assisted drug recommendation service for online pharmacies. Mobile Netw Appl 20(3):348–355

23. Doulaverakis C, Nikolaidis G, Kleontas A, Kompatsiaris I (2014) Panacea, a semantic-enabled drug recommendations discovery framework. J Biomed Semantics 5:13

Development of Intelligent Framework for Early Prediction of Diabetic Retinopathy

Adil Husain and Deepti Malhotra

Abstract Diabetes affects blood arteries throughout the body, particularly those in the kidneys and eyes. Diabetic retinopathy is a condition in which the blood vessels in the eyes get damaged as a result of diabetes (DR). Diabetic retinopathy is a severe public health problem that is one of the leading causes of blindness across the world. Diabetic retinopathy is a kind of microvascular disease that can affect diabetics. Diabetic retinopathy causes vision problems and might even result in blindness. If a diabetic person is not treated for a long time, diabetic retinopathy is more likely to develop. Diabetic retinopathy progresses to the point that it causes symptoms. In the early stages of the infection, diabetics may be unaware that they have been infected. Clinically, DR is diagnosed by looking at it directly or through imaging techniques like fundus photography or optical coherence tomography. Various methodologies were proposed by different researchers for the prediction of DR in diabetic patients, but the prediction of DR in the early stages in literature has not been visualized more broadly way. The focus of this research is to design an intelligent framework for the prediction of DR in the early stages of diabetic patients.

Keywords Machine learning · Deep learning · CNN · I-DR · etc.

1 Introduction

Diabetes is a condition that affects blood vessels all over the body, especially those in the kidneys and eyes [1]. Diabetic retinopathy is a disorder in which the blood vessels of the eye are compromised (DR). Diabetic retinopathy is a serious public health issue and one of the primary causes of blindness worldwide. Diabetic retinopathy is a microvascular problem that can develop in diabetic people [2]. Diabetic retinopathy

A. Husain (✉) · D. Malhotra
Department of Computer Science and IT, Central University of Jammu, Jammu, India
e-mail: adilhussain5057@gmail.com

D. Malhotra
e-mail: deepti.csit@cujammu.ac.in

impairs vision and can lead to blindness. If a diabetic individual is not treated for a long time, he or she is more likely to develop diabetic retinopathy [3]. Later, stages of diabetic retinopathy become symptomatic. Diabetic people may not be aware that they have been infected with the illness in the early stages. In the clinic, DR is diagnosed by examining directly at the retinal fundus or through imaging techniques such as fundus photography or optical coherence tomography [4].

The word "retinopathy" refers to damage to the retina in general. DR arises when the tiny blood arteries that supply the retina's tissue and nerve cells are damaged. There are generally no early warning signals of diabetic retinopathy [5]. It can only be identified with a thorough eye exam that searches for early indications of the illness, such as [6].

- Macular edema (swelling).
- Pale, fatty deposits on the retina.
- Damaged nerve tissue.
- Any changes to the retinal blood vessels.

By 2040, diabetes will affect an estimated 600 million people, with one-third of them having diabetic retinopathy (DR), the leading cause of vision loss in working-age adults around the world [7]. Mild non-proliferative DR (NPDR) is a kind of DR in which microaneurysms are present in the early stages. Proliferative DR (PDR) is a more advanced kind of DR that can result in significant vision loss. Regular DR tests are essential in order to receive quick treatment and avert vision loss [8]. Glycemia and blood pressure control can help to slow the progression of DR, whereas late-stage treatments like as photocoagulation or intravitreal injection can help to avoid vision loss [9]. Although many professional associations urge routine DR screening, thorough DR screening is not routinely used due to difficulties in finding human assessors [10]. Imaging investigations such as fluorescent angiography and optical coherence tomography (OCT) (see our OCT section) can aid in the identification and treatment of diabetic eye abnormalities. To address abnormalities in the eyes caused by diabetes, retina specialists employ medications, laser treatments, or surgery [11]. The earlier diabetes-related abnormalities are detected, the better the chances of keeping vision. The optometrist will have the best chance of diagnosing this and all other issues with an OCT extended eye examination. Getting your eyes tested at least once a year might help you avoid losing your eyesight due to diabetes [12]. Maintaining tight blood sugar and blood pressure management, as well as not smoking, will help lower the risk of diabetic eye abnormalities. The diagnosis of DR has been considered in the later stages of the diabetic patients, and hence, these techniques are inefficient for the detection of DR in the early stages [13]. There are two stages of DR in diabetic patients, and most of the research had been going on the later stages of DR. Both of these stages ultimately result in blindness in diabetic patients gradually [14].

If automatic DR screening is available, it relieves doctors of a significant amount of effort [15]. With the adoption of automatic DR screening, the ratio of patients to doctors will drop, saving time, and money and increasing the effective use of existing resources [16]. Another advantage of introducing automatic DR screening

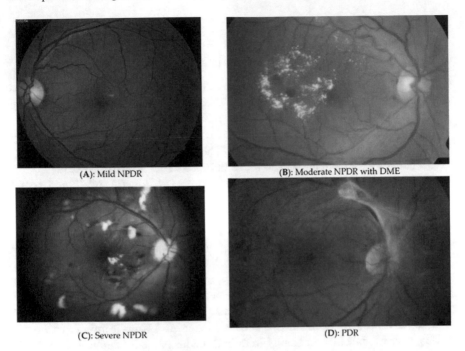

Fig. 1 Images of the retinal fundus at various stages of diabetic retinopathy. **a** stage II: mild non-proliferative diabetic retinopathy; **b** stage III: moderate non-proliferative diabetic retinopathy; **c** stage IV: severe non-proliferative diabetic retinopathy; **d** stage V: proliferative diabetic retinopathy [18]

is that patients who live in rural places and do not have access to medical facilities can be treated through telemedicine [17]. We aim to find an efficient and accurate technique for DR detection that takes less computational time and gives better results than existing methods (Fig. 1).

Depending on medical professionals

0. Normal ($\mu A = 0$) and (H = 0)
1. Mild NPDR ($0 < \mu A \leq 5$) and (H = 0)
2. Moderate NPDR ($5 < \mu A < 15$ or $0 < H < 5$) and (NV = 0)
3. Severe NPDR ($\mu \geq 15$) or (H ≥ 5) or (NV = 1)

The number of microaneurysms (μA) and the number of hemorrhages (H) and NV neovascularization are seen.

The organization of the paper is as follows: The introduction section was briefly covered in Sect. 1. Section 2 discusses numerous diabetes detection approaches that have been discussed by various researchers. The comparative analysis of several DR approaches was shown in Sect. 3. The limitations of present techniques are discussed in Sect. 4. In Sect. 5, a novel framework for detecting DR is provided. Finally, Sect. 6 brings the work to a close by highlighting a few key points.

2 Literature Review

This section summarizes the research on existing DR prediction tools.

- Amol et al. [2] established a framework that addresses MLPs. For their investigation, they only used 130 images. Backpropagation is a supervised learning approach used by MLPs to train the network.
- Kumar et al. [3] conducted a systematic investigation to identify the DR in only one type of lesion, namely hard exudates. They conducted their investigation using DIP and the RRGA method. The focus of this study is on DIARETDB1.
- Ankita et al. [4] conducted a survey to identify DR in two ways: blood vessel segmentation and lesions identification. For their survey, they used the drive, stare, and chase dataset.
- Ling et al. [5] presented a framework for DeepDR. DeepDR was trained on 466,247 fundus pictures from 121,342 diabetic patients for real-time image quality evaluation, lesion identification, and grading. A sub-network for image quality assessment, a lesion-aware sub-network, and a DR grading sub-network were created. The three deep learning sub-networks of the DeepDR system. ResNet and mask RCNN algorithms were utilized.
- Mike et al. [6] in "development and validation of a deep learning system for diagnosis of diabetic retinopathy in retinal fundus pictures" duplicated the basic strategy. The original work employed nonpublic EyePACS fundus pictures and a separate EyePACS data collection. For this study, they employed two datasets: EyePACS and MESSIDOR-2.
- Sehrish et al. [7] trained using the publicly available Kaggle dataset of retina photos, an ensemble of five deep convolution neural network (CNN) models (ResNet50, Inceptionv3, Xception, Dense121, Dense169) was used to encode the rich properties and improve categorization for different phases of DR. This ensemble model is capable of detecting many phases of DR. They concentrated on detecting all phases of DR in this investigation. This model has an accuracy rate of 80%. They made advantage of a publicly available data collection.
- Harry et al. [8] proposed using a CNN technique, and researchers were able to properly diagnose DR from digital fundus pictures and characterize its severity. They create a network with CNN architecture and data augmentation that can recognize the complex elements involved in the classification job, such as microaneurysms, exudate, and hemorrhages on the retina, and deliver a diagnosis automatically and without the need for human input. They train their network on the publically accessible Kaggle dataset with a high-end graphics processor unit (GPU) and show outstanding results, especially for a high-level classification job.
- Shirin et al. [9] to evaluate diabetic retinopathy severity stages, we combine fundus fluorescein angiography and color fundus pictures concurrently and extract 6 features using curve let transform and input them to a support vector machine. These characteristics include the number of blood vessels, the area, regularity, and number of microaneurisms in the foveal vascular zone, as well as the overall number of microaneurisms and the area of exudate.

- Shankar et al. [19] provided another method for preparing fundus pictures by utilizing histogram-based segmentation to remove areas with lesions. This paper used the synergic DL (SDL) model as a classification step, and the findings showed that the given SDL model outperforms common DCNNs on the MESSIDOR-1 database in terms of ACC, SE, and SP.
- Arcadu et al. [20] provided a method to predict DR progression defined as a 2-step worsening in early treatment diabetic retinopathy. This paper used DCNNs on 7 FOV images of the RIDE and RISE dataset with a specificity of 77% and a sensitivity of 66%.
- Wang et al. [21] to assess the severity stages of diabetic retinopathy, we integrate fundus fluorescein angiography and color fundus images simultaneously, extract 6 features using curve let transform, and feed them into a multitasking deep learning framework. The recommended approach was evaluated on two independent testing sets using quadratic weighted Cohen's kappa coefficient, receiver operating characteristic analysis, and precision–recall analysis.
- Ali et al. [22] provided a method for segmentation and classification of DR. Four types of features—histogram (H), wavelet (W), co-occurrence matrix (COM), and run length matrix (RLM)—were retrieved for texture analysis, and several ML classifiers were used to achieve a classification accuracy of 77.67%, 80%, 89.87%, and 96.33%, respectively. The data fusion technique was utilized to create a fused hybrid feature dataset to increase classification accuracy.
- Bora et al. [23] provided a deep learning system to predict the development of DR in patients.
- Gangwar and Ravi [24] provided a novel deep learning hybrid method for automatic DR detection. The suggested model was tested on the MESSIDOR-1 diabetic retinopathy dataset and the APTOS 2019 blindness detection dataset (Kaggle dataset). Our model outperformed previously reported findings. On the MESSIDOR-1 and APTOS datasets, we attained test accuracy of 72.33% and 82.18%, respectively.

3 Comparative Analysis

Table 1 summarizes the research into several DR detection approaches presented by various researchers.

As the results are measured by accuracy and specificity.

The mathematical representation are as follows:

$$\text{Specificity} = \frac{\text{Number of true negative}}{\text{No. of true negative} + \text{No. of false positive}}$$

$$\text{Accuracy} = \frac{TP + TN}{TPC + TN + FP + FN}$$

Table 1 Comparative analysis of already existing DR techniques

Author	Method used	Dataset	Accuracy	Specificity	Remarks
Amol et al. [2]	MLPNN	DIARETDB0	92%	90%	For future investigation, more color fundus photos are employed
Harry et al. [8]	CNN	Public dataset	75%	95%	Researchers may get a significantly cleaner dataset from real-world screening situations and compare this network to the five-class SVM approach
Kranthi et al. [3]	DIP/RIGA	DIARETDB1	91%	88%	DR can detect through other lesions
Ankita et al. [4]	Not applicable	DRIVE/STARE/CHASE	Not applicable	Not applicable	Scholars have only employed CNNs, however, other deep networks can be used
Mike et al. [6]	InceptionV3	EyePACS/MESSIDOR-2	89%	83.4% 67.9%	They were not able to replicate the original study
Arcadu [20]	DCNN: Inception v3	7 FOV images of RIDE and RISE datasets	Not applicable	77%	They were not able to replicate the original study
Sehrish et al. [7]	CNN	Public dataset	80%	Not applicable	We can train several models for different phases and then ensemble the results

(continued)

Table 1 (continued)

Author	Method used	Dataset	Accuracy	Specificity	Remarks
Shankar et al. [19]	DCNN: Histogram-based segmentation + SDL	MESSIDOR-1	99.28%	99.38%	They improved the efficiency and accuracy of categorization by creating architectural improvements to an existing DCNN
Wang et al. [21]	DCNN: Multitask network using channel-based attention blocks	Shenzhen, Guangdong, China	Not applicable	Not applicable	Another algorithm can be employed
Ling dia et al. [5]	ResNet/RCNN	Private dataset	82%	81.3%	Another CNN algorithm might be employed
Ali et al. [22]	ML: SMO, Lg, MLP, LMT, Lg employed on selected postoptimized hybrid feature datasets	Bahawal Victoria Hospital, Pakistan	MLP: 73.73% LMT: 73.00 SLg: 73.07 SMO: 68.60 Lg: 72.07%	Not applicable	Specificity can be added
Bora ct al. [23]	DCNN: Inception v3	EyePACS	Not applicable	Not applicable	Specificity and accuracy not mentioned
Gangwar et al. [24]	DCNN: Inception ResNet v2	APTOS 2019, MESSIDOR-1	APTOS: 82.18% MES SIDOR 1: 72.33%	Not applicable	

(continued)

where true positives (TPs), true negatives (TNs), false positives (FPs), false negative (FN)

All of the studies presented in this paper used deep learning approaches to control the diabetic retinopathy screening system. Due to a growth in the number of diabetic patients, the demand for efficient diabetic retinopathy screening equipment has lately become a major concern. Using DL for DR detection and classification solves the challenge of picking trustworthy features for ML; nevertheless, it requires a large

Table 1 (continued)

Author	Method used	Dataset	Accuracy	Specificity	Remarks
Mashal et al. [25]	DCNN: customized highly nonlinear scale-invariant network	EyePACS	85%	91%	They improved the efficiency and accuracy of categorization by creating architectural improvements to an existing CNN. DR reduces the number of steps in color fundus imaging

amount of data to train. To enhance the number of photos and overcome over fitting during the training step, most research employed data augmentation. To address the difficulty of data size and to evaluate the DL approaches, 74% of the research included in this paper used public datasets, while 26% used a mix of two or more public datasets as shown in Fig. 2.

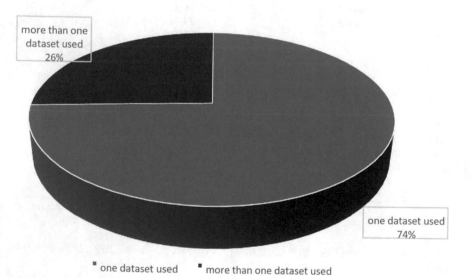

Fig. 2 Percentage of studies that used one or more datasets

4 Open Gaps and Challenges

The existing DR detection systems face several obstacles.

Early detection of signs is difficult: Traditional clinical methods do not allow for the early detection of DR. To develop a framework, it will detect DR in its early stages. Such a framework should not be time-consuming and costly.

Training models for high accuracy: To improve early-stage accuracy, researchers could train specialized models for certain stages and then ensemble the results to acquire high accuracy.

Feature reduction with maximal accuracy: To reduce the number of features taken into consideration while providing maximal accuracy which ultimately reduces complexity (time and space).

5 Proposed Framework

From Sects. 2 and 3, it is obvious that the algorithm's efficiency and accuracy varied significantly with the dataset employed. Various approaches for the prediction of DR in diabetic patients have been presented by various researchers; however, the prediction of DR in the early stages in literature has not been depicted more comprehensively. The limitation being in those techniques is that they are used to detect it in later stages of DR. So to overcome those limitations, an intelligent framework for early prediction of DR has been proposed in this paper.

Various key components involve data collection, data preprocessing, prediction model, and results.

Data collection: The data is the most important aspect of any study. We used the Kaggle dataset comprising 35,126 color fundus pictures, each with a size of 3888 X 2951 pixels. It comprises photographs from a variety of classifications based on the severity of the offense.

Dataset: distribution of different classes

Classes	Total no. of images
Class (0) normal	25,810
Class (1) mild	2443
Class (2) moderate	5292
Class (3) severe	873
Class (4) proliferative	708

Preprocessing: The dataset is first put through preprocessing step in which our aim is to enhance the images. The acquired images are enhanced so the data can help in the detection of cases of non-proliferative diabetic retinopathy (NPDR) and proliferative diabetic retinopathy (PDR), in their different stages.

To reduce the training costs, we resize each input image while keeping the aspect ratio in the initial preprocessing phase. Furthermore, we used up-sampling and down-sampling to balance the dataset. Minority classes are augmented by randomly cropping areas during the up-sampling process. To balance the samples of different classes, enhance the dataset, and avoid overfilling, flipping and 90o rotation are used. Extra instances of majority classes are deleted during down-sampling to fulfill the cardinality of the smallest class. Before flipping and rotating the images in the resulting distributions, each image is mean normalized to eliminate feature bias and reduce training them (Fig. 3).

Prediction model: CNNs will be utilized to categorize and classify images in the proposed model due to their high accuracy. It employs a hierarchical model that constructs a network, similar to a funnel, and then outputs a fully connected layer in which all neurons are coupled and the output is processed. The fundamental benefit of CNN over its predecessors is that it extracts important features automatically without the need for human intervention. As a result, CNN would be an excellent choice for the detection of diabetic retinopathy in its early stages.

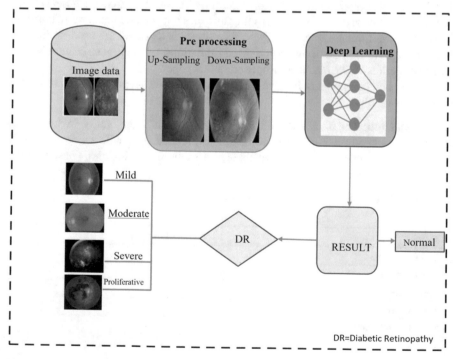

Fig. 3 I-DR framework [7]

Convolutional neural network (CNN): A convolutional neural network (CNN) is a form of artificial neural network that is specifically made to process pixel input and is used in image recognition and processing.

CNNs are effective artificial intelligence (AI) systems for image processing that employ deep learning to carry out both generative and descriptive tasks. They frequently use machine vision, which includes image and video identification, recommender systems, and natural language processing (NLP). A multilayer perceptron-like system that has been optimized for low processing demands is used by a CNN. An input layer, an output layer, and a hidden layer with several convolutional layers, pooling layers, fully connected layers, and normalizing layers make up a CNN's layers. A system that is significantly more effective and easier to train for image processing and natural language processing is produced by the removal of restrictions and increase in efficiency for image processing.

Convolutional layer, activation layer, pooling layer, and fully connected layer are four layers that work together to process and perceive data so that CNNs can classify images.

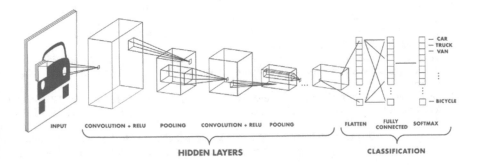

6 Conclusion

Predicting and diagnosing DR are a time-consuming process. For improved treatment of those suffering from DR, the diagnosis technique should be swift, efficient, and provide exact findings.

Diabetic retinopathy is a microvascular condition that diabetics might develop. Diabetic retinopathy is a condition that causes visual issues and can potentially lead to blindness. If a diabetic individual is not treated for a long time, diabetic retinopathy is more likely to develop. Diabetic retinopathy progresses to the point that it causes symptoms. Diabetics may be unaware that they are infected in the early stages of the virus. DR is diagnosed clinically by looking at it or by imaging techniques like fundus photography or optical coherence tomography. Various approaches for the prediction of DR in diabetic patients have been presented by various researchers, however; the prediction of DR in the early stages in literature has not been depicted more comprehensively. The primary goal of this work is to provide a short review of the existing DR detection strategies that employ retina data. As a result, the current study is an essential step toward the development of an intelligent diagnostic model that must be adopted soon for the early and accurate diagnosis of DR, which will be a great aid to the research community and health professionals in designing better medicine. The focus of this research is to design an intelligent framework for the prediction of DR in the early stages of diabetic patients.

References

1. Borys et al (2020) Deep learning approach to diabetic retinopathy detection. arXiv: 2003.02261v1 [cs. LG], 3 Mar
2. Amol et al (2015) Detection of diabetic retinopathy in retinal images using MLP classifier. In: Proc 2015 IEEE
3. Kranthi et al (2015) Automatic diabetic retinopathy detection using digital image processing. In: International Conference on Communication and Signal Processing, April 3–5, India
4. Ankita et al (2018) Diabetic retinopathy: present and past. In: International Conference on Computational Intelligence and Data Science (ICCIDS 2018)
5. Ling et al (2020) A deep learning system for detecting diabetic retinopathy across the disease spectrum. NATURE Communications
6. Mike et al (2018) Replication study: development and validation of a deep learning algorithm for detection of diabetic retinopathy in retinal fundus photographs. arXiv: 1803.04337v3[cs.CV], 30 Aug
7. Sehrish et al (2019) A deep learning ensemble approach for diabetic retinopathy detection. Received August 27, 2019, accepted October 4, 2019, date of publication October 15, 2019, date of current version October 29
8. Harry et al (2016) Convolutional neural networks for diabetic retinopathy. In: International Conference On Medical Imaging Understanding and Analysis 2016, MIUA 2016, 6–8 July, Loughborough, UK
9. Hajeb S, et al (2012) Diabetic retinopathy grading by digital curvelet transform. Hindawi Publishing Corporation Comput Mathemat Methods Med 2012(761901):11 p. https://doi.org/10.1155/2012/761901. Received 24 May 2012; Accepted 30 July
10. Wejdan et al (20020) Diabetic retinopathy detection through deep learning techniques: a review. Shalash Information Technology Department, University of King Abdul Aziz, Jeddah, Saudi Arabia, Informatics in Medicine Unlocked
11. Enrique et al (2019) Automated detection of diabetic retinopathy using SVM. Dept. de Electrica y Electr ´onica ´ Univ. de las Fuerzas Armadas ESPE Sangolqu´i, Ecuador
12. Hoda et al (2021) Automated detection and diagnosis of diabetic retinopathy: a comprehensive survey. Theoretical and Experimental Epistemology Lab, School of Optometry and Vision Science, University of Waterloo, Waterloo, ON N2L 3G1, Canada
13. Kele et al (2017) Deep convolutional neural network-based early automated detection of diabetic retinopathy using fundus image. School of Information and Communication, National University of Defense Technology, Wuhan 430019, China
14. Abhishek et al (2020) Automated detection of diabetic retinopathy using convolutional neural networks on a small dataset. Pattern Recognition Lett 135, July
15. Ratul et al (2017) Automatic detection and classification of diabetic retinopathy stages using CNN. IEEE
16. Carson et al (2018) Automated detection of diabetic retinopathy using deep learning. AMIA Jt Summits Transl Sci Proc v.2018; 2018 PMC5961805

17. Mohammad et al (2020) Exudate detection for diabetic retinopathy convolutional neural networks, Article ID 5801870. https://doi.org/10.1155/2020/5801870
18. Rishab et al (2017) Automated identification of diabetic retinopathy using deep learning 124(7):962–969, July
19. Shankar K, Sait ARW, Gupta D, Lakshmanaprabu S, Khanna A, Pandey HM (2020) Automated detection and classification of fundus diabetic retinopathy images using synergic deep learning model. Pattern Recognit Lett 133:210–216
20. Arcadu F, Benmansour F, Maunz A, Willis J, Haskova Z, Prunotto M (2019) Deep-learning algorithm predicts diabetic retinopathy progression in individual patients. NPJ Digit Med 2:92
21. Wang J, Bai Y, Xia B (2020) Simultaneous diagnosis of severity and features of diabetic retinopathy in fundus photography using deep learning. IEEE J Biomed Health Inform 24:3397–3407
22. Ali A, Qadri S, Mashwani WK, Kumam W, Kumam P, Naeem S, Goktas A, Jamal F, Chesneau C, Anam S et al (2020) Machine learning-based automated segmentation and hybrid feature analysis for diabetic retinopathy classification using fundus image. Entropy 22:567
23. Bora A, Balasubramanian S, Babenko B, Virmani S, Venugopalan S, Mitani A, Marinho GDO, Cuadros J, Ruamviboonsuk P, Corrado GS et al (2021) Predicting the risk of developing diabetic retinopathy using deep learning. Lancet Digit Health 3:e10–e19
24. Gangwar AK, Ravi V (2020) Diabetic retinopathy detection using transfer learning and deep learning. In: Evolution in computational intelligence. Springer, Singapore, pp 679–689
25. Mashal et al (2017) Detecting diabetic retinopathy using deep learning. IEEE

Healthcare Data Analysis Using Proposed Hybrid Harmony Genetic Diagnostic Model

Manju Sharma◉ and Sanjay Tyagi◉

Abstract Healthcare data analysis is a decision-based process that depends on past historical data. Analyzing such a huge volume of data as well as retaining accuracy in such sensitive data is an essential task. Henceforth, there is a necessity to design an intelligent diagnostic system that can assist humans in a proper diagnosis of disease along with making the precise results. In this paper, a new improved harmony search-based diagnostic model has been proposed for healthcare dataset analysis. Before integrating the harmony search for modeling the system, some enhancements have been done in the harmony improvement phase of the classical harmony algorithm by parameter tuning. Moreover, instead of random selection, the harmony vector is selected by a polygamous mechanism, and the crossover operator is used to increase the exploitation capability of the harmony search algorithm. The performance of proposed hybrid harmony search genetic algorithm (HHSGA)-based disease diagnostic clustering model has been analyzed based on precision, recall, accuracy, and G-measure quality indicators. The quality of the proposed algorithm is compared with other basic metaheuristic models by implementing it on four UCI repository healthcare datasets. Simulation results show that the proposed model is more robust and efficient in terms of achieving highest accuracy and correct classification of data points into clusters.

Keywords Diagnostic model · Harmony search · Clustering · Metaheuristic

1 Introduction

At present time, the medical informatics field is getting an extensive consideration among the researchers in developing an efficient and accurately predictive disease diagnostic model. The main aim of healthcare institutions is to provide quality services at reasonable costs and in minimum time. These quality services specify

M. Sharma (✉) · S. Tyagi
Department of Computer Science and Applications, Kurukshetra University, Kurukshetra, India
e-mail: manjusharmaknl@gmail.com

© The Author(s), under exclusive license to Springer Nature Singapore Pte Ltd. 2023   505
Y. Singh et al. (eds.), *Proceedings of International Conference on Recent Innovations in Computing*, Lecture Notes in Electrical Engineering 1011,
https://doi.org/10.1007/978-981-99-0601-7_39

the proper healthcare data analysis because an inaccurate medical decision some-times leads to the highly intolerable loss of lives. Analysis of the healthcare data usually helps the medical professionals in the early diagnosis of chronicle diseases. Several diagnostic models, expert systems, and medical decision systems have been described in the literature for the analysis of data from the healthcare domain as well as early diagnosis of diseases. Evolutionary and swarm intelligence algorithms are nature-inspired algorithm that mimic the process of natural phenomenon. In the biomedical field, these algorithms are emerging as a field of research for solving the optimization problems [1] (like clustering, classification, and prediction).

Clustering is an unsupervised classification or data analysis technique that assembles the data items in such a way that the data items present in a group are dissimilar from the data items present on other groups [2]. Numerous evolutionary and swarm-based diagnostic models have been proposed by researchers for the early diagnosis of a disease. Nevertheless, accuracy and precision are still challenging issues, exclusively for healthcare datasets. Hence, there is a requirement of designing an efficient intelligent system that can assist humans to make accurate judgements. The key contribution of this paper is to develop an effective diagnostic model using the concept of hybridization in harmony search using genetic algorithm operators.

Harmony search (HS) algorithm developed by Geem et al. [3] is a well-known metaheuristic technique inspired by the improvisation concept of musicians. It is gaining a wide range of popularity among researchers in different application areas like medical, agriculture, scheduling, image processing, communication system, etc. [4]. Effectiveness of a metaheuristic algorithm depends on the proper balance between its exploration and exploitation phases. Imbalance in diversification and intensification phases sometimes leads to a trap in local optima that result in the problem of premature convergence. In this paper, before employment of harmony search algorithm as a diagnostic model, some improvements have been incorporated in the traditional algorithm by incorporating genetic operators in the harmony improvisation step along with parameter tuning. The proposed diagnostic algorithm is termed as hybrid harmony search genetic algorithm (HHSGA). The efficiency of this proposed model is evaluated by implementing it on four real-life datasets of the medical domain from the UCI repository [5]. Moreover, the effectiveness of the proposed algorithm is also checked by comparing it with other state-of-the-art metaheuristic algorithms like genetic algorithm (GA), particle swarm optimization (PSO), and HS.

The remaining sections of this paper are as follows: Section 2 presents the basic harmony search algorithm and related work. Section 3 provides the application of the proposed hybrid harmony genetic search algorithm as a diagnostic model. Section 4 covers the experimentation analysis, and simulation results followed by a statistical analysis of the algorithm. Section 5 discusses the conclusion and future aspects.

2 Background

This section gives an overview of work related to diagnosis of diseases along with brief overview and various issues related to harmony search.

2.1 Related Work

Al-Muhaideb and Menai [6] discussed various approaches related to prognosis and diagnosis of disease with medical data using metaheuristic algorithms. The study includes various learning models, an appropriate selection of metaheuristic techniques as well as performance indicators. Extraction of overlapped data and structure discovery in medical datasets is a difficult process. Khanmohammadi et al. [7] designed a hybrid technique in 2017 by combining k-harmonic means and overlapping k-means algorithms (KHM-OKM) to resolve the above-mentioned problem. The resulting outcomes confirm the effectiveness of the proposed hybrid approach in extracting overlapped data and also resolving sensitivity problems. The Cuckoo search algorithm was adopted by Gadekallu and Khare [8] for feature reduction by using a rough-set-based approach for heart disease datasets. The results showed the effectiveness of the proposed approach as compared to other algorithms.

Accuracy is one of the most important metrics in medical diagnosis. Noureddine et al. [9] rectified the accuracy issues of most of the medical datasets by using a symbiotic organism search mechanism. The proposed model significantly improved the death rate count of earlier treatment of diseases. Inaccuracy in any diagnosis model is mainly due to improper parameter setting and feature selection. These disputes are resolved by Wang and Chen [10] using chaotic whale optimization algorithm for medical datasets. Kaur and Kumar [1] designed a new diagnostic model using water wave optimization algorithm. Furthermore, the premature convergence issue in the basic algorithm was also resolved using a decay operator. The proposed model achieved higher accuracy as compared to other techniques.

2.2 Harmony Search Algorithm

Harmony search algorithm is a well-known metaheuristic algorithm that imitates the search mechanism of the musicians for finding a perfect harmony. It was firstly introduced by Geem in 2001. Harmony memory and collection of pitches are similar to the candidate solutions and decision variables, respectively. Here, each decision variable provides a value for generating a global optimum solution. Just like musicians, the harmony search algorithm also undergoes a harmony improvisation phase in each iteration by changing the value of decision variables. The basic algorithm consists of the following steps:

Step 1: Initialize the objective function and parameters: Define the objective function of the problem, i.e., to minimize (or maximize) the fitness function $f(X)$. Furthermore, the values of control parameters of the algorithm like harmony memory consideration rate (HMCR), harmony memory size (HMS), pitch adjustment rate (PAR), bandwidth (BW), and maximum number of iteration (MaxIt) or stopping criteria are also initialized.

Step 2: Initialization of harmony memory: The algorithm starts with the random initialization of harmony memory (HM) through a random harmony vector (HM_{ij}), where j is the decision variable of ith harmony vector and is defined as follows:

$$HM_{ij} = LB_j + rand(0, 1) * \left(UB_j - LB_j\right) \tag{1}$$

where $i \in (1, \text{ HMS})$, $j \in (1, d)$ and d is the dimension of the problem. UB and LB are the upper and lower bounds of variables, respectively.

Step 3: Harmony improvisation: To generate a new pitch and a better harmony, HS is governed by three rules. In the first rule, a new pitch is randomly selected from the existing pitches in harmony memory. In the second rule, the algorithm selects an existing random pitch from the memory and fine-tunes it through a random bandwidth (BW) that specifies its variations. Lastly, in the third rule, the algorithm generates a new pitch. HMCR is a probabilistic parameter that decides whether to apply the first two rules or the third one for improvement of the pitch. Similarly, PAR is also a probabilistic parameter that governs the selection between the first and the second rule for pitch generation. The detailed procedure is given in Algorithm 1.

Step 4: Harmony memory updation: New harmony vector generated after the above steps is compared with the worst harmony vector. If the new one is superior to the compared harmony, then it is replaced by the worst one in harmony memory.

Step 5: Check termination criteria: Repeat steps 3 and 4 until the improvisation stopping criteria are reached. Otherwise, stop the algorithm and get the optimal solution.

Algorithm 1: Harmony Search

Begin
Define the parameters HMS, HMCR, PAR, BW, d(Dimension), MaxIt
Initialize the harmony memory randomly using Eq. 1.
Calculate the value of each harmony vector using fitness function $f(X)$.
Set Itr = 0
While(Itr < MaxIt)
    For *j = 1 to d*
        If *(rand(0,1) < HMCR)* ***then // harmony memory consideration***
            $X_j^{new} = X_j^m$ where $m \in \{1, 2, 3 \dots HMS\}$
            If *(rand(0,1) < PAR)* ***then // Pitch adjustment***
                $X_j^{new} = X_j^{new} + rand(0, 1)*BW$

        End If
      else
          $X_j^{new} = X_j^{new} + rand(0, 1)*(UB_j - LB_j)$ **//Random Selection**
      End If
    End For
      Calculate $f\left(X_j^{new}\right)$ and select X_j^{worst} harmony vector from HM

    If $f\left(X_.^{new}\right) < f\left(X_.^{worst}\right)$ ***then***

        $X_j^{worst} = X_j^{new} \& f\left(X_j^{worst}\right) = f\left(X_j^{new}\right)$

    End If
    Itr = Itr + 1
End while
Return global optimal harmony vector
End

HS has a strong exploration capability due to the presence of random selection components in harmony improvement phase. But it usually suffers from slow convergence speed and local optima problems due to the lack of global search information. It also suffers from poor intensification, due to its inability to make small moves [11]. The performance of a population-based search algorithm depends upon its ability to balance between the diversification and intensification phases. Therefore, to make HS a more efficient and effective algorithm, several modifications are incorporated into the classical HS algorithm that has been discussed in the next section.

3 Proposed Hybrid Harmony Search-Based Diagnostic Model

3.1 *Improved Harmony Search Algorithm*

This paper introduces a novel improved hybrid HS algorithm by using some operators of the GA during harmony improvisation and new harmony generation phases of the

HS. The new algorithm is named as HHSGA algorithm, which is a hybrid of both HS and GA.

Improvement 1: During the pitch adjustment stage, the value of the BW parameter is generally taken as constant by most of the researchers. However, its value affects the modification of the existing harmony vector selected from the harmony pool. In the proposed approach, initially, the value of BW is set according to maximum and minimum values of attributes in the datasets which gradually changed after each iteration as follows:

$$\text{Initialize BW}(1) = 0.02 * (\text{UB} - \text{LB}) \tag{2}$$

$$\text{BW}(\text{Itr} + 1) = \text{BW} * c \,|\, c \text{ represents a constant} \tag{3}$$

Improvement 2: In standard HS, the new harmony generation step follows the random selection mechanism for generating a new pitch. Consequently, at every iteration when a new harmony is generated, the algorithm follows the random search procedure for selecting a new harmony that enhances its exploration capability. But, at the same time, algorithm needs to deploy its exploitation capability as well. To overcome the above-mentioned shortcoming, a new hybrid approach is proposed that uses the polygamous selection [12] procedure selection of a harmony vector from the harmony memory and then improves the exploitation capability by using an arithmetic recombination operator between the selected and the worst harmony vector. If the child's harmony is better than the parent's harmony, then replace it with the worst harmonies in the harmony pool. The pseudo-code of the recombination operator is given as follows:

$$X^P = \text{Polygamous Selection}(P); \,//\, P \text{ is the harmony pool} \tag{4}$$

$$X^w = \text{Worst}(P); \tag{5}$$

$$\left(X^{c1}, X^{c2}\right) = \text{Arithmetic Crossover}\left(X^P, X^{\text{worst}}\right) \tag{6}$$

Arithmetic Crossover (X^P, X^{worst})

Input: Initialize α coefficient **Output:** Child harmonies: X^{c1}, X^{c2}

Begin

$X^{c1} = \alpha.*X^P + (1+\alpha).*X^{\text{worst}}$

$X^{c2} = \alpha.*X^{\text{worst}} + (1+\alpha).*X^P$

End

3.2 HHSGA as a Clustering Technique

In this paper, the proposed HHSGA approach is used as a clustering technique for the classification of datasets for the diagnosis of a disease. In the proposed approach, each harmony vector is a sequence of real numbers that represents the cluster centers (K). In d-dimensional space, harmony vector is represented as ($K*d$). Consider a dataset having $k = 2$ and $d = 3$. Then, each harmony vector is displayed as follows:

5.1	3.5	1.4	Cluster center 1
4.9	3.0	1.4	Cluster center 2

The classification algorithm aims to find the similarity/dissimilarity pattern between two data points. In this paper, Euclidean distance metric is used as a measurement for similarity measure, and the sum of the square error is considered as an objective function. The main aim of the clustering algorithm is to minimize the objective function that can be represented as follows:

$$f(P, C) = \sum_{m=1}^{n} \min\{||p_m - c_j||^2, \, j = 1, 2 \ldots, k\} \tag{7}$$

Here, the value min-term represents the distance measurement between data point p_m and cluster center c_j. The proposed HHSGA as a clustering technique is given as follows:

HHSGA Clustering Algorithm

Input:

  Initialize cluster centers(k),HMS, HMCR,PAR, MaxIt).
  Initialize BW as in Eq. 2.

Output:

  Cluster centers and cluster labeled data points.

(continued)

(continued)

Begin

Randomly initialize the harmony vectors from HM with k number of cluster centers. Calculate the distance of each data point from the center of cluster by using Eq. 7. Assign data objects to clusters having minimum value of fitness function

While *(Itr < MaxIt) /*Apply improved harmony search algorithm for generating the new cluster centers*/*

  For *j = 1 to d.*

   If *I(rand(0,1) < HMCR)* **then**

    $X_j^{new} = X_j^m m \in random\, cluster\, from\, the\, pool$

    If *(rand(0,1) < PAR)* **then**

     Improve X_j^{new} *by using the modified value of BW by using* Eqs. 2 and 3.

  End If

  Else

   Select X_j^P by using selection operator as in Eq. 4.

   Apply Recombination operator between X_j^P, X_j^{worst} and generate *Offspring* X_j^{c1}, X_j^{c2} using Eq. 5 and 6

  End If

 End For

Calculate fitness of X_j^{C1}, X_j^{C2}

If $f\left(X_j^{C1}, X_j^{C2}\right) < f(X_j^P, X_j^{worst})$ **then**

 Replace offspring with the parents

End if

Evaluate the centers of the new cluster and replace the existing ones

 Itr = Itr + 1

End while

Partition the data points of the dataset using the global cluster centers.
Return global best solution and labeled data points.

End

In the next section, the above-mentioned HHSGA approach is used as a basic technique for designing a disease diagnostic model.

3.3 Proposed HHSGA-Based Diagnostic Model

Many diagnostic models have been proposed by researchers for disease diagnosis. In healthcare datasets, the main aim of the diagnostic model is to maintain a high level of accuracy in the proper classification of datasets among different clusters. Henceforth, there is a requirement to design an intelligent diagnostic system that assists human in the proper diagnosis of diseases and can make precise results. In this paper, the proposed hybrid HHSGA algorithm is used as a diagnostic model for efficient classification of datasets and for generating the optimal clusters. The proposed model is shown in Fig. 1.

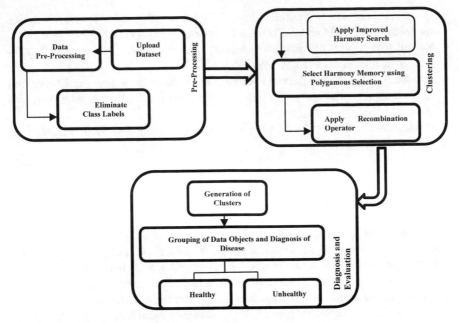

Fig. 1 Proposed HHSGA disease diagnostic model

4 Experimental Analysis

4.1 Dataset Used

See Table 1.

4.2 Cluster Quality Measures

To evaluate the performance of the model, the proposed algorithm is compared with other state-of-the-art algorithms like GA, PSO, and HS using four cluster quality measures, namely precision, recall, accuracy, and G-measure. These metrics are mathematically described as follows:

Table 1 Dataset description [5]

Dataset	Classes	Instances	Attributes
LD (Bupa)	2	345	6
Breast cancer	2	683	9
Haberman	2	306	3
Thyroid	3	215	5

- Precision: It envisions the reliability of the diagnostic model in classifying the model as positive. Mathematically, it is represented as TP/(TP + FP).
- Recall: It visualizes the model's capability to discover positive samples. Mathematically, it is represented as TP/ (TP + FN).
- Accuracy: It represents the ratio between the number of correct predictions and total number of predictions. It shows the number of times the model was overall correct [13].
- G-Measure: It is the mean of recall and precision and is represented as $\sqrt{Precision*Recall}$ [14].

4.3 Simulation Results

Results in Tables 2, 3, 4, and 5 show that the proposed HHSGA model provides more significant results as compared to other compared models. It is clear from the results that the proposed HHSGA diagnostic model provides a higher level of accuracy than the other techniques. G-measure is also an important indicator for checking the diagnostic performance of a model. It is clear from the simulation results that the proposed HHSGA model provides a better G-measure value for the proper diagnosis of a disease. It represents the true positive rate in the diagnosis of a disease. From observations, it is clear that the proposed model has better precision and recall value that makes it more efficient and robust as compared to other compared models to diagnose diseases.

Table 2 Simulation results of different metaheuristic models for the Bupa (LD) dataset

Quality metrics/algorithm	Precision mean (SD)	Recall mean (SD)	Accuracy mean (SD)	G-measure mean (SD)
GA	0.5740 (0.0028)	0.5447 (0.0020)	0.5147 (0.0020)	0.5593 (0.0022)
PSO	0.5680 (0.0158)	0.5405 (0.0116)	0.4872 (0.0153)	0.5541 (0.0137)
HS	0.5744 (0.0022)	0.5459 (0.00174)	0.4944 (0.0020)	0.5600 (0.0019)
HHSGA	0.6237 ($1.17e^{-16}$)	0.5767 ($1.17e^{-16}$)	0.5246 ($1.17e^{-16}$)	0.5998 (0.00)

Table 3 Simulation results of different metaheuristic models for the breast cancer dataset

Quality metrics/algorithm	Precision mean (SD)	Recall mean (SD)	Accuracy mean (SD)	G-measure mean (SD)
GA	0.9607 (0.0032)	0.9495 (0.0089)	0.9593 (0.0055)	0.9550 (0.0060)
PSO	0.9597 (0.0051)	0.9480 (0.0127)	0.9581 (0.0081)	0.9638 (0.0089)
HS	0.9630 (0.0013)	0.9568 (0.0023)	0.9636 (0.0016)	0.9599 (0.0018)
HHSGA	0.9640 ($1.17e^{-16}$)	0.9584 ($1.17e^{-16}$)	0.9648 ($1.17e^{-16}$)	0.9612 (0.00)

Table 4 Simulation results of different metaheuristic models for the thyroid dataset

Quality metrics/algorithm	Precision mean (SD)	Recall mean (SD)	Accuracy mean (SD)	G-measure mean (SD)
GA	0.7209 (0.0198)	0.6664 (0.0420)	0.6348 (0.0796)	0.6930 (0.0303)
PSO	0.6327 (0.1598)	0.3643 (0.0387)	0.6553 (0.1120)	0.6407 (0.0629)
HS	0.7269 (0.0655)	0.7211 (0.0215)	0.6753 (0.0737)	0.7308 (0.0425)
HHSGA	0.7385 (0.0249)	0.7236 (0.0035)	0.6883 (0.0014)	0.7239 (0.0143)

Table 5 Simulation results of different metaheuristic models for the Haberman dataset

Quality metrics/algorithm	Precision mean (SD)	Recall mean (SD)	Accuracy mean (SD)	G-measure mean (SD)
GA	0.5049 (0.0135)	0.5064 (0.0172)	0.5078 (0.0071)	0.4602 (0.0153)
PSO	0.5041 (0.0121)	0.5053 (0.0155)	0.5183 (0.0067)	0.5047 (0.0138)
HS	0.4996 (0.0157)	0.4940 (0.0189)	0.5169 (0.0250)	0. 5056 (0.0208)
HHSGA	0.5150 $(1.17e^{-16})$	0.5192 $(1.17e^{-16})$	0.5196 $(1.17e^{-16})$	0.5171 $(1.17e^{-16})$

Figures 2a–d illustrate the categories of disease related to a dataset using the HHSGA model.

The HHSGA model categorizes the liver disorder data point into two categories such as (a) patients having liver disorder and (b) patients having no liver disorder. Data objects related to the thyroid dataset have been categorized into (a) normal, (b) hypothyroidism, and (c) hyperthyroidism clusters using the proposed HHSGA model. Furthermore, the breast cancer dataset is clustered into malignant or benign categories, and the Haberman dataset is clustered into two categories, i.e., patients who survived after surgery or not.

4.4 Statistical Analysis

To analyze the performance of the proposed HHSGA model statistically, the Friedman test has been carried out on the mean value of accuracy and G-measure indicators [15]. This test signifies whether there is a difference in the performance of the HHSGA model and other compared models or not [16]. The null hypothesis (H_0) has been set that specifies that there is no difference between the HHSGA and other models. Table 6 illustrates the average ranking of models based on accuracy and G-measure indicators. The level of confidence (α) is set to be 0.05. p-value reported in Table 7 is less than α that claim the rejection of null hypothesis and indicates that there is a considerable difference between the proposed and compared models.

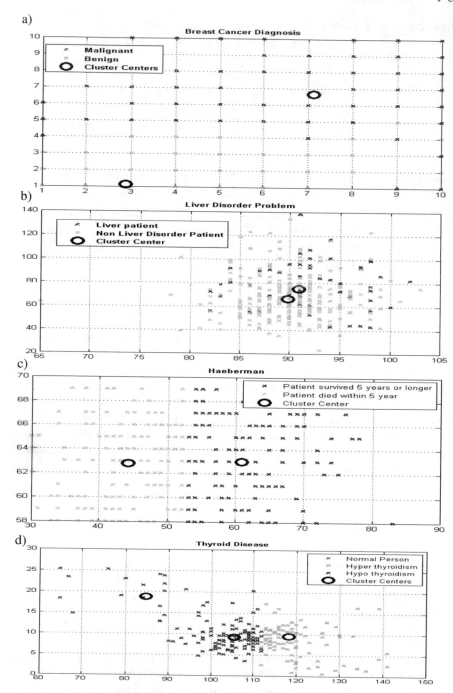

Fig. 2 Data-points classifications **a** breast Cancer **b** LD **c** Haberman **d** Thyroid datasets

Table 6 Ranking of algorithms based on accuracy (Acc) and G-measure (GM)

Datasets/algorithms	GA		PSO		HS		HHSGA	
	Acc	GM	Acc	GM	Acc	GM	Acc	GM
Bupa (LD)	2	3	4	4	3	2	1	1
Breast cancer	3	4	4	2	2	3	1	1
Thyroid	4	3	3	4	2	1	1	2
Haberman	4	4	3	3	2	2	1	1
Average ranking	3.25	3.5	3.5	3.25	2.25	2.0	1	1.25

Table 7 Friedman test summary

Measure	Statistical value	Critical value	Hypothesis	p-value
Accuracy	9.3	7.814728	Rejected	0.025557
G-measure	8.1	7.814728	Rejected	0.04399

5 Conclusion and Future Scope

In this paper, a hybrid harmony search and genetic algorithm-based diagnostic model are proposed for the classification of healthcare datasets and diagnosis of a disease. To resolve the problem of poor exploitation capability of the basic harmony search algorithm, some improvements have been made during the selection steps of the harmony vector. Instead of random selection, a polygamous selection followed by a recombination operator has been incorporated into the harmony improvisation steps. The proposed algorithm is used as a clustering technique for the classification of healthcare data points into clusters that make it applicable as a disease diagnostic model. The performance of the proposed model has been analyzed using four healthcare datasets and is compared with other state-of-the-art models. Simulation and statistical analysis using the Friedman test revealed that the proposed HHSGA model is more robust and efficient in accurate diagnosis of a disease. In future, this work can be extended for other applications like protein synthesis, software engineering, etc., using another hybrid approach by integrating the concept of multi-objective optimization in the proposed technique.

References

1. Kumar DY, Kaur A (2021) Healthcare data analysis using water wave optimization-based diagnostic model. J Info Commun Tech 20:457–488
2. Hartigan JA (1975) Clustering algorithms. Wiley
3. Geem ZW, Kim JH, Loganathan GV (2001) A new heuristic optimization algorithm: harmony search. Simulation 76:60–68

4. Hasan BHF, Abu Doush I, Al Maghayreh E, Alkhateeb F, Hamdan M (2014) Hybridizing harmony search algorithm with different mutation operators for continuous problems. Appl Math Comput 232:1166–1182

5. UCI machine learning repository: data sets. https://archive.ics.uci.edu/ml/datasets.php

6. Al-Muhaideb S, El Bachir Menai M (2013) Hybrid metaheuristics for medical data classification. Hybrid Metaheuristics, 187–217. https://doi.org/10.1007/978-3-642-30671-6_7

7. Khanmohammadi S, Adibeig N, Shanehbandy S (2017) An improved overlapping k-means clustering method for medical applications. Expert Syst Appl 67:12–18

8. Gadekallu TR, Khare N (2017) Cuckoo search optimized reduction and fuzzy logic classifier for heart disease and diabetes prediction. IJFSA 6:25–42

9. Noureddine S, Zineeddine B, Toumi A, Betka A, Benharkat A-N (2022) A new predictive medical approach based on data mining and Symbiotic Organisms Search algorithm. Int J Comput Appl 44:465–479

10. Wang M, Chen H (2020) Chaotic multi-swarm whale optimizer boosted support vector machine for medical diagnosis. Appl Soft Comput 88

11. Talaei K, Rahati A, Idoumghar L (2020) A novel harmony search algorithm and its application to data clustering. Appl Soft Comput 92:106273

12. Sharma M, Chhabra JK (2019) An efficient hybrid PSO polygamous crossover based clustering algorithm. Evol Intel, 1–19. https://doi.org/10.1007/s12065-019-00235-4

13. Verma C, Stoffová V, Illés Z, Tanwar S, Kumar N (2020) Machine learning-based student's native place identification for real-time. IEEE Access 8:130840–130854

14. Sharma M, Chhabra JK (2019) Sustainable automatic data clustering using hybrid PSO algorithm with mutation. Sustain Comp: Inform Syst 23:144–157

15. Scheff SW (2016) Chapter 8—Nonparametric statistics. In: Scheff SW (ed) Fundamental statistical principles for the neurobiologist. Academic Press, pp 157–182. https://doi.org/10.1016/B978-0-12-804753-8.00008-7

16. Verma C, Stoffová V, Illés Z (2019) Prediction of students' awareness level towards ICT and mobile technology in Indian and Hungarian University for the real-time: preliminary results. Heliyon 5:e01806

E-Learning, Cloud and Big Data

Cost Estimation Model Using Fifth Generation Language Technique for Software Maintenance Project

Mohammad Islam, Nafees Akhter Farooqui, Syed Ali Mehdi Zaidi, and Mohammad Shafeeq

Abstract Software cost estimation is a standout among the huge demanding tasks in project management for new software. In any case, the procedure of estimation is unsure as it largely relies on certain qualities that are very hazy amid the beginning periods of improvement. This exploration is to give a method to software cost estimation that performs superior to different procedures on the precision of effort estimation. A soft computing procedure has been investigated to beat the vulnerability and error in estimation. This investigation is to expand the constructive cost model by intertwining the possibility of fuzziness into the estimation of size, method of improvement projects, and the cost drivers adding to the general advancement effort. The primary goal of the explorations is to examine the job of the fuzzy inference system method in enhancing the cost estimation precision utilizing COCOMO II by describing inputs variables utilizing fifth GL systems and contrasting their outcomes. The PROMISE dataset is utilized for the assessment of the fuzzy inference system (FIS) procedures. The examinations have been completed utilizing MATLAB simulation conditions.

Keywords Fuzzy logic · Fuzzy inference system · Maintenance cost estimation model · Software cost estimation

M. Islam · N. A. Farooqui (✉)
Department of Computer Science, Era University, Lucknow, Uttar Pradesh, India
e-mail: nafeesf@gmail.com

M. Islam
e-mail: mohdislam3@gmail.com

S. A. M. Zaidi
Department of Computer Science, Shia PG College, Uttar Pradesh, Lucknow, India
e-mail: samzaididr@gmail.com

M. Shafeeq
Department of Computer Science and Engineering, BBD University, Lucknow, Uttar Pradesh, India
e-mail: shafeeq.bbd@bbdu.ac.in

© The Author(s), under exclusive license to Springer Nature Singapore Pte Ltd. 2023
Y. Singh et al. (eds.), *Proceedings of International Conference on Recent Innovations in Computing*, Lecture Notes in Electrical Engineering 1011,
https://doi.org/10.1007/978-981-99-0601-7_40

1 Introduction

Fifth generation language or fifth GL even tends to be programming languages, which contains visual tools to develop a program. It utilizes visual and graphical advancement interface device to make the source language that is assembled with a third GL or fourth GL compiler. Visual programming enables you to see object-oriented structures and drag symbols to gather program squares. There are some important points as follows:

- A fifth-generation programming language is a high level and logic language. User knowledge bases, expert systems, and less programmer control.
- Fifth GL is a programming language dependent on tackling issues utilizing limitations given to the program, as opposed to utilizing a calculation composed by a software engineer.
- Most imperatives-based, logic programming languages, and some decisive languages are fifth GL.
- Fifth GL is utilized mostly in artificial intelligence inquire about.

Artificial intelligence (AI) is growing such a mind-blowing speed; occasionally, it appears to be supernatural. There is a supposition among analysts and designers that AI could develop so monstrously solid that it would be hard for people to control. People created AI frameworks by bringing into them each conceivable they could, for which the people themselves presently appear to be compromised. AI has many other areas as soft computing technologies, neural network, genetic algorithms, fuzzy logic modeling system, etc., for finding the precise expectation of software development cost estimation. The FIS technique has been adopted to predict maintenance cost.

2 Related Work

There is basic two category of models, such as algorithmic and non-algorithmic [1]. Everyone require inputs, a precise estimate of explicit traits, for example, lines of code and other cost drivers like range of abilities which are difficult to procure amid the beginning time of software development. In 1990s, non-algorithmic was conceived to extend estimating cost. Analysts have focus toward novel methodologies that delicate registering, for example, ANN, GA, and fuzzy logic [2, 3]. A portion of early works demonstrates that fuzzy logic offers an amazing etymological interpretation that ready to speak to imprecision in sources of inputs and outputs while giving huge learning ways to deal with model's structure. It is a procedure to take care of issues, which are too complex to be in any way seen quantitatively. It depends on fuzzy-set-theory. It gives a system to speaking to semantic builds, for example, many, low, medium, and high. It gives a deduction structure that empowers fitting man reasoning limits. Unexpectedly, the binary set hypothesis depicts crisp events that do or do not happen. This experiments perspective hypothesis that clarify if the events will happen evaluating the opportunity for which the given events are required happening [4, 5].

It is the rule-based system which is the core learning containing the implied Fuzzy IF THEN guidelines in which a couple of words are portrayed by consistent part works. It can be classified into three kinds: pure, Takagi and Sugeno's fuzzy logic with fuzzifiers and defuzzifiers. The enormous bit for planning produces crisp-data as for input and envisions crisp-data as for output. This was right off the bat created by Mamdani. It has been viably associated with a collection of mechanical techniques and customer things [6, 7].

It is the initial phase in the fuzzy inference process. This includes an area change where crisp inputs sources are changed into fuzzy inputs. Crisp inputs are careful information sources estimated by sensors and go into the control system for handling, for example, temperature, weight, etc. [8–10]. Each crisp input that will be prepared by the FIS has its gathering of membership functions or sets to which they are changed. This gathering of membership functions exists inside a vast expanse of talk that holds every single important value that the crisp input can have. The accompanying demonstrates the structure of membership functions inside a vast expanse of talk for a crisp input [11, 12].

The principal of FIS is concentrated at the capacity of fuzzy logic that show characteristic. This system contains fuzzy-rules worked for expert-knowledge and called fuzzy expert systems, contingent upon their last use. Before FIS, it was at that point applied to construct expert systems for recreation objectives [13, 14]. The master frameworks depended upon the classical-boolean-logic that was not appropriate for dealing with the sequentially to the fundamental procedure wonders. Fuzzy-logic enables continuous standards that be brought into expert-knowledge-based test systems [7, 15].

The Sugeno's initial tasks, a great deal of scientists, have been engaged with structuring fuzzy systems from databases. The means of fuzzy reasoning performed by FISs are as follows:

1. Comparison of input factors with the MF on the precursor portion to get the membership estimations of each phonetic mark. (Progression is frequently said fuzzification).
2. Connection of the membership values on the reason portion to get terminating quality (level of satisfaction) of each standard.
3. Production of certified ultimately (either fuzzy or crisp) or each standard relying upon terminating quality.
4. Composite the certified consequents to deliver a crisp output. (Progression is said defuzzification).

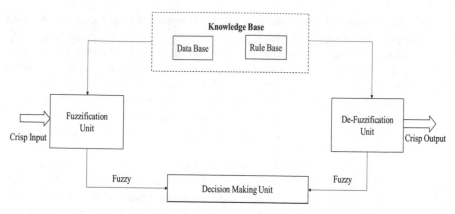

Fig. 1 Fuzzy inference system (FIS)

3 Research Methodology Used

The FIS is used to execute the differing preparing propels. Decisions were obliged making and adjusting FIS with fuzzy logic toolbox software using graphical instruments or command line capacities. The research will implement on third GL, fourth GL, and fifth GL using Mamdani FIS. Figure 1 is used as fuzzification and defuzzification.

The performance analysis and their corresponding results are compared. The results are analyzed using the criterion RMSE. Less value means that the result is more accurate.

- Select a specific kind of FIS (Mamdani).
- Define the variables for the input and output.
- Set input and output member functions.
- The data is now in rule editorial manager if-then rules.
- An explicit model structure is made, and parameters of input and output factors tuned to get the ideal output.

4 Proposed Model for Software Maintenance Cost Estimation

COCOMO II is utilized as the model-based to assess the software project cost. The model was developed by Mr. Boehm and Scattered in 1981 that utilizing aggregated data from 63 projects. It is a good manual that estimating maintenance cost for software. This proposed model implies with new factual methodologies and strategies that estimate the maintenance cost of software using fifth GL (fuzzy inference system) procedure.

The issue of software cost estimation is that everything considered and relies on single estimations of size, cost drivers, and scale factors. It is assessed dependent on recently finished projects that are fairly like the present projects [16–18]. Similarly, cost drivers and scale elements need through evaluation instead of doling out a fixed numbers value. To beat this condition, it is more brilliant to address these responsibilities to the kinds of fuzzy-sets, where the qualities of interval are utilized which is expressed through collection of membership functions like triangular MF, trapczoidal MF, and Gaussian MF [19].

The proposed fuzzy software cost estimation model is represented in the Fig. 2. Its principles contain phonetic factors identified with the undertaking. The FIS utilizes connecters "and/or" for COCOMO input factors that shape principles. The FIS incorporates many input software characteristics: seventeen cost drivers, five scale factors, one size (KDLOC), and one output as cost estimation (CE).

Fig. 2 Proposed maintenance cost estimation model using FIS techniques

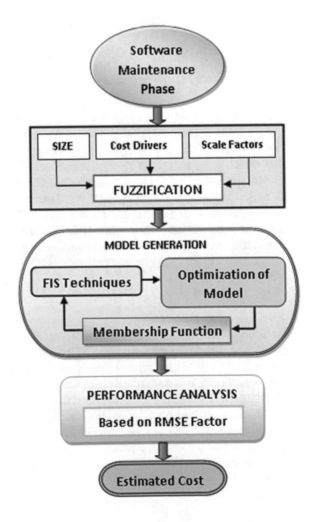

Fuzzy set takes all the input and convert into the phonetic values. For each cost driver, a different FIS is planned. Principles are created as cost drivers for forerunner parts that comparing effort-multiplier in the resulting portion. The defuzzified value for the effort-multiplier has kept for separate FIS. The scale factors are additionally fuzzified. The pcap means that programmer-capability is examined for an example. Programmer-capability for fuzzification depends on COCOMO II, calibrated model of post-architecture qualities.

Next, the model so obtained will be later subjected to optimization of its model parameters using fuzzy inference system optimization technique to arrive at better software cost estimation prediction accuracy. The fuzzy operators such union, inter-section, and complement shall be used. FIS races to create answers for progres-sive generations. Henceforth, the quality of the solutions in progressive generations improves. The procedure is ended when an ideal solution is found. The result has analyzed the criterion of root mean square error (RMSE) factor, which predicts the better software cost estimation with accuracy.

4.1 Data Used for Validation

The data used as input and output variables for ideal COCOMO II model advancement is given in Table 1. The dataset Table 2 is assembled from the examination of 40 soft-ware projects, which is adopted from Software Engineering Repository of PROMISE dataset which open access for researching reason. It comprises 26 attributes like seventeen standards COCOMO II characteristics cost drivers and five scale factors in the range that measure in thousand delivered source lines of code (KDLOC) direc-tions. The output of the model is the cost estimation (CE), measured in man-months. The estimated efforts using third GL, fourth GL, and fifth GL approaches obtained are tabulated and compared. The model equation is given as follows:

$$PM = A \times [Size]^{1.01} + \sum_{i=1}^{5} SF_i \times \prod_{i=1}^{17} EM_i \qquad (1)$$

Here, effort is indicated in terms of person-months (PM), A is a constant that is multiplicative, size is the projected-size of the software that expressed in KDLOC, EM_i ($i = 1, 2, 3, 0.17$) are effort-multipliers, and SF_i ($i = 1, 2, 3, 0.5$) are scale factors as exponent. It is a specific normal for product improvement that has impact that increments or decrements the measure for advancement effort [20]. There are team cohesion, process maturity, architecture/risk resolution, development, flexi-bility, and precedent Ness. All the effort-multipliers are gathered into 4 parts, which are project-factors, personnel, platform, and product. The items are utilized to modify the nominal effort.

Table 1 COCOMO II effort multipliers [5]

Input	Variables	Cost drivers and scale factors
1	"ACAP"	Analyst capability
2	"APEX"	Applications-experience
3	"CPLX"	Product-complexity
4	"DATA"	Size of database
5	"DOCU"	Documentations (life cycle need)
6	"FLEX"	Languages and tool-experiences
7	"PCAP"	Capability of programmer
8	"PCON"	Personnel continuity
9	"PERS"	Personnel capability
10	"PLEX"	Platform-experience
11	"PMAT"	Process-maturity-level (equivalent)
12	"PREC"	Precedentness for applications
13	"PREX"	Personal-experiences
14	"PVOL"	Volatility platform
15	"RELY"	Software-reliability (required)
16	"RESL"	Risk-resolution
17	"SITE"	Multisite-development
18	"STOR"	Main-storage (constraint)
19	"TEAM"	Team-cohesion
20	"TIME"	Time-execution (constraint)
21	"TOOL"	Software tools used
22	"SIZE"	SS
Output	Variables	Cost estimation (CE)

4.2 Fuzzy Inference System Rules Applied

Fuzzy principles for the fuzzy inference system dependent on COCOMO II are characterized with semantic factors in the fuzzification procedure. These principles depend upon connective "and" as between the input factors.

The Rules are defined as follows:
    if (rely is vl) then (effort is vl)
    if (rely is l) then (effort is l)
    if (prec is vl) then (effort is xh)
    if (pmat is vh) then (effort is l)

The following rules are used in Figs. 3, 4, and 5:
    if (pcap is very low) then (increased effort)
    if (pcap is low) then (increased effort)
    if (pcap is nominal) then (unchanged)

Table 2 Estimated-effort using that different MFs

RMSE using different generation languages approaches

P. no	Actual	COCOMO II	Third GL	Fourth GL	Fifth GL
1	2040	2215.24	2089	2075	1945
2	321	213.83	201.6	200.4	187.8
3	79	107.85	102	101.15	94.35
4	6550	7806.45	7372	7329.2	6854.7
5	724	733.16	690.9	687.14	643.9
6	121.6	149.6	166.68	165.25	125.90
7	117.6	144.9	98.87	99.39	114.13
8	33.2	42.9	42.56	42.79	36.27
9	36	40.1	48.10	48.36	36.64
10	35.2	41.5	39.79	40.01	37.38
11	8.4	18.4	6.12	5.65	5.24
12	10.8	19.3	3.63	3.65	13.81
13	352.8	369.9	444.90	447.09	340.60
14	70	81.9	64.98	65.83	63.44
15	72	79.8	45.32	45.71	61.45

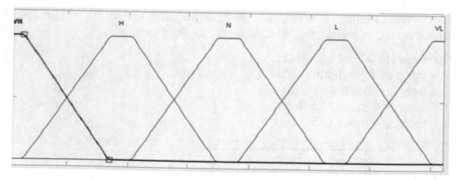

Fig. 3 PCAP fuzzification cost drivers using Gaussian MF

Fig. 4 PCAP fuzzification cost drivers using trapezoidal MF

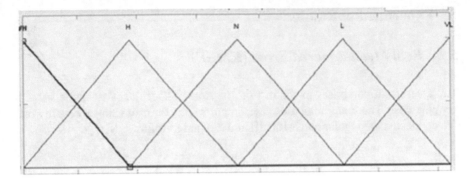

Fig. 5 PCAP fuzzification cost drivers using triangular MF

if (pcap is high) then (decreased effort)
if (pcap is very high) then (decreased effort)

Figure 6 expresses the graphical user interface that developed our model FIS. We can legitimately enter the qualities and get the relating effort. The studies have been carried out using MATLAB simulation environment.

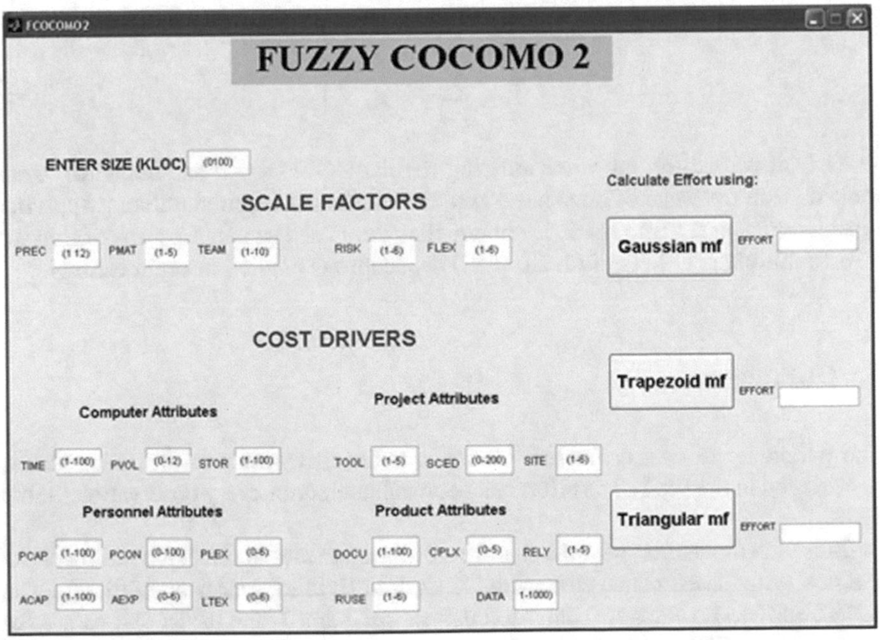

Fig. 6 Interface used for cost evaluation

5 Experimental Results

5.1 Root Mean Squared Error (RMSE)

The assessment comprises in contrasting the exactness of the calculated cost with genuine cost. There are numerous assessment scales for estimating software cost. We connected the regular one is RMSE and defined as follows:

$$\text{RMSE} = \left(\frac{1}{N} \sum_{i=1}^{N} \left(y_i - \hat{y}_i \right)^2 \right) \tag{2}$$

5.2 Mean Absolute Error (MAE)

It is a proportion of expectation precision of a forecasting method in insights, for instance, in pattern estimation. The most part communicates precision as a rate and is characterized by the equation:

$$M = \frac{1}{n} \sum_{t=1}^{n} \left| \frac{A_t - F_t}{A_t} \right| \tag{3}$$

The software effort got when utilizing regular COCOMO II and fuzzy MF were looked at. In the wake of breaking down, the outcomes acquired utilizing applying third, fourth, and fifth GL. It is shown that the cost evaluated by fuzzifying all effort-multipliers utilizing fifth GL (FIS) procedure is predicting better estimate.

6 Comparison

The parameter of cost estimation models for the assessment is the MAE that is represented in the Eq. 3. The effort has been calculated for every observation (Table 3).

Table 4 demonstrates with the chart representing to the similar examination of the real cost with the estimated cost using COCOMO II, third, fourth, and fifth GL. The RMSE and MAE values are calculated using Eqs. 2 and 3, and the RMSE values for all project for COCOMO II, third GL, fourth GL, and fifth GL are 1.2403, 1.0638, 1.075, and 0.9398, respectively. The MAE values are 0.1650, 0.1251, 0.1236, and 0.1183, respectively. This plainly demonstrates here is a reduction in the absolute errors; therefore, the proposed model is progressively reasonable for estimating cost.

Table 3 Comparison of MAE values

MAE using different generation languages approaches				
P. no	COCOMO II	Third GL	Fourth GL	Fifth GL
1	0.0859	0.024	0.0171	0.0465
2	0.3339	0.3719	0.3757	0.4149
3	0.3651	0.2911	0.2803	0.1943
4	0.1828	0.1171	0.1105	0.0385
5	0.0126	0.0457	0.0509	0.1106
6	0.2242	0.1531	0.1462	0.0746
7	0.0825	0.072	0.017	0.0451
8	0.1533	0.0866	0.0813	0.0113
9	0,4104	0.0630	0.064	0.0392
10	0.3651	0.2911	0.2803	0.1943
11	0.1828	0.1171	0.1105	0.0385
12	0.0126	0.0457	0.0509	0.1106
13	0.2242	0.1531	0.1462	0.0746
14	0.0815	0.020	0.017	0.0491
15	0.152	0.0846	0.0817	0.0121

Table 4 Comparison of RMSE and MAE factors

Comparison of cost estimation techniques	RMSE	MAE
COCOMO II model	1.2403	0.1650
Third GL model	1.0638	0.1251
Fourth GL model based on CBSD approaches	1.075	0.1236
Fifth GL model based on FIS technique	0.9398	0.1183

7 Conclusion

We conclude that the use of fuzzy logic in SCE yields more exact outcomes than the past experimental model methodology. The RMSE values of cost estimation using fifth GL based on FIS techniques give better outcomes for most extreme rules if other high-level language techniques will be used. It found that the FIS is achieving better as it shows a likely change in its intervals, and accomplished outcomes were nearer to the actual cost. FIS has the lowest MAE and highest accuracy of the three generation language software methodologies that studied.

Future work incorporates more current procedures, i.e., type-2 fuzzy-system can likewise be connected for increasingly precise forecasts of software. The above research work can be easily employed in the software industries.

References

1. Lagerström R, von Würtemberg LM, Holm H, Luczak O (2016) Identifying factors affecting software development cost and productivity. Softw Qual J 20:395–417
2. El Bajta M, Idri A, Fernandez-Alem JL, Ros JN (2015) Software cost estimation for global software development. In: 10th International Conference on Evaluation of Novel Software Approaches to Software Engineering, ENASE
3. Veeranjaneyulu N, Suresh S, Salamuddin S, Kim H-Y (2014) Software cost estimation on e-learning technique using a classical fuzzy approach. Int J Soft Eng Appl 8(11):217–222
4. Patil LV, Waghmode RM, Joshi SD, Khanna V (2017) Generic model of software cost estimation: A hybrid approach. IEEE Int Adv Comput Conf, IEEE, pp 1379–1384
5. Vu N (2010) Improved size and effort estimation models for software maintenance. University of Southern California
6. Marounek P (2012) Simplified approach to effort estimation in software maintenance. University of Economic, Prague, Faculty of Information and Statistics. J Syst Integration, 51–63
7. Wijayasiriwardhane T, Lai R, Kang KC (2011) Effort estimation of component based software development, a survey. IET Soft 5:216–228
8. Borade JG (2013) Software project effort and cost estimation techniques. Int J Adv Res Comp Sci Soft Eng 3(8):730–739
9. Chiu S (1994) Fuzzy model identification based on cluster estimation. J Intell Fuzzy Syst, 267–278
10. Farooqui NA, Ritika (2020) A machine learning approach to simulating farmers' crop choices for drought prone areas. In: Singh P, Panigrahi B, Suryadevara N, Sharma S, Singh A (eds) Proceedings of ICETIT 2019. Lecture Notes in Electrical Engineering, vol 605. Springer, Cham. https://doi.org/10.1007/978-3-030-30577-2_41
11. Aggarwal K, Singh Y, Chandra P, Puri M (2005) Measurement of software maintainability using a fuzzy model. J Comp Sci. ISSN 1549-3636
12. Sneed H (2004) A cost model for software maintenance & evolution. IEEE, 264–273
13. Patil LV, et al (2014) Develop efficient technique of cost estimation model for software applications. Int J Comp Appl 87(16):0975–8887, February
14. Maleki I, Ebrahimi L, Jodati S, Ramesh I (2014) Analysis of software cost estimation using fuzzy logic. Int J Foundations of Comp Sci Tech 4(3):27–41
15. Mukherjee S, Bhattacharya B, Mandal S (2013) A survey on metrics, models & tools of software cost estimation. Int J Adv Res Comp Eng Tech (IJARCET) 2(9):2620–2625
16. Jing QF, Zhu X-Y, Xiaoyuan XX, Baowen X, Shi Y (2017) Software effort estimation based on open-source projects: case study of Github 92:145–157, December
17. Patil PK (2015) A review on calibration factors in empirical software cost estimation (SCE) models. Int J Soft Eng Res Pract 3:1–7
18. Mohammad I, Vinodani K (2014) Development of a software maintenance cost estimation model: 4TH GL perspective. Int J Tech Res Appl, 6
19. Sehraab SK, Brara YS, Sukhjit NK, Sehra S (2017) Research patterns and trends in software effort estimation 91:1–21, November
20. Srivastava SK, Prasad P, Varma SP (2016) Evolving predictor variable estimation model for Web engineering projects. Comput Sci Eng, 68–89

Survey of Text Summarization Stratification

Arvind Jamwal, Pardeep Singh, and Namrata Kumari

Abstract The volume of data on the Internet has increased at an exponential rate during the previous decade. Consequently, the need for a method for converting this massive amount of raw data into meaningful information that a human brain can comprehend emerges. Text summarization is a common research technique that aids in dealing with a massive quantity of data. Automatic summarization is a well-known approach for distilling the important ideas in a document. It works by creating a shortened form of the text and preserving important information. Techniques for text summarizing are classified as extractive or abstractive. Extractive summarization methods reduce the burden of summarization by choosing a few relevant sentences from the original text. The implications of sentences are calculated using linguistic and statistical characteristics. This paper investigates extractive and abstract methods for text summarization. We will also explore many efforts in automatic summarization, particularly recent ones, in this article.

Keywords Text summarization · Extractive summarization · Abstractive summarization · Natural Language Processing

1 Introduction

Every day, the volume of textual data available is increasing digitally as web pages, news, academic paper, and articles. Despite having so much information available, it seems difficult to find the related information to a certain user as most of the data is not related to a certain user's need [1]. Automatic summarization helps in the extraction of needful information while rejecting the extraneous. It can also increase text legibility and cut down on the amount of time people spend searching. The ultimate goal is to

A. Jamwal · P. Singh · N. Kumari (✉)
National Institute of Technology Hamirpur, Hamirpur, India
e-mail: namrata_phd@nith.ac.in

P. Singh
e-mail: pardeep@nith.ac.in

© The Author(s), under exclusive license to Springer Nature Singapore Pte Ltd. 2023
Y. Singh et al. (eds.), *Proceedings of International Conference on Recent Innovations in Computing*, Lecture Notes in Electrical Engineering 1011,
https://doi.org/10.1007/978-981-99-0601-7_41

produce summaries that include the major themes in a clear and right manner, while not including unnecessary material [2–4]. There are multiple ways to do summarization based on input parameters and desired outputs, but there are mainly two summarization approaches in which we can divide the research, abstractive and extractive. Extractive summarization selects a few sentences or words or phrases from a source document and combines them to form a summary without modifying or changing the input sentences from the document. An abstractive summarization converts the relevant phrases collected from a text into an understandable and coherent semantic form, perhaps changing the original sentences. The combination of an extractive and abstractive summary is referred known as "hybrid text summarizing" [5].

In general, all automatic text summarizing systems have three phases in their processing design [6]. The initial stage is to identify the text's sentences, words, and other elements. In the processing phase, it transforms the input text into a summary using a text summarizing approach. Post-processing is the third phase, which entails rectifying problems in the produced draft summary [7]. Some recent reviews on automatic text summarizing have been published, with the majority focusing on extractive summarization approaches [8] since abstractive summarization is difficult and necessitates extensive Natural Language Processing (NLP). On the basis of the literature survey done in this paper, Fig. 1 is drawn. In this paper, we will be discussing the following arguments over text summarization: approaches/techniques used in summarization; the evaluation, used for text summarization; and advantages of using specific techniques.

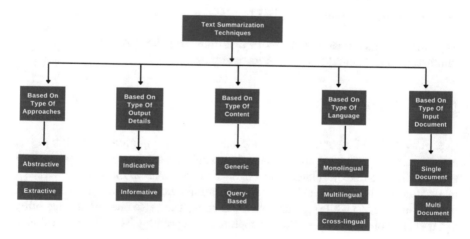

Fig. 1 Text summarization techniques

2 Types of Summarization Techniques

2.1 Based on Types of Approaches

1. In an extractive method without modifying the original content, extract notable phrases or words from the source documents and gather them to build a summary.
2. In the abstractive approach, new words are generated that are not found in the original input text document.

2.2 Depending on the Output Details

1. Indicative summary is being used to offer a fast overview for lengthy documents, which only includes points of the source text to entice the user for reading the entire work.
2. Informative might be used in place of the original document. It gives the user succinct information from the original material [8].

2.3 Based on Content

1. Generic summarizing is a system that may be utilized by any sort of user, and the summary is independent of the document's topic [8].
2. In a query-based, question-and-answer system in which the summary is decided by the user's inquiry is known as query-based summarization.

2.4 Based on the Language

1. A monolingual summarizing system accepts a single language as input and outputs a summary in that same language. For instance, consider an English language summarizing system.
2. In multilingual summarization, languages can be summed up using multilingual summarizing systems. Their input text and produced summary text, on the other hand, are written in the same language [2].
3. Multiple languages can be processed using a cross-lingual summarizing system. The output summary, on the other hand, is written in a different language from the input material, for instance, a summary of Bengali news in English [9].

2.5   Based on a Source Document

1. For single document summarizing, just one document is utilized as input.
2. Multi-document summarization accepts many documents on the same subject as input.

3   Extractive Summarization

In this form summary, just the most important and frequent terms are included from the original text. The sentences are scored, and those with the highest rating are included in the summary [10]. Figure 2 shows the architecture of the extractive text summarization as explained in [11].

3.1   TF–IDF Frequency Method

The TF–IDF is numerical insights that show how important a word is in a specific report. The word TF can relate to the repeated recurrence of the term in the report, and IDF can be defined as a metric that reduces the weight of repeating terms in the collection while increasing the weight of phrases that are encountered seldom. At this point, sentences are assessed in accordance with the things, and sentences with a high score are included in a rundown. One disadvantage of this method is that lengthier sentences generally have a higher word count since they include more words [12].

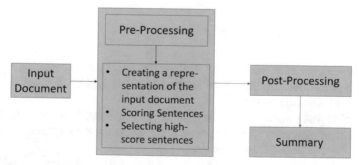

Fig. 2 Extractive text summarization techniques

3.2   Text Summarization with Fuzzy Logic

These text summarizing methods make use of a multi-valued system known as fuzzy logic. Because the logical concepts "one" and "zero" do not necessarily match the "real world," fuzzy logic [13] offers an efficient method for generating feature values for phrases that lie between these two values. Selecting a set of qualities for each sentence is the first step in sentence grading. The second step involves applying the fuzzy logic concept to each statement and assigning a score based on its relevance. This implies that each sentence is assigned a score between 0 and 1 depending on its characteristics [8].

3.3   Text Summarization with Neural Network

This method involves teaching neural networks to figure out which phrases should be contained in the summary. A 3-tiered Feed Forward neural_network is used [14].

3.4   Cluster-Based Method

It is natural to believe that summaries should cover various "themes" found in the papers. If the document collection for which the summary is being generated contains documents on completely diverse themes, document clustering becomes virtually mandatory in order to construct a relevant summary. Sentences are picked for their similarity to the theme of the cluster (C_i). The position of the sentence in the paper is the next consideration (L_i). The last criterion is it resembles the initial phrase of the text it belongs to (F_i) [15]. In this section, we have compared the summarization models based on a technique used by the model which are described in Table 1.

4   Abstractive Summarization

This type of text summarization involves the paraphrasing of the sentences and the generation of new sentences which are syntactically and semantically correct. Figure 3 shows the architecture of the abstractive text summarization.

Table 1 Comparison of models based on the technique used and their accuracy in terms of Rouge metric

Reference	Approach/technique	Remark	Dataset	Accuracy
[18]	Attentional encoder–decoder RNN	Multi-sentence summaries	CNN/DM and GigaWord	27.14
[19]	Pointer generator, coverage mechanism	Solved Inaccurate factual details	CNN/DM	31.06
[20]	Coverage mechanism	OOV words problem	XSum	24.53
[21]	Bidirectional RNN, LSTM	Solved limited awareness of the sentence context	CNN/DM	33.3
[22]	CNN	Utilize CNNs to allow parallelization over text data	CNN/DM and GigaWord	33.74
[23]	Attention mechanisms, RL	Quality summaries for long documents	CNN/DM	32
[24]	RL model	Significant speed-up in training and decoding	CNN/DM	32.41
[25]	Bottom-up attention	Training the content selection takes 1,000 sentences	CNN/DM	32.75
[26]	Reinforcement learning	Better overview than numerous baselines	CNN/DM	33.03
[27]	Transformers Encoder-Decoder part	Good semantic and context features of embeddings	CNN/DM and XSum	33.45
[27]	Transformer-based, decoder	Can produce a broad spectrum of summaries	GigaWord	36.69

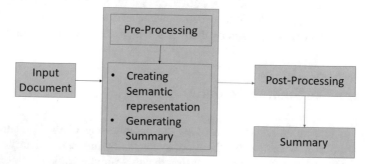

Fig. 3 Abstractive text summarization techniques

4.1 Tree-Based Method

To achieve abstractive summarization in tree-based methods, comparable phrases in the input with relevant information must be clustered and then worked with for the executive summary [14]. Similar words are grouped into trees, and dependency trees, a typical tree-based representation, are created using parsers. The trees are then created using a technique known as pruning linearization. Some of the sentence clusters in order to construct summary sentences [16].

4.2 Template-Based Method

In various fields, human summaries prefer to employ specific sentence forms. These are known as templates. Based on the style of the input document, the information in the input document is used to fill slots in applicable pre-defined templates document to conduct abstractive summarization [17]. To fill template slots, text samples can be retrieved using rules and linguistic clues [8].

4.3 Ontology-Based Method

Using an ontology, ontology-based techniques construct abstractive summarization from an input document [17]. Many texts in specialized areas are linked to and may be mapped to a domain-specific ontology. To create a summary, the mapping is traversed [12].

4.4 Graph-Based Method

Alami et al. [16] used a graph model with nodes that each represent a word and positional information while also being connected to other nodes. Directed edges indicate the structure of a sentence. Creating a textual graph that reflects the original material and producing an abstractive summary are two aspects of the graph technique. Such a strategy explores and scores multiple sub-paths in the graph in order to offer an abstractive summary [1].

5   Conclusion

Automatic text summarization produces a summary by reducing the size of the original content while maintaining important information. Despite the fact that various techniques have been presented, automatic text summarizing remains a difficult undertaking, with the results falling far short of quality human summaries. The majority of researchers concentrate on the extractive strategy. As a result, extractive summarization has a larger body of literature than abstractive summarization. This survey examined several approaches and strategies. We can say that combining two or more approaches or procedures is likely to provide positive outcomes, improving the quality of the summaries over utilizing either way alone.

References

1. Tandel A, Modi B, Gupta P, Wagle S, Khedkar S (2016) Multi-document text summarization-a survey. In: 2016 international conference on data mining and advanced computing (SAPI-ENCE). IEEE, pp 331–334
2. Young M (1989) The technical writer's handbook. University Science, Mill Valley, CA
3. Zhu J, Zhou L, Li H, Zhang J, Zhou Y, Zong C (2017) Augmenting neural sentence summarization through extractive summarization. In: National CCF Conference on natural language processing and Chinese computing. Springer, pp 16–28
4. Gupta V, Lehal GS (2010) A survey of text summarization extractive techniques. J Emerg Technol Web Intell 2(3):258–268
5. Wang S, Zhao X, Li B, Ge B, Tang D (2017) Integrating extractive and abstractive models for long text summarization. In: 2017 IEEE international congress on big data (BigData Congress). IEEE, pp 305–312
6. Kumari N, Singh P (2021) Text summarization and its types: a literature review. In: Handbook of research on natural language processing and smart service systems, pp 368–378
7. El-Kassas WS, Salama CR, Rafea AA, Mohamed HK (2020) Automatic text summarization: a comprehensive survey. Expert Syst Appl, 113679
8. Nazari N, Mahdavi M (2019) A survey on automatic text summarization. J AI Data Min 7(1):121–135
9. Gholamrezazadeh S, Salehi MA, Gholamzadeh B (2009) A comprehensive survey on text summarization systems. In: 2009 2nd international conference on computer science and its applications. IEEE, pp 1–6
10. Ko Y, Seo J (2008) An effective sentence-extraction technique using contextual information and statistical approaches for text summarization. Pattern Recogn Lett 29(9):1366–1371
11. Kumari N, Guleria A, Sood R, Singh P (2019) A supervised approach for keyphrase extraction using SVM. Int J Control Autom 12(5):469–477. Retrieved from http://sersc.org/journals/index.php/IJCA/article/view/2653
12. Kumar NV, Reddy MJ (2019) Factual instance tweet summarization and opinion analysis of sport competition. In: Soft computing and signal processing. Springer, pp 153–162
13. Oak R (2016) Extractive techniques for automatic document summarization: a survey. Int J Innov Res Comput Commun Eng 4(3):4158–4164
14. Lin H, Ng V (2019) Abstractive summarization: a survey of the state of the art. In: Proceedings of the AAAI conference on artificial intelligence, vol 33, pp 9815–9822
15. Moratanch N, Chitrakala S (2017) A survey on extractive text summarization. In: 2017 international conference on computer, communication and signal processing (ICCCSP). IEEE, pp 1–6

16. Alami N, Meknassi M, En-nahnahi N (2019) Enhancing unsupervised neural networks based text summarization with word embedding and ensemble learning. Expert Syst Appl 123:195–211

17. Gupta S, Gupta S (2019) Abstractive summarization: an overview of the state of the art. Expert Syst Appl 121:49–65

18. See A, Liu PJ, Manning CD (2017) Get to the point: summarization with pointer-generator networks. In: ACL 2017— Proceedings of the 55th annual meeting of the Association for Computational Linguistics (Long Pap.2017), vol 1, 1073–1083. https://doi.org/10.18653/v1/P17-1099

19. Narayan S, Cohen SB, Lapata M (2018) Don't give me the details, just the summary! Topic-aware convolutional neural networks for extreme summarization. In: Proceedings of the 2018 conference on empirical methods in natural language processing (EMNLP 2018), pp 1797–1807

20. Al-Sabahi K, Zuping Z, Kang Y (2018) Bidirectional attentional encoder-decoder model and bidirectional beam search for abstractive summarization. arXiv Prepr. arXiv1809.06662

21. Zhang Y, Li D, Wang Y, Fang Y, Xiao W (2019) Abstract text summarization with a convolutional seq2seq model. Appl Sci 9.https://doi.org/10.3390/app9081665

22. Paulus R, Xiong C, Socher R (2018) A deep reinforced model for abstractive summarization. In: 6th international conference on learning representations (ICLR 2018)—conference track proceedings, pp 1–12

23. Chen YC, Bansal M (2018) Fast abstractive summarization with reinforce-selected sentence rewriting. In: ACL 2018—56th annual meeting of the Association for Computational Linguistics (Long Papers, vol 1), pp 675–686. https://doi.org/10.18653/v1/p18-1063

24. Gehrmann S, Deng Y, Rush AM (2018) Bottom-up abstractive summarization. In: Proceedings of the 2018 conference on empirical methods in natural language processing (EMNLP 2018), pp 4098–4109. https://doi.org/10.18653/v1/d18-1443

25. Celikyilmaz A, Bosselut A, He X, Choi Y (2018) Deep communicating agents for abstractive summarization. In: NAACL HLT 2018 - 2018 conference of the North American chapter of the Association for Computational Linguistics: human language technologies, vol 1, pp 1662–1675. https://doi.org/10.18653/v1/n18-1150

26. Lewis M, Liu Y, Goyal N, Ghazvininejad M, Mohamed A, Levy O, Stoyanov L, Zettlemoyer L (2019) BERT: denoising sequence-to-sequence pre-training for natural language generation, translation, and comprehension. arXiv Prepr. arXiv1910.13461

27. Song K, Wang B, Feng Z, Liu R, Liu F (2020) Controlling the amount of verbatim copying in abstractive summarization. In: Proceedings of the AAAI conference on artificial intelligence, vol 34, pp 8902–8909.https://doi.org/10.1609/aaai.v34i05.6420

28. Joshi M, Wang H, McClean S (2018) Dense semantic graph and its application in single document summarization. In: Emerging ideas on information filtering and retrieval. Springer, pp 55–67

29. Modi S, Oza R (2018) Review on abstractive text summarization techniques (ATST) for single and multi documents. In: 2018 international conference on computing, power and communication technologies (GUCON). IEEE, pp 1173–1176

30. Mahajani A, Pandya V, Maria I, Sharma D (2019) A comprehensive survey on extractive and abstractive techniques for text summarization. In: Ambient communications and computer systems. Springer, pp 339–351

31. Hou L, Hu P, Bei C (2017) Abstractive document summarization via neural model with joint attention. In: National CCF conference on natural language processing and Chinese computing. Springer, pp 329–338

32. Mohd M, Jan R, Shah M (2020) Text document summarization using word embedding. Expert Syst Appl 143:112958

33. Moratanch N, Chitrakala S (2016) A survey on abstractive text summarization. In: 2016 International conference on circuit, power and computing technologies (ICCPCT). IEEE, pp 1–7

34. Hsu W-T, Lin C-K, Lee M-Y, Min K, Tang J, Sun M (2018) A unified model for extractive and abstractive summarization using inconsistency loss. arXiv preprint arXiv:1805.06266
35. Nima S (2018) Abstractive text summarization with attention-based mechanism. MS Thesis, Universitat Politecnica de Catalunya
36. Zihang D et al (2019) Transformer-xl: attentive language models beyond a fixed-length context. arXiv preprint arXiv:1901.02860
37. Khandelwal U, Clark K, Jurafsky D, Kaiser L (2019) Sample efficient text summarization using a single pre-trained transformer. arXiv preprint arXiv:1905.08836
38. Liu PJ, Saleh M, Pot E, Goodrich B, Sepassi R, Kaiser L, Shazeer N (2018) Generating Wikipedia by summarizing long sequences. In: Proceedings of the ICLR
39. Wani MA, Riyaz R (2016) A new cluster validity index using maximum cluster spread based compactness measure. Int J Intell Comput Cybern 9:179–204
40. Wani MA, Riyaz R (2017) A novel point density based validity index for clustering gene expression datasets. Int J Data Min Bioinform 17:66–84
41. Wani MA, Bhat FA, Afzal S, Khan (2020) Advances in deep learning. Springer
42. Wani MR, Wani MA, Riyaz R (2016) Cluster based approach for mining patterns to predict wind speed. In: Proceedings of the international conference on renewable energy research and applications, pp 1046–1050
43. Wani MA (2008) Incremental hybrid approach for microarray classification. In Proc. of the 7th international conference on machine learning and applications, pp 514–520
44. Riyaz R, Wani MA (2016) Local and global data spread based index for determining number of clusters in a dataset. In: Proceedings of the 15th IEEE international conference on machine learning and applications, pp 651–656
45. Lin H, Ng V (2019) Abstractive summarization: a survey of the state of the art. In: Proceedings of the 33rd AAAI conference on artificial intelligence, vol 33, pp 9815–9822. [Online]. Available: https://www.aaai.org/ojs/index.php/AAAI/article/view/5056
46. Klymenko O, Braun D, Matthes F (2020) Automatic text summarization: a state-of-the-art review. In: Proceedings of the 22nd international conference on enterprise information systems, pp 648–655
47. Shi T, Keneshloo Y, Ramakrishnan N, Reddy CK (2018) Neural abstractive text summarization with sequence-to-sequence models. arXiv:1812.02303. [Online]. Available: http://arxiv.org/abs/1812.02303
48. Devlin J, Chang M, Lee K, Toutanova K (2019) BERT: pre-training of deep bidirectional transformers for language understanding. In: Proceedings of the 2016 conference of the North American chapter of the Association for Computational Linguistics: human language technologies, pp 4171–4186
49. Radford A, Improving language understanding by generative pre-training. Open AI J, to be published
50. Rane N, Govilkar S (2019) Recent trends in deep learning based abstractive text summarization. Int J Recent Technol Eng 8(3); Sciforce (2019) Towards automatic summarization. Part 2. Abstractive methods. Sciforce Blog. [Online]. Available: https://medium.com/sciforce/towards-automaticsummarization-part-2-abstractive-methods-c424386a65ea
51. Sanad M (2019) A comprehensive guide to build your own language model in Python. [Online]. Available: https://www.analyticsvidhya.com/blog/2019/08/comprehensive-guide-language-model-nlp-python-code/. Manning CD, Schutze H, Raghavan P (2008) Introduction to information retrieval, vol 238. Cambridge University Press, Cambridge, U.K.
52. Jing K, Xu J (2019) A survey on neural network language models. arXiv:1906.03591. [Online]. Available: http://arxiv.org/abs/1906.03591
53. Mikolov T, Chen K, Corrado G, Dean J (2013) Efficient estimation of word representations in vector space. In: Proceedings of the 1st international conference on learning representations (ICLR), pp 1–12
54. Rush AM, Chopra S, Weston J (2015) A neural attention model for abstractive sentence summarization. In: Proceedings of the conference on empirical methods in natural language processing, pp 379–389. [Online]. Available: https://arxiv.org/abs/1509.00685

55. Chopra S, Auli M, Rush AM (2016) Abstractive sentence summarization with attentive recurrent neural networks. In: Proceedings of the conference of the North American chapter of the Association for Computational Linguistics: human language technologies, pp 93–98
56. Nallapati R, Zhou B, Gulçehre C, Xiang B (2016) Abstractive text summarization using sequence-to-sequence RNNs and beyond. In: CoNLL 2016—20th SIGNLL conference on computational natural language learning. Proceedings, pp 280–290. https://doi.org/10.18653/v1/k16-1028

A Systematic Study of Various Approaches and Problem Areas of Named Entity Recognition

Monica Madan, Ashima Rani, and Neha Bhateja

Abstract Named entity recognition (NER) is an essential part of the natural language processing (NLP) pipeline for the extraction of information as it finds application in many downstream applications. Many different rule-based, statistical, probabilistic, or hybrid approaches have been used to address it, the recent ones being machine or deep learning. Due to fast-changing technologies, many new developments have emerged in NER as well. This paper aims to provide a brief introduction, an overview of the new advancements, the addressed problem areas and approaches for the duration of five years from 2017 to 2021. This period has seen advances in deep learning, and some major developments are seen such as contextual embedding. The study explores different problem areas identified such as performance, handling noise or robustness, nested entities, low resources in new languages, scalability, fine-grained entities, and others being worked upon using supervised, unsupervised, reinforcement transfer learning working at character, morphological, word or sentence level, or by adding dictionaries for including knowledge. The paper can help provide a high-level input, to give an idea, of the areas that are currently being worked upon in NER along with the latest approaches with their advantages and future directions.

Keywords Deep learning · Named entity recognition · Natural language processing · Review · Survey

1 Introduction

Named entity extraction or recognition problem is part of the information extraction in NLP, also is one of the common tasks of NLP that are mostly touched directly or indirectly. One cannot miss seeing the extensive usage NER finds in biomedical and bioinformatics such as for information extraction from electronic medical records and for identification of the scientific terms in the text or identifying drug names. The other prominent area of applications of NER is language-specific implementation. To

M. Madan (✉) · A. Rani · N. Bhateja
Amity University Haryana, Gurgaon, India
e-mail: madan.monica@gmail.com

© The Author(s), under exclusive license to Springer Nature Singapore Pte Ltd. 2023
Y. Singh et al. (eds.), *Proceedings of International Conference on Recent Innovations in Computing*, Lecture Notes in Electrical Engineering 1011,
https://doi.org/10.1007/978-981-99-0601-7_42

address the different needs of various languages, extra work is required to adapt the current or to add a new method. Different languages can be character based or may be varied in morphological patterns adding to the diverse semantics and syntaxes to cater to. This chapter will be going through the introduction of NER, the standards and metrics used in general by the researchers, followed by a synopsis of the state of the art available today. For the duration of 2017 to 2021, the problem areas that the authors focused on along with the approaches taken, solution advantage and future scope are illustrated.

The motivation behind this study is that not only that it depicts the latest trends but also it has seen one of the major shifts that modified the course of solution, not only NER but also NLP in general. This shift is from feature engineering to contextual embedding, with the use of transformers, is so strong that they are naturally being adapted by the researcher to achieve state of the art. A summary of the latest trends not only helps with an overall recent picture but also facilitates the planning of the approach that one is planning to take. The papers chosen for this study are not of a particular domain or language application, but for NER in general.

The earlier approaches for the NER problem were the rule-based approaches including dictionaries and finding patterns and adding grammatical knowledge. Popular probabilistic models were used such as hidden Markov model (HMM) and conditional random fields (CRF). Machine learning approaches such as support vector machines (SVM) and decision trees require features that had to be specifically curated and were then followed by advancing deep learning. The commonly used deep learning methodologies are recurrent neural network (RNN), long short-term memory (LSTM), and convolution neural network (CNN) where (Bidirectional) Bi-LSTM-(CRF) is one of the popular models. Among the recent surveys, Li et al. [1] wrote about the deep learning advancements on NER explaining in detail, the deep learning technologies used and the resources involved. For language and domain-specific surveys, to name a few recent ones, Sharma et al. [2] presented a named entity recognition survey for the Hindi language. A survey specific to food entities was presented by Popovski et al. [3]. Georgescu Tiberiu-Marian et al. [4] worked a survey for the cybersecurity domain while Wang et al. [5] worked on a detailed survey for nested entities.

1.1 Named Entity Recognition

NER is the identification of the entities from the text that are names of a person, organization, date, or place. This work can be further subdivided into two tasks, identifying the entity with its span and then classifying it. An entity can be a single word "Amazon" or can be of multiple words such as "University of New York." Thus, it is important to identify and understand the boundary of the entity. The mentioned entity "University of New York" also represents another type of entity that is a nested entity where the system has to decide to identify "New York" as the entity in question or the "University of New York" as a single entity. Furthermore, there can be a general

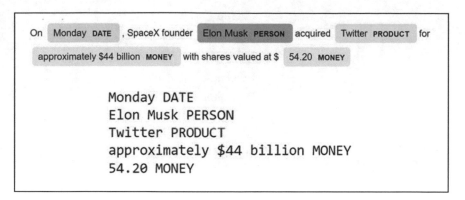

Fig. 1 Entity recognition using Spacy

number of entities such as a person, date, and organization or can be fine-grained as per the specific need of the domain. Figure 1 shows what can be achieved for NER using a commonly used library "Spacy" in Python. Here, this example, along with the correctly identified entities, also depicts an issue as SpaceX is also an entity that should have been classified as an "Organization." NER also deals with issues like Out of Vocabulary words (OOV) or emerging entities. Going forward, NER as a concept can be extended to highly visual documents often addressed as multimodal NER for improved information extraction. These were the few perspectives to view the NER problem; incoming sections will give a brief on the approaches and metrics used by earlier researchers.

1.2 Metrics

Being a multiclass classification problem, the following metrics that are universally applied are precision, recall, and $F1$ score for the evaluation of the NER models. These are widely used over accuracy as they are good measures even for class imbalance. Precision is the ratio of correct entities to all the classified entities in that class. The recall is the ratio of correct entities to actual correct entities in the class. The $F1$ score, a harmonic mean of both is a good measure and assigns equal weight to both the precision and recall. Equations 1, 2, and 3 show how the precision, recall, and $F1$ are calculated.

$$\text{Precision} = \frac{\text{True Positive}}{\text{True Positive} + \text{False Positive}} \quad (1)$$
$$\text{(Total posoitives predicted)}$$

$$Recall = \frac{True\ Positive}{True\ Positive + False\ Negative} \tag{2}$$
$$(Total\ actual\ positives)$$

$$F1 = 2 \times \frac{Precision \times Recall}{Precision + Recall} \tag{3}$$

1.3 State of the Art

NER has many datasets that are popular and used in the benchmark study serving general and specific purposes. Few popular datasets are CoNLL2003 and Ontonotes V5. On the CoNLL2003 dataset, the state of the art is with an $F1$ of 94.6 by Wang et al. [6]. The researcher has used automatic concatenation of embedding to find one for the best performance. This is very closely followed by work done by Yamada et al. [7] with the $F1$ of 94.2 based on LUKE, pretrained representations including context using transformers. For Ontonotes v5 (English), Li et al. [8] have reached an $F1$ score of 92.07 using a new loss function Dice loss in place of cross-entropy. This is very closely followed by the subject-based packed markers of Ye et al. [9] with an $F1$ of 91.9.

2 Literature Review

2.1 Methodology

The sample of the few papers used to represent this summary has been collected using the Google search engine and other reliable resources and Web sites for state of the art. The paper is an effort toward a mix of some popular and random works to represent the general work done during this time. The yearly distribution is not strict and only representational. It should be looked at subdivision for looking closely and should be taken together as the work done during these five years is presented in an easy-to-compare tabular form. Much of the work in NER areas can be seen in the applications of these models in different languages and domains. This paper is not exhaustive and does not talk about such efforts but does more about general issues faced by the NER.

2.2   Research Papers for Review from Year 2017

Table 1 shows a few of the papers from 2017. Performance is one of the major problem areas. The solution approach here is incorporating more context, sentence structure information in the training data, or a strong ensemble approach using three different algorithms particle swarm optimization (PSO), Bayesian, and CRF [10]. Performance is followed by the problem area of low availability of labeled data. Supervised learning approaches need a very big corpus to be able to have good training. To perform well in specific domains, the models need to be trained with a specific corpus. In such cases, the unavailability of the corpus can be a huge cost to the system. This brings one to the importance of unsupervised learning that does not depend on the huge amount of data for training. The details are mentioned in the table. Apart from this, scalability is another important area [11] to maintain its speed and accuracy, and Map Reduce was implemented with the model to exploit the parallelism. Also, accessibility and ease of use by the non-experts was one problem area worked upon by Dernoncourt et al. [12] using BRAT (annotation tool) and artificial neural network (ANN) for the same. The solution to the low resource leading to improved performance was by incorporating the domain knowledge using a disease dictionary and Unified Medical Language System (UMLS) along with semantic type filtering and CRF for better domain-specific results by Kanimozhi et al. [13]. In terms of models being used commonly, the Bi-LSTM-CRF, CNN, and graph convolution network (GCN) can be seen.

2.3   Research Papers for Review from Year 2018

Table 2 shows a few of the papers from 2018 where the problem areas listed are robustness, fine-grained, and nested entities addressed using summarization, using a sliding adaptive window for including the context, while the other one exploited the correlation using knowledge base and attention. Nested entities were addressed using dependencies to explore the possible span of the entities (Details in the table). Further, newly introduced popular, embedding, or vector representations of words helped the models attain significant gain in performance and are responsible for changing the course of future research in NLP. These contextual embeddings can be pertained and generated using LSTM or transformer, e.g., FLAIR, Embedding for Language Model (ELMo), and Bidirectional Encoder Representations from Transformers (BERTs). These can be computationally expensive; thus, a hybrid approach with lexical features and small embedding can be seen for saving time [19]. Apart from these, in the problem area of low resources, variations of transfer learning using graphs were used to map corpus similarities [20] using n-gram and CRF models. The same problem was also addressed using few-shot or even zero-shot learning using metric, prototypical, and transfer learning [21] but requires more work in future for the multiple classes.

Table 1 Few papers from 2017

Authors	Title	Problem area	Solution	Advantages and future scope
Cotterell et al. [14]	"Low-resource named entity recognition with cross-lingual, character-Level neural conditional random fields"	Low data resources availability	Character level for knowledge transfer from high resource language to less resource one (using CRF on 15 languages)	Prove it as a viable method, the future will need performance improvement
Centoli et al. [15]	"Graph convolution networks for named entity recognition"	Performance, dependency structure	Using graph convolutional network on Bi-LSTM & CRF	Able to show gain in performance, future needs performance improvement, can use contextual embedding
Wang et al. [16]	"Named entity recognition with gated convolutional neural networks"	Performance	GCNN—CNN using gating mechanism on CNN	SOTA training efficiency, future with deeper NN & tagging for performance improvement
Yang et al. [17]	"Neural re-ranking for named entity recognition"	Performance, structure	Sentence-level patterns for re-ranking, LSTM CNN	Performance gain, future need to distinguish similar structures using tree & experimentation with data
Liu et al. [18]	"Empower sequence labeling with task-aware neural language model"	Unsupervised	LM-LSTM-CRF, word and character level extraction of language structure features	Concise, better performance and training efficiency, no dependency on data or knowledge

2.4 Research Papers for Review from Year 2019

Table 3 lists the details from 2019. The problem areas listed here are the imbalance of data and emerging entities, apart from the previously mentioned problem areas. Cloze words objective was used to pretrain where the middle word is to be predicted using the left and the right context to achieve new state of the art. More work listed in the table is the unified approach for nested and flat entities treating NER as machine comprehension or as a query, exploiting more associated semantic information compared to NER classification general approaches which do not have a

Table 2 Few papers from 2018

Author	Title	Problem area	Solution	Advantages and future scope
Malykh et al. [22]	"Named entity recognition in noisy domains"	Noise, robust	Replace noise with the correct probable character using CNN	Achieved robustness and improved performance
Liu et al. [23]	"Fine-grained entity type classification with adaptive context"	Fine-grained	Global information using summary, adaptively window adjusting for context	Better results, future more entity types, & more than one entity in one sentence
Xin et al. [24]	"Improving neural fine-grained entity typing with knowledge attention"	Fine-grained	Correlation between context and entity using knowledge base attention	Effective and state of the art, future with more entity types, hierarchies and reducing noise
Akbik et al. [25]	"Contextual string embedding for sequence labeling"	Embedding	FLAIR embedding: contextual embedding using characters	State of the art pretrained with LSTM, future requires work on sentence-level tasks
Sohrab et al. [26]	"Deep exhaustive model for nested named entity recognition"	Nested entities	Enumerating possible spans/regions for entities	Region-level information/exploring dependencies
Peters et al. [27]	"Deep conceptualized word representations"	Embedding	Embedding for Language Models (ELMO) (Bi-LSTM)	Performance improvement using Syntax, semantics, and context

specific task in mind but extraction. To handle the imbalance of data, that is, the negative and positive examples not being in the same proportion, a new criterion "dice loss" was proposed that gives equal weightage to false positives or negatives and thus is less susceptible to being affected by the data imbalance (details in the table). Also, the problem area of the low availability of resources, the effort can be seen in the automatic generation of the corpus using bootstrapping [28] with the advantage of the possibility of adding specific information to the corpus along with being low cost, but future needs removal of incorrect labels. On the other side, phonetics was explored with the advantage of being able to work in multilingual settings by Cabot et al. [29] but with a few remaining errors of spelling and disambiguation to be removed in future.

Table 3 Few papers from 2019

Authors	Title	Problem area	Solution	Advantages and future scope
Li et al. [8]	"Dice loss for data-imbalanced NLP tasks"	Imbalance of data	A new criteria "dice loss" based on Sorensen-Dice coefficient or Tversky index	It can support a model to gain significant performance and without any architectural change
Nawroth et al. [30]	"Emerging named entity recognition on retrieval features in an affective computing corpus"	Emergent entities recognition	Based on the known query and index system using affective computing	Performs in some corpus, future needs improvement for real-world problems
Li et al. [31]	"A unified MRC framework for named entity recognition"	Unified NER & nested NER	Reformulate NER as machine reading comprehension	State of the art for both nested and flat entities
Devlin et al. [32]	"BERT: pre-training of deep bidirectional transformers for language understanding"	Embedding	Deep Representations with both left & right context using random masking	State of the art for 11 tasks in NLP with even low resources and easy fine-tuning but with high computational needs
Baevski et al. [33]	"Cloze-driven pretraining of self-attention networks"	Performance	Cloze word reconstruction for pretraining bi-directional transformer	New state of the art, future will be on the variations of the current model

2.5 Research Papers for Review from Year 2020

Table 4 lists the work in 2020. The problem areas emerge again as performance, nested entities, and class imbalance. In a certain language, the entity recognition performed better when the researcher added word position in the word vector because the entity boundary is otherwise difficult to find using Bi-LSTM + CRF and IDCNN + CRF [34]. Another approach in the direction of the low labeled data availability, thus cost-effective, was suggested by using triggers [35] with the benefit of a solution that can generalize even to the sentences that were not seen by the model, the trigger is the phrases that indicate the system toward an entity or in other words which would give a human intuition of recognizing an entity in the given text.

Table 4 Few papers from 2020

Authors	Title	Problem area	Solution	Advantages and future scope
Lou et al. [36]	"Hierarchical contextualized representation for named entity recognition"	Performance, document context	Sentence-level and document-level, contextual representation using memory network as a knowledge source	New state of the art
Thi et al. [37]	"Named entity recognition architecture combining contextual and global features"	Performance, context, and global relations	XLNet for contextual features and graph convolution network (GCN) for global features	Significant performance gains, in future, it can be tested on cross-lingual with knowledge graphs
Ye et al. [9]	"Packed levitated marker for entity and relation extraction"	Performance, span interrelation	Relation between spans—Packed Levitated Markers, subject-oriented packing	State-of-the-art and promising efficiency, future to generalize to other tasks
Osman et al. [38]	"Focusing on possible named entities in active named entity label acquisition"	Unstructured text, low resource, class imbalance	Active learning query evaluation functions, semi-supervised clustering for positive token identification using BERT, CRF	Improved performance with low data needs
Yu et al. [39]	"Developing name entity recognition for structured and unstructured"	Nested entities, global view	Global view, Biaffine model for spans, contextual embedding on multilayer Bi-LSTM	State of the art, structure information helps both flat and nested NERs

2.6 Research Papers for Review from Year 2021

The list of the latest papers in 2021 is presented in Table 5. Problem areas seen here are overfitting and unified solution to flat and nested entities. Extensive experimentation on the models by varying embedding, sentence length, and many other factors was done by researchers and also seen adding GCN and gating to the Bi-LSTM-CRF model while gaining the context and robustness. Also, seen is the use of adversarial training for robustness. The adversarial training is used to add samples to the training data using an attack algorithm; here, the data are added to fill the data gaps. Going forward to reinforcement learning, negative data and span selector were successfully utilized by researchers Peng et al. [40] to achieve comparable gain without extra

Table 5 Few papers from 2021

Authors	Title	Problem area	Solution	Advantages and future scope
Wang et al. [6]	"Automated Concatenation of Embeddings for Structured Prediction"	Performance, reinforcement learning	Automatic concatenation of embedding to find the best fit in terms of context	Good training accuracy and stronger reward function, state of the art
Xu et al. [41]	"Better feature integration for named entity recognition"	Performance	Utilizing long-distance structured information using GCN-LSTM-CRF and gating	Extensive experimentation, effectively incorporates structure, context & robust
Peng et al.[42]	"Unsupervised cross-domain named entity recognition using entity-aware adversarial training"	Unsupervised learning	Using labeled data from one domain to predict unlabeled entities using entity-aware attention & adversarial training	Performance improvement gained
Fu et al.[43]	"Exploiting Named Entity Recognition via Pre-trained Language Model and Adversarial Training"	Overfitting, poor generalization, or robustness	Better performance using BERT + Bi-LSTM-CRF, fast gradient method (FGM) to generate adversarial	Improved performance, generalization, and robustness
Yan et al. [44]	"A Unified Generative Framework for Various NER Subtasks"	One solution for flat, nested discontinuous	Span-level classification using Seq2Seq with pointer	State of the art, better results with shorter entity length representation

corpus, but optimization could be worked upon in the future for this. Figure 2 shows the areas identified in the duration in the form of a word cloud.

3 Conclusion

The future can be toward contextual embedding, i.e., ELMo, BERT, etc., as we can see the latest models achieving the state of the art using them in any NLP area. However, they have high demand in terms of computation and may need special hardware, and in the future, any effort in this direction or even the use of hybrid models will make them more available and easily accessible. One prominent problem

Fig. 2 Word cloud based on the areas identified

area identified is the need of a big corpus for the supervised approaches and points us to the unsupervised approaches that currently use transfer learning, adversarial training, etc. Efforts here in future can prove fruitful as it is yet to see more in terms of gain in results. Few problems still need effort such as OOV words due to fast emerging technologies in every field do make it difficult to catch up and can be a desirable future work area. Also, can be foreseen much more work areas in future in terms of the languages and particular domains that need corpus and specific effort, lot of work should be seen in the future for the same.

The survey aims to provide an overview of NER's latest techniques and problem areas with approaches based on recent work. With the above sample of papers, the supervised, semi-supervised, reinforcement learning, and unsupervised approaches are covered with most of the problem areas that can be seen repetitively appearing throughout this duration. The paper can help provide the future direction to researcher while summing together the latest on this front.

References

1. Li J, Sun A, Han J, Li C (2022) A survey on deep learning for named entity recognition. IEEE Trans Knowl Data Eng 34(1):50–70
2. Sharma R, Morwal S, Agarwal B (2019) Named entity recognition for Hindi language: a survey. J Discrete Math Sci Cryptogr 22(4):569–580
3. Popovski G, Koroušić Seljak B, Eftimov T (2020) A survey of named-entity recognition methods for food information extraction. IEEE Access 8:31586–31594
4. Tiberiu-Marian G, Iancu B, Zamfiroiu A, Doinea M, Boja CE, Cartas C (2021) A survey on named entity recognition solutions applied for cybersecurity-related text processing. In: Yang XS, Sherratt S, Dey N, Joshi A (ed) Proceedings of fifth international congress on information and communication technology. Springer, Singapore, pp 316–25
5. Wang Y, Zhu Z, Li Y (2022) Nested named entity recognition: a survey. ACM Trans Knowl Discov Data, 1556–4681
6. Wang X, Jiang Y, Bach N, Wang T, Huang Z, Huang F et al (2020) Automated concatenation of embeddings for structured prediction. In: Proceedings of the 59th annual meeting of the

association for computational linguistics and the 11th international joint conference on natural language processing (volume 1: long papers), pp 2643–2660

7. Yamada I, Asai A, Shindo H, Takeda H, Matsumoto Y (2020) LUKE: deep contextualized entity representations with entity-aware self-attention. In: Proceedings of the 2020 conference on empirical methods in natural language processing (EMNLP), pp 6442–6454

8. Li X, Sun X, Meng Y, Liang J, Wu F, Li J (2019) Dice loss for data-imbalanced NLP tasks. In: Proceedings of the 58th annual meeting of the association for computational linguistics, pp 465–476

9. Ye D, Lin Y, Li P, Sun M (2022) Packed levitated marker for entity and relation extraction. In: Proceedings of the 60th annual meeting of the association for computational linguistics (volume 1: long papers), pp 4904–4917, Dublin, Ireland

10. Akkasi A, Varoglu E (2017) Improving biochemical named entity recognition using PSO classifier selection and Bayesian combination methods. IEEE/ACM Trans Comput Biol Bioinf 14(6):1327–1338

11. Liu F, Xu X, Ji Z (2017) MapReduce based named entity recognition for biology literature. In: 2017 IEEE 2nd advanced information technology, electronic and automation control conference (IAEAC), pp 1268–1271

12. Dernoncourt F, Lee JY, Szolovits P (2017) NeuroNER: an easy-to-use program for named-entity recognition based on neural networks. In: Proceedings of the 2017 conference on empirical methods in natural language processing: system demonstrations, pp 97–102, Copenhagen, Denmark

13. Kanimozhi U, Manjula D (2017) A CRF based machine learning approach for biomedical named entity recognition. In: 2017 second international conference on recent trends and challenges in computational models (ICRTCCM), pp 335–342

14. Cotterell R, Duh K (2017) Low-resource named entity recognition with cross-lingual, character-level neural conditional random fields. In: Proceedings of the eighth international joint conference on natural language processing (volume 2: short papers). Asian Federation of Natural Language Processing, Taipei, Taiwan, pp 91–96

15. Cetoli A, Bragaglia S, O'harney AD, Sloan M (2018) Graph convolutional networks for named entity recognition. In: Proceedings of the 16th international workshop on treebanks and linguistic theories, pp 37–45, Prague, Czech Republic

16. Wang C, Chen W, Xu B (2017) Named entity recognition with gated convolutional neural networks. In: Sun M, Wang X, Chang B, Xiong D (ed) Chinese computational linguistics and natural language processing based on naturally annotated big data. Cham: Springer International Publishing, pp 110–121

17. Yang J, Zhang Y, Dong F (2017) Neural reranking for named entity recognition. In: Proceedings of the international conference recent advances in natural language processing (RANLP), pp 784–792, Varna, Bulgaria

18. Liu L, Shang J, Ren X, Xu FF, Gui H, Peng J et al (2018) Empower sequence labeling with task-aware neural language model. In: Proceedings of the thirty-second AAAI conference on artificial intelligence and thirtieth innovative applications of artificial intelligence conference and eighth AAAI symposium on educational advances in artificial intelligence (AAAI'18/IAAI'18/EAAI'18), Article 644. AAAI Press, pp 5253–5260

19. Ghaddar A, Langlais P (2018) Robust lexical features for improved neural network named-entity recognition. In: Proceedings of the 27th international conference on computational linguistics. Association for Computational Linguistics, Santa Fe, New Mexico, USA, pp 1896–1907

20. Sheikhshab G, Starks E, Karsan A, Chiu R, Sarkar A, Birol I (2018) GraphNER: using corpus level similarities and graph propagation for named entity recognition. In: 2018 IEEE international parallel and distributed processing symposium workshops (IPDPSW), pp 229–238

21. Fritzler A, Logacheva V, Kretov M (2019) Few-shot classification in named entity recognition task. IEEE Trans Knowl Data Eng

22. Malykh V, Lyalin V (2018) Named entity recognition in noisy domains. In: 2018 international conference on artificial intelligence applications and innovations (IC-AIAI)

23. Liu J, Wang L, Zhou M, Wang J, Lee S (2018) Fine-grained entity type classification with adaptive context. Soft Comput 22(13):4307–4318
24. Song CH, Hltcoe DL, Finin T, Mayfield J, Improving neural named entity recognition with gazetteers
25. Akbik A, Blythe D, Vollgraf R (2018) Contextual string embeddings for sequence labeling. In: Proceedings of the 27th international conference on computational linguistics. Association for Computational Linguistics, Santa Fe, New Mexico, USA, pp 1638–1649
26. Sohrab MG, Miwa M (2018) Deep exhaustive model for nested named entity recognition. In: Proceedings of the 2018 conference on empirical methods in natural language processing, pp 2843–2849. Association for Computational Linguistics, Brussels, Belgium
27. Peters ME, Neumann M, Gardner M, Clark C, Lee K, Zettlemoyer L (2018) Deep contextualized word representations. In: Proceedings of the 2018 conference of the North American chapter of the Association for Computational Linguistics: human language technologies, vol 1. Association for Computational Linguistics, New Orleans, Louisiana, pp 2227–2237
28. Kim J, Ko Y, Seo J (2019) A bootstrapping approach with CRF and deep learning models for improving the biomedical named entity recognition in multi-domains. IEEE Access 7:70308–70318
29. Cabot C, Darmoni S, Soualmia LF (2019) Cimind: a phonetic-based tool for multilingual named entity recognition in biomedical texts. J Biomed Inform 94
30. Nawroth C, Engel F, Mc Kevitt P, Hemmje ML (2019) Emerging named entity recognition on retrieval features in an affective computing corpus. In: 2019 IEEE international conference on bioinformatics and biomedicine (BIBM), pp 2860–2868
31. Li X, Feng J, Meng Y, Han Q, Wu F, Li J (2019) A unified MRC framework for named entity recognition. In: Proceedings of the 58th annual meeting of the Association for Computational Linguistics, pp 5849–5859
32. Devlin J, Chang MW, Lee K, Google KT, Language AI (2019) BERT: pre-training of deep bidirectional transformers for language understanding. In: Proceedings of the 2019 conference of the North American chapter of the Association for Computational Linguistics: human language technologies, vol 1, pp 4171–4186, Minneapolis, Minnesota
33. Baevski A, Edunov S, Liu Y, Zettlemoyer L, Auli M (2019) Cloze-driven pretraining of self-attention networks. In: Proceedings of the 2019 conference on empirical methods in natural language processing and the 9th international joint conference on natural language processing (EMNLP-IJCNLP), pp 5360–5369, Hong Kong, China
34. Du Y, Zhao W (2020) Named entity recognition method with word position. In: 2020 international workshop on electronic communication and artificial intelligence (IWECAI), pp 154–159
35. Lin BY, Lee DH, Shen M, Moreno R, Huang X, Shiralkar P et al (2020) TriggerNER: learning with entity triggers as explanations for named entity recognition. In: Proceedings of the 58th annual meeting of the Association for Computational Linguistics. Association for Computational Linguistics, pp 8503–8511
36. Luo Y, Xiao F, Zhao H (2020) Hierarchical contextualized representation for named entity recognition. In: Proceedings of the AAAI conference on artificial intelligence, vol 34, no 05, pp 8441–8448
37. Thi T, Hanh H (2021) Named entity recognition architecture combining contextual and global features. In: Ke HR, Lee CS, Sugiyama K (eds) Towards open and trustworthy digital societies. ICADL 2021. Lecture Notes in Computer Science, vol 13133. Springer, Cham
38. Osman A, Berk, Berk,sapcı B, Tastan O, Yeniterzi R (2021) Focusing on possible named entities in active named entity label acquisition
39. Yu J, Bohnet B, Poesio M (2020) Named entity recognition as dependency parsing. In: Proceedings of the 58th annual meeting of the Association for Computational Linguistics. Association for Computational Linguistics, pp 6470–6476
40. Peng S, Zhang Y, Wang Z, Gao D, Xiong F, Zuo H (2021) Named entity recognition using negative sampling and reinforcement learning. In: 2021 IEEE international conference on bioinformatics and biomedicine (BIBM), pp 714–719

41. Xu L, Jie Z, Lu W, Bing L (2021) Better feature integration for named entity recognition
42. Peng Q, Zheng C, Cai Y, Wang T, Xie H, Li Q Unsupervised cross-domain named entity recognition using entity-aware adversarial training. Neur Netw 138:68–77
43. Fu J, Liu J, Shi W (2021) Exploiting named entity recognition via pre-trained language model and adversarial training. In: 2021 IEEE international conference on computer science, electronic information engineering, and intelligent control technology, CEI 2021. Institute of Electrical and Electronics Engineers Inc., pp 665–669
44. Yan H, Gui T, Dai J, Guo Q, Zhang Z, Qiu X (2021) A unified generative framework for various NER subtasks. ACL

From Virtual World Back to Real Classrooms?

Viktória Bakonyi🄳 and Illés Zoltán🄳

Abstract In 2020 spring semester due to Covid-19, we had to quickly transform classical university courses to online ones. It was very hard, because we and the students either did not have too many experiences for using distance teaching methods and tools. During emergency situation, we got to know a lot of different IT tools and teaching practice to enhance our education quality and students used to these comfortable and modern possibilities. Now, the emergency situation is over, and we face a new problem. Is it possible to go back and continue education the same way as we did before Covid-19—which was proven to be good and effective? We made surveys just after the starting of distance education and after the first semester back to classical face-to-face teaching to collect the opinions of students. In this paper, we should like to present the experienced change in students' attitude and their need which we should follow.

Keywords Virtual classroom system · Real time · Interaction · Education · CRS · Classroom response systems

1 Introduction

To tell the truth, there were a lot of problems already with university education before Covid-19 [1–3]. Old, classical methods were not working as before. All participants in educational process were changed and maybe the most important thing, the "world" around us was changed as well. Maybe the most significant thing was noticed by many educators all over the world—students were not so active, and sometimes it seemed that they were totally boring during the lessons. Often heard, "It's a waste of time to learn such a thing being one click distance on the internet."

V. Bakonyi (✉) · I. Zoltán
Eötvös Loránd University, 1117 Pázmány s. 1/c, Budapest, Hungary
e-mail: hbv@inf.elte.hu

I. Zoltán
e-mail: illes@inf.elte.hu

© The Author(s), under exclusive license to Springer Nature Singapore Pte Ltd. 2023   559
Y. Singh et al. (eds.), *Proceedings of International Conference on Recent Innovations in Computing*, Lecture Notes in Electrical Engineering 1011,
https://doi.org/10.1007/978-981-99-0601-7_43

1.1 Problem Detection

The question is obvious: Why was the method good before some decades ago and why is it not working now? Now we experience, they forget active thinking, enough "two clicks" and go! Researchers detected two main problems:

- Prensky [4] spoke about digital natives first. It is far much than a simple word, indicator. Spending hours a day on the Internet has changed the physical activity of people's brains. Researchers state that this is the result of the must of frequent and quick decisions what we are stimulated, when reading a nice, designed homepage with a lot of inspiring links. Therefore, digital natives need bigger stimulus not to be bored.
- Surrounded by several devices parallel and making quick decisions causes not only a bigger activity in brain, but meanwhile it causes hyper-attention [5]. This problem was detected by Kathrine Hayes. Hyper-attention means that the person who use several devices parallel must switch between the stimulus, and it can be managed properly on 2% of people—at the others efficiency will decrease [6]. As IT professionals, it is easy to give an example that can highlight the problem: switching between tasks in a multitasking operating system involves a lot of steps to reset the system later on where to continue. The "resetting" of brain back to the previous thoughts needs extra efforts.

1.2 Possible Solution—Interaction

It would be impossible to stop digital development, and we do not want to do it even we would be able—but what can we do to help education? [7, 8]. We have to think of E. Dale (Cone of experience http://www.edutechie.ws/2007/10/09/cone-of-experi ence-media/) the American pedagogics measured the efficiency of learning during in different passive and active learning environment. The measured result proved his idea that activity enhance the efficiency of learning.

In the last decades, educators try to use up this result and newer and newer method-ologies appeared focusing to activate students—e.g., like developing new interac-tive course materials instead of traditional documentations, teams working, flipped classrooms or using classroom response systems (CRSs) [9–11] during lessons, etc. However, in 2014, Ferdinand von Prondzynski of Robert Gordon University in Aberdeen still said that "truly interactive lectures are still rare, and many students nowadays do not attend *these events at all.*" (https://bit.ly/37DnkNO).

1.3 E-Lection, Our CRSs (Classroom Response Systems) Solution

Some years ago, we decided to use a CRS in our teaching practice to increase inter-activity in our lectures and practices. Mostly, at large-scale lectures, presentations participated several hundreds of students, we have seen, the classical methods did not work further. If a professor asked a question, only the silent was the answer. Students did not want to raise their hands to ask when they do not understand something new or a question got to their mind—they might shamed their questions before the others. So, we recognized; their accustomed communication mode is chatting instead of classic oral communication [12]. We, as a first step, examined few possible ready-made software.

However, they are mostly well designed good solutions, but with limitations and difficulties. We think not only our students found, e.g., funny language inventions as ID names (Donald dug, No pain-No gain, etc.) causing 5–10 min laughing break. Sure such a funny situations are not fully waste of times, unfortunately they not really help the efficiency of lectures. After trial period, we decided to implement an own system avoiding unnecessary cases. See Fig. 1 [13–15].

It became a Bring Your Own Device (BYOD), bidirectional real-time Web application based on university standard authentication process (no funny ID names, no ambiguous language questions, etc.). It was written in C#, using Web Forms template, and for real-time features, we used SignalR. It was bidirectional because teacher may send quiz questions to all of the students' joint devices and students may send back answers; their questions toward the teacher or a simple do not understand signals.

The concept was working; we measured the efficiency of it and proved that the results became better using it see Table 1. In our faculty, there are 3 different training versions students may choose. That is why some of them learn the subject in autumn and others in spring. They were all Hungarian students; the lectures were given in

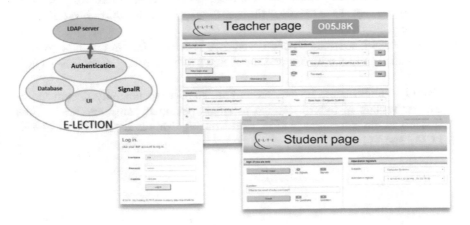

Fig. 1 E-Lection and its architecture

Table 1 Measure of efficiency [20]

		2017 without E-lection	2018 first year with E-lection
Spring	Missed exam	20.5%	17.8%
	Average	9.9	10.6
	Dispersion	4.8	5.2
	Median	10	11
Autumn	Missed exam	15.5%	14.2%
	Average	10.3	12.2
	Dispersion	4.9	5.5
	Median	10	12.2

their native language. The age group of them is between 20 and 25. They are both from the countryside and from bigger towns, mixed.

Beside the efficiency also was important for our students, that this CRS was written by their professors and in this case not true the well-known saying "Preaching water and drinking wine!"

2 Dramatic Changes—Covid-19, the Time for Distance Education

In 2020 spring from one day to another, we have to transfer our courses to virtual space without any trial period or planning. What about education which usually based on personal connections? [16–18].

- LCMS (Learning Content Management System) where educators have to upload the documents, ppt-s, video tutorials, create online tests, quizzes; on the other hand, the students have to upload the homework, performed the online tests, e.g., Moodle, Canvas, etc.
- VCS (Virtual Classroom Systems) and give lessons online in real time to keep the personal contact with our students as much as we can, e.g., Google Meet, MS Teams.
- Hybrid solution, some lessons are online, some have got only documents or another version of hybrid learning to attend personally only one third of the students at the same time, and they are rotating. This is practically mixing of LCMS and VCS.

2.1 Our Faculty Solution

In Informatics Faculty at ELTE, Hungary the dean's decision was to go on with the schedule using real-time virtual classroom system (VCS), **Microsoft Teams** with synchron online teaching method and video recording. Naturally, we started to create more documents to help student uploaded them to our LCMS system, **Canvas** preferred by our faculty. (We use Canvas for years without any problems, students like it, so we did not created any questions about it in our survey.)

At the end of the first emergency semester (and at the starting of autumn semester), we asked the students opinion about it [19–22].

2.2 The Environment of the Survey

473 Hungarian students answered the questions. Age group is between 20 and 25. They are from the country side and from bigger towns too, mixed. Filling out was not compulsory. We used Google Forms to support anonymity. Survey for Hungarian students is available at https://forms.gle/b5AMePqjmG7DvnWbA.

Now, we focus on the following 3 questions:

- Do you like online Teams courses? (Grade 1 (not at all)–5 (very much))
- What do you think, a Teams online lesson may substitute a classical live lesson, where you are personally there? (Grade 1(absolute not)–5(totally))
- What do you not like in them? (Free text).

2.3 Likeness of Online Lessons

Most of them liked it, see Fig. 2, they graded it quite well. (We must note that first semester student's likeness was lower than in case of experienced students, which average was grade 4.0. This is a little bit higher, but this is not so surprised, they are practically professionals.) 62% of Hungarian students graded 4 or 5 the likeness of Teams online courses. At about 9.5% graded it 1 or 2 others liked it. The average grade is 3.76. This value is higher than we thought before the survey.

2.4 May Substitute Classical Lessons?

The result for the second question was a little bit surprising. As we saw at previous question, they liked online lessons, but the same time they do not think it may substitute the classic face-to-face teaching see Fig. 2. The average grade is only 3.04 contrary to 3.76!

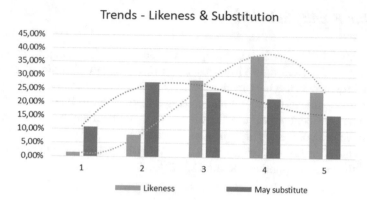

Fig. 2 May or may not substitute classical teaching

Comparing the trends of the data given in the two different questions (Hungarian survey), we can clearly see that to go into the virtual space for ever and neglect personal classes is not so acceptable for a lot of students. See the dotted trend lines, Fig. 2.

2.5 Lack of Interactivity

So, students liked very much online courses, but they can't substitute classic educational method. What my cause this controversy? What is the reason for this contradiction? We asked their opinion what they do not like.

They mentioned a lot of things; we analyzed the results and published in papers as well [19, 20]. Among others, the result proved that they need more interactivity, more personality which seemed to be lost using virtual classrooms. At Table 2, we gave the average grade of likeness of online lessons in the case the student mentioned a given problem. Naturally, if they have a problem, the likeness is lower.

Though they miss interactivity, we noticed all over distance period that students now (maybe regarding this lonely period) are less active than before. They do not like to switch on their microphone or Web camera. They do not like to share their screens. They prefer chat messages. Therefore, interaction is not easy with them.

Table 2 What students do not like

		Likeness
Reduced interaction	Mentioned	3.82
	Not mentioned	3.95
Impersonality	Mentioned	3.76
	Not mentioned	3.94
Hard to focus	Mentioned	3.58

3 We'll Be Back … to the School

Based on the epidemic regulations of the country, 2021 autumn, we tried hybrid teaching method. Half of the students were personally in the class, and half of them were online, next week they changed—to avoid crowd in the buildings. We started 2022 spring semester with distance teaching due to a strong Covid wave, but from the 1st of March, we continued with face-to-face teaching. The rules for absence were not so strict because faculty considered that may be there will be more illness among students and professors than usual. For the same reason, lessons were streamed and recorded through Teams for those who must stay at home for any reasons.

We and all over the world educators notice that something changed—again [23–25]. Yes, as we know this, one thing is constant, the change. A lot of students do not want to come into university at all; they prefer to be at home, and they neglect interactivity as many as they can—they were absent till the maximum limit. As we are involved to understand the students' needs, we again created an anonymous survey and asked their opinions.

3.1 The Environment of the Survey

It was done in April, in 2022 spring semester, after the distance period. 169 Hungarian students answered the questions. Filling out was not compulsory. The age group is between 20 and 25. They are from the countryside and from bigger towns too, mixed.

We used Google Forms to support anonymity.

Survey for Hungarian students is available at https://forms.gle/J5syCkqYXFYm uY148.

Now, we should like to focus on three questions with which we can detect how the students' needs changed toward online teaching.

- What form of teaching would you choose? (Dropdown list)
- In terms of lecture-style lessons, which do you find most effective? (Dropdown list)
- Which is the most adequate practice style? (Dropdown list).

3.2 The Teaching Mode

After two years of studying in emergency situation, they already tried fully online and hybrid teaching.

About 30% of students would like to learn fully online, and only, 10% want to go back to traditional school! A lot of students would prefer a mixed style see Fig. 3.

See, Fig. 2 again, where we collected data what they think about possible substitution of traditional education with online teaching. We can compare data (Hungarian

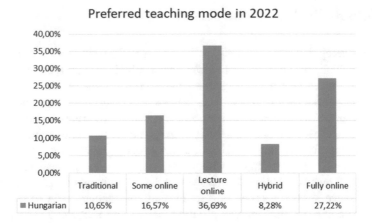

Fig. 3 Preferred teaching mode due survey in 2022

students) easily. We made equivalence between grade 1 (not possible to substitute) with fully traditional choice, grade 5 (totally substitutable), and a mixed possibility with grades 2–4. Let us see the result in Fig. 4.

We can notice that fully online mode is the winner of the last two years changes. At the beginning of emergency situation only, ~ 15% thought that online classes can substitute face-to-face lessons, now, after two years ~ 30% think that no need of personal teaching at all! It is also interesting, what is the main attractiveness of online teaching, we examined this question in a paper earlier [20].

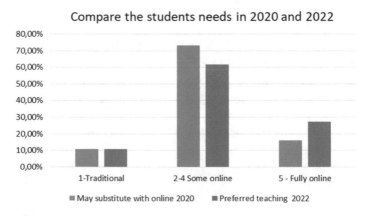

Fig. 4 Comparison of changes in students' thinking about online teaching

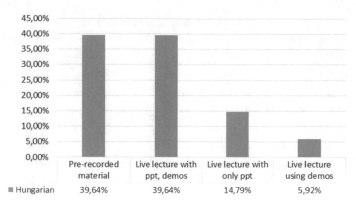

Fig. 5 Preferred lecture style

3.3 Interaction, Common Work?

One is the most precious; serious part of education is the interaction among the professors and the students, the common work, the interaction, discussion between them. Think about Dale's cone again. Nowadays, it is said, a professor should work as a tutor and not as a classical teacher who use frontal teaching tools.

3.4 Lectures

Our second question was: In terms of lecture-style lessons, which do you find most effective? The possible choices were the followings: pre-recorded material, no need for live lectures; live lecture with ppt, demos; live lecture with content outline (ppt); live lecture using demos only.

If somebody prefer pre-recorded lectures, then he/she does not want to be involved, does not want to discuss anything in real time. The possibility of real-time interaction is lost.

Let us see the results see Fig. 5. Surprisingly, at about 40% of students choose this ready-made, impersonal lecture style as most effective. But, there is another group; it is as big as the first mentioned (40%) which prefer live lectures with demos and ppt-s.

3.5 Practice Work

Our next question is: Which is the most adequate practice style?

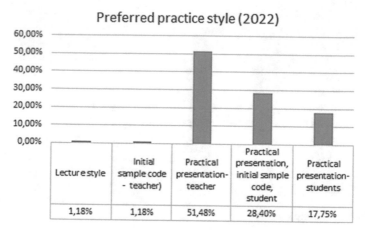

Fig. 6 Students opinions about best practice styles

The possible choices were the followings: Lecture style + small ppt; from the initial sample code—solution written by the teacher; after practical presentation, instructor writes the full solution; after practical presentation, individual solutions from initial sample code; direct practical exercise presentation—then individual solutions.

The first three choices means that professor has a bigger role—they are more classical, frontal methods. The last two choices mean, the professor role is rather a tutor role; he/she may give personal help in solutions. Meanwhile, it means a more active learning mode. But, the fact is that at more than half of the students prefers to attend practice lessons without really, individual contributions, where they might understand deeper the contexts, can ask better questions—see Fig. 6.

This kind of attitude is against Dale's experience; it is against the law of personal care and attention. Covid-19, the emergency situation, the online lessons strengthened such a strange attitude.

4 Summary

In the twenty-first century, living in a digital world, it looks changed everything, education too. We, together with other researchers, believe in active learning methods. Before Covid-19, our CRS, E-Lection was very effective to make the classical lectures more interactive, and students were activated well in it. We must note, as the saying goes, "you can have too much of a good thing," too much E-Lection made less productivity.

During emergency situation, during the last two years—when personal meeting was not possible between the professors and the students—professors and students learned and used a lot of new informatics tools to enhance learning experience. It is a very good base to increase the level of active learning and give a better motivation to students—contrary we found that a large proportion of students' priorities becomes more convenience and are less inclined to work independently. Unfortunately, this unexpected result comes at the expense of the ability to learn and the knowledge acquired.

It is clear, we cannot step into the same river twice; we must not go back simply to the teaching mode before Covid-19. The usage of new tools is required by everybody. This moment, we do not have the stone of wise; we have to work on a more successful mixed teaching mode, a new mix of tools, maybe a new E-Lection we need again, as we did before to bring back students' activity.

References

1. Zitny R et al (2016) Education using mobile technologies. In: ICETA 2016.11.24–25. Stary Smokovec IEEE, pp 115–120. ISBN 9781509046997
2. Zoltán I, Bakonyi BH, Illés Z Jr (2016) Supporting dynamic, bi-directional presentation management in real-time. In: Emil V (ed) 11th joint conference on mathematics and computer science, CEUR-WS.org, 6 p
3. Bakonyi HV, Zoltán I (2017) Real-time tool integration for lectures. In: 15th IEEE international conference on emerging elearning technologies and applications (ICETA 2017). Starý Smokovec, Slovakia, 2017.10.26–2017.10.27. IEEE Computer Society Press, Denver, pp 31–36. ISBN 978-1-5386-3294-9
4. Prensky M (2001) From on the horizon, vol 9, no 5. MCB University Press. https://bit.ly/2Ye KG7U
5. Hayes NK (2007) Hyper and deep attention: the generational divide in cognitive moods. Profession, 187–199 pp
6. Bradberry T, Multitasking damaging your brain and career, new studies suggest. https://www.forbes.com/sites/travisbradberry/2014/10/08/multitasking-damages-your-brain-andcareer-new-studies-suggest/2/#6088a80642ef
7. Shaaruddin J, Mohamad M (2017) Identifying the effectiveness of active learning strategies and benefits in curriculum and pedagogy course for undergraduate TESL students. Creative Educ 8(14). Available at https://www.scirp.org/journal/PaperInformation.aspx?PaperID=80647
8. Opre D et al (2022) Supporting students' active learning with a computer based tool. Act Learn High Educ. https://doi.org/10.1177/14697874221100465
9. Brown JL (2016) Quick, click: student response systems evolve in higher ed, New student response systems offer increased versatility. University Business. http://bit.ly/2fnJMRw
10. Dangel H, Wang C (2008) Student response systems in higher education: moving beyond linear teaching and surface learning. J Educ Technol Dev Exchange 1(1):93–104. http://www.sicet.org/journals/jetde/jetde08/paper08.pdf
11. Mader S, Bry F (2019) Audience response systems reimagined. In: Herzog M, Kubincová Z, Han P, Temperini M (eds) Advances in web-based learning—ICWL 2019. ICWL 2019. Lecture notes in computer science, vol 11841. Springer, Cham
12. Li R (2020) Communication preference and the effectiveness of clickers in an Asian university economics course. Heliyon 6(4):e03847. https://doi.org/10.1016/j.heliyon.2020.e03847
13. Bakonyi V, Zoltan I, Verma C (2021) Key element in online education to activate students with real-time tools. In: Institute of electrical and electronics engineers 2021 2nd international conference on computation, automation and knowledge management (ICCAKM) conference: Dubai, United Arab Emirates, 19 Jan 2021–21 Jan 2021. Curran Associates, Red Hook (NY), pp 326–331

14. Bakonyi V, Zoltan I, Verma C (2020) Towards the real-time analysis of talks. In: Chauhan AK, Singh G (2020) International conference on computation, automation and knowledge management Dubai. Amity University, United Arab Emirates, pp 322–327, 6 p

15. Bakonyi V, Illes Z, Verma C (2020) Analyzing the students' attitude towards a real-time classroom response system. In: Institute of Electrical and Electronics Engineers 2020 international conference on intelligent engineering and management (ICIEM), London, pp 69–73. ISBN 9781728140971

16. Kővári E, Bak G (2021) University students' online social presence and digital competencies in the COVID-19 virus situation. In: Agrati LS et al (eds) Bridges and mediation in higher distance education. HELMeTO 2020. Communications in computer and information science, vol 1344. Springer, Cham. https://doi.org/10.1007/978-3-030-67435-9_13

17. Martin F, Parker M, Deale D (2012) Examining interactivity in synchronous virtual classrooms. Int Rev Res Open Distrib Learn 13(3):227–261. https://doi.org/10.19173/irrodl.v13i3.1174

18. Bower B (2006) Virtual classroom pedagogy. In: Conference: proceedings of the 39th SIGCSE technical symposium on computer science education, SIGCSE 2006, Houston, Texas, USA, 3–5 March 3–5. https://doi.org/10.1145/1124706.1121390

19. Bakonyi V, Zoltán I (2020) Real-time and digital solutions in education during emergency situation in Hungary. In: Abonyi-Tóth A, Stoffa V, Zsakó L (eds) New methods and technologies in education, research and practice: proceedings of XXXIII. DidMatTech 2020 Conference Budapest. ELTE Informatikai Kar, Magyarország, 507 p, pp 231–240, 10 p

20. Bakonyi V, Zoltán I (2020) Real-time online courses during emergency situation in Hungary. In: ICETA 18th international conference on emerging elearning technologies and applications, Nov 12, Stary Smokovec, Slovakia

21. Bakonyi V, Zoltán I, Szabó T (2021) Real-time interaction tools in virtual classroom systems. In: Singh PK, Singh Y, Chhabra JK, Illés Z, Verma C (eds) Recent innovations in computing: proceedings of ICRIC 2021, vol 2. Springer Singapore, Singapore, pp 625–636, Paper: Chapter 47, 12 p

22. Bakonyi V, Zoltán I, Verma C (2021) Real-time education in emergency situation. In: 2021 international conference on advances in electrical, computing, communication and sustainable technologies (ICAECT), Piscataway (NJ), USA. IEEE, pp 1–6, 6 p

23. Bialystok L (2022) Education after COVID. https://doi.org/10.7202/1088373ar. Available at: https://www.researchgate.net/publication/360986155_Education_after_COVID

24. Mazzara M et al (2022) Education after COVID-19. In: Smart and sustainable technology for resilient cities and communities. https://doi.org/10.1007/978-981-16-9101-0_14. Available at https://bit.ly/3dsyzOF

25. Koopman O, Joy K, Karen K (2021) The rise of the university without classrooms after COVID-19. In: Re-thinking the humanities curriculum in the time of COVID-19. CSSALL Publishers (Pty) Ltd. Available at: https://bit.ly/3zYRf03

Aspect-Based Opinion Mining Framework for Product Rating Embedded with Fuzzy Decision

Garima Srivastava, Vaishali Singh, and Sachin Kumar

Abstract The technique of evaluating text on a subject jotted down in natural language and classifying it with polarity as neutral form, positive form, or negative form based on the sentiments, emotions, and expressions by humans in it is known as opinion mining or sentiment analysis. Analyzing and extracting opinions from such a large volume of reviews manually is probably impossible. A self-dependent automated opinion mining strategy is required to resolve this issue. It can primarily be carried out at three levels: document, sentence, and aspect level. Work proposed is an effort to rate a product and predict its demand based on features extracted from aspects of opinion expressed with fuzzy set decision boundary. Aspect identification, aspect-based opinion word identification, and inclination as positive, negative or neutral are the three major tasks in aspect-based opinion mining. Weights were assigned according to the polarity percentage of the words extracted. Polarity scores of the labeled dataset were then divided into 5 fuzzy set decision boundaries, confirmed positive, positive, neutral, negative, and confirmed negative, to accurately predict the correlation between polarity score obtained and product rating. Algorithm proposed was operated on aspect-based corpus for product rating which acquired from GitHub, comprised of 7563 rows of product aspects; the analysis suggests that maximum share 44% lies with positive range of fuzzy boundary, and this can be tapped for trend prediction of customer choice; the algorithm successfully provides a fuzzy-assisted prediction mechanism for product rating for the corpus with labeled aspects; algorithm performs at par with existing technique, and it also provides percentage share of products presence in market for future trends.

Keywords Polarity · Aspect · Correlation · Product rating · Fuzzy boundary

G. Srivastava (✉) · V. Singh
Maharishi University of Information Technology, Lucknow, India
e-mail: gsrivastava1@lko.amity.edu

G. Srivastava · S. Kumar
Amity University, Lucknow Campus, Lucknow, India
e-mail: skumar3@lko.amity.edu

1 Introduction

With the Internet's growth, an enormous amount of information is available to the public over the past decade. People worldwide digitize their diverse opinions and experiences on the Internet, making numerous pieces of information available in digital form. The Web is heading to an age where consumer opinions would be predictive and will dictate products and services. Sentiment analysis has received a lot of interest and has been recognized as an emerging study area over the past 10 years. The expressions in form of words describe a person's perception, opinion and sentiment of, or understanding of, a certain thing,

Further, we entered the age of ecommerce thanks to the Internet's and technology's rapid advancement. Ecommerce is the act of selling and purchasing goods or services over the Internet. Nowadays, many consumers express their views as for or against goods or services online. The three fundamental components of an opinion or senti-ment—opinion holder, opinion object, and opinion orientation—must be understood in order to determine opinion polarity. The person or organization that holds the opinion regarding a product or subject is known as the opinion holder. The entity that a user offers an opinion about is called an object, and an opinion's orientation determines whether that opinion is positive or negative. Figure 1 depicts architecture of a generic opinion mining framework [1–3].

Sentiment analysis can be categorized with five specific problems.

i.   Document-level sentiment analysis
ii.   Sentence-level sentiment analysis
iii. Aspect-based sentiment analysis
iv.   Sentiment lexicon acquisition
v.   Comparative sentiment analysis.

Fig. 1 Architecture of a generic opinion analysis system

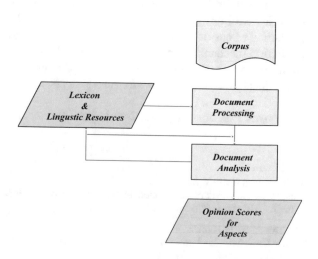

Fig. 2 Framework for opinion mining/sentiment analysis

While sentiment analysis locates and analyzes the polarity of a text, opinion mining obtains and analyzes people's views about a subject [4–6]. Opinion mining is the technique of categorizing opinions to identify if the polarity of the opinion holder as neutral, positive, or negative about a product or topic. Figure 2 illustrates the general framework or phases involved in the identification of polarity of expressions:

Data gathering. The WWW is a vast repository of data, and the process of opinion mining begins with data collection. Review sites and social networks are only two examples of diversified sources of data from which data for opinion mining can be gathered. Flipkart.com instance may be employed to accumulate data on customer opinion for any product.

Data Preprocessing. Primary step for classification involves preprocessing of data. Preprocessing helps improve the performance of classification algorithms by reducing noise. Real-time sentiment classification is accelerated with the help of preprocessing.

Preprocessing involves:

i.   Tokenization
ii. Removal of URLs, hash tags, references, and special characters
iii. Slang word translation
iv. Stemming.

The subsequent step in the classification of opinions is feature extraction or feature selection. Here, pertinent features that are used to create an efficient and precise classifier are chosen. The potential of the features selected determines how successful a classification model will be. Once features are identified various, the classification of the opinions is done using supervised, unsupervised machine learning, or lexicon-based algorithms, which is a tough problem.

Opinion Polarity, Evaluation, and Results: In this step, opinions are divided into three categories—positive, negative, and neutral—and the effectiveness of the classification methods is assessed using established performance metrics. Lastly, the opinion mining results are shown as charts or graphs. [7–10]

The paper is organized as follows: Sect. 2 presents the motivation for conducting the study; Sect. 3 describes the materials and method employed for preprocessing of the corpus and implementing the algorithm adopted for estimating the aspect score classification of the cleaned data; Sect. 4 briefly discusses the outcomes of the experiment carried out before drawing a conclusion in Sect. 5.

2   Motivation

Applying various strategies to analyze the opinions and sentiments derived from data sources is known as opinion mining [11]. Machine learning and lexicon-based techniques are two processes for classifying opinions expressed. Machine learning techniques is a commonly utilized and useful strategy in opinion mining algorithms devised because of its ability to handle enormous amounts of data and automatic implementation. The machine learning methodology uses an algorithm that enables systems to understand [12]. Algorithms for machine learning classification are a group of techniques for discovering patterns in data. Classification algorithm first creates a framework to classify the testing dataset after learning how to classify opinions from a training dataset [13]. There are various types of machine learning-based sentiment classification methods, including supervised learning approach, unsupervised learning approach, and semi-supervised learning approach [14].

Sentiment lexicons, which are collections of annotated and preprocessed sentiment phrases, are used in lexicon-based approaches [15]. Through the discovery of an opinion lexicon that analyzes the textual material, the lexicon-based technique carries out opinion mining. Lexicon-based methods can categorize opinions in two different ways [14]:

i.   dictionary-based approach and
ii. corpus-based approach.

The decision tree (DT), Naive Bayes (NB), K-nearest neighbor (KNN), support vector machine (SVM), and SentiWord-Net machine learning and lexicon-based algorithms were compared in [15], which provided a detailed assessment of various machine learning techniques. A model for combined sentiment topics (CSTs) based on unsupervised learning was introduced in [16]. According to experimental findings, this model outperformed supervised and semi-supervised methods in diversified domains.

Support vector machines and Naive Bayes supervised machine learning methods were covered in [17], which also provided a summary of the opinion mining area. They determined that the task of opinion mining is exceedingly difficult and provided a description of the metrics for performance evaluation of the opinion mining classification method: accuracy and F1-score. To examining customer feedback, researchers in [18] developed phrase-level opinion mining. Frequent itemset mining was utilized for aspect extraction, and the opinion or sentiment orientation of each aspect was determined using the supervised machine learning Naive Bayes method. In [19], a dictionary-based method for polarity analysis of Twitter corpus was presented. Combining various opinion mining classifiers is known as the hybrid classification approach for opinion mining. Shahnawaz and Astya [20] addressed many methods for sentiment analysis, including lexicon, machine learning, and hybrid sentiment classification strategies.

3   Proposed Methodology

Extracting aspect from corpus provides very useful insights into thought process of customers, which can serve as base for prospects of products and their commercial viability. Aspects are the potential entities rated by the reviewers that provide excellent information on trends of sale and customer liking [21]. An aspect for a product can be a sentence or a single word. From work studied, it can conclude that generally noun and phrases containing noun are potential aspects. Figure 3 depicts the proposed generalized flow graph for aspect-assisted opinion mining framework for product rating embedded with fuzzy decision. To segregate aspects, searching of noun and phrases containing nouns of the reviews are required. Aspect extraction is performed at sentence level for all reviews; steps can be summarized as

i.   Acquire features and split the features into the sentence, the sentence is then put up for analysis.
ii.   Evaluate the part of speech tag at the sentence level.
iii. Extract NN, NNP, NNS, etc., tagged words. Step is performed for each sentence.
iv.   Estimate acutance of words extracted.
v.   Removal of redundant words.
vi.   Work on most frequent words.
vii. Cluster synonyms and label them.
viii. Estimate subjectivity and polarity.
ix.   Classify tokens as positive, negative, and neutral.

Aspect table comprises of potential aspects and similar words. The aspect-based corpus for product rating was acquired from GitHub; aspect database comprised of 7563 rows; visible column is depicted in Fig. 4. Aspect table so formed is employed for evaluating the subjective phrases [22–24]. Stanford POS tagger was employed for tagging purpose.

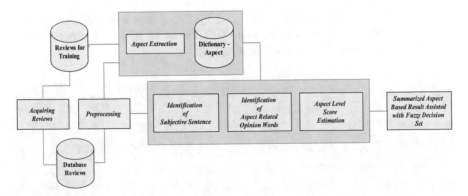

Fig. 3 Generalized flow graph aspect-assisted opinion mining framework for product rating embedded with fuzzy decision

	example_id	comment	aspect_term	term_location
0	494:1_0	buy separate ram memory rocket	RAM memory	17--27
1	311:18_1	laptop comma bass weak sound comes sounding tinny	sound	52--57
2	256:1_0	hardware seems better imac smaller	hardware	5--13
3	748:1_0	easy navigate even novice	navigate	19--27
4	128:1_0	amazing performance anything throw	Performance	8--19

Fig. 4 Aspect-based product rating database

Reviews acquired from corpus were first subjected to preprocessing steps to improve accuracy and avoid latency; steps include removal of stop words, all unnecessary non-alphabetic characters, and smileys. Python has been employed as a coding language for the work proposed. Corpus is then cleaned via:

i.   Removal of all capital letters, punctuations, emojis, links, etc. Basically, removing all that is not words or numbers.
ii. Tokenization of the corpus into words, which means breaking up every comment into a cluster of individual words.
iii. Removal of all stop words, which are words that don't add value to a comment, like "the," "a," "and," etc.

The group a, r, n, v present with database means adjective, adverb, noun, and verb, respectively. Scores are estimated and are defined by finding opinion words in the database, if present, scores are estimated for the same. In case of appearance of matched opinion word with number of occurrences more than once, then algebraic mean of polarity score is considered. Negations are to be estimated with proper justification to get the information context of the phrase. The corpus is POS tagged, and same is depicted in Fig. 5. After cleaning the corpus is lemmatized, this technique groups together the inflected forms of a word, so they can be analyzed as a single item, since they have a similar meaning. Task is performed employing lemmatize function from Python.

Finally, all empty comments from the data are removed; some people just comment emojis, punctuations, or things like that. This leaves us with 3221 comments to analyze. Figure 6 depicts the word frequency.

	example_id	comment	aspect_term	term_location	comment_t	Pos Tagged
0	494:1_0	buy separate ram memory rocket	RAM memory	17--27	[buy, separate, ram, memory, rocket]	[(buy, v), (separate, a), (ram, a), (memory, n...
1	311:18_1	laptop comma bass weak sound comes sounding tinny	sound	52--57	[laptop, comma, bass, weak, sound, comes, soun...	[(laptop, a), (comma, n), (bass, n), (weak, a)...
2	256:1_0	hardware seems better imac smaller	hardware	5--13	[hardware, seems, better, imac, smaller]	[(hardware, n), (seems, v), (better, a), (imac...
3	748:1_0	easy navigate even novice	navigate	19--27	[easy, navigate, even, novice]	[(easy, a), (navigate, r), (even, r), (novice,...
4	128:1_0	amazing performance anything throw	Performance	8--19	[amazing, performance, anything, throw]	[(amazing, a), (performance, n), (anything, n)...

Fig. 5 Corpus with POS tagging

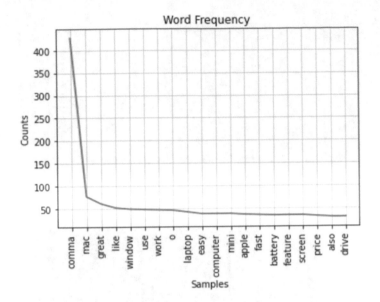

Fig. 6 Word frequency count of corpus

Figure 7 depicts the number of trigrams that appear in corpus; the graph presents information about expressions, people carry for products. For subjective phrases, identification is achieved via labeled polarity words as present in phrases defining expressions about products. Opinionated phrases so labeled are called subjective phrases, that must be analyzed, and other redundant phrases not included for expression establishment should be removed. This helps in reducing latency further.

The proposed technique takes phrases as a subjective phrase only if it has a feature present in the aspect table. The same was estimated for aspect corpus and is depicted in Fig. 8. In aspect-based opinion mining, it is of primary importance to identify words that alter the meaning of aspects, as these help in identification of opinion words that establish polarity of communication toward those aspects.

Work done employed adjectives, verbs, adverb adjective, and adverb verb combinations for polarity establishment.

The proposed technique employs adjectives, verbs, adverb adjective combinations, and adverb verb combinations as potential data for polarity estimation. Adjectives are typically very important opinion words in a phrase. Trigrams are searched forward and backward from the aspect location of the sentence to acquire opinion words. This was performed employing POS tag data; the technique doesn't provide satisfactory results if phrases have number of aspects and will not be able to accurately estimate contextual polarity scores.

Figure 9 presents the sparse diagram between subjectivity and polarity estimated from the corpus. Technique employed uses dependency relations that exist between words. The algorithm extracts opinion words for score estimation. Based on these

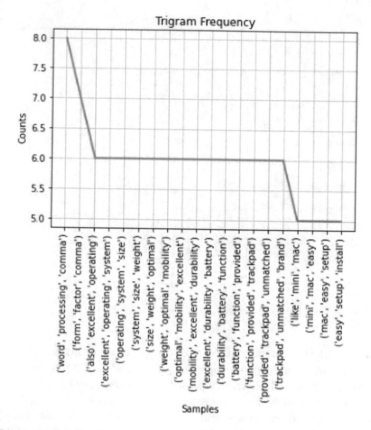

Fig. 7 Trigrams present in corpus

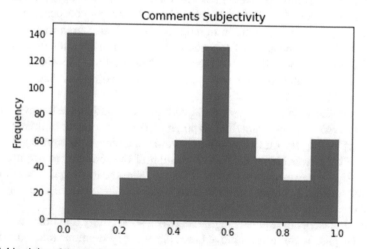

Fig. 8 Subjectivity of the comments in corpus

scores, the polarity of the aspects is estimated; values estimated are further subjected to fuzzification to predict the prospects of products.

Aspect-Level Score Calculation: To this point, the subjective phrase of a product review is estimated for aspect-related opinion words. In aspect-level score estimation step, the polarity score value of an aspect in a phrase is evaluated by algebraically adding opinion word scores in that phrase.

The proposed methodology provides polarity ratings to the opinion words algorithm before assigning priority values. Here, word sense disambiguation is not considered in the process [25–27] (Fig. 10).

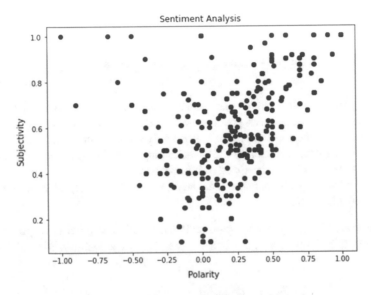

Fig. 9 Sparse diagram between subjectivity and polarity estimated from the corpus

	example_id	comment	aspect_term	term_location	comment_t	comment_l	polarity	subjectivity	sentiment
0	494:1_0	buy separate ram memory rocket	RAM memory	17–27	[buy, separate, ram, memory, rocket]	['buy', 'separate', 'ram', 'memory', 'rocket']	0.000000	0.000000	Neutral
1	311:18_1	laptop comma bass weak sound comes sounding tinny	sound	52–57	[laptop, comma, bass, weak, sound, comes, soun...	['laptop', 'comma', 'bass', 'weak', 'sound', '...	-0.041667	0.508333	Negative
2	256:1_0	hardware seems better imac smaller	hardware	5–13	[hardware, seems, better, imac, smaller]	['hardware', 'seems', 'better', 'imac', 'small...	0.250000	0.500000	Positive
3	748:1_0	easy navigate even novice	navigate	19–27	[easy, navigate, even, novice]	['easy', 'navigate', 'even', 'novice']	0.433333	0.833333	Positive
4	128:1_0	amazing performance anything throw	Performance	8–19	[amazing, performance, anything, throw]	['amazing', 'performance', 'anything', 'throw']	0.600000	0.900000	Positive

Fig. 10 Polarity, subjectivity, and sentiment of the words in corpus

4   Result and Discussion

Next step comprises of estimation of aspect score for entire review comments by customers. Total score value of the aspect from entire review is estimated by algebraically adding the phrase-wise score of the aspect. Polarity of the area separately aggregated. If the positive score is higher side, the conclusion about the product from the entire set of reviews is categorized as positive otherwise it is negative. Aspect score for the entire corpus can be estimated using following relation [28].

For each aspect j of the product:

$$\text{Positive\_ Polarity}_{\text{Aggregate}}[j] = \sum_{i} \text{Positivepol}_{i,j} \qquad (1)$$

$$\text{Negative\_ Polarity}_{\text{Aggregate}}[j] = \sum_{i} \text{Negativepol}_{i,j} \qquad (2)$$

Polarity values of the labeled dataset were then divided into 5 fuzzy set decision boundaries, confirmed positive, positive, neutral, negative, and confirmed negative, to accurately predict the correlation between polarity score obtained and product rating. Figure 11 depicts fuzzification of the polarity ranges for five different categories.

Polarity estimated was mapped on fuzzy set defined for 5 different ranges confirmed positive, positive, neutral, negative, and confirmed negative; from Fig. 12, it can be concluded that:

i.   Confirmed Positive—for the product that crossed 0.5% polarity on positive side had excellent reviews in the corpus, the percentage share is 36% of the total

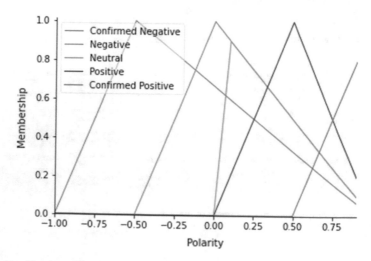

Fig. 11 Fuzzification of the polarity ranges for five different categories

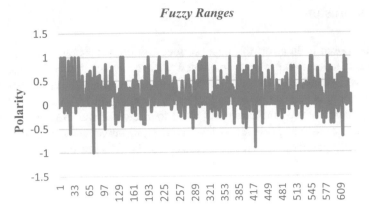

Fig. 12 Percentage share of product with fuzzy boundaries

data, and fuzzy range suggests that the products studied will be referred and bought. Product may be rated as top performing and highly saleable.

ii. Positive—product that stays between 0 and 0.5% polarity on positive side had good reviews in the aspect corpus, the percentage share is 44% of the total data, and fuzzy range suggests that products studied will be referred and has maximum percentage share. Product may be categorized as performing satisfactory and saleable.

iii. Neutral—for the product that stays at 0% polarity on axis had mixed reviews in the aspect corpus, the percentage share is 3% of the total data, and fuzzy range suggests the products studied may or may not be will be referred at all. Product under this category comes under average performing category.

iv. Negative—for the product that stays between 0 and 0.5% polarity on negative side had poor reviews in the aspect corpus, the percentage share is 12% of the total data, and fuzzy range suggests that the products studied will never be referred. Organization needs to modify or stop production of the product. Product under this category comes under below average performing category.

v. Confirmed Negative—for the product that stays between beyond 0.5% polarity on negative side had extremely poor reviews in the aspect corpus, the percentage share is 1% of the total data, and fuzzy range suggests that the products studied had no future at all and should be completely removed from list. Product under this category comes under poor performing category.

Algorithm proposed maps the polarity ranges on the fuzzy boundaries with 5 classification sets and establishes correlation between polarity and product rating through fuzzy boundary.

5 Conclusion

Work proposed is an effort to rate a product and predict its demand based on the aspect-assisted opinion mining embedded with fuzzy set decision boundary. The primary work in aspect-assisted opinion mining is aspect identification, identification and extraction of words that act as feature of aspects and its inclination detection. Polarity scores of the labeled dataset were then divided into 5 fuzzy set decision boundaries, confirmed positive, positive, neutral, negative, and confirmed negative, to accurately predict the correlation between polarity score obtained and product rating. Algorithm proposed was operated on aspect-based corpus for product rating which acquired from GitHub, comprised of 7563 rows of product aspects; confirmed positive was defined for products that crossed 0.5% polarity on positive side; the percentage share is 36% of the total data, and fuzzy range suggests that these products will be referred and bought. Positive was defined for the product that stays between 0 and 0.5% polarity on positive side; the percentage share is 44% of the total data, and fuzzy range suggests that the products studied will be referred and has maximum percentage share, Neutral was defined for the product that stays at 0% polarity on axis had mixed reviews in the aspect corpus; the percentage share is 3% of the total data; negative was defined for the product that stays between 0 and 0.5% polarity on negative side had poor reviews in the aspect corpus; the percentage share is 12% of the total data, and confirmed negative was defined for the product that stays between beyond 0.5% polarity on negative side had extremely poor reviews in the aspect corpus; the percentage share is 1% of the total data; the analysis suggests that maximum share 44% lies with positive range of fuzzy boundary, and this can be tapped for trend prediction of customer choice; the algorithm successfully provides a fuzzy-assisted prediction mechanism for product rating from the aspect-based corpus; algorithm performs at par with existing technique, and it also provides percentage share of products presence in market for future trends.

References

1. Saad S, Saberi B (2017) Sentiment analysis or opinion mining: a review. Int J Adv Sci Eng Inf Technol 7:1660. https://doi.org/10.18517/ijaseit.7.5.2137
2. Abirami AM, Gayathri (2017) A survey on sentiment analysis methods and approaches. In: Proceedings of the 2016 eighth international conference on advanced computing, pp 72–76
3. Phan HT, Tran VC, Nguyen NT, Hwang D (2020) Improving the performance of sentiment analysis of tweets containing fuzzy sentiment using the feature ensemble model. IEEE Access 8:14630–14641
4. Aung KZ, Myo N (2017) Sentiment analysis of students' comment using lexicon based approach. In: Zhu G, Yao S, Cui X, Xu S (eds) IEEE/ACIS 16th international conference on computer and information science. IEEE, pp 149–154
5. Baccianella S, Esuli A, Sebastiani F (2010) SentiWordNet 3. 0: an enhanced lexical resource for sentiment analysis and opinion mining. In: Proceedings of the seventh international conference on language resources and evaluation, 2200–2204. http://www.lrec-conf.org/proceedings/lrec2010/pdf/769_Paper.pdf

6. Abirami AM, Gayathri V (2017) A survey on sentiment analysis methods and approach. In: 8th IEEE international conference on advanced computing. IEEE Press, Chennai, India, pp 72–76. https://doi.org/10.1109/icoac.2017.7951748

7. Mishra N, Jha CK (2012) Classification of opinion mining techniques. Int J Comput Appl 56(13):1–6. https://doi.org/10.5120/8948-3122

8. Soong H-C, Jalil NBA, Ayyasamy RK, Akbar R (2019) The essential of sentiment analysis and opinion mining in social media: introduction and survey of the recent approaches and techniques. In: 9th IEEE symposium on computer applications & industrial electronics. IEEE Press, Malaysia, pp 272–277. https://doi.org/10.1109/iscaie.2019.8743799

9. Golande A, Kamble R, Waghere S (2016) An overview of feature based opinion mining. In: Corchado Rodriguez J, Mitra S, Thampi S, El-Alfy ES (eds) Intelligent systems technologies and applications. Advances in intelligent systems and computing, vol 530. Springer, Cham, pp 633–645. https://doi.org/10.1007/978-3-319-47952-1_51

10. Emam A, Alzahrani M (2017) Opinion mining techniques and tools: a case study on an Arab newspaper. In: IEEE international conference on computational science and computational intelligence (CSCI). IEEE Press, Las Vegas, NV, USA, pp 292 296. https://doi.org/10.1109/csci.2017.49

11. Sindhura, Sandeep Y (2015) Medical data opinion retrieval on Twitter streaming data. In: IEEE international conference on electrical, computer and communication technologies (ICECCT), IEEE Press, Coimbatore, India. https://doi.org/10.1109/icecct.2015.7226043

12. Bhavitha BK, Rodrigues AP, Chiplunkar NN (2017) Comparative study of machine learning techniques in sentimental analysis. In: IEEE international conference on inventive communication and computational technologies. IEEE Press, Coimbatore, India, pp 216–221. https://doi.org/10.1109/icicct.2017.7975191

13. Eshak MI, Ahmad R, Sarlan A (2017) A preliminary study on hybrid sentiment model for customer purchase intention analysis in social commerce. In: IEEE conference on big data and analytics. IEEE Press, Kuching, Malaysia, pp 61–66. https://doi.org/10.1109/icbdaa.2017.8284108

14. Rani MS, Sumathy S (2017) Analysis on various machine learning based approaches with a perspective on the performance. In: IEEE innovations in power and advanced computing technologies. IEEE Press, Vellore, pp 1–7. https://doi.org/10.1109/ipact.2017.8244998

15. Sankar H, Subramaniyaswamy V (2017) Investigating sentiment analysis using machine learning approach. In: IEEE international conference on intelligent sustainable systems. IEEE Press, Palladam, India, pp 87–92. https://doi.org/10.1109/iss1.2017.8389293

16. Usha MS, Devi MI (2013) Analysis of sentiments using unsupervised learning techniques. In: IEEE international conference on information communication and embedded systems (ICICES). IEEE Press, Chennai, India. https://doi.org/10.1109/icices.2013.6508203

17. Das MK, Padhy B, Mishra BK (2017) Opinion mining and sentiment classification: a review. In: IEEE international conference on inventive systems and control. IEEE Press, Coimbatore, India, pp 1–3. https://doi.org/10.1109/icisc.2017.8068637

18. Jeyapriya A, Selvi CSK (2015) Extracting aspects and mining opinions in product reviews using supervised learning algorithm. In: 2nd IEEE international conference on electronics and communication systems. IEEE Press, Coimbatore, India, pp 548–552. https://doi.org/10.1109/ecs.2015.7124967

19. Biltawi M, Etaiwi W, Tedmori S, Hudaib A, Awajan A (2016) Sentiment classification techniques for Arabic language: a survey. In: 7th IEEE international conference on information and communication systems. IEEE Press, Irbid, Jordan, pp 339–346. https://doi.org/10.1109/iacs.2016.7476075

20. Shahnawaz, Astya P (2017) Sentiment analysis: approaches and open issues. In: IEEE international conference on computing, communication and automation. IEEE Press, Greater Noida, India, pp 154–158. https://doi.org/10.1109/ccaa.2017.8229791

21. Kumar, Vinoth V et al (2022) Aspect based sentiment analysis and smart classification in uncertain feedback pool. Int J Syst Assur Eng Manage 13(1):252–262

22. Kumar A et al (2022) Sentic computing for aspect-based opinion summarization using multi-head attention with feature pooled pointer generator network. Cogn Comput 14(1):130–148
23. Aggarwal CC (2022) Opinion mining and sentiment analysis. In: Machine learning for text. Springer, Cham, pp 491–514.
24. Liu B (2012) Sentiment analysis and opinion mining. Synthesis Lect Human Lang Technol 5(I), 1–167
25. Hu M, Liu B (2004) Mining and summarizing customer reviews. In: Proceedings of the tenth ACM SIGKDD international conference on knowledge discovery and data mining. ACM, pp 168–177
26. Qiu G, Liu B, Bu J, Chen C (2011) Opinion word expansion and target extraction through double propagation. Comput Linguist 37(1):9–27
27. Liu Q, Gao Z, Liu B, Zhang Y (2013) A logic programming approach to aspect extraction in opinion mining. In: 2013 IEEE lWICIACM international joint conferences on web intelligence (WI) and intelligent agent technologies (IAT), vol I. IEEE, pp 276–283
28. Banjar A, Ahmed Z, Daud A, Ayaz Abbasi R, Dawood H (2021). Aspect-based sentiment analysis for polarity estimation of customer reviews on twitter. Comput Mater Continua 67(2):2203–2225

The Problems and Organization of Learning Using Distance Educational Technologies: Practical Issues

Aliya Katyetova

Abstract Since the beginning of the pandemic 2020, educational system in every country has offered and introduced new technologies and instruments for distance education. Each school independently chose an Internet platform through which distance learning was organized. Parents and children are worried about Internet connection and their speed because there are problems with it everywhere, in urban and rural areas, also regarding technology. The results of distance learning in Kazakhstan during the pandemic indicate the insufficient effectiveness of national telecommunications networks. The author of the article has experience in teaching distance learning technologies at universities and outlines the problems and basic requirements for the organization of the educational process with the use of distance educational technologies in the Republic of Kazakhstan, namely, the practical questions based on what standards and rules, whereby and how to execute them in the organization of education including a primary school. The author provides a comparative study of learning management systems used in Kazakhstan education and suggests that a future task is the identification of the student in distance learning, which can be performed using facial recognition of students during authorization. Furthermore, the author plans to conduct a questionnaire survey among teachers, which will demonstrate the situation during distance learning in computer science lessons at primary schools. These practical issues will help to avoid the main distance learning problems of schoolchildren. The present paper is useful for teachers and school management in the implementation and control of educational activities.

Keywords Distance learning · Distance educational technologies · LMS

A. Katyetova (✉)
Eotvos Lorand University, Budapest, Hungary
e-mail: akatyetova@inf.elte.hu

© The Author(s), under exclusive license to Springer Nature Singapore Pte Ltd. 2023
Y. Singh et al. (eds.), *Proceedings of International Conference on Recent Innovations in Computing*, Lecture Notes in Electrical Engineering 1011,
https://doi.org/10.1007/978-981-99-0601-7_45

1 Introduction

In the spring of 2020, every school in the world including Kazakhstan moved on to a new form of education–distance education. This required the Ministry of Education and Science to revise approaches to the educational process.

Currently, Kazakhstan's education is going through a period of reforms. The development of domestic education is focused on the best world and European standards. Among them are digitalization and automation of the educational process and the introduction of electronic textbooks in schools in Kazakhstan. According to the instructional and methodical letter "About the features of the educational process in secondary education organizations of the Republic of Kazakhstan in the 2021–2022 academic year," the subject "Digital literacy" was introduced in the first grade from January 1, 2022. The name of the subject "Information and communication technologies" in primary school has been changed to "Digital literacy" [1]. Computer science textbooks have been developed for the youngest students: Now, first-graders will be taught the basics of computer literacy.

In the national project "Quality education "Educated Nation,"" a focus places on the availability of school content from home 24/7 and the provision of Internet at least 100 Mbit/s.

During COVID-19, many schools in Kazakhstan introduced distance educational technologies. In each educational institution, there had been problematic issues to do with implementation. This article was motivated by the author's previous work in e-learning and provides some suggestions about the means and rules by which distance learning is organized.

2 Requirements for Organizations of Education

By order of the Minister of Education and Science of the Republic of Kazakhstan on November 03, 2021, № 547 "On amendments to the Order of the Minister of Education and Science of the Republic of Kazakhstan dated March 20, 2015 No. 137 "On approval of the Rules for the organization of the educational process on distance learning technologies,"" distance educational technologies are implemented independently by each educational organization, where learning by distance educational technologies follows the curricula of primary, basic secondary, general secondary and additional education, as well as of technical and vocational, post-secondary education [2].

Hereafter is a list of requirements that must be observed in the education organization.

Fig. 1 Enter page language selection [4]

2.1 Information System and Educational Portal

Each organization should have an educational portal based on the known systems LMS—learning management system (Platonus, Moodle, Prometheus, Lotus Domino/Notes, or other automated information systems) with pages containing the training-methodical, organizational, and administrative information for students.

The popular e-learning environment Moodle is used by 60% of educational institutions around the world, the United Nations, and even Google [3]. For example, many schools in Estonia and other countries use Moodle to conduct lessons and exchange information.

Below are the pages of an automated information system Platonus (see Figs. 1 and 2) and Web portal EduPage (see Fig. 3):

Any automated information system for learning at an educational institution must contain a distance learning module (unit).

The distance learning module must contain:

- authorization (authentication) system (definition of access and user rights);
- role-based access control system;
- means of protection in case of failures;
- means of protection from malicious programs;
- means of logging the user experience;
- a content management system;
- the management system movement of students;
- testing system.

The education organization should have a distributed information system of educational process management, containing a database of students, information on current progress, and providing reliable identification of the individual students. Information systems should be able to:

Fig. 2 Authorization page [4]

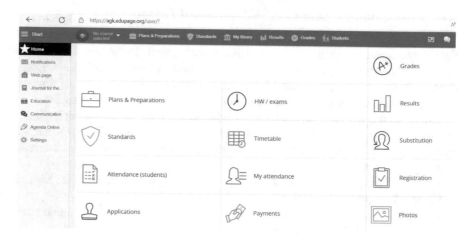

Fig. 3 Page "The main actions of the portal EduPage" [5]

- personalize the curriculum for students;
- monitor the students' progress;
- produce different kinds of reports, such as lists of students by grades, by the form of learning, by status, sheets of intermediate control knowledge, and examination lists;
- provide information about the movement of students (e.g., enrollment, exclusions, transfers); as well as allow users to enter and update the list of subjects studied in the current term/semester;
- archive data about alumni.

However, an educational portal as technology has disadvantages, one of which is the lack of a single standard for the design and content of portals, and therefore, their content varies among different educational organizations. The only common component is often user registration. The external design of the portal/site does not play a decisive role and may not be standardized, but the content and structure of the educational portal must meet certain standards. When drawing up the standard, it is necessary to use the normative documents for the organization of the educational process at the educational institution.

The main documentation for the organization of the educational process at the schools is the state standard, the standard curriculum of the subjects, methodological recommendations, and instructions. The supporting documentation includes the schedule of lessons, orders for the number of students, etc.

Technical Support and Conditions

Each participating organization should have as follows:

- equipment with access to the telecommunications network (Internet, satellite TV);
- multimedia classrooms and electronic reading rooms;
- own or purchased learning content and multimedia laboratories to create their content (local and network);
- testing complex; network learning management system—LMS and learning content management system—LCMS.

Experience shows that for successful participation in distance education technologies all staff must do specialized advanced training courses. Participating organizations should:

- deliver educational material to students through information technology or on paper;
- organize and conduct consultations in mode «online»;
- provide feedback to learners in a mode «offline»;
- control the learning achievements of students in the modes «online» and «offline»;
- create suitable conditions for teaching and academic staff to help them to design and upgrade educational resources.

As maintained by Al-araibi et al., technical readiness is the main aspect of e-learning implementation [6].

Concerning Learners

Learners (students) who want to be trained on educational technologies (DET) must write an application addressed to the head of an organization/school with a reasoned justification for the possible use of DET in training with the provision of supporting documents from the parents and/or recommendations of medical and pedagogical consultation.

The head of the education organization should specify the technologies. It should include:

- the statement of the student;
- individual curriculum of the student;
- individual schedule organization of training activities of every student;
- recommendations of medical and pedagogical consultation about the possibility of the children's participation with disabilities in the educational process by remote (distance) educational technologies.

The education organization has two days to familiarize students and their parents with the plan and timetable of an educational process used in e-learning.

E-Pedagogy in Disciplines

To provide school students with teaching materials, educational providers need to have an e-learning strategy, particularly their electronic educational methodical complexes on all subjects/disciplines of the curriculum (EEMCD), implemented using DET. The preparation of electronic teaching materials such as a set of a training package, teaching and methodical materials, handouts, visual, audio, video, and multimedia materials for educational discipline needs to be provided by the developer of the course (teacher) based on the approved work training programs developed in accordance with the curriculum.

Nowadays, many educational institutions use "case technology" where teaching materials are clearly structured and appropriately collected in a special set (case). It is therefore recommended that electronic educational methodical complex of the subject includes obligatory and optional sets of EEMCD structure which is defined below.

The compulsory suite of electronic educational and methodical complexes of disciplines should consist of:

- the working learning program including the content of the discipline, calendar thematic plan, a list of references for recommended reading (basic and supplementary), the modular partition of discipline, the schedule for distance consultations;
- electronic lecture notes;
- materials for practical works and seminars;
- tasks for students' self-study and self-study-led teacher;
- formative materials (test assignments of individual tasks);
- summative assessment materials (test examination tasks, exam questions, tickets, examinational control works).

An additional set includes methodical instructions on coursework; computer training programs: simulators, guidance materials, and multimedia versions of educational and teaching tools to help to learn the subject.

Special Structural Departments

One of the main requirements for the state and compulsory standard of education is the presence of special structural departments in the organization of education, implementing DET providing organizational, methodological, informational, and technical support for the learning process by DET, and having in its composition the following offices:

- administration of the educational process by DET;
- design of didactic means of DET;
- information and technical support.

The administration office of the educational process by DET plans and organizes the educational process by DET, shall keep such records, related to the DET, and plans to increase the training of teachers and tutors. Further, this service organizes the collection of all control materials of students (control and course works, essays, papers, written examination papers, and the like) in paper form and/or electronic form. And the transfer of these materials to tutors to assess knowledge, and provide entry of information on current progress, received from the tutors in the information database.

The DET pedagogical department works on the development, acquisition, and mastering of electronic textbooks, multimedia courses, methodical manuals, developing test systems and other means of control of knowledge, and techniques of using information resources for DET.

The information and technical support office designs, produces, and maintains software manuals, as well as informational and technical distance learning tools.

Conformity of Academic Calendar and Training Programs (Curriculum)

DET conforms to the normal academic year. All types of learning activities are carried out through:

- pedagogical communication between teaching staff or tutors and students either virtually or face-to-face at educational institutions (at the initiative of the student);
- student self-study using the materials provided.

The monitoring of progress and intermediate certification of students are performed following the working curriculum, academic calendar, and the demands of different training programs. Ongoing monitoring occurs as a result of:

- direct contact of a student with the teacher in mode "online" to use telecommunication tools;
- automated testing;
- checking of individual tasks.

The information system is protected from unauthorized access and falsifications through the use of electronic methods of limiting access and independent assessment. Each student is admitted to intermediate certification in the subject after doing and passing all control measures established by the curriculum.

The educational achievements of students for all types of training sessions are evaluated by a character of the score-rating system of knowledge assessment. Educational achievements of students are evaluated in the subjects accordingly to the curricula of general secondary education, technical and vocational, and post-secondary education.

Submission of course works (projects) by distance educational technologies is carried out:

- publicly, orally at an education institution;
- online using telecommunicational means;
- by sending to the education organization the project in paper form and electronic form and video performances after receiving reviews on course work (project).

3   IoT in Primary Education

As claimed by Dipak R. Kawade et al., "the Internet of Things (IoT) is known as the most profound technology which connects most of the digital devices with the help of the internet" [7]. IoT means the connection between people, people with things, and vice versa. Likewise, automated technology can access all information through the Internet and smart devices.

All people use computers or smartphones. By using this modern and changing world of electronics and telecommunication industry in education, we enhance the performance of primary education. IoT is a new technology that uses various methods and modern technologies to improve performance and gain higher accuracy.

Essentially, e-learning is one type of distance education. Under the legal framework for distance education, technologies are implemented by "case technology" where teaching materials are clearly structured and appropriately collected in a special set (case), that is, network, TV—technologies in online and offline modes. In distance education, the subjects are students, pedagogical staff, and those organizations that implement training programs for additional education (schools, colleges).

As highlighted by Al Rawashdeh et al., IoT helps teachers and students to provide access to course content in digital form and exchange knowledge, while increasing the effectiveness of learning by expanding interaction between teachers and students through online forums, knowledge sharing, and content sharing [8].

Each school chooses an Internet platform for the organization of the educational process in a distance learning format with the appropriate and functional infrastructure and educational content based on the principal teachers' requests [9]. These platforms stimulate learning through various functions that include online course development, assessment, and monitoring of activities for students and teachers.

In conformity with the Kazakhstan methodological recommendations on the organization of the educational process in secondary education organizations during the period of restrictive measures related to preventing the spread of coronavirus infection, one exercise in a playful form was provided for students of grades 3 and 4 on the subject of ICT in a distance format for one lesson [9]. Traditional educational methods have fulfilled the need of students in this digital era, including the pandemic time, of 2020. At the same time, there was an urgent need to acquire new technologies and techniques that would meet the students' needs. IoT is one of the best ways for this problem.

For example, during the pandemic in 2020, smartphones were used in Kazakhstan schools. These means were the main instruments for distance learning in each grade. As part of the monitoring of distance learning in schools of JSC "Information-Analytic Centre" as part of emergency distance learning in Kazakhstan, a survey was conducted, which was attended by about 20 thousand students, parents, and teachers of the country. According to the survey results in the IV term of the 2019–2020 academic year, 92% of the students used their smartphones as a learning tool. However, despite the convenience of connecting via a smartphone, this device is not the most effective for learning [10].

Teachers chose the Zoom video service for video lessons, video conferences, and online meetings. More advanced teachers used the Teams platform. Students and teachers used WhatsApp messenger daily. The WhatsApp application allowed them to connect to a video conference without spending a large amount of traffic, whereas these difficulties were when connecting through the Zoom application [10].

Besides that, the school program was fully digitalized. Nowadays, more than 12 thousand online lessons are available in two languages, Kazakh and Russian, on all electronic devices.

"Online mektep" was the most visited online school Web site during the COVID-19 (see Fig. 4) [11]. Almost the entire school curriculum is presented here: more than 24 thousand lessons under the standards of the Ministry of Education and Science of the Republic of Kazakhstan. Many resources are available in Kazakh, Russian, and English languages. What is more, the mobile application "OnlineMektep" is available on gadgets.

On this Web site, teachers prepare homework and distribute it to all students. Students can solve these tasks, and teachers can give grades. Based on the students' grades and individual characteristics of students, the teacher can update their teaching material or create new material. Teachers can prepare professional topics using smart devices that will increase student attendance and involve students in classes. It can be used to create an attractive presentation of educational programs that will help students focus on the topic, better understand the topic, and memorize information on the topic for a long time. Furthermore, it will encourage the learning process and students to have a positive attitude toward learning. IoT will increase readability for teachers, as well as reduce learning time.

As a result, the author believes that digital literacy is improved. In the book "Role of information and communication technology during COVID-19" Nidhi, S. & Meetender hold the same opinion and state that both the students and teachers

Fig. 4 Online school Web site with ICT materials and tasks for primary schoolchildren [11]

have grown up as good students to study and use digital tools during COVID-19, which has allowed them to increase their level of digital literacy [12].

Problems in the Implementation of Distance Learning

With reference to a study conducted by the research center "Paper Lab" of Kazakhstan, the most frequently noted problem in distance learning is an Internet connection and the availability of appropriate means of communication (computer, laptop, smartphone) [13]. And this minimum requirement has not been met everywhere in the country. The unavailability or unsatisfactory quality of Internet connection, especially in rural areas, was noted from the very beginning of distance learning and at all levels. The best confirmation is that the Ministry of Education was forced to admit that it was impossible to organize online school education (conducting lessons via video communication).

In 2020, the Paper Lab center conducted a study on how the first months of distance learning took place among students of primary and secondary schools. To do this, senior pupils and parents of primary school students kept diaries where they recorded how their online learning was going and what problems arose. In total, data were collected from 110 people (55 students of grades 8–10, 55 parents of primary schoolchildren of grades 2–4) from 5 regions of the country, where the responses of students from cities and villages were evenly presented. Among the most frequently encountered problems, students and parents noted the low quality and speed of the Internet, a heavy burden on teachers, but at the same time their low involvement in the learning process, low level or lack of control, knowledge gaps, maintaining the status quo in assessing students' knowledge, incomplete educational resources and applications, and other resulting problems [13].

It was not only on the Internet but in programs and Web sites that are not fully finalized. They had many disadvantages. In particular, the electronic diary Kundelik.kz, unfortunately, does not always open and is not filled properly. This Web portal is responsible for setting and viewing grades, and attendance marks, and issuing and viewing homework assignments.

A domestic platform "Sphere" was developed. It has become a platform for distance learning for schoolchildren in Almaty city. The educational content was developed through the Bilim Media Group Company. Teachers were trained to successfully conduct distance learning.

The software environment Opiq.kz allows not only to distribute, but also to create electronic textbooks. It offers students electronic textbooks from grades 1 to 11. Opiq electronic textbooks are available on any device: desktop computer, laptop, tablet, or smartphone. The interface adapts automatically. The cost of the license is 1300 tenge per month, for a set of 85 textbooks from grades 1 to 11.

Simultaneously, video lessons were shown on TV, which served as additional material for teaching schoolchildren. Unfortunately, as noted in the Ministry of Education, not all settlements of the country could receive a television picture of educational channels. These channels were unavailable in more than 600 villages, and more than 300 students live in them [14].

Below is one world example from Arizona State University (ASU), USA, how the university responded to the educational challenges presented by the COVID-19 pandemic:

ASU has ASU Prep Digital (ASUPD) which is an accredited online school for students and other users, who want to take a single online course or to learn full-time. ASUPD, constantly promoting the transition to blended learning with local schools, provided guidance and support, continuing to support more than 10,000 learners. As Carole G. Basile notes, "ASUPD was able to adapt the following initiative to support schools during the disruptions caused by the pandemic. ASU immediately responded by launching a reliable set of free online educational resources to support the transition to distance learning for students and teachers at the national level. This platform is called ASU for You and includes online classes for teaching students, access to the main materials of the ASUPD course, a library of instructional videos that help teachers and parents move to a new level, as well as full support for learning in schools. The platform ensures consistency between schools and provides teachers with tools that complement their current distance learning plans by providing metrics and assessments to make informed decisions" [15].

In addition to other problems in Kazakhstan primary education, younger students could not cope with the proposed form of education on their own, and their parents took on the function of a teacher. They received messages and sent files, explaining new topics and tasks to their children, and helping them do their homework.

Moreover, in addition to a poor Internet connection or no Internet connection at all, there was a low digital literacy of teachers, undeveloped educational infrastructure, and the unpreparedness of school management for new realities [16].

There was an additional problem with the online portal's interface. Domestic platforms have been described by many teachers as inconvenient to use. As an example, when teachers assigned tasks to students, they had to upload tasks in one window, enter a description, and send it to pupils in another window [11].

4 Contribution: Previous Work

This study contributes to the theory and practice of organizing education using distance learning technologies. It seems currently, electronic textbooks (e-books) are already familiar to everyone in the education system of Kazakhstan. They have become reliable helpers to schoolchildren and students in their independent individual learning. Their effectiveness has been proven in many schools and universities in Kazakhstan. When using e-books, the quality of academic performance increases two to three times; at the same time, the duration of training is reduced. Thus, you can learn quickly and efficiently. Additionally, e-books are useful for distance learning; this was especially noticeable during the pandemic in 2020 when all educational institutions were forced to switch from the traditional education system to a full distance learning form. Thus, we can single out Estonia's experience in the digitalization of education. Over the past three years, the use of electronic textbooks in Estonia has increased tenfold, and the educational literature industry has turned into a real provider of digital services. There, the introduction of electronic books into the educational process solved another important task—relieved children of heavy school backpack. Paper textbooks are provided to students by the school. At the end of the lessons, books remain on the shelves in school classrooms. Pupils study at home using electronic textbooks [17]. Estonia has EdTech Opiq, an interactive digital learning materials platform that replaces all old school textbooks [18].

Many universities have already realized this and have developed or are developing their distance learning systems in one form or another, including using e-books. Now, almost all universities in Kazakhstan are training personnel using distance learning technologies: Web sites and portals have been opened; special services have been created, and an educational and methodological base is being developed. For example, the following links to universities confirm this: https://satbayev.university/en/second-education, https://polytecho nline.kz, https://www.kaznu.kz/ru/17959/page/, https://www.keu.kz/ru/edu/distantsi onnoe-obuchenie.html. The Moodle and Platonus systems are taken as the basis of the portals, and their developments are also used. Portal sections allow for feedback between the participants of the process DET. Communication of training portals and automated information systems of distance learning is implemented.

In 2008, by order of the Ministry of Education and Science dated July 22, 2008, a list of basic educational organizations was determined in the priority direction of the development of higher and postgraduate education—the development of distance learning technologies—Kazakh-Russian University and Karaganda University of Kazpotrebsoyuz. However, this order was canceled in 2016 by the order of the Minister of Education and Science [19]. Despite this, higher education institutions with a special status can introduce new learning technologies, including distance learning technologies [20].

The superiority in the introduction of distance learning technology in the Republic of Kazakhstan rightfully belongs to the Kazakh-Russian University (KRU), whose educational activities since the very foundation of the university (1998–1999) have been associated with the use of information and satellite educational technology of the Modern Humanitarian Academy in Moscow [21]. This technology includes dozens of licensed automated systems that allow users to fully and cost-effectively deliver educational materials anywhere in the world, provide feedback, and monitor student progress.

International scientific and practical conferences on distance learning technologies were held annually based on KRU. In 2009, the Ministry of Education and Science confirmed the competence to conduct short-term courses on distance learning technologies with the right to issue certificates of the established sample of the Institute of Advanced Training and Retraining (IATR) of Personnel of the KRU. IATR KRU has conducted advanced training for more than 2000 teachers and staff from more than 30 national, state, and non-state educational organizations. One of the lecturers of the advanced training courses on DET was the author of the present article, who trained teachers on the development of e-books and EEMCD.

However, the announced educational portals and Web sites carried quite a bit of information intended for participants in the educational process. Most of the sites had a limited list of sections that could satisfy only a person who knew nothing about this university with information, i.e., they performed only the functions of a business card. Definitely, such information was necessary, but the fact that the bulk of universities was limited only to this opportunity did not cause much satisfaction. The main task of the university Web site is to help its students master academic disciplines, prepare for seminars, tests and exams, and contain information about the individual plan—the schedule of classes, tests, and exams, the topics of term papers, and theses, events at the university and faculty scale. The same applies to school Web sites.

There were various issues of intellectual property protection, development of high-quality content, identification of the student's identity, and other questions, which should be reflected in the regulatory framework.

Along with this, schools use their developments or a single EDUPAGE platform.

The author together with colleagues from IATR KRU helped and advised universities and other educational organizations on the development of the interface of the university portal on DET, digital textbooks, EEMCD, and introduced teachers and university staff to the rules and approaches to the successful implementation of DET in the educational process and the development of regulatory documents.

Several universities, after completing advanced training courses and using distance educational technologies, have implemented international double-degree programs for master and Ph.D. doctoral studies.

5 Discussion

Vidakis and Charitakis state that the active development of information and communication technologies (ICT) and their application in everything and everywhere, especially in the field of education, has prompted educational institutions around the world to introduce new technologies into teaching and learning processes [22].

Gupta and Bansal claim "The educational institutions and students are now ready to understand and accept the online teaching methodologies and approaches, digitalization brings with it. A number of new technologies and platforms have appeared" [23].

Thus, learning management systems (LMSs) help educational institutions to provide and implement training programs. For a software product to be included in the category of learning management software systems, the system must meet the criteria:

- allow courses and teaching materials to be contained in a centralized system accessible to students for learning purposes;
- store reports on the training progress of individual participants and the performance of training programs as a whole;
- allow customizing training programs by individual needs;
- provide opportunities for building plans and schedules and tracking the passage of training courses and disciplines.

Increasingly, Kazakhstan educational institutions use in the educational process such tools of LMS as EDUPAGE, ONLINEMEKTEP.ORG, PLATONUS, and CANVAS. If compare these tools with the most popular LMS, for instance, MOODLE, they are used for the development, management, and distribution of educational, online materials with the provision of shared access by all interested stakeholders (teachers, students, the administration of the educational institution, parents). As well, they are a repository of educational materials and students using DET have their training plans with a schedule of classes. Responsible managers and teachers too can create and add new modules, subjects, and lessons to the LMS. They set deadlines for completing homework and manage assessment criteria and grades. These systems allow for storage reports on the progress of individual participants' learning.

Table 1 provides information about LMS tools used in Kazakhstan educational organizations.

Table 1 Kazakhstan learning management systems

Description	Learning management systems			
	EDUPAGE	ONLINEMEKTEP ORG	PLATONUS	CANVAS
Used in public schools	✓	✓		
Used in colleges, universities			✓	✓
Available for parents	✓	✓		
Trilingual information system (Kazakh, Russian, English)		✓	✓	✓
Conducting e-learning	✓	✓	✓	✓
Formation of individual student curricula	✓	✓	✓	✓
Convenient and comfortable in work	✓	✓	✓	✓
Easy creation of new online courses and modules	✓	✓	✓	✓

According to the study, EduPage and Moodle systems are popular and frequently used in Kazakhstan educational organizations. At the same time, teachers and students also use the information system Platonus. However, not all tools have open source code such as MOODLE.

Each educational institution has the right to independently choose any educational platform that meets the needs of teachers and students.

All the mentioned tools allow teachers and students to download and work with e-books that motivate students for creative work. This is an environment that is familiar to a child: Gadgets and computer technologies make school classes interesting and encourage new ideas to be generated.

In addition, together with practical issues and fundamental requirements for organizing the educational process in the Republic of Kazakhstan using DET, this study can help the schools and universities in solving the future task—the student identity in distance learning. This may be the recognition of a student's face using a Webcam.

6 Conclusion

Based on the above, it can be stated that educational organizations desire to use distance learning technologies in their educational process must adhere to the regulatory framework for distance education and follow all the rules and requirements described in this paper.

At the same time, the government of Kazakhstan is working to update the subject of "ICT" in schools, the introduction of programming lessons in primary grades, and integration of information and communication technologies in humanities at universities. Additionally, this is facilitated using of the Internet of Things in education because gadgets and computer technologies make school activities interesting and encourage the generation of new ideas.

The effective work is conducted on platforms Kundelik.kz, Bilimland.kz, and Online mektep. Work is being carried out to maximize the coverage of the territory with high-speed broadband Internet, to provide schools with computer equipment, and the availability of educational Internet platforms.

In the future, the author plans to expand and continue the research and analysis of how IoT and LMS tools influence primary schoolchildren's digital literacy development. For this purpose, the author will conduct a questionnaire survey among teachers, which demonstrates to us the situation in the teaching of computer science in primary schools, including distance learning.

References

1. Instructional and Methodical Letter (2021) About the features of the educational process in secondary education organizations of the Republic of Kazakhstan in the 2021–2022 academic year, official website of the Y. Altynsarin National Academy of Education. http://uba.edu.kz/storage/app/media/IMP/IMP_2021-2022_kaz.pdf. Last accessed 27 July 2022
2. On amendments to the Order of the Minister of Education and Science of the Republic of Kazakhstan dated March 20, 2015 No. 137 "On approval of the Rules for the organization of the educational process on distance learning technologies"—Information and legal system of regulatory legal acts of the Republic of Kazakhstan "әділет". https://adilet.zan.kz/rus/docs/V2100025038
3. International Association of Universities. COVID-19: Higher Education challenges and responses, https://www.iau-aiu.net/COVID-19-Higher-Education-challenges-and-responses. Last accessed 21 July 2022
4. Distance education from the oldest domestic university. https://www.keu.kz/ru/edu/distantsionnoe-obuchenie.html
5. Humanitarian College of Astana International University. https://agk.edupage.org/user/?
6. Al-araibi, AAM, Mahrin MNb, Yusoff RCM (2019) Technological aspect factors of E-learning readiness in higher education institutions: Delphi technique. Educ Inf Technol 24:567–590
7. Dipak K, Kavita O, Poornima N (2018) IOT in primary education. In: International interdisciplinary conference on curriculum reforms in higher education: global scenario (IICCRHE-2018). Shivaji University Kolhapur
8. Al Rawashdeh AZ et al (2021) Advantages and disadvantages of using e-learning in university education: analyzing students' perspectives. Electron J e-Learn 19:107–117
9. Методические рекомендации по организации учебного процесса в организациях среднего образования в период ограничительных мер, связанных с недопущением распространения коронавирусной инфекции/Methodological recommendations on the organization of the educational process in secondary education organizations during the period of restrictive measures related to preventing the spread of coronavirus infection. The order of the Minister of Education and Science of Kazakhstan, № 548 (2020). https://www.gov.kz/memleket/entities/kdso/documents/details/63523?lang=ru
10. The National report on the state and development of the education system of the Republic of Kazakhstan as of 2020 (2021) naczionalnyj-doklad-po-itogam-2020_kaz.pdf (iac.kz), pp 201–207

11. Online classes on Bilimland.kz. https://bilimland.kz/en/courses/lower-primary-curriculum/ict/computer-abc/lesson/computer-abc. Last accessed 12 July 2022
12. Nidhi S, Meetender (2022) Online education and ICT during Covid-19: issues and challenges. In: Krishnan K, Ram C, Geeta G, Sachin G (eds) Role of information and communication technology during COVID-19, edition: Jan. 2022. Amba International Publishers and Distributors, India, pp 185–191
13. Джаксылыков С.: Дневники дистанционного обучения: какой была «дистанционная» школьная четверть глазами учеников и родителей (2020)/Dzhaksylykov S (2020) Distance learning diaries: what was the "distance" school term through the eyes of students and parents. https://drive.google.com/file/d/1a-wo91IsG_puve H2mUVCU9XliZIcC8_2/view
14. Online education in Kazakhstan. Bluescreen.kz – IT-portal about technology, games, cybersecurity, Kazakhstan and global trends. https://bluescreen.kz. Last accessed 10 July 2022
15. Basile CG (2022) Arizona State University: a learning enterprise supporting P-12 education in the COVID-19 pandemic. In: Reimers FM, Marmolejo FJ (eds) University and school collaborations during a pandemic. Sustaining educational opportunity and reinventing education, vol 8, pp 287–297. Springer, Switzerland
16. Ковязина К., Боранбай М., Бейсембаев С.: Дистанционное образование в Казахстане глазами учителей и экспертов: вызовы и возможности (2020)/Kovyazina K, Boranbai M, Beisembayev S (2020) Distance education in Kazakhstan through the eyes of teachers and experts: challenges and opportunities (2020). https://drive.google.com/file/d/1TcIzC8oSvRK yFVnuA3gfcqn5nCDwnr4N/view
17. Estonia #1 in Europe for digital learning. https://www.educationestonia.org/estonia-no-1-in-europe-for-digital-learning/. Last accessed 26 July 2022
18. Education Estonia. https://www.educationestonia.org/digital-learning-materials-by-estonian-edtech-startup-opiq/. Last accessed 26 July 2022
19. Order of the Acting Minister of Education and Science of the Republic of Kazakhstan dated July 22, 2008 No. 441 "On approval of the List of basic educational organizations in priority areas of development of higher and postgraduate education". https://online.zakon.kz/Document/?doc_id=30385785&pos=4;-108#pos=4;-108. Last accessed 28 July 2022
20. Resolution of the Government of the Republic of Kazakhstan dated June 13, 2019 No. 397. "On amendments to the Resolution of the Government of the Republic of Kazakhstan" dated February 14, 2017 No. 66 "On approval of the Regulations on the special status of higher educational institutions". https://adilet.zan.kz/rus/docs/P1900000397. Last accessed 28 July 2022
21. Kazakh-Russian University. https://dic.academic.ru/dic.nsf/ruwiki/642942. Last accessed 28 July 2022
22. Vidakis N, Charitakis S (2018) Designing the learning process: the IOLAOS platform. In: 10th international conference proceedings on subject-oriented business process management. ACM, p 15
23. Gupta M, Bansal D (2022) Role of information & communication technology (ICT) in the education sector during Covid-19. In: Krishnan K, Ram C, Geeta G, Sachin G (eds) Role of information and communication technology during COVID-19, edition: Jan.2022. Amba International Publishers and Distributors, India, pp 66–77

The Task of Question Answering in NLP: A Comprehensive Review

Sagnik Sarkar, Pardeep Singh⬤, Namrata Kumari⬤, and Poonam Kashtriya

Abstract An important task of natural language processing is a question answering (QA) (NLP). It provides an automated method of pulling the information from a given context. Thus, QA is made up of three separate modules, each of which has a core component in addition to auxiliary components. These three essential elements are answer extraction, information retrieval, and question classification. By classifying the submitted question according to its type, question classification plays a crucial role in QA systems. Information retrieval is crucial for finding answers to questions because, without the presence of the right ones in a document, no further processing can be done to come up with a solution. Last but not least, answer extraction seeks to locate the response to a query posed by the user. This paper sought to provide a comprehensive overview of the various QA methods, assessment criteria, and benchmarking tools that researchers frequently use.

Keywords Natural language processing (NLP) · QA system · Encoder · Decoder · Attention · Transformer · BERT · T5 · Knowledge graph

1 Introduction

Question answering (QA) is to provide accurate responses to questions based on a passage. In other words, QA systems enable users to ask questions and retrieve answers using natural language queries [1] and can be viewed as an advanced form

S. Sarkar · P. Singh (✉) · N. Kumari · P. Kashtriya
NIT Hamirpur, Hamirpur, HP, India
e-mail: pardeep@nith.ac.in

S. Sarkar
e-mail: 20mcs011@nith.ac.in

N. Kumari
e-mail: namrata_phd@nith.ac.in

P. Kashtriya
e-mail: poonam_phdcse@nith.ac.in

© The Author(s), under exclusive license to Springer Nature Singapore Pte Ltd. 2023   603
Y. Singh et al. (eds.), *Proceedings of International Conference on Recent Innovations in Computing*, Lecture Notes in Electrical Engineering 1011,
https://doi.org/10.1007/978-981-99-0601-7_46

of information retrieval [2]. Additionally, the QA has been utilized to create dialogue systems and chatbots designed to simulate human conversation. There are two main procedures for processing questions. The first step is to examine the structure of the user's query. The second step is to convert the question into a meaningful question formula that is compatible with the domain of QA [3]. The majority of modern NLP problems revolve around unstructured data. This entails extracting the data from the JSON file, processing it, and then using it as needed. An implementation approach categorizes the task of extracting answers from questions into one of four types:

1. IR-QA (Information retrieval based)
2. NLP-QA (Natural language processing based)
3. KB-QA (Knowledge based)
4. Hybrid QA.

2 General Architecture

The following is the architecture of the question answering system: The user asks a question. This query is then used to extract all possible answers for the context. The appropriate architecture of a question answering system is depicted in the Fig. 1.

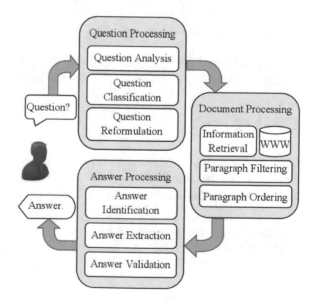

Fig. 1 Question answering systems [4]

2.1 Question Processing

The overall function of the question processing module, given a question as an input, is to process and analyze the input question so that the machine can understand the context of the question.

2.2 Document Processing

After giving the question as an input, the next big task is to parse the entire context passage to find the appropriate answer locations. The related results that satisfy the given queries are collected in this stage in accordance with the rules and keywords.

2.3 Answer Processing

The similarity is checked after the document processing stage to display the related answer. Once an answer key has been identified, a set of heuristics is applied to it in order to extract and display only the relevant word or phrase that answers the question.

3 Background

"Can digital computers think?" was written by Alan Turing in 1951. He asserted that a machine could be said to be thinking if it could participate in a conversation using a teleprinter and imitate a human completely, without any telltale differences. In 1952, the Hodgkin–Huxley model [5] showed how the brain creates a system that resembles an electrical network using neurons. According to Hans Peter Luhn [6], "the weight of a term that appears in a document is simply proportional to the frequency of the term". Artificial intelligence (AI), natural language processing (NLP), and their applications have all been influenced by these events. The BASEBALL program, created in 1961 by Green et al. [7] for answering questions about baseball games played in the American league over the course of a season, is the most well-known early question answering system. The LUNAR system [8], created in 1971 to aid lunar geologists in easily accessing, comparing, and evaluating the chemical composition of lunar rock and soil during the Apollo Moon mission, is the most well-known piece of work in this field. A lot of earlier models, including SYNTHEX, LIFER, and PLANES [9], attempted to answer a question. Figure 2 depicts the stages of evolution of the NLP models.

Fig. 2 Evolution of NLP models [10]

4 Benchmarks in NLP

Benchmarks are basically some set of some standard used for assessing the performance of different systems or models agreed upon by large community. To ensure that the benchmark is accepted by large community, people use multiple standard benchmarks. Some of the most renowned benchmarks that are used largely are as follows: GLUE, SuperGLUE, SQuAD1.1, and SQuAD2.0

4.1 GLUE (General Language Understanding Evaluation)

General Language Understanding Evaluation, also known as GLUE, is a sizable collection that includes a variety of tools for developing, testing, and analyzing natural language understanding systems. It was released in 2018, and NLP enthusiasts still find it to be useful today. The components are as follows:

1. A benchmark of nine sentence- or sentence-pair language understanding tasks constructed on well-established existing datasets and chosen to cover a wide range of dataset sizes, text genres, and degrees of difficulty;
2. A leaderboard to find the top overall model;
3. A diagnostic dataset to assess and analyze the model's performance in relation to a variety of linguistic issues encountered in the natural language domain.

4.2 SuperGLUE

General Language Understanding Evaluation, also known as GLUE, is a large collection of dataset that includes a variety of tools for developing, testing, and analysis. SuperGLUE is an updated version of the GLUE benchmark. SuperGLUE benchmark is designed after GLUE but with whole new set of improved and more difficult language understanding tasks, improved reasoning, and a new canvas of public leaderboard. It was introduced in 2019. Currently, Microsoft Alexander v-team with Turing NLRv5 is leading the scoreboard with URL score of 91.2.

4.3 SQuAD1.1 (Stanford Question Answering Dataset 1.1)

SQuAD or Stanford Question Answering Dataset was introduced in 2016 which consists of Reading Comprehension Datasets. These datasets are based on the Wikipedia articles. The previous version of the SQuAD dataset contains 100,000+ question answer pairs on 500+ articles.

4.4 SQuAD1.1 (Stanford Question Answering Dataset 2.0)

SQuAD2.0 or Stanford Question Answering Dataset combines all the 100,000 questions in SQuAD1.1 with over 50,000 unanswerable questions written so that it may look similar to answerable ones. SQuAD2.0 tests the ability of a system to not only answer questions when possible, but also determine when no answer can be found in the comprehension. Currently, the IE-NET (ensemble) by RICOH_SRCB_DML is leading the scoreboard with **EM** score of 90.93 and **F1** score of 93.21.

5 Research

In this systematic literature review (SLR), we tried to address the various steps based on the guidelines provided by the Okoli and Schabram [11], Keele [12], which emphasizing as: Purpose of the Literature Review, Searching various Literature, Practical Screen, Quality Appraisal, and Data Extraction. The amount of written digital information has increased exponentially, necessitating the use of increasingly sophisticated search tools. Pinto et al. [13], Bhoir and Potey [14]. Unstructured data is being gathered and stored at previously unheard-of rates, and its volume is growing. Bakshi et al. [15], Malik et al. [16], and Chali et al. [17], among others. The main difficulty is creating a model that can effectively extract data and knowledge for various tasks. The tendency in this situation of the question answering systems is to glean as many answers from the questions as you can. This SLR will be guided by the research questions in Table 1 in an effort to comprehend how question answering systems techniques, tools, algorithms, and systems work and perform, as well as their dependability in carrying out task.

We gathered as many journals and papers written in English in different digital libraries and reputed publications through the various keywords and tried to provide some strong evidence related to the research questions that have been tabulated earlier.

RQ_1: Fig. 3 tried to show the popularity of various models on the basis of the number of paper published in the category in every year. Here, we can observe that the BERT-based model is the most popular in this category.

Table 1 Research questions to be addressed

Question No.	Research question
RQ1	What are the popular QA techniques?
RQ2	Which domains use the question answering models?
RQ3	How it is improving the existing model?
RQ4	Contribution of other authors in the field of QA?

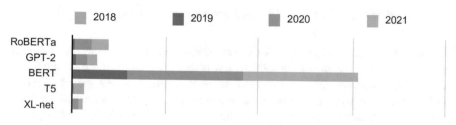

Fig. 3 A graph showing the popularity of the models

RQ_2: Fig. 4 tries to show the various question answering fields the QA models are used. We can see that general domain QA is dominantly used here.

RQ_3: The fine-tuning of different models have given rise to various improvements in the existing models. Moreover, using the different techniques over the existing model can give rise to different model which can improve the existing the model. **For Example**: The different BERT-based models like AlBERT, RoBERTa, DistilBERT with different parameters are used according to the need as shown in Table 2.

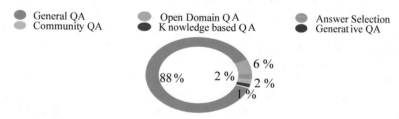

Fig. 4 Chart shows the different types of question answering area

Table 2 Different application using different models

Tasks	BERT	T5	GPT-2
Language modeling	4	3	3
Text generation	1	3	3
Question answering	7	4	3

Table 3 Table showing the area of working the models

Tasks	BERT	T5	GPT-2
Language modeling	4	3	3
Text generation	1	3	3
Question answering	7	4	3
Machine translation	2	2	1
Text classification	1		1
Text summarization	2	2	
Sentiment analysis	1	6	

RQ_4: This is the main purpose of the literature review. This question is answered in support with Table 3. Many papers have been taken into consideration for this comparison [8, 18–38]. Here, we took only three models as these are the main base models that predominate the question answering domain.

6 Conclusion

Question answering system using NLP techniques is much complicated process as compared to other type of information retrieval system. The closed domain QA systems is able to give more accurate answer than that of open domain QA system but is restricted to a single domain only. After the screening phases, we can see that the attention-based model is must preferable among the researchers. We also observed that researchers have equally turned themselves to the hybrid approaches like graph attention and applying different styles of mechanism over the base model to make their job easy. The contributions of this work are a systematic outline of different question answering systems that are able to perform better in all the different tasks. The future should try to explore the possibility of any such model that can outperform all models.

References

1. Abdi A, Idris N, Ahmad Z (2018) QAPD: an ontology-based question answering system in the physics domain. Soft Comput 22(1):213–230
2. Cao YG, Cimino JJ, Ely J, Yu H (2010) Automatically extracting information needs from complex clinical questions. J Biomed Inform 43(6):962–971
3. Joulin A, Grave E, Bojanowski P, Mikolov T (2016) Bag of tricks for efficient text classification. arXiv preprint arXiv:1607.01759
4. Allam AMN, Haggag MH (2012) The question answering systems: a survey. Int J Res Rev Inf Sci (IJRRIS) 2(3)
5. Hamed SK, Ab Aziz MJ (2016) A question answering system on Holy Quran translation based on question expansion Technique and neural network classification. J Comput Sci 12(3):169–177

6. Papineni K, Roukos S, Ward T, Zhu WJ (2002) BLEU: a method for automatic evaluation of machine translation. In: Proceedings of the 40th annual meeting of the Association for Computational Linguistics, pp 311–318
7. Hyndman RJ, Koehler AB (2006) Effect of question formats on item endorsement rates in web surveys. Int J Forecast 22(4):679–688
8. Liang T, Jiang Y, Xia C, Zhao Z, Yin Y, Yu PS (2022) Multifaceted improvements for conversational open-domain question answering. arXiv preprint arXiv:2204.00266
9. Sun Y, Wang S, Li Y, Feng S, Tian H, Wu H, Wang H (2020) Ernie 2.0: a continual pre-training framework for language understanding. In: Proceedings of the AAAI conference on artificial intelligence, vol 34, no 05, pp 8968–8975
10. Hogan A, Blomqvist E, Cochez M, d'Amato C, Melo GD, Gutierrez C et al (2021) Knowledge graphs. Synthesis Lectures on Data, Semantics, and Knowledge 12(2):1–257
11. Okoli C, Schabram K (2010) A guide to conducting a systematic literature review of information systems research
12. So D, Mańke W, Liu H, Dai Z, Shazeer N, Le QV (2021) Searching for efficient transformers for language modeling. Adv Neural Inf Process Syst 34:6010–6022s
13. Turing AM (1951) Can digital computers think? The Turing test: verbal behavior as the hallmark of intelligence, pp 111–116
14. Bhoir V, Potey MA (2014) Question answering system: a heuristic approach. In: The fifth international conference on the applications of digital information and web technologies (ICADIWT 2014). IEEE, pp 165–170
15. Bakshi K (2012) Considerations for big data: architecture and approach. In: 2012 IEEE aerospace conference. IEEE, pp 1–7
16. Malik N, Sharan A, Biswas P (2013) Domain knowledge enriched framework for restricted domain question answering system. In: 2013 IEEE international conference on computational intelligence and computing research. IEEE, pp 1–7
17. Chali Y, Hasan SA, Joty SR (2011) Improving graph-based random walks for complex question answering using syntactic, shallow semantic and extended string subsequence kernels. Inf Process Manage 47(6):843–855
18. Yao X (2014) Feature-driven question answering with natural language alignment. Doctoral dissertation, Johns Hopkins University
19. Zhang J, Zhang H, Xia C, Sun L (2020) Graph-BERT: only attention is needed for learning graph representations. arXiv preprint arXiv:2001.05140
20. Zhang X, Hao Y, Zhu XY, Li M (2008) New information distance measure and its application in question answering system. J Comput Sci Technol 23(4):557–572
21. Mozafari J, Fatemi A, Nematbakhsh MA (2019) BAS: an answer selection method using BERT language model. arXiv preprint arXiv:1911.01528
22. Sun C, Qiu X, Xu Y, Huang X (2019) How to fine-tune BERT for text classification? In: China national conference on Chinese computational linguistics. Springer, Cham, pp 194–206
23. Wang A, Cho K (2019) BERT has a mouth, and it must speak: BERT as a Markov random field language model. arXiv preprint arXiv:1902.04094
24. Wang Z, Ng P, Ma X, Nallapati R, Xiang B (2019) Multi-passage BERT: A globally normalized BERT model for open-domain question answering. arXiv preprint arXiv:1908.08167
25. Yang W, Xie Y, Lin A, Li X, Tan L, Xiong K, Li M, Lin J (2019) End-to-end open-domain question answering with BERTserini. arXiv preprint arXiv:1902.01718
26. Kale M, Rastogi A (2020) Text-to-text pre-training for data-to-text tasks. arXiv preprint arXiv:2005.10433
27. Lin BY, Zhou W, Shen M, Zhou P, Bhagavatula C, Choi Y, Ren X (2019). CommonGen: a constrained text generation challenge for generative commonsense reasoning. arXiv preprint arXiv:1911.03705
28. Ribeiro LF, Schmitt M, Schütze H, Gurevych I (2020) Investigating pretrained language models for graph-to-text generation. arXiv preprint arXiv:2007.08426
29. Agarwal O, Kale M, Ge H, Shakeri S, Al-Rfou R (2020). Machine translation aided bilingual data-to-text generation and semantic parsing. In: Proceedings of the 3rd international workshop on natural language generation from the semantic web (WebNLG+), pp 125–130

30. Moorkens J, Toral A, Castilho S, Way A (2018) Translators' perceptions of literary post-editing using statistical and neural machine translation. Translation Spaces 7(2):240–262
31. Ethayarajh K (2019) How contextual are contextualized word representations? Comparing the geometry of BERT, ELMo, and GPT-2 embeddings. arXiv preprint arXiv:1909.00512
32. Frydenlund A, Singh G, Rudzicz F (2022) Language modelling via learning to rank. In: Proceedings of the AAAI conference on artificial intelligence, vol 36, no 10, pp 10636–10644
33. Mager M, Astudillo RF, Naseem T, Sultan MA, Lee YS, Florian R, Roukos S (2020) GPT-too: a language-model-first approach for AMR-to-text generation. arXiv preprint arXiv:2005.09123
34. Qu Y, Liu P, Song W, Liu L, Cheng M (2020) A text generation and prediction system: pre-training on new corpora using BERT and GPT-2. In: 2020 IEEE 10th international conference on electronics information and emergency communication (ICEIEC). IEEE, pp 323–326
35. Puri R, Spring R, Patwary M, Shoeybi M, Catanzaro B (2020) Training question answering models from synthetic data. arXiv preprint arXiv:2002.09599
36. Wang A, Singh A, Michael J, Hill F, Levy O, Bowman SR (2018) GLUE: a multi-task benchmark and analysis platform for natural language understanding. arXiv preprint arXiv:1804.07461
37. Wang A, Pruksachatkun Y, Nangia N, Singh A, Michael J, Hill F, Levy O, Bowman S (2019) Superglue: a stickier benchmark for general-purpose language understanding systems. Advances in neural information processing systems, 32
38. Hsu HH, Huang NF (2022) Xiao-Shih: a self-enriched question answering bot with machine learning on Chinese-based MOOCs. IEEE Trans Learn Technol

Monitoring False Colour in an External Monitor for Film and Television Production: An Evaluative Study

Sambhram Pattanayak and Malini Mittal Bishnoi

Abstract Today, video production monitors are integral to television, broadcasting, and film production. In fact, because monitors accurately display what the camera sees without any changes or enhancements, they deliver a realistic image essential for video production. Though a camera offers a suggested exposure setting, images demand greater exposure accuracy. It is therefore essential to understand the camera's exposure level. Underexposure shots tend to lack clarity and overexposure causes in advertent loss of detail in image's highlights. The challenge thus is in setting the correct level of exposure to convey a message. An external monitor has certain distinguishing features that enhance picture quality in several ways. One such feature is 'false colour', which is found primarily on external monitors. False colour benefits cinematographers by transforming varied exposure levels in a frame into distinct bands. They are allowing the user to see where under or overexposure is occurring. It exposes faults that a built-in internal camera viewfinder may easily ignore while minimizing overheating issues. This paper undertakes a comparative analysis of the picture quality in terms of colour, brightness, and contrast of an external monitor with the camera's inbuilt viewfinder. This is an experimental study that determines the amount and quality of exposure that an image receives. Furthermore, this research experiments with 'false colour' viewing in terms of the IRE value chart and examines the exposure levels of all parts of the shots before recording. In this research, a stunning and perfectly exposed image was made by adjusting the false colours according to the IRE value. Underexposed areas appeared in blue, and overexposed areas in red. Checking on the skin tones faster is the key to exposing the image quickly and accurately with the help of false colour.

S. Pattanayak (✉)
Amity School of Communication, Amity University Rajasthan, Jaipur, India
e-mail: Sambhram.pattanayak@gmail.com

M. M. Bishnoi
Department of Humanities Arts and Applied Science, Amity University Dubai, Dubai, UAE
e-mail: mbishnoi@amityuniversity.ae

Y. Singh et al. (eds.), *Proceedings of International Conference on Recent Innovations in Computing*, Lecture Notes in Electrical Engineering 1011,
https://doi.org/10.1007/978-981-99-0601-7_47

Keywords External monitor · False colour · Camera viewfinder · High dynamic range · HDR · CIE · Chromaticity · Colour gamut · Cinematographer · Picture quality

1 Introduction

One of the most common reasons filmmakers choose external monitors over inbuilt cameras is the small size of the built-in screen. A production monitor is a larger version of a television set. It has a better propensity for accurately reading camera signals such as fast processing, precise linearization of an input signal using the optical-electrical transfer function, and perfect colour reproduction. Production monitors are commonly found in production trucks, studios, shooting sets, and other permanent control room sites. Like an on-camera monitor, a production monitor has monitoring tools like waveform, vectorscope, false colour, and on-screen information displays. Some also have a variety of waveforms, scopes, and histograms, ensuring that the user has all the information they need to provide proper lighting and exposure. Video-centric features such as focus peaking and zebras are becoming more common, but some cameras still lack them. Only external monitors offer these features.

Exposure is the amount of light that enters the camera's sensor, making visual information throughout time. Correct exposure is subjective, primarily based on a specific visual narrative. Aperture, shutter speed, and ISO triangulate the exposure affect and must be adjusted to take correctly exposed images. The external monitor shows the exposure values ranging from 0 to 100 of a shot under the false colour and changes them into various colours for easy viewing. So each pixel on the monitor will have an exposure value colour which is an essential weapon in a filmmaker's arsenal. It is used to convey mood and emotion, inform the viewer about the context of a scene, or provide information about characters and settings. A colour component transfer function is commonly referred to as using the 'gamma' and the chromaticity of a white point in video colour space. RGB chromaticities describe a colour gamut. These numbers specify how colour data is encoded for a specific video standard. Watching these crucial colours on an extensive professional monitor gives accurate colour fidelity.

Earlier researches have primarily focused on HDR uniform colour space, colour transfer, and colour correction methods. The current research problematizes absence of appropriate exposure and examines accurate exposure reading through the false colour technique.

2 Related Work

Since the year 2000, colour transfer has been actively researched. Reinhard et al. performed colour transfer technology that aligns the source and reference images' mean and standard deviation of colour distributions [1]. The most common format for HDR data is linear RGB, which highly correlates with the channels. RGB pixel values are frequently converted to luma-chroma colour spaces such as YCbCr or Yuv, where 'u' and 'v' denote uniform chromaticity scales to reduce the influence of pixel manipulation on one channel affecting the others [2].

Due to multi-spectral imaging techniques, many innovative multi-spectral recording devices have been developed in recent years. Zhenghao et al. solved colour bias and multicollinearity for an RGBN camera colour correcting. Although the images suffer from colour desaturation, red green blue and near-infrared (RGBN) cameras can simultaneously capture visible NIR information. The image colour bias becomes apparent when photographing outdoor situations with high NIR illumination. The ordinary least squares regression (OLSR) colour-correcting result is inadequate due to the multicollinearity in RGBN camera channels [3].

Because of time savings, modern television tales may cover much more ground than movies. By positioning each camera at a different angle and shooting the scene once, multiple cameras reduce the need for several shots of the same scene. This covers a large part of a screenplay in a considerably shorter span of time than a single camera production. Inter-camera colour consistency is an issue in multi-camera production. Chunqiu et al. developed a highbred histogram matching (HHM) algorithm to address this issue. This is accomplished by utilizing the cumulative colour histogram, which primarily involves global colour mapping and local colour straightening for uniform colour presentation among all the cameras [4].

3 Methodology

This is hands-on experimental research. A video camera and two Fresnel lamps of 800 watts each were used as the media tool. This experimental study deploys a 'false colour' technique to primarily determine and visualize wavelengths that the human eye cannot see. The camera was adjusted at various exposure levels to get the perfect image in a stable lighting condition (see Figs. 1 and 2). With the help of the false colour and its IRE values, the monitor shows that 0 is clipped out black and 100 is white. The purple colour (IRE 0) denotes the absence of all colours or black. The colours blue and blue variations (IRE 2–24) imply that it is pretty dark not much exposure to light is available. The colour dark grey (IRE 24–42) implies that light is leaving (or entering) a dimly lit place. The experiment's findings reveal that a 60–70 IRE value is an ideal choice for the shot.

The experiment technique deployed various types of exposure adjustments like f-stop, shutter speed, and gain as the primary function to get the perfect brightness.

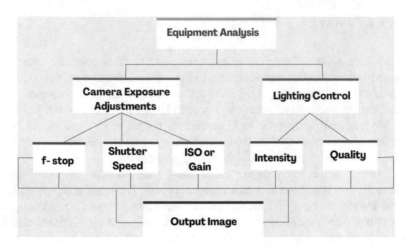

Fig. 1 Equipment analysis

Fig. 2 Pre-requisite methods

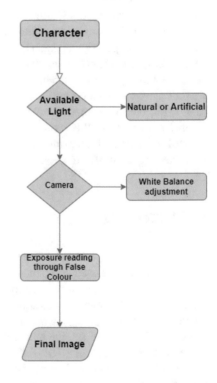

Similarly, incoming lighting control with intensity and quality was also attuned. The metre of illumination, expressed in footcandles or lux, is used to measure the amount of light present in a given area.

4 Colour Balance

The media industry is constantly improving image quality with higher frame rates, larger resolution, brighter colours, and better contrast to enhance the entire watching experience for both cinema and television. Due to the chromatic adaptation process, the human visual system perceives display colours differently depending on the environment. The trichromatic human visual system, which has three cone sensors, is used to develop digital cameras. If a camera wants to record colours like a human vision, the sensor fundamentals should possibly be similar [5]. Chromatic adaptation is the ability of the human visual system to preserve perceived colour appearance despite variations in chromaticity reflected from an object under a wide range of lighting conditions [6]. Different colour temperature cards are used to modify the colour temperatures of photos when utilizing various filming equipment. High colour temperatures produce warm and reddish images, while low colour temperatures produce incredible blue images [7]. Both resolution and colour reproduction determine a colour display's visual quality: colours, i.e. all colours with chromaticities outside the triangle associated. Because the settings of these controls choose the best image quality, it is essential to understand how different brightness and contrast levels affect image quality. Emerging display technologies can create pictures with a significantly broader colour gamut than traditional cinema and television distribution gamuts, allowing for the creation of gamut extension algorithms (GEAs) to fully utilize the colour potential of these new systems. Colour refers to all the weighted combinations of spectral wavelengths, expressed in nanometres (nm), emitted by the sun visible by the human eye (see Fig. 3). Through adaptation, the human eye can perceive a dynamic range of over 14 orders of magnitude (i.e. the difference in powers of ten between the highest and lowest luminance cvalue) in the real world [8].

Accurate colour reproduction requires chromaticity consistency, a wide colour gamut, and high brightness [9]. Colour reproduction is determined by the white point brightness level, repeatable colour gamut, and constant channel chromaticity. Most present displays are based on the trichromacy property of human vision, which creates colours by combining three carefully chosen red, green, and blue primaries in various proportions (see Fig. 4). The chromaticities of these primaries, which form a triangle in the CIE xy chromaticity diagram, determine the display's colour gamut. Colour gamut refers to the range of colours that can be reproduced on an output

Fig. 3 The whole visible colour spectrum and its associated wavelength

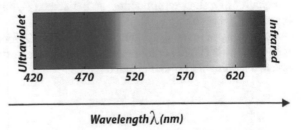

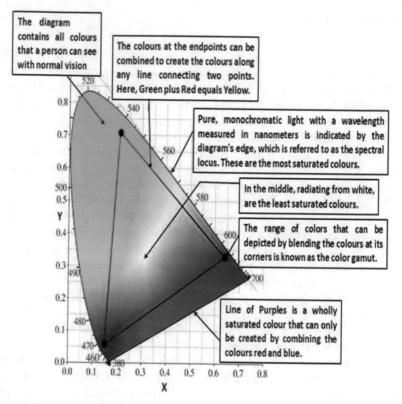

The diagram contains all colours that a person can see with normal vision

The colours at the endpoints can be combined to create the colours along any line connecting two points. Here, Green plus Red equals Yellow.

Pure, monochromatic light with a wavelength measured in nanometers is indicated by the diagram's edge, which is referred to as the spectral locus. These are the most saturated colours.

In the middle, radiating from white, are the least saturated colours.

The range of colors that can be depicted by blending the colours at its corners is known as the color gamut.

Line of Purples is a wholly saturated colour that can only be created by combining the colours red and blue.

Fig. 4 Anatomy of a CIE chromaticity diagram

device within a colour spectrum or colour space. Every screen will display different amounts of colour depending on how broad the gamut is.

There will be many colours that humans can perceive but that the display cannot produce for any given three primary colours, i.e. all colours with chromaticities outside the triangle associated with the display [10]. A wide colour gamut (WCG) is included with HDR, bringing an even more comprehensive range of colours to the table. WCG improves the quality of colour reproduction on the screen-redder reds, bluer blues, greener greens, etc. Whereas a high dynamic range improves a picture's dynamic range (with brighter brights and darker darks), a wide colour gamut enhances the quality of colour reproduction on the screen (with redder reds, greener greens, bluer blues, and so on).

A majority of imaging systems can only collect, process, and display LDR content with the ITU-R BT.709 colour gamut which is standard for high-definition television image encoding and signal characteristics.

5 High Dynamic Range (HDR)

HDR technology strives to capture, distribute, and exhibit various brightness and colour values that closely match what the human eye can see. HDR increases the contrast range between the darker and brighter sections of an image on the screen (the difference between the brightest whites and the darkest blacks that a television can display is known as contrast). While standard dynamic range (SDR) capturing and display mechanisms can only deal with 2–3 orders of brightness magnitude. The HDR has the potential to closely mimic the human visual system by increasing this range to 5 orders of brightness units and reproducing colour values that are comparable to what the human eye can perceive [8] Though wide colour gamut (WCG) seeks to improve the amount of visible colour that can be portrayed, HDR aims to express a whole range of perceptible features from shadows to highlight with sufficiently distinct tonal levels to avoid visual information loss.

WCG and high dynamic range HDR requirements should be applied to video monitors for vivid and realistic displays. However, the WCG's low transmittance rate and the HDR's high expenses for peak brightness capability are essential roadblocks [11]. Pictures that were tone-mapped to standard screens are re-rendered for high dynamic range (HDR) displays. Because these new HDR displays have a far more comprehensive dynamic range than regular monitors, a picture generated for standard monitors may likely appear overly bright when viewed on an HDR panel [12].

Different manufacturers' cameras use different interpolation algorithms to recreate the missing colour values. As a result, some systems use colour attributes to identify the camera model [13]. HDR acquisition has not been a problem. Camera manufacturers like Arri, Red, and Sony have recently released digital cinema cameras that can capture 14–16 stops of dynamic range. True HDR material, on the other hand, may capture, store, and process more than 16 stops of scene dynamic range with either ITU-R BT.709 or ITU-R BT.709 [14]. PQ-EOTF is used to build and save HDR master pictures. This guarantees that the HDR movie master makes better use of the dynamic range initially taken. Dolby algorithms can deliver a better user experience regardless of where the video was seen by adapting to the display's capabilities. Algorithms ensure that the material maintains its original creative purpose [15]. Users can enjoy a better viewing experience with HDR images because they provide the full dynamic range that the human visual system (HVS) can perceive at any level of adaptation. HDR photos have been effectively used to explore the viewing experience of presented images in the past [16].

6 Advantages of False Colour

Using an instrument to assess exposure is critical, especially when the user wants to double-check and ensure that the monitor displays the right image. The brain receives information about luminance and chrominance. Visible light, which the eye senses from a scene, consists of variable ratios of the three primary colours red, green, and blue (RGB). In terms of television, the colour white is composed primarily of red (30%), green (59%), and blue (11%) signals. The false colour is much like waveforms and histograms, in which the user can check the exposure levels of any image. Histograms give mountain shapes broad exposure, and the waveforms can see different RGB values across the horizontal values, whether its singular vectorscope or parade. False colour evaluates actual brightness values from sampling data on a coloured scale, indicating IREs ranging from 0 to 100%. False colour is beneficial because it displays each image section's Institute of Radio Engineers (IRE) value, allowing users to identify where under/overexposure occurs. It denotes a video stream's overall energy (in Mv.) by measuring an amplified electronic signal.

This is a challenging scenario for a camera's dynamic range. For instance, in a scene with a bright summer sky in the background with a character in the shade, the cinematographer must ensure that the background sky does not burn out and that the subject in the shade receives adequate exposure to expose the image appropriately. As a result, using false colours on the monitor, the sky may appear hot red. This indicates that the highlights are being sheared. Alternatively, if the cinematographer lowers the exposure to compensate for the sky, the subject may now be pushed into the purple or underexposed zone. Hence, the cinematographer may use fill lights for the character to achieve a balance.

This is commonly used in image processing to represent otherwise invisible information. It is a well-recognized technique to show photographs in varied colour schemes to highlight specific aspects. Purple, blue, black, grey, yellow, orange, and red are just a few of the hues used in these images. These colours can assist the operator in determining the amount and quality of exposure an image receives. For example, high dynamic range (HDR) photographs are frequently represented in false colour to depict the wide range of luminance in the photographed scene [17]. Mostly false colour options are available on an external monitor attached to the camera. Any digital cinema camera's raw camera files give images encoded at high colour bit depth with a native colour gamut that considerably surpasses DCI-specification standards, and in most cases, meets or exceeds Rec. 2020. The spectrum of a waveform monitor displays the same type of sampling data.

On a waveform monitor, 0% indicates that no light is transformed into an electrical signal [18]. The sensor's highest energy state is represented by 100. This is where the majority of the light from the sensor is turned into image data. The false colour matrix applies a range of reference colours to the video image (See Fig. 5) to 'fill in' parts of the image at various IREs (illustrated in Table 1). There are usually six or more colours representing the range of brightness across the entire video frame.

The false colour exposure check measures the camera's image, tints specific signal levels a particular colour, and displays the remaining as a black and white image.

A true colour image combines accurate red, green, and blue light measurements. Although at least one nonvisible wavelength is used in a false colour image, that band is nevertheless represented in red, green, or blue. As a result, the final image's colours may differ from what we expect. The inherent linkages between true colour and greyscale films relating to the same individual exist, even though true colour and

Fig. 5 False colour display on an external monitor

Table 1 IRE false colour value

Colour	Percentage value	Colour	IRE value
	110		100 to 109
	100		93 to 100
	90		84 to 93
	80		77 to 84
	70		58 to 77
	60		54 to 58
	50		47 to 54
	40		43 to 47
	30		24 to 43
	20		15 to 24
	10		8 to 15
	0		2 to 8
	−10		−7 to 2

greyscale videos are diverse [19]. White indicates sections of the scene that are 6 or more stops overexposed. Normal-neutral is 18% grey. Black is 6 or more stops underexposed. The primary distinction between a true colour image and a greyscale image is that an actual colour pixel may be considered a vector with three components (red, green, and blue). Still, a greyscale pixel can only be considered a scalar grey level [20].

7   Viewfinder Versus External Monitor

The monitor's resolution refers to how sharp a screen may be. It also has to do with the size of the screen. Because the density of the pixels makes the image appear sharper on a small screen, a lower resolution can be used. Higher resolutions are required for a larger screen. Colours are simply one facet of an image, but they usually determine how the image looks and gives a camera its identity.

To view the picture clarity on an inbuilt camera monitor while shooting in direct sunlight is a difficult chore for a camera operator. Most external displays have a 2200-nit brightness level, which is twice as bright as ordinary monitors' 1000-nit brightness. When combined with its max brightness capacity, its high pixel density, high-quality resolution, and 1200:1 contrast ratio prevent the image from being washed out by ultra-bright sunlight.

False colour employs the pixel data straight from the camera sensor to establish the integrity of the values existing in the video file, rather than just using any inspection monitor onset to determine the effectiveness of a video image. It displays them on a colourful reference scale that can be read using a colour-coded key provided by the camera or monitor manufacturer. Another compelling reason to invest in an on-camera monitor is brightness, as built-in displays frequently fall short of this criterion, or their high-brightness options quickly deplete battery life.

8   Experiment and Results

Based on the experiment with limited literature review and fewer available research, materials in the past left room for other researchers. For this exploratory research, a Panasonic video camera was used with various exposure values (dB settings—0 dB, 12 dB, and 24 dB) and two Fresnel lamps of 800 watts intensity were used for the purpose of the experiment. The key light was placed 10mts far from the subject and the backlight was 13mtr far in flood mode. False colour was monitored on a SWIT external monitor. A colour represents each exposure stop. The colours blue, cyan, green, yellow, orange, and red are presented from dark to light. The value 0 to 100, 0 is black, 100 is white, and the rest of the values in between are grey (mentioned in Figs. 6, 7, and 8).

Image Setting- 0dB	Signal Level	Colour
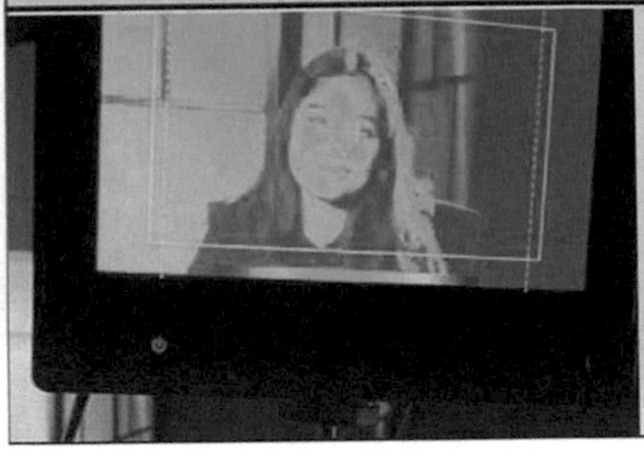	3.5%	

Fig. 6 Signal-level IRE value 3.5%

Image Setting- 12dB	Signal Level	Colour
	40%	

Fig. 7 Signal-level IRE value 40%

Image Setting- 24 dB	Signal Level	Colour
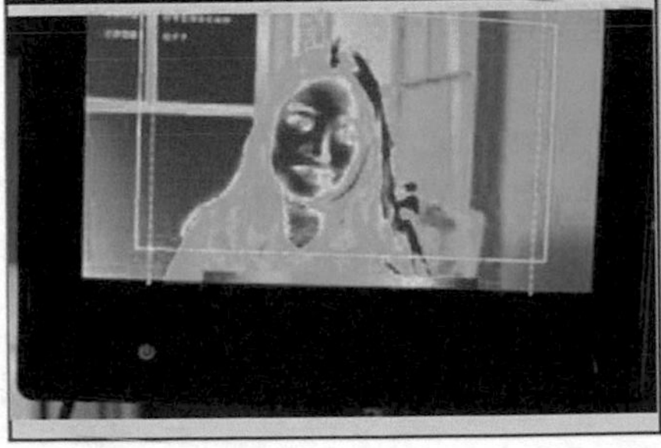	99%	

Fig. 8 Signal-level IRE value 90%

Fig. 9 Final image after false colour monitored

Adjusting false colours produced a beautiful and perfect exposed image (see Fig. 9). Overexposed elements would seem red, and underexposed parts would generally appear blue. The use of false colour on colour monitors showed what was being exposed in a frame and how much (in IER). Thus, while ensuring that there is neither underexpose nor overexpose, 'false colour' primarily achieves the perfect colour balance.

9 Conclusion and Future Work

Getting the perfect exposure in any professional camera used in film and television production is vital. This research draws out the advantages and possibilities of false colour technology to determine different luminance values represented by different colours that grasp the exposure levels of every section of the shot to see exactly where under- or overexposures are occurring. This helps the cinematographer or the camera operator to set the perfect exposure according to the available light and the scene.

Future work will focus on the external recorder, which can record Apple ProRes Raw up to DCI 4K60 directly from the sensor of some cameras. An external recorder supports essential log files from major professional camera manufacturers and a 10-bit screen with brightness to monitor log gamma footage properly.

References

1. Reinhard E, Adhikhmin M, Gooch B, Shirley P (2001) Colour transfer between images. IEEE Comput Graph Appl 21(4):34–41. https://doi.org/10.1109/38.946629
2. Mukherjee R, Debattista K, Bashford-Rogers T, Bessa M, Chalmers A (n.d.) Uniform color space based high dynamic range video compression. IEEE Trans Circ Syst Video Technol 99
3. Han Z, Jin W, Li L, Wang X, Bai X, Wang H (2020) Nonlinear Regression color correction method for RGBN cameras. IEEE Access 8:25914–25926
4. Ding C, Ma Z (2021) Multi-camera color correction via hybrid histogram matching. IEEE Trans Circuits Syst Video Technol 31(9):3327–3337
5. Finlayson GD, Zhu Y (2021) Designing color filters that make cameras more colorimetric. IEEE Trans Image Process 30:853–867
6. Choi K, Suk H-J (2014) User-preferred color temperature adjustment for smartphone display under varying illuminants. Opt Eng 53(6):61708
7. Hsu W-Y, Cheng H-C (2021) A novel automatic white balance method for color constancy under different color temperatures. IEEE Access 9:111925–111937
8. Boitard R, Pourazad MT, Nasiopoulos P (2018) Compression efficiency of high dynamic range and wide color gamut pixels representation. IEEE Trans Broadcast 64(1):1–10
9. Kim TH, Lee YW, Cho HM, Lee IW, Choi SC (1999) Optimization of resolution and color reproduction for color LCD by control of brightness and contrast levels. In: Technical Digest. CLEO/Pacific Rim '99. Pacific Rim conference on lasers and electro-optics (Cat. No.99TH8464). doi: https://doi.org/10.1109/cleopr.1999.811568
10. Bertalmio M, Vazquez-Corral J, Zamir SW (2021) Vision models for wide color gamut imaging in cinema. IEEE Trans Pattern Anal Mach Intell 43(5):1777–1790
11. Kwon KJ, Kim MB, Heo C, Kim SG, Baek JS, Kim YH (2015) Wide color gamut and high dynamic range displays using RGBW LCDs. Displays 40:9–16
12. Meylan L, Daly S, Sabine S (2007) Tone mapping for high dynamic range displays
13. Chen C, Stamm MC (2015) Camera model identification framework using an ensemble of demosaicing features. In: 2015 IEEE international workshop on information forensics and security (WIFS), 1 Nov 2015, pp 1–6
14. Parameter values for ultra-high definition television systems for production and international programme exchange (2015), [online] Available: https://www.itu.int/dms_pubrec/itu-r/rec/bt/R-REC-BT.2020-2-201510-I!!PDF-E.pdf
15. Agrawal A, Agrawal A (2021) Dolby vision: advancing the technology of cinema and home entertainment transformation of an industry. IEEE Comput Graphics Appl 41(2):96–98
16. Melo M, Bessa M, Debattista K, Chalmers A (2014) Evaluation of HDR video tone mapping for mobile devices. Signal Process Image Commun 29(2):247–256
17. Ciftci S, Akyuz AO, Ebrahimi T (2018) A reliable and reversible image privacy protection based on false colors. IEEE Trans Multimedia 20(1):68–81
18. Ma F, Jing X-Y, Zhu X, Tang Z, Peng Z (2020) True-color and grayscale video person re-identification. IEEE Trans Inf Forensics Secur 15:115–129
19. Pattanayak S, Malik F, Verma M (2021) Viability of mobile phone cameras in professional broadcasting: a case study of camera efficiency of Apple iPhone. In: International conference on computational intelligence and knowledge economy (ICCIKE), 17 March 2021, vol 11, pp 452–456
20. Liao X, Yu Y, Li B, Li Z, Qin Z (2020) A new payload partition strategy in color image steganography. IEEE Trans Circuits Syst Video Technol 30(3):685–696

Adaptive Gamification in E-Learning Platforms: Enhancing Learners' Experience

Mitali Chugh, Sonali Vyas, and Vinod Kumar Shukla

Abstract Gamification for e-learning is an approach to apply human-focused design: the systems are developed with the essentials of learners' conduct, motivation, uncertainties, and aims in mind. For instance, aspects like multiple attempts benefit learners to do away with the fear of failure, encouraging them to try again and investigate. Also, gamification triggers learners' behavior pattern that facilitates the learning process (e.g., involvement, competition, socializing, etc.). However, when implementing gamified design elements in e-learning, we must contemplate different kinds of learners, and this instigates the need for adaptive gamification. The present work intends to develop an adaptive operational gamification framework for implementation in e-learning platforms to empower learners. To propose the framework, twelve adaptive gamification research papers have been reviewed, gaming elements were identified, and an enhanced operating model is proposed. The framework is envisioned to be employed as a guide for developers and researchers who seek to develop engaging e-learning systems built on good foundational ideas. The system covers both the conceptual and operational aspects of an adaptive e-learning system to sort out the problem of disengagement.

Keywords Adaptive gamification · Conceptual model · E-learning · Game components

1 Introduction

Gamification is the utilization of game-design components and game philosophies, mechanics, and aesthetics in non-game contexts [1, 2] to support learning and resolve problems. The applications of gamification have rapidly increased in the different

M. Chugh · S. Vyas (✉)
School of Computer Science, UPES, Dehradun, India
e-mail: vyas.sonali86@gmail.com

V. K. Shukla
Information Technology, Amity University Dubai, Dubai, UAE

Y. Singh et al. (eds.), *Proceedings of International Conference on Recent Innovations in Computing*, Lecture Notes in Electrical Engineering 1011,
https://doi.org/10.1007/978-981-99-0601-7_48

627

sectors like education, training, marketing, gaming, etc., specifically focusing on the unpredictable and dynamic needs of the stakeholders [3]. Simoens et al. opine that gamification contributes to e-learning to enhance student motivation and engagement [4] and use game elements in creating worthy games that are used in the process of teaching. E-learning facilitates content delivery as a tool, web technology, and virtual learning environments making education possible everyplace and at any time [5]. Gamified e-learning systems have been designed to draw learners' attention and interest, however, have faced failures due to uniform design for all learners [6]. Schöbel and Sollner state that all learners are different, and hence, the e-learning systems must be designed to address the individual preferences of the learners, and it is the primary reason for the failure of the e-learning systems [7]. Therefore, it is essential to design tailored gamification systems that offer adaptivity of gamification components centering on special necessities [8, 9]. As gamified e-learning is a budding research area and only a few studies [10–12] have proposed and implemented the components and operational frameworks for adaptive e-learning systems, thus the present work has the following objectives to address the research gap.

- Exhaustive literature review on the structures of gamification suitable for e-learning
- To recognize the design essentials of an efficacious e-learning gamified system.
- Recommend a framework architecture for an e-learning gamification platform.

The paper is organized as follows: This section introduces the significance of adaptive gamified e-learning systems and the objectives of the study. Section 2 examines the methodology followed by the proposed adaptive operational gamified framework for e-learning systems in Sect. 3. Section 4 presents the implications of work for research and practice. Section 5 concludes the study and presents the future scope of the work.

2 Methodology

The organized literature review is conducted to address the stated objectives of the study. The research papers that are related to e-learning and gamification are included in the study. The search was conducted related to databases indexed in Google Scholar, Scopus, and Web of Science (WOS) indexes by entering keywords like e-learning gamification frameworks or gamification design frameworks in e-learning, title, abstract, and metadata, and full text includes education with the inclusion and exclusion standards as offered in Table 1.

As to the literature reviewed, the planned adaptive operational gamification framework comprises different components like informative, technical, proposal, organization, social, economic, and gamification characteristics to introduce e-learning gamification as shown in Fig. 1. In this study, we have proposed the conceptual framework along with the operational framework for implementing an e-learning platform that

Table 1 Inclusion/exclusion standards of the research

Classification	Language	Access type	Time Slot	Document type	Included content
Inclusion criteria	English	Open access	2017–2022	Articles/conference papers	Explicit discussion on gamification in E-learning platforms and adaptive/personalized/tailored gamification in e-learning platform
Exclusion criteria	Not in English	Not available as open access	–	Review, report, book chapter, etc.	–

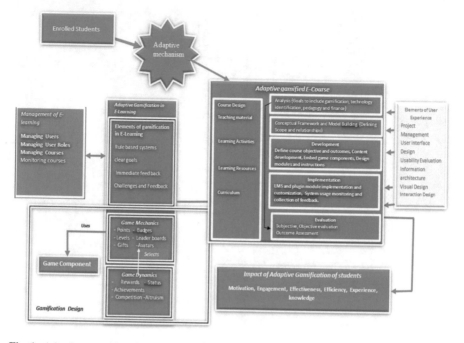

Fig. 1 Adaptive operational gamification framework

can be in the next stage be used for course implementation and evaluation (Tables 2 and 3).

3 Proposed Enhanced Operational Gamification Framework

The barriers of time and distance have been overcome using e-learning platforms. However, the drop-out ratio for many e-learning platforms is high. Among the significant reasons for the dropouts is dearth of motivation among learners. It is due the fact that all the learners do not have the same style of learning, and the e-learning platforms have same kind of learning experience. Hence, to enhance motivation for learning gamification has been introduced in the e-learning platforms.

The gamification has enhanced learner's motivation, and we have introduced the framework that specifically targets to the diverse learning styles of learners based on the student interactions with the e-learning system. Thus, the system is adaptive gamification system.

Components of the adaptive framework: The first component adaptive gamification engine presents elements of the system that are mapped with the attributes of the learners. To perform the administrative functions such as enrolling students,

Table 2 Summary of e-learning frameworks

S. No.	Reference	Game elements and adaptivity criteria	Theoretical underpinning	Gaps/challenges
1	Monterrat et al. [13]	Gamification player typology	Develop an engine to predict the profile and games elements	Implemented and evaluated mixed results
2	Santos et al. [14]	Gamification typology	Not explicitly referred to, Flow Theory	Implemented and validated to have positive impact
3	Knutas et al. [15]	Gamification player typology based on hedax element categorization	Using of DSRM, atomic lense development, and the use of ML for CSCL	Proposed
4	Böckle et al. [16]	Based on usage data, use, player or personality	Developed an adaptive framework framework with set criteria	Evaluated and shown to have an impact
5	Hassan et al. [17]	Learning styles	Learning theories and use of formulae to assign the game elements	Implemented and evaluated mixed results
6	Chtouka et al. [18]	Learner and gamification typologies	Reinforced learning and algorithm	Implemented, but not validated
7	Lavoue et al. [19]	Gamification player typology	Use of adaptation Engine based on player framework	Implemented and evaluated to showed impact
8	Kamunya et al. [10]	Motivational affordances, Psychological outcome, behavioral outcome	Self determination theory, motivational affordance theory, Flow theory, e-learning theory, Technology task fit framework and adaptivity	Proposed but not implemented
9	Jayalath and Esichaikul [12]	Motivation and engagement	ARCS framework of motivational design and game thinking	The proposed operational framework is not validated
10	Kamunya et al. [11]	Adaptive gamification	Design science research methodology (DSRM)	Implemented but not validated
11	Yamani	Gamification player typology	Instructional design framework	Implemented but not validated
12	Zineb et al.	Gamification player typology	Keller's ARCS learning framework	Proposed but not tested

Table 3 Summary of the adaptive gamification research summary

Year	2017	2018	2019	2020	2021	2022
No. of publications	1	3	3	2	1	1

access rights, etc., management of the e-learning platform exists. The adaptive game elements are stored in the repository called adaptive game techniques and dynamics which constitutes the third component of the system. The game elements initiate motivation at the time of learning. The curriculum, resources, and learning content are developed following the gamification plans contented by the adaptive device. The element influence of adaptive gamification is the desired outcome of the e-learning platform. The e-learning platform associates gamification to attain enhanced education by improving engagement, learning practices, motivation, and better knowledge.

Operational framework for adaptive framework: The operational aspect of the adaptive framework includes the use of game thinking in the non-game perspective, i.e., e-learning in our scenario. The apt usage of game components, dynamics, and mechanics results in positive effects on the learner's behavior that is in addition to course objectives and achievements. A gamified course design process (GCDP) [20] that incorporates game mechanics, elements, and dynamics is the basis of designing the proposed framework. The framework is a blend of game modules to attain the anticipated learning outcomes conferring on the student reaction. To drive applicable game dynamics, proper game mechanics are chosen that use suitable game components. At the time of course delivery, learners' marks are fed into the e-learning system to obtain the final grades. If the learner is adequately proficient, then a certificate is given, or else the learner is conveyed to attain the required skills and give a re-assessment. The operational framework facilitates diverse and dynamic learners' behavior, by different and accustomed game components, envisioned to emphasize the desired learners' behavior.

Addressing the customized and societal requirements of the learner provides a pleasing experience for the learner. Altered game mechanics, elements, and dynamics are anticipated to augment effectiveness, enthusiasm, efficiency, commitment, experience, and knowledge throughout the learning.

4 Implications of Work for Theory and Practice

In the present scenario, e-learning has acquired significance in educational pedagogy; hence, the design of e-learning platforms accustomed to the needs of users is vital. A poorly designed gamification-enabled e-learning system does not match the customized needs of the learner. To address this issue, an explicit and precise method for the design and implementation of an e-learning platform is required. The proposed enhanced adaptive operational gamification framework for e-learning platforms complements the adaptive gamification research and provides a direction to the researchers for a research plan, information gathering tools, and techniques and significantly facilitates assessment meticulousness in the present work. The proposed framework will support academicians to enhance their understanding of working of gamified e-learning systems to cater the needs of the diverse learning needs and styles of the learners. In addition, the developers of such systems will be able to comprehend the schema and operational framework of the gamified e-learning systems. As a final point, the stakeholders can apprehend about introducing adaptive gamification of e-learning, resource identification, and way to appraise the system efficiency.

5 Conclusion

Gamification triggers engagement, motivation efficiency, experience, and knowledge for the user. The present review intends to provide directions for an adaptive gamified e-learning framework including its operational model to cater to the diverse learners' needs. The gamified framework must be robust in design to be well suited for e-learning, grounded on e-learning and motivational theories and rigorous validation. The proposed framework and operational model are designed to contribute significantly to forthcoming blended e-learning agendas specifically in education. By employing the proposed framework, it is anticipated that learning outcomes will be accomplished by the students for the course, attain proficiencies through e-learning platforms and achieve higher success rates. We propose to advance this study by forming an e-learning gamified model and incorporating it into a Learning Management System (LMS). Moreover, it can also be researched which game components are chosen when compared to others in different scenarios of education and training.

References

1. Deterding S, Sicart M, Nacke L, O'Hara KD (2011) Gamification: using game-design elements in non-gaming contexts. In: Proceedings of the 2011annual conference extended abstracts on human factors in computing systems. Vancouver
2. Kapp KM (2012) The gamification of learning and instruction: game-based methods and strategies for training and education. Wiley, San Francisco
3. Nake LE, Sebastian C (2017) The maturing of gamification research. Comput Human Behav, 450–454
4. Simões J, Díaz R, Fernández A (2013) Computers in human behavior: a social gamification framework for a K-6 learning platform. Comput Human Behav 29:345–353. https://doi.org/10.1016/j.chb.2012.06.007
5. Wang HC, Chiu YF (2011) Assessing e-learning 2.0 system success. Comput Educ 57:1790–1800

6. Roosta F, Taghiyareh F, Mosharraf M (2016) Personalization of Gamification-elements in an e-learning environment based on learners' motivation. In: 2016 8th international symposium on telecommunications (IST'2016). IEEE, pp 637–642
7. Schöbel S, Söllner M (2016) How to gamify information systems—adapting gamification to individual preferences. In: Twenty-fourth European conference on information systems (ECIS), Istanbul, Turkey, pp 1–12
8. Cheng M, Lin Y, She H (2015) Learning through playing virtual age : exploring the interactions among student concept learning, gaming performance, in-game behaviors, and the use of in-game characters. Comput Educ 86:18–29. https://doi.org/10.1016/j.compedu.2015.03.007
9. Codish D, Ravid G (2014) Adaptive approach for gamification optimization. In: IEEE/ACM 7th international conference on utility and cloud computing, London, UK
10. Kamunya S, Maina E, Oboko R (2019) A gamification model for e-learning platforms. In: 2019 IST-Africa Week conference (IST-Africa), pp 1–9. https://doi.org/10.23919/ISTAFRICA. 2019.8764879
11. Kamunya S, Mirirti E, Oboko R, Maina E (2020) An adaptive gamification model for e-learning. In: 2020 IST-Africa conference (IST-Africa), pp 1–10
12. Jayalath J, Esichaikul V (2020) Gamification to enhance motivation and engagement in blended elearning for technical and vocational education and training. Technol Knowl Learn 27:91–118. https://doi.org/10.1007/s10758-020-09466-2
13. Monterrat B, Lavoué É, George S (2017) Adaptation of gaming features for motivating learners. Simul Gaming 48:625–656. https://doi.org/10.1177/1046878117712632
14. dos Santos WO, Bittencourt II, Vassileva J (2018) Design of tailored gamified educational systems based on gamer types. In: Anais dos workshops do VII Congresso Brasileiro de Informática na Educação (CBIE 2018), vol 1, p 42. https://doi.org/10.5753/cbie.wcbie.201 8.42
15. Knutas A, van Roy R, Hynninen T et al (2018) A process for designing algorithm-based personalized gamification. Multimed Tools Appl 78:13593–13612. https://doi.org/10.1007/ s11042-018-6913-5
16. Böckle M, Micheel I, Bick M (2018) A design framework for adaptive gamification applications 2. Theoretical background and related. In: 51st Hawaii international conference on system sciences, pp 1227–1236
17. Hassan MA, Habiba U, Majeed F, Shoaib M (2019) Adaptive gamification in e-learning based on students' learning styles. Interact Learn Environ 29:545–565. https://doi.org/10.1080/104 94820.2019.1588745
18. Chtouka E, Guezguez W, Amor NB (2019) Reinforcement learning for new adaptive gamified LMS. International Conference on Digital Economy. Springer, Cham., pp 305–314
19. Lavoué É, Monterrat B, Desmarais M, George S (2019) Adaptive gamification for learning environments. IEEE Trans Learn Technol 12:16–28. https://doi.org/10.1109/TLT.2018.282 3710
20. Bartel A, Hagel G (2016) Gamifying the learning of design patterns in software engineering education. https://doi.org/10.1109/EDUCON.2016.7474534

A Study on the Integration of Different DLP Systems at Different Levels

Sampath Reddy Racha and Ganesh Reddy Karri

Abstract Individuals and organizations worry more about leaking confidential information. In the past, data privacy was safeguarded by security rules and techniques like firewalls, VPNs, and IDSs. These systems aren't proactive or committed to protecting sensitive data, so rules must be implemented to do so. Confidential data can leak out in a variety of ways and through a variety of channels, which can have major impact. Therefore, there has been a push to use better systems to solve these issues. Data leakage prevention (DLP) Systems are meant to discover and halt the leaking of confidential details when the data is in use, in motion, or at rest. To detect data leaks and put a stop to them, DLP Systems employ a variety of approaches for analysing the content and context of confidential data. In this work, we have conducted a detailed survey on the various DLP Systems mechanisms that are in use. It gives a clear definition of DLP Systems and classifies the several active research directions in this area. In addition, we provide prospects for the development of more consistent DLP Systems that can compensate for some of the shortcomings of the ones that are existing now.

Keywords Loss prevention · DLP Systems · Content discovery · Content analysis

1 Introduction

DLP System is one of the most overhyped and misunderstood security tools. It could be difficult to understand the inherent worth of the tools and which products best suit those environments with at least a half-dozen different names and even more technology approaches [1] available in the market. This review will give you the background knowledge that you need on DLP Systems so you can better understand the technology, know what to look for in a product, and select the most appropriate for your cloud computing or individual company. There is no consensus on precisely what undermines the DLP solution. Some people think about encryption or DLP for

S. R. Racha · G. R. Karri (✉)
VIT-AP University, Amravati, Andhra Pradesh 522237, India
e-mail: ganesh.reddy@vitap.ac.in

© The Author(s), under exclusive license to Springer Nature Singapore Pte Ltd. 2023
Y. Singh et al. (eds.), *Proceedings of International Conference on Recent Innovations in Computing*, Lecture Notes in Electrical Engineering 1011,
https://doi.org/10.1007/978-981-99-0601-7_49

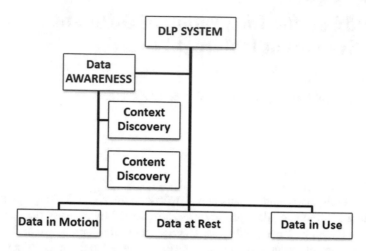

Fig. 1 DLP System view

USB ports [2], while others stick to full product suites. DLP is defined as products that use deep content analysis to identify, monitor, protection of data, in motion, and use, based on central policies [3]. DLP solutions protect sensitive data while also providing insight into how content is used within the company. Only a few businesses categorize data beyond what is publicly available and everything else. DLP enables businesses to gain a better understanding of their data and improve their content classification and management capabilities. DLP as a feature and solution both technologies is available in the market. Basic DLP functions are provided by several products, particularly email security solutions [4], but they are not complete DLP solutions. The distinction is a DLP product consists of centralized management, policy creation, and enforcement workflow for content and data monitoring and protection. The user interface and functionality are designed to address the practical and theoretical problems with content awareness-based content security (Fig. 1).

2 Literature Review

2.1 Content Versus Context

We must distinguish between content and context. Content awareness is one of the distinguishing features of DLP solutions. This is different from contextual analysis and refers to a product's capacity to do in-depth content analysis utilizing a range of methodologies. It is simple to comprehend that the content is a letter, with the envelope and surroundings serving as context. Context includes anything outside of the letter's content such as the source, header information, destination, size, recipients, sender, metadata, format, and time [5]. Context is extremely valuable, and any

DLP System should include it as a part of the overall solution. Business context analysis is a more sophisticated form of contextual analysis that takes into account the content's use at that specific time as well as its environment at the time of analysis [6]. A component of content awareness is looking into containers and assessing the contents. Content awareness allows us to utilize context without being constrained by it [7].

2.2 Content Analysis

Taking possession of the envelope and opening it is the first step in conducting content analysis [8]. Following that, the engine will need to parse the context (which is going to be necessary for the analysis) and delve into it. This is a straightforward task to complete in the case of an email containing only plain text; however, if you want to inspect the contents of a binary file, the process becomes somewhat more difficult. This problem is solved by every DLP solution available, which is file cracking. The process of file cracking [9] refers to the technology that is utilized to read and comprehend the file, regardless of how deeply buried the content may be. Crackers can sometimes read Excel spreadsheets in compressed Word documents. Unzip the file, read it, and analyse the Word document, then read and analyse the Excel data [10].

To assist in file cracking, quite a few of these programmes make use of the autonomy or verity content engines; nevertheless, all of the major tools have quite a bit of their proprietary capacity in addition to the embedded content engine. The majority of solutions can recognize standard data encryption and use it as a contextual rule to block or quarantine information [11]. Using recovery keys in conjunction with corporate data encryption, several technologies make it easier to analysis encrypted data [12].

3 Content Analysis Techniques

Once the basic information has been accessed, seven basic analysis techniques are applied on that data to discover policy breaches. Each of these techniques has its own set of advantages and disadvantages.

3.1 Rule-Based/Regular Expressions

It examines the material in search of particular rules, such as 16-digit numbers that comply with the standards of credit card checksums, medical billing codes, or some other textual analysis techniques [13]. The majority of DLP solutions improve upon fundamental regular expressions by incorporating their very own supplementary analysis criteria (for example, a name near an address near a credit card number).

3.2 Database Fingerprinting

This technique is also referred to as exact data matching at times. This method searches exclusively for exact matches within a database [14], using either a dump of the database or live data retrieved from the database (through an ODBC connection). You could create a policy, for instance, that only checks for credit card numbers in customer base, allowing you to ignore your employees when they make online purchases.

3.3 Exact File Matching

With this process, you calculate a file's hash values first and then search for additional files with the exact same fingerprint. Some people categories this technique like a contextual analysis procedure because the content of the files themselves is not studied in it. It excels in dealing with media files as well as other binaries, when textual analysis is just not viable [15].

3.4 Partial Document Matching

This technique searches for complete or partial match on the material that is being protected. You are therefore able to construct a policy to safeguard a sensitive document, and the DLP solution will search for either the full text of the document or even just a few phrases' worth's of an excerpt from the document [16]. For instance, if a worker copied and pasted even a single paragraph from a business plan for a new product into an instant message, the DLP System would alert you.

3.5 Statistical Analysis

The process of analysing a corpus of content using machine learning, Bayesian analysis, and other statistical approaches to identify policy violations in content that is similar to protected content [17]. This category comprises a large variety of statistical methods, each of which varies considerably in terms of how they are implemented and how efficient they are. Certain approaches are extremely comparable to those utilized in the prevention of spam. It can work with huge content where it might not be possible to find exact documents to match.

3.6 Conceptual/Lexicon

This technique protects information that resembles a concept. An illustration is the best way to describe something like this, so here is one: a policy that offers warnings when there is conduct that is similar to insider trading and that looks for infractions based on key phrases, word counts, and positions; this policy also issues warnings when there is an activity that is similar to insider trading [18]. Not all company rules can be described with specific examples. Conceptual analysis can find violations of policies that aren't well-defined, which other methods couldn't even begin to check for.

3.7 Pre-built Categories

Everything that can be put into one of the categories that are offered is ideal for this. Generally speaking, it is not difficult to describe the content that is associated with privacy, legislation, or industry-specific requirements. It is very easy to set up and saves substantial policy generation time. Category policies can be used as a foundation for more advanced policies that are specific to an enterprise. Most DLP products on the market are based on these seven techniques. Not every product uses every technique, and there can be big differences in how each technique is used. Most products can also use "chaining techniques", which means that they can make complex policies by combining content analysis technique with contextual analysis technique.

4 Technical Architecture

4.1 Securing Data in Motion, at Rest, and Use

A DLP System is meant to protect content throughout its entire life. In terms of DLP, this is important in three ways; scanning storage and other content repositories are part of data at rest. This is done to find where sensitive content is stored. This is what we call "content discovery". A DLP device, for instance, can search your servers for documents containing credit card details. The file may be encrypted, erased, or the owner of the file may receive a warning if the server is not permitted to store that kind of information. Data in motion is sniffing network traffic [19] (either passively or in real time with a proxy) to find out what is being sent over certain channels. One way to do this is to look for bits of sensitive source code in emails, instant messages, and web traffic. Depending on the type of traffic, tools that are already in use can often block based on central policies [15].

4.2 Data in Motion

Organizations start in DLP with network-based products that protect both managed and unmanaged systems in a wide range of ways. Most of the time, it is easier to start a deployment with network products so that you can quickly cover a large area. Early products could only do basic alerting and monitoring, but all of today's products have advanced features like that allow them to work with existing network infrastructure and offer protective controls, not just detective controls [20]. A passive network monitor forms the core of the majority of data loss prevention (DLP) solutions. In most cases, the network monitoring component is installed on a SPAN port at the gateway or in close proximity to it. Real-time full packet of data capture, session reconstruction, and content analysis is among the tasks that may be carried out using this tool [21]. However, this level of performance is not necessary except in extremely rare situations, as very few organizations are currently operating at such a high level of communications traffic [22]. Additionally, some systems restrict monitoring to pre-defined port and protocol combinations [23], as opposed to using service/channel identification based on the content of the packets being monitored [24].

The next crucial element is the incorporation of email. Due to the fact that email acts as a store and forward, a variety of features, including filtering, quarantining, and encryption integration, can be added without encountering the same difficulties as synchronous. The majority of goods already include a Mail Transport Agent, allowing you to easily add it as an additional receiver in a chain of email recipients [25]. Integration of proxies for filtering/blocking is almost inevitable that whoever deploys a DLP System will at some point want to start limiting traffic [26]. Bridge is if we use a bridge, all we need is one computer with two network cards or some kind of content analysis tools in the middle [27]. Proxy is a protocol or application-specific middleware that organizes incoming traffic into queues before forwarding it and so enables more in-depth analysis. Most commonly, HTTP, FTP, and IM are the protocols supported by gateway proxies.

Internal Networks: DLP System is rarely utilized on internal communications other than email, although being technically able to monitor internal networks. Gateways offer handy choke points, but internal monitoring presents a difficult challenge from the perspectives of cost, efficiency, and policy management. Although some DLP companies have association for internal monitoring, most organizations place less importance on this feature (Fig. 2).

Fig. 2 DLP System view of
data in motion

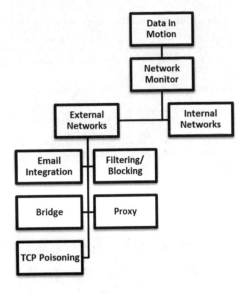

4.3   Data at Rest

Despite the fact that detecting leaks on the network is an effective strategy, it is
only one small component of the overall problem. Finding out where all of that data
is housed in the first place is proving to be of equal or even greater value to a lot
of clients these days. The technique that we are referring to here is called content
discovery. It is possible that enterprise search tools could help with this, but they
aren't very well optimized for the difficulty that is being presented here. Enterprise
data classification tools can also be helpful, but based on the feedback I've gotten
from a number of customers, I've found that they aren't very effective at locating
specific rule violations (Table 1).

4.4   Data at Rest Enforcement

The DLP tool is equipped with a variety of possible responses that it can implement
in the event that a violation of the data policy is discovered. When you receive an
alert or a report, you should handle it as though it were a breach of the network and
create an incident on the central management server. Send the user an email letting
them know that their actions may be in violation of the rules. Move the file to the
central administration server and then leave a text file with instructions on how to
make a recovery request for the file. This action is known as "quarantine and notify".
Quarantine and encrypt mean to encrypt the file in its current location while typically
leaving behind a plain text file that describes how to make a request for decryption
(Fig. 3).

Table 1 Summary of DLP System with accuracy and computational overhead

Title	Attack addressed	Accuracy (%)	Computational overhead
Legitimate data loss in email DLP [4]	Accidental data loss	94.95	High
Data leakage measures [7]	Data loss due to malicious activity	92	High
Unsupervised approach for content-based [8]	Insider attack and malicious	91.51	Low
Transformed data leaks [10]	Insider attack	92	High
Privacy-preserving detection [15]	Anomalous user	95	Low
Sensitive data exposure [6]	Unpredictable leak	87	Medium
File monitoring in cloud [2]	Sensitive data leak	88	High
Data leaks computing [18]	Data leaks caused by human mistakes	90	Low
Data leakage in cloud [20]	Guilt agent detection	97.4	High
Analysis of third-party libraries	Guilt agent detection	91.2	High

Fig. 3 Data at rest DLP System

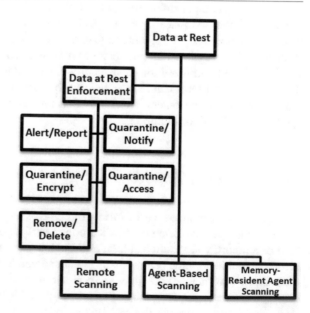

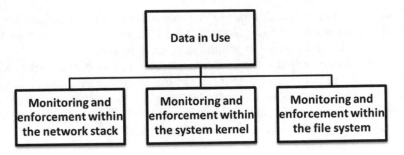

Fig. 4 View of data in use DLP System

4.5 Data in Use

DLP often starts on the network since doing so is the most effective, efficient, and cost-friendly approach to acquiring the widest possible coverage. DLP may be broken down into two categories: network-based and host-based. Network monitoring is non-intrusive and gives visibility to every system on the network, regardless of whether it is managed or unmanaged, a server or a workstation. This is because network monitoring does not interfere with the operation of the system being monitored. The exception to this is when you need to crack SSL [27]. The process of filtering is more complicated, but it is still not overly complicated on the network, and it encompasses all of the systems that are connected to the network. However, it is quite clear that this is not a comprehensive solution. If someone goes out with a laptop, you will not be able to back up your data. Nor can you prevent someone from copying data to a portable storage device such as a USB drive. The product needs to extend coverage not only to the stored data, but also to the endpoints. Control and policing are performed directly within the operating system kernel. If you plug into the kernel of the operating system, you will be able to monitor user behaviours such as copy-pasting sensitive text (Fig. 4).

4.6 Limitations of Existing DLP Systems

- Performance and storage constraints will limit the types of content analysis and the number of policies that can be imposed locally.
- It is difficult to spot and intercept confidential material that has been encrypted, as well as difficult to spot data leakage that is occurring across encrypted networks.
- When communicating sensitive information about the company to a private entity, there should be controls in place that are on the same level as the information being sent.

- Graphics files often include sensitive information about companies, such as credit card scores, academic records, and project specifications. This requires more CPU time as well as the capacity to perform more computations.
- A DLP network solution has been implemented to monitor activities, which includes email traffic, chat over IM, and communication over SSL, and more. To begin with, these solutions need to be configured in accordance with a predetermined information disclosure policy in order to distinguish private data from regular data. Unless it reveals sensitive information.

5 Conclusion and Future Scope

Since the beginning of the DLP Systems market, there have been at least one hundred different firms doing DLP evaluations. Although not all of them purchased a product and not all of them put one into operation, those who did generally found that the implementation was simpler than that of a great number of other security products. Inappropriate expectations and a lack of preparation for the business process and workflows of DLP are typically the most significant roadblocks that stand in the way of a successful DLP deployment from a purely technical point of view. Make sure that your expectations are realistic. DLP System is a very useful tool for prevent inadvertent disclosures and putting a stop to poor business processes that include the usage of sensitive data. DLP market is still a few years away from being able to halt knowledgeable bad people. Although DLP devices are still in their expanding years, they provide very high value to businesses that take the time to plan effectively and learn how to make the most of their capabilities. In future, DLP Systems pay special attention to the process of creating policies and workflows and collaborate with relevant business units to overcome existing problems. Therefore, there is a need for study that would analysis both the content and the context of data in a balanced way.

References

1. Cheng L, Liu F, Yao DD (2017) Enterprise data breach: causes, challenges, prevention, and future directions. Wiley Interdiscip Rev Data Min Knowl Discov 7. https://doi.org/10.1002/widm.1211
2. Reddy PV, Reddy KG (2021) An analysis of a meta heuristic optimization algorithms for cloud computing. In: 2021 5th international conference on information systems and computer networks (ISCON 2021), pp 2–7. https://doi.org/10.1109/ISCON52037.2021.9702376
3. Michael G (2017) ijpam.eu. 116:273–278
4. Faiz MF, Arshad J, Alazab M, Shalaginov A (2020) Predicting likelihood of legitimate data loss in email DLP. Futur Gener Comput Syst 110:744–757. https://doi.org/10.1016/j.future.2019.11.004

5. Ahmed H, Traore I, Saad S, Mamun M (2021) Automated detection of unstructured context-dependent sensitive information using deep learning. Internet of Things (Netherlands) 16. https://doi.org/10.1016/j.iot.2021.100444
6. Yang Z, Liang Z (2018) Automated identification of sensitive data from implicit user specification. Cybersecurity 1:1–15. https://doi.org/10.1186/s42400-018-0011-x
7. Tabassum SH, Naik S (2018) Detecting data leakage and implementing security measures in cloud computing, pp 111–114
8. Karim A, Azam S, Shanmugam B, Kannoorpatti K (2021) An unsupervised approach for content-based clustering of emails into spam and ham THROUGH multiangular feature formulation. IEEE Access 9:135186–135209. https://doi.org/10.1109/ACCESS.2021.3116128
9. Almeshekah MH, Gutierrez CN, Atallah MJ, Spafford EH (2015) Ersatzpasswords: ending password cracking and detecting password leakage. In: ACM international conference proceeding series, Dec 7–11, pp 311–320. https://doi.org/10.1145/2818000.2818015
10. Jiang JY, Bendersky M, Zhang M, Golbandi N, Li C, Najork M (2019) Semantic text matching for long-form documents. In: Web conference 2019—proceedings of the world wide web conference (WWW 2019), pp 795–806. https://doi.org/10.1145/3308558.3313707
11. Song H, Li J, Li H (2021) A cloud secure storage mechanism based on data dispersion and encryption. IEEE Access 9:63745–63751. https://doi.org/10.1109/ACCESS.2021.3075340
12. Ma S, Zhang Y, Yang Z, Hu J, Lei X (2019) A New plaintext-related image encryption scheme based on chaotic sequence. IEEE Access 7:30344–30360. https://doi.org/10.1109/ACCESS.2019.2901302
13. Kaur K, Gupta I, Singh AK (2017) A Comparative study of the approach provided for preventing the data leakage. Int J Netw Secur Its Appl 9:21–32. https://doi.org/10.5121/ijnsa.2017.9502
14. Periyasamy ARP, Thenmozhi E (2017) Data leakage detection and data prevention using algorithm. Int J Adv Res Comput Sci Softw Eng 7:251–256. https://doi.org/10.23956/ijarcsse/v7i4/0121
15. Alneyadi S, Sithirasenan E, Muthukkumarasamy V (2016) A survey on data leakage prevention systems. J Netw Comput Appl 62:137–152. https://doi.org/10.1016/j.jnca.2016.01.008
16. Chang D, Ghosh M, Sanadhya SK, Singh M, White DR (2019) FbHash: a new similarity hashing scheme for digital forensics. Digit Investig 29:S113–S123. https://doi.org/10.1016/j.diin.2019.04.006
17. van de Schoo R, Depaoli S, King R, Kramer B, Märtens K, Tadesse MG, Vannucci M, Gelman A, Veen D, Willemsen J, Yau C (2021) Bayesian statistics and modelling. Nat Rev Methods Prim 1. https://doi.org/10.1038/s43586-020-00001-2
18. Ali BH, Jalal AA, Ibrahem Al-Obaydy WN (2020) Data loss prevention by using MRSH-v2 algorithm. Int J Electr Comput Eng 10:3615–3622. https://doi.org/10.11591/ijece.v10i4.pp3615-3622
19. Zdonik S, Ning P, Shekhar S, Katz J, Wu X, Jain LC, Padua D, Shen X, Furht B, Subrahmanian VS, SpringerBriefs in computer science
20. Hauer B (2015) Data and information leakage prevention within the scope of information security. IEEE Access 3:2554–2565. https://doi.org/10.1109/ACCESS.2015.2506185
21. Inan HA, Ramadan O, Wutschitz L, Jones D, Rühle V, Withers J, Sim R (2021) Training data leakage analysis in language models
22. Kumari A, Game R, Kawale S (2019) Analysis of data leakage detection technique using cloud computing environment. Int J Inf Comput Sci 6:283–290
23. Xuming L, Lina C, Peng J, Xiao G, Shuo C (2019) Current status and future prospects of data leakage prevention technology: a brief review. J Phys Conf Ser 1345. https://doi.org/10.1088/1742-6596/1345/2/022010
24. Magdy S, Abouelseoud Y, Mikhail M (2022) Efficient spam and phishing emails filtering based on deep learning. Comput Netw 206:108826. https://doi.org/10.1016/j.comnet.2022.108826
25. Chen J, Paxson V, Jiang J (2020) Composition kills: a case study of email sender authentication. In: Proceedings of the 29th USENIX security symposium, pp 2183–2199
26. Alneyadi S, Sithirasenan E, Muthukkumarasamy V (2015) Detecting data semantic: a data leakage prevention approach. In: Proceedings—14th IEEE international conference on trust,

security and privacy in computing and communications, pp 910–917 (2015). https://doi.org/10.1109/Trustcom.2015.464

27. Brindha T, Shaji RS (2016) An analysis of data leakage and prevention techniques in cloud environment. In: 2015 international conference on control, instrumentation, communication and computational technologies (ICCICCT 2015), pp 350–355. https://doi.org/10.1109/ICCICCT.2015.7475303

Effect of Anime on Personality and Popularity of Japanese Culture

Kaushal Binjola◉

Abstract Anime is the unique art animation shows or movies originating from Japan. It has received widespread appreciation and recognition from the public. It is an aggregation of multiple genres and will always have something for everyone. Due to its wide popularity and ability to grasp attention, it is often the talk of the town and has led to the genesis of many communities, each based on one/multiple anime shows/movies. Observation shows that consumption of such media has led many to relate to a particular character from the show and revolve their personality around said character. They try to imitate various behavioral habits, tones, and even dialogs from these characters. Many people have begun using various Japanese terms and dialogs in natural speech. Unconsciously, anime has started to affect its watchers' natural behavior, speech, and vocabulary. This paper aims to study the range of effects of anime on an average watcher. We will also learn how anime has led to a significant increase in the knowledge and appreciation of Japanese culture and how it could affect tourism in the recent years.

Keywords Anime · Personality · Tourism · Japan · Culture

1 Introduction

This paper will explain the correlation between the increasing popularity of Japanese culture in the country and anime's role in it. The influence of anime is not restricted to Japanese culture and language but extends way above and bleeds into an individual's personality. To be closer to a character, people pick up various quirks and habits of the character, which can range from talking in a strange voice to imitating the character at all times, or performing certain habits performed by the characters, for example, Kaneki Ken, the protagonist of Tokyo Ghoul, is seeing cracking his index finger whenever he is about to power up. This is seen as cool and is also easily replicable. We also see the use of certain dialogs in everyday conversation. Expanding on the

K. Binjola (✉)
K.J. Somaiya College of Engineering, Vidya Vihar East, Mumbai, Maharashtra 400077, India
e-mail: kaushal.binjola11@gmail.com

© The Author(s), under exclusive license to Springer Nature Singapore Pte Ltd. 2023   647
Y. Singh et al. (eds.), *Proceedings of International Conference on Recent Innovations in Computing*, Lecture Notes in Electrical Engineering 1011,
https://doi.org/10.1007/978-981-99-0601-7_50

popularization of Japanese culture in India, we have seen a stark increase in people learning the Japanese language, taking an interest in consuming Japanese cuisines like ramen and sushi, and even wanting to travel to Japan to experience the world they have seen just through anime. What we would consider eccentric behavior in earlier times is turning quite normal as the popularity of shows has increased. At present, no data or paper explains the existence of a relationship between the popularity of anime and the increase of interest in Japanese culture or the behavioral changes in a person because of watching anime. In this paper, we will take data from a survey and use a Google search to evaluate the personality shift because of anime and also its impact on the increasing demand for knowledge of Japanese culture, cuisines, and even tourism to Japan. We will also see entertainment media's effect on other countries citizens. This will give us a better insight into how digital media affects its viewers and helps us understand anime watchers better. Along with its impact on Japan as a whole, it covers tourism, cuisine, and language. This will help us develop conclusions about how digital media can be used as a source to advertise one's country and even promote culture and language.

2 Literature Review

Many studies show that television shows and movies significantly impact viewers. The social learning theory provides reliable proof and states that people imitate what they observe, and a great proof of the same is children trying to imitate people around them like their parents or siblings. The same is true when people consume entertainment media. As proposed in the paper: A study on the influence of anime among anime fans in Aizawl was carried out by creating focus groups. The researchers found that the participants were keen on copying the behaviors of their favorite anime characters [1]. There is also a relation between an individual's personality and the media genre they consume. However, the media they consume is not necessarily the one they like most [2]. This postulation can mean that a genre can easily affect an individual's personality, allowing them to take an interest or like another genre they would have otherwise hated. We can also see the significant effects of media on personality, wherein a participant from the focus group started finding his younger sibling endearing after watching an anime about siblings and their adventures [1]. Participants also started picking up personality cues of their favorite characters, which changed their outlook on life [1]. There are adverse effects, such as smoking promoted in teenagers due to movies [3]. We can also see a promotion and desire for knowledge of the culture and dominant language of the country to which the media belongs. We observed this when participants of the focus group were interested in Japanese food, clothing, music, and language from watching anime [1]. Young children aged 3–8 years are learning and conversing in Hindi in Bangladesh after watching the Hindi dub of the show Doraemon [4].

The Pearson correlation coefficient is an effective method for checking the similarity between multiple data variables. Relations between multiple data variables can be formed using it. Hence, meaningful relationships or conclusions can be derived from it [5, 6]. Spearman correlation coefficient is also an important method to analyze data and form relationships, as seen when a substantial and positive relationship between the social responsibility of a club with reputation and fans' dependency on the team in the football premier league was found using Spearman's correlation coefficient [7]. Along with Spearman, it was observed that the Kendall correlation coefficient also serves as an efficient and robust method estimator [8]. Films portraying violence against women induced agreeableness in male subjects, making them accept interpersonal violence against women [9].

Data can be collected via several methods, like focus groups [1] or web scraping [2]. Technology and social media have enabled more forms of data collection as well. The use of telephones, emails, and audiovisual interviews has given a new direction to the collection of data, making data collection cheaper, efficient, and reliable [10]. The use of survey research methods is also present, which involves the use of questionnaires [11].

This paper will describe how anime, a famous Japanese media, has led to a worldwide shift in the appreciation of Japanese culture, language, personality, and behavioral changes in an individual after watching and relating to an anime.

3 Methodology

3.1 Data Collection

The data is collected via a survey using google forms. The demography of people is between the ages of 16 and 27. The respondents were asked various questions about their interest in anime, Japan, and tried to find how anime has affected their personality and lifestyle. Table 1 gives the questions asked and the possible answers respondents could have as a response.

3.2 Exploratory Analysis

On the data, we then applied various exploratory data analysis methods to understand the respondents better and form meaningful relationships between those who watch and do not watch anime to the effect anime has had on their life or their choices. An initial way is by calculating the percent responses to ordinal questions asked to people who watch anime. Time also plays an essential role in the development of bias. Critical analysis can be done by checking how prolonged exposure to anime has affected an individual's answers. We also applied different kinds of correlations like Pearson and Spearman to check the impact of various data features on each other. Meaning the degree to which the answer to a question relates to the answer to

Table 1 Survey questions and their possible answers

Question	Answer options
Name	Subjective
Age	Number
Do you watch anime?	Yes/no
When did you first start (approximate year)?	Number
Does anime affect your personality? (Made you agreeable or happy-go-lucky or cynical or pure or cold or will to never give up, etc.)	Strongly agree/agree/disagree/strongly Disagree
If it had an effect on your personality, how so? (Enter-if it didn't)	Subjective
Has anime encouraged you to learn Japanese?	Strongly agree/agree/disagree/strongly disagree
Are you interested in Japanese cuisine?	Strongly agree/agree/disagree/strongly disagree
Are you interested in Japanese culture?	Strongly agree/agree/ disagree/strongly disagree
Do you wish to travel to Japan?	Strongly agree/agree/ disagree/strongly disagree
Do you have a stereotype of Japan, based on the anime you watched?	Strongly agree/agree/disagree/strongly disagree
Are you interested in anime clothing and apparel?	Strongly agree/agree/disagree/strongly disagree
Do you ever want to cosplay?	Strongly agree/agree/disagree/strongly disagree
Do you use anime dialogs frequently?	Strongly agree/agree/disagree/strongly disagree
Has anime introduced you to Japanese music?	Strongly agree/agree/disagree/strongly disagree
Do you ever act like any favorite anime character and copy any of their characteristics?	Strongly agree/agree/disagree/strongly disagree
Have you picked up habits from anime characters? (e.g., Kaneki finger crack)	Strongly agree/agree/disagree/strongly disagree
Describe the habit you picked (if none enter-)	Subjective
Have anime characters changed the choice of liking of real-life people for you?	Strongly agree/agree/disagree/strongly disagree
Has anime introduced certain wanted characteristics in your significant other?	Strongly agree/agree/disagree/strongly disagree
On a scale of 1–10, how much do you enjoy Japanese Music?	Number
Which genres of anime do you like?	Subjective
Do you ever imagine yourself in an anime or as the main character of one?	Yes/no

another question. Meaningful conclusions can also be drawn by which features are more related to others.

Pearson coefficient is calculated as

$$\rho = \frac{\text{cov}(X, Y)}{\sigma_x \sigma_y} \tag{1}$$

where

cov(X, Y) covariance of X and Y

σ_x standard deviation of X
σ_y standard deviation of Y.

Spearman coefficient is calculated as

$$\rho = 1 - \frac{6\Sigma d_i^2}{n(n^2 - 1)} \tag{2}$$

where

d the pair-wise distances of the ranks of the variables x_i and y_i.
n the number of samples.

Kendall coefficient is calculated as

$$\tau = \frac{c - d}{c + d} = \frac{S}{\binom{n}{2}} = \frac{2S}{n(n - 1)} \tag{3}$$

where

c the number of concordant pairs
d the number of discordant pairs.

In our case, we will convert our ordinal data into numbers with Strongly Agree = 2, Agree = 1, Disagree = −1, and Strongly Disagree = −2. Following that, a correlation coefficient matrix is made, which will help calculate the correlation coefficient between every data variable with other data variables by applying the above-given formulas. The coefficients can then be extracted, and analysis can be performed on the features. The analysis will be performed on Jupyter Notebooks.

3.3   Google Trends

Google Trends is an online search tool that analyzes a portion of Google searches to compute how many searches have been done for the terms entered relative to the total number of searches done on Google over the same time. It allows the user to see how often specific keywords, subjects, and phrases have been queried over a specific period. We cannot directly check for the effect of anime on an individual's personality via Google Trends, but it helps give an accurate idea of how interest in Japan and Japanese culture has been changing over the years and how it equates to the popularization of anime. This directly relates to anime's effect on Japan and Japanese culture.

4 Analysis

4.1 Percent Analysis

A Percentage Analysis can be done by calculating the percentage of response to ordinal questions which were asked. The questions are framed in a way that shows that a high percent of Strongly Agree answers would directly correspond to a heavier impact of anime on their personality as well as their bias toward Japan and Japanese culture and vice versa is also true that a high percent of Strongly Disagree would mean no impact of Anime.

4.2 Time Exposure to Anime Analysis

A way of analyzing the effect of exposure of time on the answers given by the respondents is to take a ratio of Agree to Disagree of all questions belonging to a particular year. We have also taken an assumption as to when there are no disagreements in a year. The disagreements in the year will be given a one by default to avoid zero division error.

4.3 Correlation Coefficient Analysis

Correlation coefficient gives us how strong the relationship between two variables is. The range of correlation coefficients varies from -1 to 1. Wherein 1 signifies total positive correlation, as X increases or decreases, Y will also increase or decrease, respectively, by a scaled value of X. -1 signifies total negative correlation, as X increases or decreases, Y will also decrease or increase respectively by a scaled value of X. Spearman coefficient is used for variables that are not normally distributed and are nonlinear. For a case such as this paper, this will prove very useful. It is also useful in a rank-based system, such as one our respondents answer.

5   Result

5.1   Percent Analysis

Two-thirds of the respondents agree that anime has encouraged them to learn Japanese. Following that, 87.5% have shown an interest in Japanese cuisine. A clear indication of anime promoting tourism can be seen as 100% of all respondents agree that they are interested in Japanese Culture and have a desire to travel to Japan. The survey also shows significant marketing leverage, indicating that two-thirds of all respondents are interested in clothing and apparel displayed in anime. An exciting discovery was also made, showing anime's direct impact on personality. 59% of anime-watching respondents conveyed that they underwent a personality change or that anime has affected their personality in one way or another. Furthermore, the same has also agreed to use dialogs used in anime frequently. A more direct impact can be seen from the question, "Do you ever act like any favorite anime character and copy any of their characteristics?" 54% agree and 45% agree they also copy certain habits or personality trademarks of anime characters they are watching. 50% of all respondents also expressed the desire to be the main character of an anime.

On analyzing the subjective question of the effect of anime on personality, all the responses were positive and usually revolved around gaining the ability never to give up, being tenacious, willful, and having a strong sense of friendship.

5.2   Time Exposure to Anime Analysis

Performing the above analysis, we obtained the following distribution Fig. 1 of years to ratios. We can see that there was no determining factor as to the effect of year on the answers provided by the respondents. A notable observation was that the highest number of respondents started watching anime in 2015, and it has received one of the most balanced amounts of agreeing to disagreeing responses.

5.3   Correlation Coefficient Analysis

A moderate negative Pearson correlation of 0.544580 can be seen between the respondents' age and their want to learn a language. This can be expected as the want to learn a language decreases as the age of an individual rises. The three correlations also reaffirm our analysis of the relation of time exposure to anime and other features, as all the correlation coefficients used display a low correlation between the start year of anime and other features. A strong Pearson correlation of + 0.604790 was seen between a person using more anime dialogs and feeling that anime has affected their personality. Respondents of a higher age also imagine themselves more as the

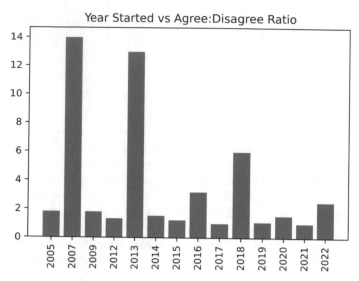

Fig. 1 Effect of time exposure to anime

main character of an anime, with a Spearman correlation between the two being +
0.497336. Following expectations, a 0.570561 positive Spearman correlation coefficient was seen between liking Japanese cuisine and wanting to travel to Japan. As
interest in Japanese cuisine increases, it will be natural to try it from an authentic
source. Those who feel that their personality has been affected by anime also show a
positive Spearman correlation of 0.560748 to imagining they are the main character
of an anime. Such a coefficient is expected. Those who had their personality affected
have been influenced by a particular character, usually the main one. A high positive correlation of 0.735706 can be seen between respondents to whom anime has
introduced specific wanted characteristics in their significant other and those who
copy specific characteristics of any characters. This is also intuitive as those who
copy characteristics of a character would also want a significant other similar to the
characters they copy. A substantial positive Pearson correlation of 0.678287 can also
be witnessed between the introduction of anime to Japanese music and the want to
learn Japanese. Lastly, according to expectations, those who enjoy Japanese music
are more likely to have Japanese music introduced to them via anime, as shown by
a powerful positive Spearman correlation coefficient of 0.736528.

5.4 Google Trends Analysis

Google Trends gives us a great look at how interest in the Japanese language changed
to anime. According to Fig. 2, which queries through Google web searches from 2004
to the present, we see a similar shape shared by the two graphs with Japanese language

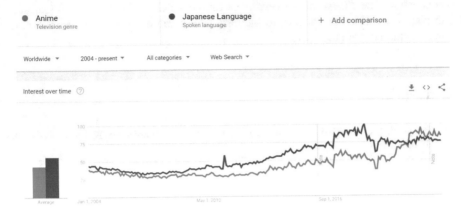

Fig. 2 Web searches of anime and Japanese language from 2004 to present

being offset by some value. Both line charts rise and fall at the same time and can be seen to have similar slopes. The two slopes also show a very high positive Spearman correlation coefficient of 0.885203. Spearman is used here as the distributions are not normal.

An outlier point can be seen in the Japanese graph around 2011, which can be associated with the Japan Tohoku earthquake and tsunami 2011. The spike of searches in January of 2017 and fall of searches in March of 2019 cannot be attributed to any concrete point or theory.

6 Conclusion and Future Work

Using the analysis work presented in this paper, we can say that anime has substantially affected the viewer's personality and has introduced various positive changes to their mindset. Japanese culture has also received a massive boost from anime, with people wholeheartedly supporting Japanese cuisine and songs and wanting to learn the Japanese language. A significant take from the survey can be the impact of anime on tourism. Japan can use anime as a marketing strategy to invite the youth to explore Japan firsthand. They can also extend the use of anime to popularize their education and work sector to invite people from other countries to pursue education and work in Japan, increasing their economy.

Further analysis can also be done by considering the economic change in Japan's industry due to anime and tourism over the years. Sales of anime merchandise and profits due to tourism can also serve as validation. Furthermore, data can be collected on the increase in travel to Japan by foreign nationals for a vacation to experience Japan firsthand. Social media can be used to collect data on the increase in the number of posts relating to anime and Japan and can be used to perform sentiment analysis to understand the change of personality of anime watchers. Unsupervised learning can also be explored to understand the data better, and meaningful interpretations can be gathered from the same.

References

1. Muankimi ML (2017) A study on the influence of Anime among Anime Fans in Aizawl. IJCHSSR 1(1)
2. Jak˘sa S (2020) What anime to watch next? The effect of personality on anime genre selection.
3. Heatherton TF, Sargent JD (2009) Does watching smoking in movies promote teenage smoking? Curr Dir Psychol Sci 18(2):63–67
4. Islam NN, Biswas T (2012) Influence of Doraemon on Bangladeshi children: a CDA perspective. Stamford J English 7:204–217
5. Benesty J, Chen J, Huang Y, Cohen I (2009) Pearson correlation coefficient. In: Noise reduction in speech processing. Springer, Berlin, Heidelberg, pp 1–4
6. Adler J, Parmryd I (2010) Quantifying colocalization by correlation: the Pearson correlation coefficient is superior to the Mander's overlap coefficient. Cytometry A 77(8):733–742
7. Javaran SH, Sajadi SAN, Karamoozain M (2014) The relation- ship between the social responsibility of club with reputation and fans' dependency on the team in the football premier league
8. Croux C, Dehon C (2010) Influence functions of the Spearman and Kendall correlation measures. Stat Methods Appl 19(4):497–515
9. Malamuth NM, Check JV (1981) The effects of mass media exposure on acceptance of violence against women: a field experiment. J Res Pers 15(4):436–446
10. Aborisade OP (2013) Data collection and new technology. Int J Emerg Technol Learn (iJET) 8(2):48–52
11. Morgan GA, Harmon RJ (2001) Data collection techniques. J Am Acad Child Adolescent Psych 40(8):973–976

Synthetic Time Series Data Generation Using Time GAN with Synthetic and Real-Time Data Analysis

Tanya Juneja, Shalini Bhaskar Bajaj, and Nishu Sethi

Abstract Synthetic time series data generation is a wide area to research, and lot of attention has drawn recently. The assumption for generating multivariate time sequenced data is well proportioned and continuous without missing values. A productive time series data model should retain momentum, such that the new sequence maintains the actual relationship between different variations occurrence over time. Different existing approaches that put forward generative adversarial networks into the sequence system that prohibits care for temporary interactions that differ from sequential time data. Simultaneously, monitored sequence prediction models—allowing for fine control over network dynamics—are naturally determined. Time series data generation has problems like informative missing data values leading to untraceable challenge and long sequences with variable length. These problems are biggest challenges in making of a powerful generative algorithm. Herein, we are using an innovative structure to produce real-time sequence which associates the adaptability of different unattended prototype through controls provided during the supervised model training. Privacy data analysis safeguards data privacy and data sharing. Generation of accurate private synthetic data is a NP hard problem from any considered scenario. This paper discusses about privacy parameter and data analysis of real and synthetic data. The synthetic data is enclosed in the boundaries of real data and have similar behavior.

Keywords Data privacy · Dynamic bayesian network · Synthetic data · Time series

T. Juneja (✉) · S. B. Bajaj · N. Sethi
Amity University Haryana, Manesar, Gurugram, India
e-mail: tanya.juneja@ymail.com

S. B. Bajaj
e-mail: sbbajaj@ggn.amity.edu

© The Author(s), under exclusive license to Springer Nature Singapore Pte Ltd. 2023
Y. Singh et al. (eds.), *Proceedings of International Conference on Recent Innovations in Computing*, Lecture Notes in Electrical Engineering 1011,
https://doi.org/10.1007/978-981-99-0601-7_51

1 Introduction

The temporal model poses a unique challenge on productive models. The model not only gives the function of apprehend the distribution of parameters within various instances of time, but it should also seize the potential for intricate variables of variables present at that period. Specifically, in modeling multivariate data y_1: $T = (y_1 \ldots y_T)$, we wish to do accurately download conditional distribution $\pi p(y_t | y_1 : t - 1)$ and temporary changes. Introducing the time series generative adversarial networks (Time GANs), an ordinary structure aimed at producing real-time serialized data for diverse domains. Furthermore, the unpredictable losses of both original and practical contradictions, respectively, we present the tracked losses that follow the steps by means of the real data as monitored, obviously to encourage models to seize uncertain distribution following steps within data points. It yields to take benefits of the fact that more information acquired on training data rather than every data point is original or artificial so, to acquire from transformation since actual follow-up. Importantly, supervised losses are jointly reduced training both embedded networks and generators, so that the hidden area is not just functional promote efficiency of parameter setting to expedite generator learning temporary relationships. Eventually, the framework for dealing hybrid data preparation with together discrete plus continuous time series data generated simultaneously.

Our approach is the first to integrate the unstructured GAN structure in autoregressive models with control given to supervise training. The benefits in a sequence of analysis with different realistic plus virtual datasets scenarios. Preferably, we drive TSNE [1] and PCA [2] analyze to visualize similarity in distribution produced actual distribution. In addition, using the "train on synthetic, test on real (TSTR)" framework [3, 4] in the subsequent predictive function, we test fitness of the facts produced keeps original speculative features. We find that Time GAN accesses consistently and significant developments beyond modern standards in producing real-time series.

Also using Gretel for different comparisons between real and synthetic data. Gretel Synthetics called privacy filter, which provides the kind of weapons to cover data for the making of weak points that often exploited by enemies to attack. For example, synthetics that are remarkably like the original data can lead attacks on membership ideas and attribute disclosure. Another major privacy risk appears when you have "external" records, especially if they are like external one's training records. Combat both conditions, we created filters like outlier, which are both can dial to a certain extent based on the desired level of privacy.

2 Related Work

Time GAN, a productive time sequence model, taught to fight and collaborate with capabilities to learn embeddings. A gap consisting supervised loss and uncontrolled loss. Therefore, route is in middle of the road of a wide range of research areas, including themes from automated models to predict sequencing, GAN-based approaches to generating sequence, and learning to represent a time series.

Automatic repetitive networks trained with a high probability system [5] tend to be errors can be large guesses when sampling multiple steps, due to differences between close-loop training and open-loop inferences. Based on curriculum learning [6], planned sample firstly aimed as a solution, in which the models trained to produce output embedded in a mixture of both previous speculation and basic truth data [7]. Encouraged to adapt to the adversarial domain [8], professor forcing has been instrumental in training apartheid facilitator to differentiate between free running and hidden conditions imposed by teachers, thus promoting network training and sample power to combine [9]. The methods of criticizing the character [10] have proposed, introducing the character to the character in targeted results, trained to measure the next worth proposed functions which guide player's free operation predictions [11].

Contrastingly, different research indicates directly benefited from the GAN framework within temporary setting. Firstly, (C-RNN-GAN) [12] clearly applied on GAN structure on sequence data, by LSTM networks to generate generators and discriminators. Data points reproduced repeatedly; it takes as includes audio data and data created with prior timeline. GAN with normal conditions (RCGAN) [3] took the same approach, introducing a small structural difference as a drop relying on previous output while preparing additional inputs [13]. The crowd of the studies since then has used these frameworks to produce a sequence of actions in such a variety domain such as text [14], finance, bio signals, sensor and intelligent grid data [15], and regenerative conditions [16]. Recent work [17] suggested making a timeline information to treat samples infrequently. However, unlike our proposed method, these methods are only dependable regarding the contradictory learning response, it itself may not be enough to confirm especially since the network is effectively holding temporary power over training data.

Finally, learning to represent the process of time sequences deals principally on benefits of acquiring codes combined to benefit sub-functions such as prediction [18], prediction [19], and separation [20]. Different activities study helps to read hidden presentations for purposes of previous training [21], segregation [22], and interpretation [23]. Meanwhile in a stable state, various operations seem to discover benefits of combining autoencoders along with enemy training, accompanied by objectives that have same learning paradigms [24], which allow for effective thinking [25], as well improving production capacity [26]—the method used later in production different structures by coding and producing complete sequence of discrimination [27]. On the contrary, our proposed method incorporates time series data randomly, incorporating the intensity of the other measurement of time, as well as the use of embedded network to identify space with a lower side of a productive model to learn the step-by-step distribution and subtle flexibility of data.

3   Proposed Work

Generative adversarial networks (GANs) exist as a category of machine learning structure accustomed to train productive models. These networks are interesting sufficient for its past implementation to operate in key areas of the same infrastructure energy systems, agriculture, and other industries [28]. As a consequence, it is so important to study utilizations of GAN on a large scale. By means of productive opposition networks, we can generate performance data points that cannot separate from the original data according to the data traits by transferring the sequence of real data and arbitrary sound as response to the frame. Conventional GAN structure composed of two neural networks connected to one another, specifically, generator network along with discrimination network. Mutually, the generator and the discriminator systems are set up with the first bias of the appropriate tiers. Generator network considers random audio as capture and generate data sequences, which transmitted to the discrimination network.

3.1   Time GAN

Time GAN contains four structure modules: embedding function, retrieval function, sequencing generator, and subsequent discrimination. Important understanding that autoencoding parts trained jointly with opposing components, such as Time GAN concurrently acquires coding features, making presentations, and repeating over time. Embedding network provides hidden space, the opposing structures operates containing space, and hidden power your real and virtual dataset synced with supervised losses. We describe each one in sequence.

Architecture: The GAN time variant intended to seize characteristics of real data and the complex flexibility of those features over time. The newly launched embedding network provides a rewind map between features and hidden representation. The network embedding and the recovery network maintains the relationship between the hidden vectors and elements in the hidden area, whereas the function of the generator and the racist network remains unaffected. Like traditional GAN, generator and discrimination are associated with successive losses, also known as unsupervised loss [29]. Inclusive to unattended losses, the model repaired with two other losses activities, identified as supervised losses and reconstruction losses, coupled with it automatic installers.

Four parts are as: (I) Generator—Produces data sequences. (II) Discrimination—Distinguishes sequence of data as actual or duplicate. (III) Embedding network—Consists of a rewind map amid features and hidden features illustration. (IV) Restore—Consists of a map between the feature and the hidden space.

Three types of losses can describe as: (1) Uncontrolled Loss—Loss of work in relation to electricity generation and racism (minimum size). (2) Monitored Loss—How well the generator accurately calculates future data in a hidden location. (3) Reconstruction Loss—Compares reconstructed and original data, referring to automated embedding

Embeddings and recovery function: The embedding and recovery functions offer mappings among features and latent spaces, permitting to acquire basics of temporal dynamics of the data through lower-dimensional representation in adversarial network. Let GS: GX denotes the latent vector spaces resultant characteristics space $S:X.$GX includes constant and progressive time-based characteristics to the latent codes gS; g_1: $T = e\,(s;\, \times\, 1$: $T)$. Herein, we execute e through repetitive structure, where $eS : S- >$ GS is an embedding network for unchanging characteristics, and $eX : GS * GX * X- >$ GX a recurrent embedding network for progressive time-based features. t X takes constant and progressive time-based codes back to their feature representations s; $\times$ $1 \cdot T = r\,($GS; g_1; $T)$. Implementations of r across a feedforward network at each step , $rS : GS- > S$ and $rX : GX- > X$ are recovery structure for constant and progressive time-based embeddings [30]. The embedding and recovery functions characterized with the help of several structure of selection, along with one perquisite that it adopts autoregressive plus spontaneous sequencing.

3.2 Datasets

In this paper, we have taken three multivariate time series datasets. These datasets described below:

1. Power: Database characterized by sound repetition, high magnitude, and related features. The UCI appliances' power forecasting database contains various estimates, with continuous value that includes different temporal features that measured soon.
2. Stock: This database taken from Yahoo finance contains six variables. Open, Up, Down, Close, Adj Close, and Volume.
3. Air quality: The site contains 9358 times the mid-hour responses from a series of five chemical oxide sensors embedded in a high-quality chemical device. The device found in a stadium in a highly polluted area, road level, in the heart of an Italian city. Data recorded from March 2004 to February 2005 (one year) represents the longest record available for cooling chemical sensors. Low true hours CO hours, non-methane hydrocarbons, benzene, total nitrogen oxides (NO_x), and nitrogen dioxide (NO_2) provided by a certified reference reviewer.

4 Experiment and Evaluation

Quality testing metrics often referred to observable examination. Subjected on the function and structure of the model results. Distinctive viewing method used. Duplicate visualization data is modest, but when we consider higher-degree time series

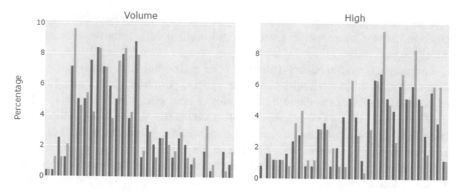

Fig. 1 Field data comparison between training data (violet) and synthetic data (sea green) on dataset stock field volume and high

data, visualization turns difficult. It is difficult to detect any minor inconsistencies between actual and artificial data points. That is, it must be impossible considering model was real distribution. In Fig. 1, training data and synthetic data at various instances can be seem close to with difference less than 1. As at time interval in high at time interval 10–11 value seems to be 3.3 for synthetic while 3.4 for training data.

4.1 PCA and TSNE

TSNE [1] and PCA [2] analyzes on both the original and synthetic datasets (flattening the temporal dimension). These visualization helps in estimating the closeness of the distribution of created data points resemblance to real in 2D plot, resulting in qualitative assessment. In Fig. 2, there is PCA and TSNE analysis of the datasets. In Fig. 2, PCA analysis shows that all data points are within the original data points, and similarly for TSNE, original data encapsulates synthetic data. The descriptive analysis of data points shows that data is close to original data and varies between lowest and highest real data points.

4.2 Correlation

The term autocorrelation or correlation refers to the degree of similarity between (A) and a given time series, and (B) its late version, over (C) consecutive intervals. In other words, autocorrelation intends to measure the relationship between the current value of a variable and any past values you can reach. Therefore, the time series of automatic integration attempts to estimate the current values of the variance by comparing the historical data for that variance (Fig. 3).

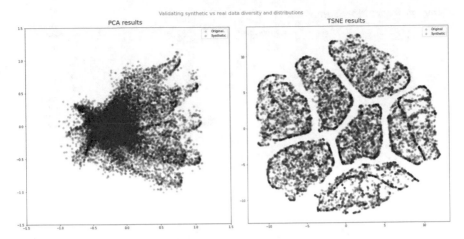

Fig. 2 PCA and TSNE result for synthetic (red color) and real data (black color) for energy data

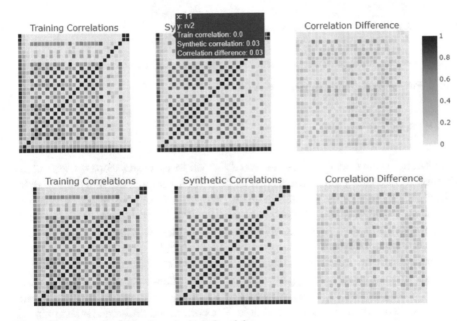

Fig. 3 Correlation analysis of synthetic and real data

4.3 Privacy Protection

Privacy protection [31] is some on basis of four characteristics: (I) outlier filter, (II) similarity filter, (III) overfitting prevention, and (IV) differential privacy. Figure 4 gives us details about privacy of datasets.

Privacy Protection Summary

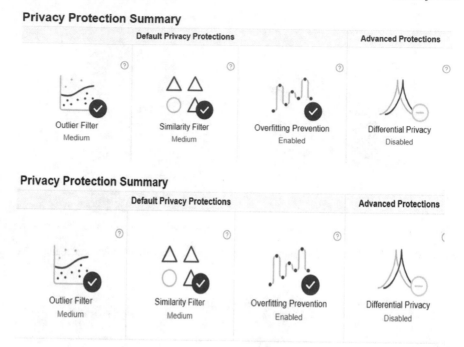

Fig. 4 Privacy protection of synthetic and real data

5 Result

The synthetic data generated is not exactly close to real data values. Data values duplicated depending on datasets such as zero values duplicated in synthetic data, while 130 data values duplicated in energy datasets. In the worst-case generation of synthetic data, Boolean of linear statistical is NP hard problem [32]. Differential privacy disabled for all three datasets which is very efficient to prevent duplicate data generation. Synthetic quality score is above 85 which indicates all the privacy factors and analysis are favorable and synthetic data is unique and in structured format. PCA analysis of the dataset overlapped properly for the taken dataset as seen in Fig. 2. Observation from Fig. 3 various parameters in the datasets are correlated while certain parameter is distant in that instance of time. The correlation difference is less than 0.2 in most parameter considered for comparison. Differential privacy disabled indicates that no privacy compromised during synthetic data generation. Overall accuracy and quality of data ranges between 80 and 90. The privacy outliners enable data to be secure for usage and (Table 1) encourages to obtain high accuracy also enabling synthetic data to fully capture characterstics of original data.

Table 1 Parameter analysis on three datasets

Parameter	Energy	Stock	Air quality
Synthetic quality data score	90	80	88
Privacy	Good	Exceptionally good	Exceptionally good
Field correlation stability	94	70	88
Deep structure stability	88	85	88
Field distribution stability	99	88	91
Outlier filter	Disabled	Medium	Medium
Similarity filter	Disabled	Medium	Medium
Overfitting prevention	Enabled	Enabled	Enabled
Differential privacy	Disabled	Disabled	Disabled

6 Conclusion and Future Scope

Herein, we present Time GAN, a unique framework for a generation of serialized time data that includes variation of the unmanaged GAN method and conditional variable duration controls provided by autoregressive supervised models. Using donations for supervised losses as well as collaboratively trained embedding network, Time GAN demonstrates consistent and critical development in addition to high-quality benchmarks in the production of real-time series data. We also investigated privacy parameters on the generated time series data using Time GAN with the help of Gretel. The privacy of data encourages to preserve original data with high accuracy enabling to fully capture characteristics during synthetic data generation. Data generation can be used in different fields like IOT and health care by developers to develop strong data model with improved hyperparameter adjustments embedding network creation.

References

1. Bermperidis T, Schafer ST, Gage FH, Sejnowski T, Torres EB (2022) Dynamic interrogation of stochastic transcriptome trajectories using disease associated genes reveals distinct origins of neurological and neuropsychiatric disorders. Front Neurosci 703

2. Dogariu M, Ştefan L-D, Boteanu BA, Lamba C, Ionescu B (2021) Towards realistic financial time series generation via generative adversarial learning. In: 29th European signal processing conference (EUSIPCO), pp 1341–1345
3. Esteban C, Hyland SL, Rätsch G (2017) Real-valued (medical) time series generation with recurrent conditional gans. arXiv preprint arXiv:1706.02633
4. Yoon J, Jordon J, van der Schaar M, PATE-GAN (2019). Generating synthetic data with differential privacy guarantees. In: International conference on learning representations
5. Liu X, Li L (2022) Prediction of labor unemployment based on time series model and neural network model. In: Computational intelligence and neuroscience
6. Persaud D (2022) Arts education in a time of crisis: COVID-19 in Los Angeles, 2020–2022. Doctoral dissertation, UCLA
7. Raval M, Dave P, Dattani R (2021) Music genre classification using neural networks. Int J Adv Res Comput Sci 12(5)
8. Ganin Y, Ustinova E, Ajakan H, Germain P, Larochelle H, Laviolette F, Marchand M, Lempitsky V (2016) Domain-adversarial training of neural networks. J Mach Learn Res 17(1):2096–2030
9. Lamb AM, Parth Goyal AGA, Zhang Y, Zhang S, Courville AC, Bengio Y (2016) Professor forcing: a new algorithm for training recurrent networks. In: Advances in neural information processing systems, pp 4601–4609
10. Sajja RK, Killari V, Nimmakayala SA, Ippili V (2022) Machine learning algorithms for asl image recognition with lenet5 feature extraction. Int J Adv Res Comput Sci 13(3)
11. Bahdanau D, Brakel P, Xu K, Goyal A, Lowe R, Pineau J, Courville A, Bengio Y (2016) An actor-critic algorithm for sequence prediction. arXiv preprint arXiv:1607.07086
12. Mogren O (2016) C-rnn-gan: continuous recurrent neural networks with adversarial training. arXiv preprint arXiv:1611.09904
13. Mirza M, Osindero S (2014) Conditional generative adversarial nets. arXiv preprint arXiv:1411.1784
14. Zhang Y, Gan Z, Carin L (2016) Generating text via adversarial training. In: NIPS workshop on adversarial training, p 21
15. Zhang C, Kuppannagari SR, Kannan R, Prasanna VK (2018) Generative adversarial network for synthetic time series data generation in smart grids. In: IEEE international conference on communications, control, and computing technologies for smart grids (SmartGridComm). IEEE Publications, pp 1–6
16. Chen Y, Wang Y, Kirschen D, Zhang B (2018) Model-free renewable scenario generation using generative adversarial networks. IEEE Trans Power Syst 33(3):3265–3275. https://doi.org/10.1109/TPWRS.2018.2794541
17. Ramponi G, Protopapas P, Brambilla M, Janssen R (2018) T-cgan: Conditional generative adversarial network for data augmentation in noisy time series with irregular sampling. arXiv preprint arXiv:1811.08295
18. Dai AM, Le QV (2015) Semi-supervised sequence learning. In: Advances in neural information processing systems, pp 3079–3087
19. Lyu X, Hueser M, Hyland SL, Zerveas G, Raetsch G (2018)Improving clinical predictions through unsupervised time series representation learning. arXiv preprint arXiv:1812.00490
20. Srivastava N, Mansimov E, Salakhudinov R (2015) Unsupervised learning of video representations using lstms. In: International conference on machine learning, pp 843–852
21. Miuccio L, Panno D, Riolo S (2022) A Wasserstein GAN autoencoder for SCMA networks. IEEE Wirel Commun Lett 11(6):1298–1302
22. Li Y, Mandt S (2018) Disentangled sequential autoencoder, p10. arXiv preprint arXiv:1803.02991
23. Hsu W-N, Zhang Y, Glass J (2017) Unsupervised learning of disentangled and interpretable representations from sequential data. In: Advances in neural information processing systems, pp 1878–1889
24. Larsen ABL, Sønderby SK, Larochelle H, Winther O (2015) Autoencoding beyond pixels using a learned similarity metric. arXiv preprint arXiv:1512.09300

25. Dumoulin V, Belghazi I, Poole B, Mastropietro O, Lamb A, Arjovsky M, Courville A (2016) Adversarially learned inference. arXiv preprint arXiv:1606.00704
26. Makhzani A, Shlens J, Jaitly N, Goodfellow I, Frey B (2015) Adversarial autoencoders. arXiv preprint arXiv:1511.05644
27. Kim Y, Zhang K. Rush AM, LeCun Y et al (2017) Adversarially regularized autoencoders. arXiv preprint arXiv:1706.04223
28. Asre S, Anwar A (2022) Synthetic energy data generation using time variant generative adversarial network. Electronics 11(3)
29. Pei H, Ren K, Yang Y, Liu C, Qin T, Li D (2021) Towards generating real-world time series data. In: IEEE international conference on data mining (ICDM), pp 469–478. https://doi.org/10.1109/ICDM51629.2021.00058
30. Yoon J, Jarrett D, van der Schaar M (2019) Time-series generative adversarial networks. Curran Associates, Inc.
31. Boedihardjo M, Strohmer T, Vershynin R (2021) Privacy of synthetic data: a statistical framework

How Big Data Analytical Framework has Redefined Academic Sciences

Imran Rashid Banday, Majid Zaman, S. M. K. Quadri, Muheet Ahmed Butt, and Sheikh Amir Fayaz

Abstract Academic institutions generate a large amount of heterogeneous data and academic leaders want to make the most of that data by evaluating it for improved decision-making. The volume is not the only difficulty; the organization's data format (structured, semi-structured, and unstructured) creates complication in academic functioning and decision-making on a daily basis. Academic data sets have expanded in amount and proportion to the point that standard data processing and analytics cannot provide satisfying results; moreover, it is not only about volume but also about what is to be done with data. In this paper, Big Data is used in an academic institution (University of Kashmir) to analyse data (structured, semi-structured, and unstructured) from multiple sources for smart decision-making, and machine learning models on Big Data are used to predict academic behaviour such as student performance, academic bias, and job market viability.

Keywords Big Data · Data mining · Heterogeneous sources · Big Data in higher education

1 Introduction

Higher education institutions cannot avoid the effect of Big Data. Massive volumes of educational data are gathered and created every day in the higher education system from many sources and in various forms [1]. In higher education institutions, information technology has been employed to automate the majority of the manual procedures (from student admissions to tests to result declaration or physical file tracking

I. R. Banday · M. A. Butt · S. A. Fayaz
Department of Computer Sciences, University of Kashmir, Srinagar, India

M. Zaman (✉)
Directorate of IT and SS, University of Kashmir, Srinagar, India
e-mail: zamanmajid@gmail.com

S. M. K. Quadri
Department of Computer Science, Jamia Millia Islamia, New Delhi, India

records to administrative activities), resulting in massive repositories of data sources. Higher education institutions already have massive amounts of data, both physical and digital, housed in a variety of repositories [1]. Furthermore, there is a massive surge in digital data, which nearly doubles each year. The student life cycle is the primary driver of this data increase. One of the most obvious advantages of Big Data is its ability to handle enormous volumes of data at once and discover trends. The potential benefits of Big Data analysis in higher education are still mostly unknown [2]. Accrediting organizations (such as NIRF and QS BRICS), governments, and other stakeholders investigate and audit universities in order to discover innovative ways to enhance and monitor student progress and other institutional practices [3]. People are very interested in this data, and they want to acquire more and more information about students in order to better understand students and programmes. Previously, if a student wanted to follow a programme at the University of Kashmir (UoK), he would just enrol in the course based on the already mentioned statues, with no regard for the University. However, with the implementation of the choice-based credit system in higher education, we can now recommend course combinations to students, or even if a candidate is already enrolled in a course, we can recommend subject combinations for him/her in the following semester and or programme for higher education. In this research, we provided a methodology that demonstrates how institutions (specifically the University of Kashmir) can handle large amounts of data [4].

1.1 Big Data Introduction

Big Data basically refers to huge volume of data that cannot be stored and processed using the traditional approaches within the given time frame. How huge this data needs to be in order to be classified as Big Data. There is lots of confusion while referring to the Big Data. Usually, Big Data term is used to data that is either in gigabytes or terabytes or petabytes or exabytes or anything that is larger than this size. However, this does not define the term Big Data completely, even a small amount of data can be referred as big data depending on the context it is being used. The size of the data will vary between megabytes and petabytes, depending on the domain [5, 6]. Big Data is therefore context-specific and can apply to various sizes and types from domain to domain, but the common challenge facing all of these domains is to be able to make sense of the data by analysing it at a high analytical level.

1.2 Big Data Definition

Big Data concerns large volume, structured or unstructured, complex, growing datasets with multiple, independent heterogeneous, and or homogenous sources. SAS defines Big Data as follows: "Big Data is a term that describes the large volume of data—both structured and unstructured—that inundates a business on a day-to-day basis [7]. But it is not the amount of data that is important. It is what organizations do with the data that matters. Big Data can be analysed for insights that lead to better decisions and strategic business moves".

IBM definition for Big Data is: "Big Data is a term applied to data sets whose size or type is beyond the ability of traditional relational databases to capture, manage and process the data with low latency [8]. Big Data has one or more of the following characteristics: high volume, high velocity, or high variety".

1.3 Big Data Source

Primarily data is obtained from following types of sources.

(1) Internal Sources: Internal sources are those data sources that provide generally organized data that originates from within the enterprise. Data from internal sources (like student life cycle, customer management, resource management, etc.) is generally operational data. This type of data is used by enterprises for their daily operations like OLTP [9].

(2) External Sources: External sources are those data sources that provide unorganized data that originates from the external environment of an enterprise. Data from these sources (like Internet, industry partners, government, weblogs, etc.) is analysed to understand the external entities (students, customers, and competitors) [10].

1.4 Big Data Types

Structured data: Structured data can be defined as the data that has predefined format and usually stored in tabular format [11]. The sources for this type of data are flat files (delimiter separated values), relational databases.

Unstructured data: Unstructured data is the data that may or may not have and predefined format or repeating patterns. Unstructured data consists of data having different formats like text, images, audio, video, emails, etc. [11]. Sources for this type of data are documents, logs, survey results, feedbacks, social networking platforms, and mobile data.

Semi-structured data: Semi-structured data can be defined as the data that does not follow the proper structure of the data models as in relation databases [11]. This type of data contains labels or mark-up components in order to separate elements and generate hierarchies of records and fields in the given data.

Figure 1 shows the percentage of structured, semi-structured, and unstructured data [7].

Fig. 1 Percentage structure
of Big Data

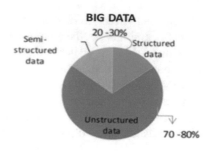

2   Literature Survey

There is a lot of research on Big Data in Academia, and we have highlighted some of the most significant publications in this study. We believe there is not a single research employing models on big data for Artificial Intelligence solutions for predicting academic behaviour including student performance, academic bias and job market viability.

A Big Data-based novel knowledge teaching assessment method in universities is reviewed by Xin [12]. The discussion of knowledge teaching evaluation systems in colleges and universities over the last ten years is summarized in this paper. It also examines the factors that are preventing further advancements in teaching evaluation and draws on performance management theory. The evaluation system offers a structured framework for integrating assessment methods used in university knowledge teaching theory and practice.

Brock et al. [13] suggested that the research model adds elements of organizational learning capacities to the classic technology adoption paradigm. This study polled 359 information technology workers from 83 countries who are studying at the University of Liverpool Online and work in a variety of sectors. This study combines two technology adoption paradigms to give meaningful academic and practical information.

Wen et al. [14] examine works on increasing Hadoop cluster energy efficiency and group them into five categories: energy-aware cluster node management, energy-aware data management, energy-aware resource allocation, energy-aware job scheduling, and alternative energy-saving strategies. Authors briefly explain each category's logic and compare and contrast the relevant works in terms of their benefits and drawbacks. Furthermore, they present our findings and suggest future research directions, including energy-efficient cluster partitioning, data-oriented resource classification and provisioning, resource provisioning based on optimal utilization, EE and locality aware task scheduling, machine learning-assisted job profiling, elastic power-saving Hadoop with containerization, and efficient big data analytics on Hadoop.

Rehman et al. [15] looked into the latest BDA technologies, methods, and methodologies that can lead to the development of intelligent IIoT systems into their research. Authors created a taxonomy by classifying and categorizing the literature based on key factors (e.g. data sources, analytics tools, analytics techniques, requirements, industrial analytics applications, and analytics types). The foundations and case studies of the many businesses that have profited from BDA were presented. Authors in this research also go through the several advantages that BDA brings to the IIoT. Furthermore, they highlight and debate the critical concerns that must be addressed as research directions in the future.

Williamson et al. [16] emphasis the paper on a large ongoing data infrastructure project in UK Higher Education. It looks at the infrastructure's sociotechnical networks of organizations, software programmes, standards, dashboards, and visual analytics technologies, as well as how these technologies are integrated with governmental market reform imperatives. The study highlights how higher education is being remade through the utopian goal of a "smarter university" while also being transformed through the political agenda of marketization.

Cantabella et al. [17] present a case study conducted at the Catholic University of Murcia, in which student behaviour was analysed over the course of four academic years according to learning modality (on-campus, online, and blended), taking into account the number of LMS accesses, tools used by students, and events associated with them. Due to the difficulties of managing the vast number of data created by users in the LMS (up to 70 GB in their research), statistical and association rule approaches were used in conjunction with a Big Data framework to speed up statistical analysis of the data. The gathered findings were shown and reviewed using visual analytic tools in order to uncover patterns and shortcomings in student's use of the LMS.

3 Big Data at University of Kashmir

Higher educational institutions like universities and their affiliated colleges hold very huge quantity of data related to students, courses, and staff. Analysing this data can allow us to obtain insights which can enhance the operational effectiveness of educational organization. By doing statistical analysis on this educational big data, variables like student course selection, examination results, and career prediction of each student can be processed [18].

Before we talk about this educational Big Data, let us first examine how higher education institutions look like without Big Data. Let us take the example of University of Kashmir (UoK) (Fig. 2), but it can be very similar at any other universities. In University, lots of operations take place during life cycle of a student. Actually these operations are referred as processes that take place at different stages right from registration of student in a particular programme to completion of degree. In 2001, the first process, which was automated in UoK, was result compilation. For this purposes, database management systems (dBase) were used which were having ".dbf" file as underlying file format. As of now, this system is still in use for result preparation of few small programmes. In year 2004, registration section was automated. Registration section is the back bone of university, where in they keep data

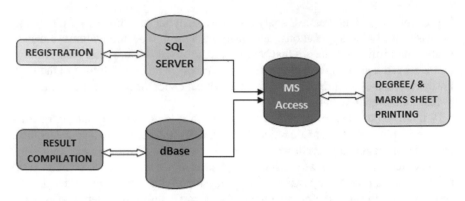

Fig. 2 Different process at University of Kashmir (UoK)

like student name, DOB, college, address, course opted, etc., in short all the information related to the student is available here. This process was having platform which was built using SQL SERVER at back end and VB as front end tool. Simultaneously, another process was automated which was handling degree preparation and printing, with MS access at backend DBMS.

This is how organizations have been working for quite some time. In University of Kashmir, it was working in similar way before automation of every process of student life cycle at UoK that took place, i.e. after year 2008. Now, problem is that we have data everywhere. Educational institutions have changed today, all the operations/ processes in student life cycle as shown in below figures are online, and each process has its own database thus huge volume of data at multiple locations is being generated rapidly.

The identification of Big Data and the sources from which it originates is the reference point from where one has to start. It is assumed that we are dealing with University of Kashmir (UoK) atmosphere, but it can be very similar at any other universities. This data is available in physical files (non-electronic format) at the university, data from multiple databases of different processes as illustrated in the figures above, data in the form of logs from various processes and university servers, data from external educational boards such as the JK State Board of Education, data from affiliated colleges, and data from the feedback forum (the university's feedback/grievance portal for students). Today all the processes involved in student life cycle (at University of Kashmir) is happening online and each process as is having its own data source. Below (Fig. 3) mentioned pictures show the detailed explanation of these processes.

The student life cycle as demonstrated in the Fig. 3 above starts with the admission of a student in the University. The next important step is the student verification and registration process. Once the student is enrolled in the university, he becomes eligible for appearing in the internal exams conducted by the concerned college/department and the external exams conducted by the university. The colleges/departments after

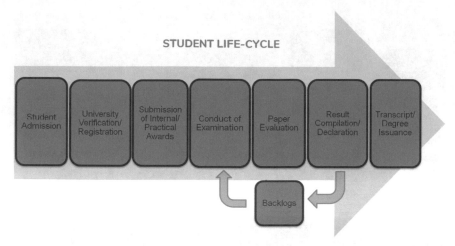

Fig. 3 Detailed process of students life cycle

conducting the internal exams/practical submit the awards to the university examination system where these are awards recorded for each student. Next, university conducts the examinations, and the papers written by students of different colleges/departments are being evaluated by University examination system. The results for both internal and external exams are finally compiled and declared. For students who pass all the subjects, university awards the degree. The students who fail in one or multiple subjects are considered as backlogs and for these students' examination process is repeated [19].

The examination conduct process is the core process in the student life cycle. As demonstrated in Fig. 4, this process is all about the conduct of examination. There is a repository of the student academic history.

From this repository, the internal/practical marks for the regular students are pushed into the examination conduct system where they get recorded for the compilation of the final results. Through the examination conduct system, there are other tasks that get initiated. These tasks are as listed below.

(1) The generation of admit cards for all students who are eligible for sitting in the exams.
(2) The centre statements for the conduct of exams. In these statements', centres are allocated for all the colleges and departments for the conduct of their exams.
(3) Attendance sheets for all designated centres to track the record of present and absent students for different exams.
(4) Centre-wise/paper statements that is taken care by conduct and paper setting section.

The backlog students are also included in the same examination conduct process after due process of examination form submission and fee payment by the concerned students.

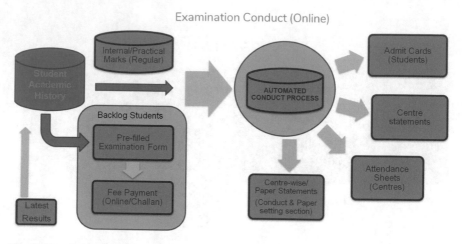

Fig. 4 Examination conduct process

The result automation process is the culmination of entire examination conduct and evaluation process. As demonstrated in Fig. 5, the automated result compilation mechanism gets several inputs that are necessary for the result compilation and declaration. This mechanism gets inputs like examination conduct enrolment data, e-awards (Practical exams) from departments and colleges, external e-awards from departments/nodal colleges and OMR scanning data. All these inputs get compiled, and the output is the result of the examination. The results are uploaded to the university website and a hard copy in the form of result gazette is send to the concerned department/college. The result of all the students gets recorded in the student academic history repository.

Also, the students are given option to submit the re-evaluation form within a stipulated time in case they are not satisfied with their exam results. The re-evaluation request for a paper invokes again the entire paper evaluation and result compilation process for the concerned student/s. In case of any change in the result, student academic history repository gets updated [20].

4 Technical Implementation

Traditional approaches are having extremely inefficient architecture for processing Big Data as it is time-consuming, costly, and complex. The data is enormous, and the systems are small. The processing of the large component must be transferred to the small component thus resulting in demand of parallel computing, in which several calculations are performed at the same time by dividing big problems into smaller chunks and solved in parallel. Emerging technologies provide new and interesting

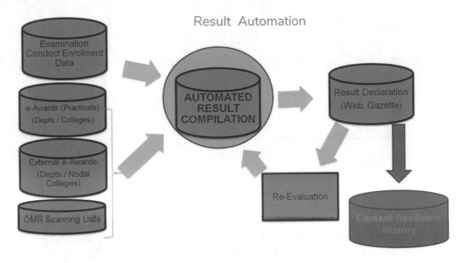

Fig. 5 Result automation and declaration process

approaches to procedure and remodel huge data, described as complex, unstructured, or huge quantities of data, into significant knowledge. There are several key frameworks in Big Data area such as:

4.1 MapReduce

MapReduce is a distributed computing processing technique and application model based on Java. It is one of the most widely used systems for solving parallel processing problems involving large datasets and a large number of computers (nodes). The Map and Reduce [21] tasks are the two key components of the MapReduce algorithm. Map converts a set of data into tuples (key/value pairs), which are then used to generate another set of data. Second, the reduce task takes a map's output as input and merges the data tuples into a smaller set.

4.2 Hadoop

Hadoop software library is a system that uses simple programming models to allow for the distributed processing of large data sets across clusters of computers [22]. Hadoop is designed in such a way that it can be scaled from a single server to thousands of computers with very high fault tolerance.

4.3   *NoSQL*

NoSQL also known as not only SQL are databases that store data in a format other than relational tables.

5   Technical Implementation

For academic enterprise analysis, we need to design a framework for that we have to first identify the problem for which framework will be designed. For instance, suppose university authorities or government officials are looking for information regarding how many students of science stream of a particular district have received their degrees that have completed their programme without any backlog, of particular years. So, to generate that information, we actually need the data from all the identified sources and data processing algorithm that will make it ready for analysis. An algorithm for this will collect data from all known data sources and clean it up as needed. The next stage will be to move unstructured data to Hadoop and perform any required data transformations. Following data transformations, the algorithm will create a data warehouse using the transformed data as well as other structured data. Finally, we can analyse the data that has already been placed in the data warehouse.

To design analytics architecture using above-mentioned algorithm, there are few steps that need be taken after the data from physical files, multiple databases, and other sources (external sources) has been identified. The crucial step is to get all the data from these sources/databases integrated or accessible at one place/platform. The architecture will be having minimum four layers (Fig. 6). The data sources are covered by the lowest layer. Structured data could be stored in relational databases, or all of the academic enterprise's relational databases can even be explicitly integrated at this layer. NoSQL databases can be used to store unstructured data. Apache Hadoop will be the next layer. The data from the lowest layer is processed by Hadoop. This tool has the advantage of being able to work with both relational data and NoSQL. The Apache Hive tool can be used at the same layer and can be used for working with NoSQL data and converting it to RDBMS. In next layer, we now have a data warehouse. Once we have all data in data warehouse, we can use any business intelligence tool/platform (like Tableau) and create visualizations.

6   Conclusion and Future Studies

Big Data analytics is employed less frequently in higher education than in other industries, and the need for it is growing in the education sector as well. We are approaching a new era in which Big Data mining in education sector will assist us in discovering learning in modern times. Big Data analytics is critical in this era of data

Fig. 6 Academic enterprise
architectural framework

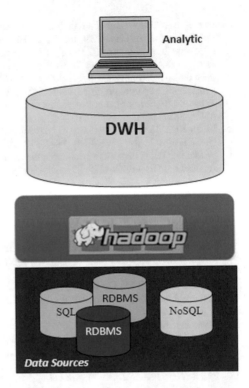

floods because it can bring unexpected insights and aid in better decision-making in the education sector. Big Data analysis of educational data is not as easy as it may appear. We highlighted a very important asset in this research, which is the identification of heterogeneous data sources in the University of Kashmir's (educational setting), as well as the concerns associated with big data mining and the design of a framework. The design of the framework is only one of the challenges associated with it. The approach given in this paper incorporates Hadoop into existing systems and data warehouse to provide a solution for educational institutes' expanding data. This suggested Big Data architecture for education will allow for collecting data, storage, and analysis, which will all have potential benefits for future initiatives. The academic challenges will be well addressed by Big Data because unforeseen ideas such as job vacancies, course recommender systems for students, and student retention can only be well designed and integrated with the existing system when big data is implemented in the education setup.

References

1. Weiss SM, Indurkhya N, Introduction to big data in education and its contribution to the quality improvement processes. https://doi.org/10.5772/63896

2. Daniel BK (2017) Big data in higher education: the big picture. Springer International Publishing Switzerland, p 19. In: Daniel BK (ed), Big data and learning analytics in higher education. https://doi.org/10.1007/978-3-319-06520-5_3
3. Prinsloo P, Archer E, Barnes G, Chetty Y, Van Zyl D (2015) Big(ger) Data as better data in open distance learning. Int Rev Res Open Distrib Learn 16(1). https://doi.org/10.19173/irrodl.v16i1.1948. https://id.erudit.org/iderudit/1065938ar
4. Tulasi B (2013) Significance of big data and analytics in higher education. Int J Comput Appl (0975–8887) 68(14)
5. Imran RB, Majid Z, Quadri SMK, Muheet AB (2018) Big data mining: a literature review. IJRECE 6(3). ISSN: 2393-9028 (PRINT) I ISSN: 2348-2281 (ONLINE)
6. Williams P (2017) Assessing collaborative learning: big data, analytics and university futures. Assess Eval High Educ 42(6):978–989
7. Shamsuddin NT, Aziz NI, Cob ZC, Ghani NL, Drus SM (2018) Big data analytics framework for smart universities implementations. In: International symposium of information and internet technology. Springer, Cham, pp 53–62
8. Wu X, Zhu X, Wu G-Q, Ding W (2014) Data mining with big data. IEEE Trans Knowl Data Eng 26(1)
9. Chen H, Roger HL, Veda C (2012) Business intelligence and analytics: from big data to big impact. MIS Q 36(4). Eller College of Management, University of Arizona, Tucson, AZ 85721 U.S.A.
10. Diebold F (2012) On the origin(s) and development of the term "Big Data". Pier working paper archive, Penn Institute for Economic Research, Department of Economics, University of Pennsylvania
11. Fayaz SA, Altaf I, Khan AN, Wani ZH (2019) A possible solution to grid security issue using authentication: an overview. J Web Eng Technol 5(3):10–14
12. Xin X, Shu-Jiang Y, Nan P, ChenXu D, Dan L (2022) Review on A big data-based innovative knowledge teaching evaluation system in universities. J Innov Knowl 7(3):100197
13. Brock VF, Khan HU (2017) Are enterprises ready for big data analytics? A survey-based approach. Int J Bus Inf Syst 25(2):256–277
14. Wu WenTai, Lin WeiWei, Hsu C-H, He LiGang (2018) Energy-efficient hadoop for big data analytics and computing: a systematic review and research insights. Futur Gener Comput Syst 86:1351–1367
15. ur Rehman MH, Yaqoob I, Salah K, Imran M, Jayaraman PP, Perera C (2019) The role of big data analytics in industrial Internet of Things. Future Gener Comput Syst 99:247–259
16. Williamson B (2018) The hidden architecture of higher education: building a big data infrastructure for the 'smarter university.' Int J Educ Technol High Educ 15(1):1–26
17. Cantabella M, Martínez-España R, Ayuso B, Yáñez JA, Muñoz A (2019) Analysis of student behavior in learning management systems through a Big Data framework. Future Gener Comput Syst 90:262–272
18. Rasmitadila R, Humaira MA, Rachmadtullah R (2022) Student teachers' perceptions of the collaborative relationships between universities and inclusive elementary schools in Indonesia. F1000Research 10:1289
19. Kenett RS, Prodromou T (2021) Big Data, analytics and education: challenges, opportunities and an example from a Large University unit. Big Data Educ Pedagogy Res, pp 103–124
20. Durbin P, A framework for turbulence modeling using Big Data: Phase II final report Karthik Duraisamy (PI) University of Michigan, Ann Arbor, MI Juan J. Alonso (Co-I) Stanford University, Stanford, CA
21. Phan A-C, Phan T-C, Cao H-P, Trieu T-N (2022) Comparative analysis of skew-join strategies for large-scale datasets with MapReduce and spark. Appl Sci 12(13):6554
22. Zhang J, Wang F, Zhou J (2022) Research on the construction of university data platform based on hybrid architecture. In: International conference on wireless communications and applications. Springer, Singapore, pp. 104–111

Software Complexity Prediction Model: A Combined Machine Learning Approach

Ermiyas Birihanu, Birtukan Adamu, Hailemichael Kefie, and Tibebe Beshah

Abstract The need for computers increased quickly. As a result, the program is utilized in a significant and intricate manner. More complex systems are being developed by software businesses. Additionally, customers expect great quality, but the market requires them to finish their assignment faster. Different measuring methods are employed by software firms. Some of these include customer feedback after it has been given to customers, software testing, and stakeholder input. The objective of this project is to use a combination of machine learning techniques to predict software bug states using the NASA MDP dataset. The research process considered data preprocessing methods and applied singular and combination machine learning algorithms. To create the model, the single classifiers were combined using the voting method. Accuracy, precision, and recall were used to evaluate the model's effectiveness, along with tenfold cross-validation. The promising result was recorded by a combination of J48 and SMO classifiers. Before attempting to test the software product, the researcher retrieved attribute data from the source code; the complexity of the software product will then be ascertained using the constructed model. The main contribution of this study is to improve software quality by incorporating a machine learning framework into the present software development life cycle between implementation and testing.

Keywords Software complexity · Prediction model · Machine learning · Software metrics · Software faults

E. Birihanu (✉) · H. Kefie
Software Engineering, Wolkite University, Wolkite, Ethiopia
e-mail: ermiyas.birhanu@wku.edu.et

B. Adamu
Applied Informatics, Silesian University of Technology, Gliwice, Poland
e-mail: birtbir170@student.polsl.pl

T. Beshah
School of Information System, Addis Ababa University, Addis Ababa, Ethiopia

© The Author(s), under exclusive license to Springer Nature Singapore Pte Ltd. 2023 681
Y. Singh et al. (eds.), *Proceedings of International Conference on Recent Innovations in Computing*, Lecture Notes in Electrical Engineering 1011,
https://doi.org/10.1007/978-981-99-0601-7_53

1 Introduction and Problem Definition

Software companies are building various complex systems from time to time. Usually, software businesses construct a variety of complicated systems. Additionally, customers want great quality while the market expects that projects be finished quickly. Most frequently, the quantity of faults or defects discovered in software products can be used to assess the software's quality. Customer reviews may be one technique to determine or judge the software quality of a product. However, there are at least two issues with consumer feedback metrics. In the first place, it does not provide specific feedback regarding potential areas for improvement, and in the second, it occurs too late to implement any changes. Software fault classification is used to both discover and prevent problems.

The need for computers increased quickly. Therefore, the usage of software grows into enormous and complex. Software is computer programs that are used to instruct the computer to do a given task. Software Development Life Cycle (SDLC) [1] describes the methodology by which the development of any software takes place. SDLC includes feasibility analysis, requirement analysis and specification, designing, implementation, testing, and maintenance. In this study, the researchers were focused only on software testing and maintenance. Software testing [2] provides a way to reduce maintenance and overall software costs, as well as faults. Various software techniques are applied for testing the software; to reduce error, but, it cannot possible be error-free. The systems that are taking an input such as characters numbers, and files; it is easy to test for any small input, but what if the line of code is complex and large?. Software testing is a means of determining the quality of the software. It is the most important and expensive stages in SDLC. Project managers must understand "When to stop test?" and "Which part of the code to be tested?" the answers to these queries will have a direct impact on resource allocation (i.e., the experience of test staff, how many people to allocate for testing), cost, and defect rates as well as product quality. Manual software inspection gets more difficult as software grows and complexity. Software defect prediction is an alternate strategy that is utilized in this context to predict potential impacts clearly. Testing is an important component of software engineering since it often consumes between 40% and 50% of development efforts and requires more work for systems that need higher levels of reliability [2–4]. Software defects have an impact on the price and time of development in addition to the software's quality. Various factors, most of which are connected to human mistake, design or coding issues, data input errors, documentation faults, and communication breakdowns, could also cause the system to be faulty. We will use software fault prediction to account for the most common human variables prior to putting the software systems via testing and maintenance.

Now a day, there not available several datasets that could be mined to discover useful knowledge regarding software defects. We were used a variety of machine learning approaches to the publicly accessible datasets CM1, JM1, KC1, KC2, and KC3 from the National Aeronautics and Space Administration (NASA) software repository. The goal is to divide the software modules into modules that are prone to faults and not-faults. To extract meaningful information from massive datasets, machine learning is one of the most important and inspiring fields. Therefore, it is advised to use this method when developing software applications. The fundamental goal of machine learning is to uncover patterns in data that can be useful. Data mining methods include organized formats (for instance, many databases) and text mining, which uses unstructured data (for example, natural language documents). Machine learning tasks can be classified into two categories, descriptive and predictive. To predict the future, the predicative task uses the dataset to infer the current data. The main goal of this project is to create and refine a software complexity predictive model.

2 Related Works

Machine learning has been developing quickly, resulting in a wide range of learning algorithms for various purposes. Machine learning techniques also used for software engineering. These techniques overall worth is mostly determined by how well they work to solve issues in the actual world. As a result, for the field to advance, algorithm reproduction and application to new tasks are crucial. However, several machine learning experts have just lately released work on the creation of models for predicting software defects.

In this section, we examined the research that has been conducted to date divided into three categories: software defect prediction using classification approaches [5–9], clustering approaches [10, 11], and ensemble approaches [12, 13]. According to these categories assessments, we have seen combining two or more classifiers produces better performance; hence, we used the voting method to combine two machines learning techniques.

3 Research Methodology

This study is based on data collected from NASA MDP; a dataset created by NASA for research purposes. To fully comprehend the data, it is necessary to have a close working relationship with domain experts like software testers and developers.

3.1 Design of Software Complexity Prediction Model

The design and development of the software complexity prediction model were the main emphases of this research study. We started by using the chosen data from the NASA Promise MDP, which contains training data such as software metrics and associated values. Second, the input data received preparation through the handling of

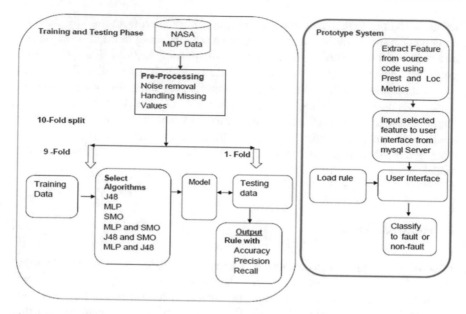

Fig. 1 Proposed architecture of software complexity prediction model

missing values and the removal of noise. Third, we applied a 10-fold cross-validation test in the devloped model while using the selected techniques (J48, MLP, SMO, and Vote). Finally, a performance report was examined, and software complexity predictions are in line with fault or non-fault one. The researcher creates a prototype system with an input form that automatically feeds from the database when the user clicks on the load data button to assess if it has flaws or not to ensure the design and development of a model. The architecture of the software complexity prediction model is given in Fig. 1.

3.2 Data Exploration

Data on software faults are difficult to find and even while commercial software companies do not make the data available to the public, for measuring software faults. Due to its initial version and the fact that 60% of software defect studies chose it as a priority, in this study, we collected data from NASA MDP Promise [14]. Before using machine learning techniques, we considered several factors, which are covered in the following subsections.

There are now 13 datasets targeted for software metrics based on NASA Promise repositories [15]. It includes the distinct programming languages, code metrics (Code size, Halstead's complexity, and McCabe's Cyclomatic complexity), and time limits. We have selected five datasets (CM1, JM1, KC1, KC2, and KC3) out of the 13

Table 1 Dataset information

	CM1	JM1	KC1	KC2	KC3
Programming language	C	C	C++	C++	Java
Line of code (LOC)	20k	315k	43k	18k	18k
Modules	498	10,885	2109	522	458

available.[1] Tables 1 and 2 illustrate the attributes we utilized and the description of the dataset.

3.3　Feature Extraction

In this section, we took two sample codes that are measured by software metrics tools (Prest and Loc metrics). We used Java source code to extract features using Prest and Loc metrics. The researcher extracted the feature from the source code based on mathematical methods which is several operands and operators.

1. **Prest Based Extracting Attributes**: It can simultaneously parse all files written in C, C++, Java, JSP, and SQL using several parsers. We were used remote method invocation for scientific calculator Java source code. Table 3 provides a full set of extracted attributes with corresponding values.
2. **Loc Metrics-Based Extracting Attributes**: We used source code for remote method invocation to extract attributes. Loc metrics were used to calculate physical line, executable logical, blank lines, total lines of code, McCabe VG complexity, comment, and header comment. Since Prest cannot support these properties, we only considered IO code, comment, Halsted comment, and Halsted blank lines. In the Loc metrics tool first the user's browser the source code to be extracted attributes (useful information of the source code), then click on count Loc.

4　Experimental Results

The main goal of this study was to predict software fault using machine learning approach by single classifiers and combing of two classification methods. We used Weka 3.6.9 machine learning software to develop a predictive model.

---

[1] http://promise.site.uottawa.ca/SERepository/datasets-page.html.

Table 2 List of features and their descriptions

Attribute name	Data type	Description
Loc	Numeric	McCabe's line count of code
V(g)	Numeric	McCabe "Cyclomatic complexity"
ev(g)	Numeric	McCabe "essential complexity"
iv(g)	Numeric	McCabe "design complexity"
N	Numeric	Halstead total operators + operands
V	Numeric	Halstead "volume"
L	Numeric	Halstead "program length"
D	Numeric	Halstead "difficulty"
I	Numeric	Halstead "intelligence"
E	Numeric	Halstead "effort"
B	Numeric	Halstead
T	Numeric	Halstead's time estimator
lOCode	Numeric	Halstead's line count
lOComment	Numeric	Halstead's count of lines of comments
lOBlank	Numeric	Halstead's count of blank lines
lOCodeAndComment	Numeric	Line of code and comment
uniq_Op	Numeric	Unique operators
uniq_Opnd	Numeric	Unique operands
total_Op	Numeric	Total operators
total_Opnd	Numeric	Total operands
branchCount	Numeric	The flow graph
Defects	String	Module reported to defect or not defect

4.1 Model Building Using Single Classifiers

The main objectives of this study were to enhance the performance and accuracy of single classifiers. To achieve this, we first measured the accuracy and performance of single classifiers and recorded it using the method and performance measure that we chose. Three algorithms—decision tree, multi-layer perceptron, and support vector machine were selected by the researchers. Table 3 which discussed below is an experimental result summary for selected machine learning methods.

Table 3 Result summary on single classifiers

Dataset	Learning method																
	J48			MLP				SMO			
	Correctly classified	Precision	Recall	Correctly classified	Precision	Recall	Correctly classified	Precision	Recall
CM1	87.95	0.90	**0.96**	87.55	0.901	0.969	89.56	0.903	0.993
JM1	79.7	0.83	0.93	80.95	0.841	**0.99**	80.73	0.807	**1**
KC1	84.54	0.88	0.93	85.91	0.872	0.978	84.78	0.996	0.85
KC2	81.41	0.87	0.89	84.67	0.873	0.945	82.76	0.828	0.98
KC3	**89.3**	**0.92**	**0.96**	**90.61**	**0.919**	0.983	**90.82**	**0.952**	0.90

Tenfold cross-validation

4.2 *Model Building Using Combined Methods*

Based on the researchers showed that combining of two methods, improve the performance and accuracy of single classifiers [16]. We combined two classifiers in order to evaluate the performance and accuracy of combined classifiers using a vote algorithm with the average probability combinational method. Table 4 which discussed below is an experimental result summary for selected machine learning methods.

5 Performance Evaluation

In Sects. 3.1 and 3.2, the researcher observed the following basic research work outcomes:

- J48: As the dataset size increased, the number of leaves, the size of the tree, and the time required to create the model all increased, while the model's recall, precision, and accuracy all decreased.
- When the size of the dataset is increased, the models for MLP, SMO, J48 & MLP, J48 & SMO, SMO & MLP perform worse in terms of recall, precision, and accuracy, and take more time to develop.

6 Experimental Discussion

Single and combination classifiers are employed to create the software fault predicting model, and six independent experiments are conducted for each classifier and dataset. Based on all criteria, 30 experiments were performed. The dataset was classified into two sections, i.e., training (90 %) and testing (10%) using ten cross-validation. The experiment was created to determine if growing the size of the dataset enhances or degrades the performance of algorithms, to assess the impact of algorithms on performance, and to compare the effectiveness of algorithms in predicting software faults. The model produced by the J48 classifier was chosen as the most effective one for predicting software defects. J48 decision tree generated 25 rules for predicting software complexities data from which the researchers considered only 7 rules. The graphical user interface was designed and developed using Java NetBeans IDE 8.0.2 on these seven selected rules. The developed model points out that extracting of the source code information before software testing then applying the developed model to determine whether the software product has faults or not.

Table 4 Result summary on combined method

Dataset	Learning method											
	J48 and MLP			J48 and SMO				MLP and SMO				
	Correctly classified	Precision	Recall	Correctly classified	Precision	Recall		Correctly classified	Precision	Recall		
CM1	88.5	0.903	0.978	89.5	0.901	0.993		89.5	**0.901**	0.998		
JM1	80.7	0.808	**0.999**	80.72	0.807	**1**		80.72	0.807	**1**		
KC1	85.4	0.881	0.957	84.6	0.848	0.997		84.6	0.845	**1**		
KC2	81.8	0.874	0.901	82.7	0.828	0.998		79.5	0.795	**1**		
KC3	**90.6**	**0.921**	0.981	**90.8**	**0.908**	**1**		**90.82**	0.906	**1**		

Tenfold cross-validation

Table 5 Effect of pruning

	Number of instances	J48 with all attributes and pruned parameter			J48 with all attributes and Un-pruned parameter		
		Accuracy (%)	Precision	Recall	Accuracy (%)	Precision	Recall
CM1	498	**87.95**	0.904	0.969	**88.35**	0.919	0.955
KC3	458	**89.3**	0.924	0.961	**88.2**	0.925	0.947
KC2	522	**81.4**	0.873	0.896	**80.2**	0.88	0.87
KC1	2109	**84.5**	0.885	0.939	**83.8**	0.887	0.927
JM1	10,885	**79.7**	0.837	0.919	**79.0**	0.837	0.919

Table 6 Effect of dataset size J48 classifiers model

J48 with all attributes	Instance number	Number of rules	Size of the tree (size of complexity)	Recall	Precision	Accuracy
CM1	498	5	9	0.969	0.901	87.95
KC3	458	7	13	0.983	0.919	**89.3**
KC2	522	26	51	0.945	0.873	81.41
KC1	2109	56	111	0.978	0.872	84.54
JM1	10,885	340	679	0.99	0.841	79.7

6.1 Effect of Dataset Size

The outcomes of the experiments demonstrate that increasing the data size affects how well machine learning algorithms work. To understand why machine learning algorithm performance declined as dataset size increased, the researchers ran additional experiments. We used J48 classifiers for the CM1, JM1, KC1, KC2, and KC3 datasets with pruned and unpruned parameters. Table 5 illustrates how pruning affects the effectiveness of prediction models as data size grows.

According to the findings of the experiments and additional study by the researcher, machine learning algorithms perform worse because datasets grow since doing so complicates models, produces more rules, and lowers model recall, accuracy, and precision. The example in Table 6 is very apparent.

6.2 Effect of Methods on Performance

Out of all the approaches, the combination of the J48 and SMO method produced the results for KC3 data with the highest accuracy (90.82). Figure 2 depicts the impact of the algorithm on performance for the CM1 dataset; when compared to other

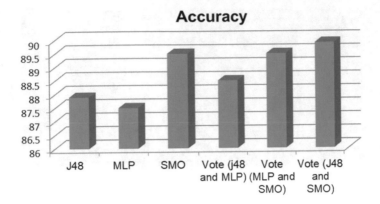

Fig. 2 Model comparison for CM1 dataset

techniques, J48 and SMO showed higher performance; the second performance was recorded for MLP with SMO and SMO methods for the CM1 dataset.

6.3 Extracted Rules from Decision Trees

When compared to the other five classifiers in this study, J48 classifiers achieved relatively high performance based on performance evaluators like specificity and execution time. As can be seen in the results of Experiment 1 using the J48 decision tree, 25 (twenty-five) rules for predicting software defect were produced. We chose 7 (seven) criteria that account for the majority of instances in the provided dataset, and then, the researchers held in depth discussions with domain experts to ensure that the chosen rules actually apply to all instances.

7 Software Complexity Classification System: A Prototype

In this study, an attempt was made to design and develop an operational application prototype named software complexity classification system that uses the classification rules generated from J48 classifiers. The prototype is used to classify a software product into one of the software labels (fault or not-fault). The software complexity classification system includes the user to load extracted source code data from the database and predict the software product based on the trained into defect or not defected one if we click on predict button. From Fig. 3, there are 22 inputs (attributes), and output can be defect or not defect. Inputs are given from extracted source code data. Exit button is used to close out from the graphical user interface. The prototype was developed based on the rules generated by the J48 classifier. To demonstrate the predicted outcome of software defects as TRUE based on rule 7 in

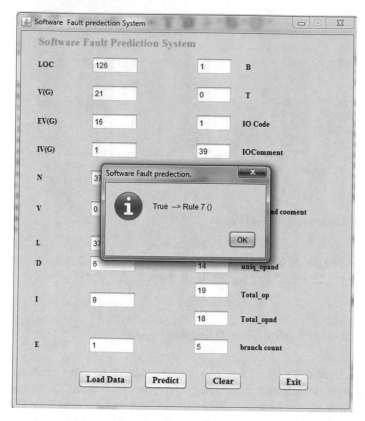

Fig. 3 Software complexity prediction prototype user interface with sample result

experiment one, we put the extracted source code and data into the prototype as we saw in Fig. 3. We used Java source code to predict software faults according to the selected rule generated from J48 classifiers.

8 Conclusion and Future Works

Due to its advantages, software-based system development is now growing more than in past years. Before the software system is made available to end users, quality assurance is necessary. We have several quality measurements, including software testing, CMM, and ISO standards, to improve the quality of the software. Software testing is currently becoming more and more crucial to the dependability of software. Using machine learning techniques, software defect prediction can significantly increase the effectiveness of software testing and direct resource allocation.

This study main objective is to use a mixed machine learning method for software complexity prediction in software product. This study attempted to generate practical predictive models from the NASA MDP dataset and develop a novel graphical user interface for effective utilization of the developed model.

In the data exploration section, we had a total of 498, 10885, 2109, 522, and 458 records and 22 attributes. To get the dataset ready for experiments, we did noise removal and handled missing values. To build the model, we used machine learning algorithms such as a decision tree, support vector machine, multi-layer perceptron, and a combination of single classifiers using the voting method. We also introduced a novel, visual representation method (Fig. 3) which visually conveys information about the extracted software attributes from the source code using Prest and LOC. The performance of the model was measured by accuracy, precision, and recall. This study helps the software project managers in predicting and fixing bugs before the product is given to clients, ensuring the software's quality.

In the future, we plan to extend the model to more software repositories (Eclipse, JEdit, Open-source software, and AR datasets). Additionally, we also plan to perform to examine the effect of attribute reduction on the performance of machine learning algorithms and apply other software metrics, i.e., resource metrics.

References

1. Saini M, Kaur K (2014) A review of open source software development life cycle models. Int J Softw Eng Appl 8:417–434
2. Zhang S, Zhang C, Yang Q (2003) Data preparation for data mining. Appl Artif Intell 17:375–381
3. Sharma C, Sabharwal S, Sibal R (2014) A survey on software testing techniques using genetic algorithm. arXiv preprint arXiv:1411.1154
4. Shivaji S (2013) Efficient bug prediction and fix suggestions. University of California, Santa Cruz
5. Pal S, Sillitti A (2021) A classification of software defect prediction models. In: 2021 International conference nonlinearity, information and robotics (NIR), pp 1–6
6. Liu C, Sanober S, Zamani AS, Parvathy LR, Neware R, Rahmani AW (2022) Defect prediction technology in software engineering based on convolutional neural network. In: Security and communication networks, vol 2022
7. Ha TM, Tran DH, Hanh LT, Binh NT (2019) Experimental study on software fault prediction using machine learning model. In: 2019 11th International conference on knowledge and systems engineering (KSE), pp 1–5
8. Peng X (2022) Research on software defect prediction and analysis based on machine learning. J Phys Conf Ser, p 012043
9. Jorayeva M, Akbulut A, Catal C, Mishra A (2022) Machine learning-based software defect prediction for mobile applications: a systematic literature review. Sensors 22:2551
10. Marjuni A, Adji TB, Ferdiana R (2019) Unsupervised software defect prediction using median absolute deviation threshold based spectral classifier on signed Laplacian matrix. J Big Data 6:1–20
11. Park M, Hong E (2014) Software fault prediction model using clustering algorithms determining the number of clusters automatically. Int J Softw Eng Appl 8:199–204

12. Elahi E, Kanwal S, Asif AN (2020) A new ensemble approach for software fault prediction. In: 2020 17th International Bhurban conference on applied sciences and technology (IBCAST), pp 407–412
13. Li R, Zhou L, Zhang S, Liu H, Huang X, Sun Z (2019) Software defect prediction based on ensemble learning. In: Proceedings of the 2019 2nd international conference on data science and information technology, pp 1–6
14. Malhotra R (2015) A systematic review of machine learning techniques for software fault prediction. Appl Soft Comput 27:504–518
15. Gray D, Bowes D, Davey N, Sun Y, Christianson B (2012) Reflections on the NASA MDP data sets. IET Softw 6:549–558
16. Wang T, Li W, Shi H, Liu Z (2011) Software defect prediction based on classifiers ensemble. J Inf Comput Sci 8:4241–4254

Work with Information in an Outdoor Approach in Pre-primary Education

Jana Burgerová and Vladimír Piskura ⓘ

Abstract Outdoor education, due to its characteristics and accessibility, currently represents one of the innovative approaches in the field of pre-primary education. In recent years, this approach has also been implemented in the content of tertiary education. One of the compulsory areas of the curriculum that should also be covered in the outdoor approach is mathematics and work with information. The existing generation of students, future teachers, is referred to as the alpha generation and is specific in terms of the use of technology. The present research aimed to analyse the knowledge and attitudes towards outdoor education and towards information literacy, to analyse their interrelation with future pre-primary teachers in the first and last year of their studies by comparing the selected items. The research was conducted on a sample of 153 respondents using the questionnaire as the main research tool. We present findings and conclusions regarding outdoor education, which is familiar to students but is an unexplored and relatively unknown area in terms of its connection with information work. The conclusions correspond with current trends that note the shortcomings in using technology for effective and practical solutions despite the digital skills of the current generation.

Keywords Pre-primary education · Work with information · Outdoor approach

1 Introduction

One of the innovative (although not new) approaches to education, especially in pre-primary education, is outdoor education. "For the kindergarten teachers, the most handy and most accessible environment is the nature. Children love nature, they are interested in everything that surrounds them, and have an unstoppable curiosity"

J. Burgerová · V. Piskura (✉)
University of Presov, 17 Novembra 15, 08001 Presov, Slovakia
e-mail: vladimir.piskura@unipo.sk

J. Burgerová
e-mail: jana.burgerova@unipo.sk

[4]. Contemporary globalized society is characterized by, among other things, lack of exercise, being in nature, loss of physical activity, increasing stress, and overall demands on the psyche, which is manifested by a decrease in physical fitness, an increase in various civilizational diseases, and thus a deterioration in the health status of the entire population. Physical activity is an essential biological expression of life and necessary for a child's healthy growth and development [11]. To explain outdoor education in a simple way, we can generally imagine education in outdoor spaces outside the classroom. Outdoor education can also be understood as the transfer of education from the traditional classroom to the schoolyard. The question is, why would we do it? There are different opinions. There are several opinions. Whether people are discovering something new about nature, developing skill, or simply learning how to be comfortable in the outdoors, they often experience a wide range of emotion [6]. When teachers realize the teaching outside the class, they often observe the improvement of the children's behaviour, and the whole class enjoys learning; children with specific needs are often more successful in the outdoor environment.

In pre-primary education, the national curriculum for pre-primary education document, which sets out the state's essential requirements for providing institutional education in Kindergarten, is the most important document and the content of education. It is structured into areas, out of which two are mathematics and work with information, describing the standards of informatics education and digital technologies. "Outdoor" and technology seem to represent two incompatible domains at first glance. This challenged our interest; particularly, we were interested in the perception of existing students on the interconnection or intersection of those two domains.

There is a generation of young people, currently studying in higher education referred to by the terms as Millennial Generation, Howe and Strauss [7], the Net-Generation surrounded by digital media [10], the Digital Natives by Prensky [8] and considered to be "native speakers" of the digital language of computers and the Internet. They are also called and described as the Y and Z Generation replacing Generation X [13], the iGeneration [9], or Digital Learners, which offers a more global vision of the twenty-first century learner [2]. Net-Generation generally refers to the generation that has grown up in the Internet age. The concept of Net-Generation or Digital Natives was first proposed by American educator Prenskyin [8] and he has made pro-and-con distinctions between Digital Natives and Digital immigrants [5]. Regardless the label, the young people of this era are generally considered universally able to and skilled in working with digital technologies. Still, there is research that speaks against this claim and which demonstrates that there is no automatic ability for this generation to transfer digital competencies and digital skills into, for example, academic settings. In fact, there is no evidence that today's students want to use these technologies for educational purposes either. Our experience gained in training the future primary and pre-primary education teachers at the Faculty of Education of the University of Prešov also sounds against the generally overestimated digital skills of today's students. The results of testing conducted in 2016 using the IT Fitness Test on a sample of 532 students showed the significant gaps in the digital skills of our students [3]. Also, the direct experience of teachers technology enhanced subjects teaching in which it is possible to directly observe the work of students, points to the absence of skills in working with spreadsheets, graphic editors, presentation software, the use of cloud computing, but especially the use of technology for problem solving. Alghamdi et al. [1] realize study with aims to explore kindergarten student teachers' readiness to integrate technology into their future classrooms and factors affect their integration. The first-phase results showed that participants were ready to implement technologies while having positive attitudes towards technology integration. The second-phase results confirmed all participants were able to transfer their technical skills into professional practice. However, few were ready to practically apply their pedagogies. The results indicate three main factors, including technological resources, the school infrastructure, and the number of students in their classrooms. It is recommended to improve teacher preparation programme to develop teacher technology readiness. In a study titled primary school teachers' attitudes towards technology use and stimulating higher-order thinking in students: a review of the literature. Wijnen et al. [12] present interesting results focused on teachers' attitudes towards technology. They conducted two separate literature reviews on teachers' attitudes towards (1) using technology (78 articles) and (2) stimulating higher-order thinking in students (18 articles). To structure the potential underlying constructs constituting teachers' attitudes in these two contexts, they used the Theory of Planned Behaviour. They identified nine factors related to primary school teachers' attitudes towards using technology in their teaching and four factors related to primary school teachers' attitudes towards stimulating higher-order thinking. Furthermore, they found that it was not always possible to establish the impact of each factor on teachers' intended or actual use of technology and behaviours stimulating higher-order thinking, respectively.

Considering the above-mentioned characteristics of the students, and all mentioned researches, in the present research, we focused our attention on the students' perception of the interconnection between the outdoor and computer science fields and the possibility of implementing computer science activities in the MOE through an outdoor approach.

2 Participants

We electronically surveyed 419 students. All participants were students of the Faculty of Education, University of Prešov, in the following fields of study: pre-school and elementary pedagogy and pre-school and elementary pedagogy and pedagogy of psychosocially disturbed in the academic year 2021/2022. The target group consisted of students in the first (Y1) and third year (Y3) of bachelor's studies. We received 153 completed questionnaires, representing a return rate of 37%. Of the 153 students, 99 (65%) were first-year students, and 54 (35%) were third-year students. The participants were deliberately selected; they were future educators who will work in pre-primary settings. Therefore, the survey was administered at the beginning and end of their studies.

3   Measurement

The research aimed to analyse the knowledge and attitudes towards outdoor education, knowledge about information literacy, and to analyse their interrelationship in the groups of future kindergarten teachers in the first and the last year of their studies. For this purpose, we constructed a non-standardized research instrument, a questionnaire, which contained a total of 6 items. The first three items (p1–p3) in the form of a Likert scale (ranging from 1 to 6) focused on outdoor education. The items p4–p6 focused on information literacy education in kindergartens, with two Likert scales items and 1 (p6) multiple choice item.

The following hypotheses were formulated:

H1: There is a statistically significant difference in the scores of the two study groups between items p1 and p2.

H2: There is a statistically significant difference between items p2 and p3 in the ratings of the two treatment groups.

H3: There is a statistically significant difference between items p4 and p5 in the scores of both treatment groups.

H4: There is a statistically significant difference in scores between items p3 and p5 in both treatment groups.

The data collection was done electronically through MS Forms tool. The link to the research tool was provided to the students electronically. Data collection was conducted between 3 and 18 May 2021. Each Likert-type item contained a six-item scale, and a summary score was calculated for each item.

Statistical analysis was performed in the freely available programmes Jasp and Jamovi. Comparison of scores between the selected items was done by paired t-test. The above frequentist techniques were used in order to assess whether the difference between the selected items was statistically significant. Therefore, Bayesian statistics were used to assess the relative degree of empirical evidence in favour of the null (H0) or alternative hypothesis (Ha). Thus, for each of the items tested, a so-called Bayes factor was calculated to indicate the extent to which our beliefs in favour of one of the hypotheses need to be revised in the light of the observed data. A Bayes factor greater than 3 can already be considered as significant evidence in favour of a given hypothesis. At link 1 (bit.ly/38wPbCP) is available research tool and at link_2 (bit.ly/3taKYMj) the complete data.

3.1   Results

At the beginning of the analysis, we were interested in the summary results of the Likert-type questionnaire items. Items p1, p2, and p3 focused on awareness and attitudes towards outdoor education and their overall mean rating in both study groups is high, at 5.29 on a 6-point scale, with the lowest rating in item p3 in Y1: 4.96 and the highest rating in item p2, also in Y1 5.65. Items p4 and p5 were focused informatively,

Table 1 Basic descriptive indicators for all liquor-type items

	Group	N	Mean	Median	SD	SE
p1	A	99	5.12	6.00	1.127	0.1133
	B	54	5.20	6.00	1.088	0.148
p2	A	99	5.65	6.00	0.644	0.0647
	B	54	5.59	6.00	0.901	0.123
p3	A	99	4.96	5.00	1.019	0.1025
	B	54	5.28	5.50	0.811	0.110
p4	A	99	4.71	5.00	0.824	0.0828
	B	54	4.85	5.00	1.035	0.141
p5	A	99	4.20	4.00	1.160	0.1166
	B	54	4.39	5.00	1.338	0.182

Note A-first year (Y1), B-third year (Y3)

and their overall mean rating in both treatment groups is moderately high at 4.53 out of a 6-point scale. Comparing all the outdoor (Øp1, p2, p3) and informative (Øp4, p5) Likert-type items, we detect a difference in the ratings of all participants at 0.76 points. Items focused on outdoor education had more positive ratings than those focused on computer science. All descriptive statistics for each item are presented in Table 1. The summary ratings show that students rated the items more positively than negatively overall, averaging 4.99 out of 6 on the 6-point scale.

Hypothesis H1

To begin our analysis, we tested for differences in scores between the p1 and p2 items. Item p1 read: I know what outdoor education/outdoor access is in kindergarten. In terms of the evaluation level, both monitored groups (R1 and R3) are at an almost identical level, with a difference of only 0.08 points. In terms of distribution, the data in both groups are similarly inconsistent. In the first year, a larger number of participants evaluated the highest level, while the same number is scattered on levels 5, 4, and 3, the lowest levels are occupied only sporadically. In R3, the situation is similar, but the lowest level is not occupied at all. In item p2, we asked the students whether they consider spending more time outside to be beneficial for a child in kindergarten. Of all the items, the students rated it the most positively, at an average level of 5.62. The evaluation of both monitored groups is almost at an identical level (difference 0.05b). Based on the descriptive indicators (Table 2), the data in R1 are consistent. The majority of students rated grade 6, while no one chose the first 3 grades. In R3, the data are more scattered, but still consistent in the positive region. A paired t-test was used due to the nature of the research data and the research population, the results of which are presented in Table 3.

The calculated *p*-value is at the level of statistical significance with a medium effect size. Based on this fact, we reject the null hypothesis. The performed Bayesian analysis demonstrates that assuming the validity of Ha, i.e. p(D|Ha), the observed

Table 2 Basic descriptive indicators for items p1 and p2

Item	N	Mean	Median	SD	SE
p1	153	5.15	6.00	1.11	0.0898
p2	153	5.63	6.00	0.742	0.0600

Table 3 Paired t-test between item p1 and p2 for all students

	Statistic	±%	df	p	Cohen's d
Student's t	−4.84		152	<0.001	−0.391
Bayes factor10	4001	1.57e−10			

data are 5220 times more likely than the data that can be expected in the case of validity of H0, i.e. p(D|H0). Thus, with robust empirical support, we can conclude that item p2 was rated significantly better by students compared to item 2.

In the next part of the evaluation, we statistically compare items p2 and p3 and in both treatment groups to see whether students who know that being outdoors more often is beneficial for the child in the Nursery School as a future educator will implement outdoor learning in the Nursery School and how this changes over the course of the study.

It is evident from the data visualisation in Graphs 1 and 2 that first-year students rated items p2 and p3 diametrically differently. This fact is confirmed by the descriptive indicators in Table 4. The observed difference is 0.69 points from the 6-point scale. Given the calculated standard deviations, the data for item p2 are consistent, while the data for item p3 are scattered. Based on the visualized data on Graphs 1 and 2, for item p2, first-year students are unanimous in their belief that being outdoors more often is beneficial for the child. For item p3, the data no longer suggest this is the case, and we find a spectrum of ratings. Students in the first year are not unanimous in their approach to the subsequent implementation of outdoor education in teaching practice.

Hypothesis H2

A paired t-test was calculated to test the statistical significance of the differences between items p2 and p3 in the first year, which are presented in Table 5.

The p-value reaches the level of statistical significance, and thus, we reject the null hypothesis with a large effect size (Cohen's d = 0.663). The calculated Bayes factor probability values clearly tend (BF10 = 4.75) to favour the alternative hypothesis (BF10 > 3), and the observed level of evidence in favour of Ha is sufficient to declare its validity. Thus, it can be concluded that the difference in scores between p2 and p3 and in Y1 was confirmed, respectively. The observed data constitute adequate empirical evidence to decide on the validity of Ha.

The same statistical analysis was carried out in the third, final year of the Bachelor's degree. The visualized data in Graphs 3 and 4 show similar characteristics in this case as well. They are consistent for item p2 and scattered in the last three

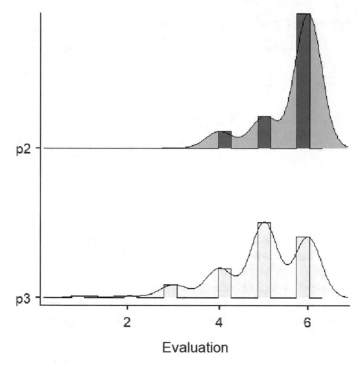

Graph 1 Observed distribution in p2 and p3 for Y1(A)

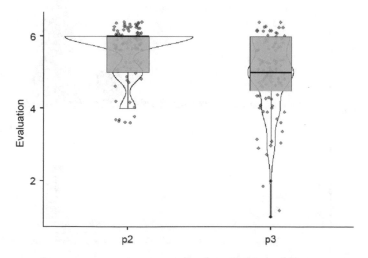

Graph 2 Density distribution of the data in item p2 and p3 for Y1(A)

Table 4 Basic descriptive variables for items p2 and p3 in Y1

Item	N	Mean	Median	SD	SE
p2	99	5.65	6.00	0.644	0.0647
p3	99	4.96	5	1.019	0.1025

Table 5 *t*-test between item p2 and p3 for Y1

	Statistic	±%	df	p	Cohen's d
Student's *t*	6.59		98.0	<0.001	0.663
Bayes factor10	4.75e+6	2.09e−15			

grades for item p3, a slight change from the first year. The difference in mean scores between items p2 and p3 in Y3 compared to Y1 is half of that in Y1, 0.32 points on a 6-point scale. From the descriptive indicators (Table 6), we can see that compared to the first year, this item was rated slightly more positively.

To test the statistical significance of the differences between p2 and p3 in Y3, a paired t-test was calculated, and the results of which are presented in Table 7. The calculated p-value reaches the level of statistical significance, and we are thus able to

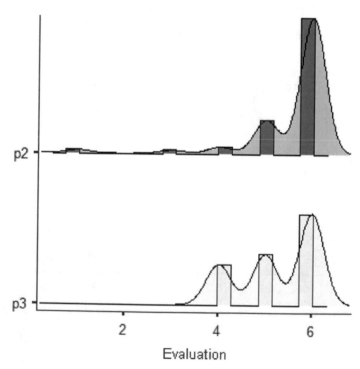

Graph 3 Observed distribution in item p2 and p3 for Y3(B)

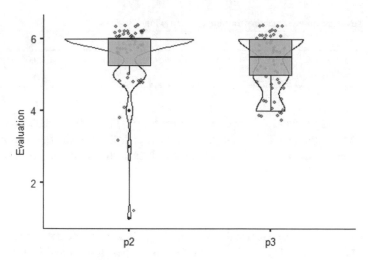

Graph 4 Density distribution of the data in item p2 and p3 for Y3(B)

reject the null hypothesis with a small effect size (Cohen's d = 0.384). Furthermore, the calculated Bayes factor likelihood values show that Ha is a model that describes the observed data better than H0 (BF10 = 5.14), and the observed level of evidence in favour of Ha is sufficient to declare its validity.

The difference between p2 and p3 in Y3 is confirmed as the observed data provide sufficient empirical evidence for the validity of Ha.

Hypothesis H3

In our analyses, we further investigated whether there is a statistically significant difference between the scores of items p4 and p5, and thus whether there is a correlation between the level of computer literacy represented by item p4 and the idea of implementing computer science education activities in the MOE through an outdoor approach. We investigated this separately in the first and third years and compared the results.

Table 6 Basic descriptive indicators for items p2 and p3 in Y3

Item	N	Mean	Median	SD	SE
p2	54	5.59	6.00	0.901	0.123
p3	54	5.28	5.50	0.811	0.110

Table 7 t-test between item p2 and p3 for Y3

	Statistic	±%	df	p	Cohen's d
Student's t	2.82		53.0	0.007	0.384
Bayes factor10	5.14	3.96e−9			

Table 8 Key descriptive indicators for items p4 and p5 in Y1

Item	N	Mean	Median	SD	SE
p4	99	4.71	5.00	0.824	0.0828
p5	99	4.20	4	1.160	0.1166

Table 8 gives the basic descriptive statistics for the items. From the mean values, we can see that students in the first year rated item p4 more positively than item p5, with a difference of 0.505 points. From the values of the standard deviations, but also the visualization of the data in Graphs 5 and 6, we can see that there is also a difference in the variance of the data. While the data for item p4 look consistent, in item p5, it is diversified. Thus, in item p4, more students rated positively, while the first stages of the evaluation are not occupied at all.

A paired t-test was used to test the statistical significance of the detected difference in scores between items p4 and p5 for first-year students, the results of which can be found in Table 9. As the table gives, the p-value reaches the significance level, on the basis of which we are able to reject the null hypothesis and thus confirm the existence of a certain difference.

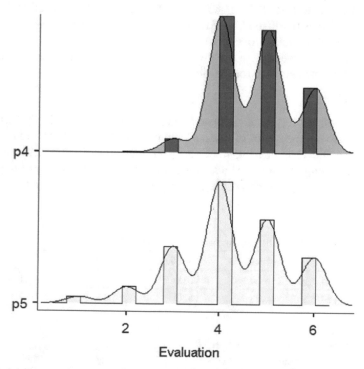

Graph 5 Observed distribution in item p4 and p5 for Y1(A)

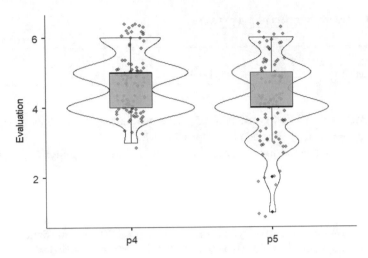

Graph 6 Density distribution of the data in item p4 and p5 for Y1(A)

Table 9 *t*-test between item p4 and p5 for Y1

	Statistic	±%	df	p	Cohen's d
Student's *t*	4.29		98.0	<0.001	0.431
Bayes factor10	407	3.49e−10			

The performed Bayesian analysis shows that assuming the validity of Ha, i.e. p(D|Ha), and the observed data are 407 times more probable than the data that can be expected in the case of validity of H0, i.e. p(D|H0). Thus, with robust empirical support, we can conclude that there is a statistically significant difference between the first-year scores of item p4 and p5. Students in their first year of study who have some level of information literacy (item p4 = 4.76/6) think they can master information technology, but they cannot imagine using this technology to carry out outdoor activities.

The same statistical analysis was carried out in the third year of study. The basic descriptive statistics are presented in Table 10.

Already from the mean values, we can conclude that the difference between these items in year 3 is almost at the same level as that of the freshmen of 0.463 (the difference of differences is only 0.042 points). A slight difference occurred in the density distribution of the data. Overall, the data density distribution shows similar

Table 10 Basic descriptive variables for items p4 and p5 in Y3

Item	N	Mean	Median	SD	SE
p4	54	4.85	5.00	1.04	0.141
p5	54	4.39	5.00	1.34	0.182

Table 11 t-test between item p4 and p5 for Y3

	Statistic	±%	df	p	Cohen's d
Student's t	2.54		53.0	0.014	0.345
Bayes factor10	2.70	4.34e−9			

characteristics to Y1 (p4 consistent, p5 scattered), but at p4 students already rate more heterogeneously than in Y1 (Graphs 7 and 8).

Again, we subjected the detected difference to statistical analysis. As given in Table 11, the p-value reaches the significance level with a medium effect size (Cohen's d intervals: small 0.2–0.5 medium), on the basis of which we are able to reject the null hypothesis and thus confirm the existence of some difference.

The performed Bayesian analysis does not have sufficient evidential power (BF10 < 3) to confirm the alternative hypothesis Ha (BF10 = 2.70), and the data are also not conclusive in favour of H0 (BF01 = 0.370). We do not find a statistically significant difference in this case; we conclude from the data that there is a difference between items p4 and p5, but the data are ambiguous in this case, and so we are unable to obtain empirical evidence of its magnitude/probability in favour of Ha or H0.

Hypothesis H4

Another area of interest was the correlation between *p3*, where students were asked to express a preference for implementing outdoor education in the MOE *(p3: As a future educator, I will implement outdoor education in the MOE)*, and p5, where students were asked to express whether they can imagine implementing educational activities in the information domain outdoors (p5: *I can imagine implementing educational activities in the information domain outdoors/outdoors—e.g. I can imagine implementing educational activities in the information domain outdoors/outdoors—e.g. I can imagine I can think of ways to implement educational activities in the outdoor environment (e.g. in a park or a schoolyard (digital animation, digital games, elementary programming).* We tested these items separately for the first and third years and compared the analysis results.

We report the basic descriptive statistics in Table 12. The difference in item p3 and p5 scores is 0.758 in the first year, with item p3 being more positively rated. As can be seen from the visualization in Graphs 9 and 10, the density distribution of the data is very similar for each item, with p3 being slightly more consistent in the positive region.

A paired t-test was used to test the statistical significance of the detected difference in scores between items p3 and p5 for first-year students, and the results of which

Table 12 Baseline descriptive indicators for items p3 and p5 in Y1

Item	N	Mean	Median	SD	SE
p3	99	4.96	5.00	1.02	0.102
p5	99	4.20	4	1.16	0.117

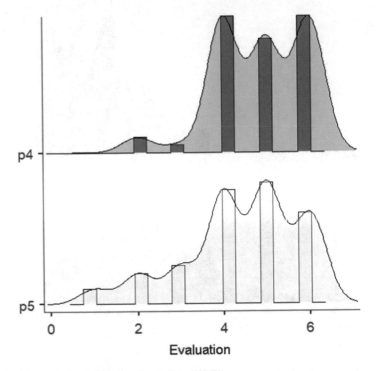

Graph 7 Observed distribution in p4 and p5 for Y3(B)

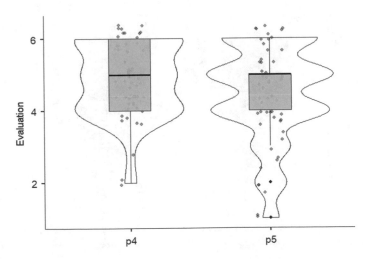

Graph 8 Density distribution of data in item p4 and p5 for Y3(B)

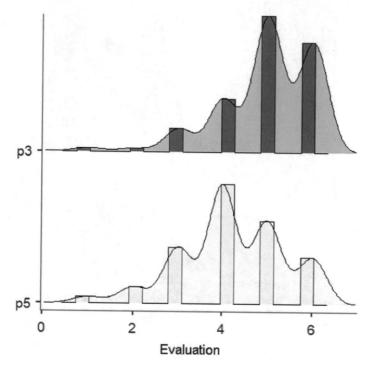

Graph 9 Observed distribution in item p3 and p5 for Y1(A)

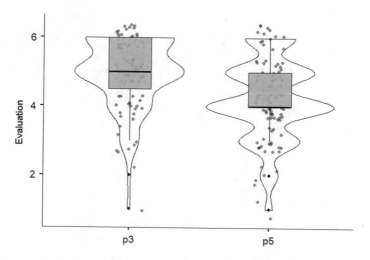

Graph 10 Density distribution of the data in item p3 and p5 for Y1(A)

Table 13 t-test between item p3 and p5 for Y1

	Statistic	±%	df	p	Cohen's d
Student's t	5.82		98.0	<0.001	0.585
Bayes factor10	165,751	2.88e−13			

Table 14 Baseline descriptive indicators for items p3 and p5 in Y3

Item	N	Mean	Median	SD	SE
p2	54	5.28	5.50	0.811	0.110
p3	54	4.39	5.00	1.338	0.182

can be found in Table 13. The calculated p-value reaches the level of significance, on the basis of which we are able to reject the null hypothesis and thus confirm the existence of a certain difference.

The implemented Bayesian analysis provides robust empirical support for the Ha model. Assuming the validity of Ha, i.e. p(D|Ha), the observed data are 165,751 times more likely than the data that can be expected in the case of the validity of H0, i.e. p(D|H0). Thus, with robust empirical support, we can conclude that there is a statistically significant difference between the first-year scores of item p3 and p5.

We performed the same statistical analysis for the final year Y3. Table 14 reports the basic descriptive statistics for items p3 and p5 in the third year. Compared to Y1, we observe more positive evaluations for both items under study. The detected difference between the items is at 0.899 (the difference in differences is + 0.141 for year 3).

In terms of the scatter in the data visualized in Graphs 11 and 12 for item p3 in Y3, the data look consistent compared to Y1. Comparing p3 and p4 in Y3, we observe a significant difference in the density distribution of the data. Item p3 is rated significantly more consistently in the positive domain by third parties, with the first 3 rating levels not occupied at all, whereas item p5 is rated significantly more diversely by third parties and all rating levels are occupied comparatively.

The calculated p-value is at the level of statistical significance with a large effect size, formally demonstrating the existence of a statistically significant difference between p3 and p5 in Y3 and thus formally rejecting H0 (Table 15).

The implemented Bayesian analysis also in Y3 provides robust empirical support for the Ha model. Assuming the validity of Ha, i.e. p(D|Ha), the observed data are 5487 times more likely (with Bayes facts greater than 3 being sufficient as formal evidence for the validity of the hypothesis) than the data that can be expected in the case of the validity of H0, i.e. p(D|H0). Thus, with robust empirical support, we can conclude that there is a statistically significant difference between the item scores of p3 and p5 in the third year.

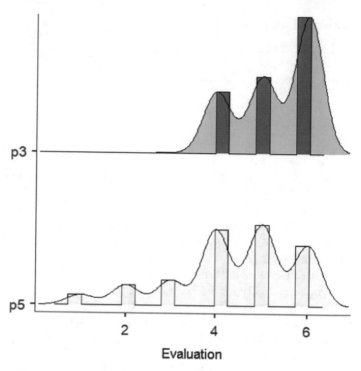

Graph 11 Observed distribution in item p3 and p5 for Y3(B)

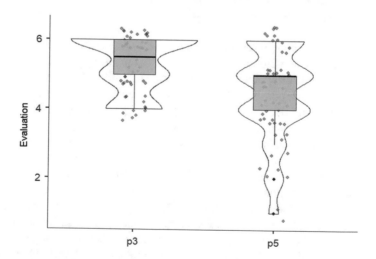

Graph 12 Density distribution of data in item p3 and p5 for Y3(B)

Table 15 t-test between item p3 and p5 for Y3

	Statistic	±%	df	p	Cohen's d
Student's *t*	5.21		53.0	<0.001	0.709
Bayes factor10	5487	4.22e−12			

4 Conclusions and Discussion

The study aimed to analyse the knowledge and attitudes towards outdoor education and information literacy and to find out their interrelationship among prospective pre-service teachers in the first and final years of their studies in the MOE. In general, we can conclude that students rated the items overall more positively than negatively.

Outdoor education represents a new theoretical area that several students have encountered for the first time at university. Being outdoors is generally beneficial for people physically and psychologically, and it is also a stably implemented part of the day that has been in the organization of the day in kindergartens since their inception, which in our opinion, may have caused the significantly higher score of the p2 item dealing with the dependence of the frequency of being outdoors on the child's well-being.

It was interesting to compare items p2 and p3, where, on the one hand, students declare that they know what outdoor education is and that being outdoors more often is beneficial for the child in kindergarten (and with a higher rating in the first year), on the other hand, they rate significantly lower the item in which they, as future educators, should implement this type of education, which we observe in both groups studied.

On the one hand, students are proactive in implementing outdoor activities; on the other hand, when technology enters into it a problem arises in outdoor activities for students. The observed difference was confirmed in both study groups and did not change during the study. Also, the future teachers in the third, i.e. final year, plan to implement outdoor education in the MoE; however, they already rate the implementation of outdoor educational activities in the computer field in the MoE significantly lower.

From the results of the conducted questionnaire research and statistical analysis, we can conclude that students are familiar with the concept of outdoor education and generally consider more frequent staying outdoors as beneficial for the child. However, students know this concept more theoretically and applying the acquired knowledge to the practical level is a potential problem for them. A similar problem is observed in the field of technology. On the one hand, students take the initiative in the implementation of outdoor activities; on the other hand, if they have to use technology in a practical way, a problem arises in outdoor activities. Our task is, therefore, to provide opportunities to link these seemingly incompatible areas.

References

1. Alghamdi J, Mostafa F, Abubshait A (2022) Exploring technology readiness and practices of kindergarten student-teachers in Saudi Arabia: a mixed-methods study. Educ Inf Technol. https://doi.org/10.1007/s10639-022-10920-0
2. Bullen M, Morgan T, Qayyum A (2011) Digital learners not digital natives. In: La Cuestión Universitaria, č 7, s 60–68. ISSN 1988-236x
3. Burgerová J, Adamkovičová M, Piskura V, Kochová H (2017) Pilot testing of it skills of preschool and elementary education students by the it fitness test. ICTE J [online], roč 6, č 2, s 35–43. 1805 3726. Dostupné na: https://periodicals.osu.eu/ictejournal/dokumenty/2017-01/ictejournal-2017-1-article-4.pdf
4. Campan A-S, Bocos M (2019) The outdoor activities in preschool education. In: Chis V, Albulescu I (ed) ERD 2018—education, reflection, development , 6th edn, pp 1–6. https://doi.org/10.15405/epsbs.2019.06.1
5. Zhu F, Villanueva LE (2021) Cognitive schema of net-generation college students: an empirical study from three universities. Converter 2021(6):413–424. Retrieved from http://converter-magazine.info/index.php/converter/article/view/405
6. Gilbertson K, Ewert A, Siklander P, Bates T (2022) Outdoor education: methods and strategies. Human Kinetics
7. Howe N, Strauss W (1991) Generations: the history of America's future, 1584 to 2069. Quill, New York
8. Prensky M (2001) Digital natives, digital immigrants part 1. On the horizon [online], roč 9, č 5, s 1–6. ISSN 1074-8121. Dostupné na: https://doi.org/10.1108/10748120110424816
9. Rosen DL, Teaching the iGeneration. [online].[cit 22-4-2016]. Dostupné na: http://www.ascd.org/publications/educational-leadership/feb11/vol68/num05/Teaching-the-iGeneration.aspx
10. Tapscott D (1998) Growing up digital: the rise of the net generation. McGraw-Hill Book, New York. ISBN 0-07-134798-4
11. Volkovová T (2015) Outdoorové aktivity vo výchove mimo vyučovania. Bratislava: Metodicko-pedagogické centrum v Bratislave, s 62. ISBN 978-80-565-1386-6
12. Wijnen F, Walma van der Molen J, Voogt J (2021) Primary school teachers' attitudes toward technology use and stimulating higher-order thinking in students: a review of the literature. J Res Technol Educ, pp 1–23. https://doi.org/10.1080/15391523.2021.1991864
13. Zhao E, Liu L (2008) China's generation Y: understanding the workforce. In: 2008 4th IEEE international conference on management of innovation and technology [online]. IEEE, pp 612–616. ISBN 978-1-4244-2329-3. Dotupné na: https://doi.org/10.1109/ICMIT.2008.4654435

Cyber Security and Cyber Physical Systems and Digital Technologies

Human Versus Automatic Evaluation of NMT for Low-Resource Indian Language

Goutam Datta⬤, Nisheeth Joshi, and Kusum Gupta

Abstract Machine Translation (MT) is an important application of natural language processing that converts a source language to a target language automatically with the injection of a parallel corpus. Researchers from academia and industry are currently very active in designing high-performance translation systems. The MT system has witnessed a big paradigm shift, and a recent neural network-based translation system, i.e., the Neural Machine Translation system, has almost replaced the statistical Machine Translation (SMT) system. This paper mainly focused on two things: firstly, the design of the NMT model and then evaluating its performance with human and automatic metrics on a low-resource Bengali-to-English language pair. Secondly, we have checked the performance of two popular online translators: Google Translate and Bing. We have evaluated the performance of these two translators with the most popular and widely used automatic evaluation metrics, Bilingual Evaluation Understudy (BLEU) and Word Error Rate (WER). BLEU's evaluation process is primarily based on the n-gram matching approach; hence, sometimes scores are not reliable. WER computes the Levenshtein distance between hypothesis words and reference words. Human evaluation is considered to be the best in MT evaluation. Hence, we have also computed the translation score with the gold-standard human evaluation metric. This research will be helpful as a part in evaluating the performance of various MT engines, especially in domain-specific low-resource language pairs, and also the performance of MT engines can be judged with human evaluators.

G. Datta (✉) · N. Joshi · K. Gupta
School of Mathematical and Computer Science, Banasthali Vidyapeeth, Banasthali, Rajasthan, India
e-mail: gdatta1@yahoo.com

N. Joshi
e-mail: jnisheeth@banasthali.in

K. Gupta
e-mail: gupta_kusum@yahoo.com

G. Datta
Informatics Cluster, University of Petroleum and Energy Studies, Dehradun, India

© The Author(s), under exclusive license to Springer Nature Singapore Pte Ltd. 2023 715
Y. Singh et al. (eds.), *Proceedings of International Conference on Recent Innovations in Computing*, Lecture Notes in Electrical Engineering 1011,
https://doi.org/10.1007/978-981-99-0601-7_55

Also, these human evaluators are used to verify the accuracy of available automatic evaluation metrics.

Keywords Neural Machine Translation · BLEU · Word Error Rate

1 Introduction

One of the initial objectives of computers was the automatic translation of text between languages. Being flexible with human language, MT is considered to be one of the most challenging jobs in Artificial Intelligence. Historically, rule-based systems were employed for this task, but in the 1990s, statistical methods superseded them. Deep neural network models have recently achieved state-of-the-art performance in the field of Neural Machine Translation [1, 2]. There are different models proposed by the researchers in the NMT domain. Some of the popular NMT systems as proposed by the researchers in different phases are as follows: Bahadanu's Attention model, transformer model, etc. [3, 4]. Google helped solve many of NMT's issues, including issues with handling rare words and a lack of robustness, among others, with their NMT system [5]. At Google, in the research paper published in 2017 and titled "Attention Is All You Need," the transformer model was presented for the first time. Its purpose was to transform one sequence into another without making use of any recurrent neural networks (RNNs). The transformer models are state of the art, and their design is based on a self-attention mechanism that allows them to function (Fig. 1). The simple transformer model is a sequence-to-sequence model having encoder and decoder blocks. Each encoder and decoder block has a stack of identical blocks. Also, each encoder block has two important components such as self-attention and a position-wise feedforward network.

There is another challenge of MT, i.e., evaluating the performance of MT systems [6–8]. Although human evaluation is considered to be the best, it is a time-consuming process. There are many automatic evaluation metrics proposed by the researcher, such as BLEU [9], precision, recall, F-measure, cosine similarity, METEOR [10]. There is another human-like automatic metric proposed by many researchers [11–13].

In this paper, we have experimented with a transformer-based model on a low-resource Bengali-to-English language pair (tourism dataset). Bengali is a language that is mostly spoken in India and Bangladesh. English and Bengali are both morphologically rich languages. We have evaluated the performance of our model and two more MT engines, such as Google and Bing Translate, with automatic metrics BLEU, WER, and human evaluation. We also compared the scores of BLEU with gold-standard human evaluation. The rest of the paper is organized as follows: Sect. 2 describes some previous work on NMT and its performance evaluation with automatic metrics. Section 3 highlights our overall methodology. Section 4 outlines our experimental setup; in Sect. 5, we have presented some analysis and discussion. Finally, we put conclusions and future direction in Sect. 6.

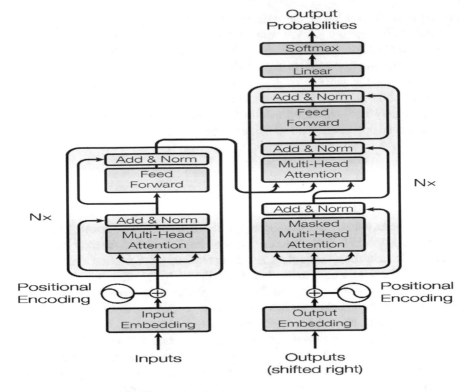

Fig. 1 Transformer model [4]

2 Previous Work

In the paper [12], the authors primarily focused on evaluating the performance of MT systems. They designed a framework for different types of errors and categorized them. They performed a comparative analysis of the types of errors available in the raw text generated by the MT engine and then performed some post-editing processes to remove them. The aim was to find out the recurrent errors that are produced by the translation system. This may eventually be helpful for future researchers to produce robust translation systems. Human evaluation is considered to be the best in evaluating the performance of MT systems, but it is time-consuming and cannot be reproducible. In bilingual translation systems, if the human evaluator understands both languages, then he can correctly judge the quality of the translation generated by MT engines. Automatic evaluations are dependent on human evaluation because they require one or more reference translations to correlate the MT outputs. Sometimes, even automatic metrics fail to generate correct scores when the languages are morphologically rich. In the paper, the authors address this issue of automatic metrics by proposing a human-like evaluation metric where they have exploited the use of scalar and multidimensional quality metrics. Their transformer-based NMT model was trained on low-resource English-to-Irish language pairs. In this study, they presented a comparative analysis with human evaluation of the outputs generated by RNN-based and transformer-based models on English-to-Irish language pairs [6].

In another work, researchers during the development of MT evaluation mainly focused on the semantics of the language. They proposed an approach where the adequacy of the language is the prime concern rather than fluency. This evaluation scheme mainly tries to find the semantic similarity between the source and the target languages and does not depend on the reference translation [13].

3   Methodology

We have used a tourism dataset collected from TDIL (https://tdil-dc.in/). Firstly, we swapped the parallel corpus and took Bengali to English. That is, the source language is Bengali and the target is English. After the required preprocessing of the dataset, we trained the NMT model (transformer) with default hyperparameters. After model convergence, we randomly picked three samples from the test set. The same test dataset was given to online translation engines: Google and Bing. The translation qualities of our tuned model, Google and Bing translate models, were verified with gold-standard human evaluators. Human evaluation metrics were calculated using that questionnaire. For the purpose of our MT engines' (our transformer model, Google Translator, and Bing) qualitative performance, we have mainly focused on the following evaluation metrics: BLEU, WER, and human evaluation.

3.1   Human Evaluation

As stated before, to generate a human score, we have carried out a survey that is based on the questionnaires. Five individuals with a satisfactory to excellent level of linguistic expertise in both of these languages (English and Bengali) are participated in the survey. Questions primarily addressing the adequacy and fluency of the translation were developed as part of the questioners' inquiries. The adequacy and fluency of the response were evaluated using a scale with a value that ranges from 0 to 5. The concept of adequacy ensures that the meaning of the source text and the translated output is the same; that the source sentence and the target sentence are both complete; and that there is no distortion between the source sentence and the target sentence, among other things. In a similar vein, fluency guarantees that both the source language and the target language have the correct syntax. A score of 5 is given for adequacy when all meanings between the source and target languages are correct; a score of 4 is given when most of the meaning is correct; a score of 3 is given when much meaning is given; a score of 2 is given when little meaning is given, and a score of 1 is given when no meaning is given. When evaluating fluency, a score of 5 indicates flawless language, a score of 4 indicates good language, a score of 3 indicates non-native language, a score of 2 indicates diffluent language, and a score of 1 indicates an incomprehensible language. We have shown the translated results of three randomly picked sentences from a test set that were translated from Bengal to English by our transformer model and freely available online translators, Google Translator and Bing Translator, from Microsoft. This translated text's quality was evaluated with the human evaluator and the automatic metrics BLEU and WER.

3.2 BLEU is an Automatic Metric

Bilingual Evaluation Understudy (BLEU) is used to measure the quality of MT output by comparing the MT output with reference text. It tries to match MT output with reference text with an n-gram-based approach, where n is the number of tokens; in a unigram, n equals to one; in a bigram, n equals to two; and so on.

Word Error Rate (WER).

The Word Error Rate (WER) is exploited in evaluating the performance of MT systems and speech recognition systems [14]. Researchers in the field of text and the speech translation system used WER to evaluate the quality of translations of their model [15]. As stated before, WER computes the Levenshtein distance between hypothesis words and reference words divided by the total number of reference words.

WER is calculated with the help of the following expression:

$$\text{WER} = \frac{S + D + I}{N}, \tag{1}$$

where S is the total number of substitutions in the target sentence, D is the total number of deletions in the target sentence, I is the number of insertions, and N is the total number of words in the reference.

The overall methodology is represented in Fig. 2. With the help of processed dataset and default set of hyperparameters, we trained our model. Hyperparameter tuning is essential in any machine learning model [16–18]. We then tested our model with the test set. The model's performance was evaluated with two automatic metrics such as BLEU and Word Error Rate (WER). As stated before, we also evaluated our model with human evaluators. Their scores were averaged to compute the final score.

4 Experimental Setup

We used open NMT's transformer model and trained the model on a Bengali–English dataset. Our model was trained in the NVIDIA Tesla V100 environment. We trained our model up to 10,000 training steps. After that, model starts converging. We split our entire dataset as follows: 70% training set, 15% validation set, and 15% test set.

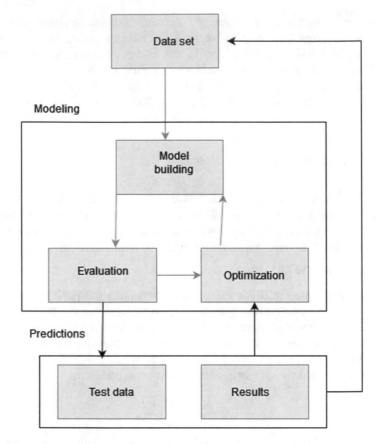

Fig. 2 Schematic representation of the model building and quality evaluation process

The model took around 3 hours for the entire training process. Proper hyperparameter tuning is very important in any machine learning model. Few of the selected hyperparameters of our transformer model are represented in Table 1.

With the help of selected hyperparameters, we trained our model.

The purpose of the selected hyperparameters is as follows:

Table 1 Snapshot of some selected hyperparameters of the transformer model

Hyperparameters	Values
Optimizer	Adam
Learning rate	2
Dropout	0.1
Heads	8
Decay_method	Noam

Table 2 Randomly picked test sets and their translation scores as generated by human and automatic evaluation metrics

Test set	Translation engine	Human evaluation (averaged) (in scale 0 to 100)	BLEU (scale 0 to 1)	WER (in %age)
1	Our model	30	0.14	88
2	Our model	25	0.12	91
3	Our model	35	0.11	95

Table 3 Randomly picked test sets and their translation scores as produced by human and automatic evaluation metrics (in Google translation engine)

Test set	Translation engine	Human evaluation (averaged) in scale (0–100)	BLEU (scale 0 1)	WER (in %age)
1	Google	90.5	0.11	73.3
2	Google	90.5	0.67	81.81
3	Google	100	1.0	0

Optimizer: It helps the model by selecting suitable parameters so that loss can be reduced and accuracy can be increased.

Learning rate: It is a tunable parameter that selects step size in each iteration during its movement to the minimum loss function.

Dropout: It is a regularization technique used in machine learning and deep learning.

Heads: It is the number of heads in the transformer model.

Decay_method: In machine learning, decreasing the learning rate is a type of decay. Similarly, increasing the learning rate is a warm-up process. Noam is a type of decay method used in deep learning where warm-up and decay both exist.

We noticed after around 10,000 epochs there was no further improvement in training and validation accuracy. We verified the performance of our model with randomly picked sentences from the test dataset. And, the same sentences were fed to other online engines Google and Bing. Performance evaluation of all translators was carried out with automatic and human evaluation metrics. The performance of our model, Google Translate, and Bing Translate, is reported in Tables 2, 3, and 4, respectively. The graphical representations of all these results are reported in Figs. 3, 4, and 5, respectively.

5 Analysis and Discussion

From our above experimentation, we can see that our model's performance is not satisfactory (Table 2 and Fig. 3). WER, BLEU, and human evaluation scores are

Table 4 Randomly picked test sets and their translation scores as produced by human and automatic evaluation metrics (in Bing translation engine)

Test set	Translation engine	Human evaluation (Averaged) in 0–100	BLEU in a scale (0–1)	WER (in %age)
1	Bing	90	0.12	66.6
2	Bing	90	0.14	68.18
3	Bing	60	0.75	50

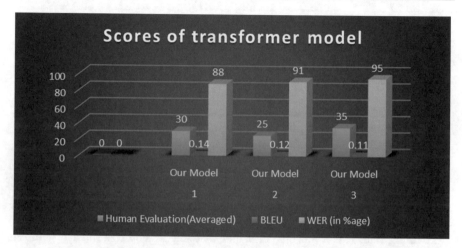

Fig. 3 Translation quality scores as generated by human evaluation and automatic evaluation metrics (BLEU and WER) for the transformer model

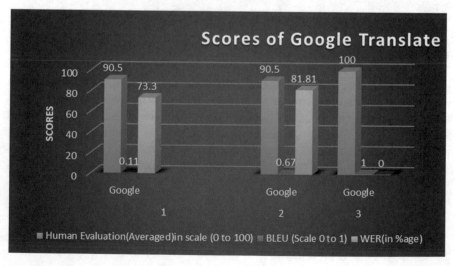

Fig. 4 Translation quality scores as generated by human evaluation and automatic evaluation metrics (BLEU and WER) for Google Translate

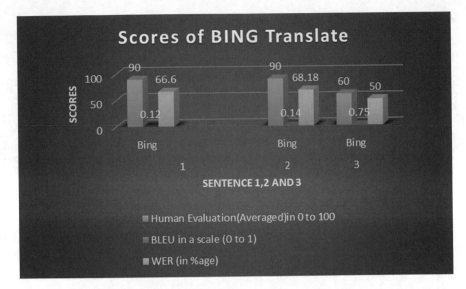

Fig. 5 Translation quality scores as generated by human evaluation and automatic evaluation metrics (BLEU and WER) for Bing Translate

presented in Table 2 and Fig. 3. There are possibly many factors for this. We have used tourist datasets that are too small for deep learning models. Even the dictionary vocabulary was very small for our model. In the case of morphologically rich languages, the handling of rare words is a serious problem if the vocabulary is not large enough. Another reason for model underperformance is that our model was tuned with the default hyperparameter setting. Next, in the evaluation part, we have evaluated the performance of Google Translate and Bing Translate. We have picked three sentences randomly from the test set. As we can observe from the tables and graphical representations in Section 4: Table 3 and Fig. 4, we can see that in some instances, the BLEU score is less than the human score. The WER is also significantly higher, whereas, for the same test sentence, the human score is quite high. The same thing we can notice in Bing Translate; also, results are presented in Table 4 and Fig. 5. In some cases, the scores generated by the automatic metric and the human score differ significantly. The reason is that the BLEU is based on n-gram matching between translated texts and one or more reference texts. It does not capture the overall semantics of the translated text. Similarly, in WER also, if there are many substitutions of the words (although the substituted words have the same meaning) and many deletions, the WER percentage will be higher. These are some of the reasons; though automatic metrics are faster and reusable, their scores are sometimes not satisfactory.

6 Conclusion and Future Work

From the above experimentation, we can conclude that proper tuning of hyperparameter is important in designing any machine learning model. The use case we considered is an NMT system. We have seen that NMT exploits deep learning, and hence, it requires a huge corpus to develop a better model. Furthermore, computing the scores of MT systems with automatic evaluation metrics such as BLEU, WER, and others is useful for evaluating the overall translation quality of MT models. However, MT evaluation is a challenging task since the same word can convey different meanings in some other contexts. Most of the automatic metrics are precision and recall based and fail to capture the semantics of the words. Hence, MT evaluation is still an interesting research area, and as discussed before, sometimes intervention of human evaluation is required because human evaluation is the best [19], though it is time-consuming.

In the future work, researchers are focusing on the strategies to capture the semantics of the hypothesis and reference texts along with the n-gram-based matching to enhance the accuracy of the automatic metric scores.

References

1. Softmax G, Softmax G, Comparative study of neural machine translation models for Turkish Language
2. Stahlberg F (2020) Neural machine translation: a review. J Artif Intell Res 69:343–418. https://doi.org/10.1613/JAIR.1.12007
3. Bahdanau D, Cho KH, Bengio Y (2015) Neural machine translation by jointly learning to align and translate. In: 3rd International conference on learn represent ICLR 2015—conference track proceedings. Published online 2015, pp 1–15
4. Vaswani A, Shazeer N, Parmar N et al (2017) Attention is all you need. In: Advances in neural information processing systems, pp 5999–6009
5. Wu Y, Schuster M, Chen Z et al (2016) Google's neural machine translation system: bridging the gap between human and machine translation. Published online 2016, pp 1–23. http://arxiv.org/abs/1609.08144
6. Fomicheva M, Specia L (2019) Taking MT evaluation metrics to extremes: beyond correlation with human judgments. Comput Linguist 45(3):515–558. https://doi.org/10.1162/coli_a_00356
7. Lin T, Wang Y, Liu X, Qiu X (2021) A survey of transformers. 1(1). http://arxiv.org/abs/2106.04554
8. Datta G, Joshi N, Gupta K (2021) Empirical analysis of performance of MT systems and its metrics for English to Bengali: a black box-based approach. In: Paprzycki M, Thampi SM, Mitra S, Trajkovic L, El-Alfy E-SM (eds) Intelligent systems, technologies and applications. Springer Singapore, pp 357–371
9. Papineni K, Roukos S, Ward T, Zhu W-J (2002) {B}leu: a method for automatic evaluation of machine translation. In: Proceedings of the 40th annual meeting of the association for computational linguistics. Association for Computational Linguistics, pp 311–318. https://doi.org/10.3115/1073083.1073135
10. Banerjee S, Lavie A (2005) METEOR: an automatic metric for mt evaluation with improved correlation with human judgments. In: Proceedings of the Acl workshop on intrinsic and extrinsic evaluation measures for machine translation and/or summarization, ACL 2005, pp 65–72

11. Lankford S, Afli H, Way A (2022) Human evaluation of English—Irish transformer-based NMT. Published online 2022, pp 1–19
12. Escribe M (2019) Human evaluation of neural machine translation: the case of deep learning. Published online 2019, pp 36–46. https://doi.org/10.26615/issn.2683-0078.2019_005
13. Licht D, Gao C, Lam J, Guzman F, Diab M, Koehn P (2022) Consistent human evaluation of machine translation across language pairs, p 1. http://arxiv.org/abs/2205.08533
14. He X, Deng L, Acero A (2011) Why word error rate is not a good metric for speech recognizer training for the speech translation task? In: 2011 IEEE international conference on acoustics, speech and signal processing (ICASSP) ICASSP, Published online 2011, pp 5632–5635. https://doi.org/10.1109/ICASSP.2011.5947637
15. Min DJ, Pérez-Rosas V, Mihalcea R (2021) Evaluating automatic speech recognition quality and its impact on counselor utterance coding. In: Proceedings of the seventh workshop on computational linguistics and clinical psychology: improving access CLPsych 2021—NAACL 2021. Published online 2021, pp 159–168. https://doi.org/10.18653/v1/2021.clpsych-1.18
16. Yang L, Shami A (2020) On hyperparameter optimization of machine learning algorithms: theory and practice. Neurocomputing 415:295–316. https://doi.org/10.1016/j.neucom.2020.07.061
17. Agnihotri S (2019) Hyperparameter optimization on neural machine translation. Creat Components, p 124. https://lib.dr.iastate.edu/creativecomponents/124/
18. Lim R, Heafield K, Hoang H, Briers M, Malony A (2018) Exploring hyper-parameter optimization for neural machine translation on GPU architectures. Published online 2018, pp 1–8. http://arxiv.org/abs/1805.02094
19. Joshi N, Mathur I, Darbari H, Kumar A (2013) HEval: yet another evaluation metric. IJNLC 2(5).arXiv:1311.3961v1

Framework for Detection of Malware Using Random Forest Classifier

Mohsin Manzoor and Bhavna Arora

Abstract Malware poses a challenging threat to this digital world of communication, as it can manipulate or perform any baffling activity inside a computer by corrupting important files and disabling the network system with malicious attacks. Since the malware writers have evolved in their malware designing technique over the past decade, so the detection mechanism should be robust enough to identify the latest malware and also should curb its propagation inside a computer. The traditional. method of malware detection like signature-based. methods fails to detect the new complex malware as the developers now develop variations of these and camouflage the malware. Hence, the malware goes undetected using the traditional techniques. So, there is a dire need for an effective and an efficient malware detection system that could identify and detect the obscured and encrypted malware with maximum accuracy so that the big enterprise or any other institution that relies on the digital exchange of information should not sustain any kind of financial and information loss. The utilization of machine learning in malware detection has been found effective in detecting hidden and obscured malware. This study focuses on a quick and concentrated assessment of malware detection utilizing several machine learning approaches and recommends a framework that uses the Random Forest classifier-based strategy for detecting malware. The maximum accuracy obtained during this research was about 98.5%.

Keywords Malware · Malware detection techniques · Machine learning

M. Manzoor (✉) · B. Arora
Department of Computer Science and Information Technology, Central University of Jammu
Samba India (181143), Samba, India
e-mail: Mohsinbeigh03@gmail.com

B. Arora
e-mail: Bhavna.csit@cujammu.ac.in

1 Introduction

Malware, sometimes known as "harmful software," is a catch-all term for any malicious program or code that causes computer damage. Malware is malicious software created to infiltrate, damage, or disable a computer system by obtaining partial control over the operations of the information and communication (ICT)-enabled devices. The malware that has the power to spread is the most hazardous since there is no central control, making it difficult to defend them [1]. The types of malware that are usually found in any kind of malware attack are enlisted as worms, spywares, viruses, Trojan viruses, ransomware and adwares [2]. Malware is designed to attack internet-based applications as practically every aspect of life. Malware can be detected using different methods. The most popular method is using the classifier to determine whether the file is a malware or not. In order to detect the malware, several issues and concerns arise. There are various types of malwares that differ from each other through the propagation modes, functionality, and performance. The malware writers encrypted their malware so that the former detection mechanism should fail to identify and detect the malware. Key limitations of signature-based detection methods are that they vary from different antivirus vendors; furthermore, the signature-based malware detection technology only works with known malware. Malware authors employ a variety of ways to enable their produced malware to readily evade and fool signature-based detection methods, which include mutations like encryption, oligiomorphism, polymorphism, and metamorphism, usually done by mutation engines [3]. Mutation engines are computer program that can change one program into another with different codes by encrypting the target software using various keys and a decryption module that may be customized widely [4]. These mutations are described as follows:

(a) *Encryption*: The malware developers did not want their malicious code to be noticed and detected by any detection mechanism, and for that purpose, encryption is very simple way. This type of mutation consists of the phenomena of encrypting and decrypting the malwares while escaping from the detectors and infecting the target, respectively [3].

(b) *Oligiomorphism*: It is another important technique used for achieving the camouflage behavior of malwares, and the oligomorphic malwares were first emerged during 1990's under the name "whale" (as a DOS virus). Unlike encryption technique where the decryptor remains uniform or same for each malware, the pleomorphism supports the unique decryptor for each malware infection. This technique is considered as an advancement in the encryption.

(c) *Polymorphism*: This technique is the blend of encryption and oligiomorphism, but these malwares are more infectious than the others. The first polymorphic virus was first designed by "Mark Washburn" during late 1990's and was labeled as virus 1260 [3]. It is very difficult and hard for the antiviruses to detect such type of malwares as these malwares change their behavior and appearance upon the generation of new copies.

(d) *Metamorphism*: This type of malware obfuscation technique holds the phenomena of altering the syntax of the malware while the semantics remains the same, which means that newly generated malware copies apparently look different, but deep inside, their working or execution remains the same [3].

The paper is organized in different sections: The first section discusses about the basic malware and the advanced malwares; it also discusses different techniques for malware detection and different machine learning approaches for the same. Section 2 summarizes the research work carried out by different researchers. The proposed framework has been detailed out in Sect. 3. Section 4 discusses the results followed by conclusion in Sect. 5.

1.1 Malware Detection Techniques

The detection methods for malware are classified into several groups from various perspectives, such as the signatures they produce, their behavior upon executing under controlled scenarios, and the heuristic analysis; based on these perspectives, the different techniques of malware detection are depicted in Fig. 1 [5].

1. *Signature-Based Malware Detection*:

A file's signature is a one-of-a-kind property, similar to an executable's fingerprint. Signature-based methods are more efficient and faster than other methods since they leverage patterns taken from diverse malware to identify them [5].

2. *Behavior-Based Detection*:

Malware detection techniques based on behavior examine a program's way of executing inside a machine to determine if the given piece of code is malicious or not. Behavior-based techniques are immune to the flaws of signature-based techniques since they watch what an executable file performs [5].

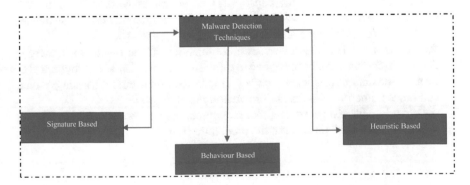

Fig. 1 Malware detection methods [5]

3. *Heuristic-Based Detection*:

Heuristic evaluation is a virus-spotting approach that includes looking for suspicious features in code; the heuristic model was designed to discover suspicious qualities in unknown, new viruses, modified versions of current threats and known malware samples [5].

1.2 Machine Learning (ML) Approaches for Detection of Malware

Malware is difficult to detect because it is comprised specialized programs designed to achieve a certain aim and executed in a predetermined manner [6]. Over time, various researchers have used various machine learning techniques for observing the obscured and encrypted malware. These techniques are enlisted as follows [6]:

1. *K-means*: The purpose of this iterative approach for classifying or grouping items is to divide the dataset into K groups or clusters, where K is a positive integer number. This strategy is most effective when it comes to detecting polymorphic malware.
2. *Naïve bays*: The Naive Bayes probabilistic classifier is created using the Bayes classifier. When compared to other methods, the Naive Bayes approach is more suited for more competent results. This method works effectively for identifying worms and viruses on mobile devices.
3. *SVM*: In SVM, the decision boundaries are defined by the decision planes. It is a binary classifier that is not probabilistic. It is fantastic at detecting malware on Android.
4. *Decision tree*: For classification and regression, it is a supervised learning strategy that works like a flowchart, starting from the root node and bifurcating the data to the leaf nodes till the entropy is decreased. This technique is also effective in detecting polymorphic malware.
5. *Neuro-fuzzy*: To improve the capabilities of neural networks, this approach combines the benefits of fuzzy logic and artificial neural networks. This strategy works well for detecting malware that has been disguised [6].
6. *Random Forest*: This classifier uses a combination of the results of numerous decision trees that have been applied to different subsets of a dataset for improvement in the dataset predicted accuracy. It is based on ensemble learning, which combines numerous classifiers to solve complicated problems.
7. *ANN*: Unlike the K-nearest neighbor technique, the all nearest neighbor (ANN) classifier selects "k" throughout the procedure (Fig. 2).

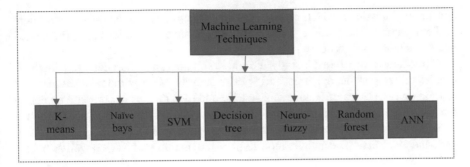

Fig. 2 Taxonomy of different machine learning techniques for malware detection [6]

2 Literature Survey

Based on the various machine learning approaches, various authors have proposed different methods, along with the objectives and the accuracy for the detection of malware. A brief description of some of the research is as follows.

Souriet et al. [5] The paper give out an analytical and elaborated survey of malware detection approaches by employing machine learning. The authors had reviewed more than 50 papers, and according to their detailed survey, the detection techniques are bifurcated into two classifications, including signature and behavior-based detections. The research specifies an appropriate category for malware detection techniques rather than scanning and statistical analysis.

Tahir [3] The paper discusses in detail about the different malwares in addition to different malware detection techniques. According to the author, malware is the biggest threat to this digital flow of information as it can manipulate or perform any obfuscation activity inside a computer by corrupting important files and disabling the network system with malicious attacks. The author classified the malware as Trojans, worms, rootkit, adwares, spywares, sniffers, robot-networks (botnets), keyloggers, spamware, ransomware, etc. The author also briefed about how malware creators are advanced in their developing malware approach, as the malware creators had developed new tools to make the malware be unidentified inside a computer or a network, and the authors labeled the technique as camouflage behavior.

Stahlbock et al. [7] Malware is an ever-changing hazard in the internet era. The author presented a deep learning architecture as well as an intelligent malware model. Detection: to show how a deep learning architecture is utilized, this article employs extracted Windows API calls from portable.executable (PE) files, and autoencoders (SAEs) are a two-part deep learning approach for intelligent malware identification. Phases: For intelligent malware detection, unsupervised pretraining and supervised backpropagation can be utilized.

Liu et al. [8] This study presented a malware analysis system based on machine learning with three modules: data processing, decision-making, and new virus detection. The data processing module is in charge of grayscale pictures, opcode features, and import functions, which are all used to extract malware features. Then, there is the detection. The module searches for new malware families using the shared nearest neighbor (SNN)-based clustering method.

Damodaran et al. [9] In this study, the authors compare malware.detection methodologies utilizing static, dynamic, and hybrid analysis. Markov models that are used to train on both types of feature sets—static and dynamic, as to compare detection rates among virus families (HMMs). They also include hybrid cases, such as when dynamic analysis is utilized during training, but static approaches are used for detection, and vice versa. The authors look static and dynamic analyses, as well as hybrid methodologies.

Shijo et al. [10] This research proposes a static and dynamic analyses-based approach for assessing and identifying an unknown executable file. Machines are used in the technology. With training data consisting of known viruses and benign applications, learning is possible. After inspecting both the binary code and the dynamic, the feature vector has determined behavior. The proposed technique employs both static and dynamic analyses, resulting in improved efficiency and classification accuracy. The results of the tests show that the static approach is 95.8% correct, the dynamic technique is 97.1% correct, and the combined method is 98.7% correct. In the experimental setup and assessment, static analysis was done on 997 viral files and 490 clean files, with each file assessed using the strings program.

Saeed et al. [11] The current status of malware infection and the effort being done to build anti-malware or malware detection systems are examined in depth in this study. As a result, it provides malware detection system developers with an up-to-date comparison reference. The major purpose of this review article is to look into the present state of malware and detection technologies. In addition, the study examines the approaches and technology utilized to develop anti-malware. The authors in this paper provide a thorough comparison of major malware families along with a summary of malware detection systems.

Garcia et al. [12] The authors of this work employed a way of converting a malware binary into an image and then utilizing Random Forest to categorize malware families. The method's usefulness in identifying malware is demonstrated by the accuracy of 0.9562. The authors used the Malimg Dataset, which comprises 9342 malware samples from 25 distinct malware families, to evaluate their ideas.

Roseline et al. [13] This study presents a complete machine learning-based anti-malware solution that employs a visualization method that depicts malware as 2D graphics. The proposed method employs a layered ensemble approach that resembles the primary features of deep learning while excelling them. The proposed system requires no hyperparameter change or backpropagation and requires reduced model complexity.

Chen et al. [14] The researchers look at the peculiarities of API execution and offer this double retrieval approach based on semantics and structural data. Based on the API data features and attention method, researchers also designed and built a sliding local attention detection system. The authors divided their work into three phases, i.e., in phase first, the authors did analyze the characters or features of the API execution sequence. In the phase second, the researchers proposed a new novel feature extraction model that is totally based on API execution sequence, and in the last phase, based upon the API data characteristics and attention mechanism feature, the authors framed a model on detection framework scenario called sliding local attention detection system (SLAMS) that is totally based on local attention mechanism and sliding Windows method, and the result shows the accuracy of 97.23%.

Anderson et al. [15] The authors of this paper describe the many assaults against machine learning models that have been demonstrated in the field of information security. The researchers used a gradient-boosted decision tree model with a 0.96 area under the receiver operating characteristic score that was trained on 100,000 malicious and benign data (ROC AUC).

Aafer et al. [16] Researchers performed a rigorous study to extract significant characteristics from malware behavior gathered at the API level and then used the feature set to test multiple classifiers. Their results show that using the KNN classifier, researchers can attain an accuracy of 99 percent and a false positive rate of only 2.2 percent. In this research, researchers attempt to address the limitations of permission-based warning methods by developing a robust and lightweight classifier for Android applications that can be used to detect malware. The authors use a generic data mining strategy to develop a classifier for Android applications in this research.

Galen et al. [17] The authors of this paper investigate how machine learning-based models perform when it comes to detecting malware that is in the portable executable (PE) format. Given the extensive use of PE format executables (which include.exe files) on Windows-based machines, when results for these files are available, they may be of substantial practical use. Many machine learning-based malware detection studies train and test malware detection models on a large dataset of malware and good ware samples' models.

Raman [17] The author of this study proposed a list of seven key characteristics for differentiating between malware and clean software. To discover these traits, we used the assumption that attributes from various regions of a PE file would be linked less to one other and more to the file's class, dirty or clean. These characteristics can be utilized as raw data or as input to malware classification algorithms. The categorization data can be used by antivirus software to increase detection rates.

Bekerman et al. [18] The authors did offer a supervised technique for detecting malware by analyzing network data in great detail. At the network layer, the proposed technique removes 972 behavioral characteristics from various protocols. The researchers then utilize a feature selection strategy to highlight the most important or relevant features. According to the authors, who based their findings on an experimental investigation of real network traffic from various circumstances (Table 1).

3   Proposed Framework

In this section, the methodology and the framework that are being implemented are discussed. The technique that is used along with the dataset is given in detail.

Table 1 Comprehensive overview of the review of recent papers

S. no.	References	Author/year	Objectives	Method	Dataset	Accuracy
01	[7]	Galen et al./2020	Detecting the malware in executables that are in portable executable (PE) format	Random Forest	Private	99.55%
02	[8]	Chen et al./2020	Sliding local attention mechanism (SLAM)	CNN	Public dataset (the dataset of Alibaba 3rd Security Algorithm Challenge)	97.23%
03	[1]	Kumar R et al./2019	ScaleMalNet to detect new malware attacks	Deep learning	"Ember" dataset is containing 70,140 benign and 69,860 malicious files	98.1%
04	[9]	Liu et al./2017	Feature extraction using grayscale photos, the n-gram opcode, and key functions	SNN, backpropagation, classification	Kingsoft, ESET NOD32, and Anubis	For known malware, 98%, and for unknown malware, 86%
05	[10]	Anderson et al./2017	A rudimentary reinforcement learning-based architecture for attacking static PE anti-malware engines	Reinforcement learning technique	Private	96%
06	[11]	Stahlbock et al./2016	Deep learning framework for malware detection (short for DL4MD)	Neural network and deep learning	"Comodo Cloud Security Centre"	96%

(continued)

Table 1 (continued)

S. no.	References	Author/year	Objectives	Method	Dataset	Accuracy
07	[12]	Garcia et al./2016	Converting malware binaries into the grayscale images	Random Forest method	Malimg dataset	95%
08	[13]	Bakerman et al./2015	Network traffic packet analysis	Naïve Bayes, j48 and Random Forest	Network traffic capture collected by Verint and Emerging threats	90%
09	[14]	Aafer et al./2013	Malware activity recorded at the API level	KNN classifier	McAfee and Android Malware Genome Project	99%
10	[15]	Raman/2012	Using machine learning approaches. The resulting algorithms classify malware	IBK, J48, PART, J48 Graft, Ridor, and Random Forest	Clean files (benign files) from Windows XP and 7 and dirty files (malware) from VX Heaven's archive	j48 Graft and Random Forest with an accuracy of 98.55% and 98.21%, respectively

3.1 Technique

The focus of this research is to propose a method that could detect the latest malwares with maximum accuracy by using the behavior analysis. The technique that this research uses is the Random Forest classifier technique for malware detection. Random Forest employs the supervised learning approach algorithm for learning. It can be utilized for classification as well as for regression issues. It is based on ensemble learning, which means that it is a strategy for solving complicated problems by merging numerous classifiers as well as enhancing the model's performance. Random Forest is a machine learning approach that unites the number of decision trees formed on various subsets of a dataset and combines the results and provides the average result to improve the model's predicted accuracy. The important features or advantages of using a Random Forest classifier are that it predicts output behavior with high accuracy and runs efficiently, and it keeps track of accuracy even when a large chunk of data is missing. Next, we need to develop a model that can classify the output into two output classes, namely "Malware" and "Benign."

3.2 Dataset

Datasets are crucial in determining the requirement for malware detection depending on performance. The dataset "Malware Detection" used in this research has been collected from "Kaggle" repository, the entire dataset consists of about 216,352 instances, in which 75,503 files are malware files and 140,849 files are benign (normal) files. In our experiment, we extracted 36 characteristics or parameters that may be used to determine if a file is authentic or malicious, and the full dataset was split into training and testing phases in a 75:25 ratio. During the experiment, the greatest accuracy attained was 98.5%.

3.3 Methodology

The methodology is shown in Fig. 3. The following steps are performed to achieve the results:

1. Importing dataset: The dataset mentioned above is imported into the model.
2. Data preprocessing: After collecting the mixed data containing both malware and benign files from the dataset, the entire data are preprocessed. The preprocessing includes removal of null values, and most importantly, the dataset is preprocessed in such a way that it does not overfit the proposed designed model. The outputs from this process play a vital role for analyzing the data.
3. Splitting dataset: The dataset is splitted into the training and testing sets in the ratio of 75% and 25%, respectively.
4. Fitting Random Forest algorithm: We have used Random Forest classifier with default hyperparameter settings except the use of Gini as criterion in place of

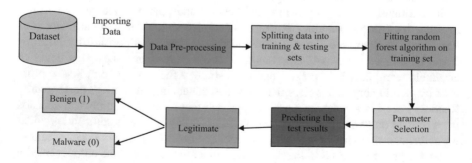

Fig. 3 Proposed framework of the model

Table 2 Parameter description for the model

Parameter label	Value range
1. Id	1–216,352
2. Size of optional header	224 or 240
3. Major linker version	8–48
4. Section alignment	4096 or 8192
5. File alignment	512 or 4096
6. Major operating system version	4–10
7. Minor operating system version	0–2
8. Size of header	512–4096
9. Section min entropy	512–35,840
10. Minor image version	0–20,512
11. Subsystem version	2–9
12. Minor subsystem version	0–20
13. Size of header	512–4096
14. Size of stack commit	0–8196
15. Loader flags	0 or 1
16. Legitimate	**0 or 1**

entropy. After the splitting phase, the Random Forest classifier is applied on the training set.

5. Feature selection: The important parameters or characteristics that are used by the model to label any sample as malware or benign are depicted in Table 2 along with the range of values. The parameters ranging from 1 to 15 are the important input features for the model and the parameter with label "Legitimate" is the output parameter; the model will learn the trends for the classification of the samples based on enlisted parameters.

6. Predicting the results: So, with the final output of these algorithms, we can determine whether the given file taken into consideration is malware, i.e., 1 or benign, i.e., 0.

4 Results and Discussion

According to the literature survey, different researchers have proposed different models for malware detection, but the Random Forest classifier has shown the maximum accuracy and minimum false-positive and false-negative rate as compared to the other proposed models. The work on this project is still going on in Central University of Jammu (Jammu and Kashmir). The results obtained from the project are depicted below as (Table 3; Figs. 4 and 5).

Table 3 Evaluation parameters of the results

	Precision	Recall	F1-score	Support
Benign files (0)	0.99	0.99	0.99	35,129
Malware files (1)	0.98	0.98	0.98	18,959
Macro average	0.98	0.98	0.98	54,088
Weighted average	0.99	0.99	0.99	54,088

Fig. 4 Confusion matrix

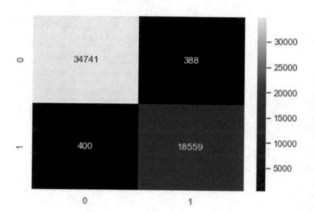

Fig. 5 ROC curve

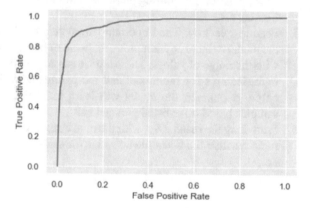

5 Conclusions

Prior identification of malware is a critical task in order to reduce the frequency of harmful malware assaults. In certain cases, the Random Forest algorithm is the best strategy for detecting camouflage malware. This study presents a detailed examination of detecting malwares using machine learning-based algorithms. It is obvious by this investigation that Random Forest-based approaches have a greater accuracy in malware detection strategies. As a result, future research should focus on developing strong and current models that can readily detect encrypted malware. In this comprehensive study, Random Forest is found to be the most promising technique in the detection mechanism of malware; other notable characteristics of the Random Forest classifier are its ability to anticipate output behavior with high accuracy and efficiency, even when a substantial piece of data is missing, and its ability to maintain track of correctness even when a huge chunk of data is missing.

References

1. Vinayakumar R, Alazab M, Soman KP, Poornachandran P, Venkatraman S (2019) Robust intelligent malware detection using deep learning. IEEE Access 7(c):46717–46738. https://doi.org/10.1109/ACCESS.2019.2906934
2. Pirscoveanu RS, Hansen SS, Larsen TMT, Stevanovic M, Pedersen JM (2015) Analysis of malware behaviour classification from ML
3. Tahir R (2018) A study on malware and malware detection techniques. Int J Educ Manag Eng 8(2):20–30. https://doi.org/10.5815/ijeme.2018.02.03
4. Li X, Loh PKK, Tan F (2011) Mechanisms of polymorphic and metamorphic viruses. In: Proceedings of European intelligence and security informatics conference EISIC 2011, pp 149–154. https://doi.org/10.1109/EISIC.2011.77
5. Souri A, Hosseini R (2018) A state-of-the-art survey of malware detection approaches using data mining techniques. Human-centric Comput Inf Sci 8(1). https://doi.org/10.1186/s13673-018-0125-x
6. Sharma N, Arora B (2021) Data mining and machine learning techniques for malware detection. Adv Intell Syst Comput 1187:557–567. https://doi.org/10.1007/978-981-15-6014-9_66
7. Stahlbock R, Weiss GM (2016) Data mining : the 2016 WorldComp international conference proceedings. In: Proceedings of international conference on data mining, pp 61–67, [Online]. https://search.proquest.com/openview/a090ba95404b143e4bbfbb4e0b6bebab/1?pq-origsite=gscholar&cbl=1976357
8. Liu L, Sheng Wang B, Yu B, Xi Zhong Q (2017) Automatic malware classification and new malware detection using machine learning. Front Inf Technol Electron Eng 18(9):1336–1347. https://doi.org/10.1631/FITEE.1601325
9. Damodaran A, Di Troia F, Visaggio CA, Austin TH, Stamp M (2017) A comparison of static, dynamic, and hybrid analysis for malware detection. J Comput Virol Hacking Tech 13(1):1–12. https://doi.org/10.1007/s11416-015-0261-z
10. Shijo PV, Salim A (2015) Integrated static and dynamic analysis for malware detection. Procedia Comput Sci 46(Icict 2014):804–811. https://doi.org/10.1016/j.procs.2015.02.149
11. Saeed IA, Selamat A, Abuagoub AMA (2013) A survey on malware and malware detection systems. Int J Comput Appl 67(16):25–31. https://doi.org/10.5120/11480-7108
12. Garcia FCC, Muga FP (2016) Random forest for malware classification, pp 1–4, [Online]. http://arxiv.org/abs/1609.07770
13. Roseline SA, Geetha S, Kadry S, Nam Y (2020) Intelligent vision-based malware detection and classification using deep random forest paradigm. IEEE Access 8:206303–206324. https://doi.org/10.1109/ACCESS.2020.3036491
14. Chen J et al (2020) SLAM: a malware detection method based on sliding local attention mechanism. In: Security and communication networks, vol 2020. https://doi.org/10.1155/2020/6724513
15. Anderson HS, Filar B, Roth P (2017) Evading machine learning malware detection. In: BlackHat DC, p 6, [Online]. https://github.com/EndgameInc/gym-malware, https://www.blackhat.com/docs/us-17/thursday/us-17-Anderson-Bot-Vs-Bot-Evading-Machine-Learning-Malware-Detection-wp.pdf

16. Aafer Y, Du W, Yin H (2013) DroidAPIMiner: mining API-level features for robust malware detection in android. In: Lecture Notes of the institute for computer sciences, social informatics and telecommunications engineering. LNICST, vol 127 LNICST, pp 86–103. https://doi.org/10.1007/978-3-319-04283-1_6
17. Galen C, Steele R, Performance maintenance over time of random forest-based malware detection models
18. Raman K, Selecting features to classify malware
19. Al-Sammarraie NA, Al-Mayali YMH, Baker El-Ebiary YA (2018) Classification and diagnosis using back propagation Artificial Neural Networks (ANN) algorithm. In: 2018 International conference on smart computing and electronic enterprise ICSCEE 2018, pp 1–5. https://doi.org/10.1109/ICSCEE.2018.8538383
20. Bekerman D, Shapira B, Rokach L, Bar A (2015) Unknown malware detection using network traffic classification. In: 2015 IEEE conference on communications and network security, CNS 2015, pp 134–142. https://doi.org/10.1109/CNS.2015.7346821

Machine Learning-Based Intrusion Detection of Imbalanced Traffic on the Network: A Review

S. V. Sugin and M. Kanchana

Abstract Cyber threats are a very widespread problem in today's world, and because there are an increasing number of obstacles to effectively detecting intrusions, security services, such as data confidentiality, integrity, and availability, are harmed. Day by day, attackers discover new sorts of threats. First and foremost, the type of attack should be carefully assessed with the aid of Intrusion Identification Methods (IIMs) for the prevention of these types of attacks and to provide the exact solution. IIMs that are crucial in network security have three main features: first, they gather data, then they choose a feature, and finally, they choose an engine. As the amount of data produced grows every day, so does the number of data-related threats. As a result of the growing number of data-related attacks, present security applications are insufficient. In this research, the Modified Nearest Neighbor (MNN) and the Technique for Sampling Difficult Sets (TSDS) are two machine learning techniques that have been suggested to detect assault in this research. It is intended to employ an IIM technique based on a machine learning (ML) algorithm by comparing literature and giving expertise in either intrusion detection or machine learning algorithms.

Keywords IIM · Imbalanced traffic network · Technique for Sampling Difficult Sets · ML · DL

1 Introduction

The use of the internet has been steadily expanding recently. It offers a lot of possibilities in applications, considering education, business, healthcare, and a variety of other industries. Everyone has access to the internet. This is where the primary issue arises. The information we obtain from the internet must be protected. This Intrusion

S. V. Sugin (✉) · M. Kanchana
Department of Computing Technologies, School of Computing, SRM Institute of Science and Technology, Kattankulathur, Chennai, India
e-mail: sugin.sv@gmail.com; ss9372@srmist.edu.in

M. Kanchana
e-mail: kanchanm@srmist.edu.in

Y. Singh et al. (eds.), *Proceedings of International Conference on Recent Innovations in Computing*, Lecture Notes in Electrical Engineering 1011,
https://doi.org/10.1007/978-981-99-0601-7_57

Identification (IIM) ensures data security over the network and system. Firewalls and other traditional ways of implementing, for the sake of security, authentication procedures have been implemented [1]. The first level of protection for data was considered, and the second level of protection was studied.

IIM is used to detect illegal or aberrant conduct. An attack is initiated on a network that is exhibiting unusual activity. Attackers take advantage of network flaws such as poor security procedures and practices, as well as program defects such as buffer over-flows, to cause network breaches [2]. It is possible that the attackers are less accessible component services on the lookout to get more control of access or black hat attackers looking to check on regular internet users for critical information. Methods for identifying intrusion can be centered on detecting misuse or based on detecting anomalies. Misuse-based IIM examines traffic on the network and compares it to a set of criteria in a database of predefined malicious activity signatures. Attacks are identified in the identification of anomalies method.

2 Intrusion Identification Methods (IIMs)

Access to the network or a hacker's use of a resource is referred to as an intrusion. An intrusion is used to diminish the integrity, confidentiality, and availability of a resource. In the current world, an intruder tries to obtain entry to illegal metrics and causes harm to the hacker actions that are identified [3] (Fig. 1).

Intrusion Identification Methods (IIMs) detect all of these types of harmful actions on a network and alert the network administrator to secure the information needed to defend against these attacks [2]. The development of IIM has increased security in a network and the protection of service data.

As a result, an Intrusion Identification Method (IIM) is a network and computer security solution that keeps track of network traffic [4]. Firewall security is provided by an IIM. A firewall protects an enterprise by detecting dangerous internet activity, whereas an IIM detects attempts to breach firewall protection or gain access, and

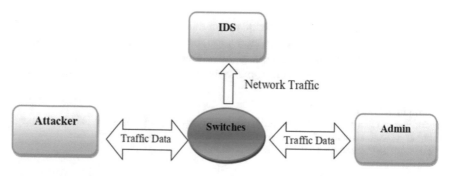

Fig. 1 Intrusion identification methods (IIMs)

it quickly notifies the administrator that something needs to be done. As a result, IIMs are security systems that detect various attacks on the network and ensure the security of our systems.

3 Network Intrusion Identification Model Framework

Faced with this unbalanced traffic on the internet, we suggested the Technique for Sampling Difficult Sets (TSDS) algorithm, which compresses the majority class samples, while in tough situations, enhancing the quantity of minority samples is a must to decrease the training set's imbalance and allow the Intrusion Identification Method to improve category performance [5]. For classification models, as classifiers, employ RF, SVM, k-NN, and Alex Net.

The intrusion identification model presented in Fig. 2 was proposed. Data preprocessing such as processing of duplicates, incomplete data, and missing data is done first in our intrusion identification structure [6]. The test and training sets were then partitioned, with the sets of practice being treated for metrics balance with the help of our suggested TSDS algorithm. We utilize StandardScaler to normalize and digitize the sample labels and analyze the data before modeling to speed up the convergence [7]. Likewise, the practice set is processed and utilized for the training data to be constructed, which is then evaluated using the test set.

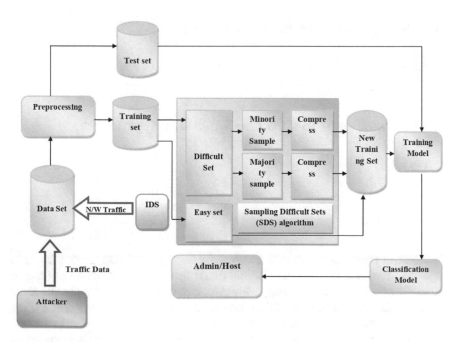

Fig. 2 Network intrusion identification system model framework

Several traffic data types have comparable patterns in imbalanced network traffic, and minority attacks, in particular, might be hidden within a significant tough for the classifier to understand the distinctions between them during the training phase because there is a lot of typical traffic [8]. The redundant noise data is the majority class in the unbalanced training set's comparable samples. Because the majority class's number is substantially greater than the class of the minority predictor, who is not able to understand the minority class's spread, the majority level is compact. Discrete traits in the minority class remain constant, but constant attributes change [9]. As a result, the continuous qualities of the minority class are magnified to provide data that adheres to the genuine distribution. As a result, we propose the TSDS algorithm as a means of redressing the imbalance.

First, using the Modified Nearest Neighbor (MNN) technique, the near-neighbor and far-neighbor sets were created from an unbalanced set of data [10]. Because the samples from the collection of near-neighbors are so similar, the classifier has a hard time recognizing the distinctions between the groups. In the identification process, we refer to them as "exhausting instances and extracts." Then, in the tough set, they move in and out of the samples from the minority. Likewise, the augmentation samples from the easy set and the toughest set's minorities are merged to make a new set of exercises. In the MNN method, the K-neighbors are used as the availability aspect for the complete algorithm [11]. The number of problematic samples grows as the scaling factor K increases, as does the compression.

3.1 Comparison of Accuracy on Datasets

See Table 1 and Fig. 3.

3.2 Comparison of Various ML-Based IDS Approaches

See Table 2.

4 Discussions

The research trends in benchmark datasets for evaluating NIDS models are also graphically illustrated. The KDD Cup '99 dataset is shown to be the most popular, followed by the NSL-KDD dataset. However, the KDD '99 dataset has the issue of being quite old and not resembling present traffic data flow. Other datasets are accessible as well, but the research trend in these datasets is quite low due to the new dataset's lack of appeal in research. It is suggested that researchers can be encouraged

Table 1 Comparison of accuracy on datasets

S. No.	Author	Attack	Dataset	Accuracy (%)
1	L. Liu, IEEE Access [1]	Denial of Service (DoS)	NSL-KDD	78.24
2	J. Alikhanov, IEEE Access [3]	Distributed Denial of Service (DDoS)	NSL-KDD,AWSCIC-IDS	84.61
3	T. Kim, IEEE Access[2]	Distributed Denial of Service (DDoS)	CSE-CIC-IDS2018	88.97
4	Z. K. Maseer, IEEE Access [12]	Denial of service (DOS)	CIC-IDS2017	85.88
5	M. Wang, IEEE Access[8]	Neptune	NSL-KDD	89
6	A. Kavousi, IEEE Transactions[10]	Havex Malware	LUBE-SOS	82.83
7	Z. Chkirbene, IEEE Systems [13]	Denial of service (DOS)	NSL-KDD	80
8	M. A. Siddiqi, IEEE Access[6]	Botnet	ISCX-IDS2012	96.51
9	G. De Carvalho Bertoli, IEEE Access [14]	Malware	AB-TRAP	54
10	Y. Uhm, IEEE Access [9]	Denial of service (DOS)	CIC-IDS2017	97.78
11	D. Han, IEEE [4]	Botnet, Distributed Denial of Service (DDoS)	Kitsune	81.65, 79.55
12	L. Jeune, IEEE Access[7]	Botnet, Distributed Denial of Service (DDoS)	DARPA1998	86.34, 80
13	S. Wang, IEEE Access[15]	Distributed Denial of Service (DDoS)	UNSW-NB15	90
14	M. Injadat, IEEE Transactions [16]	Distributed Denial of Service (DDoS)	UNSW-NB2015	74
15	W. Seo, IEEE Access[17]	Distributed Denial of Service (DDoS)	UNSW-NB15	95.8
16	D. Gumusbas, IEEE Journal [11]	Denial of service (DOS)	AWID2018	78.4

(continued)

Table 1 (continued)

S. No.	Author	Attack	Dataset	Accuracy (%)
17	C. Liu, IEEE Access[18]	Distributed Denial of Service (DDoS)	NSL-KDD, CIS-IDS2017	99.87
18	Y. Li, IEEE Access[19]	Denial of service (DOS)	NSL-KDD	94.25
19	Y. Tang, IEEE Access [20]	Denial of service (DOS)	UNSW-NB15	88.53
20	G. Siewruk, IEEE Access [21]	Denial of service (DOS)	NSL-KDD	98
21	W. Xu, IEEE Access[22]	Denial of service (DOS)	NSL-KDD	90.61
22	A. G. Roselin, IEEE Access [23]	Distributed Denial of Service (DDoS)	NSL-KDD	81.82
23	A. R. Gad, IEEE Access[24]	Distributed Denial of Service (DDoS)	NSL-KDD, KDD-CUP99	80.65
24	Z. Li, IEEE Journal [5]	Denial of service (DOS)	NSL-KDD, CIC-IDS2017	93.12
25	L. Le Jeune, IEEE Access [7]	Distributed Denial of Service (DDoS)	NSL-KDD	94.7
26	Y. D. Lin, IEEE Access [25]	Denial of service (DOS)	CSE-CIC-IDS2018	97
27	M. D. Rokade, (ESCI) [26]	Denial of service (DOS)	NSL-KDD-CUP-1999	88.50
28	P. F. Marteau, IEEE Transactions [27]	Denial of service (DOS)	CIDDS	80
29	W. Wan, Z. Peng, (ICCEA) [28]	Denial of service (DOS)	NS-KDD	80.49
30	M. Lopez-Martin, IEEE Access[29]	Distributed Denial of Service (DDoS)	UNSW-NB15	91

to use modern datasets with more detailed attributes that are more relevant to today's environment.

5   Conclusion

In this review, we studied the dataset assault through machine learning techniques. It reviewed ML models from different assaults available in the dataset. As a result of

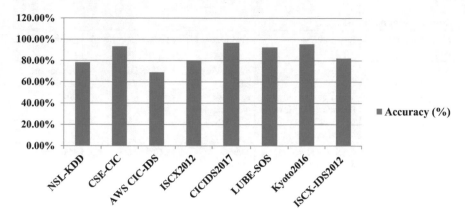

Fig. 3 Comparison of classifier accuracy on datasets

Table 2 Comparison of the related works

S. No.	Authors	Key findings	Techniques used	Dataset	Limitations
1	L. Liu, IEEE Access [1]	Demonstrating advantages over existing methods and the high potential for usage in emerging NIDS	To present a novel Difficult Set Sampling Technique (DSSTE) method	NSL-KDD, CSE-CIC	Intrusion detection systems have a hard time predicting the distribution of malicious attempts
2	J. Alikhanov, IEEE Access [3]	On the NIDS detection rate, different extraction strategies are applied	Sketch-Guided Sampling (SGS) techniques are used	NSL-KDD, AWS CIC-IDS	The impact of sampling on NIDS based on anomalies should be less evaluated
3	T. Kim, IEEE Access[2]	Through pattern matching with incoming packets, the NIDS attacks and detects intrusions very efficiently	The classification detection rate and classification speed may both be increased by using ML-NIDS	ISCX2012, CSE-CIC-IDS2018	The ML-NIDS defects may be exploited to dramatically enhance prediction
4	Z. K. Maseer, IEEE Access [12]	Anomaly-based IDS (AIDS) can identify malware and violent attacks by analyzing the sent data in depth	Implementing anomaly-based IDS (AIDS) dataset	CIC-IDS2017	Increase the vulnerability of AIDS

(continued)

Table 2 (continued)

S. No.	Authors	Key findings	Techniques used	Dataset	Limitations
5	M. Wang, IEEE Access [8]	An Improved Conditional Variational Autoencoder (ICVAE) with a enhance detection rates	Framework uses SHapley Additive exPlanations (SHAP)	NSL-KDD	Framework not in real time
6	A. Kavousi, IEEE Transactions [10]	Anomaly Detection Model based on LUBE and SOS	The use of prediction intervals (PIs) is used to develop an intelligent anomaly detection approach	LUBE-SOS	Malicious attacks with different severities, data can attack easily
7	Z. Chkirbene, IEEE Systems [13]	Unsupervised and supervised learning approaches are used to create triangle area-based closest neighbors (TANN)	The Euclidean distance map (EDM) is a novel method for detecting anomalies using sequential algorithms	UNSW-NB, NSL-KDD	In compared to modern system procedures, the EDM technique has a lower warning rate
8	M. A. Siddiqi, IEEE Access [6]	The detection rate of intrusion detection is high when guided ML methods are used	IDS approaches based on a random forest were utilized	CIC-IDS2017, ISCX-IDS2012	The reinforcing procedure provided less efficiency
9	G. De Carvalho Bertoli, IEEE Access [14]	The AB-TRAP is used to identify attackers in both local (LAN) and global (internet) aspects	AB-TRAP organizes the process of designing and implementing NIDS systems	AB-TRAP	Applying machine learning algorithms to give fresh techniques is a key point in favor of not recycling old datasets
10	Y. Uhm, IEEE Access [9]	To reduce the minority class problem, a service-aware partitioning method was developed	Random forest (RF) and decision tree (DT), as well as deep neural networks (DNNs), are used to build NIDS	CIC-IDS2017, Kyoto2016	Improve the real-time intrusion prevention algorithm that has been presented

(continued)

Table 2 (continued)

S. No.	Authors	Key findings	Techniques used	Dataset	Limitations
11	D. Han, IEEE [4]	Network Intrusion Identification Methods based on anomaly also use machine learning (ML) techniques	Particle Swarm Optimization (PSO) based on algorithm for traffic mutation	Kitsune	The scalability of ML-focused NIDS is being improved
12	L. Jeune, IEEE Access [7]	Intrusion Detection Expert System (IDES) and HIDS	The botnet was utilized in a large-scale (DDoS) effort on the (DNS)	DARPA1998, NSL-KDD	Real-world scenario is not synthesized in the datasets
13	S. Wang, IEEE Access [15]	To protect networks against malicious access	Used firewalls, deep packet inspection systems and intrusion detection systems	NSL-KDD, UNSW-NB15	The performance validated by UNSW-NB15 cannot be clearly categorized
14	M. Injadat, IEEE Transactions [16]	SMOTE is done to increase the training model's performance and decrease network traffic data class imbalance	In order to apply Z-score normalization and SMOTE, data preprocessing is required	CIC-IDS2017, UNSW-NB2015	When compared to the CBFS approach, the IGBFS method had a higher detection accuracy
15	W. Seo IEEE Access [17]	In signature-based detection and anomaly detection, cyberattacks have made significant progress	Convolutional neural networks' (CNNs) algorithm is used	UNSW-NB15	To develop real-time IPSs and identify current network system vulnerabilities
16	D. Gumusbas, IEEE Journal [11]	Artificial Neural Networks (ANNs) and Deep Belief Networks	Packet CAPture (PCAP) and the NetFlow protocol	AWID2018, CIC-IDS2017	To do classification, another ML model is required
17	C. Liu, IEEE Access [18]	Adaptive Synthetic Sampling (ADASYN)	Convolutional Neural Network (CNN), Long Short-Term Memory (LSTM)	NSL-KDD, CIS-IDS2017	It takes a long time and has a low efficiency
18	Y. Li, IEEE Access [19]	Domain Generation Algorithm (DGA)	Hidden Markov model (HMM)	NSL-KDD	DNN model classification should be improved

(continued)

Table 2 (continued)

S. No.	Authors	Key findings	Techniques used	Dataset	Limitations
19	Y. Tang, IEEE Access [20]	Randomly initializing weights and deviations increases the speed of an extreme learning machine (ELM)	Improved particle swarm optimized online regularized extreme learning machine (IPSO-IRELM)	NSL-KDD, UNSW-NB15	To increase IRELM's capacity to classify data
20	G. Siewruk, IEEE Access [21]	Context-aware software vulnerability classification system	Continuous Integration and Continuous Deployment (CICD)	NSL-KDD	Improve the vulnerability performance
21	W. Xu, IEEE Access [22]	The network is recreated using Mean Absolute Error (MAE)	Autoencoder (AE)-based deep learning approaches	NSL-KDD	Improve the performance of the dataset
22	A. G. Roselin, IEEE Access [23]	To identify malicious network traffic, BIRCH clustering technique is used	Optimized Deep Clustering (ODC)	NSL-KDD	ODC technique has a lower detection rate of anomalies
23	A. R. Gad, IEEE Access [24]	Synthetic minority oversampling technique (SMOTE)	The Chi-square (Chi^2) approach was used to pick features. ODC technique has a lower detection rate of anomalies	NSL-KDD, KDD-CUP99	Less complexity
24	Z. Li, IEEE Journal [5]	Gated Recurrent Unit and Long Short-Term Memory	Broad Learning System	NSL-KDD, CIC-IDS2017	Less accuracy BLS algorithms
25	L. Le Jeune, IEEE Access [7]	PCCN-based approaches are used	Intrusion Detection Expert System	NSL-KDD	IDES performance should be improved
26	Y. D. Lin, IEEE Access [25]	Variational autoencoder and multilayer perception model are used	Range-based sequential search algorithm	CSE-CIC IDS2018	Improve the categorization of segmentation
27	M. D. Rokade, (ESCI) [26]	SVM-IDS approach based on deep learning	Artificial Neural Network algorithm	KKDDCUP99, NLS-KDD	Classification and detection of high-class objects should be improved

(continued)

Table 2 (continued)

S. No.	Authors	Key findings	Techniques used	Dataset	Limitations
28	P.F.Marteau IEEE Transactions [27]	One-class SVM classifier (1C-SVM) is used	Semi-supervised DiFF-RF algorithm	CIDDS	Inaccurate datasets
29	W. Wan, Z. Peng,(ICCEA) [28]	All single DNN classifiers are integrated using the AdaBoost technique	Generative Adversarial Networks (GAN)	KDD99, NS-KDD	Increase the sample deduction accuracy rate
30	M. Lopez-Martin, IEEE Access [29]	Radial Basis Function (RBF) is implemented	Radial Basis Function Neural Networks (RBFNNs)	NSL-KDD, UNSW-NB15	Improve the suggested dataset's performance metrics

the growing number of data-related assaults, present security applications are insufficient. In this research, the Modified Nearest Neighbor (MNN) and the Technique for Sampling Difficult Sets (TSDS) are two machine learning techniques that have been suggested to detect assault in this research. More recent and updated datasets must be utilized in future research in order to assess deployed algorithms in order to deal with more current harmful intrusions and threats.

References

1. Liu L, Wang P, Lin J, Liu L (2021) Intrusion detection of imbalanced network traffic based on machine learning and deep learning. IEEE Access 9:7550–7563. https://doi.org/10.1109/ACC ESS.2020.3048198
2. Kim T, Pak W (2022) Robust network intrusion detection system based on machine-learning with early classification. IEEE Access 10:10754–10767. https://doi.org/10.1109/ACCESS. 2022.3145002
3. Alikhanov J, Jang R, Abuhamad M, Mohaisen D, Nyang D, Noh Y (2022) Investigating the effect of traffic sampling on machine learning-based network intrusion detection approaches. IEEE Access 10:5801–5823. https://doi.org/10.1109/ACCESS.2021.3137318
4. Han D et al (2021) Evaluating and improving adversarial robustness of machine learning-based network intrusion detectors. IEEE J Sel Areas Commun 39(8):2632–2647. https://doi.org/10. 1109/JSAC.2021.3087242
5. Li Z, Rios ALG, Trajkovic L (2021) Machine learning for detecting anomalies and intrusions in communication networks. IEEE J Sel Areas Commun 39(7):2254–2264. https://doi.org/10. 1109/JSAC.2021.3078497
6. Siddiqi MA, Pak W (2021) An agile approach to identify single and hybrid normalization for enhancing machine learning-based network intrusion detection. IEEE Access 9:137494–137513. https://doi.org/10.1109/ACCESS.2021.3118361
7. Le Jeune L, Goedemé T, Mentens N (2021) Machine learning for misuse-based network intrusion detection: overview, unified evaluation and feature choice comparison framework. IEEE Access 9:63995–64015. https://doi.org/10.1109/ACCESS.2021.3075066

8. Wang M, Zheng K, Yang Y, Wang X (2020) An explainable machine learning framework for intrusion detection systems. IEEE Access 8:73127–73141. https://doi.org/10.1109/ACCESS.2020.2988359
9. Uhm Y, Pak W (2021) Service-aware two-level partitioning for machine learning-based network intrusion detection with high performance and high scalability. IEEE Access 9:6608–6622. https://doi.org/10.1109/ACCESS.2020.3048900
10. Kavousi-Fard A, Su W, Jin T (2021) A machine-learning-based cyber attack detection model for wireless sensor networks in microgrids. IEEE Trans Industr Inf 17(1):650–658. https://doi.org/10.1109/TII.2020.2964704
11. Gumusbas D, Yıldırım T, Genovese A, Scotti F (2021) A comprehensive survey of databases and deep learning methods for cybersecurity and intrusion detection systems. IEEE Syst J 15(2):1717–1731. https://doi.org/10.1109/JSYST.2020.2992966
12. Maseer ZK, Yusof R, Bahaman N, Mostafa SA, Foozy CFM (2021) Benchmarking of machine learning for anomaly based intrusion detection systems in the CICIDS2017 dataset. IEEE Access 9:22351–22370. https://doi.org/10.1109/ACCESS.2021.3056614
13. Chkirbene Z et al (2021) A weighted machine learning-based attacks classification to alleviating class imbalance. IEEE Syst J 15(4):4780–4791. https://doi.org/10.1109/JSYS.2020.3033423
14. De Carvalho Bertoli G et al (2021) An end-to-end framework for machine learning-based network intrusion detection system. IEEE Access 9:106790–106805.https://doi.org/10.1109/ACCESS.2021.3101188
15. Wang S, Balarezo JF, Kandeepan S, Al-Hourani A, Chavez KG, Rubinstein B (2021) Machine learning in network anomaly detection: a survey. IEEE Access 9:152379–152396. https://doi.org/10.1109/ACCESS.2021.3126834
16. Injadat M, Moubayed A, Nassif AB, Shami A (2021) Multi-stage optimized machine learning framework for network intrusion detection. IEEE Trans Netw Serv Manage 18(2):1803–1816. https://doi.org/10.1109/TNSM.2020.3014929
17. Seo W, Pak W (2021) Real-time network intrusion prevention system based on hybrid machine learning. IEEE Access 9:46386–46397. https://doi.org/10.1109/ACCESS.2021.3066620
18. Liu C, Gu Z, Wang J (2021) A hybrid intrusion detection system based on scalable K-means+ random forest and deep learning. IEEE Access 9:75729–75740. https://doi.org/10.1109/ACCESS.2021.3082147
19. Li Y, Xiong K, Chin T, Hu C (2019) A machine learning framework for domain generation algorithm-based malware detection. IEEE Access 7:32765–32782. https://doi.org/10.1109/ACCESS.2019.2891588
20. Tang Y, Li C (2021) An online network intrusion detection model based on improved regularized extreme learning machine. IEEE Access 9:94826–94844. 10.1109/ ACCESS. 2021.3093313
21. Siewruk G, Mazurczyk W (2021) Context-aware software vulnerability classification using machine learning. IEEE Access 9:88852–88867. https://doi.org/10.1109/ACCESS.2021.3075385
22. Xu W, Jang-Jaccard J, Singh A, Wei Y, Sabrina F (2021) Improving performance of auto encoder-based network anomaly detection on NSL-KDD dataset. IEEE Access 9:140136–140146. https://doi.org/10.1109/ACCESS.2021.3116612
23. Roselin AG, Nanda P, Nepal S, He X (2021) Intelligent anomaly detection for large network traffic with optimized deep clustering (ODC) algorithm. IEEE Access 9:47243–47251. https://doi.org/10.1109/ACCESS.2021.3068172
24. Gad AR, Nashat AA, Barkat TM (2021) Intrusion detection system using machine learning for vehicular ad hoc networks based on ToN-IoT Dataset. IEEE Access 9:142206–142217. https://doi.org/10.1109/ACCESS.2021.3120626
25. Lin YD, Liu Z-Q, Hwang R-H, Nguyen V-L, Lin P-C, Lai Y-C (2022) Machine LEARNING with variational autoencoder for imbalanced datasets in intrusion detection. IEEE Access 10:15247–15260. https://doi.org/10.1109/ACCESS.2022.3149295
26. Rokade MD, Sharma YK (2021) MLIDS: a machine learning approach for intrusion detection for real time network dataset. In: 2021 International conference on emerging smart computing and informatics (ESCI), pp 533–536. 10.1109/ ESCI50559.2021. 9396829

27. Marteau PF (2021) Random partitioning forest for point-wise and collective anomaly detection-application to network intrusion detection. IEEE Trans Inf Forensics Secur 16:2157–2172. https://doi.org/10.1109/TIFS.2021.3050605

28. Wan W, Peng Z, Wei J, Zhao J, Long C, Du G (2021) An effective integrated intrusion detection model based on deep neural network. In: 2021 International conference on computer engineering and application (ICCEA), pp 146–152. 10.1109/ ICCEA53728. 2021.00037

29. Lopez-Martin M, Sanchez-Esguevillas A, Arribas JI, Carro B (2021) Network intrusion detection based on extended RBF neural network with offline reinforcement learning. IEEE Access 9:153153–153170. https://doi.org/10.1109/ACCESS.2021.3127689

A Novel Approach to Acquire Data for Improving Machine Learning Models Through Smart Contracts

Anuj Raghani, Aditya Ajgaonkar, Dyuwan Shukla, Bhavya Sheth,
and Dhiren Patel

Abstract Despite the Big Data Revolution, critical aspects to improve machine learning models have been overlooked concerning the benchmarked datasets and their nature that are available for the model. In this work, we propose a blockchain-based decentralised, trustless platform using "Smart Contracts" that is tailored exclusively for data collection from proficient developers and machine learning model improvement. Elastic Weighted Consolidation is used to update this model to take into account the characteristics of the incoming dataset(s) in order to avoid catastrophic forgetting, which occurs when a model only learns from fresh data and ignores its existing knowledge. A rewarding mechanism has been discussed, which ensures that low-quality data is not rewarded and good information is compensated fairly based on the improvement made to the model. It fosters a favourable environment for competition. This platform is conceived as a marketplace that provides monetary incentives for developers to partake in improving and contributing to model development.

Keywords Machine learning · Datasets · Blockchain · Smart contracts · Crowdsourcing

1 Introduction

According to IBM, machine learning is a subfield of artificial intelligence (AI) and computer science which focuses on using data and algorithms to simulate how humans learn while constantly increasing its accuracy [1]. Machine learning (ML) is a subset of artificial intelligence (AI) that gives computers the capacity to learn using their "own" intuition rather than being explicitly programmed by humans.

A. Raghani (✉) · A. Ajgaonkar · D. Shukla · B. Sheth
VJTI, Mumbai, India
e-mail: adraghani_b18@it.vjti.ac.in

D. Patel
SVNIT, Surat, India

755

Y. Singh et al. (eds.), *Proceedings of International Conference on Recent Innovations in Computing*, Lecture Notes in Electrical Engineering 1011,
https://doi.org/10.1007/978-981-99-0601-7_58

The development of computer algorithms that can adapt to new data is the focus of machine learning.

At the very heart of ML lies Data. Datasets are collections of instances that all share particular attributes, where each sample is a single row of data. Machine learning algorithms are given training datasets to help them "learn" and then validation datasets to ensure that the model interprets the data as it should. Datasets govern the development, utility, and success of machine learning models. They are the foundation for developing cutting-edge solutions but also serve as inhibitors of the potential of machine learning models [2]. Machine learning has experienced exponential growth, which can be attributed to three factors: (a) algorithms with high efficiency, (b) improved computational power, and (c) the availability of vast amounts of labelled data [3].

The ability of models to represent data and the processing power of GPUs have significantly improved over the past few years, yet data is still being overlooked in the quest to improve machine learning. Simple examples are the ImageNet benchmark for visual object recognition and the GLUE benchmark for English textual understanding [4, 5]. It is evident that the vastly increased availability of data has been a key driver of AI/ML, coupled with faster processors, less expensive storage, and theoretical advancement. We have produced and catalogued data exponentially more quickly in recent years than ever before.

To support the notion of the importance of data for improving our results vastly, we need to look no further than well-known and established benchmark datasets. The direction of the aims, ideals, and research agendas of machine learning advancement has been greatly influenced by benchmark datasets. According to reports, machine learning algorithms perform astonishingly well when put to the test against these benchmark datasets. Quality datasets are, therefore, crucial to the construction and assessment of models in the field of machine learning. The desire to acquire newer and more varied datasets is motivated by the realisation that a single group of developers or institutions cannot, on their own, create or amass a dataset that best serves their intended purpose, i.e. has a dataset that is highly representative of the data the systems would encounter in real-world use.

However, recent developments reinforce the understanding that datasets limit the capability of machine learning and artificial intelligence along with the emergence of concerns regarding biases and societal prejudices creeping into the realm of computing and AI as a result of the shortcomings in the underlying data leading to disturbing and unfavourable trends in the domain. Particularly, current data processes have a propensity to abstract away human labour, arbitrary assessments and biases, and variable circumstances involved in dataset development [6]. To elaborate on one facet of data concerns, Paullada et al. [7] imply that the relationship between inputs and target labels seen in datasets is not necessarily significant and that the way objectives and data are gathered might cause models to rely on unreliable heuristics. The issues this raises go beyond false conclusions drawn from benchmarking studies. When machine learning models can use fictitious cues to predict outcomes well enough to surpass a baseline on test data, the resulting systems may give the impression that fictitious tasks that do not correspond to real-world capabilities are legitimate. Formally speaking, certain tasks have the ability to be specified but cannot have an adequate extensional realisation frequently because the task's underlying theory is flawed.

The type of unique, skilled, and methodical annotation used in dataset collection was found to be "slow and expensive to acquire," so developers turned towards the unrestricted collection of increasingly large amounts of data from the web, alongside increased reliance on crowd workers and contributing developers. Recently, the machine learning field has turned to approaches with much more robust data requirements.

In order to address this issue, which is at the heart of data science and machine learning, our solution—which is covered in more detail in Sect. 3—sets out to develop better machine learning models and solutions. To do this, it will tap into the world's machine learning talent and enhance machine learning solutions. It will also create a decentralised platform on the blockchain. In order to increase the performance of current machine learning models, we thus make use of breakthroughs in the field of blockchain technology and the theoretical underpinnings of federated machine learning.

The remainder of the paper is structured as follows: the backdrop of present systems is given in Sect. 2, along with background information on the fundamental ideas upon which our solution is built. The summary of the proposal and the justification for the system's architecture are included in Sect. 3. Key components and the architecture are highlighted in Sect. 4. The proposal is restated and submitted for confirmation in Sect. 5. Section 6 contains the paper's conclusion as well as a list of references.

2 Background

At the outset, we acknowledge prior research that shares the objectives of our proposal and provides the groundwork for the combination of blockchain technology, smart contracts, and machine learning. The most insightful work is presented by Harris and Waggoner [8], who lay out their idea for a system that would allow users to collaborate to build a dataset and host a continuously updated model using smart contracts. We acknowledge their proposal and have worked to refine and build on its foundations to create a functioning marketplace. We also give credit to the DanKu protocol proposal, which was the first to advocate using blockchain technology to establish contracts that provide compensation for a machine learning model that has been trained for a specific dataset [9]. This proposal is a derivative of their initial proposal, from soliciting machine learning models as solutions for certain datasets to soliciting datasets to refine existing machine learning models. Similarly, Marathe et al., in their proposal DInEMMo [10], suggest a complete marketplace for both MI, model and data sharing, which provides theoretical insight for this proposal.

For conceptual similarity, we also note Waggoner et al. [11], who put forward a technique for buying data from a set of individuals. Participants are encouraged to participate if they feel that their data points are representative or that the information they provide will help the mechanism make more accurate predictions in the future using a test set.

The goal to cultivate, encourage, and increase the engagement of experienced developers in machine learning creation and refinement to advance the field has served as the main inspiration for this idea. At the same time, the Open-Source movement has greatly enabled and inculcated the values of learning, sharing, and cooperation. However, it is still limited because this strategy depends on the other contributors' willingness to participate. Hence, we propose a blockchain-powered marketplace designed to facilitate the improvement of machine learning models by adequately incentivising collaboration. Before elaborating on this system, we briefly examine the core technologies and concepts which serve as its foundation.

2.1 Blockchain

Blockchain is a decentralised and distributed ledger system that promotes trust and dependability since data and transactions are only added to the chain once participants have come to an agreement. We use its services in our system to make sure that such agreements will happen smoothly, securely, and reliably [12].

2.2 Smart Contracts

In essence, smart contracts are computer programmes that are kept on a blockchain and linked to a specific blockchain address that contains the contract's source code [13]. A smart contract's code cannot be modified when it is published, and anybody can interact with it [14]. Transactions are traceable and irreversible, and the code governs their execution. Transactions are traceable and irreversible, and the code regulates their execution. Without the need for a centralised authority, a legal system, or an external enforcement mechanism, smart contracts enable trusted transactions and agreements to be made between dispersed, anonymous parties. Simple "if/when...then" phrases that are typed into code and placed on a blockchain are how smart contracts operate. When predefined circumstances have been verified to have been met, a network of computers will carry out the actions.

2.3 Dataset Sourcing

The main concept is that supplemental data contracts are provided, where the best model currently available is improved with additional data. The best model currently available can be improved with new data points through a data contribution contract that the organisers can design. Participants are compensated if their data contribution improves the performance of the best model, i.e. fills in the gaps in the training data already available. For instance, newly developed data can include fresh texts, movie titles, images, etc. In a study, Chen et al. [15] make the claim that if we have a test dataset, we can pay data providers based on how well the model trained on their provided data performs on the test dataset. When a data buyer lacks access to a test dataset, they also investigate the creation of incentive systems for obtaining high-quality data from multiple data sources.

2.4 Federated Machine Learning

A machine learning technique called federated learning [16] enables machine learning models to learn from several datasets located in various locations (such as local data centres or a central server) without exchanging training data. This lowers the likelihood of personal data breaches by allowing personal data to stay on local sites. Without transferring data, federated learning is used to train machine learning algorithms utilising a variety of local datasets. Without storing training data in a centralised location, this enables businesses to develop a shared global model. Our method, which receives data from contributors and then trains the current model on the newly obtained data, is inspired by federated machine learning. On the decentralised storage system, the updated model weights are kept.

3 Proposed Solution

As acknowledged earlier, further strides in the development of machine learning solutions are severely limited by the availability of quality data. This provides the motive of improving the quality of data used for training machine learning models. As a result, the development process would be substantially accelerated, and more reliable models would be produced using the newer, better data. The easiest method to try to do this is to collect data manually. However, collecting data manually can be very expensive and time-consuming, not to mention fraught with considerations, as explained in the introduction.

A web platform serves as our solution. The problem is solved by an "Organiser" or contract creator who wants to increase the accuracy of an already existing machine learning model. They build a blockchain-based smart contract [17] with the following specific details:

1. ML model description,
2. validation dataset description,
3. current machine learning model file,
4. maximum total reward that they offer to contributors.

The submitted machine learning model is tested on the validation dataset, and base accuracy is evaluated on the backend server. The validation dataset and submitted machine learning model are uploaded onto the IPFS [18]. The hashes from the validation dataset and machine learning model uploading, as well as the base accuracy assessed earlier, are then added to the smart contract. The Ethereum blockchain is where the smart contract is stored and deployed. Initiation of the contract results in creating a new competition on the application.

The "Developer" or "Data Contributor" views the contract and reads the model and data definitions. The developer may upload the data if doing so will enhance the model's accuracy. The existing model is trained on the new data provided; after training is completed, the new model is tested on the validation data uploaded earlier, and a new accuracy is calculated. If the new accuracy is more than the base accuracy, the submission is considered successful. Multiple such developers can participate in the contract, and if the accuracy obtained by evaluating new models trained on their datasets is more than the base accuracy, then those developers will be eligible for some portion of the reward stated by the organiser.

When the organiser downloads the final model, which is an aggregation of all the models with accuracy more than the base model on the validation data, then the rewards will be distributed to all the eligible developers who contributed data to the competition. A continual learning approach is proposed for averaging the weights when the final model is created. The architecture diagram of the solution is illustrated in Fig 1.

4   Functional Overview

We propose a solution where the participants collaboratively improve a model and use smart contracts to store the submissions and incentivise contributors that improve the model and apply federated learning with a continual learning approach. The reward mechanism is designed to receive good data only. The basic outline of the flow of the application is illustrated in Fig 2. Our framework has three phases:

An **initiation phase** in which an organiser stakes a token to be awarded to contributors and shares the validation data and base model,

A **contribution phase** in which participants submit training data samples and train their data on the base model,

A **remuneration phase** in which the provider pulls the model and the best models are averaged, and the reward is distributed based on the improvement made by individual submissions. The specifics needed to explain the proposal's intricacies are covered in the parts that follow.

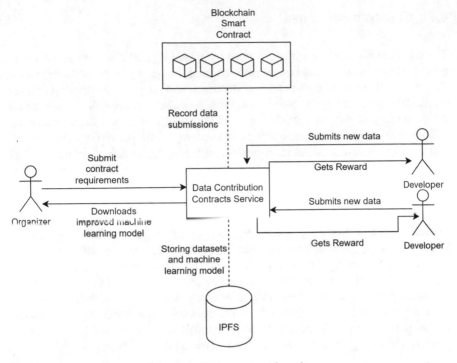

Fig. 1 Architecture diagram of data contribution contracts' service

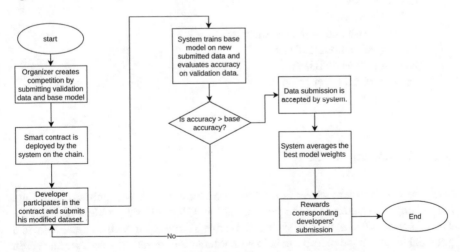

Fig. 2 Flowchart for data-sourcing contracts

4.1 Commitment Phase

The organiser initialises the smart contract by uploading the validation data and base model. The validation data is used for assessing the quality of the data submitted. This is done by first calculating the base accuracy, and once the base model is trained on the data submitted by the contributor, the new accuracy score will act as feedback on the quality of the data. Additionally, if the validation data is kept private, the dataset given may be biased, which may encourage data providers to report data that is biased in favour of the validation set, reducing the usefulness of the acquired data for subsequent learning or analysis activities. The organiser then specifies the reward he is ready to stake. Mechanisms exist for regulating the reward price set by the organiser [19]. This is to make sure that the contributors are adequately compensated. In case the contract fails, the reward is refunded back. All the computations in these phases are done off-chain to reduce gas costs on Ethereum and the incapability to handle floating-point calculations. Once the organiser is satisfied with the contract, the contract is submitted to be deployed. The Data Contribution Contracts Service (DCCS) then deploys the contract in the organiser's account, and the model file and validation data are uploaded on IPFS to ensure privacy and availability. The Hash generated from it is then stored in the contract as a struct instance, and the contract is now publicly available on the portal for contributors. The gas cost of the whole process is borne by the organiser. The code below is the Solidity snippet of a contract entity.

```
struct Contract{
    uint 256 reward;
    address payable organiserAddress;
    string validationDataHash;
    string baseModelHash;
    string model description
}
```

4.2 Participation Phase

Once the commitment phase is over, the contract is available publicly now, and contributors can understand the problem at hand through the problem description. The organiser offers metadata, such as the range of values for each characteristic and potential labels on the records, to help with interaction. In recent years, we have seen the emergence of marketplaces like WorldQuant and Xignite, where data is available as a commodity. These marketplaces' main objective is to facilitate communication between data contributors and data consumers when they need data to carry out certain tasks, such as developing new machine learning models, enhancing the precision of existing ones, or doing statistical estimation. Recent works have demonstrated methods that strike a balance between exploration (learning more about the data that the provider has) and exploitation (putting that knowledge to use in allocating the limited data acquisition budget) [20]. Concerns like data privacy, repeating submissions to steal rewards, and data security are not addressed here. Various data acquisition techniques are used by contributors, like data discovery, data acquisition, and data generation (crowdsourcing or synthetic data) [21]. Once the user is satisfied with the data collected, the user uploads the data. The data is received by the DCCS service, and the base model is trained on this data. After the training is finished, the validation data hash is fetched from the contract and accuracy is computed, and the data contributed is stored on the IPFS. Since EVM cannot handle floating-point numbers, we propose storing the decimal part and integer part separately. This will ensure fairness even in close-case scenarios. Again, the objective of storing on IPFS and not EVM is to reduce gas costs. All off-chain API calls are handled by Oracalize, a service that allows smart contracts to access data from other blockchains and the World Wide Web [22]. The submission details are stored on the contract. Code below is the Solidity snippet of a Submission entity:

```
struct Submission{
    address payable contributor;
    uint 256 accuracyInteger;
    uint256 accuracyDecimal;
    string contributedDataHash;
    string baseModelHash;
    string improvementInAccuracy;
}
```

4.3   Reward Phase

Once the organiser is satisfied with the improvement in the metrics so far, the contract can be closed. Here, the organiser will receive a newly trained model with a better accuracy score on the validation data. Optionally, the organiser can also acquire the data points that contributed to the improvement. This will be useful when the organiser wants to periodically again create a new contract in the future. We take inspiration from the federated learning approach for combining the learnings from all submissions that improved the base model. This phase is further divided into two sub-phases.

4.3.1   Reward Mechanism

Incentives are required to encourage people to contribute new data that will help improve the model's performance. Data contributors can earn points and badges when other contributors validate their contributions, just like on websites like Stack Overflow. Incentives must not be distorted or misaligned. Because the prize structure is winner-take-all, second place is equivalent to not competing at all. This eventually leads to an equilibrium, in which only a few teams compete, and potential new teams never form because catching up appears highly unlikely. A structure like this is anticollaboration. Competitors are strongly encouraged to keep their techniques confidential. This is in stark contrast to many other crowdsourced projects, such as Wikipedia, where participants must build on the work of others [23].

Willingness of the contributors to improve the model is the one thing that the proposal is dependent on. So, there is a need to automate this process. Since we are storing the new accuracy scores of all submissions, it makes sense to reward only those who contribute "good data." As proposed in our previous work [24], we put forward using a fair and equitable incentive system. Let's assume the base metric score is B, and we have "N" contributors and their metric scores are $S_1, S_2,..., S_N$, so the incentive received by the ith contributor will be R_i which will be:

$$R_i = \left((S_i - B) / \sum_{(k=0)}^{n} (S_k - B) \right) * R, \tag{1}$$

where R denotes the total reward specified in the contract. The mechanism described above ensures that low-quality data is not rewarded. Because there is no winner-take-all scheme, the reward mechanism encourages more people to contribute. It fosters a healthy level of competition. The whole mechanism for reward distribution is shown in Algorithm 1.

4.3.2 Combining the Best Submissions

Continual Learning

The ability to learn tasks sequentially is critical for artificial intelligence development. In general, machine learning models are incapable of this, and it has long been assumed that forgetting of weights learned is an unavoidable property of models dependent on previous input. This process is known as continual learning [25]. Since, in our approach, we want the models to improve upon the existing learned parameters, we have used the Elastic Weighted Consolidation (EWC) algorithm [26]. The basic idea behind this algorithm is a quadratic penalty, introduced to constrain the network parameters to stay within the low error region for task A when learning to perform B. The quadratic penalty acts as a "spring" of sorts to anchor the parameters to previously learned solutions, hence the name Elastic Weight Consolidation. This iterative process is executed over all the submissions that improved the base model.

Federated Averaging

For all the submissions that have improved the base accuracy, we now have to combine the learnings from all models into one global model.

There are two types of aggregation for federated learning:

1. parameter aggregation: aggregate according to parameters or intermediate calculation parameters of the client-side models,
2. model aggregation (M): aggregate according to the client-side models.

In the original study by McMahan et al., the weighted average of the local learning parameters on each client is used to maintain the global model after they have been aggregated by the central server. The federated matched averaging (FedMA) algorithm compares elements with similar feature extraction signatures and matches and averages hidden components, such as channels for convolution layers, hidden states for LSTM, and neurons for fully connected layers [27]. FedMA builds the shared global model in a layer-by-layer fashion. Precision-weighted federated learning takes into account the second raw moment (uncentered variance) of the stochastic gradient when computing the weighted average of the parameters of independent models trained in a federated learning setting [28]. It handles the heterogeneity of data across various clients. They also make sure that data privacy is not violated in any form. This aspect of the system is evolving, and research is going on to improve the ability to accumulate the weights learned into one ensemble model.

Algorithm 1: Reward Transfer

1:   $subs \leftarrow$ getAllDataSubmissions()

2:   $N \leftarrow$ length(submissions)

3:   $totalImprov \leftarrow 0$

4:   **for** $i = 1$ **to** $N -$ **do**

5:       **if** $subs_i[accuracyImproved] > 0$ **then**

6:          $totalImprov \leftarrow totalImprov + subs_i[accuracyImproved]$

7:       **end if**

8:   **end for**

9:   $reward \leftarrow$ getRewardValue()

10:   **for** $i = 1$ **to** $N - 1$ **do**

11:     $reward\ Awarded \leftarrow \frac{subs_i[accuracyImproved]}{totalImprov} * reward$

12:     $transferRewarded(rewardedAwarded, subs_i[walletAddress])$

13:   **end for**

---

5 Validation

We attempt to validate our proposal by simulating the envisioned marketplace by way of a contract organiser and a set of developer accounts, all of which interact and provide data to the open contract. The organiser initiates the contract by giving the details mentioned earlier and posts the contract on the marketplace. However, no deduction is made from the organiser's wallet at this time. Published contracts are available on the marketplace and can be viewed by all the users, except the organiser, from their developer dashboard, as shown in Fig. 3. Developers can upload new datasets from their dashboards for a particular contract. The base model is trained on the newly submitted data on the backend server using the elastic weight consolidation (EWC) so that the model learns new data and also remembers old data. If the performance of the new model on the validation dataset uploaded by the organiser is better than the old model, the weights of the new model are stored on the IPFS network and the hash for these weights is stored in the smart contract along with the contributor address. The organiser can view the performance of the submissions, as shown in Fig. 4. On contract termination, the organiser downloads the model, and for all the models that have performed better than the base model, their weights will be averaged. The amount of the reward is deducted from the organiser's account and deposited into the account of all the contributors whose data was accepted into the model development process.

This proposal, however, is subjected to certain technical limitations similar to some of those presented by Harris and Waggoner [9], namely, Bad intent and response, Illusionary and Ambiguous Data, and Overwhelming the Network, which are briefly explained as follows:

Bad intent and response: this essentially refers to the ability of certain developers to "game" the system in a manner, where in bad faith, they can block other submissions by filling up the submission limit without making any meaningful contribution. A wealthy and determined agent can corrupt the contract. The incentive mechanism should make it costlier for the contributor with every wrong submission or set submission limits.

Illusionary and Ambiguous Data: on the other hand, the "organisers" who use this framework must carefully evaluate the type of model and how providing unclear data can affect the system, as doing so may merely give the impression of greater

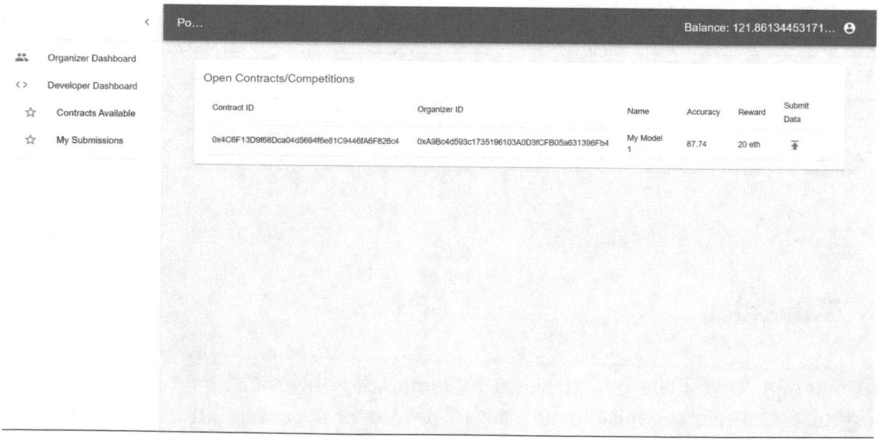

Fig. 3 Viewing available contracts as a developer

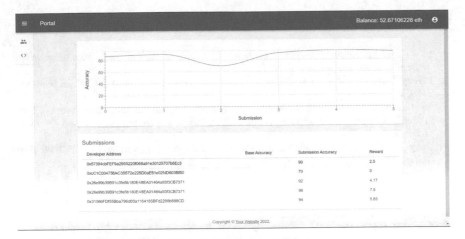

Fig. 4 Viewing published contract submissions from organiser dashboard

performance. To avoid a bad submission penalty, the contributors should make sure that they meet the data quality standards mentioned in the contract.

Overwhelming the Network: public blockchain-based applications have had dependability problems as a result of network congestion. While we circumvent this by using a private blockchain, its scalability is somewhat constrained, and resources are relatively scarce. When adding data that necessitates making new transactions, this can be problematic for this framework.

6 Conclusion

In this paper, we have introduced a system that intends to source better quality datasets and improve the efficiency of existing machine learning models by incentivising contributors. Contributed datasets are stored securely on IPFS, which also makes the system fault-tolerant to data storage failures and highly scalable. The gas costs of keeping enormous amounts of data on-chain are reduced by off-chain interactions; instead, the data is kept on IPFS, and only the hashes of the data will be saved in the smart contract. We use smart contracts to maintain a record of all contributors and ensure fair and unbiased distribution of rewards among the contributors. Our novel rewarding mechanism filters out bad-quality data. Since the model has to improve on the existing benchmark, the EWC algorithm is used to avoid catastrophic forgetting of prelearned weights of submissions that improve baseline metrics. Our system creates a platform to improve machine learning models using blockchain technology. With advances in technology, the solution can be scaled for the improvement of more complex machine learning models. Also, we anticipate the use of blockchain in AI not only for improving machine learning models but also in various other aspects of machine learning.

References

1. IBM Cloud Education (2020) Machine learning, 15 July 2020. Retrieved from https://www. ibm.com/cloud/learn/machine-learning
2. Halevy A, Norvig P, Pereira F (2009) The unreasonable effectiveness of data. IEEE Intell Syst 24(2):8–12
3. Sun C, Shrivastava A, Singh S, Gupta A (2017) Revisiting unreasonable effectiveness of data in deep learning era. In: Proceedings of the IEEE international conference on computer vision, pp 843–852
4. Deng J et al (2009) Imagenet: a large-scale hierarchical image database. In: 2009 IEEE conference on computer vision and pattern recognition. Ieee
5. Wang A, Singh A, Michael J, Hill F, Levy O, Bowman S (2018) GLUE: a multi-task benchmark and analysis platform for natural language understanding. In: Proceedings of the 2018 EMNLP workshop BlackboxNLP: analyzing and interpreting neural networks for NLP, Association for Computational Linguistics, pp 353–355
6. Scheuerman MK, Denton E, Hanna A (2021) Do datasets have politics? Disciplinary values in computer vision dataset development. In: Computer Supported Cooperative Work, CSCW
7. Paullada A et al (2021) Data and its (dis) contents: a survey of dataset development and use in machine learning research. Patterns 2(11):100336
8. Harris JD, Waggoner B (2019) Decentralized and collaborative AI on blockchain. In: 2019 IEEE international conference on blockchain (Blockchain). IEEE
9. Kurtulmus AB, Daniel K (2018) Trustless machine learning contracts; evaluating and exchanging machine learning models on the ethereum blockchain. arXiv preprint arXiv:1802. 10185
10. Marathe A et al (2018) DInEMMo: decentralized incentivization for enterprise market-place models. In: 2018 IEEE 25th international conference on high performance computing workshops (HiPCW). IEEE
11. Waggoner B, Frongillo R, Abernethy JD (2015) A market framework for eliciting private data. In: Advances in neural information processing systems, vol 28
12. NIST Blockchain Technology Overview Draft NISTIR8202, 23 Jan 2018
13. Pinna A, Ibba S, Baralla G, Tonelli R, Marchesi M (2019) A massive analysis of Ethereum smart contracts empirical study and code metrics. IEEE Access 7:78194–78213
14. Bartoletti M (2020) Smart contracts. Front. Blockchain, 4 June 2020
15. Chen Y, Shen Y, Zheng S (2020) Truthful data acquisition via peer prediction. Adv Neural Inf Process Syst 33:18194–18204
16. McMahan B, Moore E, Ramage D, Hampson S et al (2016) Communication efficient learning of deep networks from decentralized data. arXiv preprint arXiv:1602.05629
17. Wood G (2014) Ethereum: a secure decentralized generalized transaction ledger. Ethereum Project Yellow paper 151(2014):1–32
18. Benet J (2014) IPFS—content addressed, versioned, P2P file system
19. Chen L, Koutris P, Kumar A (2018) Model-based pricing for machine learning in a data marketplace. arXiv preprint arXiv:1805.11450
20. Li Y, Yu X, Koudas N (2021) Data acquisition for improving machine learning models. arXiv preprint arXiv:2105.14107
21. Roh Y, Heo G, Whang SE, A survey on data collection for machine learning: a big data-ai integration perspective. IEEE
22. Liu X, Chen R, Chen YW, Yuan S-M (2018) Off-chain data fetching architecture for Ethereum smart contract, pp 1–4. https://doi.org/10.1109/ICCBB.2018.8756348
23. Abernethy JD, Frongillo R (2011) A collaborative mechanism for crowdsourcing prediction problems. In: Advances in neural information processing systems, vol 24
24. Ajgaonkar A et al (2022) A blockchain approach for exchanging machine learning solutions over smart contracts. In: Science and information conference. Springer, Cham
25. Ring MB (1998) Child: a first step towards continual learning. In: Learning to learn. Springer, pp 261–292

26. Kirkpatrick J et al (2017) Overcoming catastrophic forgetting in neural networks. Proc Natl Acad Sci 114(13):3521–3526
27. Wang H et al (2020) Federated learning with matched averaging. arXiv preprint arXiv:2002.06440
28. Reyes J et al (2021) Precision-weighted federated learning. arXiv preprint arXiv:2107.09627

Blockchain Framework for Automated Life Insurance

Vaishali Kalsgonda, Raja Kulkarni, and Shivkumar Kalasgonda

Abstract The insurance process is done manually in India. The insurance client has to depend on an insurance agent from buying to the claim firing process which leads to wrong entry, fraudulent claims, and cost overhead on the agent's commission. An insurance client has to maintain documents and make them available at the time of claim firing, which is very cumbersome. Blockchain smart contracts will provide a secure, automatic, cost-cutting, paperless, and real-time insurance process. This paper aims to provide a blockchain solution for life insurance claim processing, as it is the most purchased and needy insurance. An architectural illustration of a blockchain prototype is contributed to this paper. As a result, the authors found that using Hyperledger Composer, we can develop fine-grained insurance applications.

Keywords Blockchain · Hyperledger · Insurance · Smart contracts

1 Introduction

Blockchain development for insurance is in its early stage. Many researchers are interested in adopting blockchain for the insurance process. Using blockchain for insurance will provide secure and automatic claim processing. In most cases, family members have no idea about any life insurance purchased by a dead person, and delay is made for claim processing or collecting all required documents for an insurance

---

Raja Kulkarni and Shivkumar Kalsgonda—These authors contributed equally to this work.

V. Kalsgonda (✉)
Department of Computer Science, Shivraj College, Gadhinglaj, Maharashtra 416502, India
e-mail: vpkalsgonda@gmail.com

R. Kulkarni
Department of Computer Studies, Chhatrapati Shahu Institute of Business Education Research, University Road, Kolhapur, Maharashtra 416004, India
e-mail: drrvkulkarni@siberindia.edu.in

S. Kalasgonda
E-Commerce, Distributed Systems, Seattle, USA

Y. Singh et al. (eds.), *Proceedings of International Conference on Recent Innovations in Computing*, Lecture Notes in Electrical Engineering 1011,
https://doi.org/10.1007/978-981-99-0601-7_59

claim; sometimes, beneficiary does not get any benefits of insurance purchased. If a blockchain solution is used for the insurance process, there is no need to maintain documents and make them available at the time of claim processing by the client, and also, the documents will be kept secure in the blockchain. The insurance process will become automatic by use of smart contracts, and the smart contract is a small program written on top of the blockchain and responsible for transaction processing under particular conditions. For implementing blockchain solutions for the insurance industry in India, a consortium blockchain network has to be used as multiple organizations are involved in this network and only authorized nodes have to give access permissions, and this approach does not require the use of cryptocurrencies. So, implementing blockchain for the insurance industry in India, Hyperledger Fabric is the best solution. In this paper, the authors aim to design a blockchain solution for life insurance using the Hyperledger Composer tool of the Hyperledger Fabric framework.

2 Literature Review

In India, banking, insurance, and card industries are coming together to form a consortium to realize the benefits of blockchain at an industry level. Insurers are focusing on using blockchain solutions to speed up claim processing; also, blockchain is the best solution for avoiding errors, which will introduce while manual entry [1]. Integrating blockchain with IoT devices can be used such as supply chain [2], Unmanned Aerial Vehicle, and [3] also for accessing and managing IoT devices [4, 5]. Using existing blockchain frameworks, we can develop applications that are solved only by binary conditions [6]. The papers dealing with implementation details are fewer in number, and in most of the papers, researchers are interested to use Ethereum [7]. Blockchain technology has a lot of benefits for an insurance company, but we cannot fully automate all the steps in insurance processing. Certain validations by manual step are required in the current claim process. Blockchain can be adopted for limited use cases where there is no requirement for complex regulatory processing [8]. Nath [9] reported that blockchain technology to share fraud intelligence will make it harder for any fraudulent activity by criminals and further suggested adopting this technology step by step to avoid the big bang. Integrated use of both Hyperledger Fabric and Ethereum is used for implementing blockchain and for transportation insurance is used for getting advantage of both the private and public blockchains. Hyperledger Fabric is used for storing data and Ethereum is used for modeling payments [10]. As the insurance industry suffers from fraud [11], researchers are going to use AI for fraud detection and permanently save the result in blockchain to minimize claim refund losses. Mayank [12] proposed a blockchain framework for insurance to offer fine-grained access control and experimented by scaling up the network to test the robustness of the system and finally concluded that the network size is directly proportional to the confirmation time; the more the number of nodes, the more will be the confirmation time, and in short, the slower will be the network. Blockchain solutions to the insurance industry will speed up the insurance processing with reduced cost [13]; as per the architectural concern, consortium blockchain network will be preferable for automatic processing and the public blockchain will be the solution for payment purposes. Though the blockchain solution has many advantages, there are issues like scalability, network lighting, and lack of skill, to write bug-free smart contracts. Once these drawbacks are overcome, blockchain solutions will successfully automate insurance processing. According to a systematic literature study [14], more improvements are needed to accept blockchain technology in the insurance sector such as forming a consortium, prototyping the use cases, reaching the average users, etc., but the insurance industry will get more benefits than other industries. MIStore [15] is a blockchain solution to store medical insurance records securely in a distributed environment implemented using the Ethereum platform. The efficiency of MIStore is dependent on the efficiency of the Ethereum platform if another platform is used for implementing the same it might provide better throughput. Blockchain-based crop insurance [16] will result in automatic payouts in case of natural disasters, and farmers will not have to worry about climate change. Manual entry in medical insurance results in fraud of ten billion per year in the USA [17], because of a lack of endorsement of stakeholders. Blockchain solutions to the health insurance sector will provide security, immutability, and transparency.

3   Background Work

Blockchain is the most disruptive technology in today's world, as its decentralized working system makes it free from fault tolerance, and its p2p system and consensus mechanism remove the unwanted cost of the third party in a trustless environment. Blockchain's cryptographic feature makes it secure. A smart contract written on top of the blockchain provides atomicity. In short, blockchain applications are secure, minimize cost, and work fast and automatically. So, most of the enterprise applications are interested in adopting blockchain solutions.

3.1   Blockchain

The primary objective of blockchain is to avoid the double spending problem and provide security. In blockchain, every completed transaction is recorded in an immutable ledger in a verifiable, secure, transparent, and permanent way, with a time stamp and other details [18]. Each blockchain employs a consensus mechanism and the choice of mechanism varies among networks to resolve different states, or "forks" in the network [19] detail a variety of consensus mechanisms. In this paper, we propose to add blockchains as a mechanism to design a smart contract application for life insurance.

3.2 Life Insurance Process

In life insurance or assurance, according to the contract between an insurance policyholder and an insurer or assurer, the beneficiary gets the amount including funeral expenses from the insurer after the death of an insured person based on the premium amount paid by the assurer. The beneficiary also gets benefits in other situations like terminal illness or critical illness. The policyholder can pay a premium as either one term payment or periodically. Life policies apply terms and conditions to limit the liability of the insurer, and death claims relating to suicide, fraud, war, riot, and civil commotion are excluded in the contract of policies.

3.3 Death Registry Process in India

In India, it is mandatory under the law to register every death with the concerned State/UT Government within 21 days of its occurrence. If the death has taken place in hospitals, nursing homes, or medical institutions, such deaths are to be reported by the head of the institutions within 21 days of the death to the concerned registrars. If the death has taken place at home, it is the responsibility of the family member to report the same within 21 days to the sub-registrars. A death certificate is then issued after proper verification [20].

4 Gap Analysis

Most of the researchers provide novel architecture, models, and frameworks using blockchain, but there is a lack of technical details about used blockchain elements. There are several papers present on blockchain framework using Hyperledger Composer for health care, supply chain, and banking sector, but there is only one paper available related to the insurance sector, which provides only general framework for the insurance process and only two participants, as insurance company and insurance client [12], but by further studies, it is found that there will be different participants, endorsement policies, consensus algorithms, ordering peers, and different algorithms for smart contracts for a different type of insurance policies. Implementing blockchain solutions for the insurance industry is a very challenging job [21]; firstly, insurers have to train or hire staff having proper technical knowledge for implementing blockchain solutions, and second insurer has to invest in adopting new practices. The proposed system in this paper aims to design a blockchain framework for life insurance having four types of participants as insurance company, death registry, verifier, and insurance client. The development language of the proposed work is Hyperledger Composers modeling language which supports JavaScript API.

5   Material and Methods

The designing of the proposed framework is done according to Hyperledger Composer which runs on top of Hyperledger Fabric.

5.1   Hyperledger Fabric

Hyperledger Fabric is a project led by IBM, under the Linux Foundation. It provides a platform for modular architecture to develop Enterprise Blockchain solutions. Hyperledger Fabric supports building private (permissioned) business networks [22]. Composer Playground is a web-based tool for modeling and testing business networks. Playground communicates with the local Fabric runtime directly [23].

5.2   Development Environment of Hyperledger

Figure 1 shows the development environment of Hyperledger Composer. The different files such as Network Model, Transaction Logic, Access Control Rules, and Query File are packaged and archived in a single file called a Business Network Archive file. The BNA file is then deployed onto a Fabric network. Composer REST Server is a tool that allows us to generate a REST API server based on our business network definition. Using the Yeoman configuration tool makes it easy to create an angular front-end against the REST API.

6   Proposed Framework

In the proposed framework, details of design methodology, blockchain participants, assets, transactions, transaction logic, access control rule, and query file are discussed. Transaction logic will be implemented in JavaScript; considering this, the algorithm of transaction logic is written. Transaction file contains business logic of claim processing.

6.1   Methodology

For the proposed framework, phases such as analysis, design, and implementation are used. In the analysis phase, the authors analyze the requirements of the proposed work and identify participants, assets, and transactions. Depending on the analysis,

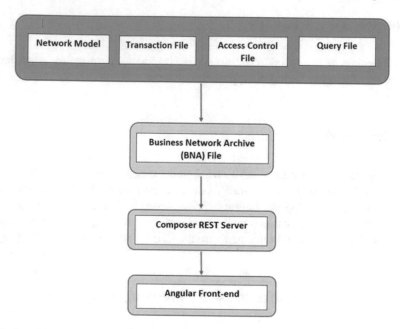

Fig. 1 Development environment overview for Hyperledger [18]

author designs the architecture of the proposed system which is shown in Fig. 2. The blockchain network consists of three different organizations such as insurance organization, a death registry, and a verifier organization. In this architecture, death registry workers, insurance workers, claim verification workers, and insurance customers interact with a proposed blockchain network with their respective web apps, through a Nodejs server. Hyperledger framework provides Fabric SDKs for interacting with clients on the blockchain network. Here, the client is referred to as a worker. Each organization's participating node or clients are called peer nodes. Transactions made by these peers are submitted to the blockchain network only after validating it by ordering service. The detailed design of the smart contract is discussed in the next section. The authors will implement the design in the implementation phase.

6.2 Design of Proposed Work

In our proposed work for the network model, author designed participants, asset, and transactions as follows.

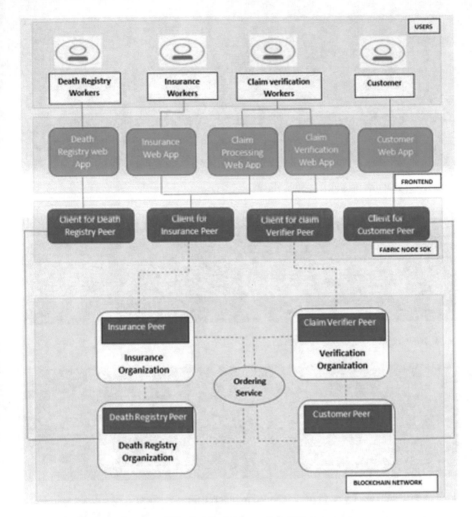

Fig. 2 Architecture of proposed blockchain framework for life insurance

6.2.1 Participants

Insurers or insurance industry, insurance policyholders or customers, death registries, and validators are the participants of the proposed system. The main transaction occurs between customers and the insurance industry. The endorsement peers are the insurance industry, customers, and the death registry office. We require some nodes as validators for validating the state of the blockchain. The testing for creating new participants is shown in Figs. 3, 4, 5 and 6. These records are stored in the form of a participant registry.

```
1    {
2      "$class": "org.insurchain.network.client",
3      "cid": "c1",
4      "adharNo": 345678902345,
5      "pan": "AAAAPZ1234C",
6      "DOB": "1990-03-07"
7    }
```

Fig. 3 New customer is added to blockchain

```
1    {
2      "$class": "org.insurchain.network.insurer",
3      "insId": "i1",
4      "address": "Mumbai"
5    }
```

Fig. 4 New insurer is added to blockchain

```
1    {
2      "$class": "org.insurchain.network.deathReg",
3      "dRegid": "d1",
4      "address": "Kolhapur"
5    }
```

Fig. 5 New death registry participant is added to blockchain

```
1    {
2      "$class": "org.insurchain.network.verifier",
3      "vid": "v1",
4      "address": "Mumbai"
5    }
```

Fig. 6 New verifier participant is added to blockchain

```
1  {
2      "$class": "org.insurchain.network.Policy",
3      "policyNo": "l1",
4      "type": "lifePolicy",
5      "Discription": " this is life policy"
6  }
```

Fig. 7 New policy asset is added to blockchain

6.2.2 Assets

In our proposed work, asset is a policy and attributes of assets are policy number, start date, status, premium, and beneficiary. Every insurance policy has a unique policy number, so this is our prime attribute using this we can make transactions. Figure 7 demonstrates adding a new asset to blockchain. The record of the new asset is stored in the asset registry.

6.2.3 Transactions

From buying a policy to refunding of policy, there will be transactions like client registration, buy policy, claim, pay premium, refund, and history; these all are listed in Table 1. The most important transaction is automatic claim processing when an authorized person makes an entry for life status as death. Therefore, the design of claim processing is the main objective of our proposed work.

Table 1 Transaction with description

Transaction	Description
Client registration	The client has to register for insurance processing
Buy insurance	When clients purchase insurance, new information such as policy number, start and end date, Aadhaar number, and status is added. Life status will be "Alive" at the time of buying the insurance policy
Pay premium	This transaction will be used for paying a premium
Claim processing	When any authorized person in the death registry office changes the life status as "dead" and cause of death, then depending upon the cause of death automatically claims fire
Policy expiry	When the policy expires automatically, it will give notification and saves the record with the status as "Expired"

6.2.4 Transaction Logic

As discussed previously, the researcher's main objective is to design a smart contract for the automatic claim processing. When an insurance client purchases an insurance policy, a smart contract registers life status as alive (Algorithm 1) along with client id (Cid), insurance company id (Iid), and Policy number (Policy No). Here, Cid and Iid records are fetched from the participant registry and Policy No from the asset registry as these are nothing but prime attributes. Figure 8 demonstrates the transaction, for buying the new policy. When life status of insurance client is changed as dead by authorized person in the death registry office (Algorithm 1), it gets verified by authorized validator. Once validator approves the status as "valid", system automatically files the insurance claims (Algorith 2). Here, terms used for participant registry and asset registry are already discussed in section participant and asset, respectively. The sequence diagram shown in Fig. 9 shows the process of automatic claim processing, and as per Algorithms 1 and 2, there is no need to apply for claim processing manually by the beneficiary and wait for benefits.

Algorithm 1 Policy: buyPolicy

Input: Cid, Iid, PolicyNo, Premium
Output: status=alive
    1: *Cid* ←Discover (Client Id from ParticipantRegisty)
    2: *Iid* ←Discover (Insurer Id from ParticipantRegisty)
    3: *PolicyNo* ← Discover (Policy Number from AssetRegisty)
    4: *Status* ←" alive"
    5: *Add* ← AssetRegistry

```
1  {
2      "$class": "org.insurchain.network.BuyPolicy",
3      "policy": "resource:org.insurchain.network.Policy#l1",
4      "ins": "resource:org.insurchain.network.insurer#i1",
5      "clnt": "resource:org.insurchain.network.client#c1",
6      "premium": 1000,
7      "st": "Alive"
8  }
```

Fig. 8 Transaction for buying policy

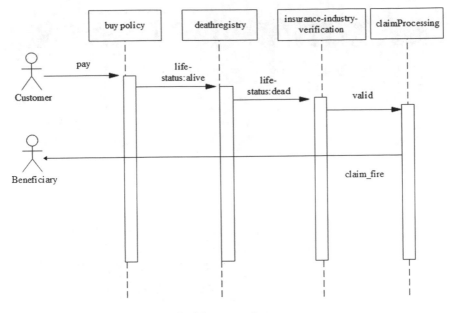

Fig. 9 Sequence diagram of automatic claim processing

Algorithm 2 Policy: claimPolicy

Input: *Cid, Iid, PolicyNo, Did, Vid*
   1: *Cid ← Discover(ClientId from ParticipantRegisty)*
   2: *Iid ← Discover(Insurer Id from ParticipantRegisty)*
   3: *PolicyNo ← Discover(PolicyNumber from AssetRegisty)*
   4: *Vid← Discover(Verifier Id from ParticipantRegisty)*
   5: **if** status=Dead **then**
  6:             Verify death cause
   7:                **if** cause = valid **then**
   8:                   Refund (Cid, Iid, PolicyNo, amount)
   9:               **end if**
 10: **end if**
 11: Update AssetRegisty

6.2.5   Access Control Rule

Access control rules define which participant can view which part of the system. In our proposed design, insurance clients can only view details of their own policies. In Fig. 10, you can view that the admin can access the data of all clients and Fig. 11 shows that the client can access its own data only; Fig. 12 shows that client does not have the authority to create a new participant.

Likewise, in our proposed system, authorized persons in the death registry can only change the life status of policyholders and cannot access another part of the

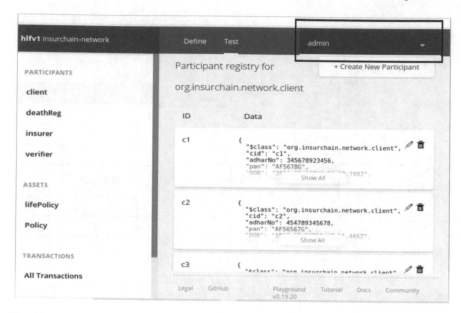

Fig. 10 Admin can access all client's data

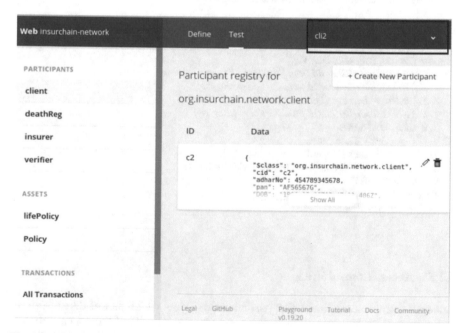

Fig. 11 Client can access only its own data

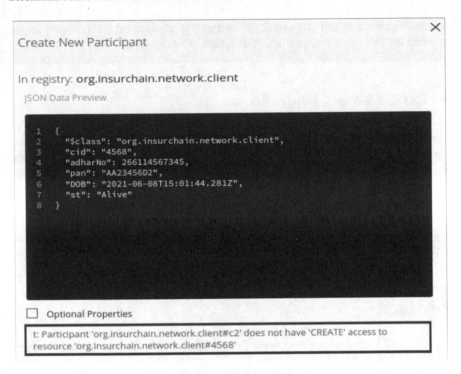

Create New Participant

In registry: org.insurchain.network.client

JSON Data Preview

```
1   {
2       "$class": "org.insurchain.network.client",
3       "cid": "4568",
4       "adharNo": 266114567345,
5       "pan": "AA23456D2",
6       "DOB": "2021-06-08T15:01:44.281Z",
7       "st": "Alive"
8   }
```

☐ Optional Properties

t: Participant 'org.insurchain.network.client#c2' does not have 'CREATE' access to resource 'org.insurchain.network.client#4568'

Fig. 12 Client does not have the authority to create a new participant

system. A person from the insurance industry verifies the death reason, whether the reason for death is valid or not for the claim processing. After the confirmation, the person who verifies has the authority to make appropriate entries in verification form.

6.2.6 Query File

In query files, we can write queries like SQL queries. In the proposed work, the author designs query for admin and policyholders. In our framework, we design queries to perform all admin activities, also queries for the customer for accessing own activities.

7 Test Result

In this paper, simple data is used for demonstration purposes. The authors show all implementation details of assets, participants, and transactions. Here, Hyperledger Composer-based implementation is used for testing purposes and found that the system works well according to ACL. Only authorized persons can access or update specific data. Thus, the result shows that Hyperledger Fabric is the best-suited solution for building a fine-grained permissioned blockchain.

8   Conclusion and Future Scope

This paper proposes a design of a blockchain-based framework for implementing the life insurance process which will provide a fine-grained solution for the automatic claim processing. The proposed framework is based on Hyperledger Fabric. Insurance clients need not have to maintain the required documents and provide them at the time of the claim process and wait to get benefits.

The author's purpose is to develop smart contracts based on the proposed design, using raft orderer service with a permeant database, user interface, and all required networking artifacts in the future.

References

1. Kalsgonda V, Kulkarni R (2020) Applications of blockchain in insurance industry: a review. PIMT J Res (PIMT JR)
2. Caro MP, Ali MS, Vecchio M, Giaffreda R (2018) Blockchain-based traceability in Agri-Food supply chain management: a practical implementation. In: 2018 IoT vertical and topical summit on agriculture Tuscany (IOT Tuscany), Tuscany, Italy
3. Jensen IJ, Selvaraj DF, Ranganathan P (2019) Blockchain technology for networked swarms of unmanned aerial vehicles (UAVs). In: 2019 IEEE 20th international symposium on a world of wireless, mobile and multimedia networks (WoWMoM), Washington, DC, USA
4. Ding S, Cao J, Li C, Fan K, Li H (2019) A novel attribute-based access control scheme using blockchain for IoT. IEEE Access 7:38431–38441
5. Huh S, Cho S, Kim S (2017) Managing IoT devices using blockchain platform. In: 2017 19th International conference on advanced communication technology (ICACT), PyeongChang, Korea (South)
6. Hans R, Zuber H, Rizk A, Steinmetz R (2017) Blockchain and smart contracts: disruptive technologies for the insurance market. In: AMCIS 2017 Proceedings
7. Fekih L (2020) Application of blockchain technology in healthcare: a comprehensive study. In: International conference on smart homes and health telematics
8. Gatteschi V, Lamberti F, Demartini C, Pranteda C, Santamaria V (2018) To blockchain or not to blockchain: that is the question. In: IT professional, pp 62–74
9. Nath I (2016) Data exchange platform to fight insurance fraud on blockchain. In: 2016 IEEE 16th international conference on data mining workshops (ICDMW), Barcelona, Spain
10. Li Z, Xiao Z, Xu Q, Sotthiwat E, Goh RSM, Liang X (2018) Blockchain and IoT data analytics for fine-grained transportation insurance. In: 2018 IEEE 24th international conference on parallel and distributed systems (ICPADS), Singapore
11. Dhieb N, Ghazzai H, Besbes H, Massoud Y (2020) A secure AI-driven architecture for automated insurance systems: fraud detection and risk measurement. IEEE Access, pp 58546–58558
12. Raikwar M, Mazumdar S, Ruj S, Gupta SS, Chattopadhyay A, Lam KY (2018) A Blockchain framework for insurance processes. In: 2018 9th IFIP international conference on new technologies, mobility, and security (NTMS), Paris, France

13. Gatteschi V, Lamberti F, Demartini C, Pranteda C, Santamaría V (2018) Blockchain and smart contracts for insurance: is the technology mature enough? Future Internet. 10(2):20. https://doi.org/10.3390/fi10020020

14. Kar AK, Navin L (2021) Diffusion of blockchain in insurance industry: an analysis through the review of academic and trade literature. Telematics Inform 58:101532. https://doi.org/10.1016/j.tele.2020.101532. ISSN 0736-5853

15. Zhou L, Wang L, Sun Y (2018) MIStore: a Blockchain-based medical insurance storage system. J Med Syst 42:149. https://doi.org/10.1007/s10916-018-0996-4

16. Jha N, Prashar D, Khalaf OI, Alotaibi Y, Alsufyani A, Alghamdi S (2021) Blockchain based crop insurance: a decentralized insurance system for modernization of indian farmers. Sustainability 13(16):8921. https://doi.org/10.3390/su13168921

17. Ismail L, Zeadally S (2021) Healthcare insurance frauds: taxonomy and blockchain-based detection framework (Block-HI). IT Professional 23(4):36–43. https://doi.org/10.1109/MITP.2021.3071534

18. Leka E, Lamani L, Selimi B, Deçolli E (2019) Design and implementation of smart contract: a use case for geo-spatial data sharing. In: 42nd International convention on information and communication technology, electronics and microelectronics (MIPRO), Opatija, Croatia

19. Tschorsch F, Scheuermann B (2016) Bitcoin and beyond: a technical survey on decentralized digital currencies. IEEE Commun Surv Tutor, pp 2084–2123

20. News: Information, 20 May 2019. [Online]. Available: https://www.indiatoday.in/information/story/death-certificateprocedure-india-delhi-1529341-2019-05-20

21. Grima S, Spiteri J, Romānova I (2020) A STEEP framework analysis of the key factors impacting the use of blockchain technology in the insurance industry. Geneva Pap Risk Insur Issues Pract 45:398–425. https://doi.org/10.1057/s41288-020-00162-x

22. hyperledger-fabric, 2019. [Online]. Available:https://hyperledgerfabric.readthedocs.io/en/release-1.4

23. Welcome to Hyperledger Composer, [Online]. Available: https://hyperledger.github.io/composer/latest

Blockchain for IoT: A Comprehensive Review for Precision Agricultural Networks

Abhiudhaya Upadhyaya, Yashwant Singh, and Pooja Anand

Abstract Internet of Things (IoT) has gained popularity in recent years due to the services it offers and its wide usage in the field of science and technology such as smart agriculture, smart home devices. We now live in a world surrounded by smart gadgets and utilize it almost on daily basis. As the usage is increasing, the amount of data being disseminated is also increasing due to which a lot of data is being exposed to threats. So, it must be made sure that the channels through which data is being transferred should reach securely to the endpoint without compromising integrity, confidentiality, and authentication, ensuring that data reached should not be modified or tampered. In recent years, an important technique has been developed called "blockchain", which can help in improving security, trust, speed, visibility, immutability, and traceability. Motivated by these facts, this work explores the potential of the blockchain and the efficacy of using it to deal with growing security and performance challenges in IoT associated within precision agriculture.

Keywords Internet of Things (IoT) · BIoT · Security · Cryptography · Blockchain · Global Unique Identifier (GUID) · Public Key Infrastructure (PKI) · IoT challenges

1 Introduction

We live in a world surrounded by smart devices and gadgets. One of the revolutionary innovations of the twentieth century is Internet of Things (IoT). IoT is an

A. Upadhyaya (✉) · Y. Singh · P. Anand
Department of Computer Science and Information Technology, Central University of Jammu, J&K 181143 Bagla, India
e-mail: abhiudhayu@gmail.com

Y. Singh
e-mail: yashwant.csit@cujammu.ac.in

P. Anand
e-mail: poojaanand892@gmail.com

© The Author(s), under exclusive license to Springer Nature Singapore Pte Ltd. 2023   787
Y. Singh et al. (eds.), *Proceedings of International Conference on Recent Innovations in Computing*, Lecture Notes in Electrical Engineering 1011,
https://doi.org/10.1007/978-981-99-0601-7_60

area of active research and is playing an important role in advancing technical sector. Internet of Things is defined as a new technology paradigm envisioned as a global network of machines and devices which are capable of communicating with each other [1]. It is also a smart network that connects everything to the internet for the purpose of exchanging data and interacting via information detecting devices such as sensors, gateways [1]. These devices are resource constraint and have very less power and storage capacity [2]. The term IoT was coined by Ashton in 1998; since then, significant progress has been done in IoT domain [3]. The number of smart connected devices is increasing every year; by this rate of growth, on forecasting, the number of devices connected would reach 23.5 billion by 2029 [4]. IoT offers plethora of applications in science and technology. Smart industry, smart logistics, smart agricultural, intelligent transportation, smart grid [5], smart environmental protection, smart safety, smart medical, smart wearables, home gadgets are some of the use cases [6, 7].

Considering the execution of numerous approaches to secure IoT devices, new types of attacks will continually emerge due to variances in security standards. Even though IoT devices are secure on their own, they become vulnerable to a variety of threats when linked to an insecure network such as device authentication, DoS/DDoS attack [8], intrusion detection malware detection [9]. As per studies, 70% of the smart devices are vulnerable to cyberattacks. Furthermore, attacks such as Mirai (2016), Persirai (2017), and BrikerBot (2017) in the past have exploited IoT devices security [10]. Today, a large number of devices communicate with one other on a daily basis, resulting in a large amount of data transfer. Many crucial IoT systems such as Internet of Drones (IoD) and Internet of Robotic Things (IoRT) are used in surveillance and warfare activities [11]. As a result, there can be security and privacy breach while transferring important data, which is a major concern.

In such a scenario, blockchain has quickly emerged as an important technology in recent years. Satoshi Nakamoto is widely regarded as the inventor of blockchain. He also introduced the electronic cash currency "Bitcoin", which sparked interest in blockchain research [12]. Blockchain is a public, trusted, shared, and immutable ledger of all transactions, has decentralized storage, offers high transparency and security [10]. Blocks are formed from transactions and are linked together in a chain utilizing the information of adjacent blocks called hash. These records are protected from tampering and change using a variety of cryptographic techniques [12]. Blockchain has been integrated with a range of domains since its inception to handle a variety of problems and provide better solutions to existing ones; one such domain is Internet of Things. In the realm of the Internet of Things, blockchain has the ability to tackle issues of privacy, security, and lack of trust. Thus, the use of blockchain in the IoT domain has given rise to a new blockchain domain in the IoT known as blockchain Internet of Things (BIoT) [13].

The organization of the paper is as follows: Sect. 1 consists of the introduction. Section 2 consists of related work. Section 3 includes the blockchain and its requirements. Section 4 and 5 includes the IoT security challenges in precision agriculture. Lastly, Sect. 6 depicts the IoT performance challenges (Fig. 1).

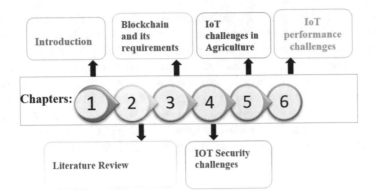

Fig. 1 Graphical layout of manuscript

2 Literature Review

Elhoseny et al. [14] presented a hybrid security model in healthcare services for securing the diagnostic text data in medical images during transmission. The model was developed using a technique known as 2D Discrete Wavelet Transformation 1 Level (2D-DWT-1L) steganography and then integrated with a hybrid encryption scheme which consists of both AES and RSA algorithms. The model was applied on two datasets, i.e., DME and DICOM datasets. The results exhibited that the model has a PSNR value of 57.02, and the value of MSE was 0.1288 which reveals higher performance than the existing one. In another work, Khari et al. [15], the work in this paper focused on the security of data in IoT devices using cryptography and steganography techniques for secure data transmission using an Elliptic Galois Cryptography (EGC) protocol which includes cryptography to encrypt data, then embed this data into low complexity images using steganography. The proposed work (EGC) showed 86% efficiency, when compared to the existing techniques. Aggarwal et al. [16] proposed a model for internet of drones, which provide secure communication, data collection, and transmission among drones and users as well by utilizing a public blockchain having Ethereum platform which included selection of forger node, creating blocks and validation, applying proof-of-stake mechanism. The results revealed that the proposed model covered all security aspects such as authentication, authorization, accountability (AAA), data integrity (DI), identity anonymity (IA), verification and validation (VV) which made the system more scalable, reliable, and superior by evaluating the computation cost and time. Another work done by Nikooghadam et al. [17] proposed a safe and lightweight authentication and key management protocol for IoT-based WSNs to provide a secure communication link between user and sensor nodes. The method for forward verification is Automated Validation of Internet Security Protocol and Applications (AVISPa) tool. The results of the research revealed that communication cost and storage cost showed better performance when compared to the existing protocols.

Truong et al. [18] have proposed a framework named SASH that integrates IoT platform with blockchain which provides plethora of advantages including security and data transmission among IoT devices. In SASH, basically, blockchain technology is utilized which consists of data marketplace, two sharing schemes, and prefix encryption. Firmware and Hyperledger have been used as platform to implement SASH. The proposed work showed a moderate overhead. In another work, Karati et al. [19] proposed a new generalized CLSC (gCLSC) cryptography, certificateless signcryption technique to assure secure communication of data through IoT devices. It functions as a digital signature and encryption both and can be adopted where authenticity, confidentiality, and lightweight are instrumental factors. The performance metric of proposed work revealed that the computational cost was way better, and also, the gCLSC observed minimal storage almost 50% less when compared to CLSC.

ASeyfollahi et al. [20] proposed a reliable data dissemination for IoT using an algorithm called Harris Hawks Optimization. The mechanism is equipped with fuzzy hierarchical network model for Wireless Sensor Network (WSN) to provide reliable and secure data aggregation. The results showed that RDD improved the energy consumption, packet forwarding distance, and packet delivery ratio by 3.12%, 17.5%, and 43.5%, respectively, when compared to other methods. Gochhayat et al. [21] proposed a novel distributed key management scheme for IoT devices which provides security and privacy to user sensitive data. The method includes delegating the resource-consuming cryptographic to local entity, and this entity coordinates with other peer entities to provide a authentication mechanism. The scheme also exploits the merits of mobile agents by deploying them in subnetworks when required. The results of the research manuscript revealed that it reduced communication overhead.

Li et al. [22] proposed a node-oriented secure data transmission (NOSDT) algorithm for securing the transmission in social networks based on IoT as nodes are highly vulnerable to attacks by malicious nodes. The scheme analyzes the behavior of malicious node and provides secure data transmission by selecting new reliable nodes. Influence model is designed to reduce transmission in malicious node. Upon evaluation, NOSDT improved the performance of social networks by reducing the transmission impact of malicious nodes. In another work, Khan et al. [23] introduced a blockchain-based solution for secure and private IoT communication in smart home network. Smart home architecture based on blockchain is empowered with Deep Extreme Learning Machine. DELM learns quickly, and extreme learning machine is a feedforward neural network. In this proposed system, they used backpropagation approach during learning phase and the network adjusts weight to achieve accuracy. The proposed approach achieved accuracy of 93.91%, outperforming other algorithms.

Mahdi et al. [24] proposed a new stream cipher procedure called Super ChaCha over standard ChaCha for securing data communication in IoT devices. A change in rotation is done as a modification in the procedure. The input cipher is also altered from column to diagonal to zigzag to alternate form. The results showed that Super ChaCha successfully passed all five benchmarking and NIST tests, and with a small increase in time, memory, and power and a small drop in throughput, the complexity level rises considerably. By brute-force attack, Super ChaCha requires 2512 likely keys to break. Also, Pampapathi et al. [25] proposed an efficient and reliable data distribution and secure data transmission using improved adaptive neuro fuzzy inference system (IANFIS) and modified elliptical curve cryptography (MECC) in IoT. The proposed work consists of methods including registration and data communication, sending, data broker, security analysis, and local calculations. On comparing ANFIS with IANFIS, the performance of IANIFS was better; on the other hand, the MECC was 96% better at security when compared to the existing ECC.

Manogaran et al. [26] have proposed a blockchain-assisted secure data sharing (BSDS) model in industrial IoT. The goal is to maximize response rate by reducing false alarm progression (FAP). This model is in charge of data capture and dissemination security, both inbound and outbound. The BSDS was found to achieve a 5.67% high response rate with FAP of 4% and reduced the failure rate by 2.11% rate with FAP of 2%, respectively. Further, it achieved 3.12% less FAP with the response rate of 0.95, maximized response rate by 6.63% with the failure rate of 0.06, and reduced delay by 11.91% with the FAP of 5.2%. In another work, Naresh et al. [27] have proposed an identity-based online/offline signcryption (OOSC) scheme suitable to provide secure message communication among IoT devices, gateway, and server. This approach is divided into two phases: online and offline, with the offline phase performing heavy mathematical computations and the online phase performing minor computations. On experimenting, the results revealed that the proposed scheme takes less computational time and provides security against IND-CC2.

Miao et al. [28] proposed a Federated Learning-Based Secure Data Sharing mechanism for IoT, named FL2S for secure data sharing in IoT. Federated learning (FL) framework is developed based on the sensitive task decomposition. In addition, to improve data sharing quality, deep reinforcement learning (DRL) technology is utilized. The results revealed that FL2S achieved better privacy protection and data quality in secure data sharing. Table 1 shows the various techniques used in IoT devices for securing data along with results and challenges.

From the above analysis, it is clear that many of the techniques utilized are based on cryptography, which can be potentially breached for IoT networks. So, it is clear that a robust solution such as blockchain is needed for several issues described in Sect. 4.

3 Blockchain and Its Requirements

A growing list of records and a decentralized, immutable, distributed ledger for keeping time-stamped transactions between multiple computers in a peer-to-peer network is what blockchain is. Data in blockchain is stored in blocks which are linked cryptographically. Every block contains a cryptographic hash of the previous block. As a result of forming a chain, these blocks altogether in a network are called blockchain. Whenever a new block is added into a blockchain, consensus mechanism

Table 1 The table presents various security techniques utilized for securing data communication with results and challenges in different IoT domains in recent years

Author	Year	Technique/method	Result/challenges	Domain
Elhoseny et al. [14]	2018	2D-DWT-1L Steganography integrated with AES and RSA	Better performance in hiding confidential data into transmitted cover image and securing	Healthcare-based IoT
Khari et al. [15]	2019	Elliptic Galois Cryptography	EGC showed 86% better efficiency	IoT
Aggarwal et al. [16]	2019	Public blockchain on Ethereum platform	Better computational cost and time It can also be performed on private blockchain	Internet of Drones (IoD)
Nikooghadam et al. [17]	2019	Authentication and key management protocol (AVISPa) tool	Better communication and storage cost No machine-to-machine security protocol in industrial IoT	IoT
Truong et al. [18]	2019	SASH, integrating blockchain with IoT platform	Showed moderate overhead. SASH is yet to be implemented in global policies in network	IoT
Karati et al. [19]	2019	gCLSC Cryptography, certificateless signcryption	Minimal storage, i.e., 50% less on comparing CLSC Can be enhanced by incorporating revocation and discarding the bilinear pairing efficiently	IoT
Seyfollahi et al. [20]	2020	Reliable Data Dissemination for the Internet of Things (RDDI) using Harris Hawks Optimization (HHO)	Improved energy consumption, packet forwarding distance, and packet delivery ratio by 3.12%, 17.5%, and 43.5%, respectively Yet to be evaluated on jamming attack	IoT

(continued)

Table 1 (continued)

Author	Year	Technique/method	Result/challenges	Domain
Gochhayat et al. [21]	2020	Delegating the resource-consuming cryptographic to local entity	Reduced the communication overhead, generation of extra certificates	IoT
Li et al. [22]	2020	Node-oriented secure data transmission (NOSDT)	Improved performance of social networks by reducing the transmission impact of malicious nodes More work can be done on forwarding path, better methods for detection of malicious nodes	Social network in IoT
Khan et al. [23]	2020	Blockchain empowered with Deep Extreme Learning Approach (DELM)	Accuracy = 93.91% Exploring extensions through the application of further datasets and architectures	Smart home IoT
Mahdi et al. [24]	2021	Super ChaCha	Increased complexity time requires 2512 keys to break brute-force attack	IoT
Pampapathi et al. [25]	2021	Improved Adaptive Neuro Fuzzy Inference System (IANFIS) and modified elliptical curve cryptography (MECC)	Better performance than ANFIS; also, MECC is 96% better when compared to ECC	IoT
Manogaran et al. [26]	2021	Blockchain-assisted secure data sharing (BSDS)	5.67% high response rate with FAP of 4% and reduced the failure rate by 2.14% rate with FAP of 2%, respectively	Industrial IoT
Naresh et al. [27]	2021	Identity-based online/offline signcryption scheme (OOSC)	Less computational time and secure against IND-CC2 Yet to be implemented in healthcare monitoring systems and in industrial systems	IoT

(continued)

Table 1 (continued)

Author	Year	Technique/method	Result/challenges	Domain
Miao et al. [28]	2021	Federated Learning-Based Secure Data Sharing (FL2S)	Better privacy protection and data quality in secure data sharing	IoT

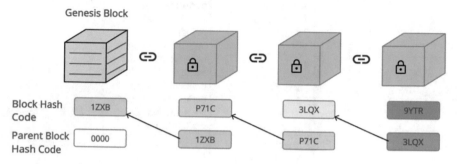

Fig. 2 Blockchain consisting of blocks linked via hash codes [29]

ensures that it is one and only version of truth that is agreed upon by all nodes of blockchain [29]. A visualization of blockchain is given in Fig. 2.

Blockchain technology comprises several components on which blockchain process depends and operates. There are protocols that must be followed for creation, addition, and storing data in blocks. Some of the major requirements of blockchain are:

- Smart contracts: It is a program that is stored on blockchain when certain conditions are met, makes transaction transparent and irrevocable.
- Tokenization: It enables the digital representation of rights, goods, and services. Tokenization enables users to exchange values and trust without engaging any centralized authority.
- Data security: Data security is an important aspect and necessary need of blockchain technology so as to protect data from tampering.
- Decentralized data storage: It is a prerequisite for a distributed or scattered system.
- Immutability: In distributed ledger, all the transactions/records stored should not be modified. This ensures that data is safe and secure.
- Consensus: It makes sure that every new block added should only be integrated in blockchain when all the legit users in the network agree.
- Typed blocks: It is necessary for smart contracts as well as high-speed business transactions. So, formatting is different for the different kinds of blocks which includes time, consensus algorithm, transactions per blocks, and data types of content it has.
- Sharding: It is required for dividing the content over subset of nodes so that there is less burden on each node.

- Access rights management: For managing access privileges and assigning purpose, cryptography techniques such as private and public key encryption cryptography as well as distributed databases with user identity are needed.
- Standards used to manage permissioned blockchains: Due to the obvious immutability of the blockchain network, data can only be viewed in a precise order. Though public certificates are accessible on the public blockchain, those without the private key cannot be granted authority. As a consequence, the data should always be organized according to data aspects such as the user's IP address, name, code, and XML schema. As part of the communication process, all of these information are shared with the consortium.
- Standard data formatting: Standardizing data formats while considering Application Programming Interfaces (APIs) is also required in blockchain systems. To communicate within the similar network, any organization in the blockchain network must utilize similar data formatting.
- Updatability: Updating data in distributed ledger is an essential component for records for every node that carries a transaction inside a peer-to-peer network, and data must be formatted and updated in a methodical manner.
- P2P encryption between blockchain nodes: Blockchain requires encryption to protect transactions among end nodes which might join together.
- UX: The user interface or UX is among the most important aspects of a system since it offers users with a simple and convenient application environment [29].

4 Security Challenges in IoT

IoT security is vital for smooth functioning of the system. There are several security parameters that need to be taken into consideration. Hence, to make IoT network more secure and reliable, we must meet the security challenges. Some of the parameters that we must focus on are as below [30]:

- Confidentiality: It essentially means hiding information from people who are not authorized to read it. IoT collects various sensitive information from devices such as in medical healthcare equipment, heart rate, temperature, and pressure detectors. The important details from these devices can be traced. So, every bit of information must be transmitted confidentially from sensor devices.
- Integrity: It is defined as preventing modification of data by unauthorized person in communication process. In IoT, a single change in data bit could potentially alter the entire meaning from NO to YES.
- Availability: It can be defined as the services running the system should be within reach at any time to anyone who is an authenticated entity. Availability in smart home, like turning off the smart lights via remote control (RC), should be able to perform anytime as per user's request [31].
- Trust and privacy: Trust and privacy are important aspects in IoT security. In IoT, trust management consists of data gathering that is reliable, data fusion and mining that is reliable, and user privacy that is enhanced. Frequency of answers,

consistency of replies, physical proximity, common aims, common ecosystem, history of engagement, and availability are some of the characteristics used to calculate trust. Privacy is also an important aspect. It is defined as no information would be shared with anyone without the consent of the person. In IoT, data collection is vital in public services; hence, regarding what and to whom the information should be shared, the decision must be made by the person himself.

- Authentication and access control: Authentication means that the two (both) communication parties are confident that they are conversing with the genuine counterpart. The confidentiality, integrity, and availability of data are all ensured by a successful authentication system [32].
- Accountability: In a heterogeneous IoT context, accountability assurance can help determine which device produced which data and also which device processed which data. It has the ability to hold people accountable for their acts.
- Auditability: It basically means that we must be able to monitor each event that occurs in an IoT system in order to provide auditability.
- Non-repudiation: In the Internet of Things, non-repudiation means "the system must assure whether the event occurred or not" [33].

For the various security issues in IoT networks, blockchain can perform really well as blockchain is centralized, immutable, traceable, has high speed, can store records, and can provide security and privacy. Hence, it can serve better than the traditional methods used. These were some important terms that should be considered while strengthening IoT security. There are many issues that can be solved via blockchain. Also, the aforementioned security parameters can be made robust by blockchain.

5   IoT Challenges in Precision Agriculture: Case Study

Blockchain technologies can minimize the security challenges in IoT networks and along with that provide solutions to some more major challenges in precision agriculture described below:

- Extending address space: The communication protocol IPv6 address space constraint is a significant scalability issue when it comes to addressing of IoT devices. Blockchain has 160-bit address space, whereas IPv6 consists of 128-bit address space. Considering 160-bit address space, blockchain can create and allot address spaces about $1.46*10^{48}$ for IoT nodes. Hence, the possibility of address collision is very unlikely. This is secure and sufficient enough to provide IoT devices a Global Unique Identifier (GUID). Thus, no necessary requirement is there for a central body to provide and generate limited Internet-Assigned Number Authority with blockchain technology.
- Managing identity of things: Blockchain can be used to create trustworthy identities, track ownership of commodities, products, and services. Another advantage of blockchain is data translucency and end-to-end process traceability. For

instance, the TrustChain protocol has been proposed for validating and administering trustworthy transactions in distributed IoT networks while retaining integrity. Each and every block in the TrustChain displays a transaction among two IoT participants, plus the hash codes of earlier transactions are used to create new transactions. Aside from security, the fundamental benefit of TrustChain is that each and every agent in the system monitors the interaction of others and gathers records to calculate trustworthy levels. In addition, blockchain can provide control functions for trustworthy and for decentralized transactions as well. It also enables remote asset management and rapid end-to-end data verification among IoT devices.

- IoT transaction verification: The network of blockchain has the potential to play a significant role in the authentication as well as authorization of IoT systems. Entire IoT transactions made by devices are recorded on the distributed or shared ledger and may be monitored and traced safely using blockchain. The valid sender, who has a unique private key (PK) and GUID, will always cryptographically confirm each and every IoT transaction communicated with the blockchain system. As a result, it would be easier to confirm authentication and integrity of the triggered or activated transaction. Blockstack is a popular blockchain approach that makes use of JSON Web Token (JWT) to easily authenticate IoT transactions. One of the applications of blockstack is that it can be utilized in smart greenhouse for authenticating access.

- Securing IoT communications: Many classical protocols such as DTLS/TLS protocols have some limitations particularly in computation time or memory needs. Furthermore, using the common PKI protocol, these methods have certain issues with centralized governance and control of key production and distribution. By providing each and every device, a unique pair of GUID and PKI: once installed and connected to the blockchain network, the blockchain could solve these challenges and improve key management among IoT devices. With aid of blockchain, additional secure communication enhancements can be imagined, such as eliminating the need for a handshake phase in the DTLS or TLS protocols to exchange PKI certificates. As a result, for creating secure interactions between IoT devices, the ideal solution for covering runtime computing and memory management needs would be blockchain. Furthermore, IoT device firmware can be continuously hashed into a blockchain for identifying IoT malware and alerting device owners to take necessary security actions against the identified malicious bot. The transmitter node hashes a message to send to another IoT node and stores the hash code in a blockchain network. The recipient node, on the other hand, hashes the identical message. The verification protocol states that if the hash value on the received message matches the hash value on the blockchain, then the message has not been modified or tampered during transmission [34].

6 IoT Performance Challenges and Blockchain Solutions

Because of the rising number of linked IoT gadgets in precision agricultural networks, IoT systems will need to coordinate a large variety of network topologies in the future and evaluate massive amounts of data at a rapid rate. As a result, the performance of IoT networks in precision agriculture systems portrays some more major challenges. Five obstacles might be characterized as IoT performance issues in precision agriculture networks as shown in Fig. 3. Blockchain technology can help with such issues as well.

- Blockchain and sensing problem: Mainly, this issue arises in perception layer of IoT model. Many agricultural apparatuses such as tractors, irrigation machines, smart greenhouses, farming devices consist of sensors embedded, which constantly generate data about operating status and allow IoT nodes to send and receive data via IoT cloud. In this scenario, blockchain can be utilized to define communication protocols among these sensors in addition to keep track of all M2M transactions. For example, IOTA, which is a new update of blockchain platform, is designed in a way to perform large transactions in IoT devices using IOTA as well as DAG. Application of IOTA is that it can eliminate the scalability issue in precision agriculture.
- Blockchain and energy consumption problem: Prominently, this issue is related to the network layer in the IoT layer paradigm. Generally, IoT devices are low constraint, i.e., they are expected to be low-power devices. As IoT gadgets in precision agriculture become more widespread, wireless devices have to be used, which consumes much greater energy than that wired one. However, because of the decentralization aspect of Blockchain, it may be possible to implement certain solutions to the energy consumption challenge. Blockchain such as private blockchain could be used to ensure that the ratio of high compute power to high bandwidth connection for the IoT node is maintained. Blockchain will help in

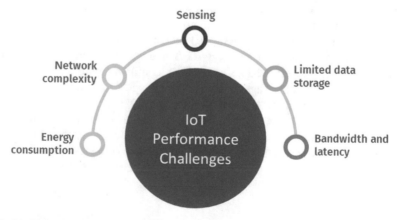

Fig. 3 Five IoT performance challenges

maximizing electrification as blockchain has decentralized ledgers, so it can help by establishing decentralized energy ledgers for things utilized in precision agriculture network, for instance, monitoring a variety of sensors and batteries. We will also be able to measure the amount of energy consumed by IoT devices and sensors in real time, all because of blockchain [35].

- Blockchain and networks' complexity problem: Again, this issue affects the network layer in the IoT layer paradigm. Various heterogeneous network topologies in precision agriculture lead to complicated communications in IoT network system. In precision agriculture, farming equipment's have to communicate through various platforms and infrastructures so as to interact flawlessly. It is achievable; however, it will be tough, costly, moreover time-consuming. So, to address these issues, blockchain can assist in the acquisition and management of data through secured standard-based and decentralized networks. Through design patterns, blockchain can handle communications among IoT devices, resulting in less complication in IoT communications and a reduction in data transmission delay within precision agriculture networks [34].

- Blockchain—bandwidth and latency problem: One more IoT performance challenge is bandwidth along with latency in device communication. In IoT, data traffic emerges outside the data center. As a result, communication among an enormous number of scattered IoT devices must be enabled as soon as possible. Every short time period, a significant number of various IoT devices must be updated. Furthermore, device communication requires a large number of route options and several levels of packet inspection. These issues can be mitigated if the data center is replaced with blockchain. Because of blockchain's decentralization, the burden will be spread closer to the endpoints. As a result, IoT connectivity will be built with low latency and inefficient bandwidth. Due to the increasing scale of blockchains, the requirement for computational power, storage, and bandwidth has become critical. These constraints, on the other hand, can be rectified with a private blockchain, which can execute over 1000 transactions per second on Ethereum or Bitcoin.

- Blockchain and limited data storage problem: Massive amounts of data must be saved and handled using adaptable archives due to the quick rise of IoT precision agricultural networks. Conventional cloud-based storage systems are limited in their ability to manipulate vast amounts of varied types of IoT data. So, this restriction is based on real-time data monitoring, high availability, scalability, security, and low latency requirements. For the solution to these mentioned constraints of cloud-based storage, IoT endpoints would be able to perform greater real-time data analysis and manipulation with blockchain-based storage. Blockchain is also a great choice for data security and availability as in blockchain, data cannot be modified. Moreover, unauthorized data exchange can be a risk in the cloud, whereas blockchain allows customers the option to create access protocols without relying on a third party [34].

7 Conclusion

The IoT network is growing at a rapid speed in terms of its size and wide implementation in different sectors of society. The usage nowadays has become enormous, and a large amount of data is being transferred, which has made data more vulnerable to threats. The need of the hour is securing IoT devices to make communication safer and more reliable. Blockchain has an instrumental role in securing IoT device communication. It is an immutable and distributed ledger, transparent, and traceable providing enormous benefits. In this paper, we took an overview of blockchain by defining it conceptually and its required parameters. Considering blockchain over traditional methods has several benefits. Some IoT issues were also discussed such as addressing address space, managing object identification, and transaction verification. Blockchain has also played a crucial role in providing solutions for IoT performance challenges in IoT precision agriculture such as in sensing, network complexity, energy consumption, limited data storage, bandwidth, and latency issues. Further, a reliable precision agricultural infrastructure framework can be developed. Also, the implementation of a blockchain-based technique to secure data dissemination in smart homes will be carried out in the future.

References

1. Lee I, Lee K (2015) The Internet of Things (IoT): Applications, investments, and challenges for enterprises. Bus Horiz 58(4):431–440. https://doi.org/10.1016/j.bushor.2015.03.008
2. Bodkhe U, Tanwar S (2021) Secure data dissemination techniques for IoT applications: research challenges and opportunities. Softw Pract Exp 51(12):2469–2491. https://doi.org/10.1002/spe.2811
3. Bhuvaneswari V, Porkodi R (2014) The internet of things (IOT) applications and communication enabling technology standards: an overview. In: Proceedings—2014 international conference on intelligent computing applications, ICICA 2014, pp 324–329. https://doi.org/10.1109/ICICA.2014.73
4. Estimated data on number of IoT devices connected worldwide. https://www.statista.com/statistics/1183457/IoT-connected-devices-worldwide/. Accessed 17 Jan 2022
5. Anand P, Singh Y, Selwal A, Singh PK, Felseghi RA, Raboaca MS (2020) IoVT: internet of vulnerable things? threat architecture, attack surfaces, and vulnerabilities in internet of things and its applications towards smart grids. Energies (Basel) 13(18). https://doi.org/10.3390/en13184813
6. Chen S, Xu H, Liu D, Hu B, Wang H (2014) A vision of IoT: applications, challenges, and opportunities with China perspective. IEEE Internet of Things J 1(4): 349–359. Institute of Electrical and Electronics Engineers Inc. https://doi.org/10.1109/JIOT.2014.2337336
7. Real world examples of IoT. https://www.edureka.co/blog/iot-applications/. Accessed 17 Jan 2022
8. Malhotra P, Singh Y, Anand P, Bangotra DK, Singh PK, Hong WC (2021) Internet of things: evolution, concerns and security challenges. Sensors 21(5):1–35. https://doi.org/10.3390/s21051809
9. Wu H, Han H, Wang X, Sun S (2020) Research on artificial intelligence enhancing internet of things security: a survey. IEEE Access 8: 153826–153848. Institute of Electrical and Electronics Engineers Inc. https://doi.org/10.1109/ACCESS.2020.3018170

10. Anand P, Singh Y, Selwal A, Alazab M, Tanwar S, Kumar N (2020) IoT vulnerability assessment for sustainable computing: Threats, current solutions, and open challenges. IEEE Access 8:168825–168853. https://doi.org/10.1109/ACCESS.2020.3022842
11. Shah Y, Sengupta S (2020) A survey on classification of cyber-attacks on IoT and IIoT devices. In: 2020 11th IEEE annual ubiquitous computing, electronics and mobile communication conference, UEMCON 2020, pp 0406–0413. https://doi.org/10.1109/UEMCON51285.2020.9298138
12. Kamran M, Khan HU, Nisar W, Farooq M, Rehman SU (2020) Blockchain and Internet of Things: a bibliometric study. Comput Electr Eng 81. https://doi.org/10.1016/j.compeleceng.2019.106525
13. Bhushan B, Sahoo C, Sinha P, Khamparia A (2021) Unification of Blockchain and Internet of Things (BIoT): requirements, working model, challenges and future directions. Wireless Netw 27(1):55–90. https://doi.org/10.1007/s11276-020-02445-6
14. Elhoseny M, Ramírez-González G, Abu-Elnasr OM, Shawkat SA, Arunkumar N, Farouk A (2018) Secure medical data transmission model for IoT-based healthcare systems. IEEE Access 6:20596–20608. https://doi.org/10.1109/ACCESS.2018.2817615
15. Khari M, Garg AK, Gandomi AH, Gupta R, Patan R, Balusamy B (2020) Securing data in Internet of Things (IoT) using cryptography and steganography techniques. IEEE Trans Syst Man Cybern Syst 50(1):73–80. https://doi.org/10.1109/TSMC.2019.2903785
16. Aggarwal S, Shojafar M, Kumar N, Conti M (2019) A new secure data dissemination model in internet of drones. In: ICC 2019 - 2019 IEEE International Conference on Communications (ICC), Shanghai, China, pp 1–6. https://doi.org/10.1109/ICC.2019.8761372
17. Ostad-Sharif A, Arshad H, Nikooghadam M, Abbasinezhad-Mood D (2019) Three party secure data transmission in IoT networks through design of a lightweight authenticated key agreement scheme. Fut Gener Comput Syst 100:882–892. https://doi.org/10.1016/j.future.2019.04.019
18. Truong HTT, Almeida M, Karame G, Soriente C (2019) Towards secure and decentralized sharing of IoT data. In: Proceedings - 2019 2nd IEEE International Conference on Blockchain, Blockchain 2019, pp 176–183. https://doi.org/10.1109/Blockchain.2019.00031
19. Karati A, Fan CI, Hsu RH (2019) Provably secure and generalized signcryption with public verifiability for secure data transmission between resource-constrained IoT devices. IEEE Internet Things J 6(6):10431–10440. https://doi.org/10.1109/JIOT.2019.2939204
20. Seyfollahi A, Ghaffari A (2020) Reliable data dissemination for the Internet of Things using Harris hawks optimization. Peer-to-Peer Netw Appl 13(6):1886–1902. https://doi.org/10.1007/s12083-020-00933-2
21. Gochhayat SP et al (2020) Reliable and secure data transfer in IoT networks. Wireless Netw 26(8):5689–5702. https://doi.org/10.1007/s11276-019-02036-0
22. Li X, Wu J (2020) Node-oriented secure data transmission algorithm based on IoT system in social networks. IEEE Commun Lett 24(12):2898–2902. https://doi.org/10.1109/LCOMM.2020.3017889
23. Khan MA et al (2021) A machine learning approach for blockchain-based smart home networks security. IEEE Netw 35(3):223–229. https://doi.org/10.1109/MNET.011.2000514
24. Mahdi MS, Hassan NF, Abdul-Majeed GH (2021) An improved chacha algorithm for securing data on IoT devices. SN Appl Sci 3(4). https://doi.org/10.1007/s42452-021-04425-7
25. Pampapathi BM, Nageswara Guptha M, Hema MS (2021) Data distribution and secure data transmission using IANFIS and MECC in IoT. J Ambient Intell Human Comput. https://doi.org/10.1007/s12652-020-02792-4
26. Manogaran G, Alazab M, Shakeel PM, Hsu CH (2021) Blockchain assisted secure data sharing model for internet of things based smart industries. IEEE Trans Reliab. https://doi.org/10.1109/TR.2020.3047833
27. Naresh VS, Reddi S, Kumari S, Divakar Allavarpu VVL, Kumar S, Yang MH (2021) Practical identity based online/off-line signcryption scheme for secure communication in internet of things. IEEE Access 9:21267–21278. https://doi.org/10.1109/ACCESS.2021.3055148
28. Miao Q, Lin H, Wang X, Hassan MM (2021) Federated deep reinforcement learning based secure data sharing for Internet of Things. Comput Netw 197. https://doi.org/10.1016/j.comnet.2021.108327

29. Bodkhe U et al (2020) Blockchain for Industry 4.0: a comprehensive review. IEEE Access 8:79764–79800. https://doi.org/10.1109/ACCESS.2020.2988579

30. Malik R, Singh Y, Sheikh ZA, Anand P, Singh PK, Workneh TC (2022) An improved deep belief network IDS on IoT-based network for traffic systems. J Adv Transp 2022:1–17. https://doi.org/10.1155/2022/7892130

31. Anand P, Singh Y, Selwal A, Singh PK, Ghafoor KZ (2021) IVQFIoT: intelligent vulnerability quantification framework for scoring internet of things vulnerabilities. Expert Syst. https://doi.org/10.1111/exsy.12829

32. Anand P, Singh Y, Selwal A (2021) Internet of things (IoT): vulnerabilities and remediation strategies. In: Singh PK, Singh Y, Kolekar MH, Kar AK, Chhabra JK, Sen A (eds) Recent innovations in computing. ICRIC 2020. Lecture notes in electrical engineering, vol 701. Springer, Singapore. https://doi.org/10.1007/978-981-15-8297-4_22

33. Patel C, Doshi N (2019) Security challenges in IoT cyber world, pp 171–191. https://doi.org/10.1007/978-3-030-01560-2_8

34. Torky M, Hassanein AE (2020) Integrating blockchain and the internet of things in precision agriculture: analysis, opportunities, and challenges. Comput Electron Agric 178. Elsevier B.V. https://doi.org/10.1016/j.compag.2020.105476

35. Georgiou K, Xavier-De-Souza S, Eder K (2018) The IoT energy challenge: a software perspective. IEEE Embedded Syst Lett 10(3):53–56. Institute of Electrical and Electronics Engineers Inc. https://doi.org/10.1109/LES.2017.2741419

The Proof of Authority Consensus Algorithm for IIoT Security

Sonali B. Wankhede and Dhiren Patel

Abstract Blockchain-based technologies have been created across a range of businesses for improving data security and integrity. A network chain is the most remarkable blockchain applications as part of the Industrial Internet of Things (IIoT). Sensor data in real time aids industrial equipment and infrastructures in making decisions and taking specified actions. Without sufficient security, an IIoT system can cause operational disruption and financial loss and lead to malicious activities and system manipulation. Detecting anomalies in IIoT networks is critical for network security. In this paper we discuss the Proof of Authority consensus algorithm. The Proof of Authority (PoA) depends on the reputation of trusted parties in blockchain network. The PoA consensus mechanism is based on the values of identities on a network, and validators stake their own identities and reputation. So, the validating nodes that are randomly picked as trustworthy safeguard the Proof of Authority Blockchain network. In Proof of Authority model, transactions are reviewed by already approved network users, and it operates with a fixed number of block validators. Since the identities of the nodes are trusted and known, the process can be used in applications like supply chains and trade networks.

Keywords Blockchain · Industrial internet of things · Trust evaluation · Proof of authority

S. B. Wankhede (✉) · D. Patel
VJTI Mumbai, Mumbai, India
e-mail: sbwankhede_p21@ce.vjti.ac.in

D. Patel
e-mail: dhiren@coed.svnit.ac.in

SVNIT Surat, Surat, India

1 Introduction

Machine to machine (M2M) connections are predicted to increase from 5.6 billion in 2016 to 27 billion by 2024 [1]. The increase in number shows that Internet of Things (IoT) is important emerging markets that will be a cornerstone of the growing digital economy. The gadgets will be connected to the Internet and local devices but at the same time will be able to communicate with other devices on Internet, so the security and privacy of this IoT applications plays a major role. Some countries like Western Europe, North America, and China are the primary driving countries [1]. Upcoming IoT applications cannot reach high demand and may lose all of the potential without a trusted and interoperable IoT ecosystem. Internet of things has its own unique security challenges including the issues of Privacy, Authentication, Management, and information storage. The immutable and tamperproof data security characteristics of blockchain make it a significant tool in a variety of fields, including health care, military, banking, and networking.

The Industrial Internet of Things (IIoT) is a subcategory of Internet of Things (IoT). IIoT networks serve as a platform for a variety of applications and enable us to respond to customer needs, particularly in industrial settings like smart factories [2]. Due to the benefit of blockchain technology, it is widely used in smart factories, smart homes, smart cities, and healthcare systems [2, 3].

Numerous items in current smart factories are connected to public networks and smart systems such as temperature monitoring systems, Internet-enabled lights, IP cameras, and IP phones assist many activities. These gadgets store personal and sensitive information and provide life-saving services [2, 4]. The key difficulty will be securely storing, collecting, and sharing data. Through rigorous authentication, Blockchain Technology assures data integrity in the smart factories, as well as the availability of communication backbones. User's data privacy should be protected during transmission, consumption, and storage in smart factories [5].

Fraudsters intending to access, edit, or use stored data for harmful purposes are vulnerable to tampering. Such attacks can be classified as anomalous events, as they deviate significantly from normal behavior [3, 6].

The major goal of this article is to identify suspicious parties and transactions in a blockchain-based IIoT network targeted at smart factories. Abnormal conduct can also be used as a proxy for suspicious behavior [5].

Rest of the paper is organized as follows: Sect. 2 discusses related work in this area. Section 3 discusses foundations and methodology.

Section 4 discusses challenges in IIOT. Section 5 discusses blockchain to address challenges in IIoT with conclusions and references at the end.

2 Related Works

The smart factory has become the Center of interest as the IIoT develops. The research on smart factory for securing access to industrial data is very crucial, and it relies heavily on developing technologies like AI and machine learning [7]. There are basically three categories of related study. Various surveys on the security of IoT and privacy issues are already available.

Yuchen and colleagues compiled a list of security concerns in IoT applications. The authors of [8] talked on the security concerns with services based on locations. Their focus was on the specific issues of IoT device localization and positioning. In [9], Anne et al. focus on IoT middleware security challenges and present a full assessment of related protocols and security issues. M. Guizani et al. examined several trust management strategies for IoT, as well as their benefits and drawbacks, in [10].

Jung et al. [10] proposed a smart factory web based on the IIoT concept. Shin et al. [11] merged the intelligent computing technology for gathering information utilizing ICT technology in industry and presented a network topology based on edge computing to satisfy the real-time needs of IIoT.

Domova et al. [12] proposed a set of data mining alarm algorithms for alert processing in smart factories.

Lee et al. [13] presented the re-industrial architecture to boost intelligent production, sharing and equipment management in smart factories, as well as to construct a reliable cloud platform. The use of blockchain as a trusted entity to ensure the security of IIoT has become widespread.

Wan et al. [14] built and reshaped a distributed network using traditional IIoT architecture and blockchain. The flexibility and security capabilities of blockchain for creating a credit mechanism based on consensus and the proof of work mechanism based on the credit to deal with IIoT devices.

3 Methodology

Blockchain and the Internet of Things are significant technologies that will have huge impact on the IT and telecommunication industries. These technologies are focused at increasing transparency, visibility, comfort, and trust. The IIoT devices collect real-time data from sensors and blockchain. A distributed, decentralized, and shared ledger is the key to data security [15]. The entries in blockchain are time stamped and chronologically arranged. By using the cryptographic hash keys, the entries in the ledger are firmly tied with the prior entries. We discuss here a foundation block consensus algorithm called as Proof of Authority to be used for security in smart factories.

3.1 Proof of Authority

Proof of Authority (PoA) is a new family of consensus algorithms with great performance and fault tolerance. Blockchain platforms can be classed as permissionless or permissioned or restricted to a specific number of authority nodes. The PoA protocol is a novel BFT algorithm that targets only approved nodes and permits to join consortiums and submit transactions [16].

The scalability and performance differences between Byzantine Fault Tolerant (BFT)-like algorithms and PoA algorithms include that PoA algorithms require few message exchanges, and therefore, they give greater performance, and they can also be implemented at a wider scale than standard BFT methods [15]. The crucial fact is that the PoA protocol can work normally with opponents accounting for half of the total players, whereas Byzantine tolerates only third of the total [16]. In the worst-case scenario, the actual performance of PoA protocol lacks comprehensive examination.

PoA algorithm's consensus mechanism is reliant on group of trusted nodes known as authorities. Every authorized node has a unique ID, and at least $N/2 + 1$ of them must be trusted for the network. The nodes demonstrating their authority in PoA are given the rights of generating new blocks. Validators are the nodes that execute the software that allows them to insert transactions into blocks. The process does not require validators to continually monitor the systems, but it does necessitate keeping the machine secure. Proof of Authority is suitable for private as well as public networks where trust is spread, such as the POA Network. Because the PoA consensus mechanism relies on the value of identities, block validators stake their own reputation rather than coins. The faith in the identities chosen secures PoA.

PoA consensus may vary depending on the implementation; however, they are commonly applied under the following conditions:

1. Validators must verify their true identity: A candidate must be prepared to invest money and risk his or her reputation. A rigorous approach decreases the danger of choosing shady validators and encourages long-term commitment to the blockchain.
2. Validators must be chosen in the same way for all applicants:

    To keep the blockchain's integrity, validators' identities must be validated. PoA consensus has the following benefits:

1. Risk tolerance is high as long as 51% of the nodes are not malicious.
2. The duration between fresh blocks being generated is predictable.
3. Transaction volume is high.
4. Far more sustainable than computationally intensive algorithms like Proof of Work.

Permissionless (Bitcoin, Ethereum) and permissioned consensus techniques can be found on blockchain platforms like Apla or Ethereum. The nodes in a permissioned blockchain are pre-authenticated. This enables the employment of consensus types in addition to other benefits, giving a higher transaction rate.

Table 1 Comparison of PoA, PoS, and PoW algorithms

Proof of authority	Proof of stake	Proof of work
Take a more human approach to security by integrating validators who stake their reputation	It benefits affluent users: the more your investment, the more likely a block will be validated	If the number of validators is small, block validators can produce a lot of power
Avoids decentralization	It is a decentralized system	It is a decentralized system
Secure against replay and 51% attacks	51% attacks	51% attacks
It was proposed in 2017	It was proposed in 2011	It was proposed in 1993
Takes up less time and energy	The issue of high levels of energy resource waste has been addressed	Energy consumption is crucial to the network's security since it lets it to preserve accurate records of transactions

Table 1 shows the comparisons of some of the consensus algorithms.

3.1.1 PoA Consensus in Apla Platform

Only validating nodes have the ability of generating new blocks. The blockchain registry keeps track of validating nodes. The order of generating new blocks are determined by the order in which they appear in this list.

The current leader node generates new block at the current moment and is determined by the formula below.

$$\text{Leader} = ((\text{Time} - \text{First})/\text{Step})$$

where

Leader—current leader node, Time—current time

First—generation time of first block

Step—numberofseconds, Nodes—number of current nodes.

- **Generation of new blocks**:

The current time interval's leader node generates the new block. The leader's role is given to the next validating node from the list during each time interval.

The new block is created by the leader node as follows:

1. Accept the new block and verify the following:

    i. New block was created by current interval's leader node.
    ii. This leader node has generated no further blocks.
    iii. The block has been produced successfully.

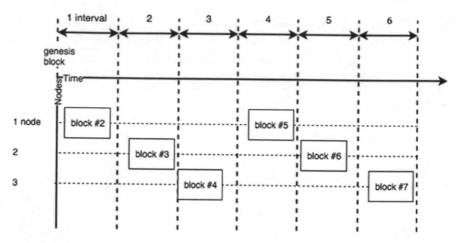

Fig. 1 Generation of new blocks—Apla platform

2. Execution of transactions from the block in sequential order and checking if all transactions are completed appropriately and that block production limits are not exceeded.

3. Accept or reject the block:

   i.   If there is successful block validation then add new block to the blockchain of the node.
   ii. If the validation fails then reject the block and provide an error transaction of block.
   iii. If the validating node that generated erroneous block continues to do so, it may be blocked or terminated.
   iv. Transaction queue is used to collect new transactions.

- **Verification of new blocks**:

   1. Accept new block and verify the following:
      i. The block was created by the current interval's Leader node.
      ii. The Leader node has generated n further blocks, and the block has been produced successfully.
   2. Execute transactions from the block one by one. Check that all transactions are completed appropriately and that block production limits are not exceeded.
   3. Accept or reject the block based on the previous step:
      i. If block validation is successful, add the new block to the node's blockchain.
      ii. If block validation fails, reject the block and provide a faulty block transaction. It may be blocked or removed if the validating node that generated this incorrect block continues to do so.

Fig. 2 Industrial internet of things

4 Challenges in IIoT

IIoT devices being low cost and constrained, there are number of challenges to address.

1. Device privacy: IIoT devices are vulnerable to disclosing personal data.
2. Cost and traffic: To keep up with the exponential increase of IIoT devices.
3. Cloud service overload and insufficiency: Cloud services are unavailable due to attacks, software faults, power outages, and other issues.
4. Defective architecture: The components of an IIoT device have a single point of failure that affects the device and the network.
5. Data manipulation: Data from IIoT devices is extracted and then manipulated before being used in an inappropriate way.

5 Benefits of Blockchain in IIoT

Let us apply blockchain to address some of the challenges discussed earlier.

1. **To store data from devices**: this arrangement is cloud-connected and allows IIoT apps to be used from anywhere.
2. **Secure data storage**: The distributed design of blockchain avoids the risk of becoming a single point of failure that many cloud-based IIoT applications face. The data created by the devices can be simply saved on the blockchain in a secure manner regardless of the distance between them [17].
3. **Data encryption**: The original data can be saved in the cloud, and hash key can be linked to it. Since, only 256 bit hash key can be saved in blockchain. Any change in the data will result in the change of hash of the data, keeping the information safe.

The hash allows only authorized parties to access the data from the cloud. Probability of storing faulty data from devices is reduced by using blockchain as a solution.

4. **To prevent unwanted access**: IIoT systems require regular connectivity between different nodes. The communication in blockchain is based on public and private keys. Even if the data is accessed by an unwanted party, the contents will be incomprehensible by encrypting the data using keys. Therefore, blockchain data structure addresses the security challenges of IIoT applications.

5. **Blockchain-based proxy architecture for resource-constrained devices**: Despite the fact that blockchain provides a variety of security characteristics for a distributed environment, IIoT has a unique resource limitation.

    IoT devices can't keep big ledgers due to the limited resources. Various efforts have been made to make the usage of blockchain in IIoT easier. One of the possible methods is proxy-based architecture. Proxy servers can be set up in the network to store data in an encrypted format. The client can download the encrypted resources from the proxy servers.

    Since smart factories handle sensitive data, storing it on the blockchain is both monetarily and computationally burdensome. As a result, the smart factory stores the actual smart device and sensor data. The smart factory data also provides information on the data type and control states.

6   Conclusion

In this paper, we have discussed the Proof of Authority consensus algorithm addressing the challenges in IIoT devices. The consensus conditions for the Proof of Authority and the working mechanism are explained. It has been observed that Proof of Authority is an appealing alternative to Proof of Work-based protocols for many blockchain applications due to its excellent performance and energy efficiency. This survey is considered to be a significant resource for improving security in IIoT applications.

References

1. Rajesh Kandaswamy DF Blockchain-based transformation. https://www.gartner.com/en/doc/3869696-blockchain-basedtransformation-a-gartner-trend-insight-report/, online; accessed June 5, 2018
2. Wan J, Li J, Imran M, Li D, Amin FE (2019) A blockchain-based solution for enhancing security and privacy in smart factory. IEEE Trans Ind Inform 15(6):3652–3660
3. Scicchitano F, Liguori A, Guarascio M, Ritacco E, Manco G (2019) Blockchain attack discovery via anomaly detection. Con-siglio Nazionale delle Ricerche, Istituto di Calcolo e Reti ad Alte Prestazioni (ICAR)

4. Xu Q, He Z, Li Z, Xiao M, Goh RSM, Li Y (2020) An effective blockchain-based, decentralized application for smart building system management. In: Real-time data analytics for large scale sensor data. Elsevier, pp 157–181
5. Podgorelec B, Turkanovic M, Karakati S (2020) A machine learning-based method for automated blockchain transaction signing including personalized anomaly detection. Sensors 20(1):147
6. Saad M, Spaulding J, Njilla L, Kamhoua C, Shetty S, Nyang D, Mohaisen D (2020) Exploring the attack surface of blockchain: a comprehensive survey. IEEE Commun Surv Tutor 22(3):1977–2008
7. Tan L, Xiao H, Yu K, Aloqaily M, Jararweh Y (2021) A blockchain empowered crowdsourcing system for 5g-enabled smart cities. Comput Stand Interf 76:103517
8. Tan L, Xiao H, Yu K, Aloqaily M, Jararweh Y A blockchain empowered crowdsourcing system for 5g-enabled smart cities. Comput Stand Interf 76:103
9. Ngu AH, Gutierrez M, Metsis V, Nepal S, Sheng QZ (2017) Iot middleware: a survey on issues and enabling technologies. IEEE Internet Things J 4(1):1–20
10. Din IU, Guizani M, Kim B-S, Hassan S, Khan MK (2019) Trust management techniques for the internet of things: a survey. IEEE Access 7:29 763–29 787
11. Shin M, Woo J, Wane I, Kim S, Yu H (2019) Implementation of security mechanism in iiot systems. In: Hwang S, Tan SY, Bien F (eds) Proceedings of the sixth international conference on green and human information technology. Springer Singapore, pp 183–187
12. Domova V, Dagnino A (2017) Towards intelligent alarm management in the age of iiot. In: 2017 Global internet of things summit (GIoTS), pp 1–5
13. Huh S, Cho S, Kim S (2017) Managing iot devices using blockchain platform. In: 2017 19th international conference on advanced communication technology (ICACT), pp 464–467
14. Wan J, Li J, Imran M, Li D, Fazal-e-Amin (2019) A blockchain-based solution for enhancing security and privacy in smart factory. IEEE Trans Ind Inform 15(6):3652–3660
15. Miller D (2018) Blockchain and the internet of things in the industrial sector. IT Prof 20(3):15–18
16. Tseng L (2016) Recent results on fault-tolerant consensus in message-passing networks. In: International colloquium on structural information and communication complexity. Springer, pp 92–108
17. Alphand O, Amoretti M, Claeys T, Dall'Asta S, Duda A, Ferrari G, Rousseau F, Tourancheau B, Veltri L, Zanichelli F (2018) Iotchain: a blockchain security architecture for the internet of things. In: 2018 IEEE wireless communications and networking conference (WCNC). IEEE, pp 1–6

Machine Learning for Security of Cyber-Physical Systems and Security of Machine Learning: Attacks, Defences, and Current Approaches

Ruxana Jabeen, Yashwant Singh, and Zakir Ahmad Sheikh

Abstract The scope of cyber-physical systems (CPS) is extending and is currently being utilized in many critical infrastructures. Any sort of security reach in such infrastructures could trigger human as well as financial loss. So, the security of these systems is of utmost importance. The feasibility of many approaches has been explored by researchers to ensure the security of CPS, one such approach is based on intelligent methods like machine learning (ML) and deep learning (DL). The involvement of intelligent methods also extends the attack surface of CPS, so there is a need to make them secure as well because they are also vulnerable to many types of adversarial attacks. Thus in this work, we review the existing approaches of adversarial attacks like Momentum-based, GAN-based, Fast Gradient Sign Method, Momentum Iterative Fast Gradient Sign Method, the variant of Natural Evolution Strategies, zero-order (ZOO) random position ascent method, Deep fool, and many more. Moreover, we also explored some defence strategies like Modifying Data, Modifying Models, Using Auxiliary Tools, etc., that could be utilized to prevent adversarial attacks. Moreover, we have proposed a methodology for ensuring the security of CPS and the security of defender (intelligent model, i.e. ML and DL) itself.

Keywords Cyber-physical systems · Cyber-attacks · Adversarial attacks · Adversarial defence · Adversarial machine learning · Adversarial threat model

1 Introduction

The Internet of Things (IoT) is a rapidly evolving technology that offers a wide range of services, making it the most rapidly evolving technology with a significant

R. Jabeen (✉) · Y. Singh · Z. A. Sheikh
Department of Computer Science and Information Technology, Central University of Jammu, Rahya Suchani Bagla, Jammu and Kashmir 181143, India
e-mail: jabeenruxana@gmail.com

Y. Singh
e-mail: yashwant.csit@cujammu.ac.in

© The Author(s), under exclusive license to Springer Nature Singapore Pte Ltd. 2023
Y. Singh et al. (eds.), *Proceedings of International Conference on Recent Innovations in Computing*, Lecture Notes in Electrical Engineering 1011,
https://doi.org/10.1007/978-981-99-0601-7_62

influence on society and corporate networks. IoT has developed an important section of modern social life, with applications in knowledge, industry, and health, where it stores sensitive data about businesses and individuals, economic transaction data, product development, and marketing.

As a result of its increased usage in critical systems, it invoked a greater focus on cyber-criminals. The success of IoT cannot be overlooked; nonetheless, assaults and threats against IoT devices and facilities are increasing by the day. Cyber-attacks have become a part of IoT, harming users' lives and societies; hence, proactive measures to guard against cyber-attacks must be done. Cybercrime is a global threat to government and business infrastructure, and it can harm people in a variety of ways. Cybercrime is expected to cost the global economy up to $6 trillion every year, according to estimates. Cyber-attacks can be caused by several factors, including (a) inadequate cyber security in some countries and (b) cyber-criminals using new technology to attack. (c) Services and other business plans can be used to commit cybercrime. As the Internet of Things is evolving at such a rapid pace, it is vital to identify IoT security imitators, problems and incursions on IoT. [1] Infrastructure, as well as the consequences of these risks and assaults, should be extensively investigated. Criminalizing, investigative authorities and processes, digital evidence, threat and norms, and international monitoring and collaboration are all concepts and materials covered by transmission methods and local legal systems and national laws. The scope (multilateral or regional) and application of these contracts varied [1]. We can see the rise or technological advancements from the timeline in Fig. 1. These advancements also grab the attention of malicious users.

Various traditional definitions have been offered to the field of machine learning. In his founding work, Arthur Samuel defines device knowledge as an "area of training aimed at giving supercomputers the potential to study without having to be taught". Depending on the nature of the data labelling, machine learning is classed as supervised, unsupervised, or semi-supervised. The output is labelled and supervised learning predicts an unknown (input, output) mapping using known (input, output) examples (regression and classification). Only input models are sent to the system in unsupervised learning (e.g. estimation of probability density function and clustering). Semi-supervised learning (e.g. image/text recovery systems) is a blend of supervised and unsupervised learning in which a part of data is partially labelled and that portion is used to understand the unlabelled part [2]. Deep learning is a branch of machine learning that deals with artificial neural networks (ANN) which are based on the structure and function of the brain [3]. Deep learning allows deep networks to be placed in a parameter space region where supervised fine-tuning prevents local minima. For jobs where a huge large data is available, even if classification is done for a small number of instances, deep learning approaches obtain very good accuracy [4]. Deep learning has since produced multiple state-of-the-art successes in fields like speech recognition, image recognition, and language translation and is now applicable for a wide range of Artificial intelligence (AI) applications. Deep learning, as it is commonly known, is a statistical technique for categorizing patterns based on sample data utilizing multiple-layer neural networks. Corporations have spent billions of dollars to recruit deep learning experts.

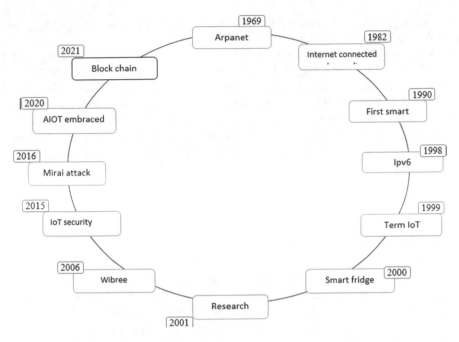

Fig. 1 IoT timeline

Currently, intelligent methods based on machine learning and deep learning have shown encouraging outcome in dealing with cyber-attacks. Specifically, the deep learning models possess the greater capability to deal with cyber-attacks considering the larger flows of data. Also, ML and DL can deal with known as well as unknown attacks. Artificial intelligence (AI) approaches that deal largely with deductive inference can be contrasted to machine learning's inductive inference, i.e. generalizations from a set of observed cases. One of the major applications of machine learning techniques is ontology learning. Ontologies can be learned from scratch using machine learning techniques, or they can be enhanced using existing ontologies. Learning data comes from a variety of sources: linked data, social networks, tags, and textual data. Another common machine learning application is the learning of mappings from one ontology to another (for example, using association rules or similarity-based approaches). Machine learning algorithm is a computational approach that achieves a target without being expressly stated (i.e. "difficult") to do so. These algorithms are "soft programmed" in the sense that they automatically adjust or adapt their design as a result of repetition (i.e. experience) to get better and better at doing the target objective. Training is the process of adapting, and it entails supplying samples of input data as well as desired outcomes.

Manipulation of the system. One of the most common attacks on machine learning systems is to force high-volume algorithms to generate incorrect predictions. Researchers created a platform that integrates human–computer interaction, analytics, gamification, and deception to entice harmful individuals into specific traps while piquing their curiosity.

When an adversary can feed erroneous data into your model's training pool, causing it to learn something it shouldn't, this is known as a poisoning attack. The most typical symptom of a poisoning attack is a shift in the model's border. The flaws with DL's privacy protection have been exposed. According to privacy and security issues, the DL model may be copied or reverse engineered, personal data set can be deduced, and even a recognized face picture of the victim can be recreated. Furthermore, recent research has discovered that the deep learning model is susceptible to adversarial samples altered by undetectable noise, causing the DL model to forecast incorrectly within elevation confidence.

1.1 Architecture of CPS

Many architectural frameworks have been proposed by researchers for CPS as shown in Fig. 2. Among all, one well-known architecture is based on layers namely the perception execution layer, application control layer, and data transmission layer. The perception execution layer is made up of physical components such as sensors and actuators. The primary responsibility of the application control layer is to provide services to users. The transmitting data layer serves as a link between the application control layer and the perceptual execution layer, allowing data to be sent [5]. The other architectures are four-layer, five-layer, and seven-layer architectures. In seven-layer architecture, there are various layers like the physical layer, application layer, data accumulation layer, connectivity layer, edge computing layer, data abstraction layer, collaboration, and process layer. The connectivity layer guarantees that data is transmitted reliably. The analysis and processing of data is the emphasis of the edge computing layer. Data is stored in a variety of ways in the data abstraction layer to construct performance applications. The application information is shared with people and business processes through the collaboration and process layer [6].

Application layer: Researchers in [7] to accomplish varied smart IoT systems, the application layer connects IoT with various types of users (individuals or systems), as well as their specialized requirements.

Data Abstraction Layer: This layer is aware of the many languages used to express data where the information is stored. It'll be able to manage network requirements to appropriate data resources as a result. Layer allows multi-agent systems entities to access knowledge via Java calls regardless of the true information representation language. The data abstraction layer is made up of many application programming interfaces (API) as well as a new component is known as Data Accessing Layer. The application programming interfaces are just a set of Java techniques that connect stored data in one area to the remainder of the network [8].

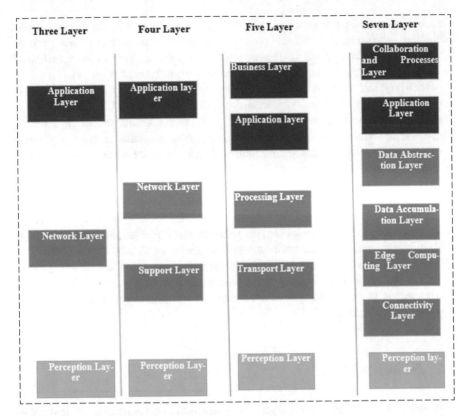

Fig. 2 Various layered architectures for CPS

Connectivity layer: Virtual connections acquired from infrastructure providers are used by the connectivity layer to run virtualization with the required geographic footprint, dependability, and efficiency for telecom operators. Researchers show how wide-area IP multicast, which operates on the highest of a dependable effective network, may be used to provide IPTV distribution at a low cost [8].

Perception layer: This bottom-most layer comprises various robotic things like sensors, vehicles, smartphones, home equipment, and actuators. The intelligent IoT develops a multi-robotic system and delivers innovative features through distributed activities by contacting and integrating. This layer is responsible to operate in the environment, sensing the data, acquiring, and transmitting it to the upper layer [9].

Edge computing layer: The evaluation, filtering, and transformation of information are the emphasis of the edge of the network layer. Users connect with the data centre to get software, hardware, and other computer means, as well as to store information, in models of cloud computing [10].

Accumulation layer: Mobile data is converted to static data via the data gathering layer [10].

Collaboration and processes layer: To get ideas, it consumes and distributes application data with business owners' processes. It provides responses or action to be taken against the data provided. This action for instance can even be an actuation of an electromechanical device upon the instruction from the controller [9].

Cyber-physical systems are devices that combine computer, memory, and interconnection to manipulate and connect with physical processes. A CPS is a device that is managed or monitored by application algorithms and connected to the Internet, with physical and software components that are tightly integrated. From the perspective of the IoT, cloud technology, portability, Big Data, and networks of networked devices and sensors, software architectural concepts must function in an open and highly interactive world [11].

Cyber-physical systems and the IoT are utilized in smart manufacturing to constantly monitor operations and automate tasks that were previously handled by people. Several kinds of services may be established, publicized, and given to clients in the industrial and corporate environment that focuses on these notions. The motive of Industry 4.0 is to improve the effectiveness of design and manufacturing processes, as well as the number and quality of services while lowering prices. Industry 4.0 is a step forward in establishing developed schemes by utilizing new technology and organizational structures [12].

1.2 Security of CPS

Cyber-attacks on the application layer, data transmission layer incursions, and application control layer security problems are all types of network assaults, as per the framework [5, 13–19]. A cyber-attack is an attempt to disable computers or breached a computer system to launch additional attacks or steal data. Attackers use different techniques to launch a cyber-attack. Various cyber and physical attacks are shown in Fig. 3.

1. **Physical Attack**: A physical attack can occur if an IoT device is physically accessed. This type of assault can be carried out by a single employee of the organization who has contact with the IoT device.
2. **Denial of Service (DOS)**: This sort of assault is difficult to steal information from websites or other online services. Hackers take advantage of the service with a big number of botnets by flooding it with requests, causing it to collapse and become unreachable.
3. **Botnet Attack**: It is an assault that occurs when an IoT system is transformed into distantly operated bots that may be utilized as a part of a botnet. Cyber-attackers can access the Internet and convey sensitive and confidential data.
4. **Ransomware Attack**: It is a form of cyber-attack in which a criminal encrypts data and prevents others from accessing it. The decoding is subsequently sold for a profit by the hacker. This sort of attack will have a considerable impact on daily operations.

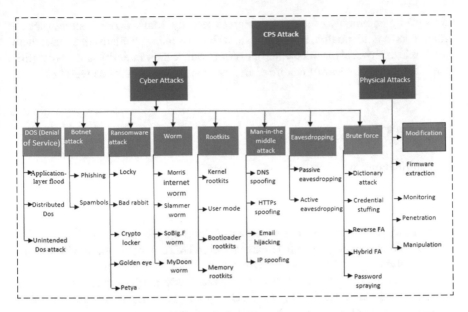

Fig. 3 Various categories of the cyber-physical system

5. **Eavesdropping Attack**: It is a sort of attack where an attacker hijacks traffic with an IoT device and a server through an unsecured connection to get sensitive and confidential data.
6. **Man-in-the-Middle Attack**: A man-in-the-middle attack occurs when an attacker intercepts information between two systems, by monitoring the discussions of two parties.

Because the operations conducted by some CPS are so vital, we want security that is both effective and unique. In CPS, there are basic system distinctions that will push us to think about security in terms that are more directly related to the physical application. We can enhance the security of both cyber-physical systems, where such an approach is most obviously merited, and cyber-only systems, where such a method is more readily overlooked, by including security as required fields design [20].

Information technology systems are vulnerable to a broader spectrum of assaults and design faults than CPS. A full threat model involves both the attacker's environment and the device under attack. Any danger, whether from a hacker or a bug, takes advantage of the program's failure to completely comply with an implicit or explicit definition of its attributes. The physical plant may fail if the timing of real-time processes is disrupted. False data being entered into a control scheme can result in poor choices and damaging actions [21]. Unrestricted information flow can compromise the privacy of data in a system. When physical components are introduced to a cyber system, it becomes much more difficult to determine information flow and mitigate the resulting confidentiality issue. The phrases "availability, integrity, and secrecy" are commonly used to characterize security. Many cyber systems place a premium on confidentiality. Physical systems possess many problems that make them vulnerable to physical assaults which can compromise their integrity and availability. Confidentiality is a security feature of the cyber-physical systems as well [22].

1.3 Contribution

The contributions of our paper entail the following:

The paper reviews existing security concerns in cyber-physical systems (CPS).

- The paper reviews adversarial exploitability of ML and DL techniques.
- The paper reviews adversarial attack and defence methods.
- Discussion about Adversarial Threat Model (ATM) as a prevention methodology and its use case is also included in this paper.
- The paper also discusses the use of generative adversarial networks (GAN) in general and Conditional GAN (C-GAN) in particular for generating fake data samples to perform adversarial attacks.
- The paper also proposes a security methodology for CPS ensuring resilience against adversarial attacks on ML- and DL-based defender.

1.4 Paper Roadmap

The road map of the article is as follows. In Sect. 2, we introduce the security of machine learning and discuss the adversarial attack methods, defence methods, adversarial threat model, and its use case. In Sect. 3, we discuss the use of generative adversarial networks (GAN) for generating fake data samples to perform adversarial attacks. In Sect. 4, we propose a security methodology for CPS based on adversarial machine learning. Section 5 researches challenges and gaps. Finally, the paper is concluded in Sect. 6.

2 Security of Machine Learning

The word "cyber security" refers to a set of methods and technology for restricting access to, attacks on, changes to, or demolition of networks, computers, data, and applications. Network security measures and computer (server) security systems make up cyber security systems. Each of these entities must have at least one antivirus program, firewall, and intrusion detection system (IDS) [23]. IDSs detect and assess unlawful use, alteration, duplication, and damage to information systems in addition to identifying them. Security breaches are defined as internal intrusions (intra-organizational assaults) and exterior intrusions (inter-organizational attacks). Artificial intelligence and machine learning-based approaches for identifying security threats have become indispensable in our lives because dispensable in detecting security threats. The protection of computer systems from cyber-attacks is one of the most important challenges for national and international security. This article presents a survey on security concerns using a variety of machine learning approaches [24]. Machine learning applications will undoubtedly increase in quantity. We were only able to address a handful of the most important features due to space constraints. Machine learning has grown to be a significant research topic with many theories, techniques, and algorithms. We believe that on the Semantic Web, there will not be a single dominant method for machine learning, but that we may expect unique solutions from various machine learning research areas [25].

Oppositional approach of machine learning tries to mislead prototypes by using unreliable input. Phases based on required framework adversarial assaults may be categorized into three groups: training phase, testing phase, and model implementation stage attacks. By manipulating the training data set, changing input features, or altering data labels, adversaries assault the target model's learning stage. As a result, the attacks used at the model project level are quite similar to the attack patterns used in the test phase. Various types of adversarial attacks exist as shown in Fig. 4. Abstraction-based adversarial attacks are of two forms, i.e. white-box and black-box attacks. In white-box attacks, attackers gain entry to the algorithm of the target framework, attributes, layout, and other data and use it to produce adversarial samples. As a result, white-box assaults are more successful than black-box attacks, allowing the target model to attain an error rate of 84–96%. Even though black-box attacks have a lower success rate than white-box attacks, with an error rate of around 84–96%, these attacks can be performed by applying the transferability on data sets, inversion of the framework, and extraction on the framework, despite the fact that attacks of dark type have a lower chance than white-box attacks, with a failure rate of around 84–96%, disputes relating can be carried out using the generalizability of adversarial sets of data, model inversion, and prototype extraction, although black-box risks don't ask for any data from the target image [26].

Authors in [27] have classified DNN adversarial attacks into gradient-based and non-gradient-based. In a gradient-based system, the key idea of which is that many neural network classification systems are end-to-end differentiable. In a non-gradient-based environment, this group's assaults do not necessitate any understanding of the DNN gradients. One Pixel Attack and Zeroth Order Optimization (ZOO) are two of the most well-known individuals in the group.

2.1 Adversarial Attack Methods

There is an increasing concern about adversarial attacks on associated apps. Recent findings reveal that increasing a modest amount of perturbations to a picture (invisible to humans) can force a classifier to make targeted errors. With a few exceptions, the

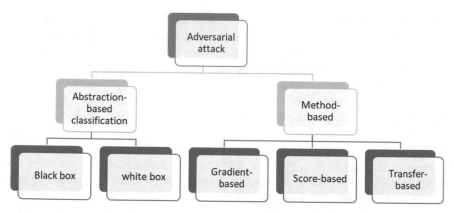

Fig. 4 Various types of adversarial attacks

vast bulk of existing work has concentrated on constructing adversarial cases in the digital realm. The main idea is to mimic the physical-to-digital transition that was presented by (1) building the physical things (for example, picture printing) and (2) the network resource digitalizes physical results (camera). Our task is to make those types of noises that are incompatible with each other. To construct more robust adversarial situations, the D2P approach to replicate the complicated impact produced by physical devices is proposed. As per our observations, the modelled transformation boosts the assault efficacy [26].

Many machine learning algorithms are subject to input perturbations that are practically unnoticeable. Because the majority of methods for creating such perturbations rely on either precise details (gradient-based attacks) or also confidence rankings like probabilities of class type (score-based attacks), none of which are available in most real-world machine learning, it was unclear how much a risk adversarial perturbations pose for the safety of actual machine learning techniques. The Boundary Assault is a choice attack that begins with a big adversarial perturbation and attempts to diminish it while remaining hostile. In basic computer vision tasks like ImageNet, the approach is conceptually straightforward, involves almost little hyper-parameter adjustment, does not rely on replacement models, and is comparable with the best gradient-based methods. They used the attack on two Clarifai.com black-box algorithms. The Boundary Attack in particular, as well as the decision-based attack class as a whole, brings up new paths for studying machine learning model resilience, as well as new concerns about the safety of machine learning systems in use [10].

The author in [28] to assess the resilience of an adversarial assault, employ momentum and proxy. Low chance black-box system can be fooled by adversarial assaults. They propose a large family of iterative techniques that are momentum-based to boost adversarial attacks. They used motion iterative techniques on an ensemble of models to increase the success chances of black-box assaults and demonstrate that adversarial models trained with heavily defended skills are at risk of our black-box attacks. Although genuine black-box attacks are difficult to execute, utilize dynamic questions to develop a proxy model that perfectly mimics the characteristics of a particular model, essentially converting black-box assaults to white-box assaults. Adversarial training is the commonly researched way of increasing the robustness of DNN [29]. Adverbially trained framework learns to withstand perturbations in the gradient direction of the error function after several tries. Training multimodal adversarial augmented the training set with hostile samples generated by other hold-out models as well as the model being evaluated. As a result, the ensemble adversarial trained models are resistant to single-step and black-box assaults. A set of dynamic incremental gradient-based algorithms is an offer to create adversarial instances that can deceive both box models. Among numerous approaches, adversarial training is the most well-researched method for improving the resilience of DNNs [28].

For attacking ensemble models, the authors in [28] suggest the application of the logits technique to numerous models with fused logit activation functions. The logarithm connections between probability predictions are captured by an array of logits. The finely detailed outputs are aggregated using models fused by logits of all models which flaws can be found quickly to target an ensemble of models in particular. We prove that models that were trained adversarial are susceptible to our black-box assaults after analysing three ensemble methodologies.

An adversarial machine learning approach should be used to look at the vulnerabilities of machine learning. In the IoT data fusion and aggregation, the procedure they describe is a partial modelling defence based on adversarial machine learning. The Wireless Sensor Network (WSN) is frequently regarded as the foundation of an IoT system. Although data interchange between multiple sensor smart objects and connector centres is required, many Internet of Things (IoT) systems have tight speed, privacy, and reliability of the system. In a wireless context, ensuring that the true information collected by sensor systems is safely acquired by the hub/data centre is critical for IoT system security and dependability. As the scope of IoT systems that increases rapidly with more sensors added, also machine learning has started to play a critical role in the parsing and learning from massive data sets produced by IoT devices. Although machine learning improves IoT systems running more efficiently, it may also be used by adversaries to conduct attacks against IoT technology. This research looks at the safety of data condensation procedures in the IoT. Actuators, RFID, switches, and sensors are some examples of IoT devices that collect original data or information. In wireless architectures like waveform design, analysis of signals and security, deep learning has initiated to be utilized. Machine learning is a method of extracting characteristics from data and making decisions based on previous and real-time streaming data [30] (Table 1).

2.2 Adversarial Defence Methods

Several adversarial attack approaches, including image classification, object recognition, semantic segmentation, face identification, and injecting perturbations into input images decrease CNN performance on a wide range of vision tasks. The defence

Table 1 Existing works in adversarial attacks and their methodologies

Methods	References	Black box	White box	Descriptions	Adversarial mechanism equation
Tiny perturbation based	Szeged et al. [9]	✓	✗	Add tiny perforated to evade non-label	...
Momentum based	Dong et al. [28]	✓	✗	Iterative technique based on momentum to improve adversarial attack success rates	$ar g_x$·max $\bar{x}(x^*, y)$, S.t. $||x^* - x||$ $\omega \leq \epsilon$,
GAN-based	Adate et al. [31]	✓	✗	GAN is used to generate adversarial noise and categorize each picture	...
Active learning	Penchant Li et al. [32]	✓	✓	Query generation using white-box technique patterns learning technique to decrease the required queries	...

(continued)

Table 1 (continued)

Methods	References	Black box	White box	Descriptions	Adversarial mechanism equation
Momentum iterative fast gradient sign technique (MI-FGSM)	Jinping Dong et al. [14]	✔	✔	Dynamic incremental gradient-based techniques for improving the chances of creating adversarial cases	⋮
The variant of natural evolution strategies (NES)	Andrew Ilyas et Al. [33]	✔	✘	NES version to present the adversarial example generation strategy under search limits, restricted information,	⋮
ZOO (zero order)	Chen et al. [34]	✔	✘	Use the zero-order (ZOO) Arbitrary position climbing method in conjunction image compression, hierarchy attack, and sampling methods to conduct black-box attacks	⋮

(continued)

Table 1 (continued)

Methods	References	Black box	White box	Descriptions	Adversarial mechanism equation
ATNs (adversarial transformation network)	Baluja et.al. [35]	✖	✖	Adversarial transformation network (ATN) generated adverse samples using a feed-forward neural network	…
Deep fool	Moosavi-Dezfooli et al. [36]	✖	✖	The minimal norm adversarial perturbation iterative approach is used to calculate the minimum norm. The minimal norm adversarial perturbation iterative approach is used to calculate the minimum norm	$\widehat{K}(x) = \arg_k \max f_k(x)$
One-step methods of a target class	Kurakin et al. [37]	✔	✖	Getting adversarial samples *from* Formula [5] calculation	$X^{adv} = X - \varepsilon \, \mathrm{Sign}\,(\bigtriangledown x \, J(X, y \, \mathrm{target}))$

(continued)

Table 1 (continued)

Methods	References	Black box	White box	Descriptions	Adversarial mechanism equation
IGS (iterative gradient sign method)	Wang et al. [38]	✘	✔	The huge data set's high dimensionality and complex data manifolds make it difficult for the defence to characterize adversarial samples	$x_{adv}^{i} = x_{adv}^{i-1} - \text{clip } \varepsilon$ $(\alpha . \text{sgn } (\bigtriangledown \times L(F(x_{adv}^{i-1}), y \text{ true})))$
HOUD-INI	Cisse et al. [39]	✔	✘	Creating adversarial samples that may be modified for task loss to deceive the gradient-based learning engine	$\nabla_{g}[H(\hat{y}, y)] = \begin{cases} -C.e^{-\|\delta g(y,\hat{y})\|^{2}/2}l(y,\hat{y}) & g = g\theta(x,\hat{y}) \\ C.e^{-\|\delta g(y,\hat{y})\|^{2/q}}l(y,\hat{y}) & g = g\theta(x,\hat{y}) \\ 0, otherwise \end{cases}$
Antagonistic network for generating rogue images (ANGRI)	Sarkar et al. [40]	✔	✔	Antagonistic network for Image feature perturbation was calculated for generating rogue images (ANGRI)	$\hat{x} = A(X, t)$

(continued)

Table 1 (continued)

Methods	References	Black box	White box	Descriptions	Adversarial mechanism equation
Universal perturbation	Moosavi-Dezfooli et al. [41]	✘	✘	For a variety of models, creating image-independent general adversarial perturbation	$\hat{k}(x+v) \neq \hat{k}(x) \, for \, "most" \, x \sim \mu$
ILCM (iterative least likely class method)	Kurakin et al. [37]	✔	✘	The most implausible labels identified by the classification are utilized to change the objective label within the BIM approach	$X_0^{adv} = X, \ X_{N+1}^{adv} = \text{Clip}X, \epsilon\{X_N^{adv} - \alpha \text{Sign}(\nabla X J(X_N^{adv}, yLL))\}$
Jacobian based saliency map attack (JSMA)	Paper [42]	✘	✘	Gradients of the information gathering are applied to every input component, and the resultant data is then used to construct adversarial samples using sophisticated saliency map techniques	...
One pixel	Su et al. [43]	✔	✘	To create adversarial samples, only one pixel per image was changed	$x_i(g+1) = (x_{r1}(g) + F(x_{r2}(g) - x_{r3}(g)), r1 \neq r2 \neq r3$

approaches in light of CNN susceptibility. Black-box and white-box assaults are the two basic types of adversarial attacks. In white-box attacks, CNN models are believed to be known, whereas, in black-box attacks, they are unknown. Physical assault methods, in addition to algorithmic attacks, generate real-world objects. CNN's models were misclassified as a result of this. Existing tracking-by-detection technologies entail two basic stages in terms of online updates. While using an offline pre-trained CNN model, trackers do not update online in step1. Trackers collect data in step2 to update the model, samples from prior frames were found online. It's important to keep in mind that trackers aren't model adaptability that is improved by gathering samples incrementally with online updates, which may assist protect against hostile perturbations. They deploy on two state-of-the-art trackers, and an adversarial assault and defence were presented. On the OTB100 data set, we first assess the baseline performance. For all of the experiments, we build our approaches on top of DaSiamRPN and RT-MDNet [44].

A wide number of domains, including computer vision, natural language processing, and anomaly detection, have seen tremendous success with machine learning systems. The machine learning technologies being suggested for detecting cyber-attacks, these attackers are a major source of concern. The security community is particularly concerned about the emergence of adversarial machine learning, particularly concerning the applicability of these bouts to intrusion detection systems (IDS) [45, 46] (Table 2).

2.3   Adversarial Threat Model ATM)

Any machine learning model's security is evaluated in terms of hostile intentions and capacities, as well as vulnerability assessments in machine learning techniques, considering the adversary's capability. It is described in terms of the attack surface, adversarial capabilities, and adversarial aims (adversary categorization) [60].

(i)   **The Attack Surface**

A machine learning-based system may be compared to a data processing pipeline in general. The system's primitive sequence of operations may be summarized as follows during testing: (a) gathering data gathered from sensors or stored in databases, (b) moving the data into the digital realm, (c) processing the converted data using a machine learning model to create an output, and (d) acting on the outcome. The system collects sensor data (camera pictures) from which modelling characteristics (a tensor of image pixels) are generated and employed in the calculations [30]. Based on the attack surface, the following attacks are possible:

Table 2 Adversarial defence methods

Defence name	References	Defence method	Description
Adversarial training	Good fellow et al. [47]	Modifying data	It entails creating adversarial examples on the target model utilizing various attack strategies. These hostile instances are combined with the original training set to create an enhanced training set, which is then used to retrain the target model
Gradient hiding	Papernot et al. [48]		Adversarial attack tactics rely on the elevation information given by the classifier hence masking or altering the gradients will fool the adversaries
Blocking the transferability	Chen et al. [49]		It's a three-step NULL Labelling strategy that prevents hostile samples from spreading from one to another network. It basically adds a new NULL label to the information and classified it as such using a training model that can survive adversarial attacks
Data compression	Dziugaite et al. [50]		It was revealed that employing the JPG method for compression can improvise a significant set of network model identification efficiency interrupted due to FGSM attack interruption Using a comparable JPEG compression strategy, a protective system against FGSM and DeepFool attacks was established

(continued)

Table 2 (continued)

Defence name	References	Defence method	Description
Data randomization	Xie et al. [51]		Randomness defence approaches add randomness to the input (such as resizing or padding) or to the model parameters (such as adding randomness to the model parameters) to reduce the impact of adversarial examples
Regularization	Lyu et.al[52]	Modifying model	Carefully regulating the networks global Lipschitz constant, is a layer-wise regularization strategy for decreasing the network's sensitivity to tiny perturbations
Defensive distillation	Soll et al. [53]		The idea behind this strategy is that soft-label training delivers more information than hard-label training since they encode the relative distinctions across classes. As a result, it has been recommended that robust classifiers be trained using this strategy, known as defensive distillation

(continued)

- **Evasion Attack**: In an antagonistic situation, this is the most prevalent sort of attack. During the testing phase, the adversary tries to circumvent the system by altering harmful samples. This option assumes that the training data is unaffected.
- **Poisoning Attack**: Contamination of the training data is a form of attack that occurs during the machine learning model's training period. An enemy seeks to contaminate the training data by introducing precisely planned samples, ultimately jeopardizing the entire learning process.
- **Exploratory Attack**: The training data set is unaffected by these attacks. Black-box access to the model attempt to understand as much as possible

Table 2 (continued)

Defence name	References	Defence method	Description
Feature squeezing	Xu et al. [54]		Reduce the colour bit complexity of each pixel and use spatial levelling as a feature squeezing approach. Adversarial perturbations can be reduced to a certain extent via variance minimization. It randomly selects a small pixel value and reconstructs the simplest picture that is compatible with those pixels
Deep contractive network	Wu et al. [55]		To reduce adversarial noise the network for compression which is a deep network actually employs a noise reduction automated encoder as a result of this, DCN applied a softening penalty same as the convolutional auto encoder (CAE) in the process of training, which was proven to offer some protection against assaults like L-BGFS
Mask defence	Reddy Kalavakonda et al. [56]		By training the original pictures and related adversarial samples the mask layer captures the disparities between the actual pictures and the output attributes of the prior network model layer

(continued)

Table 2 (continued)

Defence name	References	Defence method	Description
Defence-GAN	Esmaeilpour et al. [57]	Using auxiliary tools	Defence-GAN is a type of defence that use a WGAN that has been trained on valid training samples to learn how to remove perturbations from adversary examples
MagNet	SAYED et al. [58]		To distinguish between legal and adversarial data MagNet employs a detector. Any deviation in a sample for testing and the manifold exceeds a threshold the detector rejects that one
High-level representation guided denoiser (HGD)	Liao et al. [59]		High-level representation guided denoiser (HGD) is used for creating a strong adversarial model that can with stand both black-box and white-box attacks

about the underlying system's learning mechanism and patterns in training data.

(ii) **Adversarial Classification**

Adversary classification is defined as a game between Classifier and Adversary in which the Classifier tries to learn a feature $yc = C(x)$ that will predict exactly the classes of instances in the training set and the Adversary attempts to make the Classifier predict true positives of the training set as negative by changing them from x to $x' = A(x)$. The opponent and the classifier are constantly seeking to surpass one another by increasing their payoffs. The classifier uses an expense Bayes learner to lower its anticipated cost, assuming that the adversary always takes the best strategy, while the opponent tries to adjust features to lower its own estimated value. The existence of adaptive opponents in a system can considerably reduce the performance of the classifier, especially if the classifier is unaware of the adversary's existence or kind. The Classifier aims to develop a classifier C that promotes its expected utility (UC), whereas the Adversary's goal is to find a feature modification technique A that optimizes its own expected utility:

$$U_C = \sum\nolimits_{(x,y)\epsilon xy} p(x, y)[u_c(c(A(x), y))] - \sum\nolimits_{X_i\epsilon X_c(x)} V_i \qquad (1)$$

$$U_A = \sum\nolimits_{(x,y)\epsilon xy} p(x, y)[U_A(CA(x), y) - W(x, A(x))] \qquad (2)$$

2.4 ATM Case Study

Recently, the cyber realm has begun to investigate adversarial machine learning. The developers used the DREBIN Android malware data set to produce adversarial samples. Using a modified Jacobian saliency map approach, only 20 antagonistic characteristics could be added by the writers. The authors were able to fool a classifier that is actually neural network-based 50–80% of the time based on a target network from a normal beginning performance of over 95%. That research, however, required white-box access to the detector. When the authors could only query malware classifiers, they revealed how to attack them. An approximate training data set for a replacement malware classifier was labelled by repeated queries. The substitute detector was then employed as a stand-in for the real target, which was unknown. To create the stealthy attacks GAN was used which functioned very well. A random forest, which had a 0.19% accuracy versus the adversarial data, was the best-performing classifier [45].

Attacker Modelling: When bearing in mind how attackers in virtual systems can act, it's important to keep the following in mind, it needs to establish a consistent model for describing an intruder that can be used to inspire adversarial research principles. Such principles do not hold the same weight in the cyber world; hence, alternate attacker modelling is required. The following are some of the qualities we recommend.

Levels of Perturbation: Certain fields in network traffics are not modified either the attack will fail because a faulty packet will be formed, or specific areas will be encoded, causing the attack to fail.

Attacker Knowledge: The attacker's understanding of the destination system can be measured in terms of data about the targeted neural network and computer vision protections. It must also indicate how much of an IT system the invader is familiar with in the cyber realm. In systems that are cyber-physical nature such as systems of industrial control, the attacker's grasp of dynamic behaviour is also crucial. If the attacker has no idea how their changes will affect the evolution of the system, all they can do is optimize mercilessly for another time step.

Timing: Defining the invader's capacity to decide the strike's initial point is important. Some systems will become more susceptible to a sneaky attacker in future, making it much easier to hack the system.

Human in the Loop: Finally, see if a person is required in extra to the IDS [61]. Depending on the assault, the level of disturbance necessary to deceive an intrusion detection system may be abundant to be perceptible to a man watching a human–machine interaction. The significance of these changes is based on the real system and human response times: for instance, if the system under assault is a control source, variations might happen too rapidly for a person to behave.

3 GAN-Based Adversarial Attacks

A generative adversarial network (GAN) is a type of construct in neural network technology that offers a lot of potential in the world of artificial intelligence. A generative adversarial network is composed of two neural networks: a generative network and a discriminative network as shown in Fig. 5. Generative adversarial networks (GAN) are an emerging technology for semi-supervised and unsupervised learning [62].

Generative (G): It is also known as a counter feeder. It constantly tries to produce fake data set samples. It keeps improvising or improving the samples until it generates a perfect one that completely fools the intelligent model [63].

Discriminative (D): It is another pair of the neural network and it constantly tries to catch the counter feeder. Since its job is to catch the counter feeder that's why it is called a Discriminator.

Typically, the GAN is based on equation as shown below [64].

$$\text{Min max } V(G, D) = E_{X \sim \text{Pdata}(x)}[\log(D(X)] + E_{z \sim Pz(z)}[\log(1 - D(G(z))) \quad (3)$$

where E_x: Expected value overall real data instances.

E_z: Expected value over all random inputs to the generator.

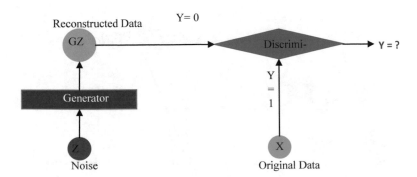

Fig. 5 Structure of a generative adversarial network (GAN)

Pdata(x): Represented the probability distribution of our original data.

$Pz(z)$: Represented the distribution of the noise.

$D(X)$: Discriminators estimate the probability of real data instances.

$D(G(Z))$: Discriminators estimate the probability of a fake data instance.

3.1 Conditional GAN (C-GAN)

There are various types of GANs such as deep convolutional (DC GAN), Conditional GAN, Least Square GAN (LSGAN), Auxiliary Classifier GAN (ACGAN), Dual Video Discriminator GAN, Tab GAN, etc. Conditional GAN is a widely used GAN, and Tab GAN is used to generate tabular synthetic data samples. The Conditional GAN, these networks may be built by simply feeding the extra auxiliary information (e.g. label) into the GAN, transforming it into extra auxiliary information (e.g. label) into the GAN, transforming it into a CGAN as shown in Fig. 6.

The CGAN generator uses the extra auxiliary information y (label, text, or images) and a latent vector z to generate conditional real-looking data (G (z|y)), and the CGAN discriminator uses the extra auxiliary information y (label, text, or images) and real data x to distinguish between generator generated samples-D (G (z|y)) and real data x. The data creation may be controlled using CGAN. This isn't doable with the standard GAN. [65]. The following is the CGAN updated loss function [66]:

$$\text{Min max } V(G, D) = E_{x \sim \text{pdata}(x)}[\log D(x|l)] + E_{z \sim pz(z)}[\log(1 - D(G(z|l)))] \quad (4)$$

where E_x: Expected value overall real data instances.

E_z: Expected value over all random inputs to the generator.

Pdata(x): Represented the probability distribution of our original data.

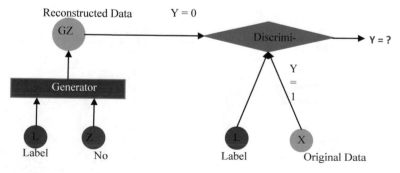

Fig. 6 Structure of a conditional GAN (C-GAN)

$Pz(z)$: Represented the distribution of the noise.

$D(x|l)$: Discriminates an image given a label L.

$D(G(z|l))$: Generates an image given a label L.

4 Proposed Security Method

Our methodology is based on the evaluation of learning model (ML and DL) performance in non-adversarial scenarios and adversarial scenarios with and without adversarial learning. Our proposed methodology is depicted in Fig. 7. We evaluate the performance of ML and DL models in our framework in three different scenarios, i.e. on a normal data set, on an adversarial data set, and adversarial data set with adversarial learning. Various adversarial attacks will be considered such as data modification, perturbation, corruption, etc., [67] as assessed thereby for attack severity. We initially select a model and its structure by performing hyper-parameter tuning. Also, the data set selected is pre-processed and split into the training set and testing set for performance evaluation purposes. Our methodology consists of three phases. In Phase 1, the model is trained and tested on a normal data set without adversarial attack consideration and the performance is measured as P1. In Phase 2, we perform adversarial attacks on the testing set, and the model trained in Phase 1 is tested for performance, and its resulting performance is indicated as P2. In Phase 3, we consider the adversarial learning of the model and train it to show resilience against adversarial attacks. The adversarial learned model is then tested for performance on the Phase 2 testing set containing adversarial attacks. The performance is measured and indicated with P3. Consequently, we compare the performances of P1, P2, and P3 in general and P2, and P3 in particular to assess the improvement achieved through adversarial learning.

5 Research Challenges and Gaps

(a) **Effective and efficient defence against white-box attacks**: To the authors' knowledge, no defence has been developed that can strike a compromise between efficacy and efficiency. Adversarial training is the most successful in terms of efficacy, but it comes at a high computational time. In terms of capability, several randomization and information and communications technology defences/detection methods may be configured in a matter of seconds. Although certificated defences suggest a path to theoretically assured privacy, their high efficiency falls well short of the actual requirements.

(b) **Causality behind adversarial samples**: Although many adversarial assaults have been developed, the cause of adversarial examples is still a mystery. The pervasiveness of adversarial data was attributed to model structures and based

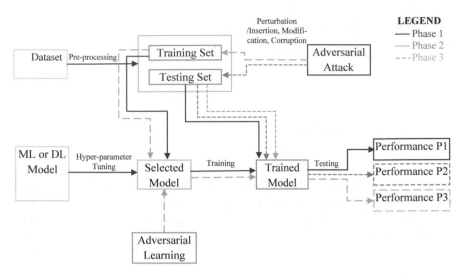

Fig. 7 Proposed performance evaluation methodology

learning in the early research on the topic, which assumed that suitable tactics and network design considerably improved adversarial resilience. However, such efforts—particularly those that result in obscured gradients—create a false sense of safety. Recent research, on the other hand, has discovered that adversarial susceptibility is more likely to be caused by data transfer geometry and a lack of training information.

(c) **Existence of a general robust decision boundary**: Because there are so many distinct adversarial assaults described by different metrics, it's logical to wonder: is there a basic strict disciplinary boundary that a specific type of DNN with a specific training technique can learn? The answer to this question is now "no". Even though PGD adversarial training shows excellent resilience to a wide variety of L_∞ assaults demonstrate that it is still subject to adversarial assaults as evaluated by other L_p standards like EAD and CW2. Establish that for a two-concentric-sphere database, the optimum L_2 and L_∞ decision bounds are distinct and that the discrepancy rises with the similarities of the data sources.

(d) **Model privacy**: Model extraction threats pose a major danger to the security of learnt models through unlawful duplication. According to Behzadan and Hsu, one way to mitigate this is to raise the cost of such assaults or to timestamp the regulations. We may see some randomization in the medium to protect against such threats, but this will result in an unreasonable loss of performance. A potential subject of research is creating approaches that might induce restricted randomization in the system to protect against such assaults.

6 Conclusion

In this survey, we discuss the security issues in cyber-physical systems (CPS) and the use of ML and DL for its security. We discussed the difficulties faced by traditional security approaches to dealing with complex attacks. The ML and DL security methods are vulnerable to adversarial attacks, and there is a need to ensure their security before their utilization for security in CPS domains. Their security is evaded by adversarial attacks both at a testing time (poisoning attack) and training time (evasion attack) to misclassify the data by performing data corruption, data perturbation, and data modification. Depending on the knowledge of the attacker about the attacking model, the capabilities of the attacker vary from black box to white box. The white one contains, and the attacker possesses greater data about the model and its parameters, while in the case of the black box, the attacker possesses no understanding of the model being attacked and its parameters. We also discuss the general Adversarial Threat Model (ATM) which could be used as a risk prevention technique. Moreover, we have covered the use of generative adversarial networks (GAN) for creating fake data samples to perform adversarial attacks. Conditional GAN (C-GAN) which considers the labels in supervised data set is also discussed in great detail. Based on adversarial attacking and defence methods, we propose a methodology to evaluate the performance of ML and DL models. We consider three scenarios of evaluating performance, i.e. normal performance, performance against adversarial attacks, and performance against adversarial attacks with adversarial learning.

References

1. Kagita MK, Thilakarathne N, Gadekallu TR A review on cyber crimes on the internet of things
2. El Naqa I, Murphy MJ What is machine learning ? pp 3–11
3. Jordan MI, Mitchell TM (2015) Machine learning: trends, perspectives, and prospects, vol 349, no 6245
4. Arnold L et al (2016) An introduction to deep learning to cite this version : HAL Id : hal-01352061 an introduction to deep learning
5. Cao L, Jiang X, Zhao Y, Wang S, You D, Xu X (2020) A survey of network attacks on cyber-physical systems. IEEE Access 8:44219–44227
6. Rana B (2020) A systematic survey on internet of things : energy efficiency and interoperability perspective, no. August, pp 1–41
7. Zhong C, Zhu Z, Huang R (2017) Study on the IOT architecture and access technology
8. Zhu Y, Sampath RZ, Jennifer R (2008) Cabernet : connectivity architecture for better network services, no. December
9. Ray PP (2017) Internet of robotic things: concept, technologies, and challenges. IEEE Access 4:9489–9500
10. Benchmarking EEC, Hao T, Huang Y, Wen X, Gao W, Zhang F Edge AIBench: towards comprehensive end-to-end edge computing benchmarking, pp 1–8
11. Garcia-rodriguez J, Azorin-lopez J, Tom D, Fuster-guillo A, Mora-mora H (2021) IA-CPS : intelligent architecture for cyber-physical systems management, vol 53, no June
12. Sacala IS, Pop E, Moisescu MA, Dumitrache I, Caramihai SI, Culita J (2021) Enhancing CPS architectures with SOA for industry 4.0 enterprise systems enhancing CPS architectures with SOA for industry 4.0 enterprise systems, no January 2022

13. Fung WW et al Protection of keys against modification attack HKUST theoretical computer science center research report HKUST-TCSC-2001–04
14. Wilhelm M, Schmitt JB, Lenders V Practical message manipulation attacks in IEEE 802.15.4 wireless networks, pp 2–4
15. Mcdermott JP Attack net penetration testing, pp 15–21
16. Ashok A, Govindarasu M, Wang J (2017) Cyber-physical attack-resilient wide-area grid, no i, pp 1–17
17. Luckett P (2016) Neural network analysis of system call timing for rootkit detection, pp 1–6
18. Mohammad AH (2020) T. World, and I. Science, Ransomware Evolution, growth and recommendation for detection, no February
19. Tyagi AK, Aghila G (2011) A wide scale survey on botnet, vol 34, no 9, pp 9–22
20. Neuman C Challenges in security for cyber-physical systems
21. I.- Things safety and security in cyber—physical systems and internet-of-things systems, vol 106, no 1, pp 9–20 (2018)
22. Akella R, Tang H, Mcmillin BM (2010) Analysis of information flow security in cyber—physical systems. Int J Crit Infrastruct Prot 3(3–4):157–173
23. C. Security and T. Monitoring (2011) Importance of intrusion detection system (IDS), vol 2, no 1, pp 1–4
24. Salloum SA, Alshurideh M Machine learning and deep learning techniques for cybersecurity : a review, vol 2. Springer International Publishing
25. Introducing machine learning, no November (2014)
26. Jan STK, Messou J, Lin Y, Huang J, Wang G Connecting the digital and physical world : improving the robustness of adversarial attacks
27. Taran O, Rezaeifar S, Voloshynovskiy S Bridging machine learning and cryptography in defence against adversarial attacks, no 200021
28. Dong Y et al Boosting adversarial attacks with momentum, pp 9185–9193
29. Li J, Zhao R, Huang J, Gong Y, M. Corporation, and O. M. Way (2014) Learning small—size DNN with output—distribution—based criteria, no September, pp 1910–1914
30. Luo Z, Lu Z (2020) Adversarial machine learning based partial-model attack in IoT, pp 13–18
31. Adate A, Saxena R (2017) Understanding how adversarial noise affects single image classification. In: Proceedings of the international conference on intelligent information technologies, Chennai
32. Li P Query-efficient black-box attack by active learning
33. Ilyas A, Engstrom L, Athalye A, Lin J (2018) Black-box adversarial attacks with limited queries and information
34. Chen P (2017) ZOO : zeroth order optimization based black-box attacks to deep neural networks without training substitute models, pp 15–26
35. Baluja S, Fischer I (2016) Adversarial transformation networks : learning to generate adversarial exampl arXiv:1703.09387v1 [cs.NE] 28 Mar 2017, no 2013
36. Fawzi A, Frossard P (2016) DeepFool : a simple and accurate method to fool deep neural networks, pp 2574–2582
37. Goodfellow IJ (2017) Amls, pp 1–17
38. Wang DD, Li C, Wen S, Xiang Y (2020) Defending against adversarial attack towards deep neural networks via collaborative multi-task training, vol 5971, no AUGUST 2019, pp 1–12
39. Cisse M, Adi Y, Keshet J Houdini : fooling deep structured prediction models
40. Sarkar S, Mahbub U UPSET and ANGRI : breaking high performance image classifiers, vol 20742, no 1, pp 1–9
41. Fawzi O, Frossard P Universal adversarial perturbations, pp 1765–1773
42. Papernot N, Mcdaniel P, Jha S, Fredrikson M, Celik ZB, Swami A (2016) The limitations of deep learning in adversarial settings
43. Su J, Vargas DV, Sakurai K One pixel attack for fooling deep neural networks, pp 1–15
44. Jia S, Ma C, Song Y, Yang X Robust tracking against adversarial attacks
45. Zizzo G, Hankin C, Maffeis S, Jones K (2019) INVITED : adversarial machine learning beyond the image domain

46. Li J, Yang Y, Sun JS, Tomsovic K, Qi H (2021) ConAML: constrained adversarial machine learning for cyber-physical systems, vol 1, no 1. Association for Computing Machinery
47. Goodfellow IJ, Shlens J, Szegedy C (2015) Explaining and harnessing, pp 1–11
48. Papernot N, Mcdaniel P, Goodfellow I Practical black-box attacks against machine learning, pp 506–519
49. Chen Y Blocking transferability of adversarial examples in black-box learning systems
50. Dziugaite GK, Roy DM (2016) A study of the effect of JPG compression on adversarial images arXiv:1608.00853v1 [cs.CV] 2 Aug 2016, no Isba
51. Xie C, Wang J, Zhang Z, Zhou Y, Xie L, Yuille A Adversarial examples for semantic segmentation and object detection, vol 1, pp 1369–1378
52. Lyu C (2015) A unified gradient regularization family for adversarial examples
53. Soll M, Hinz T, Magg S, Wermter S (2019) Evaluating defensive distillation for defending text processing neural networks against adversarial examples
54. Xu W, Evans D, Qi Y (2018) Feature squeezing : detecting adversarial examples in deep neural networks, no February
55. Wu EQ, Zhou G, Zhu L, Wei C, Ren H, Sheng RSF (2019) Rotated sphere Haar wavelet and deep contractive auto-encoder network with fuzzy Gaussian SVM for pilot's pupil center detection. IEEE Trans Cybern 1–14
56. Kalavakonda RR, Vikram N, Masna R, Bhuniaroy A (2020) A smart mask for active defense against coronaviruses and other airborne pathogens, vol 2248, no c
57. Esmaeilpour M, Cardinal P, Koerich AL (2021) Multi-discriminator Sobolev defense-GAN against adversarial attacks for end-to-end speech systems, vol X, no X, pp 1–10
58. Sayed E, Member S, Yang Y, Member S (2019) A comprehensive review of flux barriers in interior permanent magnet synchronous machines. IEEE Access 7:149168–149181
59. Liao F, Liang M, Dong Y, Pang T, Hu X Defense against adversarial attacks using high-level representation guided denoiser, pp 1778–1787
60. Ren K, Zheng T, Qin Z, Liu X (2020) Adversarial attacks and defenses in deep learning. Engineering
61. Duque S, Nizam M (2015) Using data mining algorithms for developing a model for intrusion detection system (IDS). Procedia Comput Sci 61:46–51
62. Hoang Q, Nguyen TD, Le T, Phung D (2017) Multi-generator generative adversarial nets, no August
63. Wang K, Gou C, Duan Y, Lin Y, Zheng X, Wang F (2017) Generative adversarial networks : introduction and outlook, vol 4, no 4, pp 588–598
64. Creswell A et al (2017) Generative adversarial networks : an overview, no April, pp 1–14
65. Jabbar A, Li X, Omar B A survey on generative adversarial networks : variants, applications, and training, pp 1–38
66. Wang Y et al (2018) Accepted Manuscript
67. Chakraborty A, Alam M, Dey V, Chattopadhyay A, Mukhopadhyay D (2018) Adversarial attacks and defences: a survey, vol x, no x

Legal Challenges of Digital Twins in Smart Manufacturing

Ridoan Karim⬛, Sonali Vyas⬛, and Ahmed Imran Kabir⬛

Abstract Digital twins are interconnected systems in which modifications to one data item would affect other sections of the model. As more and more parties can use and rely on data that could include errors, the essential question remains relating to data ownership, liabilities, and intellectual property. It may be difficult to regulate the technological system with existing laws. It can lead to trust issues as various parties rely on the exactness of the information supplied by each other in this technological platform. Hence, this study particularly analyse the digital twins existing challenges related to legal, ethical, and policy in the smart manufacturing industry and recommends possible solutions for the future utilization of digital twins.

Keywords Law · Technological change · Digital twins · Smart manufacturing · Ethics · Policy

1 Introduction

The influence of digital twins is becoming more important as digitalization and the Internet of Things (IoT) grow more ubiquitous. Nevertheless, the complicated nature of digital twins paves legal, ethical, and social issues. Most notably, digital twins are intangible concepts that connects several issues relating to intellectual property and data protection. As a result, it is critical to first comprehend what digital twins are, how they are used, and what the legal ramifications are in the real-life scenarios.

R. Karim (✉)
School of Business, Department of Business Law and Taxation, Monash University, Subang Jaya, Malaysia
e-mail: ridoan.karim@monash.edu

S. Vyas
School of Computer Science, UPES, Dehradun, India

A. I. Kabir
School of Business and Economics, United International University, Dhaka, Bangladesh

At its most basic level, a digital twin may simply serve as a central store of data, including information on how a certain asset—such as a building—was planned and created [1]. The technology might be utilized for building's creation, administration, operation, and maintenance. Nevertheless, the technology utilization is not so straight-forward; rather it is complex and diverse, incorporating virtual promontories of practical beings, at the multiple state-of-the-art statuses. This is the unique nature of the technology—as it provides a precise benchmark of frame developments to impact choices and provide solutions and establish a sustainable future for the smart manufacturing industries. Nevertheless, issues pertaining to scalability are providing a hard time governing a model of this complexity. Additionally, issues pertaining to data ownership and privacy protection are all potentially ambiguous in this early phrase of the technology.

As a result, there are certain legal, policy, and ethical constraints on using this technology. Digital twins are linked systems in which changes to one data item have an impact on other parts of the model. As more parties are able to access and depend on data that may include inaccuracies, the fundamental issue of data ownership, liability, and intellectual property remains unanswered. Existing laws may make it difficult to govern the technological system. As multiple parties depend on the accuracy of information provided by each other on this technology platform, it might lead to trust difficulties. As a result, this chapter will examine the current legal, ethical, and regulatory difficulties of digital twins in the smart production business and potential solutions for their future use.

2 Digital Twins: Philosophy or Reality?

Numerous industries are particularly interested in the "digital twins" technology on a larger scale. PREDIX, a digital dual platform created by General Electric, can accurately assess and forecast the effectiveness of the industrial base [2]. The SIEMENS company's unique intelligent platform aids in virtually every manufacturing process control, including design, production, and operation [3]. ABB concentrates on developing fresh possibilities for decision-making based on information gathered through comparable technologies [4]. By providing a flexible platform built on the Internet of Things for modelling and analysing interactions between people, locations, and gadgets, Microsoft has also increased the scope of its digital twin product line [5]. These industry leaders' have significantly expanded the use of "digital twin" technologies within different sectors.

The National Aeronautics and Space Administration's (NASA) aircraft design through digital twin was one of the first significant and successful efforts in this realm of technology [6]. With the help of this system, we may now test the compatibility and dependability of different parts used in the aviation sector and simulate different scenarios for their further development [7]. The utilization of this digital model assists in resolving a number of difficulties pertaining to the safety features.

Worldwide health care and life sciences are also utilizing digital twins to deliver tailored therapies [8]. This involves giving doctors the freedom to use clinical support tools within the digital health care system [9]. However, there is still more possibility. Digital medicines, virtual reality-based treatments, enhanced human body comprehension, hospital management, and disease modelling are other sectors where research into the usage of digital twins should be further enhanced.

Digital twins are also being used to create digital cities [9]. Dynamism is a characteristic of large cities that makes managing urban infrastructure more challenging. One simply needs to consider the millions of people that reside in big cities, as well as the numerous hospitals, schools, workplaces, and retail establishments—that need the ongoing care of local government [10]. The National Research Foundation's (NRF) "Virtual Singapore" project offers 3D semantic modelling of the urban environment, allowing users to directly relate the data to the real world by displaying actual locations, the characteristics of various modes of transportation, or elements of buildings and urban infrastructure [11]. The platform also provides a range of dynamic real-time indicators, as well as data on demographics, temperature, and traffic, in addition to the standard map [12].

Digital twins also started playing a great role in the automotive industry [7]. Concepts for brand-new automobiles are created digitally. In contrast to a practical physical depiction, this offers an excellent visual portrayal of a future vehicle. Things could eventually need to alter as that vehicle advances in the development process based on how it responds or performs in the actual world. Digital twins are being employed more frequently to expedite and enhance the development process of autonomous vehicles [13]. Vehicle testing has also gotten exceedingly complicated in recent times, and this complexity increases when we are talking about autonomous cars. Through the use of real-world data, digital twins facilitate that process and aid in the elaboration of AI judgments. As a result, automakers now have the technology and equipment to evaluate car designs before they are put on a testing track or real roads.

Digital twins also hold the promise to transform global supply chain [14]. There may not be an appropriate way to test out new ideas, goods, or modifications to the present supply chain because it is continuously in motion, at least not without causing significant delays. The solution to this issue is digital twins, which let logistics teams experiment while also gaining a precise picture of how it might impact operations [15]. They are an integral part of the ever-evolving digital supply chain, which is vastly improving operations through technologies like advanced analytics, robotics, automation, rapid manufacturing, and more.

To discover possible issues before they are introduced to the market, new product packaging or revised designs, for instance, can be virtualized by putting them through a digital twin model. Teams may utilize the same technologies to optimize inventory and positioning, evaluate environmental or transportation conditions, and more. Imagine a prediction model that can be used to analyse almost any situation and produce answers that are both extraordinarily accurate and realistic.

Digital twins are useful tools in the area of architectural design as well [17]. Most of the time, designers constantly attempt to construct or produce model according to certain criteria. However, there might occasionally be unexpected circumstances that require designers to make changes to a model or design that go against their wishes or preferences. By coming up with innovative ways to understand past certain constraints, digital twins can assist preserve some of those original designs while also supplying real-world data to realize practical notions. Building information modelling (BIM) data is being combined with scans and operational data by designers to produce extraordinarily precise digital representations of individual structures or large communities [18].

Digital twins are influencing our future, whether we are talking about improved building processes, or more precise and well-designed architectural models, whether virtual research for new health care items, or many other possibilities. Utilizing digital twins to optimize different operations and procedures, several industrial sectors are experiencing outstanding outcomes.

3 Digital Twins and Smart Manufacturing: Usage and Practices

The notion of digital twins is a progression of modelling and simulation technologies. Conventional simulator approaches are only capable of analysing system performance to a limited extent [19]. By using IoT technology, digital twins provides a breakthrough in simulation modelling and engineering analysis capabilities [20]. Digital models are used to describe virtual smart manufacturing. There are three types of digital models: digital shadow, digital replica, and digital twin [21]. The automated projection of system architectures is emphasized in a digital duplicate. The digital shadow stresses mathematical modelling to characterize a system's physical and chemical properties. A digital twin is diverse simulations where automated systems represent multi-dimensional factors. It improves a material product or design through sensing and by analysing collected information [22].

Digital twins is also known as digital surrogates, as stated by Shao and Kibira [23] and Valckenaers [24]. Digital twins and its other intelligent agents make up real-world smart systems. Manufacturing, aviation, health care, and medicine are among the industries where digital twins is being explored [25]. It's widely acknowledged as one of the technology cornerstones for new industry developments [26].

3.1 Digital Twins-Based Smart Manufacturing

Digital twin utilizes multi-dimensional information of a product to create its twin [27]. Conceptualization and formation of comparable digital twins are all part of the design process. For example, there are three characterization of utilizing digital twins:

- Digital twin idea stage (Functional model): Utilizing multi-dimensional simulator to evaluate the concept. In this stage, both simulation's precision and pace are important [28].
- Digital twin formation stage (Behavioural model): Simulations are used to establish key technical design characteristics. This level focuses on semi-physical system commissioning [29].
- Digital twin fine-tuning stage (Intelligence model): The digital twins are utilized to assist in the construction of the prototype smart manufacturing simulation (SMS). The digital twin technique might help with SMS validation by reducing test durations and increasing design efficiency [30]. The biggest immediate benefit of using digital twins is that it eliminates the need for expensive physical contracting and testing. Nevertheless, the final purpose of using digital twins is design innovation, not just confirmation. In the following lifespan stages of SMS, the developed digital twin from the design phase may be utilized for observing, optimizing, diagnosis, and forecasting [31].

3.2 Digital Twins Architecture in Smart Manufacturing

The integration of complicated and interrelated functional models of smart manufacturing are conceptualized at the first stage of digital twins. Smart manufacturing is defined as the creation of entities utilizing AI-driven smart tools and incorporating engineering effective technologies altogether [32]. The individualized needs from the client and product domains, as well as the customized industrialized intelligence models are transferred to the SMS configuration to create identical implementation domain in the smart production process. To effectively implement digital twin, the framework of the Function Structure Behaviour Control Intelligence Performance (FSBCIP) (See Fig. 1) is implemented throughout the process [33]. As stated by Wu et al. [34], a functional model is an organized description of the actions that makes a product. This approach can be considered as conceptual design phase which emphasizes the manufacturer demand on product quality, assembly, maintainability, and safety. The SMS's functions are dictated by the product characteristics. Several planning processes, including selecting tools and designing mechanics, are frequently included in this paradigm [35]. Once the functional model is prepared, the engineers work on the structural model. In a structure model, the integration and communication between mechanical configurations that perform production utilities, principles, and behaviours are described [36]. The mechanical structure's interconnection provides the basis for the manufacturing system's material, which contains the information that need to be transferred and transformed in the motion behaviour stage. A behaviour prototype is a logical representation of mechanical broadcast, motion-form conversion, and their interrelationships [37] that contains the information and material links to create the SMS process. In SMS, layout planning, carrier design, topology description, and buffer design are frequently included [25]. Afterwards, a control model is formed to address process structure, performance, and calculation using mathematical or engineering methods to control process output. Control model is basically the design of the electrical and pneumatic system, including several network of sensors, which forms mechanical controls and multi-process communication system [38]. As per Peruzzini and Pellicciari [39], from the standpoint of control design, an intelligence model is designed to characterize, create, and evaluate an SMS's learning capacity, optimization ability, and autonomous ability. For obtaining long-term success in SMS, it contains a systematic procedure and a series of reasoning codes, computational optimization methods and machine learning algorithms [1] (See Fig. 1).

The assessment of system performance, such as competence, robustness, adaptivity, extensibility, and flexibility, is frequently analysed in a performance model. Personalized criteria and specifications are included in the SMS input. The SMS produces the final digital twins, which include all replicas, control systems, implementation systems, and intelligence algorithms. The FSBCIP framework is an extension of the traditional function–behaviour–structure paradigm [40]. The primary purpose of FSBCIP's framework is to meet each customer's unique needs, like location space constraints, industrialized volume, product eminence, and fabrication competence. Step by step, the product domain's individuality criteria are reassigned to the SMS area [41].

Rule-based cognitive systems and case-based cognitive systems are other two types of traditional knowledge-based systems of digital twins. In the digital intelligence era, discovering veiled information and patterns in comprehensive case data to expand design competence is a superior approach to learn [42]. In a complicated SMS, there are frequently too many underlying connected aspects for subject specialists to

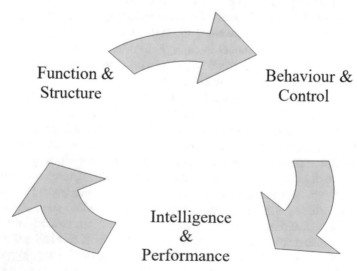

Fig. 1 Framework of the function structure behaviour control intelligence performance (FSBCIP)

consider, which may be gathered by deep learning and big data analytics approaches. In summary, a digital twin model with intelligent capabilities may help smart manufacturing become smarter [43]. Most modern digital twin models are limited in the understanding and basic thinking of SMS design options. Integrating state-of-the-art performance algorithms into digital twin models in future could result in the ability to conceive and produce new design solutions automatically.

3.3 Digital Twins-Based Blockchain System

Because smart manufacturing requires multi-field knowledge and expertise, a reliable and interconnected process is required to allow engineers from all groups to collaborate on investigation, and planning to create a single model. The present digital twin solutions are generally centralized. Under cyber-attack, securing data in the runtime network is a major concern [44]. To counter the assaults that endanger the collaborative network, new cybersecurity approaches must develop [45]. Blockchains provide a new approach to collaborative smart manufacturing that is both safe and efficient. Cryptographic technology and distributed consensus methods are used in blockchains to secure the network communication and therefore to improve cyber-security among varied members [46]. There has been a lot of study on blockchain-enabled smart industrial applications [25], including blockchain-secured data exchange in combined DTSMSD. To offer availability and traceability of design information, a blockchain-enabled digital twin resolution is required [47]. However, practical concerns like flexibility and scalability of the system are still there.

3.4 Web Services and Cloud Computing for Distributed Digital Twins

Apart from blockchain, web services and cloud computing for assisting design, analysis and simulation in a distributed digital twin situation are another kind of crucial enabling technology for realizing collaborative smart manufacturing. Both designers and computer resources are spread globally in the context of collaborative smart manufacturing [48]. Cloud computing and web service technologies enable several comparable units in far-away places to share design and work jointly [49]. Another reason for adopting cloud computing technology is the necessity for high-performance computing for replication work in smart manufacturing [50], which might enable less costly computing resource allocation, dynamic distribution, and growth. By combining online services and cloud computing, designers may work on SMSD projects from anywhere.

4 Legal Issues of Digital Twin

Other than real estate and business, there are many digital twin use cases related to healthcare sector today. In 2023, the marketplace for digital twin is predictable to be worth $15.6 billion, by a yearly progress rate of above 40%. The possibilities of digital twins for smart manufacturing are endless, and they are reshaping the real estate industry's future. However, the digital twin's inventive potential should not overwhelm the legal framework's value. Indeed, the creation of the digital twin necessitates the removal of certain legal stumbling blocks. The digital twin is built on contextualized data that is then translated into knowledge. Data, algorithms and AI systems are therefore at the centre of the digital twin. As a result, the legal framework for these notions is crucial to the growth of the digital twin. The digital twin creates legal concerns about data, liability, algorithmic reliability, and intellectual property. As a result, the digital twin becomes a scientific as well as a legal concern.

4.1 Lack of Laws and Regulations

The legal ramifications of digital twins are many and may be difficult to navigate, particularly if there is no regulatory framework that can establish norms and provide a consistent, secure basis for which we can rely. Consequently, well-documented contracts are essential to control communication with digital twin's technology users. Cyber-security is an essential part of digital twin technology. An uncontrolled digital twin resolution can have serious real-world significances, especially if it is used to control composite systems or to make or adapt a physical creation and use it or sell it in the market. Laws must thus control the unique characteristics of digital twin resolutions and must be detailed to address the most significant dangers and possible ramifications, while also being flexible enough to adapt to technological and real-world linked counterpart evolutions.

4.2 Intellectual Property and Data Privacy

Data ownership, in particular, is a kind of intellectual property that has yet to be fully applied to digital twins. The legal ramifications of data sharing and secrecy create a legal quandary. Digital twins would be meaningless if data is not shared. Most intellectual property and legal standards, on the other hand, do not allow for non-essential data exchange, particularly in complicated situations when ownership is already ambiguous. Hence, legal considerations deriving from dangers related with data sharing must be considered. To preserve confidentiality while exchanging sensitive data, appropriate safeguards must be considered. Non-disclosure agreements and confidentiality terms in contracts may be used to enforce this. There has not been any legal framework devoted to the digital twin until now. However, data protection requirements are similarly valid in the usage of personal data in digital twins [51].

The smart manufacturing industry is made up of ideas, which is why intellectual property generates value and deserves legal rights under patent protection. A patent may be filed depending on how the technology is created, utilized, and how new ideas are produced. In general, intellectual property rights and patents are difficult to monitor. The use of intellectual property rights is conceivable if the algorithm is included in a patentable invention [52]. For example, consider the scenario of a discovery made possible by the modelling of a digital twin. The innovation might therefore be legally protected, as long as the standards for novelty are encountered: the origination needs to be original and capable of industrial application.

4.3 Digital Twin and the Liability

The digital world's actors are putting a lot of pressure on the extent of liability. The traditional principles of civil and criminal culpability are being utilized for robots and autonomous technologies. Machines were formerly thought to be incapable of acting on their own, but in the new virtual world, they have overcome the gap and become autonomous. Do we need to adapt our paradigms and enable robots to represent us in an authorized structure that governs virtual communities [53]? Is it possible to hold the actual person liable for the harm produced by his or her digital twin? Who is to blame: the computer programmer, the industrialist, the owner of robot, or the machine itself? The legal application of digital twin in its current stage does not answer the above questions.

The legal effect of digital twins on liability is likely to be the most complicated. To hold someone legally accountable for any kind of loss, you must be able to demonstrate that the liable party failed to execute their obligation, which resulted in the loss or harm for which they are liable. Because digital twins are a network of interconnected technologies, proving responsibility is challenging. As a result, changes in one system have an impact on the whole model. Errors may be difficult to see and track when there are several users and data sources involved. One option to address this issue is to establish a centralized command centre that monitors and approves data updates. Even yet, mistakes may happen and are tough to track down. Errors might also have come from outside the digital twin ecosystem. For example, if the physical twin's sensors are not operating properly, then it is normal for the whole system to fail. To assist the best functioning of digital twins, it is necessary to specify goals, qualities, and effective functions of a task. To build trust and responsibility, these concepts might be expressed in the application of digital twins.

4.4 Algorithmic Integrity and Digital Twins

The data quality will determine the digital twin's virtual picture, which will mirror the actual product. Alternatively, data reliability is a significant and difficult problem. The AI research group has just recently started to focus on approaches to identify and eliminate prejudice in supervised automated learning systems' training data sets. It has been shown that the computerized and automatic evaluation of large data increases the danger of prejudice [54]. As a result, data quality is critical: it should be impartial, and comprehensive, that is, reflective of diverse developments and consistent across time.

5 Conclusion

Because of the complexity, digital twins have substantial legal significances. The usage and models of digital twins will rise in tandem with digitalization. The legal effect of digital twins on liability is likely to be the most complicated. To hold someone legally accountable for any kind of loss, you must be able to demonstrate that the liable party failed to execute their obligation, which resulted in the loss or harm for which they are liable. The use of digital twins will grow and become more complex in the imminent times. Consequently, it is important that all groups involved in the use of digital twins form a pure individuality, function, and accountability from the beginning. If the models are prepared appropriately, this will ensure that all groups are legitimately secured.

References

1. Lu Q, Xie X, Parlikad AK, Schooling JM (2020) Digital twin-enabled anomaly detection for built asset monitoring in operation and maintenance. Autom Constr 118:103277
2. Bundin M, Martynov A, Shireeva E Legal issues on the use of "digital twin" technologies for smart cities. In: International conference on electronic governance and open society: challenges in Eurasia. Springer, pp 77–86
3. Butt J (2020) Exploring the interrelationship between additive manufacturing and Industry 4.0. Designs 4:13
4. Jones D, Snider C, Nassehi A, Yon J, Hicks B (2020) Characterising the digital twin: a systematic literature review. CIRP J Manuf Sci Technol 29:36–52
5. Semeraro C, Lezoche M, Panetto H, Dassisti M (2021) Digital twin paradigm: a systematic literature review. Comput Ind 130:103469
6. Millwater H, Ocampo J, Crosby N (2019) Probabilistic methods for risk assessment of airframe digital twin structures. Eng Fract Mech 221:106674
7. Grieves M, Vickers J (2017) Digital twin: mitigating unpredictable, undesirable emergent behavior in complex systems. Transdisciplinary perspectives on complex systems. Springer, pp 85–113
8. Elayan H, Aloqaily M, Guizani M (2021) Digital twin for intelligent context-aware IoT healthcare systems. IEEE Internet Things J 8:16749–16757

9. Erol T, Mendi AF, Doğan D The digital twin revolution in healthcare. In: 2020 4th international symposium on multidisciplinary studies and innovative technologies (ISMSIT). IEEE, pp 1–7
10. Deng T, Zhang K, Shen Z-JM (2021) A systematic review of a digital twin city: a new pattern of urban governance toward smart cities. J Manage Sci Eng 6:125–134
11. Ketzler B, Naserentin V, Latino F, Zangelidis C, Thuvander L, Logg A (2020) Digital twins for cities: a state of the art review. Built Environ 46:547–573
12. Shahat E, Hyun CT, Yeom C (2021) City digital twin potentials: a review and research agenda. Sustainability 13:3386
13. Yun H, Park D (2021) Virtualization of self-driving algorithms by interoperating embedded controllers on a game engine for a digital twining autonomous vehicle. Electronics 10:2102
14. Augustine P (2020) The industry use cases for the digital twin idea. Adv Comput 117:79–105. Elsevier
15. Barricelli BR, Casiraghi E, Fogli D (2019) A survey on digital twin: definitions, characteristics, applications, and design implications. IEEE Access 7:167653–167671
16. Bécue A, Maia E, Feeken L, Borchers P, Praça I (2020) A new concept of digital twin supporting optimization and resilience of factories of the future. Appl Sci 10:4482
17. Ruohomäki T, Airaksinen E, Huuska P, Kesäniemi O, Martikka M, Suomisto J Smart city platform enabling digital twin. In: 2018 International conference on intelligent systems (IS). IEEE, pp 155–161
18. Sepasgozar SM, Hui FKP, Shirowzhan S, Foroozanfar M, Yang L, Aye L (2020) Lean practices using building information modeling (Bim) and digital twinning for sustainable construction. Sustainability 13:161
19. Nujoom R, Mohammed A, Wang Q (2018) A sustainable manufacturing system design: a fuzzy multi-objective optimization model. Environ Sci Pollut Res 25:24535–24547
20. De Paolis LT, Bourdot P (2019) Augmented reality, virtual reality, and computer graphics: 6th international conference, AVR 2019, Santa Maria al Bagno, Italy, June 24–27, 2019, Proceedings. Springer, Part II
21. Zakoldaev D, Korobeynikov A, Shukalov A, Zharinov I Digital forms of describing industry 4.0 objects. In: IOP conference series: materials science and engineering. IOP Publishing, p 012057
22. Negri E, Fumagalli L, Macchi M (2017) A review of the roles of digital twin in CPS-based production systems. Proc Manuf 11:939–948
23. Shao G, Kibira D Digital manufacturing: requirements and challenges for implementing digital surrogates. In: 2018 winter simulation conference (WSC). IEEE, pp 1226–1237
24. Valckenaers P (2020) Perspective on holonic manufacturing systems: PROSA becomes ARTI. Comput Ind 120:103226
25. Leng J, Wang D, Shen W, Li X, Liu Q, Chen X (2021) Digital twins-based smart manufacturing system design in industry 4.0: a review. J Manuf Syst 60:119–137
26. Tao F, Zhang H, Liu A, Nee AY (2018) Digital twin in industry: State-of-the-art. IEEE Trans Industr Inf 15:2405–2415
27. Tao F, Qi Q, Wang L, Nee A (2019) Digital twins and cyber–physical systems toward smart manufacturing and industry 4.0: correlation and comparison. Engineering 5:653–661
28. Nativi S, Mazzetti P, Craglia M (2021) Digital ecosystems for developing digital twins of the earth: the destination earth case. Remote Sensing 13:2119
29. Haag S, Anderl R (2018) Digital twin—proof of concept. Manuf Lett 15:64–66
30. Konstantinov S, Ahmad M, Ananthanarayan K, Harrison R (2017) The cyber-physical e-machine manufacturing system: virtual engineering for complete lifecycle support. Proc CIRP 63:119–124
31. Hasan HR, Salah K, Jayaraman R, Omar M, Yaqoob I, Pesic S, Taylor T, Boscovic D (2020) A blockchain-based approach for the creation of digital twins. IEEE Access 8:34113–34126
32. Zheng P, Sang Z, Zhong RY, Liu Y, Liu C, Mubarok K, Yu S, Xu X (2018) Smart manufacturing systems for industry 4.0: conceptual framework, scenarios, and future perspectives. Front Mech Eng 13:137–150

33. Sanna A, Giacalone G (2021) Digital twin and machine learning solutions for the manufacturing environment
34. Wu C, Zhou Y, Pessôa MVP, Peng Q, Tan R (2021) Conceptual digital twin modeling based on an integrated five-dimensional framework and TRIZ function model. J Manuf Syst 58:79–93
35. Glaessgen E, Stargel D The digital twin paradigm for future NASA and US air force vehicles. In: 53rd AIAA/ASME/ASCE/AHS/ASC structures, structural dynamics and materials conference 20th AIAA/ASME/AHS adaptive structures conference 14th AIAA, pp 1818
36. Qi Q, Tao F, Hu T, Anwer N, Liu A, Wei Y, Wang L, Nee A (2021) Enabling technologies and tools for digital twin. J Manuf Syst 58:3–21
37. Bao J, Guo D, Li J, Zhang J (2019) The modelling and operations for the digital twin in the context of manufacturing. Enterprise Inf Syst 13:534–556
38. Friedland B (2012) Control system design: an introduction to state-space methods. Courier Corporation
39. Peruzzini M, Pellicciari M (2017) A framework to design a human-centred adaptive manufacturing system for aging workers. Adv Eng Inform 33:330–349
40. Scacchi A, Catozzi D, Boietti E, Bert F, Siliquini R (2021) COVID-19 lockdown and self-perceived changes of food choice, waste, impulse buying and their determinants in Italy: QuarantEat, a cross-sectional study. Foods 10:306
41. Taylor C, Murphy A, Butterfield J, Jan Y, Higgins P, Collins R, Higgins C (2018) Defining production and financial data streams required for a factory digital twin to optimise the deployment of labour. Recent Adv Intel Manuf 3–12. Springer
42. Moyne J, Iskandar J (2017) Big data analytics for smart manufacturing: case studies in semiconductor manufacturing. Processes 5:39
43. Gao RX, Wang L, Helu M, Teti R (2020) Big data analytics for smart factories of the future. CIRP Ann 69:668–692
44. Danilczyk W, Sun Y, He H Angel: an intelligent digital twin framework for microgrid security. In: 2019 North American power symposium (NAPS). IEEE, pp. 1–6
45. Gupta N, Tiwari A, Bukkapatnam ST, Karri R (2020) Additive manufacturing cyber-physical system: supply chain cybersecurity and risks. IEEE Access 8:47322–47333
46. Leng J, Jiang P, Xu K, Liu Q, Zhao JL, Bian Y, Shi R (2019) Makerchain: a blockchain with chemical signature for self-organizing process in social manufacturing. J Clean Prod 234:767–778
47. Spellini S, Chirico R, Lora M, Fummi F Languages and formalisms to enable EDA techniques in the context of industry 4.0. In: 2019 Forum for specification and design languages (FDL). IEEE, pp 1–4
48. Leng J, Jiang P (2018) Evaluation across and within collaborative manufacturing networks: a comparison of manufacturers' interactions and attributes. Int J Prod Res 56:5131–5146
49. Avventuroso G, Silvestri M, Pedrazzoli P (2017) A networked production system to implement virtual enterprise and product lifecycle information loops. IFAC-PapersOnLine 50:7964–7969
50. Cohen Y, Faccio M, Pilati F, Yao X (2019) Design and management of digital manufacturing and assembly systems in the Industry 4.0 era, vol 105. Springer, pp 3565–3577
51. Lkhagvasuren G Ensuring rights of the data subject in non-EU countries. In: Proceedings of the 12th international conference on theory and practice of electronic governance, pp 465–467
52. Mulligan DK, Kluttz D, Kohli N (2019) Shaping our tools: contestability as a means to promote responsible algorithmic decision making in the professions. Available at SSRN 3311894
53. Bourcier D (2001) De l'intelligence artificielle à la personne virtuelle: émergence d'une entité juridique? Droit et société 847–871
54. Schmid PC, Amodio DM (2017) Power effects on implicit prejudice and stereotyping: the role of intergroup face processing. Soc Neurosci 12:218–231

LSTM-Based Encoder–Decoder Attention Model for Text Translation and Simplification on the Constitution of India

Meith Navlakha, Russel Lobo, Rishabh Bhargava, and Ruhina B. Karani

Abstract Natural language processing techniques can be used on judicial and legislative documents like the Constitution for making it more accessible to the general audience. Various approaches such as Natural Machine Translation (NMT), text simplification of complex sentences can be performed on the Constitution. The model proposed in this paper can be used for the purpose of translation and simplification of the Constitution of India. The model is a LSTM variant forming an encoder–decoder network integrated with an attention layer and is trained using teacher forcing method. The data set used for the task of translation consists of Parallel Corpus from IIT Bombay English–Hindi concatenated with our own curated data set of 300 English articles of the Constitution of India translated to Hindi, whereas for simplification task it consists of our own curated data set of complex 300 English articles of the Constitution of India translated to simplified sentence. The proposed model for translation has a BLEU score of 15.34, and the research has elaborated on the performance analysis of the generated outputs and the BLEU score. For simplification, the results show some inconsistencies which can be improved by increasing the data set for simplification task.

All authors have contributed equally.

M. Navlakha (✉) · R. Lobo · R. Bhargava · R.B. Karani
Department of Computer Engineering, Dwarkadas J. Sanghvi College of Engineering, Mumbai, India
e-mail: meithnavlakha@gmail.com

R. Lobo
e-mail: russellobo0014@gmail.com

R. Bhargava
e-mail: rishabhb1403@gmail.com

R.B. Karani
e-mail: ruhina.karani@djsce.ac.in

Keywords NLP · Translation · Text simplification · Cognitive linguistics · Long- and short-term memory · Seq2Seq · Teacher forcing · Attention · Indian languages · COI, constitution of India

1 Introduction

Natural language processing (NLP) has witnessed significant advancements over the few years, helping the machine to understand and interpret human language across many disciplines, in the fields of computational linguistics, language modelling, language translation and simplification, and being capable of analysing and extracting linguistic context from a substantial volume of text. Generally, the legislative and judicial records of various countries incorporate technical terminologies which are not comprehensible to the layman. Moreover, some of the jargon used in judicial terms may have significant meaning in that field, while having inconsequential meaning in the normal day-to-day conversation context. As a result, most of the state-of-the-art (SOTA) models on the English–Hindi NMT pair fail to capture the appropriate context when it comes to legislative and judicial translation and simplification. Through the paper, we propose using NLP approaches to bridge the gap by developing a model that focuses especially on the legislative and judicial context of the sentences.

The proposed model will help citizens better understand the law and order and other legislative documents, making these records more accessible to the common man. Most people find the language used in the Constitution of India complicated and complex leading them to eschew it. The simplification of the articles in the constitution may help a citizen to get an easier and more limpid understanding of the laws in the constitution. Also, in India, we have a large diversity present in the amounts of the language spoken across various states. The Constitution of India is written in English (and some other native languages) where less than half of the population can even read and understand English. Thereby, using our language translation model, we can help bridge the gap and make the constitution of the people easily comprehensible to all. Machine translation is one of the most important fields in NLP and can prove to be helpful to remove the language barrier by translating the constitution to their native languages.

The scope of the research is to develop an LSTM-based encoder–decoder teacher forcing and attention model for NMT between the English and Hindi language pair focusing on the judicial and legal context of the sentences. However, the inadequacy of pertinent data sets and hardware resource constraints have been the major hurdles in the field. Our contributions through this research:

1. A model that focuses on the legal and judicial context while translating from English to Hindi.
2. A novel data set having English to Hindi translation of the articles in the Constitution of India.

In the following Sects. 2 and 3, we discuss the related works on English to Hindi machine translation. Section 4 elaborates on the proposed methodology of our research. Section 5 gives a brief description of the corpus used, and Sect. 6 expounds on the experimental designs and set-up for the model. Finally, in Sect. 7, we analyse the results and performance of the model and conclude the paper in Sect. 8.

2 Literature Review

As there is a lot of potential in using NLP models for machine translation and sentence simplification, recent research has shown a myriad of methods that provide a cohesive end-to-end structure, rather than conventional methods that utilise various submodels and long pipelines. Cho et al. [1] propose a neural machine translation model consisting of encoder and decoder. Neural machine translation models usually employ an encoder and a decoder. Using two models, the RNN encoder–decoder and newly created gated recursive convolutional neural networks, the features of this technique were investigated. But with a rise in sentence length and an increase in unknowns, the performance started to degrade.

The model suggested by He et al. [2] improve translation diversity and quality by using a council of specialised translation models rather than a single translation model. Each mixture component chooses its own training data set leading to soft clustering of the parallel corpus. Encoder–decoder architectures have dominated the field of sequence modelling in NLP since Sutskever et al. [3] released their "sequence to sequence" learning model. The resulting translations from Seq2Seq models sometimes lack diversity. The reason for this is the differences in styles, genres, subjects, or ambiguity in the translation process which causes semantic and syntactic changes in a corpora that SEQ2SEQ models are unable to detect.

Other approaches include DNNs that have proven to be powerful models. However, they cannot be used to map sequences to sequences which is a significant limitation. Sutskever et al. [3] present a sequence to sequence (seq2seq) model that uses a multilayered LSTM and another deep LSTM. It was observed that LSTM did not have difficulty with long sentences and LSTM outperformed the phrase-based SMT system. Escolano et al. [4] suggest an alternative approach where encoders–decoders which is language specific will be flexible to learn new language by learning their corresponding modules was proposed. The method was successful to add new language, without the need of retraining, thus outperforming universal encoder–decoder by 3.28 points on average. Reference [5] introduces mT5, a multilingual variant of T5 which is pretrained on a new Common Crawl-based data set covering 101 languages.

Mahmud et al. [6] implemented a GRU-based unidirectional RNN model with an encoder–decoder attention which is a neural machine translation (NMT) model. It is focused on English–Bangla translation and a small-sized balanced data set is used for the same. A huge BLEU score of 50.07 was achieved. The Statistical Machine Translation (SMT) proposed by Brown et al. [7] makes use of the rules of probability for language translation.

Martin et al. [8] build upon the Seq2Seq model. Additional inputs have been added to the original sentences at the time of training, in the form of control tokens. The aim is to make the user in charge of how the model simplifies sentences on four important features of sentence simplification. Four control tokens have been introduced namely NbChars, LevSim, WordRank and DepTreeDepth. ACCESS scores best with SARI score of 41.87, a substantial advancement over previous SOTA (40.45), and third to best FKGL (7.22). With respect to the SARI score, the second and third models, DMASS+DCSS (40.45) and SBMT+PPDB+SARI (39.96), both make use of the external resource Simple Paraphrase Database (Pavlick and Callison-Burch, 2016) that was derived from data that was 1000 times larger than the training set used in this paper.

Narayan et al. [9] present a hybrid approach to sentence simplification. It combines a machine translation module that handles reordering and replacement with a model that encodes probabilities for splitting and deletion. The approach is based on semantics. The SARI score is 28.61, 31.40, 30.46 on Newsela, Turk Corpus, PWKP/WikiSmall data sets, respectively.

A unique end-to-end neural programmer–interpreter (Reed and de Freitas 2016) that learns to generate edit operations directly, similar to a human editor, was introduced by Dong et al. [10]. The suggested architecture consists of a programmer and an interpreter; the programmer predicts an edit operation that will simplify things, such as add, delete and keep; the interpreter performs the edit operation while keeping track of a pointer. It gives a SARI score of 38.22, 32.35, 31.41 on WikiLarge, WikiSmall, Newsela data sets, respectively.

As opposed to prior work that uses source-side similarity search for retrieving memory and bilingual corpora as translation memory (TM), in their proposed paradigm, Cai et al. [13] suggest using monolingual memory and crosslingual learnable memory retrieval. In the proposed framework, abundant monolingual memory can be TM due to the crosslingual memory retriever. Also, for the ultimate translation objective, the memory retriever and NMT model can be jointly optimised.

The randomness of neural networks leads to the existing neural network machine translation models unable to effectively reflect the linguistic dependencies and having unsatisfactory results when dealing with long sentence sequences. Xu et al. [14] propose a machine translation model with entity tagging improvement. It implements LSTM with attention mechanism to tune the extent to which the context at source language influences the target language sequence. The mean BLUE value achieved is 24.7%.

3 Neural Machine Translation (NMT)

Recurrent Neural Networks were initially proposed for NMT by Medsker, Larry et al. [11]. RNNs relies on the idea of repeated units which makes it easier to learn and maintain context owing to their cyclic structure. However, in a sentence like: "Goalkeepers have to always look at the attackers footwork and on-field position to determine whether the ball will be passed or shot. So, I believe in a game of football instinctive play is best suited for ___ position", the blank here can have a variety of answers :- Goalkeeper, RightFullback, LeftFullback, CenterBack, CenterBack. RNNs find it difficult to learn the dependencies and relationships and preserve context for long sentences . Therefore, in general for language with such complex contexts, RNN is generally avoided for the use of encoder and decoder design. Long- and Short-Term Memory (LSTM) models were introduced for encoding and decoding to solve the drawbacks of RNNs with better context preservation capabilitites.

3.1   Long- and Short-Term Memory (LSTM)

Long- and Short-Term Memory is proposed by Hochreiter et al. [12] is a type of RNN that is used to solve issues with long-term temporal dependence. The structure of LSTMs is similar to that of a chain. In a module of LSTM, there are four layers of neural networks. These layers interact with one other as well as with other modules in order to facilitate learning. Figure 1 depicts a typical LSTM module structure.

3.2   Encoder and Decoder

LSTM can be used to solve sequence-to-sequence prediction issues utilising the encoder–decoder approach. The method makes use of the encoder, which encodes the input sequence, and the decoder, which decodes the encoded input sequence into the target sequence.

3.3   Bahdanau Attention Method

The performance of the seq2seq LSTM model declined with the increase in the length of the input sequence. A Deep Neural Network could focus just on a few important topics by using the attention mechanism instead of storing the entire

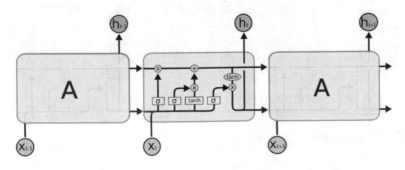

Fig. 1   LSTM internal structure

context and then passing it to the decoder. An attention mechanism that simultaneously learnt to position and translate was introduced by Bahdanau et al. [15] and depicted in Fig. 2. The Bahdanau attention could alleviate the performance bottleneck of traditional encoder–decoder systems. Cho et al. [1] and Sutskever et al. [3] used a RNN encoder–decoder framework for neural machine translation along with the attention layer. The encoder and decoder states were combined linearly using this additive technique. Contrary to the seq2seq concept without attention, the context vector was created by all concealed states of the encoder (forward as well as backward) and decoder. To help attention, focus on the most crucial information, the input and output sequences were aligned using an alignment score parameterized by a feed-forward network based on the context vector connected to the source position and previously generated target words, the model would predict the target word.

The main components in the Bahdanau encoder–decoder architecture:

1. The hidden decoder state at time step $t - 1$ is represented by s_{t-1}.
2. The context vector at time step t is denoted by c_t. Each decoder step generates it in a different way to produce the target word, y_t.
3. h_i is an annotation that concentrates heavily on the ith word out of T total words while capturing the information found in the words making up the complete input sentence, $\{x_1, x_2, ..., x_T\}$.

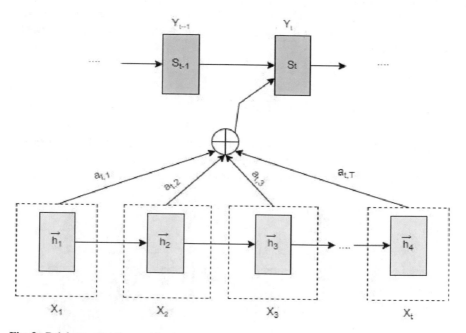

Fig. 2 Bahdanau attention architecture

4. Each annotation, h_i, is assigned a weight value at i at the current time step, t.
5. An alignment model, $w(.)$, produces an attention score called $e_{t,i}$ that rates how well s_{t-1} and h_i match.

The Bahdanau attention algorithm works as followed:

1. A collection of annotations, h_i, are produced by the encoder from the input sentence.
2. These annotations and the prior hidden decoder state are put into an alignment model. This data is used by the alignment model to produce the attention scores, $e_{t,i}$.
3. The attention ratings are normalised into weight values, $a_{t,i}$, in a range between 0 and 1, by applying a softmax function to them.
4. A context vector, c_t, is produced by adding these weights to the previously computed annotations in order to create a weighted total of the annotations.
5. To compute the final output, y_t, the context vector is supplied to the decoder together with the previous hidden decoder state and the previous output.
6. Up until the end of the sequence, steps 2 through 6 are repeated.

$$a_{t,s} = \frac{\exp(\text{score}(h, \overline{h_s}))}{\sum\limits_{s'} \exp(\text{score}(h, \overline{h_{s'}}))}$$

$$c_t = \sum_s a_{t,s} \overline{h_s}$$

$$s_t = f(c_t, h_t) = \tanh(W_c[c_t; h_t])$$

4   Proposed Methodology

Figure 3 depicts the various steps involved in the proposed methodology for both text translation and simplification. Translation and simplification models have been trained separately but have the same underlying methodology. First, we pre-process the original sentences from our corpus by removing the special characters, punctuations and extra white spaces. Then, the pre-processed text is tokenized and fed to the input layer which transforms the tokenized text to 256 embedding vectors. The embeddings are passed to the LSTM layer of the encoder. The decoder receives the weights from the attention layers connecting the encoder and the decoder. The decoder works at the word-level generating one word at a time which stops when the <END> special token is generated or maximum length is reached.

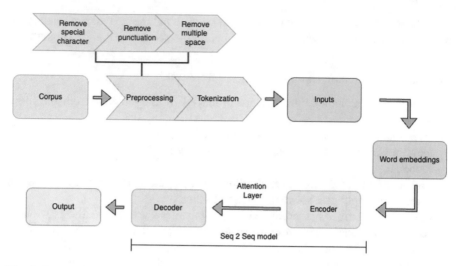

Fig. 3 Proposed methodology of the proposed Seq2Seq model for training

4.1 Pre-processing and Tokenization

Firstly, for an English data set, we convert all the letters in the data set to lowercase letters. Then, the punctuation marks (!.,) and special characters($#) were replaced by a single space character. Later the multiple spaces were replaced by a single space, to indicate the beginning and conclusion of the statement when training a "start" and "end" token is inserted into the front and back of the text, respectively. After performing these actions, we now finally tokenise the corpus. Similarly, for the training process, we perform the above sequence of actions on the parallel Hindi data set.

4.2 Word Embedding

Every sentence is transformed to a fixed-sized dimensional vector representation using the Keras embedding layer. Our proposed model uses word-level embedding, leveraging the learned vector representation for the sentences with similar meaning words that have closer orientation.

4.3 Model Architecture

The LSTM-based encoder–decoder network is used for mapping the input and target sentences in the sequence-to-sequence model. The proposed model leverages teacher forcing in order to have a faster convergence. Initially, learning is fairly

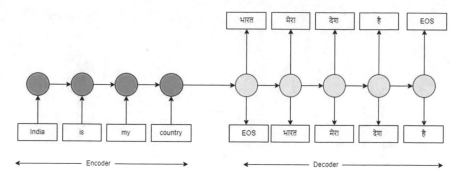

Fig. 4 Word-level Seq2Seq model using LSTM for text translation with English tokens as input to the Encoder and predicted Hindi output tokens by the Decoder

low for training, so using teacher forcing we would feed back the actual expected results rather than what the model has predicted, leading to a faster updation of the weights in the correct direction. Figure 4 depicts our base seq2seq model for English to Hindi translation.

4.4 Attention

Our proposed model stacks an attention layer bridging all the encoder units to each decoder unit to improve the existing seq2seq models. Attention offers a learning method where the decoder can learn in a more weighted context from the encoder and interpret which part of the subsequent encoding network requires more "attention" when predicting the output. Although the attention-based seq2seq model requires higher computational resources, the results are quite convincing compared to the traditional seq2seq. The attention model creates a context vector that is filtered for each output time step instead of just encoding the input sequence into a single fixed context vector.

Figure 5 depicts the proposed model with the attention layer. Both the encoder and decoder consist of a LSTM layer of 256 units with a dropout of 0.2. Initially, the input layer feeds in the input text and passes it to the embedding layer which transforms the tokenized sentence into a 256 dimensional embedding. One embedding layer for each encoder and decoder is initialised at the start. All the encoder units are connected to Bahdanau attention layer passing the context vector to decoder unit one at a time.

5 Corpus Description

5.1 COI Translation Corpus

The prerequisite for a machine translation model is the availability of parallel corpora for source and target languages. In terms of the availability of parallel data, Hindi is a less resourceful and atypical language when compared to its European

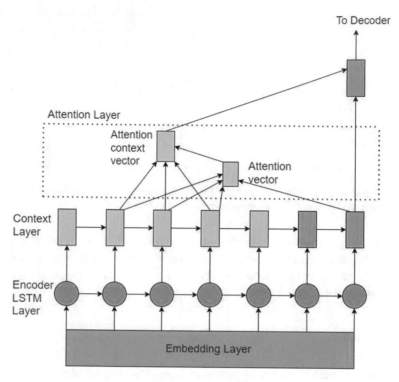

Fig. 5 Seq2seq model architecture using LSTM with attention

analogues. Also, our major focus is on the Judicial and Legal domain narrowed down the choices. Thus, we had to create our own data set for the Constitution of India. Our proposed translation parallel corpus includes these two sources:

1. Parallel Corpus from IIT Bombay English–Hindi corpus. It contains data from multiple disciplines out of which only the Judicial domain and different Indian Government website sources were filtered out.
2. Curated our own custom English to Hindi data set on the Constitution of India with reference to the official Indian Constitution Hindi version.

The corpus is exhaustive with an eclectic diversity in the vocabulary. Table 1 illustrates the sentence distribution with respect to each domain in the data set.

90% of the data set's sentences were used for training, 9% for testing, and 1% for validation of the translation model.

5.2 COI Simplification Corpus

Alongside, another parallel data set for text simplification was curated, consisting of articles from the Constitution of India illustrated in Table 2. Original articles

Table 1 Domain-wise distribution of a number of sentences in translation data set

Dataset	Domain	No. of sentences
IITB Corpus	Judicial domain corpus-I and II	8734
	Different Indian Government websites	2,89,615
	Gyaan-Nidhi Corpus	2,27,123
	Mahashabdkosh: administrative domain	1,59,822
	Hindi–English wordnet linkage	1,75,175
Constitution of India	All the articles	300

Table 2 Dataset distribution for simplification data set

Dataset	Domain	No. of sentences
Constitution of India	Articles	300

and its corresponding simplified sentences were clubbed to form the dataset records. The simpler versions of sentences are easily comprehensible to a layman's understanding.

6 Experimental Set-Up

6.1 Experimental Resources

The model is trained on Google Colab having the following hardware configurations:

- 12 GB RAM
- 16 GB GPU
- Nvidia K80/T4 GPU.

6.2 Online Training

Online Training technique has been implemented as it is computationally infeasible to train the entire corpus due to resource constraints. We have fragmented the COI translation data set into four fragments and trained the model in a cyclic manner each fragment at a time. In this way the model learns in an incremental manner without taxing high computational requirements.

7 Results and Analysis

7.1 Evaluation Metrics

BLEU-4 [14] is a metrics used to evaluate the similarity between the predicted sentence to its corresponding reference parallel. In our case, it is the translated or simplified sentence compared to its original sentence. BLUE score scales from 1.0, indicating an absolute similarity to 0 indicating an absolute mismatch.

7.2 Analysis

BLUE score of our model with different configurations, on the validation dataset (1% holdout from the corpus), has been presented in Table 3.

As observed in Table 4, our proposed model performed well for shorter sentences. The translated sentence must be similar to the reference sentence in the data set. The model is capable of capturing the context behind the judicial and legislative jargon and translating them correctly.

However, as observed in Table 5, the model's performance degrades as the sentence length increases. It starts off correctly but then starts repeating the words till it gets cut at the max length. This is mainly due to the fact that our model uses only a single LSTM layer. The model's performance for longer sentences can be enhanced by stacking more LSTM layers or even using bidirectional LSTM layers. The simplification model isn't added to the tables due its low performance, resulting into repeated words. This is majorly because the current simplification corpus only has 300 records.

Table 3 BLUE score on validation set

Configuration	BLUE score
Seq2seq lstm with teacher forcing	12.53
Seq2seq model with attention (Bahdanau)	15.34

Table 4 Performance of the model on short sentences

Source	India is my country
Seq2seq model with attention (Bahdanau)	भारत मेरा देश है
Google	भारत मेरा देश है।
Reference	भारत मेरा देश है।

Table 5 Performance of the model on long sentences

Source	Provision for just and humane conditions of work and maternity relief: the state shall make provision for securing just and humane conditions of work and for maternity relief
Seq2seq model with attention (lBahdanau)	न्याय और मानवीय परि स्थि ति यों के लि ए प्रबन्ध: है है है है है है है है है है है है है है है है
Google	कार्य की न्यायोचि त एवं मानवीय दशाओं तथा प्रसूति सहायता का प्रावधान: राज्य कार्य की न्यायोचि त एवं मानवीय दशाओं को सुनि श्चि त करने तथा प्रसूति सहायता के लि ए प्रावधान करेगा।
Reference	काम की न्यायसंगत और मानवोचि त दशाओं का तथा प्रसूति सहायता का उपबंध—राज्य काम की न्यायसंगत और मानवोचि त दशाओं को सुनि श्चि त करने के लि ए और प्रसूति सहायता के लि ए उपबंध करेगा।

8   Conclusion and Future Works

In the current work, we have focused on different NLP techniques like text translation and text simplification on the legislative and judicial text records in order to reduce the complexity and make it more comprehensible to the common people. Performance and results of the proposed LSTM-based seq2seq model incorporating teacher forcing have been expounded in Table 3. Both the translation and simplification models have trained on our custom curated data set on the Constitution of India. The seq2seq lstm model with teacher forcing gives a BLUE score of 12.53, however adding an attention layer enhances the model. It helps the model better understand the context of the sentences. The NMT model for English–Hindi language pairs having a BLUE score of 15.34 works well for short sentences as seen in Table 4. The NMT model successfully captures the judicial and legislative context in the given sentence. However, we observe a decrease in the performance for longer sentences as seen in Table 5. The simplification model has a relatively lower performance mainly due to the fact that we only have around 300 records for simplification, and the accuracy can future be increased by working on the data set. Moreover, the performance can further be improvised by increasing the epochs and the encoder–decoder units which are currently limited due to hardware constraints.

The research work can further be extended by increasing the simplification and translation data set with more judicial and legislative sentences. Our proposed model uses teacher forcing, other methods like beam-search, and state-of-the-art transformers can also be experimented with to increase the performance. The accuracy of the text translation model can further be improved by augmenting the data set with more judicial specific English–Hindi pairs.

References

1. Cho K, Merriënboer BV, Bahdanau D, Bengio Y (2014) On the properties of neural machine translation: encoder-decoder approaches. arXiv:1409.1259
2. He X, Haffari G, Norouzi M (2018) Sequence to sequence mixture model for diverse machine translation. arXiv:1810.07391
3. Sutskever I, Vinyals O, Le QV (2014) Sequence to sequence learning with neural networks. Adv Neural Inf Process Syst 27
4. Escolano C, Costa-jussà MR, Fonollosa JAR, Artetxe M (2020) Multilingual machine translation: closing the gap between shared and language-specific encoder-decoders. arXiv:2004.06575
5. Xue L, Constant N, Roberts A, Kale M, Al-Rfou R, Siddhant A, Barua A, Raffel C (2020) mT5: a massively multilingual pre-trained text-to-text transformer. arXiv:2010.11934
6. Mahmud A, Al Barat MM, Kamruzzaman S (2021) GRU-based encoder-decoder attention model for English to Bangla translation on novel dataset. In: 2021 5th international conference on electrical information and communication technology (EICT). IEEE (pp 1–6)
7. Brown PF, Cocke J, Della Pietra SA, Della Pietra VJ, Jelinek F, Mercer RL, Roossin P (1988) A statistical approach to language translation. In: Cooling Budapest 1988 volume 1: international conference on computational linguistics
8. Martin L, Sagot B, de la Clergerie E, Bordes A (2019) Controllable sentence simplification. arXiv:1910.02677
9. Narayan S, Gardent C (2014) Hybrid simplification using deep semantics and machine translation. In: The 52nd annual meeting of the association for computational linguistics, pp 435–445
10. Dong Y, Li Z, Rezagholizadeh M, Cheung JCK (2019) EditNTS: an neural programmer-interpreter model for sentence simplification through explicit editing. arXiv:1906.08104
11. Medsker L, Jain LC (eds) (1999) Recurrent neural networks: design and applications. CRC press
12. Hochreiter S, Schmidhuber J (1997) Long short-term memory. Neural Comput 9(8):1735–1780
13. Kunchukuttan A, Mehta P, Bhattacharyya P (2017) The IIT Bombay English-Hindi parallel corpus. arXiv:1710.02855
14. Papineni K, Roukos S, Ward T, Zhu WJ (2002) BLEU: a method for automatic evaluation of machine translation. In: ACL
15. Bahdanau D, Cho K, Bengio Y (2014) Neural machine translation by jointly learning to align and translate. arXiv:1409.0473

Compiling C# Classes to Multiple Shader Sources for Multi-platform Real-Time Graphics

Dávid Szabó and Illés Zoltán

Abstract We are implementing a multi-platform real-time rendering library for .NET C#. This library supports low-level programming of Graphical Processing Units in our computers on the abstraction level of the graphics APIs, therefore it makes it possible to develop our own rendering algorithm or graphics engine in C#. With our library we can use shaders, buffers, textures, pipelines and other well-known low-level graphical resources to implement a graphical application in C#. This library is an abstraction layer above the low-level graphics APIs (currently supporting Vulkan and multiple versions of OpenGL). However, every graphics API supports different shading languages. To provide a C# abstraction layer for these APIs the layer needs to abstract their shader languages as well. In this paper, we are presenting our methods of using the .NET Roslyn Compiler API to make an Embedded Domain Specific Language in C# for implementing shader code. We propose the methods used in our library (SharpGraphics) to develop C# classes as shaders. These "shader classes" and their implementation are parsed and compiled into shader source codes in multiple shader languages during development time.

Keywords C# · Source generator · Real-time · Shaders · Vulkan

1 Introduction

Nowadays, with the .NET Ecosystem, especially with the newly arrived .NET 5 and 6 [1] we can reach multiple platforms and operating systems from a single source code. A standard C# class library can be compiled for Windows, Linux, Mac, to mobile platforms like Android or iOS and even to the Web as a webservice or webpage. Using multi-platform packages and organizing our code into libraries

D. Szabó (✉) · I. Zoltán
Eötvös Loránd University, Budapest, Hungary
e-mail: sasasoft@inf.elte.hu

I. Zoltán
e-mail: illes@inf.elte.hu

© The Author(s), under exclusive license to Springer Nature Singapore Pte Ltd. 2023 869
Y. Singh et al. (eds.), *Proceedings of International Conference on Recent Innovations in Computing*, Lecture Notes in Electrical Engineering 1011,
https://doi.org/10.1007/978-981-99-0601-7_65

we can accelerate our development for multiple systems. Only the system specific features must be reimplemented for every platform, our application specific models can be reused for all platforms. In our research we are merging this multi-platform C# area and technology with the world of low-level real-time graphics and rendering.

To implement a rendering algorithm or graphics engine You need to use graphics APIs to access the Graphical Processing Unit (GPU) in Your device. Using these APIs You can program GPUs, allocate resources on it and issue state modification and drawing commands to render sequence of images in real-time. Such graphics APIs are OpenGL, Vulkan, DirectX and Metal. These are C/C++ interfaces which can be used to communicate with the GPU's Device Driver, therefore most applications using these APIs are implemented in C or C++. While the APIs have so-called binding libraries for C# (extern PInvoke static methods and structures generated from the C/C++ interfaces) these libraries do not fit the previously presented multi-platform C# environment perfectly. Every graphics API support different platform. DirectX supports only Windows, Vulkan supports only modern devices, OpenGL with the combination of OpenGL ES and WebGL supports every platform (excluding Mac or iOS), but these are older APIs. The combined use of most these APIs is needed to support as many platforms and devices as possible. Our goal is to provide a possibility for .NET C# developers to use graphics APIs through our low-level abstraction layer.

Our library presents a way for developing multi-platform real-time rendering applications using C#. The library will provide a set of classes and functions for implementing low-level rendering algorithms in a cross-platform .NET application. The focus is to support implementing the rendering code only once in a shared library and all supported platforms can reference and consume this single codebase. Therefore, the library is an abstraction of Graphics APIs (currently Vulkan with vk .net [2] and OpenGL with OpenTK [3]). When the application initializes, the most fitting Graphics API can be selected to be used based on the current platform, device and capabilities. Furthermore, the real-time rendered images can be embedded into existing .NET applications' user interfaces, just like any other regular UI component.

In this paper, after the brief introduction of our library's structure the focus is on the GPU sided code implementation, the shader programming with the library. The custom algorithms that are executed by the GPU during our drawing commands are implemented in shader languages and these differ between graphics APIs. To hide this difference, we have implemented C# abstractions not just for the Graphics APIs, but for their supported shading languages as well. The goals, advantages and implementation are detailed in later sections of this paper.

2 Real-Time Graphics Library in C#

Our library is a structure of libraries, not just a single package. There is a base library which is the graphics API abstraction which the user will use for development most of the times. There are the implementation layer packages which are the individual

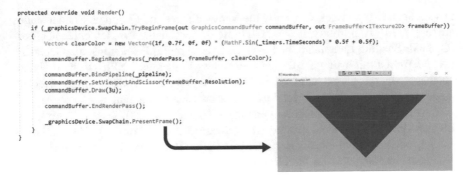

```
protected override void Render()
{
    if (_graphicsDevice.SwapChain.TryBeginFrame(out GraphicsCommandBuffer commandBuffer, out FrameBuffer<ITexture2D> frameBuffer))
    {
        Vector4 clearColor = new Vector4(1f, 0.7f, 0f, 0f) * (MathF.Sin(_timers.TimeSeconds) * 0.5f + 0.5f);

        commandBuffer.BeginRenderPass(_renderPass, frameBuffer, clearColor);

        commandBuffer.BindPipeline(_pipeline);
        commandBuffer.SetViewportAndScissor(frameBuffer.Resolution);
        commandBuffer.Draw(3u);

        commandBuffer.EndRenderPass();

        _graphicsDevice.SwapChain.PresentFrame();
    }
}
```

Fig. 1 CPU-side C# implementation of rendering the hello triangle sample application using our library (SharpGraphics)

Graphics APIs (currently Vulkan, modern OpenGL and OpenGL ES 3.0) implementing the base library [4]. Lastly there are the view layer packages which adds support for embedding and presenting the rendered images into multiple .NET UI frameworks like WPF, Xamarin, Avalonia, etc.

Using the base library, it is possible to implement a custom rendering algorithm in a multi-platform C# class library. All the rendering features like using shaders, buffers, pipelines, etc. can be coupled in this module, regardless of what platform or which graphics API will be targeted. The decoupled architecture makes it possible to reuse most of the code implemented with our library, because the only platform specific part is the initialization of the Management, Device and UI surface. The rest of the rendering application can be shared between platforms.

2.1 Example of CPU-Side Rendering Code

To render something, a Render Pass must be created with a single Step and a Pipeline. The Render Pass specifies the output images' types and their role in the rendering, when and how it wants to render into them. A Pipeline is a set of configurations for drawing commands. A Pipeline specifies the used shaders, the type of shader inputs and resources and other drawing specifications.

For more complex solutions it is possible utilize Buffers, Pipeline Resources (Uniforms in OpenGL or Descriptor Sets in Vulkan) and Textures [5] (Fig. 1).

2.2 GPU Shaders

As stated previously, the focus of this paper is the implementation of shaders using our library. Shaders are the short programs that are executed by the GPU in specific drawing stages. Currently the Vertex Shader and Fragment Shader stages are supported by our library. In OpenGL these shaders shall be implemented in GLSL (OpenGL Shading Language), however GLSL has multiple versions, also OpenGL ES and WebGL are using a different versioning. Vulkan uses SPIR-V as its shader language which is an extended GLSL 460 compiled into an intermediate bytecode language. Both DirectX and Metal uses their own shader languages.

Currently our library supports generating GLSL 420, 430, 440, 450, 460, 300 ES and SPIR-V bytecode. For modern hardware GLSL 460 or SPIR-V would been enough, but the current role of OpenGL in our library is to maintain compatibility with older and simpler platforms therefore we're supporting older GLSL code generation as well.

2.3 Shader Classes in C#

In our library it is possible to promote a C# class into a "shader class". This will make the build system run our Source Generator (see in the next section) on the class and it will generate the shader sources and bytecodes from the class in the correct format to let us use it during Pipeline creation. To promote a class to shader class it shall be inherited from the proper ShaderBase class from our library and a Shader Attribute must be added to the class (presented on Fig. 2). This way we can also specify what kind of shader (Vertex of Fragment) we are implementing with this class. Inside this class the regular C# variables and programming constructs can be used, just like in the implementation a normal C# class. We can even use types or functions from outside of the class, those will be generated into the shader as well. However, there are some rules to be careful about, for example it is not allowed to use any reference types (shaders support only value types). Also, the class must be a partial class because our generator will generate additional partial classes storing the compiled shader source codes in fields.

In the class the abstract main method must be overridden. This will be the entry point of the shader which will be called for every vertex/fragment. Let us start with a simple Passthrough Vertex and Fragment Shader combination to understand the semantics of shader development in our library. These shaders are just receiving input and sending it forward without any modification. Implementation of much more complex shaders is possible of course, but this is a great starting point.

To implement a simple Passthrough Vertex Shader (sets the incoming position from the vertex input data as the final position for the vertex without any transformations or modifications) we must create input variables to receive the vertex input data. We can declare a local private variable in our shader class as a Vector4 (Vector4 and other mathematics structures are in the official C# System.Numerics namespace, our shader generator is using these types) and we shall add the in attribute from our library to this variable. This will indicate that during shader generation this will be generated with an in storage qualifier into the shader source (similarly for output variables we have an out attribute). In the Vertex Shader we have some built-in

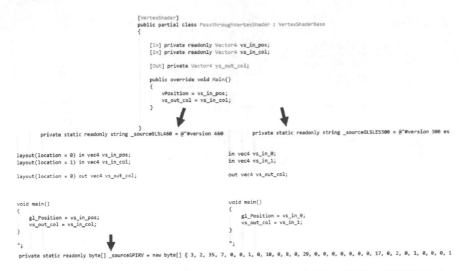

Fig. 2 Vertex passthrough shader implemented in C# and compiled into GLSL and SPIR-V using our library

variables like vPosition which stands for the built-in gl_Position variable in GLSL or vID which is gl_VertexID in GLSL. To finish our Vertex Shader class, we shall assign the input position to vPosition stating that this value is going to be the final vertex position. Also, we can create an in and an out Vector4 variable for sending the incoming color value of the vertex to the Fragment Shader.

The Fragment Shader can be implemented analogue to the Vertex Shader. We shall declare an in Vector4 variable for receiving the color from our Vertex Shader and an out Vector4 for stating the final fragment color. In the overridden main function, we shall assign the input color to the output color (Fig. 3).

2.4 Complex C# Shaders

Vertex Shader for 3D Transformations. To do 3D transformations, the transformation matrices are needed in the Vertex Shader. The final vertex position must be transformed with the MVP matrix (model/world, view and projection transformations multiplied into a single matrix). In the Fragment Shader we could require the world position of the fragments and the normal vector as well. To get the world position the world transformation matrix is needed on its own and to calculate the transformed normal we need the inverse transpose of the world transformation matrix. These matrices can be stored in a structure (just a regular C# struct). In our Transform Vertex Shader there shall be a Uniform variable of this structure to be able to receive the current object's transformation during the draw command.

Fig. 3 3D transformation vertex shader implemented in C# and compiled into GLSL and SPIR-V using our library

Uniform variables are constants inside the shader, but before the drawing command is issued on the CPU sided code that will eventually execute our shader, we can set the values of these Uniform variables. Conveniently, in our library the same C# structures can be used both in the CPU sided code when we are setting the value and in the GPU shaders where we use it (mind the StructLayout attribute! We will explain memory layouts later in this section). In the C# shader class, we can use the Uniform attribute from our library to indicate that a field is a uniform variable. We need to assign the set and binding inside this set to this variable, because these are the ids which we can use on the CPU sided code to select which of the Uniform variables (in case we have multiple) we want to assign a value.

For vector and matrix operations we are consuming the C# System.Numerics namespace with its types and functions. By implementing a valid calculation with this API, the formula will be compiled into the correct GLSL formula.

Fragment Shader for Phong Shading. Phong Shading [6] is a straightforward way of simulating light effects in 3D graphics. An example of a Phong shader using our library could be implemented like on Fig. 4.

In this shader we are using multiple Uniform variables for multiple purposes. Material stores the current object's material information (how it reacts to light, how

```
[FragmentShader]
public partial class NormalsFragmentShader : FragmentShaderBase
{

    [In] private readonly Vector3 vs_out_pos;
    [In] private readonly Vector3 vs_out_norm;
    [In] private readonly Vector2 vs_out_tex;

    [Out] private Vector4 fs_out_col;

    [Uniform(set: 0u, binding: 1u)] private MaterialData material;
    [Uniform(set: 0u, binding: 2u)] private LightData light;
    [Uniform(set: 0u, binding: 3u)] private SceneData scene;
    [Uniform(set: 0u, binding: 4u)] private TextureSampler2D textureSampler;

    public override void Main()
    {
        Vector4 ambient = light.ambientColor * material.ambientColor;

        Vector3 normal = Vector3.Normalize(vs_out_norm);
        Vector3 toLight = -light.direction;
        float diffuseIntensity = MathHelper.Clamp(Vector3.Dot(normal, toLight), 0f, 1f);
        Vector4 diffuse = light.diffuseColor * material.diffuseColor * diffuseIntensity;

        Vector3 reflect = Vector3.Reflect(light.direction, normal);
        Vector3 toCamera = Vector3.Normalize(scene.cameraPosition - vs_out_pos);
        float specularIntensity = MathF.Pow(MathHelper.Clamp(Vector3.Dot(reflect, toCamera), 0f, 1f), light.specularPower);
        Vector4 specular = light.specularColor * material.specularColor * specularIntensity;

        Vector4 textureColor = textureSampler.Sample(vs_out_tex);

        fs_out_col = (ambient + diffuse + specular) * textureColor;
    }

}
```

```
[StructLayout(LayoutKind.Explicit)]
public struct LightData
{
    [FieldOffset(0)]
    public Vector3 direction;

    [FieldOffset(16)]
    public Vector4 ambientColor;
    [FieldOffset(32)]
    public Vector4 diffuseColor;
    [FieldOffset(48)]
    public Vector4 specularColor;

    [FieldOffset(64)]
    public float specularPower;
}
```

Fig. 4 Phong shading fragment shader implemented in C# using our library

shiny it is, etc.), Light stores the virtual world's sun's, the directional light's information (its color and direction) and Scene presents other scene dependent data like the camera (viewer) position.

When using structures as uniform variables in a shader language we must consider the memory layout of the fields inside a structure. Every device (GPU) favors a different memory layout as an optimal layout. While we would expect that the fields of a structure are stored in the memory by the order of the declarations the device may expecting it in a different order, simply because that order is more optimal for its internal architecture. Also, fields may have padding between them, leaving unused bytes among the bytes of our actual variables in the memory. When we see a structure with two float fields, we would expect that it should be stored on 8 bytes of memory (2 * 4, because a float is stored on 4 bytes). The device may introduce padding between the fields, to match a particular memory layout, for example organizing the structure to have all its fields starting at every 16th byte [7]. To overcome this inconsistency shader languages introduced standard layouts which are device independent memory layout options for declaring structures. Our library is using the GLSL std140 layout for Uniform variables which means the C# structures must obey the rules of this layout. Like in the example of our LightData structure, You may need to explicitly set the memory offset (location within the structure) of all fields to match the shader language's expectations. Currently our shader generator has no means of checking or enforcing the correct layout, but this is something that we are planning to add later.

In the Phong shader we are using a Texture Sampler (a resource to read pixels from an image in memory) to colorize our geometry. For these purposes we have introduced some types in our library which You can use in the C# shader classes to access certain shader features like texture samplers and sampler operations.

3   C# Source Generators

Source Generators are part of the Roslyn Compiler Platform SDK [8] that is a C# development library that enables developers to access the syntax tree and model of their compiled C# application either during development time or during runtime. Using a Source Generator, we can implement custom code checking and refactoring tools or we can generate C# source files as well which will be added to the compilation during the compilation. For our Shader Generators we have used the C# source file generation feature of the API.

While it is possible to use Source Generators to create a standalone application, like a command-line tool, we have created a Source Generator project which can be referenced and used from any C# projects. These kinds of projects are called Analyzers, .NET Standard 2.0 class library projects where the <IsRoslynComponent> true </IsRoslynComponent> option must be added to the project file and it needs references to the Microsoft.CodeAnalysis.Csharp and Microsoft.CodeAnalysis.Analyzers packages. For both references we need to add PrivateAssets = "all" attribute, to indicate that the reference of our analyzer should not see these references, otherwise it may try to make that project as an Analyzer project as well and that would result in inconsistency errors with the project type. An Analyzer project created this way can be referenced from another C# project of the solution, by adding OutputItemType = "Analyzer" to the referencing project.

During compilation of the referencing project the compiler system will invoke the Analyzer project's compiled dll. The Analyzer is executed, and the implemented custom code will traverse the current state of compilation, optionally add additional C# sources to it. All of this is happening completely during compilation time, no runtime reflection or other runtime performance penalties will be applied.

Visual Studio itself uses the Analyzer's dll as well, but it only loads the dll when Visual Studio launches and loads the project. This means that every time when You modify Your Analyzer project's source code and recompile it, You need to restart Visual Studio forcing it to use the newly compiled dll.

While an Analyzer project can be used for a lot of purposes, we are currently focusing on source generation. To get started we need a class in the Analyzer project implementing the IIncrementalGenerator interface (ISourceGenerator is available as well, but Incremental Generators are more optimal to use). In the implemented Initialize method we can access the context of the compilation of the project that is referencing our Analyzer. We can filter the classes, structures, functions, variable declarations, or all other programming constructs of this compilation to collect only the parts of the code which we want to work on. It is important to filter early and to filter well, because on larger projects a non-performant Analyzer can greatly slow down the compilation and the responsiveness of Visual Studio.

```
<ItemGroup>
  <!-- Package the generator in the analyzer directory of the nuget package -->
  <None Include="$(OutputPath)\$(AssemblyName).dll" Pack="true" PackagePath="analyzers/dotnet/cs" Visible="false" />
    <!-- Package the base SharpGraphics dependency alongside the generator assembly -->
  <None Include="$(PkgSharpGraphics)\lib\netstandard2.0\*.dll" Pack="true" PackagePath="analyzers/dotnet/cs" Visible="false" />
</ItemGroup>

<PropertyGroup>
  <GetTargetPathDependsOn>$(GetTargetPathDependsOn);GetDependencyTargetPaths</GetTargetPathDependsOn>
</PropertyGroup>

<Target Name="GetDependencyTargetPaths">
  <ItemGroup>
    <TargetPathWithTargetPlatformMoniker Include="$(PkgSharpGraphics)\lib\netstandard2.0\*.dll" IncludeRuntimeDependency="false" />
  </ItemGroup>
</Target>
```

Fig. 5 Project file of a Roslyn source generator using other package references

After successful filtering we can schedule the execution of our code generator method with the filtered data. The filtered data is usually some kind of Syntax information (ClassDeclarationSyntax in our Shader Generator's case because we are filtering for the classes which are promoted with our Shader Attribute). Using these Syntax classes, we can travel the whole syntax tree if needed. We can query the fields and methods inside the found classes. We can get the statements inside a method and the expressions that are constructing those statements. To get more detailed type information about expressions we can get the semantic model for syntax trees. Using these models, the type of the fields and the declaration or other information of these types can be found. Based on this information it is possible to issue warnings, refactoring or to generate completely new C# sources and add these to the compilation. All this parsing and generation happens during development and compile time.

We are creating a library for developers to use; therefore, we need to be able to package this analyzer project into a NuGet Package. Any .NET Standard project can be issued for packing using the <GeneratePackageOnBuild> true </GeneratePackageOnBuild> option in the project file, however we need additional options for properly packing the Analyzer and its references and DLLs. For all other nuget package references we shall add the PrivateAssets = "all" and GeneratePathProperty = "true" attributes. Thanks to the latter the path to the compiled DLL of the reference can be accessed in the project file later for packaging. On Fig. 5 we present the modifications to the project file that are required to pack the DLLs of referenced packages into our Analyzer's package.

4 Shader Generation

After the compilation has been filtered for C# classes annotated with the Shader Attribute our generator prepares the GLSLangValidator (explained later), then it decides about the type and stage of the shader based on the Shader Attribute. Currently only Vertex and Fragment Shaders are supported, so the generator tries to find either of these two Attributes. We collect information about our C# shader class into a ShaderClassDeclaration class. The fields of the class are filtered by our in, out and uniform attributes and the main function of the class is searched for as well. On

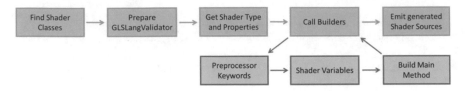

Fig. 6 Structure and compilation steps of our shader generator module

this prepared information we start our Builders. We have a Builder for all supported shader and their task is to traverse the syntax tree of the C# shader class and generate the corresponding shader source as a string. Lastly the generated strings are added to the compilation as fields in a C# class, therefore the library will be able to get these shader source strings during Pipeline creation (Fig. 6).

4.1 Shader Generator Preparation

We can determine the stage of the shader by the type of our Shader Attribute, our library has a separate VertexShaderAttribute and FragmentShaderAttribute for this role (both are inheriting from a ShaderAttribute base class so after filtering we can do further analysis based on the subclass types). Later this system can be extended with support for Compute Shaders or other Graphics Shader stages as well. We collect all this information to the previously mentioned ShaderClassDeclaration class which is provided to the Builders.

The generated shader sources are emitted into the compilation as partial classes. The ShaderBase class which our C# classes needs to inherit from, provides properties for getting the generated shader source for every supported shader language and version. For each shader language and version, a partial class is generated, holding the generated source in a private field, and providing access to it using a property getter. To properly emit these partial classes, we also need to get the namespace of the C# shader class which the generator is currently parsing. First, we collect all the generated sources into a List, with the name of their shader language (to know which property and field it should emit the source into) and the source string itself. For the string-based GLSL the source string is the multi-line string of the generated string itself. For the bytecode-based SPIR-V the source string is the instantiation of a byte array with the content of the compiled bytes. We are not utterly satisfied with emitting the compiled shaders into the code this way. We are experimenting with ways to emit the generated shaders as Embedded Resource files into the compilation.

4.2   Shader Builders

Builders are the soul of our whole Shader Generation package because Builders are responsible for deeply inspecting the C# shader class's syntax-tree and generate the source string statement by statement. To be able to extend the system in the future for supporting more shader languages (for supporting DirectX and Metal), we have defined a ShaderBuilderBase class which is responsible for exploring the syntax-tree. Specialized Builders for GLSL or other shader languages are inherited from this base class to implement the language specific generation steps. Multiple GLSL builders are also inheriting from each other. While the basic keywords and syntax of the most modern GLSL 460 are the same in the older GLSL ES 300 on mobile hardware, some modern features are not supported, and those parts of the source generation must be overridden to fall back to the older feature set.

ShaderBuilderBase defines the main structure of the generated shader source, by first generating the Preprocessor keywords, then shader level variables and lastly the main function. Generation is done using a StringBuilder, which is an optimal way of concatenating multiple strings into a single string in C#. The base class keeps track of the current indentation to always emit the proper number of tabulators at the beginning of the lines. It also stores the line number for adding structure definitions or functions, which are discovered during the exploration of the main function.

Variables. We generate the in, out, uniform and local shader level variables in this order. The declarations have been grouped into these storage categories before the Builders are started, during the preparation. For every category we generate the declaration for each variable in declaration order. First, we need to get the shader type of the declared variable. Our library differentiates the C# types in the implemented C# shader classes and the types that are supported in the shader language. We need to map these types to each other. For primitive types, the mapping is simple (int to int, float to float, etc.), but for more complex types we need to emit a different string into the shader code (Vector3, which is System.Numerics.Vector3 is vec3 in the shader language). Structures are even more complex (see later). When the type of the variable is known, we can begin the generation of the declaration. We need to generate the proper storage qualifier (in, out, uniform) based on the Attributes attached to the declaration. This is specialized for the GLSL Builders to generate the qualifier keyword before the type.

For uniform variables struct types are supported, so this behavior must be handled at the declaration to generate the struct inlined to the declaration. If the uniform variable is an array of structs, then the inline struct will have only a single struct array field with the actual struct instances inside. Older GLSL versions do not support explicit location settings for the uniform variables so for GLSL ES 300 we need to override this generation with a specialized behavior to rename these uniform variables during shader generation consistently using the location settings. This way the CPU side code can still deterministically get the location of these variables in our library. The generator supports renaming variables during the generation, therefore every time when we are using a variable in the shader code that is renamed, the new name will be used.

Statements. The main function (or any function) is built up from statements. After generating the interface of the function (void main() {…}) the statements of the function must be parsed and generated in order. For every supported statement kind (Expression, Declaration, If, For, etc.) we have functions to build the corresponding source for that statement kind in the shader language. Statements are built up from expressions which are also built up from other expressions. We have builder methods expressions too. To better understand, we can observe the syntax-tree of an Expression Statement which assigns the transformed input vertex position to the output position in the Vertex Shader on Fig. 3. The Expression Statement is a Simple Assignment Expression which have a Left Expression (output position which is an Identifier Name Expression) an operator (equals) and a Right Expression that is an Invocation Expression (calling the static Transform extension function of Vector4). The parameters of this invocation are also Expressions and so on, until we reach the bottom of the Syntax-tree where we usually find an Identifier Name (name of a variable) or a Literal (like a float token).

Our generator is built up with the structure presented on Fig. 7 to be able to discover the Syntax-tree and generate the equivalent shader source based on it. All the methods are virtual; therefore, their behavior can be overwritten in the derived GLSL or other shader builders if special behavior is needed for a shader language.

Types. Whenever the generator encounters the need to find the type of an expression (variable declarations, invocations, member access) it checks if that type has a corresponding type in the shader language (int for int, vec3 for Vector3, etc.). If there is no corresponding type, then it assumes that it must be a custom structure and tries to get its declaration syntax for generating the struct. If there is no such structure defined in the C# compilation, then it is an unknown type and it will result in a generator error.

Generating structures are done in a separate StringBuilder, because we need to add the declaration at the beginning of the shader source code (after preprocessor keywords and before variable declarations). Generating the struct is straightforward, like generating the variables of the shader. We store the fact into a Dictionary, that this struct has been generated, so the next time when the generator encounters the same struct type it will realize that it is already generated.

4.3   Shader Validation and SPIR-V Compilation

So far, the C# compiler can detect errors made against the C# specification and our Generator can detect some of the errors that are rendering the implemented shaders incompatible with the target shader languages. Before linking the build into a final compilation, we are using the GLSLangValidator tool to catch any other error that passed the first two lines of defense.

```
protected virtual void BuildMethodBody(string name, IEnumerable<StatementSyntax> statements)
{
    foreach (StatementSyntax statement in statements)
        BuildMethodBodyStatement(statement);
}
protected virtual void BuildMethodBodyStatement(StatementSyntax statement)
{
    _sb.AppendIndentation(_indentation);
    switch (statement)
    {
        case ExpressionStatementSyntax expression: BuildExpression(expression.Expression); _sb.AppendLine(';'); break;
        case LocalDeclarationStatementSyntax localDeclaration: BuildStatement(localDeclaration); _sb.AppendLine(';'); break;
        case ReturnStatementSyntax returnStatement: BuildStatement(returnStatement); _sb.AppendLine(';'); break;
        case BreakStatementSyntax breakStatement: _sb.AppendLine("break;"); break;
        case ContinueStatementSyntax continueStatement: _sb.AppendLine("continue;"); break;
        case IfStatementSyntax ifStatement: BuildStatement(ifStatement); break;
        case ForStatementSyntax forStatement: BuildStatement(forStatement); break;
        case BlockSyntax block: BuildStatement(block); break;
    }
}
protected virtual void BuildExpression(ExpressionSyntax expression)
{
    switch (expression)
    {
        case AssignmentExpressionSyntax assignment: BuildExpression(assignment); break;
        case ArrayCreationExpressionSyntax arrayCreation: BuildExpression(arrayCreation); break;
        case OmittedArraySizeExpressionSyntax omittedArraySize: break;
        case InitializerExpressionSyntax initializer: BuildExpression(initializer); break;
        case ObjectCreationExpressionSyntax objectCreation: BuildExpression(objectCreation); break;
        case PrefixUnaryExpressionSyntax prefixUnaryExpression: BuildExpression(prefixUnaryExpression); break;
        case PostfixUnaryExpressionSyntax postfixUnaryExpression: BuildExpression(postfixUnaryExpression); break;
        case LiteralExpressionSyntax literalExpression: BuildExpression(literalExpression); break;
        case SimpleNameSyntax simpleName: BuildExpression(simpleName); break; //Base of IdentifierName
        case ElementAccessExpressionSyntax elementAccess: BuildExpression(elementAccess); break;
        case InvocationExpressionSyntax invocation: BuildExpression(invocation); break;
        case MemberAccessExpressionSyntax memberAccess: BuildExpression(memberAccess); break;
        case BinaryExpressionSyntax binaryExpression: BuildExpression(binaryExpression); break;
        case CastExpressionSyntax castExpression: BuildExpression(castExpression); break;
        case ParenthesizedExpressionSyntax parenthesizedExpression: BuildExpression(parenthesizedExpression); break;
    }
}
protected virtual void BuildExpression(AssignmentExpressionSyntax assignment)
{
    BuildExpression(assignment.Left);
    switch (assignment.Kind())
    {
        case SyntaxKind.SimpleAssignmentExpression: _sb.Append(" = "); break;
        case SyntaxKind.AddAssignmentExpression: _sb.Append(" += "); break;
        case SyntaxKind.SubtractAssignmentExpression: _sb.Append(" -= "); break;
        case SyntaxKind.MultiplyAssignmentExpression: _sb.Append(" *= "); break;
        case SyntaxKind.DivideAssignmentExpression: _sb.Append(" /= "); break;
        case SyntaxKind.ModuloAssignmentExpression: _sb.Append(" %= "); break;
    }
    BuildExpression(assignment.Right);
}
```

Fig. 7 Internals of our shader generator. For all statements in a method, it builds the statement and the expressions inside the statement

GLSLangValidator is a command-line tool from Khronos (developers of OpenGL and Vulkan) for parsing GLSL code and detecting errors in it. This tool can be also used for compiling an extended version of GLSL 460 to SPIR-V bytecode to support Vulkan, therefore we must use this tool in our Generator for Vulkan compatibility. The Validator application must be executed for all generated shader strings to be validated. In the case of SPIR-V, we must also read the compiled bytecode file.

Preparing GLSLangValidator in our Generator. Our Generator needs to be able to run the Validator on the generated sources during compilation time. Currently, we are using GLSLangValidator by writing the shader string into a file, then passing the path of the file to the Validator for parsing. All of this is happening during compilation or development time; therefore, a lot of shaders and supported languages may slow down Visual Studio. We are experimenting with other ways for speeding up the process.

We can get the executable for the Validator from its official GitHub repository [9]. We have added executables (for multiple platforms) to the Generator project as Embedded Resources. This way we can access the executable files while the Generator is executing. During the preparation phase of our Shader Generator, we are copying the Validator executable to the Temp folder in the system, if it is not already there (on Mac and Linux we need to execute the chmod command from our Generator to add executable rights to this file). Visual Studio or build systems may launch multiple instances of a Generator/Analyzer at once so we need to handle accessing the Temp folder from multiple threads simultaneously. We create a temporal folder for the current session, and we write the completely generated shader string into a text file in this folder.

Validation. After the Validator executable and the shader file to validate is in place we need to launch the Validator as a new process with the path to the shader file in its launch arguments. To validate the only argument the Validator needs is the path to the shader file. For SPIR-V compilation we need to add -V to compile and -o {output_file_path} for the compiled file's path.

We are starting the Validator process with the built-in System.Diagnostics.Process class in C#. In Windows, we can get the output of the Validator so we can throw a compile time error with the error received from the Validator.

After starting the Validator Process, we need to wait for the process to end. We are doing an active pooling wait on the resource, because the async waiting on processes is not available on .NET Standard 2.0. If the compilation has been canceled (manually by the user or automatically by the IDE) we shall terminate the process to finish our generator as soon as possible.

If the generator has a non-zero exit code, then errors occurred during the compilation. Otherwise, the shader is correct. For SPIR-V we can read the compiled binary file into our Generator to emit the byte array into the source code.

5 Conclusion

With our forthcoming library we are providing a possibility for .NET C# developers for implementing low-level real-time rendering algorithms and solutions in C# and to embed the rendered frames into the surfaces of .NET UI frameworks, like WPF and Xamarin. Using our library, both the CPU and GPU graphics algorithms can be implemented in C#. The CPU rendering code is an abstraction layer for Graphics APIs and the native GPU shaders are generated from C# classes during compilation time.

We are presenting a new kind of development environment for both regular C# programmers and graphics programmers. We would like to recommend this paper and our approach both to .NET and graphics developers .NET developers could power their application with multi-platform real-time graphics modules and graphics developers could reach other platforms and new interesting development environments with our approach.

Acknowledgements EFOP-3.6.3-VEKOP-16-2017-00001: Talent Management in Autonomous Vehicle Control Technologies—The Project is supported by the Hungarian Government and co-financed by the European Social Fund.

References

1. Announcing .NET 6, https://devblogs.microsoft.com/dotnet/announcing-net-6/, last accessed 2022/06/30
2. vk.net, https://github.com/jpbruyere/vk.net, last accessed 2022/06/30
3. OpenTK, https://github.com/opentk/opentk, last accessed 2022/06/30
4. Szabó D, Illés Z (2022) Real-time rendering with OpenGL and Vulkan in C#. In: Recent innovations in computing, vol 2. Springer, pp. 599–611
5. Sellers G, Kessenich J (2016) Vulkan programming guide, 1st edn. Addison-Wesley
6. Bui Thong P (1975) Illumination for computer generated pictures. In: Communications of the ACM 18(6):311–317
7. Szabó D, Illés Z (2020) Vulkan in C# for multi-platform real-time graphics. In: 18th international conference on emerging e-learning technologies and applications, pp 687–692
8. .NET Compiler Platform SDK, https://docs.microsoft.com/en-us/dotnet/csharp/roslyn-sdk/, last accessed 2022/06/30
9. glslang, https://github.com/KhronosGroup/glslang, last accessed 2022/06/30

An Enhanced GLCM and PCA Algorithm for Image Watermarking

Pratyaksha Ranawat, Mayank Patel, and Ajay Kumar Sharma

Abstract Digital image watermarking is a technique to provide safety and authentication to sensitive data in the form of images. It prevents unauthorized access and misuse of the data. In today's world of the Internet, where there is a high risk of sensitive data being misused, it is essential to take some measures to ensure data safety. In the proposed technique, features are extracted and analyzed from the image that must be protected by watermarking. Then the watermark is embedded into this cover image to get the final watermarked image. We have used GLCM and PCA algorithms to improve the efficiency of the watermarking scheme to attain maximum security. This technique is less complex and performs well. With this technique, a blind watermark is generated, which is more robust against attacks and thus improves the security of confidential data. We have used the parameters PSNR and MSE to test the efficiency of the proposed technique. We have used MATLAB for simulation.

Keywords Image processing · PSNR · MSE · Watermarking

1 Introduction

1.1 Introduction to Image Processing

In image processing, we first convert the image to digital form, and after that, various image processing operations are performed. Some operations include visualization, recognition, sharpening, restoration, retrieval, etc. These facilitate improving the quality of the image or extracting useful information from the image. The input of an image processing algorithm is an image, and the output can be an image or the features of that image.

P. Ranawat (✉) · M. Patel · A. K. Sharma
Geetenjali Institute of Technical Studies, Udaipur, Rajasthan, India
e-mail: pratyaksharanawat@gmail.com

© The Author(s), under exclusive license to Springer Nature Singapore Pte Ltd. 2023   885
Y. Singh et al. (eds.), *Proceedings of International Conference on Recent Innovations in Computing*, Lecture Notes in Electrical Engineering 1011,
https://doi.org/10.1007/978-981-99-0601-7_66

1.2 Digital Image Watermarking

Digital image watermarking is a technique to embed a watermark into a multimedia product to protect the data from unauthorized access and tampering. It prevents illegal copying of data. Audio, video, text, or images [1] are protected from unauthorized users with the technique of digital watermarking. We can classify watermarking as visible watermarking and invisible watermarking.

Visible watermarks are visible to the human eye. They are not very secure as anyone can resize the image and crop the watermark. When embedding a visible watermark, we need to ensure it does not occlude the critical portions of the image.

In invisible watermarking [2], the watermark gets concealed in the cover image. It makes use of the technique of steganography. It is instrumental in proving the ownership of digital data. The authorized user can only identify the invisible watermark.

We have two ways of extracting the watermark from the transmitted image. In the case of a blind watermark, information about the host image or watermark is not required for extraction. In the case of a semi-blind watermark, information about the watermark is required, but no information about the cover image is required. In the case of a non-blind watermark, we require all the information about the host image and watermark.

1.3 Steps in Digital Image Watermarking

There are three essential components in the generic model of digital image watermarking:

i.   Watermark generation
ii. Watermark embedding
iii. Watermark extraction (Fig. 1).

1.4 Requirements for Digital Image Watermarking

The requirements for digital image watermarking are as follows [3]:

i.   Robustness: A robust watermark can sustain attacks like compression, noise filtering, shearing, etc. It performs various other operations like conversion of digital to analog and analog to digital, cutting, and image enhancement. All the watermarking algorithms do not have the same level of robustness; some are robust against some manipulation and failed against other powerful attacks. However, every watermark does not need to be robust; in some cases, it can be more fragile.

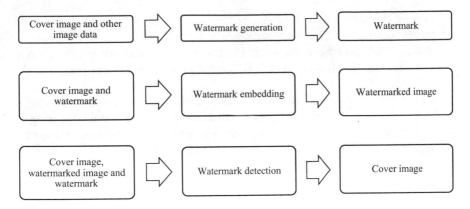

Fig. 1 Steps in digital watermarking

ii. Imperceptibility: Imperceptibility, also called invisibility or fidelity, is an essential requirement of watermarking system. The watermark should not be easily visible to the human eye. The original image and the watermarked image should look similar. When the impermeability is achieved, the robustness [4] and the capacity will reduce and vice-versa. So, the impermeability is reduced by increasing the robustness and the capacity. The watermark is not always invisible; sometimes, it is preferred to be visible per the user's demand.

iii. Capacity: It is often referred to as Payload and is the number of bits embedded into the images. The capacity of every image varies as per the watermarked design. The capacity tells about the limit of watermark information that can be embedded and also satisfies the impermeability and robustness.

iv. Security: The image can resist attacks. These attacks are responsible for modifying the [5] embedded watermark. There are three parts of attacks: unauthorized removal, unauthorized embedding, and unauthorized detection. According to watermarking, specific characteristics are present in the watermark, which help resist attacks.

v. Low Complexity: Watermarking is very costly because it is very complex. As it is very complicated and readily available for business purposes, the economics can be described using watermark embedders and detectors. The high speed of embedding and detector are two significant issues of complexity.

1.5 Applications of Digital Image Watermarking

In image processing watermarking has proved to be a promising technique. It has several applications of its own, and some of them are described below:

a. Copyright Protection: We use the copyright information to embed the information into new production. We extract the watermark in case of any ownership dispute, and the owner has to provide evidence of his ownership.

b. Content Authentication: It is required [6] to authenticate the content. It will prevent the access of unauthorized users in the watermark.
c. Broadcast Monitoring: It is used more frequently in advertising that the content is broadcasted as the contract between the advertisement company and the customer.
d. Owner Identification: Traditional methods verify that the owner has a visual mark. We can overcome this by using advanced software to modify the images.
e. Fingerprinting: It is used to safeguard customers. When the customer gets a legal copy of any product and starts distributing it illegally in the market, he is punished by using fingerprinting copyright [7]. In this, the transaction is traced by embedding every recipient's unique, robust watermark. So, the owner can quickly identify who redistributed the product and extract the watermark from further illegal copies.
f. Medical Applications: Watermarking plays a significant role in the [8] medical industries. It protects the confidential reports of the patient from getting accessed by any unauthorized person. It is used to find all the information and reports of any patient in case they lose them. It protects the copyright of medical images. The widespread use of the Internet makes it very easy to share images publicly. Thus, the protection of confidential data becomes very important.

1.6 Watermarking Properties

Following are a few of the properties desired in the watermarked image:

i. Effectiveness: It is the most essential and desired property of the watermark. According to this, the watermark should not get detected easily during transmission, whereas it should be easily extracted at the receiver end.
ii. Host signal quality: We embed the watermark in the host signal, so we need to make sure that there are minimum changes made to the host signal and that the changes made are not visible easily. It must be ensured that the quality of the host signal is maintained.
iii. Watermark size: It is imperative to check the size of the watermark. The watermarked data usually transmits over insecure channels, and if the watermark size is more, it will be visible by all and be detected easily, thus hampering security.
iv. Robustness: The watermark must be robust at all times to sustain any attacks or threats during transmission and maintain safe data transfer.

1.7 Threats for Watermarking

There are various threats to degrade the watermarking and hinder data security. With the advancement in watermarking techniques, the attackers are also coming up with new ideas and techniques to ruin the watermark. Some of the attacks are:

a. Image Compression—The lossy compression of the images is responsible for the destruction of watermarking process.
b. Geometric transformations—Includes rotation, translation, sheering, and resizing of the images.
c. Image Enhancements—It includes sharpening, color calibration, and contrast change.
d. Image Composition—It involves the addition of text, windowing with another image, etc.
e. Information Reduction—It includes cropping.

2 Present Work

2.1 Problem Definition

With the growing Internet usage, there has been a serious concern regarding the security of confidential content. For this, many new techniques have emerged, and one of them is digital watermarking. The two major categorizations of the watermarking techniques are the spatial domain and frequency domain. In the spatial domain watermarking technique, we insert the watermark into the cover image by modifying the lower-order bits of the cover image. This technique minimizes the complexity and computational values. However, the robustness of this technique is low against a specific type of attack.

Frequency domain watermarking includes techniques like Discrete Cosine Transform (DCT), Discrete Fourier Transform (DFT), and Discrete Wavelet Transform (DWT), which make use of some inverse transformations. These techniques provide security against more attacks, but this technique's complexity and computational values are very high. Therefore, we need to design a robust technique against a maximum number of attacks while having minimum complexity and computational values.

2.2 Objectives of Research

- Examine the properties of the Discrete Wavelet Transform technique.
- Design a novel enhanced technique to generate a blind watermark.
- Compare the designed technique with DWT based on PSNR, BER, and MSE.

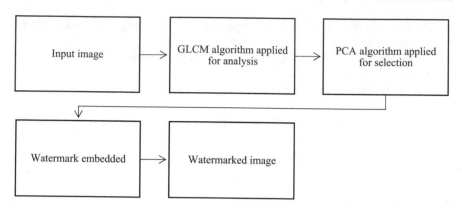

Fig. 2 Process of generating a watermarked image

2.3 Research Methodology

In the base paper, we observed that the OS-ELM approach is applied to generate a semi-blind watermarked image. DWT algorithm is deployed for extracting and analyzing the features of the image.

The technique suggested in this paper uses GLCM and PCA algorithms for generating a blind watermark. This technique is less complex, and the watermarked image is more robust against various security attacks (Fig. 2).

Steps followed in this research methodology

1. Initially, all the confidential and non-confidential images are considered. The confidential image is hidden within the non-confidential image. The keys used for encryption are generated from non-confidential images.
2. In the next step, we apply the GLCM algorithm to extract features such as co-relation and homogeneity from the confidential image.
3. Then we apply the PCA algorithm to select features from the already extracted features. The similarity between pixels is analyzed, and the image is compressed using various mathematical formulae and statistics.
4. Finally, the watermark created using non-confidential data is embedded into the cover image using OS-ELM, a machine technique.

Gray-level co-occurrence matrix (GLCM)

It is a [9] statistical method for understanding the spatial relationship among pixels in an image. It calculates how often pairs of pixels with specific values and in a specified spatial relationship occur in an image. The various statistics derived using GLCM are:

1. Contrast: It measures the local variations.
2. Co-relation: It is a measure to calculate the probability of occurrence among specific pixel pairs.

3. Energy: It provides the sum of squared elements in GLCM. It is also known as the angular second moment.
4. Homogeneity: It measures the closeness of the distribution of elements in the GLCM to the GLCM diagonal.

The various steps of the GLCM algorithm are outlined below [10]:

1. The first step in the GLCM algorithm is to quantize the image data. We treat each sample as a single image pixel; the sample's value is the pixel's intensity. These intensities are quantized into a specified number of discrete gray levels.
2. Create the GLCM, a $N \times N$ matrix where N is the number of levels specified under quantization.

The matrix is created as follows:

i. Let s be the sample under consideration.
ii. We denote the set of samples surrounding s by W, which fall within a window centered upon s, of a size that is specified under window size.
iii. Consider only the samples in set denoted by W. We define the value of i, j by the number of occurrences of two samples of intensities i and j in a specified spatial relationship.
iv. Make the GLCM symmetric.

    a. Make a transposed copy of GLCM.
    b. Add this copy to the GLCM itself.

This will produce a symmetric matrix in which the relationship i to j is indistinguishable from the relation j to i.

i. Normalize the GLCM. Divide each element by the sum of elements.

    1. Calculate the selected features. The calculations will only use the values of GLCM.

$$\text{Entropy} = \sum_{i,j=0}^{N-1} -\ln(P_{i,j}) P_{i,j} \tag{1}$$

$$\text{Energy} = \sum_{i,j=0}^{N-1} P_{ij}^2 \tag{2}$$

$$\text{Contrast} = \sum_{i,j=0}^{N-1} P_{ij}(i-j)^2 \tag{3}$$

$$\text{Homogeneity} = \sum_{i,j=0}^{N-1} \frac{P_{ij}}{1+(i-j)^2} \tag{4}$$

$$\text{Correlation} = \sum_{i,j=0}^{N-1} \frac{P_{ij}(i - \mu)(j - \mu)}{\sigma^2} \tag{5}$$

Principal Component Analysis (PCA)

Principal Component Analysis, or PCA, is a dimensionality-reduction [11] method used to reduce the dimensionality of large data sets by transforming a large set of variables into a smaller one that still contains most of the information in the large set.

Following are the steps involved in PCA [12]:

1. Standardization: It is imperative to standardize variables initially so that all of them are on the same scale. This will enable each variable to contribute equally to the analysis. Mathematically, standardization can be done by subtracting the mean from each variable's value and then dividing it by the standard deviation.

$$z = \frac{\text{value} - \text{mean}}{\text{standard deviation}} \tag{6}$$

2. Covariance matrix computation: We calculate the covariance matrix to see how the variables in the input data vary from the mean with respect to each other. It is a $p \times p$ symmetric matrix consisting of covariances associated with all possible pairs of initial variables.

3. Eigenvectors computation and eigenvalues computation: To calculate the principal components of the data, we compute Eigen values and Eigen vectors, a concept of linear algebra. This helps reduce the dimensionality without much loss of information, and this can be achieved by discarding the components with low information.

4. Feature vector: It is a matrix with eigenvectors that we choose to keep and consider relevant. This is the first step toward reducing the dimensions. We arrange the eigenvectors in decreasing order of their eigenvalues to find the most significant principal components.

5. Recast the data along the principal component axes: In the previous steps, we have learned that we did just one change on the data set, i.e., standardization. Apart from those, we calculate only principal components and perform operations on them. The input data always remains in terms of initial variables. In this step, we use the feature vector formed using eigenvectors of the covariance matrix to reorient the data from the original axes to the ones reoriented by the principal components.

$$\text{FinalDataSet} = \text{FeatureVector}^T * \text{StandardizedOriginalDataSet}^T \tag{7}$$

3 Result

We have used MATLAB for the simulation of our proposed algorithm. MATLAB is a programming language based on matrix that is very user-friendly, thus, making the workflow fast and easy. It is prevalent in the various fields of engineering and science as it provides an interactive environment for developing algorithms, analysis and visualization of data, and numerical computations. First, the user enters a scale factor ranging from 0 to 7, as the input image is 8 bits. The watermarked image is generated according to this entered scale factor value (Figs. 3 and 4).

PSNR—Peak signal to noise ratio is the ratio of the maximum power of an image to the corrupting noise power that affects the quality of that image. In image processing, PSNR is generally used to determine the error created by compression in the form of noise in the original signal.

MSE—Mean square error demonstrates the cumulative squared error between the compressed and original image. It is mainly used in statistical models to measure the difference between the observed and predicted values (Fig. 5 and Table 1).

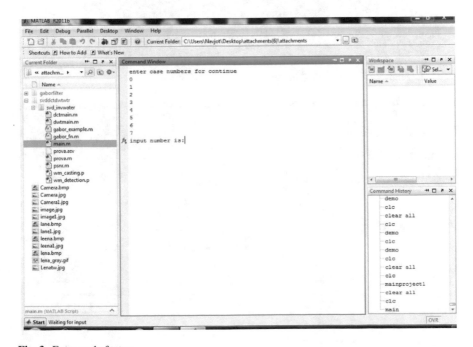

Fig. 3 Enter scale factor

Fig. 4 Generation of watermarked image

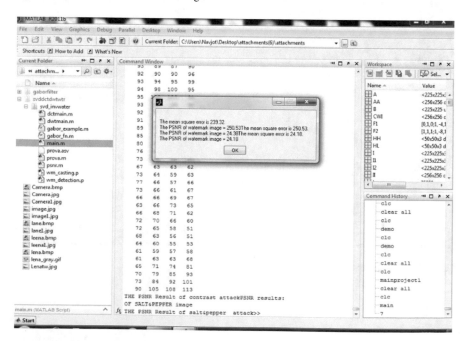

Fig. 5 Parametric values of watermarked image

Table 1 Result comparison

	Parameter values	Base	Proposed
Watermarked image	PSNR	13.3917	18.0129
	MSE	3001.26	2874.83
	Correlation cocfficient	0.01	0.01
	Entropy	7.9990	7.9989
Contrast attack	PSNR	20.0542	26.0537
	MSE	647.22	547.30
	Correlation coefficient	0.96	0.01
	BER	4.2319	4.2200
Sharpened attack	PSNR	23.6209	29.4842
	MSE	284.70	243.80
	Correlation coefficient	0.97	0.98
	BER	7.003	6.9047
Salt and pepper attack	PSNR	22.4476	27.484
	MSE	373.00	293.80
	Correlation coefficient	0.96	0.91
	BER	7.9012	7.9036
Decrypted image	PSNR	13.3848	18.0130
	MSE	3006.02	3274.75
	Correlation coefficient	0.01	0.00
	BER	7.6833	3.4237
Elapsed time (s)		0.011795	0.011994

4   Conclusion

The proposed method described in this paper helps in securely embedding the watermark into the cover image to obtain a more robust watermarked image. GLCM and PCA algorithms generate a blind watermark that can safely transmit over any secured or unsecured channel. In today's world, there is wide internet usage, thereby increasing the risk of unauthorized access. Attackers can easily tamper with data over digital channels, posing a severe security problem.

Unauthorized sources can access sensitive data, and they can also make changes to this data and distribute it illegally. Therefore, it is vital to have proper security measures to make sensitive data transmission secure. Watermarking is one such measure to induce security in digital data.

This paper achieves it by deploying a combination of GLCM and PCA algorithms. It is simulated on MATLAB, and parameters such as PSNR and MSE are calculated to verify results. The effectiveness of this proposed technique is also verified against specific image processing attacks, and it is found to perform well.

5   Future Work

There is always a chance to improvise and improve any work that is done. Following are a few points that need to be taken into consideration to improve the proposed algorithm in the future:

1. This algorithm can be improved in terms of elapsed time, and it can be minimized.
2. The efficiency of the proposed algorithm can be verified in the presence of other attacks to see how well it performs or needs to be improved.
3. The complexity of the algorithm can be reduced further.

References

1. Vleeschouwer CD, Delaigle JF, Macq B Invisibility and application functionalities in perceptual watermarking. In: 2002 an overview proceedings of the IEEE, vol 90, pp 64–77
2. Jun X, Ying W Towards a better understanding of DCT co-efficient in watermarking. PACIIA 2008, pp 206–209
3. Jun S, Alam MS Fragility and robustness of binary phase only filter based fragile/semi-fragile digital image watermarking instrumentation and measurement. 2008 IEEE Trans 57:595–606
4. Gonzalez RC, Woods RE (2009) Digital image processing, 3rd ed. Prentice Hall of India
5. Zeki AM, Abdul Manaf A A novel digital watermarking technique based on ISB. In: 2009 world academy of science, engineering and technology, vol 50, pp 989–996
6. Bamatraf A, Ibrahim R, Salleh MNBM Digital watermarking algorithm using LSB. 2010 in ICCAIE, pp 155–159
7. Charles Fung AG, Walter G A review study on image digital watermarking. In: 2011 presented at the 10th international conference on Networks St. Maarten The Netherlands Antilles
8. Zeki AM, Manaf AA, Foozy CFM, Mahmod SS A watermarking authentication system for medical images. 2011 presented at CET Shanghai China
9. Sheikh R, Patel M, Sinhal A Recognizing MNIST handwritten data set using PCA and LDA. In: 2020 international conference on artificial intelligence: advances and applications, pp 169–177
10. Khan S, Sharma AK, Patel M Performance analysis on modified hybrid DB scan clustering technique to enhance average total execution time. In: 2020 3rd International conference on intelligent sustainable systems (ICISS), pp 1603–1605. https://doi.org/10.1109/ICISS49785.2020.9316086

11. Agrawal S, Patel M, Sinhal A An enhance security of the color image using asymmetric RSA algorithm, 2021. In: Purohit S, Singh Jat D, Poonia R, Kumar S, Hiranwal S (eds) Proceedings of international conference on communication and computational technologies. Algorithms for intelligent systems. Springer
12. Shekhawat VS, Tiwari M, Patel M A secured steganography algorithm for hiding an image and data in an image using LSB technique, 2021. In: Singh V, Asari VK, Kumar S, Patel RB (eds) Computational methods and data engineering advances in intelligent systems and computing, vol 1257. Springer

Author Index

Printed in the United States
by Baker & Taylor Publisher Services